COMPOSITION AND PETROLOGY OF THE EARTH'S MANTLE

McGRAW-HILL INTERNATIONAL SERIES IN THE EARTH AND PLANETARY SCIENCES

AGER: Principles of Paleoecology
BERNER: Principles of Chemical Sedimentology
BROECKER and OVERSBY: Chemical Equilibria in the Earth
CARMICHAEL, TURNER, and VERHOOGEN: Igneous Petrology
DE SITTER: Structural Geology
DOMENICO: Concepts and Models in Groundwater Hydrology
EWING, JARDETZKY, and PRESS: Elastic Waves in Layered Media
GRANT and WEST: Interpretation Theory in Applied Geophysics
GRIM: Applied Clay Mineralogy
GRIM: Clay Mineralogy
HOWELL: Introduction to Geophysics
HYNDMAN: Petrology of Igneous and Metamorphic Rocks
JACOBS, RUSSELL, and WILSON: Physics and Geology
KRAUSKOPF: Introduction to Geochemistry
KRUMBEIN and GRAYBILL: An Introduction to Statistical Models in Geology
LEGGET: Geology and Engineering
MENARD: Marine Geology of the Pacific
MILLER: Photogeology
OFFICER: Introduction to the Theory of Sound Transmission
RAMSAY: Folding and Fracturing of Rocks
RINGWOOD: Composition and Petrology of the Earth's Mantle
ROBERTSON: The Nature of the Solid Earth
SHROCK and TWENHOFEL: Principles of Invertebrate Paleontology
STANTON: Ore Petrology
TOLSTOY: Wave Propagation
TURNER: Metamorphic Petrology
TURNER and VERHOOGEN: Igneous and Metamorphic Petrology
TURNER and WEISS: Structural Analysis of Metamorphic Tectonites

**McGRAW-HILL
BOOK COMPANY**
New York
St. Louis
San Francisco
Auckland
Düsseldorf
Johannesburg
Kuala Lumpur
London
Mexico
Montreal
New Delhi
Panama
Paris
São Paulo
Singapore
Sydney
Tokyo
Toronto

A. E. RINGWOOD, FAA, FRS
Professor of Geochemistry
Research School of Earth Sciences
The Australian National University

Composition and Petrology of the Earth's Mantle

This book was set in Times Roman by Textbook Services, Inc.
The editors were Robert H. Summersgill and James W. Bradley;
the production supervisor was Judi Frey.
Kingsport Press, Inc., was printer and binder.

Library of Congress Cataloging in Publication Data

Ringwood, A E date
Composition and petrology of the earth's mantle.

(McGraw-Hill international series in the earth and planetary sciences)
Includes bibliographies.
1. Earth—Mantle. 2. Petrology. I. Title.
QE511.R5 551.1'1 74-10721
ISBN 0-07-052932-9

COMPOSITION AND PETROLOGY OF THE EARTH'S MANTLE

1 2 3 4 5 6 7 8 9 0 KPKP 7 9 8 7 6 5

Dedicated to my wife GUN and children KRISTINA and PETER,
particularly for their Cheerful Acceptance of many
holidays lost during preparation of this book

CONTENTS

PREFACE

The purpose of this book as originally conceived was twofold—firstly, to review some relatively young fields of experimental petrology and secondly, to discuss the bearing of information from these fields upon the composition and constitution of the mantle.

A considerable literature in experimental petrology has developed during the last 20 years, much of it of a highly specialized nature. In consequence, many significant developments have become somewhat inaccessible to students and to researchers in neighbouring earth science fields. It seemed that there was a need for a review-synthesis of some of these developments which would be intelligible to a wide spectrum of advanced students and professionals in the earth sciences. This volume represents an attempt to meet part of this need.

Because of the broad scope of the subject and the enormous amount of new data which has become available, a complete, or even a representative, coverage of experimental petrology seemed out of the question. The general experimental topics I have chosen to discuss are those with which I am most familiar since they embrace research projects which have been pursued by my colleagues and myself at the Australian National University. Experimental work by other laboratories on these topics has also been reviewed. In a situation where one is attempting to evaluate the results of other laboratories in relation to one's own, it is difficult to be entirely objective, particularly where differences have arisen. In these cases, I have stated my preferred interpretations and have attempted to justify them in terms of relevant data and evidence.

In applying the experimental results to the composition and constitution of the mantle, I have attempted to introduce a broad array of evidence from several fields—particularly, natural petrology, mineralogy, geochemistry, cosmochemistry, and geophysics—at a level which I hope will be widely intelligible. The objective has been to summarize the relevant evidence which complements the experimental petrology and bears upon the present petrological and chemical composition of the principal regions of the mantle. It seems to me that we are approaching the stage where a broadly self-consistent understanding of the chemical and physical nature of the mantle is in sight, and the present volume represents an attempt to document this thesis.

Whilst the predominant emphasis is on the evidence relating to the mantle's present composition and constitution, its development in time is not entirely ignored. The chemical and petrological evolution of the mantle is discussed in terms of plate tectonics in Chapter 8, and the specific role of phase transformations in influencing mantle motions and plate movements is treated in Chapter 15. In view of the availability of some excellent recent texts and the copious recent literature, further discussion of the revolution wrought upon the earth sciences by the new global tectonics was not considered necessary.

Although not originally intended, the book developed into something of a dichotomy during writing, leading to the decision to divide it into two parts. Part 1 represents a relatively self-contained summary of the composition and petrology of the upper mantle-crust system combined with a discussion of the petrogenesis of the two principal classes of magmas which have been erupted from the earth's interior—basaltic and andesitic (orogenic). This part draws upon a wide array of observational data from natural petrology and integrates this with experimental results on phase equilibria at high pressures and temperatures displayed by natural rock systems—particularly basaltic and peridotitic compositions. Most of this work was carried out in collaboration with Dr. D. H. Green and, to a smaller extent, with Dr. T. H. Green and Dr. I. A. Nicholls. In addition, I have been very fortunate to have had access at an early stage to experimental work carried out by these scientists alone and in collaboration with others. I should acknowledge at this point the great debt I owe to these colleagues, particularly David Green, for innumerable discussions over the years which have greatly influenced my viewpoints on many topics, as well as for the fact that without their collaboration, the experimental work discussed in Part 1 would never have seen the light of day. To a substantial degree, Part 1 presents an overview of experimental work carried out by a strongly interacting research group at the Canberra laboratory.

Part 2, covering the deep mantle, differs from Part 1 in its approach as well as in its content. The vast amount of data provided by direct studies of the petrology and geochemistry of rocks from the upper mantle is no longer available in dealing with the nature of the deep mantle. Moreover, the petrogenesis of magmas which plays such an important role in studies of the composition and evolution of the upper mantle does not appear to contribute directly to our understanding of the deep mantle.

Whereas in Part 1 much of the interdisciplinary evidence complementary to the experimental investigations was drawn from petrology and mineralogy, in Part 2 there is a greater emphasis on geophysics, geochemistry, and crystallography. Constraints upon the chemical composition of the deep mantle are set by the overall composition of the upper mantle (Part 1) combined with geophysical, geochemical, and cosmochemical considerations. The primary emphasis in the present treatment of the deep mantle is placed upon the occurrence of high-pressure phase transformations and the bearing of these on the depth-variation of physical properties of the deep mantle. To these ends, a detailed review of experimental data on high-pressure phase transformations and relevant silicate and model systems is presented.

The final chapter, on the origin of the earth, was added some time after the remainder of the book had been completed. It represents an attempt to utilize the conclusions of earlier chapters on the composition and constitution of the mantle as boundary conditions for theories of the earth's origin. Many current theories in this area are based heavily upon evidence supplied by meteorites. I have long been convinced that much key evidence relating to the origin of the earth is to be found only in the direct study of the earth. This point of view is developed at some length in Chapter 16.

Some idiosyncrasies of style should be explained. Frequent references have been made to primary sources, and inevitably, in view of the scope of the book, a large number of these have been to papers by my colleagues and myself. To lessen the effects of repetition and to minimize interruptions to the continuity of text, I have numbered most of the citations and placed them in footnotes. References so cited have been assembled at the ends of chapters, rather than combined into a single bibliography at the end of the book. Although this procedure results in a good deal of repetition it leads to chapters which are more integral and self-contained. For the same reason, three diagrams appear twice in different chapters. An author index has been omitted since I am unconvinced that the labour of compilation is offset by its use to the general reader. The combination of reference lists attached to chapters with footnote citations within chapters should make it possible to locate the contributions of individuals without undue effort. Except for Chapter 16, the manuscript of this book was delivered to the publishers between January and April of 1973.

It should be admitted, though, that the above referencing system caused problems both in typesetting and in the making up of pages in order to maintain the integrity of the system. Unfortunately, these difficulties are reflected in the cost of the book as well as in the time taken to produce it. Offsetting this, however, is the advantage given the reader of being able to see immediately and on the same page not only the particular references being cited but their chronological order as well, thereby providing a developmental account of ideas on the earth's mantle.

Finally, I wish to express my sincere appreciation to colleagues who have read and commented upon the material in this volume—particularly to Drs. D. H. Green, R. C. Liebermann, I. Nicholls, and V. Oversby of the Research School of Earth Sciences, Australian National University, and to Dr. T. H. Green of the School of Earth Sciences, Macquarie University.

A. E. RINGWOOD

COMPOSITION AND PETROLOGY OF THE EARTH'S MANTLE

INTRODUCTION

0-1 SOME PERSPECTIVES

The past 25 years have witnessed major advances in the solid earth sciences. Some of these, for example, the hypothesis of sea-floor spreading and plate tectonics, have enjoyed a spectacular and immediate impact. In this particular case, the impact has arisen from the novelty of the hypothesis and from its capacity to provide an integrated working explanation of a remarkably diverse array of geophysical and geological observations. In other fields, for example, the constitution of the earth's interior, major progress has also been made, although the impact has not been as dramatic. It is only after reading review papers and books written on this subject more than 25 years ago that the extent and significance of recent progress in this field become fully apparent. Frequently, the progress has resulted not from the introduction of fundamentally new concepts but rather from a greatly improved ability to discriminate among a multitude of competing hypotheses. Thus, the hypothesis that the earth's core consists mainly of metallic iron is of ancient lineage. Nevertheless, it is only comparatively recently that it has been possible to elevate this hypothesis to the status of scientific theory and to relegate to states of low or negligible probability the competing hypotheses that the core consists of a high-pressure metallic form of olivine[1] or of primordial solar material.[2] Likewise, the suggestion that phase transformations play an important

[1]Ramsey (1948, 1949).
[2]Kuhn and Rittman (1941).

role in the mantle was made many years ago. Nevertheless, during the early 1950s this hypothesis was considered controversial, and several alternative hypotheses not requiring phase changes to explain the physical properties of the mantle were proposed. It is only during the last 10 years that the case for phase transformations has been proven beyond all reasonable doubt.

Prior to 25 years ago, most[1] of the significant papers and books on the solid earth were written by geologists and seismologists. There was only limited interaction between these disciplines. The geologists in general were preoccupied with the nature and evolution of the crust and, with some notable exceptions,[2] declined to speculate seriously about the constitution and evolution of the mantle. The widespread usage of the nonspecific terms *sima* and *ultrasima* for the upper mantle is an indication of this state of mind. Knowledge of the earth beneath the crust was derived primarily from seismology. The variation of *P*- and *S*-wave velocities with depth determined the major subdivisions of the earth's interior. Knowledge of these velocities as determined by Jeffreys[3] and Gutenberg,[4] together with constraints imposed by the earth's mean density and moment of inertia, enabled Bullen[5] to determine the distribution of density, pressure, and other elastic properties with depth. Such data on the elastic properties of the mantle, in the hands of Jeffreys,[6] Bullen,[7] Gutenberg,[8] and Birch,[9] enabled the construction of working hypotheses relating to the properties and nature of the materials in the earth's interior.

These scientists placed particular emphasis upon the tentative and uncertain nature of many of the inferences drawn ultimately from fundamental seismic data. A footnote by Birch[10] has been widely quoted in this respect. A more extensive discussion of the nature and uncertainties of geophysical data and inferences by Gutenberg[11] merits attention. The cautious appraisal by these and other scientists of the results and limitations of geophysical methods was entirely reasonable and justified at a time when it was not possible to test hypotheses relating to the earth's

[1]Among the important exceptions to this statement are the geochemical papers of Goldschmidt and the studies of the earth's gravity field by Vening Meinesz.
[2]E.g., Daly, Holmes.
[3]Jeffreys (1939).
[4]Gutenberg (1951, 1958, 1959b).
[5]Bullen (1936, 1940).
[6]Jeffreys, (1939, 1959).
[7]Bullen (1940, 1947).
[8]Gutenberg (1951, 1959a, b).
[9]Birch (1952).
[10]Birch (1952, p. 234):
Unwary readers should take warning that ordinary language undergoes modification to a high-pressure form when applied to the interior of the earth, e.g.,

High-pressure form	*Ordinary meaning*
certain	dubious
undoubtedly	perhaps
positive proof	vague suggestion
unanswerable argument	trivial objection
pure iron	uncertain mixture of all the elements

[11]Gutenberg (1959a, pp. 1-7).

deep interior by direct experiments under controlled laboratory conditions. However, recent developments in the earth sciences entitle us to take a more optimistic view of future prospects than is implicit in the references cited.

This optimism is based upon two developments. First is the application of experimental high pressure-high temperature techniques to study the properties and stability fields of rocks and minerals which may be present within the earth's deep interior and the physical chemistry of magma genesis in this region. It is now possible to reproduce statically in the laboratory the *P-T* conditions existing in the mantle down to a depth of about 800 km, and it is likely that this equivalent depth range will be extended substantially in the near future. The application of crystal chemistry, particularly the study of germanate model systems within the available experimental pressure range, has provided a great deal of information regarding the probable stability and behaviour of the corresponding silicates throughout the entire pressure range of the mantle. Measurements of the compressibilities of minerals can now be carried out under static pressures equal to those at depths of about 800 km, whilst precise determinations of the pressure and temperature derivatives of *P*- and *S*-wave velocities in minerals at quite modest pressures provide the data needed for the formulation of equations of state which are applicable over the entire pressure range of the mantle. Static high-pressure techniques are now being supplemented by dynamic (shock-wave) methods which enable the generation of pressures exceeding those in the centre of the earth (about 3.7 Mbars) for a few microseconds. This interval is sufficient to permit the determination of key equation-of-state variables and other important properties such as electrical conductivity. These techniques, as yet in their infancy, have already permitted the testing of many hypotheses formerly based upon seismic data. The stage is being reached at which high-pressure investigations are suggesting hypotheses which can be resolved only by increasingly sophisticated seismic techniques. The interaction between these fields, already close, must increase greatly in the years to come and should enable the construction of widely self-consistent earth models capable of explaining a comprehensive range of observations.

The second cause for optimism comes from the increasingly broad and interdisciplinary nature of the earth sciences. The past 25 years has seen a vast expansion of the numbers and range of scientists from different disciplines who have opted to apply their individual skills and techniques to problems of the earth sciences. As a result, there has been a spectacular development in such fields as marine geology and geophysics, geochemistry and cosmochemistry, isotope geochemistry, experimental petrology, as well as in many branches of geophysics. These developments are yielding and will continue to yield solutions to some of the most fundamental problems of the earth sciences. Many of these new developments have arisen from cooperative projects between specialists from different disciplines, and the resultant cross-fertilization of ideas has clearly played an important role.

Within any single discipline, however, progress is increasingly hindered by the problem of nonuniqueness of explanations which can be advanced to satisfy a given set of observations. This was recognized many years ago, for example, in the interpretation of gravity observations. Such interpretations are of limited value unless they are closely coordinated with seismic studies. In an analogous manner, earth models obtained by inversion of surface-wave or free-oscillation data are seldom unique, and it is often possible to satisfy the data with a range of models which differ only marginally in the probabilities by which they should be preferred. In order to constrain the choices further, the seismologist must consider evidence derived from other fields—e.g., by the application of equations of state of presumed mantle materials or by the use of constraints of a petrological nature. The geochemist and petrologist is in a precisely analogous position and must be prepared to consider and evaluate constraints derived from fields other than his own if he wishes to propose hypotheses which have any claim to generality. Thus, future progress in the earth sciences is likely to depend more and more upon the extent to which specialists in single fields are able to communicate with and interpret the results of workers in neighbouring disciplines. It is only by such interdisciplinary investigations that it will be possible to formulate models of the earth's interior which offer the highest degree of internal consistency in explaining observations from many fields and which, accordingly, have the highest probability of being correct.

The interdisciplinary approach to earth science problems is an ideal which is beset by some obvious practical difficulties. The enormous growth in the volume of published literature in recent years renders it very difficult to a specialist in one field to be adequately conversant with important developments in related fields. The range of fundamental knowledge and techniques which are employed in the earth sciences is far too broad to be mastered by individuals. Accordingly, when a specialist ventures outside his field and attempts to draw upon knowledge from other fields, there is an increased likelihood that errors of fact and interpretation will be introduced. Very frequently, the specialist will find that results from other fields appear to contradict theories which have developed within his own field. In this event, if a synthesis is to be achieved, he will be obliged to weigh the conflicting evidence and to make judgements which are necessarily partly subjective. Faced with this situation, it is almost inevitable that the synthesizer will give greater weight to evidence obtained within his own field than to conflicting evidence arising from fields with which he is less familiar. Examples are to be found within this volume.

These problems are fundamental and unlikely to disappear, despite the broader educational background which earth science students are receiving. Nevertheless, despite its inherent limitations, the interdisciplinary approach remains as an essential prerequisite for progress.

0.2 PRINCIPAL SUBDIVISIONS OF THE EARTH'S INTERIOR

These are based upon the depth distribution of seismic wave velocities as determined by Jeffreys[1] and Gutenberg[2] and later workers (Fig. 0-1). Although signifi-

[1]Jeffreys (1939).
[2]Gutenberg (1951, 1958, 1959b).

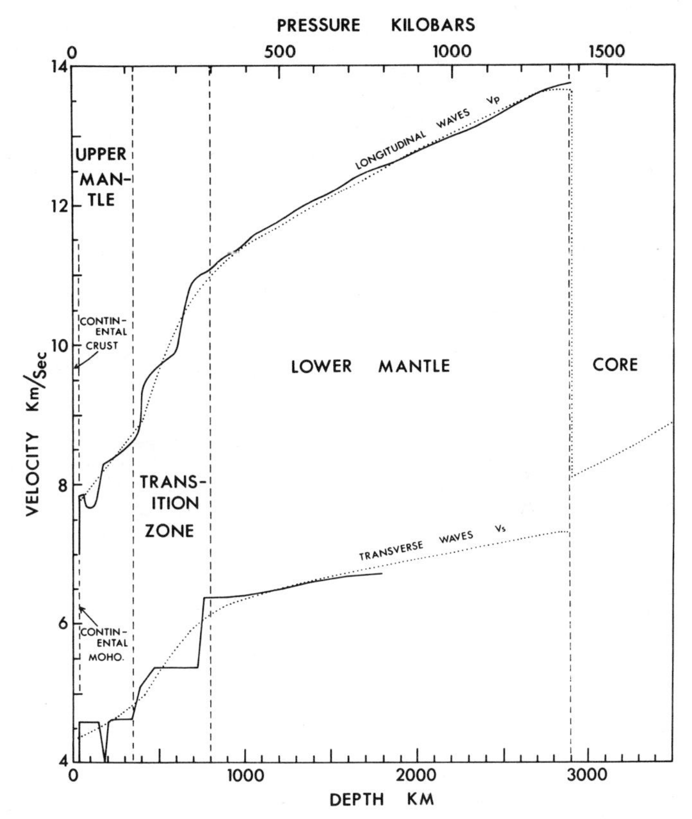

FIGURE 0-1
Seismic velocity distributions in the mantle. *P* waves—solid line: Johnson (1967, 1969); *S* waves—solid line: Nuttli (1969); broken lines—Jeffreys (1939).

cant modifications to these velocity distributions have since been proposed, the broad features remain unchanged. The velocity-depth profiles divide the earth into a number of well-defined regions, initially termed by Bullen[1] Regions A, B, C, D, E, F, and G. In the present case, we shall prefer a more descriptive terminology, as follows (Figs. 0-1 and 0-2).

The *crust* is defined as the region extending from the surface to the *Mohorovicic discontinuity*, which lies at a depth of 30 to 50 km under most continental regions and about 10 to 12 km under most oceanic regions. Below the crust and extending to a depth of about 400 km lies the *upper mantle*. This region is characterized by regional variations in velocity distributions, by the widespread presence of a velocity minimum (low-velocity zone), particularly for *S* waves, and by generally low velocity gradients, except perhaps at the boundaries of the low-velocity zone. The *transition zone* between 400 and 1000 km is characterized by

[1]Bullen (1947).

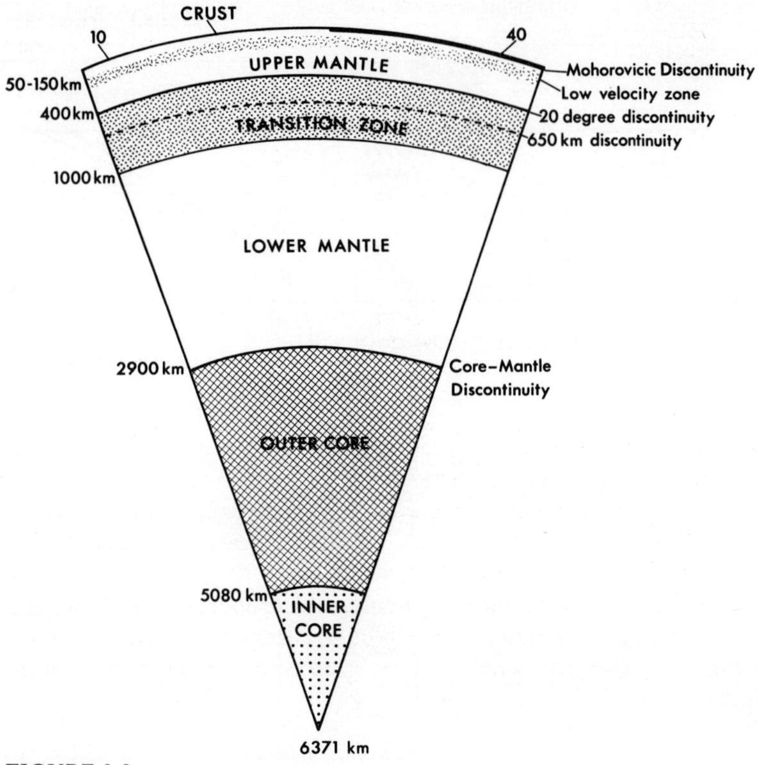

FIGURE 0-2
Principal subdivisions of the earth.

high velocity gradients on the average. Recent investigations[1,2] have shown that much of the total velocity increase in this region is concentrated in two restricted depth intervals around 400 and 650 km. The *lower mantle*, from 1000 km to 2900 km, is characterized by a moderate and relatively uniform increase of velocities with depth, except between 2700 and 2900 km, where velocities may decrease slightly. It will sometimes be convenient to refer to the combined regions of the transition zone and lower mantle as the *deep mantle*, 400 to 2900 km. The boundary of the earth's *core* is reached at a depth close to 2900 km. This is marked by a major first-order seismic discontinuity for *P* waves and by cessation of transmission of *S* waves. The core is divided into two regions, the *outer core* and the *inner core*, as shown in Fig. 0-2. The dimensions and masses of the individual regions are given in Table 0-1.

The present definition of the crust in terms of seismological parameters is in common usage. Nevertheless, confusion has resulted from the practice of some workers to define the crust more literally in terms of rheological properties. Ac-

[1]Niazi and Anderson (1965).
[2]Johnson (1967).

Table 0-1 DIMENSIONS AND MASSES OF THE INTERNAL LAYERS*

Region	Depth to boundaries (km)	Fraction of volume	Mass (10^{25} g)	Fraction of total mass	Fraction of mass of mantle
Crust †	0–Moho	0.008	2.4	0.004	0.006
Upper Mantle	Moho–400	0.16	62	0.10	0.15
Transition Zone	400–1000	0.22	100	0.17	0.24
Lower Mantle	1000–2900	0.44	245	0.41	0.6
Outer Core	2900–5100	0.154 }	189	0.32	
Inner Core	5100–6371	0.008 }			

*Based on Bullen (1947), Birch (1952).
†Based on Poldervaart (1955).

cording to this usage the crust refers to an outer, relatively rigid shell, possessing a substantial long-period strength and overlying a weaker mobile shell. Since the rheological properties of common mafic and ultramafic rocks are highly temperature-dependent, but comparatively insensitive to changes of (solid) phases and chemical composition, the Mohorovicic discontinuity does not in general correspond to a discontinuity in rheological properties. Following Daly[1] we will use the terms *lithosphere* for this outer rigid region and *asthenosphere* for the underlying weaker region which is widely believed to be identical with the low-velocity zone.

REFERENCES

BIRCH, F. (1939). The variation of seismic velocities within a simplified earth model in accordance with the theory of finite strain. *Bull. Seis. Soc. Am.* **29**, 463–479.

——— (1952). Elasticity and the constitution of the earth's interior. *J. Geophys. Res.* **57**, 227–286.

BULLEN, K.E. (1936). The variation of density and the ellipticities of strata of equal density inside the earth. *Mon. Not. Roy. Astron. Soc., Geophys. Supp.* **3**, 395–400.

——— (1940). The problem of the earth's density variation. *Bull. Seis. Soc. Am.* **30**, 235–250.

——— (1947). "*An Introduction to the Theory of Seismology.*" Cambridge, New York. 276 pp.

DALY, R.A. (1940). "*Strength and Structure of the Earth.*" Prentice-Hall, Englewood Cliffs, N.J. 434 pp.

GUTENBERG, B. (1951). PDDP, p¹p¹ and the earth's core. *Trans. Am. Geophys. Union* **32**, 373–390.

——— (1958). Velocity of seismic waves in the earth's mantle. *Trans. Am. Geophys. Union* **39**, 486–489.

———(1959a). "*Physics of the Earth's Interior.*" [International Geophysics Series, vol. 1, J. V. Mieghem (ed.).] Academic, New York. 240 pp.

——— (1959b). The asthenosphere low-velocity layer. *Ann. Geofis. Rome* **12**, 439–460.

[1]Daly (1940).

JEFFREYS, H. (1939). The times of P, S and SKS and the velocities of P and S. *Mon. Not. Roy, Astron. Soc., Geophys. Supp.* **4,** 498–533.

——— (1959). "*The Earth,*" 4th ed. Cambridge, New York. 420 pp.

JOHNSON, L.R. (1967). Array measurements of P velocities in the upper mantle. *J. Geophys. Res.* **72,** 6309–6325.

——— (1969). Array measurements of P velocities in the lower mantle. *Bull. Seism. Soc. Am.* **59,** 973–1008.

KUHN, W., and A. RITTMAN (1941). Über den Zustand des Erdinnern und seine Enstehung aus einem homogenen Urzustand. *Geol. Rundschau* **32,** 215–256.

NIAZI, M., and D.L. ANDERSON (1965). Upper mantle structure of western North America from apparent velocities of P waves. *J. Geophys. Res.* **70,** 4633–4640.

NUTTLI, O.W. (1969). Travel times and amplitudes of S waves from nuclear explosions in Nevada. *Bull. Seism. Soc. Am.* **59,** 385–398.

POLDERVAART, A. (1955). Chemistry of the earth's crust. In: A. Poldervaart (ed.), "*Crust of the Earth,*" pp. 119–144. *Geol. Soc. Am. Spec. Paper* 62.

RAMSEY, W.H. (1948). On the constitution of the terrestrial planets. *Mon. Not. Roy. Astron. Soc.* **108,** 406–413.

——— (1949). On the nature of the earth's core. *Mon. Not. Roy. Astron. Soc., Geophys. Supp.* **5,** 409–426.

PART ONE

Crust and Upper Mantle

1

MINERAL STABILITY FIELDS IN SOME KEY CRUSTAL ROCK TYPES

1-1 INTRODUCTION

Information on the constitution of the crust has been derived principally from geology and petrology, on the one hand, and from seismology and gravity studies, on the other. The combination of results from these fields has provided a reasonably direct understanding of the nature of the upper continental crust to depths of 15 to 20 km. However, our understanding of the nature of the lower continental crust is less direct. In recent years, supplementing the traditional sources of data, investigations in the field of experimental petrology have provided further constraints upon the range of compositions which are permissible in this region. In the present section, we shall review evidence bearing upon the subsolidus *P-T* stability fields of mineral assemblages displayed by rock compositions of particular significance in the deeper levels of the crust. The results in this field are of general importance and bear directly not only upon the nature of the lower crust but also upon such problems as the nature of the Mohorovicic discontinuity, the possible role of the gabbro-eclogite transformation in tectonic phenomena, and the quantitative characterization of metamorphic facies. The present review will concentrate principally upon investigations of complex natural rock compositions of particular relevance to broad geologic and geophysical problems, rather than upon simplified systems containing few components. The latter have been widely discussed in the literature. Although they yield rigorous data which are fundamental to the interpretation of more complex systems and provide a close approximation to the behaviour of some natural systems, there remain other important natural systems, particularly basalts, which can be studied more profitably by direct methods. For reasons of economy of space, the discussion is limited to the stability fields displayed by anhydrous rocks, for which relevant experimental data have become available only relatively recently. There is a vast literature dealing with the natural and experimental petrology of hydrous metamorphic rock systems, and several excellent texts already exist.[1]

We will commence by discussing experimental investigations on basaltic rocks. Pioneering work in this field was carried out by Boyd and England (1959) and Yoder and Tilley (1962). The former workers first succeeded in transforming a natural basalt to eclogite under experimental conditions, whilst the latter demonstrated that, for a number of basaltic compositions, the eclogitic mineral assemblage appeared stable above about 20 kbars (1200°C), whereas the gabbroic assemblage appeared stable up to about 10 kbars. Only a small number of experimental runs was carried out however, and no attempt was made to study the detailed phase relationships involved in the gabbro-eclogite transformation and the nature of the transitional mineral assemblages.

The first systematic experimental study of these topics was carried out by Ringwood and Green (1964, 1966) and Green and Ringwood (1967).[2] This was followed shortly by the work of Cohen et al. (1967) and Ito and Kennedy (1968, 1970). Some differences existed at first between results obtained by the two laboratories, and these caused confusion in applying the results to important geophysical and petrologic problems. A factual and well-balanced review by Wyllie[3] conveys the confused state of the subject which existed at the time. Fortunately, however, the reasons for the experimental discrepancies have since been largely clarified, [4,5,6] and it is now possible to present a consistent synthesis.

Because this topic has been somewhat controversial and because of their considerable importance to many petrological and geophysical problems, the experimental results from the two laboratories are discussed separately in some detail, particularly where differences have arisen.

1-2 EXPERIMENTAL INVESTIGATIONS ON THE GABBRO-ECLOGITE TRANSFORMATION

Results from the Canberra Laboratory

Over 300 experimental runs at high pressures and temperatures have been undertaken in order to elucidate the nature of the gabbro-eclogite transformation in

[1]E.g., Turner (1968), Winkler (1965).

[2]These results were also communicated in 1966 via the widespread circulation of "*Petrology of the Upper Mantle*," Publication 444, Department of Geophysics and Geochemistry, Australian National University, 252 pp.

[3]Wyllie (1971), chap. 5.

[4]Ito and Kennedy (1971).

[5]Green and Ringwood (1972).

[6]Kennedy and Ito (1972).

basaltic[1,2,3,4] and related[3,5,6,7] systems. The emphasis in these experiments[8] was to obtain a detailed knowledge of the petrological equilibria by means of closely spaced series of runs on a wide spectrum of basaltic compositions, with the further objective of elucidating the effects of changes of bulk chemical composition upon the equilibria. The compositions of the basalts which were investigated are shown in Table 1-1. Results on nonbasaltic compositions are discussed in Sec. 1-5.

The results of these experiments are shown in simplified form in Fig. 1-1, in which the principal phase assemblages found in the typical basalts as a function of pressure at 1100°C are displayed. Although the pressures required for appearance and disappearance of phases vary between individual basalts, there is an important qualitative resemblance between the sequence of phase assemblages displayed by all rocks. For each basalt there are three principal mineral stability fields corresponding closely with naturally observed mineral assemblages. The low-pressure assemblage is that of a gabbro or pyroxene granulite. It is characterized by the presence of pyroxene and plagioclase ± olivine ± quartz ± spinel, according to the particular bulk chemistry. Garnet is not present.

In each basaltic composition, as pressure increases, a point is reached at which garnet occurs in the mineral assemblage. With further increase in pressure, the proportion of plagioclase correspondingly decreases. Thus, we have a field of coexisting garnet, pyroxene, and plagioclase which may be termed the *garnet granulite field.* The pyroxenes in the garnet granulite field are rich in Al_2O_3 in the form of Mg and Ca Tschermak's molecules.[9] With further increase in pressure, these components are rejected and crystallize as garnet, and accordingly, the garnet/pyroxene ratio rises. Plagioclase becomes more sodic as it decreases in abundance. At the highest pressures, sodic plagioclase breaks down to form jadeite which enters into solid solution to form an omphacitic pyroxene. Free quartz may or may not be formed, depending upon the silica saturation of the rock. A further increase in the garnet/pyroxene ratio occurs as the Tschermak's molecule component of the pyroxene is reduced to low levels. These transformations mark the beginning of the eclogite mineral assemblage, characterized by the coexistence of pyrope-rich garnet and omphacitic pyroxene ± quartz.

Thus we see that, in all the basaltic compositions studied, the transformation

[1]Ringwood and Green (1964, 1966).

[2]Green and Ringwood (1967, 1972).

[3]T. Green (1967).

[4]Ringwood and Essene (1970).

[5]Green and Lambert (1965).

[6]T. Green and Ringwood (1968).

[7]T. Green (1970).

[8]The boundaries of the principal mineral reactions—appearance of garnet and disappearance of plagioclase—were successfully reversed at 1100°C. Changes in oxidation state of the charge and loss of iron to the platinum container during runs were investigated by chemically analyzing numerous runs after completion for FeO and Fe_2O_3. The loss of iron from normal basalt averaged 1.4% FeO during runs. An iron-loss of this magnitude is incapable of affecting the mineral stability fields of the rock to a significant degree. Likewise, the changes in oxidation state during runs were found to be of minor proportions.

[9]These are the pyroxene-type molecules $MgAl(AlSi)O_6$ and $CaAl(AlSi)O_6$ which are used to represent the solid solution of alumina in pyroxenes.

Table 1-1 CHEMICAL COMPOSITIONS AND CIPW NORMS OF BASALTIC COMPOSITIONS INVESTIGATED BY GREEN AND RINGWOOD (1967, 2—6), T. GREEN (1967, 1), AND RINGWOOD AND ESSENE (1970, 10)

	(1) High-alumina basalt	(2) Quartz tholeiite	(3) Alkali-poor quartz tholeiite	(4) Alkali olivine basalt	(5) Oxidized alkali olivine basalt	(6) Alkali-poor olivine tholeiite	(10) Apollo 11 basalt
SiO_2	49.9	52.2	49.9	45.4	45.4	46.2	41.6
TiO_2	1.3	1.9	2.1	2.5	2.5	0.1	10.3
Al_2O_3	17.0	14.6	13.9	14.7	14.7	14.5	10.6
Fe_2O_3	1.5	2.5	2.8	1.9	9.8	0.5	—
FeO	7.6	8.6	9.7	12.4	4.2	11.8	17.2
MnO	0.2	0.1	0.2	0.2	0.2	0.3	0.3
MgO	8.2	7.4	8.5	10.4	10.4	12.5	8.0
CaO	11.4	9.4	10.8	9.1	9.1	13.0	10.6
Na_2O	2.8	2.7	1.8	2.6	2.6	0.8	0.5
K_2O	0.2	0.7	0.1	0.8	0.8	—	0.2
P_2O_5	—	0.2	0.2	—	—	—	0.1
Cr_2O_3	—	—	—	—	—	0.2	0.6
Total	100.0	100.0	100.0	100.0	99.7	100.0	100.0
$\frac{100 \times Mg}{Mg + Fe^{++}}$	68	61	61	60	82	66	45
norms: CIPW							
Qz	—	2.5	2.8	—	—	—	—
Or	1.0	4.8	0.5	4.5	4.5	—	1.3
Ab	23.5	22.1	15.4	18.0	22.0	6.8	4.7
Ne	—	—	—	2.2	—	—	—
An	33.4	25.5	29.3	26.2	26.2	35.9	28.3
Di	18.9	17.1	19.6	15.7	15.0	23.5	25.7
Hy	9.4	20.6	23.7	—	6.8	9.6	21.9
Ol	9.3	—	—	25.8	8.4	23.0	17.0
Ilm	2.5	3.6	4.2	4.8	4.8	0.2	(rutile) 9.7
Mt	2.2	3.6	4.2	2.9	6.7	0.7	—
Ap	—	0.4	0.5	—	—	—	—
Haem	—	—	—	—	5.2	—	—
Chrom	—	—	—	—	—	0.3	1.0
Normative plagioclase	An_{59}	An_{52}	An_{64}	An_{58}	An_{54}	An_{83}	An_{82}

from gabbro or pyroxene granulite to eclogite proceeds through an intermediate mineral assemblage characterized by coexisting garnet, pyroxene(s), and plagioclase. This possesses a stability field ranging from 2 to 12.5 kbars at 1100°C in width and is identical with the natural garnet-clinopyroxene granulite subfacies recognizable in some metamorphic terrains.[1,2] The results plotted in Fig. 1-1 show that rather modest changes in chemical composition cause large changes in

[1]De Waard (1965).
[2]Green and Ringwood (1967).

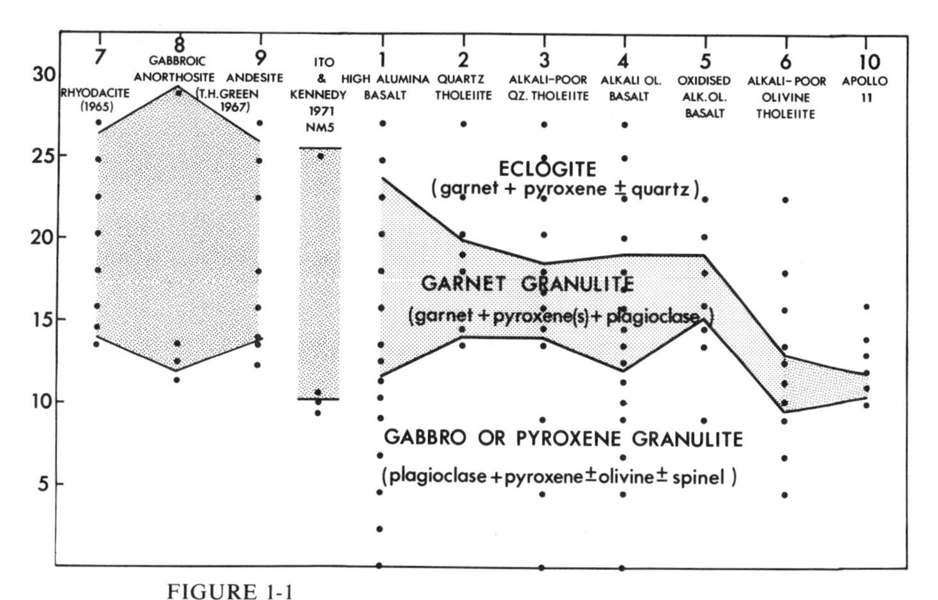

FIGURE 1-1
Diagram illustrating the effect of chemical composition on the pressure required for the incoming of garnet (lower boundary) and the outgoing of plagioclase (upper boundary) for some basaltic and nonbasaltic rock compositions. All data are at 1100°C. Each point represents an experimental run. Compositions 1 to 6 and 10 are from Table 1-1. Compositions 7, 8, and 9 are from references shown. (*From Ringwood and Green, 1966, and Green and Ringwood, 1972, with permission.*)

the pressures and width of the gabbro-eclogite transformation. Thus the pressures at which garnet first appears vary between 9.6 and 15.2 kbars (1100°C), whilst the pressure required to cause the final disappearance of plagioclase varies between 12.0 and 23.3 kbars. These wide variations in transition parameters are explicable in terms of the chemical and mineralogical equilibria involved in the transformation (Sec. 1-3). They are also of considerable geophysical significance (Sec. 2-4) and emphasize the hazards in attempting to draw far-reaching geophysical or petrological conclusions on the basis of an extensive study of a single rock composition.[1]

In order to apply the experimental results to the petrology and properties of the lower crust, it is necessary to extrapolate to lower temperatures, and for this the temperature gradient of the transformation must be established. Ringwood and Green (1966) determined the pressures of the gabbro-eclogite transformation in quartz tholeiites 2 and 3 over the temperature interval 1000 to 1250°C (Fig. 1-2). The slope of the plagioclase-out boundary in Fig. 1-2 is 27.5 bars/°C, and the garnet-in slope is 15 bars/°C. It was recognized that these gradients possessed substantial experimental uncertainties,[2] and accordingly, it

[1]E.g., Ito and Kennedy (1970).
[2]Ringwood and Green (1966, pp. 399-400).

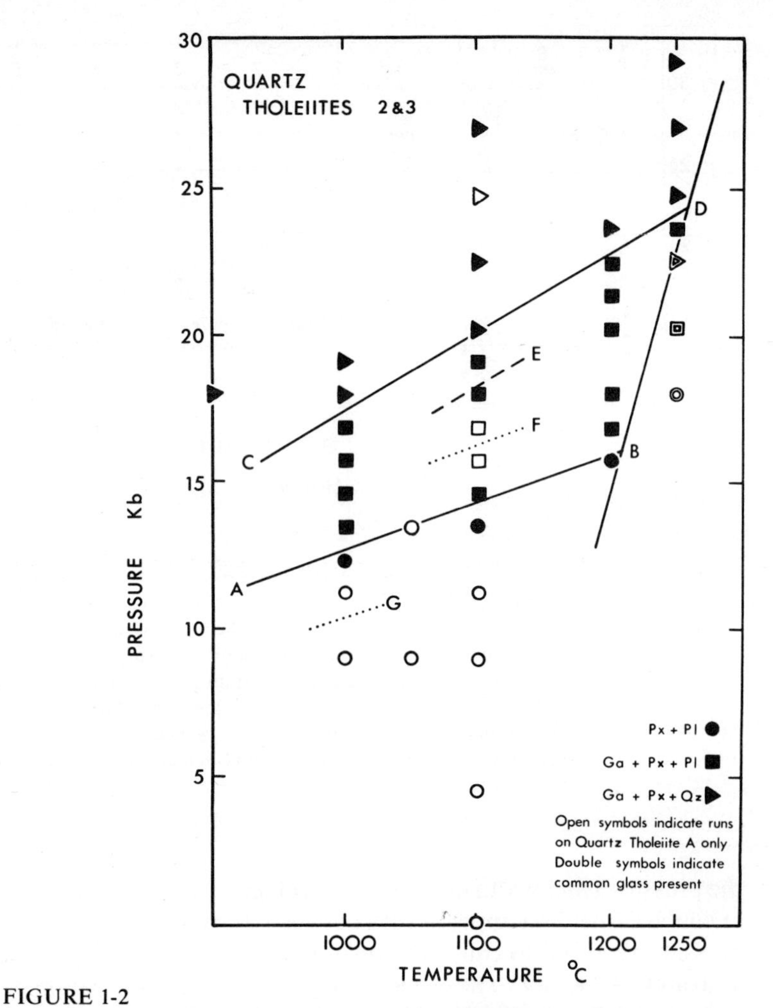

FIGURE 1-2
Mineral assemblages observed in quartz tholeiites [Table 1-1, (2) and (3)] over a range of temperature and pressure conditions. The eclogite field lies above CD and the pyroxene granulite field lies below AB. Area between AB and CD represents the garnet granulite stability field. Boundary BD is the solidus. Px = pyroxene, Pl = plagioclase, Ga = garnet, and Qz = quartz. Plagioclase is absent on the high-pressure side of CD in quartz tholeiite 2 and on the high-pressure side of the line E in quartz tholeiite 3 composition. The line G marks the appearance of garnet in highly iron-enriched quartz tholeiite. F marks the first appearance of garnet in magnesian quartz tholeiite. (*From Green and Ringwood, 1967, with permission.*)

was proposed to use the *average* of the two gradients for downward extrapolation, as well as the average of slopes of several simple mineralogical equilibria involved in the gabbro-eclogite transformation (Table 1-3, page 24). With these considerations in mind, the experimental transition intervals determined at

1100°C were extrapolated to lower temperatures, using the mean gradient of 21 bars/°C and a gradient for the plagioclase-out boundary of 24 bars/°C.

Experimental Results from the UCLA Laboratory

A study of the gabbro-eclogite transformation in a high-alumina basalt with a composition resembling an oceanic tholeiite was reported by Cohen, Ito, and Kennedy (1967). The composition and norm of this basalt (NM5) are given in Table 1-2. The composition is seen to compare most closely with the high-alumina basalt of Table 1-1. Most of the experimental runs were concerned with solid-liquid equilibria. A small number, however, was stated to represent subsolidus conditions, and it was claimed[1,2] that the transition interval between gabbro and eclogite assemblages was only 3 to 4 kbars wide near 1200°C compared with the 12-kbar interval found for the high-alumina basalt by the Canberra laboratory (Fig. 1-1). In a subsequent, more detailed study of NM5, however, Ito and Kennedy (1970) concluded that their earlier results had been plagued by failure to achieve equilibrium, and they concluded that the garnet-granulite transition interval was 10 kbars wide on the basis of reversed experimental determinations of phase boundaries.

Still more recently, Ito and Kennedy (1971) returned to NM5 to discover that nearly all their previously stated subsolidus runs on the gabbro-eclogite transformation had been carried out above the solidus. In the vicinity of the transition interval, the key runs reported earlier had been carried out about 70°C above their newly determined solidus. As a result, the first appearance of garnet was greatly delayed and plagioclase disappeared much too early, accounting for the small tran-

[1]Cohen, Ito, and Kennedy (1967).
[2]Ito and Kennedy (1968).

Table 1-2 CHEMICAL COMPOSITION AND CIPW NORM OF BASALT NM5 INVESTIGATED BY COHEN, ITO, AND KENNEDY (1967) AND ITO AND KENNEDY (1970, 1971)

Composition		CIPW norm	
SiO_2	49.9		
TiO_2	1.3	Qz	
Al_2O_3	16.8	Or	2.2
FeO	11.4	Ab	24.6
MnO	0.2	An	31.4
MgO	7.6	Di	11.5
CaO	9.3	Hy	13.2
Na_2O	2.9	Ol	14.1
K_2O	0.4	Ilm	2.6
P_2O_5	0.2	Ap	0.4
$\frac{100\ Mg}{Mg + Fe}$	54		

sition width reported by Cohen et al. (1967). The new transition interval at 1100°C was found to be 15 kbars. The newly determined solidus was found to be 150°C below the earlier solidus at 30 kbars. Their new solidus, in fact, agreed closely with the solidus determined by Ringwood and Green (1966) for the quartz tholeiite (Fig. 1-2) and with solidi determined by T. Green and Ringwood (1968) for several natural garnet-pyroxene-quartz systems.

The results (Fig. 1-3) of Ito and Kennedy (1971) appear to be definitive in establishing the position of the gabbro-eclogite transformation in NM5 and supersede their previous results on this rock. An important contribution was their success in obtaining reversals of equilibria at temperatures down to 800°C, thereby yielding a more accurate gradient (20 bars/°C for the plagioclase-out reaction) than was used by Ringwood and Green (24 bars/°C).

In addition to the results on NM5 discussed above, the UCLA laboratory has provided some reconnaissance results on the gabbro-eclogite transformation in some highly alkalic and olivine-rich rocks[1,2] and in compositions rich in Al_2O_3.[2,3]

[1] Ito and Kennedy (1968).
[2] Godovikov and Kennedy (1969).
[3] See also T. Green (1969).

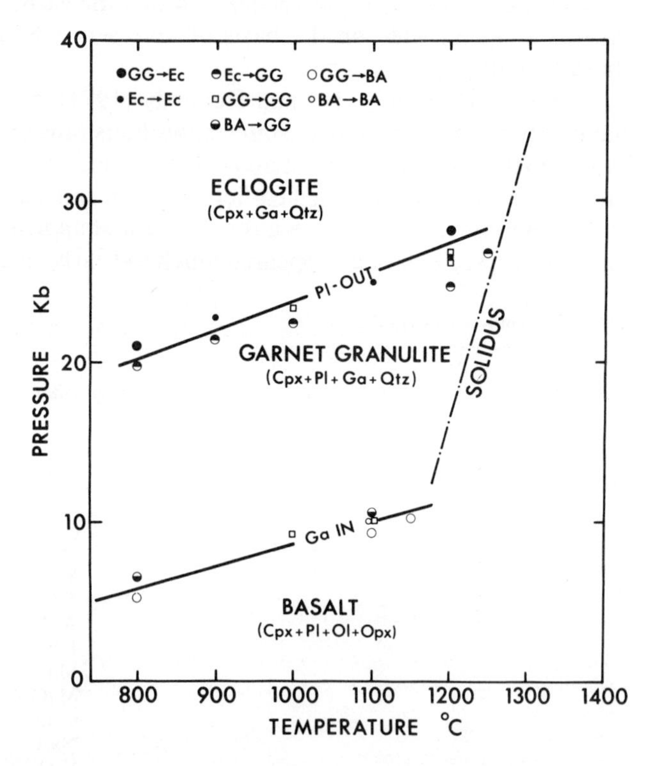

FIGURE 1-3
Phase relationships displayed by basaltic composition NM5 over a range of pressures and temperatures. Abbreviations: BA = basalt, GG = garnet granulite, Ec = eclogite, Ga = garnet, Cpx = clinopyroxene, Opx = orthopyroxene, Pl = plagioclase, Ol = olivine, Qtz = quartz. (*After Ito and Kennedy, 1971.*)

Discussion

The latest results of Ito and Kennedy on NM5 are compared in Fig. 1-1 with Canberra results on a range of compositions. The transitional interval in NM5 is seen to be generally similar although somewhat wider than in the Canberra high-alumina basalt. This is explicable in terms of the mineralogical equilibria involved in the transition.[1] The entry of garnet at slightly lower pressure is a consequence of the higher iron content of olivine in NM5. In this rock, garnet is formed by the reaction of olivine + plagioclase, and as expected, the pressure for this equilibrium is similar to that found in the corresponding reaction in the olivine tholeiite (No. 6, Fig. 1-1). Likewise, the slightly higher pressure required for elimination of plagioclase in NM5 is the result of compositional differences[1] and is seen to be similar to the pressure required for disappearance of plagioclase in the andesite composition (No. 9, Fig. 1-1). It is instructive to compare the results on NM5 with those of the quartz tholeiite (No. 2, Fig. 1-1) at 1100°C and 20 kbars. Kennedy and Ito (1971) found that NM5 crystallized to a "plagioclase eclogite" under these conditions, containing 9.3% plagioclase, whereas Ringwood and Green found that the quartz tholeiite crystallized to a quartz eclogite lacking plagioclase. When 9.3% plagioclase is subtracted from the composition of NM5, the residual composition is found to be very similar to that of quartz tholeiite No. 2. Thus the *eclogite component* of NM5 at 1100°C, 20 kbars is essentially identical with the quartz eclogite formed from quartz tholeiite 2 under the same *P-T* conditions.[1]

There can be no reasonable doubt that, despite Kennedy and Ito's (1971) statements to the contrary, their latest results on NM5 are entirely concordant with the earlier Canberra work on a wide range of basalt compositions. The most important result of the recent work on NM5 is the evidence that a gradient of 20 bars/°C is more appropriate in extrapolating the garnet granulite-eclogite boundary to lower temperatures than the 24 bars/°C used by Ringwood and Green (1966). This revised gradient, supported by data on simple systems, is adopted in subsequent discussions.

1-3 PETROLOGICAL APPLICATIONS

Mineralogical Equilibria Involved in the Gabbro-Eclogite Transformation

The general characteristics of the transition interval between gabbro and eclogite fields have already been discussed (Sec. 1-2, Fig. 1-1), and the changes in mineralogy across the transformations in two basaltic compositions are illustrated in Figs. 1-4 and 1-5. We consider now some of the detailed mineralogical equilibria and the reasons for the wide variation in transition parameters as a function of bulk composition.

The principal reactions in the pyroxene granulite field of Fig. 1-1 can be

[1]Discussed in detail by Green and Ringwood (1972).

represented by the following simplified equations.[1,2] (In these equations, Fe^{++} may be substituted in varying degrees for Mg^{++}.)

1 Oversaturated compositions (characterized by the presence of free quartz):

$$\underset{\text{anorthite}}{CaAl_2Si_2O_8} + m\underset{\text{enstatite}}{MgSiO_3} \rightleftharpoons$$

$$\underset{\text{diopside}}{CaMgSi_2O_6} + \underset{\text{aluminous enstatite}}{(m-2)MgSiO_3 \cdot MgAl_2SiO_6} + \underset{\text{quartz}}{SiO_2} \qquad (I)$$

$$\underset{\text{anorthite}}{CaAl_2Si_2O_8} + m\underset{\text{diopside}}{CaMgSi_2O_6} \rightleftharpoons \underset{\text{aluminous diopside}}{mCaMgSi_2O_6 \cdot CaAl_2SiO_6} + \underset{\text{quartz}}{SiO_2} \qquad (II)$$

2 Undersaturated compositions (no free quartz):

$$2\underset{\text{olivine}}{Mg_2SiO_4} + 2\underset{\text{anorthite}}{CaAl_2Si_2O_8} \rightleftharpoons$$

$$\underset{\text{aluminous diopside}}{CaMgSi_2O_6 \cdot CaAl_2SiO_6} + \underset{\text{aluminous enstatite}}{2MgSiO_3 \cdot MgAl_2SiO_6} \qquad (III)$$

$$2\underset{\text{olivine}}{Mg_2SiO_4} + \underset{\text{anorthite}}{CaAl_2Si_2O_8} \rightleftharpoons \underset{\text{diopside}}{CaMgSi_2O_6} + 2\underset{\text{enstatite}}{MgSiO_3} + \underset{\text{spinel}}{MgAl_2O_4} \qquad (IV)$$

These reactions may be used to subdivide the pyroxene granulite field into two regions termed the *low-pressure granulite field* (gabbroic field) and the *intermediate-pressure granulite field.* The low-pressure granulites are characterized by the stable coexistence of plagioclase and olivine, whereas in the intermediate-pressure granulites, these minerals are incompatible—reactions (III) and (IV).[2]

In each basaltic composition, as pressure increases, a point is reached at which garnet appears in the mineral assemblage. With further increase of pressure, the proportion of garnet steadily increases, whilst the proportion of plagioclase correspondingly decreases. Thus we have here a field of coexisting garnet, pyroxene, and plagioclase.

The principal reactions occurring in the garnet granulite field of Fig. 1-1 are:

1 Oversaturated compositions:
Garnet forms by two different types of reaction:

(i) $$4\underset{\text{enstatite}}{MgSiO_3} + \underset{\text{anorthite}}{CaAl_2Si_2O_8} \rightleftharpoons \underset{\text{pyrope garnet}}{Mg_3Al_2Si_3O_{12}} + \underset{\text{diopside}}{CaMgSi_2O_6} + \underset{\text{quartz}}{SiO_2} \qquad (V)$$

(ii)

$$\underset{\text{aluminous enstatite}}{mMgSiO_3 \cdot MgAl_2SiO_6} \rightleftharpoons \underset{\text{pyrope garnet}}{Mg_3Al_2Si_3O_{12}} + \underset{\text{enstatite}}{(m-2)MgSiO_3} \qquad (VI)$$

$$\underset{\text{aluminous diopside}}{mCaMgSi_2O_6 \cdot CaAl_2SiO_6} + 2\underset{\text{enstatite}}{MgSiO_3} \rightleftharpoons$$

$$\underset{\text{pyrope-grossular}}{CaMg_2Al_2Si_3O_{12}} + m\underset{\text{diopside}}{CaMgSi_2O_6} \qquad (VII)$$

[1]Kushiro and Yoder (1966).
[2]Green and Ringwood (1967).

Owing to selective reaction of anorthite, the residual plagioclase becomes more albite-rich with increasing pressure. The final disappearance of Na-rich plagioclase is represented by:

$$\underset{\text{albite}}{NaAlSi_3O_8} + \underset{\text{diopside}}{mCaMgSi_2O_6} \rightleftharpoons \underset{\text{omphacite}}{mCaMgSi_2O_6 \cdot NaAlSi_2O_6} + \underset{\text{quartz}}{SiO_2} \quad \text{(VIII)}$$

2 Undersaturated compositions:

$$\underset{\text{anorthite}}{CaAl_2Si_2O_8} + \underset{\text{enstatite}}{4MgSiO_3} + \underset{\text{spinel}}{MgAl_2O_4} \rightleftharpoons \underset{\text{garnet}}{2Ca_{0.5}Mg_{2.5}Al_2Si_3O_{12}} \quad \text{(IX)}$$

$$\underset{\text{albite}}{NaAlSi_3O_8} + \underset{\text{enstatite}}{2MgSiO_3} + \underset{\text{spinel}}{MgAl_2O_4} \rightleftharpoons \underset{\text{jadeite}}{NaAlSi_2O_6} + \underset{\text{garnet}}{Mg_3Al_2Si_3O_{12}} \quad \text{(X)}$$

We see from Fig. 1-1 that garnet appears at lower pressures in the undersaturated olivine basalts. This is caused in part by the reaction (IX) of spinel and plagioclase and also by the fact that the reaction (III) of olivine with plagioclase at relatively low pressures results in the formation of pyroxenes which are more highly aluminous in the case of the undersaturated basalts than for oversaturated basalts at the same *P-T* conditions. With increasing pressure, the aluminous pyroxenes of undersaturated compositions yield garnet according to reactions (VI) and (VII) before this reaction becomes important in the oversaturated basalts.

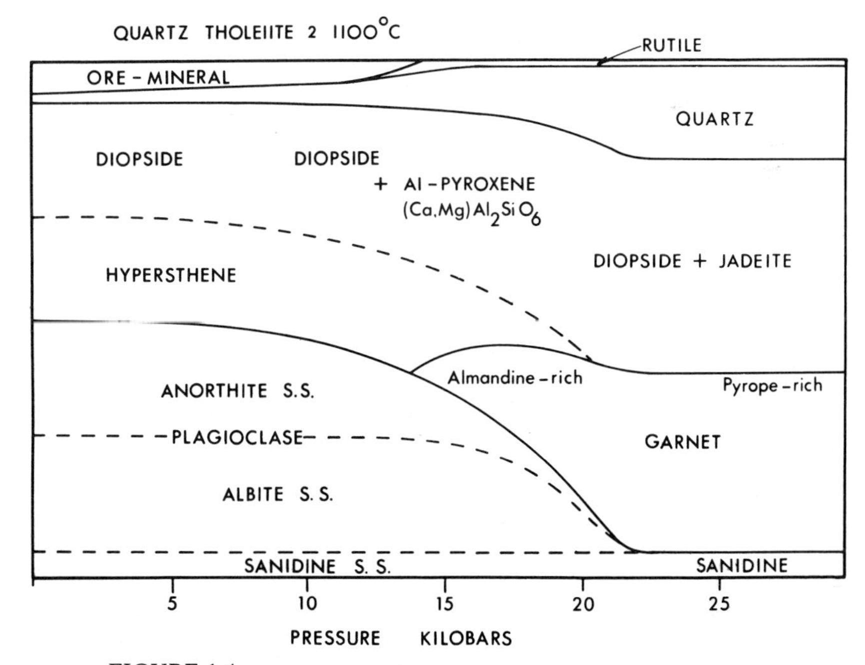

FIGURE 1-4
Approximate proportions of mineral phases present in quartz tholeiite Table 1-1 (2) in the pressure range 0 to 30 kbars at 1100°C. (*From Green and Ringwood, 1967, with permission.*)

With further increase of pressure, all rocks (Fig. 1-1) reach the eclogite stability field, characterized by coexisting garnet, omphacite ± quartz. In this field the omphacite possesses a high ratio of jadeite to Tschermak's molecule,[1] whilst plagioclase is absent in rocks of normal basaltic composition. (However, it may be present in rocks of nonbasaltic composition in eclogite facies.[2])

The results plotted in Fig. 1-1 show that rather modest changes in chemical composition cause large changes in the positions of the field boundaries. The reasons for the variation are readily explicable in terms of the chemical and mineralogical equilibria involved[3]—reactions (I) to (X). The pressures required for the incoming of garnet are smaller in undersaturated rocks (Fig. 1-1, Table 1-1, Nos. 1, 4, 6, and 10)[4] than in oversaturated rocks (Table 1-1, Nos. 2 and 3), whilst the pressure required for the final disappearance of plagioclase is decreased in basaltic compositions which are poorer than usual in soda (Fig. 1-1, Table 1-1, Nos. 3, 6, and 10). Changes in the FeO/MgO ratio have an important influence over the pressure required for incoming of garnet,[3] these pressures being smaller the higher the FeO/MgO ratio of the rock. Changes in oxidation state do not affect the pressure at which plagioclase disappears. However, the entry of garnet requires higher pressure in highly oxidized rocks (Table 1-1, No. 5).

For many purposes connected with geophysical applications the simplified threefold subdivision of basalt stability fields (Fig. 1-1) into pyroxene granulite, garnet granulite, and eclogite is convenient. However, in petrological applications of the results, it is more appropriate to use a subdivision based upon particular mineral reactions as follows.[5,3]

1 Low-pressure granulites Characterized by the stable association of olivine and plagioclase.

2 Intermediate-pressure granulites Characterized by the stable association of orthopyroxene + plagioclase and incompatibility of olivine + plagioclase.

3 High-pressure granulites Characterized by the stable association of garnet + clinopyroxene + plagioclase + quartz and incompatibility of hypersthene and plagioclase. This field is equivalent to the garnet-clinopyroxene-granulite subfacies of de Waard.[5]

4 Eclogites Characterized by the association of garnet, omphacite ± quartz, in rocks of basaltic composition, the omphacite having a high ratio of jadeite to Tschermak's molecules.

These fields are clearly defined by the experimental data between 1000 and 1250°C. The mineral assemblages observed experimentally in this temperature range correspond closely to the mineral assemblages observed in naturally occur-

[1]White (1964).

[2]Green and Lambert (1965).

[3]A more detailed discussion is given by Green and Ringwood (1967).

[4]The degree of saturation of Apollo 11 basalt (No. 10, Fig. 1-1) depends upon whether titania is calculated as rutile or ilmenite.

[5]De Waard (1965).

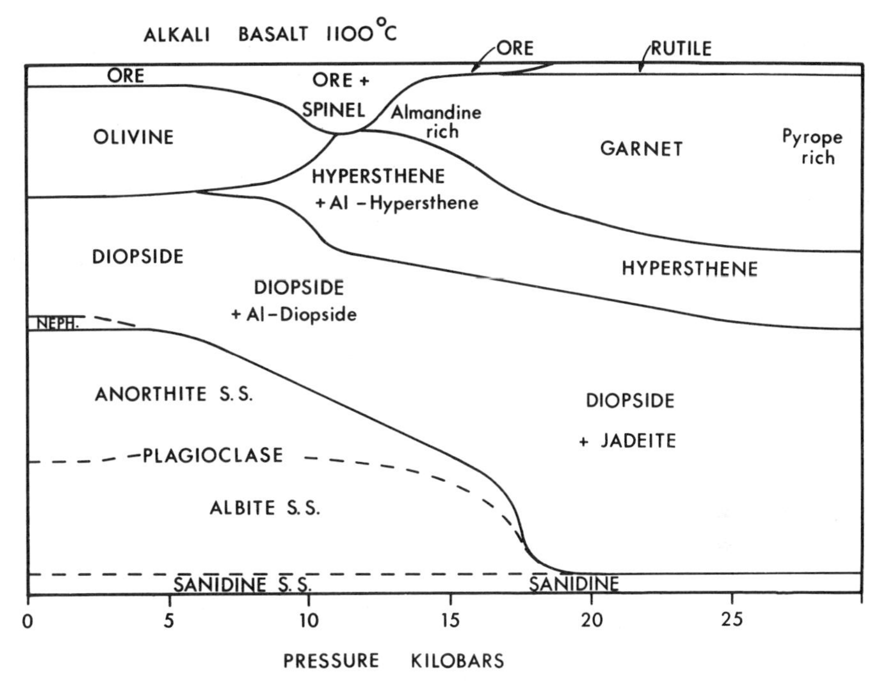

FIGURE 1-5
Approximate proportions of mineral phases present in alkali basalt Table 1-1 (4) in the pressure range 0 to 30 kbars at 1100°C. (*From Green and Ringwood, 1967, with permission.*)

ring granulites and eclogites which were probably established at temperatures in the vicinity of 500 to 800°C. A detailed review of the comparison between experimental and natural assemblages is given elsewhere.[1] The close correspondence between experimental and natural equilibria justifies extrapolation of experimental results to lower temperatures in order to interpret the conditions of formation of natural granulites and eclogites.

Extrapolation of High-Temperature Gabbro-Eclogite Transformation Boundaries to Low Temperatures

The average gradient of the garnet granulite-eclogite boundary in NM5 obtained by Ito and Kennedy (1971) over the temperature interval 800 to 1250°C was 20 bars/°C (Fig. 1-3). Support for the general applicability of this gradient comes from a study of gradients observed in simple systems which are closely related to the equilibria involved in the complex basaltic systems—e.g., the breakdown of sodic plagioclase to form jadeite + quartz (jadeite enters into solid solution in pyroxene) and the breakdown of aluminous pyroxenes to form garnet and low-

[1]De Waard (1965).

alumina pyroxenes. In many cases these gradients have been determined over a wider range of temperatures than for the basaltic system and are therefore probably more reliable. A list of gradients of other equilibria which are closely related to the granulite-eclogite transformations is given in Table 1-3. The average gradient of the 10 equilibria is also 21 bars/°C, which is almost identical with the gradient obtained for the natural system (Fig. 1-3). Thus the use of a gradient in the vicinity of 20 bars/°C for extrapolating garnet granulite-eclogite boundaries in the complex system to lower temperatures is strongly supported. The gradients of the gabbro (pyroxene-granulite) to garnet granulite boundary determined by Ringwood and Green (1966)—15 bars/°C—and Ito and Kennedy (1971)—14 bars/°C—are in close agreement.

The granulite field boundaries observed in quartz tholeiite have been extrapolated to lower temperatures and pressures using these gradients in Fig. 1-6. It is seen that the temperature on the garnet granulite-eclogite boundary at a pressure corresponding to the base of the normal continental crust is 600°C. If the temperature at the base of the crust is lower than 600°C, then eclogite would be

Table 1-3 GRADIENTS OF SIMPLE EQUILIBRIA RELATED TO THE GARNET GRANULITE-ECLOGITE TRANSITION*

No.	Equilibrium	*dp/dt* (bars/°C)†
1	Albite = jadeite + quartz	20
2	Albite + diopside = omphacite (40% Jd) + quartz	19
3	Aluminous enstatite (6% Al_2O_3) = enstatite (< 6% Al_2O_3) + pyrope	34
4	Albite + nepheline = 2 jadeite	18
5	Clinopyroxene + orthopyroxene + anorthite + spinel = garnet + clinopyroxene (Overall composition: 1 forsterite + 1 anorthite)	16
6	Clinopyroxene + orthopyroxene + forsterite + spinel = garnet + forsterite + clinopyroxene (Overall composition: 1 forsterite + 2 anorthite)	25
7	Clinopyroxene + orthopyroxene + quartz = garnet + quartz + pyroxene (Overall composition: 2 enstatite + anorthite)	10?‡
8	4 enstatite + spinel = forsterite + pyrope	17
9	Anorthite + wollastonite = grossularite + quartz	26
10	Anorthite = grossularite + kyanite + quartz	24
11	Tschermak's pyroxene = grossularite + corundum	64

*References for reactions:

1 Birch, F., and P. Le Comte (1960). *Am. J. Sci.* **258**, 209.
2 Kushiro, I. (1965). *Carnegie Inst. Washington Yearbook* **64**, 112.
3 Boyd, F., and J. England (1964). *Carnegie Inst. Washington Yearbook* **63**, 157.
4 Robertson, E.C., F. Birch, and G. J. F. MacDonald (1957). *Am. J. Sci.* **255**, 115.
5,6 Kushiro, I., and H. S. Yoder (1965). *Carnegie Inst. Washington Yearbook* **64**, 89.
7 Kushiro, I., and H. S. Yoder (1964). *Carnegie Inst. Washington Yearbook* **63**, 108.
8 MacGregor, I. D. (1964). *Carnegie Inst. Washington Yearbook* **63**, 156.
9–11 Hays, J. F. (1967). *Carnegie Inst. Washington Yearbook* **65**, 234.

†Average gradient for reactions 1 to 10 is 21 bars/°C. (Inclusion of the extreme gradient No. 11 would only serve to strengthen the conclusions stated in the text.)

‡ This gradient is based on very limited experimental data and has a large uncertainty.

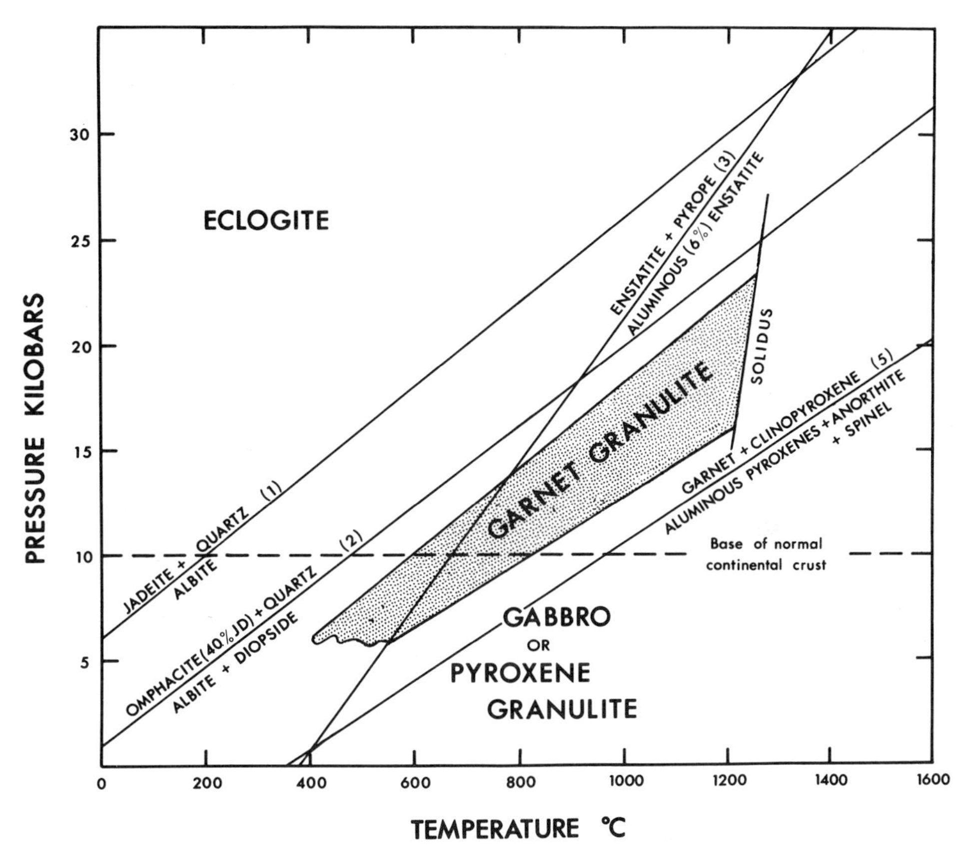

FIGURE 1-6
Extrapolated stability fields of eclogite, garnet granulite, and pyroxene granulite-gabbro for the quartz tholeiite (2) composition. The gradient of the eclogite to garnet granulite boundary is 20 bars/°C, and the gradient of the gabbro to garnet granulite boundary is 15 bars/°C.

the stable mineral assemblage for a basalt of this composition throughout the continental crust.

The temperature distribution in the crust is determined principally by the surface heat-flux and the depth distribution of radioactivity. Studies of numerous models[1,2,3,4,5] (Fig. 2-3) show that, in stable continental regions of normal crustal thickness characterized by mean heat-flows of 0.8 to 1.2 μcal/cm^{2}sec, the temperature at the base of the crust is usually less than 600°C for most reasonable assumptions concerning radioactivity distribution. This conclusion is particularly

[1]Birch (1955).
[2]Clark (1961, 1962).
[3]Clark and Ringwood (1964).
[4]Hyndman, Lambert, Heier, Jaeger, and Ringwood (1968).
[5]Roy, Blackwell, and Birch (1968).

applicable to Precambrian shields characterized by mean heat-flows of $< 1.0\ \mu cal/cm^2$ sec, where the temperature at the base of the crust may be less than 450°C[1,2,3] (Fig. 2-3). Such temperature distributions, taken in conjunction with the experimental results discussed above, indeed imply that eclogite is the stable modification of quartz tholeiite throughout very large regions of continental crust, under anhydrous conditions. In five out of seven cases (Fig. 1-1), experiments on other basalts showed that they required *smaller* pressures to reach the eclogite stability field than were required in the case of the quartz tholeiite. Thus the conclusion that eclogite is the stable mineral assemblage throughout much of the normal continental crust for many rocks of basaltic composition is generalized and reinforced. This result may be regarded as particularly well established in the cases of undersaturated basaltic rocks possessing low contents of alkalis. For example, extrapolation of the eclogite-garnet granulite boundary of the olivine tholeiite (Fig. 1-1, No. 6) at 20 bars/°C yields a temperature of 950°C at pressure of 10 kbars. This temperature is believed to be higher than the temperature at a depth of 35 km in almost all continental regions, and accordingly, this rock composition would crystallize as eclogite under almost all conceivable ranges of P-T conditions in the (anhydrous) continental crust.

The above considerations may be somewhat surprising since they imply that the basalts, dolerites, and gabbro which are commonly observed in the upper crust may be highly metastable relative to the P-T conditions of their environment.

It is important to consider the bearing of geological field evidence upon these topics. Eclogites occurring in the crust clearly have more than one mode of origin. For example, those occurring in diamond pipes appear to be derived from the mantle. However, there is another important class of eclogite occurrences as small lenses, inclusions, and sometimes as conformable bodies in regional metamorphic rocks of amphibolite, greenschist, and glaucophane-schist facies. Many petrologists[4,5,6,7,8,9,10,11] have interpreted these eclogites as being formed in situ in the crust, by metamorphism of basaltic rocks under conditions of low water-vapour pressure, and in some cases, by localized shear stress. This is in complete harmony with the experimental evidence.

The survival of gabbroic mineral assemblages in the crust is attributed, on the other hand, to kinetic considerations. Mafic rocks emplaced in the crust originally as magmas complete their crystallization above 1000°C to stable gabbroic or pyroxene granulite assemblages. On cooling in a normal crustal environment,

[1]Clark and Ringwood (1964).
[2]Hyndman, Lambert, Heier, Jaeger, and Ringwood (1968).
[3]Roy, Blackwell, and Birch (1968).
[4]Backlund (1936).
[5]Kozlowski (1959).
[6]Smulikowski (1960).
[7]Bearth (1959).
[8]Korzhinsky (1937).
[9]Coleman, Lee, Beatty, and Brannock (1965).
[10]Bryhni, Green, Heier, and Fyfe (1970).
[11]Green and Mysen (1972).

such mafic rocks would pass through the garnet granulite stability field and often into the eclogite stability field. Whether or not a given mafic rock will actually undergo these transformations depends upon kinetic factors, particularly the pressure-temperature-time path followed and the degree of shear stress or deformation encountered. This question is discussed further in Sec. 2-1.

Referring to Fig. 1-1 we see that the high-alumina basalts (1) and NM5 require substantially higher pressures to eliminate plagioclase than are found for other basalts. This characteristic is also shared by other compositions high in Al_2O_3, such as the gabbroic anorthosite and andesite.[1,2] Actually, from the strict petrological point of view, these compositions attain the eclogite facies (characterized by coexisting garnet and omphacite low in Tschermak's molecule ± quartz) at a pressure of about 20 kbars (1100°C), and the mineral assemblage at higher pressures but preceding elimination of plagioclase is best called a *plagioclase eclogite*.[3,4] This mineral assemblage [omphacite + garnet + felspar(s) ± quartz] would also be stable under anhydrous conditions throughout the normal continental crust in regions possessing low heat-flows.

Large areas of continental crust are known to possess heat-flows in the range 1.5 to 2.0 hfu. Geothermal calculations show that the temperature at a depth of 35 km in such regions may often be much higher than 600°C. In such regions, the stable anhydrous mineral assemblage for most basaltic compositions would be that of garnet granulite. These topics are taken up again in Chap. 2.

1-4 CHANGES IN PHYSICAL PROPERTIES ASSOCIATED WITH THE TRANSFORMATION

Green and Ringwood (1967) studied the changes in mineralogy as several basalts transformed to eclogite. For a series of closely spaced runs, the proportions of phases present were estimated by comparison of powder patterns and diffractometer records with specially prepared standards (Figs. 1-4 and 1-5). In 4 out of 5 basalts studied, they found that, as pressure increased through the garnet granulite field, the proportion of garnet steadily increased, whilst the proportion of plagioclase steadily decreased, and they concluded "to a first approximation, the changes in mineralogy across the garnet granulite zone occur at a uniform rate and accordingly the corresponding changes in seismic velocity and density will also change regularly throughout the garnet granulite zone." This latter prediction was checked by Green (1967), who measured the densities of a series of runs as the alkali-olivine basalt transformed to eclogite (Fig. 1-7). Because of the linear relationship between P velocity and density,[5] seismic wave velocity can be expected to vary similarly across the transition region.

[1]T. Green (1967, 1970).
[2]Godovikov and Kennedy (1969).
[3]Green and Ringwood (1967).
[4]Ito and Kennedy (1971).
[5]Birch (1961).

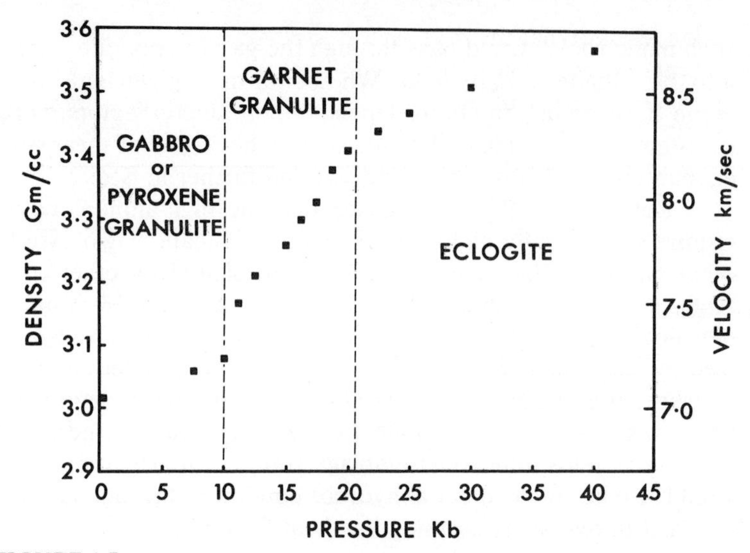

FIGURE 1-7
Densities of experimental charges of alkali basalt (4, Fig. 1-1) over a range of pressures at 1100°C. Possible errors are similar to those noted in Fig. 1-8. (*From Green, 1967.*)

The low-alkali olivine tholeiite (Fig. 1-1) constituted a partial exception[1] to this generalization. The incoming of garnet and the disappearance of plagioclase occurred within a pressure interval of only 3.4 kbars, with much of the change occurring within 2.3 kbars.[2] However, at the point where plagioclase disappeared, the rock contained only 30% garnet and the pyroxene was highly aluminous. With increase of pressure between 12.4 and 22.5 kbars, the aluminous components of the pyroxene exsolved to form further garnet, which increased to 55%. The density of this rock thus increased relatively rapidly from 3.10 to 3.38 g/cm^3 between 10.1 and 12.4 kbars, followed by a further, more gradual increase to 3.61 g/cm^3 between 12.4 and 22.5 kbars.

The density variations through the gabbro-eclogite transformation in NM5 and in a variant of NM5 containing 2% additional Al_2O_3 were also investigated by Ito and Kennedy (1970, 1971). They concluded that "the transition consists of two sharp density discontinuities. The discontinuities are so sharp that seismic wave reflections might well be detected from the phase boundaries."

The actual experimental data of Ito and Kennedy (1971) are plotted in Fig. 1-8, together with possible error bars.[3] There does indeed appear to be a significant inflection of the density-pressure relationship between 14 and 18 kbars. (A corresponding indication appears on Fig. 1-7.) Nevertheless, it is a gross exagger-

[1]The Apollo 11 basalt (Fig. 1-1) behaves generally similarly to the low-alkali olivine tholeiite. This is to be expected since the composition of the Apollo 11 basalt is equivalent to that of the low-alkali olivine tholeiite plus 10% rutile.
[2]Ringwood and Green (1966, table 4).
[3]Green and Ringwood (1972).

ation to state that the data of Fig. 1-8 depict the existence of two abrupt density discontinuities. The first claimed "discontinuity" is seen to be spread out over a pressure interval of at least 5 kbars (equivalent to a depth interval of 17 km), whilst the second "discontinuity" is at least 8 kbars wide (28 km).

The bearing of these experimental results on the nature of the Mohorovicic discontinuity is discussed in Sec. 2-4.

1.5 STABILITY FIELDS IN ACID-INTERMEDIATE ROCKS

By comparison with mafic rocks, much less work has been carried out in determining the *P-T* stability fields of mineral assemblages in acid and intermediate

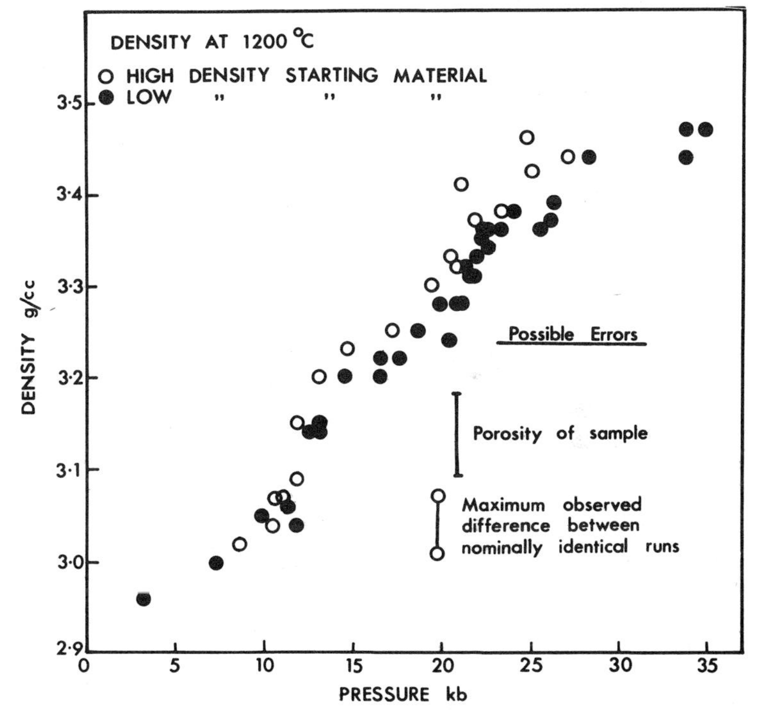

FIGURE 1-8
A replot of Ito and Kennedy's (1971) data (fig. 3) on densities of experimental charges of NM5 at 1200°C under various pressures. The data points are shown with a size representing an uncertainty of ±0.01 in density. Possible errors are due to measurement using very small samples and to the estimated porosity (Ito and Kennedy, 1971) and unknown permeability of the samples. Additional errors may have resulted from the necessity to grind off the outer iron-enriched layers of the sample (Ito and Kennedy, 1970) and from partial melting (the solidus lies below 1200°C at 15 kbars). The maximum observed difference between nominally identical runs (1200°C, 12 kbars) is also illustrated. (*From Green and Ringwood, 1972, with permission.*)

rocks. Fortunately, this work has shown that many of the results obtained upon the quartz tholeiites are relevant in a variety of more acid rocks.

Green and Lambert (1965) investigated the mineral stability fields in an anhydrous granite over a wide range of pressures at 950 and 1100°C. The composition and norm of this rock are given in Table 1-4. The granite crystallized under low pressures to the assemblage quartz, alkali felspar, plagioclase, orthopyroxene, and rutile. With increasing pressure,[1] garnet was first observed at 12 kbars at 950°C and at 14 kbars at 1100°C. These pressures are similar to the pressures required for incoming of garnet in the quartz tholeiites (Fig. 1-1). The reaction which is responsible for the formation of garnet in the granite is similar to the reaction responsible for formation of garnet in the quartz tholeiite [Sec. 1-3, reaction (V)]. Likewise, the gradient for the incoming of garnet in the granite is 17 bars/°C, which is in good agreement with the corresponding gradient in the tholeiites, considering the various sources of error.

With increasing pressure, the proportion of garnet increases, orthopyroxene reacts away, and it is replaced by clinopyroxene which becomes increasingly rich in jadeite. At 1100°C and at 21 kbars, which is similar to the pressure required for the formation of eclogite in quartz tholeiite, the mineral assemblage observed is alkali felspar (albite 30, sanidine 70), quartz, omphacite, and garnet. (This mineral assemblage is similar to the mineral assemblage of the quartz tholeiite in eclogite facies, except for the presence of alkali felspar.) With increasing pressure, the proportion of alkali felspar in the granite decreases, and omphacite plus quartz increase, as an increasing proportion of albite in the alkali felspar breaks down to yield quartz and jadeite (which enters the omphacite).

A detailed study of mineral stability fields at high pressure in diorite (andesite) and gabbroic anorthosite[2] compositions has been carried out by T. Green (1970), see Table 1-4, and reconnaissance results on basaltic andesite (56% SiO_2) and dacite (65% SiO_2) also exist.[3] These rocks display analogous behaviour to the granite and basalts. The low-pressure assemblage consists of plagioclase, pyroxene, and quartz. With increasing pressure, garnet appears (at 8 kbars, 900°C in the diorite), and there is a field of garnet + pyroxene + plagioclase + quartz (garnet granulite). When the pressure has reached a value which would convert basaltic rocks to quartz eclogite, the mineral assemblage displayed by these acid-intermediate rocks consists of garnet, omphacite, quartz, and alkali felspar. (The survival of alkali felspar at high pressures is due to the stability of orthoclase, which does not transform, i.e., dry, until about 100 kbars[4] and hence is able to stabilize some albite by solid-solution formation.)

There is thus a certain unity about the response of the important crustal rock series varying from gabbro to granite in composition to high pressures and temper-

[1]These results include a friction correction not applied in the original paper of Green and Lambert.

[2]See also T. Green (1967) for a study of kyanite eclogite.

[3]T. Green and Ringwood (1968).

[4]Ringwood, Reid, and Wadsley (1967).

Table 1-4 COMPOSITION AND CIPW NORMS OF THE GRANITE INVESTIGATED BY GREEN AND LAMBERT (1965) AND THE DIORITE AND GABBROIC ANORTHOSITE INVESTIGATED BY T. GREEN (1970)

	Granite*	Diorite	Gabbroic anorthosite
SiO_2	69.6	59.9	53.5
TiO_2	0.6	0.7	1.0
Al_2O_3	14.7	17.3	22.5
Fe_2O_3	1.7	—	0.9
FeO	1.7	6.3	4.7
MnO	0.1	—	0.1
MgO	1.0	3.4	2.1
CaO	2.5	7.1	9.9
Na_2O	3.4	3.7	3.7
K_2O	4.6	1.6	1.1
CIPW norms			
Qz	25.7	9.2	2.1
Or	27.2	9.4	6.5
Ab	28.5	31.3	31.3
An	10.9	25.8	41.5
Diop	—	7.8	6.3
Hyp	3.4	15.0	8.5
Mt	2.5	—	1.3
Ilm	6.5	1.3	1.9

*Plagioclase $Ab_{72}An_{28}$.

atures. The mineral stability fields (Fig. 1-1) exhibited by the quartz tholeiite are observed also in the more acid rocks at similar pressures, with the principal difference[1] that alkali felspar remains a stable phase in the more acid rocks at pressures beyond those required for the formation of quartz eclogite. The *P-T* stability fields of mineral assemblages exhibited by oversaturated basaltic rocks thus provide a framework for classification and quantification of mineral facies in anhydrous rocks. These considerations are useful when it is necessary to make predictions about the properties and behaviour of acid-intermediate rocks under the *P-T* conditions which exist in the lower crust (Sec. 2-1).

Examples of granitic rocks in the eclogite facies (i.e., composed of quartz, alkali felspar, garnet, and omphacite) have not yet been recorded, possibly due to insufficient attention being paid to the nature of the pyroxenes in natural occurrences of so-called granulites. However, Kozlowski (1959) has provided detailed descriptions of rocks of andesitic composition which display this mineral assemblage and are thus to be regarded as belonging to the eclogite facies.[2]

[1]Kyanite may also be a significant phase in some compositions (T. Green, 1967).
[2]Green and Ringwood (1967).

REFERENCES

BACKLUND, H. G. (1936). Zur genetischen Deuten der Eklogite, *Geol. Rundschau* **27,** 47–61.

BEARTH, P. (1959). Uber Eklogite, Glaucophanschiefer und metamorphe Pillolaven. *Schweiz. Mineral. Petrog. Mitt.* **39,** 267–286.

BIRCH, F. (1955). Physics of the crust. In: A. Poldervaart (ed.), "*Crust of the Earth,*" pp. 101–118. *Geol. Soc. Am. Spec. Paper* 62.

——— (1961). The velocity of compressional waves in rocks to 10 kbar, 2. *J. Geophys. Res.* **66,** 2199–2224.

BOYD, F. R., and J. L. ENGLAND (1959). Experimentation at high pressures and temperatures, *Carnegie Inst. Washington Yearbook* **58,** 82–89.

BRYHNI, I., D. H. GREEN, K. S. HEIER, and W. S. FYFE (1970). On the occurrence of eclogite in western Norway. *Contrib. Min. Petrol.* **26,** 12–19.

CLARK, S. P. (1961). Geothermal studies. *Carnegie Inst. Washington Yearbook* **60,** 185–190.

——— (1962). Temperatures in the continental crust. In: C. M. Herzfeld (ed.), "*Temperature, Its Measurement and Control in Science and Industry,*" vol. 3, pp. 779–790. Reinhold, New York.

——— and A. E. RINGWOOD (1964). Density distribution and constitution of the mantle. *Rev. Geophys.* **2,** 35–88.

COHEN, L. H., K. ITO, and G. C. KENNEDY (1967). Melting and phase relations in an anhydrous basalt to 40 kilobars. *Am. J. Sci.* **265,** 475–518.

COLEMAN, R. G., D. E. LEE, I. B. BEATTY, and W. W. BRANNOCK (1965). Eclogites and eclogites, their differences and similarities. *Bull. Geol. Soc. Am.* **76,** 483–508.

DE WAARD, D.(1965). A proposed subdivision of the granulite facies. *Am. J. Sci.* **263,** 455–461.

ENGEL, A., and C. ENGEL (1962). Progressive metamorphism and amphibolite, Northwest Adirondack Mountains, New York. In: A. E. J. Engel, H. L. James, and B. F. Leonard (eds.), "*Petrologic Studies,*" Buddington Volume, pp. 37–82. Geological Society of America, New York.

ERNST, W. G. (1968). "*Amphiboles,*" pp. 1–125. Springer-Verlag, New York.

FYFE, W. S., F. J. TURNER, and J. VERHOOGEN (1958). Metamorphic reactions and metamorphic facies. *Geol. Soc. Am. Memoir* **73,** 1–259.

GODOVIKOV, A. A., and G. C. KENNEDY (1969). Eclogites. In: S. P. Korykobskii (ed.), "*Problems of Petrology and Genetic Mineralogy,*" vol. 1, pp. 48–61, Nauka (in Russian).

GREEN, D. H. (1967). Effects of high pressure on basaltic rock. In: H. H. Hess and A. E. Poldervaart (eds.), "*Basalts: The Poldervaart Treatise on Rocks of Basaltic Composition,*" Interscience, a division of Wiley, New York.

——— and I.B. LAMBERT (1965). Experimental crystallization of anhydrous granite at high pressures and temperatures. *J. Geophys. Res.* **70,** 5259–5268.

——— and B. O. MYSEN (1972). Genetic relationship between eclogite and hornblende and plagioclase pegmatite in Western Norway. *Lithos* **5,** 147–161.

——— and A. E. RINGWOOD (1967). An experimental investigation of the gabbro to eclogite transformation and its petrological applications. *Geochim. Cosmochim. Acta* **31,** 767–833.

——— and A. E. RINGWOOD (1972). A comparison of recent experimental data on the gabbro-garnet granulite-eclogite transition. *J. Geol.* **80,** 277–288.

GREEN, T. H. (1967). An experimental investigation of sub-solidus assemblages formed at high pressure in high-alumina basalt, kyanite eclogite and grosspydite compositions. *Contr. Mineral. Petrol.* **16,** 84–114.

——— (1970). High pressure experimental studies on the mineralogical constitution of the lower crust. *Phys. Earth Planet. Interiors* **3,** 441–450.

——— and A. E. RINGWOOD (1968). Genesis of the calc-alkaline igneous rock suite. *Contr. Mineral. Petrol.* **18,** 105–162.

———, A. E. RINGWOOD, and A. MAJOR (1966). Friction effects and pressure calibration in a piston-cylinder apparatus at high pressure and temperature. *J. Geophys. Res.* **71,** 3589–3594.

HYNDMAN, R. D., I. B. LAMBERT, K. S. HEIER, J. C. JAEGER, and A. E. RINGWOOD (1968). Heat flow and surface radioactivity measurements in the Precambrian shield of West Australia. *Phys. Earth Planet. Interiors* **1,** 129–135.

ITO, K., and G. C. KENNEDY (1968). Melting and phase relations in the plane tholeiite-lherzolite-nepheline basanite to 40 kilobars with geological implications. *Contr. Mineral Petrol.* **19,** 177–211.

——— and ——— (1970). The fine structure of the basalt eclogite transition. *Mineral. Soc. Am. Spec. Paper* **3,** 77–83.

——— and ——— (1971). An experimental study of the basalt-garnet granulite-eclogite transition. In: "*The Structure and Physical Properties of the Earth's Crust,*" pp. 303–314. *Am. Geophys. Union, Geophys. Monograph* **14.**

KENNEDY, G.C., and K. ITO (1972). Comments on: "A comparison of recent experimental data on the gabbro-garnet granulite-eclogite transition." *J. Geol.* **80,** 289–292.

KORZHINSKY, D. S. (1937). Dependence of mineral stability on depth. *Zap. Vses. Mineral. Obshestva* **66** (2), 369–396 (in Russian).

KOZLOWSKI, K. (1959). On the eclogite-like rocks of Stary Gieraltow (east Sudetan). *Bull. Acad. Polon. Sci., Ser. Sci. Chim., Geol., Geograph.* **6** (11), 723–728.

KUSHIRO, I., and H. S. YODER (1966). Anorthite-forsterite and anorthite-enstatite reactions and their bearing on the basalt-eclogite transformation. *J. Petrol.* **7,** 337–362.

LAMBERT, I. B., and P. J. WYLLIE (1968). Stability of hornblende and a model for the low velocity zone. *Nature* **219,** 1240–1241.

RINGWOOD, A. E., and E. ESSENE (1970). Petrogenesis of Apollo 11 basalts, internal constitution and origin of the moon. "*Proc. Apollo 11 Lunar Science Conference.*" *Geochim. Cosmochim. Acta* **34,** suppl. 1, 769–799.

——— and D. H. GREEN (1964). Experimental investigations bearing on the nature of the Mohorovicic Discontinuity. *Nature* **201,** 566–567.

——— and ——— (1966). An experimental investigation of the gabbro-eclogite transformation and some geophysical implications. *Tectonophysics* **3,** 383–427.

———, A. F. REID, and D. H. WADSLEY (1967). High pressure $KAlSi_3O_8$, an aluminosilicate with 6-fold coordination. *Acta Cryst.* **23,** 1093–1095.

ROY, R., D. D. BLACKWELL, and F. BIRCH (1968). Heat generation of plutonic rocks and continental heat flow provinces. *Earth Planet. Sci. Letters* **5,** 1–12.

SMULIKOWSKI, K. (1960). Petrographical notes on some eclogites of the East Sudetes. *Bull. Acad. Pol. Sci., Ser. Sci. Geo. Geograph.* **8** (1), 11–19.

TURNER, F. J. (1968). "*Metamorphic Petrology.*" McGraw-Hill, New York.

WHITE, A. J. R. (1964). Clinopyroxene from eclogites and basic granulites. *Am. Min.* **49**, 883–888.

WINKLER, H. G. F. (1965). "*Petrogenesis of Metamorphic Rocks.*" Springer-Verlag, New York. 220 pp.

WYLLIE, P. J. (1971). "*The Dynamic Earth.*" Wiley, New York. 416 pp.

YODER, H. S., and C. E. TILLEY (1962). Origin of basalt magmas: an experimental study of natural and synthetic rock systems. *J. Petrol.* **3**, 342–532.

2

COMPOSITION OF THE CRUST

2-1 THE CRUST IN STABLE CONTINENTAL REGIONS[1]

The seismic velocity distributions within the continental crust vary widely in different geologic provinces, and broad generalizations about continental structure are difficult to sustain. In this section, discussion is concentrated upon the crust beneath relatively stable continental regions such as Precambrian shields and platforms with low relief. Many platforms consist of comparatively undeformed, shallow sedimentary basins resting upon basement complexes of crystalline igneous and metamorphic rocks. Within such regions, some valid generalizations can be made. In contrast, the crust beneath continental orogenic regions appears to possess a more complex character, varying widely from region to region, and generalizations are more difficult to establish.

The thickness of the crust beneath stable continental regions is usually between 35 and 45 km, with an average close to 40 km.[2] Seismic *P* velocities in the

[1]Much of the discussion in Secs. 2-1 and 2-2 is based on papers by Ringwood and Green (1966a,b).
[2]Pakiser and Robinson (1966).

upper crust are commonly between 5.8 and 6.4 km/sec,[1,2] and it is widely believed that they increase with depth, velocities in the vicinity of 6.5 to 7.2 km/sec being commonly inferred for the lower crust. However, the nature of the downward increase is not settled; in some regions there is evidence for a discontinuity—the Conrad[3,4]—whereas in others, the increase is probably continuous.[5] At the base of the crust, seismic *P* velocity is believed to increase rapidly to about 8.2 km/sec, marking the Mohorovicic discontinuity (Moho). It is widely believed that a large part of the velocity increase from ~7 km/sec in the lower crust to ~8 km/sec in the mantle occurs within a narrow depth interval of a few kilometres or less, although there is a great deal of debate about the detailed nature of this transition. Once again, regional variations are almost certainly involved. The above generalizations are probably broadly correct for much of the stable continental crust. However, they are not of universal applicability.[6] Sections of a typical shield region are shown in Fig. 2-1.

[1]Gutenberg (1959a).
[2]Steinhart and Meyer (1961).
[3]Richards and Walker (1959).
[4]Clowes, Kanasewich, and Cumming (1968).
[5]Tatel and Tuve (1955).
[6]In some relatively stable continental regions, a layer with a velocity of 7.0 to 7.4 km/sec occurs in the lower crust—ref. Figs. 2-6, 2-7, and Everingham (1967).

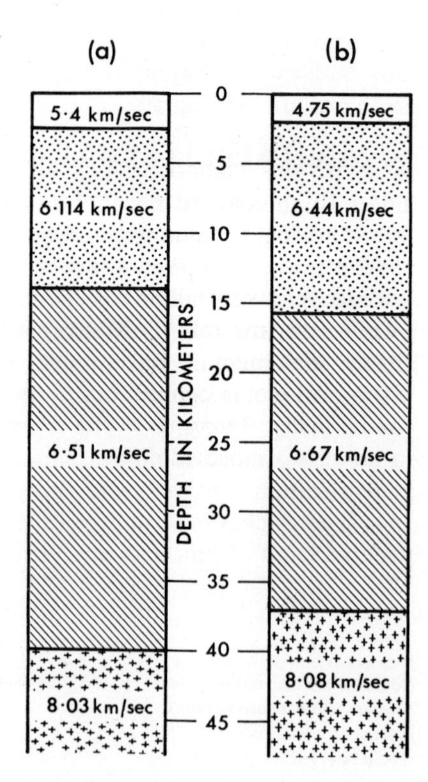

FIGURE 2-1
Crustal sections for two typical stable regions—the Precambrian shields in (*a*) central Wisconsin and (*b*) upper Michigan. (*From Steinhart and Meyer, 1961.*)

Table 2-1 AVERAGE COMPOSITION OF CONTINENTAL SHIELD CRYSTALLINE SURFACE ROCKS. (*After Poldervaart*, 1955)

Component	Wt %
SiO_2	66.4
TiO_2	0.6
Al_2O_3	15.5
Fe_2O_3	1.8
FeO	2.8
MnO	0.1
MgO	2.0
CaO	3.8
Na_2O	3.5
K_2O	3.3
P_2O_5	0.2

Density (estimated): 2.69 g/cm^3
V_P (estimated): 6.25 km/sec

The Upper Crust

Seismic *P* velocities of 5.8 to 6.4 km/sec in the upper crust correspond to those of acid igneous and metamorphic rocks and have led to the inference that the upper crust is of "granitic" composition. This view is supported by geological observations and by geochemical estimates of the mean chemical composition,[1,2,3,4,5] which are in close agreement. The average chemical composition of continental shield crystalline rocks as estimated by Poldervaart[5] is shown in Table 2-1. A rock with this composition would consist principally of quartz, felspars, mica, and/or amphibole but is closer to a granodiorite than a granite.

Woollard's[6,7] studies of the density of the upper crust generally support this model but lead to a slightly higher density. He obtained an average density of 2.74 g/cm^3 for crystalline basement rocks from 1158 samples distributed widely over North America. This mean value is supported by analyses of the negative gravity anomalies which exist over granitic intrusions[7] (density 2.65 to 2.70 g/cm^3) in the upper crust and which therefore imply a slightly higher upper crustal density than 2.70 g/cm^3. A density of 2.74 matches that of a mafic granodiorite or quartz diorite.

[1]Sederholm (1925).
[2]Grout (1938).
[3]Vogt (1931).
[4]Nockolds (1954).
[5]Poldervaart (1955).
[6]Woollard (1966, 1968).
[7]Woollard (1962).

The Lower Crust under Anhydrous Conditions

Whilst the composition of the upper crust can be established reasonably well from a combination of observations based upon petrology, geology, seismic velocities, and densities, the nature of the lower crust in shield and platform regions is less clear. Large regions of the lower crust are characterized by *P* velocities in the range 6.5 to 7.2 km/sec. These velocities are similar to those of basalts, dolerites, and gabbros,[1] and it has been widely accepted that the lower crust is indeed composed of these rocks.[e.g.,2,3,4,5] Thus the general concept of a chemically layered crust, the upper layer being granitic and the lower gabbroic, has been widely retained, although the layering is not generally considered to be regular and continuous over large regions, thus leaving plenty of room for crustal heterogeneity.

The inference that the lower crust is composed of gabbroic or basaltic rocks is not supported, however, by studies of the stability fields of mineral assemblages in mafic rocks. The experimental investigations discussed in Chap. 1 showed that the gabbroic mineral assemblage was not thermodynamically stable under the P-T-P_{H_2O} conditions existing in the lower crust. If the lower crust is of mafic composition, in thermodynamic equilibrium, and is relatively dry, it should be composed largely of eclogite ($V_p \approx 8.4$ km/sec, $\rho \approx 3.5$ g/cm^3) in regions of low heat-flow or, in regions of high heat-flow, of garnet granulites possessing velocities of 7.3 to 8.0 km/sec and densities of 3.2 to 3.4 g/cm^3 (Secs. 1-3 and 2-4). These velocities do not match those of the lower crust in most nonorogenic regions. We must conclude that, if the crust is relatively "dry" and approaches chemical equilibrium, it cannot be of mafic composition.

The extent to which mafic rocks in the crust approach chemical equilibrium deserves comment. Basalts, dolerites, and gabbros are common rocks in the upper crust, yet the above considerations imply that they are mostly metastable relative to eclogites and granulites. To what extent might this occur also in the lower crust? Clearly, this depends on kinetic considerations.

The preservation of metastable gabbroic mineral assemblages in the upper crust appears to require cooling to temperatures of 200 to 300°C on a time scale of up to 10^6 to 10^7 years under conditions of low water-vapour pressure and low shear stress. Studies of the response of dry mafic rocks to metamorphism in the crust, however, have revealed innumerable examples where in situ transformation to granulites and eclogites has occurred.[6] The transformation appears to require temperatures higher than 300°C, often combined with solid-state deformation incurred during orogenesis. This latter effect is probably of major significance at lower temperatures.

[1]Birch (1960, 1961).
[2]Birch (1958).
[3]Gutenberg (1955).
[4]Gutenberg (1959a).
[5]Pakiser and Robinson (1966).
[6]Reviewed by Green and Ringwood (1968).

Large regions of the lower crust have probably been subjected to temperatures of 300 to 700°C for periods of 10^8 to 10^9 years. Moreover, the complex tectonic development of the crust over such periods has probably resulted in widespread and repeated deformation. It appears most unlikely that gabbros would remain in their highly metastable state under these conditions. By comparison with the observed response of gabbros to metamorphism in exposed regions of the crust, it would be expected that gabbros in the lower crust would transform to granulite and eclogite mineral assemblages more closely in equilibrium under the prevailing *P-T* conditions. This is not to state that complete equilibrium will always be maintained. This will depend on the *P-T* path in time and on the deformation history. Thus it is conceivable that transformations in a given rock might effectively cease when a granulite mineral assemblage has been attained, even though subsequent changes in the *P-T* conditions might take the rock into the eclogite stability field. The transformation of garnet granulites to eclogite with falling temperature under the pressures in the lower crust occurs at comparatively low temperatures (Fig. 1-6) and may require accompanying deformation and shearing in order to proceed at geologically significant rates. On the other hand, the transformation of gabbro to garnet granulite under the pressures in the lower crust occurs at higher temperatures (Fig. 1-6) and probably proceeds spontaneously by thermal activation without needing activation by deformation.

Accordingly, in regions where the lower crust is relatively "dry," we conclude that the occurrence of seismic velocities in the vicinity of 6.6 to 7.2 km/sec implies the presence of large amounts of minerals of relatively low seismic velocity—e.g., quartz and alkali felspars to counterbalance the seismically fast garnet and pyroxene.[1] An estimate of the mineral assemblages, densities, and seismic velocities which would be displayed by typical acid and intermediate rocks in the eclogite facies, based upon recent experimental results discussed in Sec. 1-5, is given in Table 2-2. It appears that rocks of mean composition approaching quartz diorite in the eclogite facies may provide the best match to the physical properties of the lower crust. Alternatively, rocks of average dioritic composition in the garnet granulite facies would also provide a satisfactory match. If the lower crust has an average composition in this range, whereas the upper crust is taken to be acid, the mean composition of the entire crust in such regions is found to be rather more acid than is currently assumed in some geochemical models.[2,3] On the other hand, in some regions where a lower crustal layer possessing a velocity of 7.2 to 7.8 km/sec is found, this layer could well consist of mafic rocks in the garnet granulite facies,[4] and the average composition of the crust would approximate that of an intermediate igneous rock (andesite).

An alternative crustal model has been proposed by T. Green (1970) and is

[1] See also Green and Lambert (1965).

[2] Poldervaart (1955).

[3] Pakiser and Robinson (1966).

[4] Section 2-2. See also Ito and Kennedy (1971).

Table 2-2 COMPOSITIONS, MINERAL ASSEMBLAGES, AND ESTIMATED DENSITIES AND SEISMIC VELOCITIES OF TYPICAL ACID-INTERMEDIATE ROCKS AT LOW PRESSURES AND AT HIGH PRESSURES

	Chemical composition (wt %)			
Component	**Diorite**		**Granodiorite**	
SiO_2	60.4		65.6	
TiO_2	0.7		0.7	
Al_2O_3	17.3		16.2	
Fe_2O_3	2.0		1.4	
FeO	4.3		3.6	
MgO	3.5		1.8	
CaO	7.2		4.9	
Na_2O	3.3		3.6	
K_2O	1.3		2.2	
	Mineral assemblage (wt %)			
Mineral	**Low pressure**	**High pressure**	**Low pressure**	**High pressure**
Quartz	14.8	21.6	21.7	26.2
Alkali felspar ($Ab_{50}Or_{50}$)	—	15.2	—	25.3
Orthoclase	7.7	—	13.0	
Plagioclase	56.8	—	52.0	
Omphacite	—	34.6	—	25.0
Clinopyroxene	5.8	—	2.3	
Hypersthene	11.0	—	7.7	
Garnet	—	19.0	—	13.8
Kyanite	—	8.9	—	9.0
Ore minerals	4.2	0.7	3.3	0.7
Density (g/cm^3)	2.83	3.13	2.78	3.00
V_p (km/sec) (approx.)	6.6	7.3	6.4	7.2

probably applicable in some regions. He demonstrated that rocks of gabbroic anorthosite composition in a high-grade metamorphic facies (plagioclase + orthoclase + garnet + pyroxene ± kyanite) would provide a close match to the seismic velocities of the lower crust in many regions. A lower crust of this type might evolve by partial melting and differentiation of a pre-existing andesitic crust of the kind now forming in some island-arc regions.[1] Experimental results showed that partial melting under conditions of relatively low water pressure would produce granite-granodiorite melts which would migrate into the upper crust, leaving behind a refractory residuum with a composition generally resembling gabbroic anorthosite.[2]

[1]Taylor and White (1965).
[2]T. Green (1970).

The Lower Crust under Hydrous Conditions

The above conclusions are applicable to regions of the lower crust where the partial pressure of water is low. However, in regions where the temperature is less than 600°C and where a substantial partial pressure of water vapour exists, the studies of metamorphic petrologists show that basaltic rocks would occur dominantly as amphibolites, probably in the almandine amphibolite facies. They would be composed mainly of the minerals amphibole, plagioclase, epidote, and almandine-rich garnet. Such rocks would possess P velocities of 7.0 to 7.6 km/sec and densities of 3.0 to 3.25 g/cm^3 (Table 2-3).

This range of velocities is somewhat higher than the values which are usually assigned to the lower crust in stable regions. Drake and Nafe[1] have pointed out that layers with velocities between 7.0 and 7.5 km/sec are comparatively rare beneath Precambrian shields, although they do exist.[2] Somewhat lower velocities, matching those most commonly inferred, would be produced by rock complexes consisting dominantly of amphibolite but also containing an appreciable proportion of interbedded granitic rocks. The mean chemical composition of the lower crust in such regions would then be basic-intermediate. On the other hand, in less evolved platform regions and in young orogenic belts, lower layers with velocities in the range 7.0 to 7.5 km/sec are commonly observed,[1,3] and amphibolite could well be the principal constituent.

[1]Drake and Nafe (1968).
[2]Everingham (1967).
[3]Section 2-2.

Table 2-3 *P* VELOCITIES (MEASURED AT 4 kbars) AND DENSITIES OF SOME AMPHIBOLITES AND HORNBLENDE-RICH ROCKS

Rock	Density (g/cm^3)	Velocity (km/sec)	Ref.	Mineralogy
Hornblende gabbro	3.11	7.02	1	Amph 68, Plag 31
Diopside hornblendite	3.20	7.08	1	Amph 57, Px 43
Epidote amphibolite	3.13	7.56	2	Amph 50, Ep 34, Plag 11, Qz 5
Epidote amphibolite	3.26	7.52	2	Amph 57, Ep 47, Plag 3, Qz 3
Amphibolite	3.04	7.08	2	Amph 73, Plag 19, Qz 3
Amphibolite	3.03	7.04	2	Amph 68, Plag 26, Qz 4
Amphibolite	3.12	7.27	3	Amph 86, Serp 6, ore 8

Abbreviations: *Amph* = amphibole, Plag = plagioclase, Qz = quartz, Ep = epidote, Px = pyroxene, Serp = serpentine, ore = ore minerals.
References:
1 Kanamori and Mizutani (1965).
2 Christensen (1965).
3 Birch (1960, 1961)

Petrological Evolution of the Lower Crust

Thus it appears that there are two types of rocks—amphibolites and intermediate rocks in granulite-eclogite facies—which offer the best match to the properties of the lower crust. The occurrence of these two types of rocks in a given crustal environment will be strongly influenced by the partial pressure of water, and this in turn will be determined by the geologic evolution of the crust. It is possible to distinguish two distinct geological environments that might give rise to high and low water-vapour pressures, respectively.

1 An initial high water content is probable in sediments and volcanics formed in a thick geosynclinal pile. Furthermore, a high proportion of basaltic material is typical of a eugeosynclinal environment. Such water-rich sequences, if subjected to burial and particularly to penetrative shearing during folding, would probably be converted to hydrated mineral assemblages (zeolite, glaucophane schist, and greenschist facies). As temperatures at the base of the geosyncline rose to about 400°C, metamorphism to amphibolites and mica schists would occur. Such sequences, if not subjected to higher grades of metamorphism, and if incorporated into the lower crust by the particular crustal evolution processes operating, could yield seismic *P*-wave velocities of 6.8 to 7.2 km/sec. A limit to the persistence of such hydrated assemblages in the lower crust is provided by their instability relative to anhydrous granulitic assemblages at higher temperature. This transformation may occur at temperatures between 500 and 700°C and at rather moderate pressures.[1]

Thus the possibility of a largely amphibolitic lower crust seems to require eugeosynclinal sedimentation and basaltic volcanism on oceanic-type crust followed by deep burial but with temperatures remaining below about 650°C.

2 Whilst the previous model appears applicable in some regions,[2] it does not appear appropriate for other more extensive regions of the stable continental crust, characterized by the occurrence of crystalline rocks ("basement") at shallow depth. These terrains of igneous granitic rocks, migmatites, and varied metamorphic rocks imply that, at an earlier stage of crustal evolution, temperatures in these rocks must have been within the range 400 to 600°C. At the time when these temperatures were generated, the temperatures in the subjacent rocks now forming the lower crust must have been still higher and probably in the range 600 to 1000°C. Under such conditions, amphibolites would have been converted into granulites, and it is probable that the lower crust would have become rather thoroughly dehydrated. Subsequently, after the period of active evolution had proceeded to completion and general cooling of the lower crust to its present state occurred,

[1]Engel and Engel (1962).

[2]E.g., Lake Superior (Ringwood and Green, 1966b).

hydrous mineral assemblages would be unable to reform because of lack of water; hydration by downward access of water from the upper crust appears unlikely, although it may occur on a local and restricted scale. As discussed before, in such regions of low water-vapour pressure, the occurrence of seismic velocities in the 6.5 to 7.2 km/sec range points towards the occurrence of acid-intermediate rocks in the granulite and eclogite facies as the principal constituents of the lower crust.

The preceding discussion regarding the nature of the deeper crust has been strongly influenced by recent developments in experimental petrology (Chap. 1). Nevertheless, it should be recalled that similar conclusions have been reached by many petrologists on the basis of more traditional arguments. Few metamorphic petrologists have ever paid more than lip service to the geophysical hypothesis that the lower crust consists of basaltic or gabbroic rocks. It is generally recognized that the most likely candidates for the lower crust are high-grade metamorphic rocks in the granulite and eclogite facies, possessing mineral assemblages very different from that of gabbro. A cogent statement of this view is given by den Tex (1965), who concludes that the lower crust is composed of a heterogeneous mixture of acid and intermediate granulites and charnockites (55%), basic granulites and eclogites (40%), and ultrabasic rocks (5%). The mean chemical composition of this mixture would correspond to that of intermediate rock types. From the petrologic viewpoint, a heterogeneous model such as this is more plausible than a chemically homogeneous lower crust. From the geophysical point of view, however, average properties are the more important, and there is little difference in this respect between den Tex's model and the model which was arrived at earlier in this section.

Geochemical and Geothermal Evidence upon Crustal Zoning

Similar conclusions regarding the nature of the lower continental crust have been reached by Heier and coworkers,[1,2,3] arguing from geochemical grounds. They found that the abundances of uranium, thorium, and rubidium in high-grade granulite-facies rocks are strongly depleted, by a factor up to 10, compared to the average abundances of these elements in upper crust amphibolite-facies rocks possessing otherwise similar major element compositions. These depletions are believed to have been caused by the dehydration processes involved in metamorphism to granulite-facies conditions. Much of the uranium and thorium occurring at grain boundaries and associated with hydrous minerals such as amphibole and

[1]Heier (1965a,b).
[2]Heier and Adams (1965).
[3]Lambert and Heier (1967, 1968).

biotite has apparently migrated upward and out of the lower crust as water is expelled. It also observed that, in high-grade granulite-facies terrains, highly acid rocks are not nearly as abundant as in lower grade amphibolite-facies terrains. These authors plausibly argue that partial melting of extensive regions of the lower crust is attained during the highest stage of metamorphism, resulting in the upward migration of the low-melting-point fraction, highly enriched in silica, alkalis, water, and especially, uranium and thorium. These processes ultimately result in the formation of a lower crust of generally intermediate chemical composition, highly depleted in U and Th, overlain by a more acid upper crust, highly enriched in these elements.

This model is strongly supported by recent investigations of surface heat-flow integrated with studies of the abundances and distribution of heat-producing elements in near-surface rocks.[1,2,3] It is known that Precambrian shields are characterized by relatively low heat-flows; yet the upper crust of shields contains "normal" abundances of the heat-producing elements U, Th, and K. These observations can be satisfied only by models[2,3] in which the radioactive elements in the shield crust are concentrated in a thin surface layer (5 to 10 km thick) and the crust below this layer contains very low abundances of these elements, consistent with those observed in granulite facies rocks of intermediate composition. Geothermal profiles which have been inferred[2] for the West Australian Archaean shield are shown in Figs. 2-2 and 2-3 and Table 2-4. These are compared to corresponding profiles inferred for the crust in the Snowy Mountains area of southeas-

[1]Lambert and Heier (1967).
[2]Hyndman, Lambert, Heier, Jaeger, and Ringwood (1968).
[3]Roy, Blackwell, and Birch (1968).

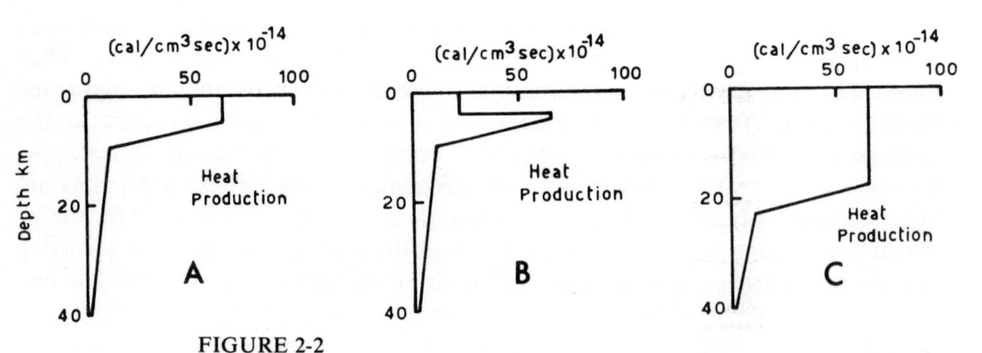

FIGURE 2-2
Estimated heat-production profiles for (*a*) a West Australian Archaean shield area with mainly granitic surface rocks; (*b*) a shield area with greenstone surface rocks; and (*c*) a Palaeozoic region in Eastern Australia with a high heat-flow and granitic surface rocks. (*From Hyndman, Lambert, Heier, Jaeger, and Ringwood, 1968, with permission.*)

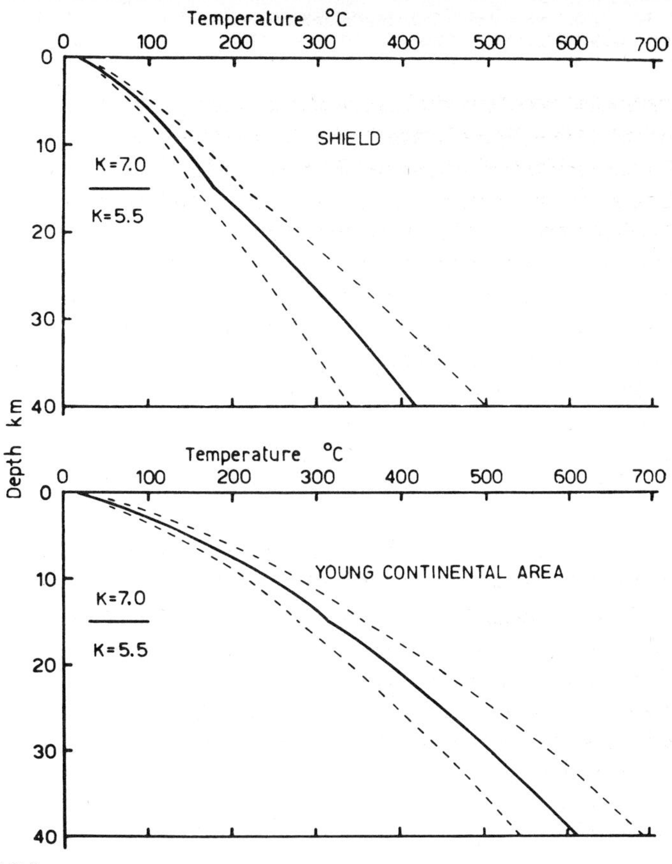

FIGURE 2-3
Crustal temperature profiles for a low-heat-flow shield area and for a younger high-heat-flow area. Parameters used in calculations are given in Fig. 2-2 and Table 2-4. (*From Hyndman, Lambert, Heier, Jaeger, and Ringwood, 1968, with permission.*)

Table 2-4 DATA FOR RADIOACTIVITY AND TEMPERATURE PROFILES OF FIG. 2-2. (*After Hyndman, Lambert, Heier, Jaeger, and Ringwood,* 1968)

	West Australian Precambrian Shield		**Eastern Australia (Snowy Mountains, Palaeozoic)**
	Granitic surface rocks	**Greenstone surface rocks**	**Granitic surface rocks**
Surface heat-flow (μcal/cm^{2}sec)	1.15	0.98	2.00
Heat-flow at 40 km	0.45	0.45	0.55
Near-surface heat-production (cal/cm^{2}sec $\times 10^{-13}$)	6.4	2.2	6.2

tern Australia characterized by extensive outcrops of Palaeozoic granitic rocks and a relatively high surface heat-flow.

An extensive investigation[1] of the relationship between surface heat-flow and the radioactivity of near-surface rocks in the United States produced further interesting results. The data implied the existence of clearly defined heat-flow provinces in each of which the heat-flow Q and the near-surface heat-production A are related by an equation of the form $Q = a + bA$. An example of this relationship displayed by plutons in the central and northeastern regions of the United States is shown in Fig. 2-4. The simplest explanation of this linear relation is that the radioactivity measured at the surface is constant from the surface to depth b but varies from place to place. Thus the fraction of heat-flow from the lower crust and upper mantle, a, remains constant within a heat-flow province, whilst the variable upper crustal radioactivity generates the variable heat-flow observed at the surface.

The data in Fig. 2-4 are derived from a stable region which has been subjected to only mild disturbances since the Precambrian. The slope of the line yields a depth of 7.5 km for the heat-producing layer, whilst the intercept gives 0.8 μcal/cm^{2}sec as the heat-flow originating from the lower crust and mantle. Studies of other provinces yielded similar thicknesses for the heat-producing layer but very different values for the heat-flux below this layer. Roy et al. argue that the heat-flow relationship (Fig. 2-4) for the central-eastern province of the United States is likely to be generally applicable in stable regions of the continental crust. The geothermal-geochemical arguments lead generally to a model for the upper and lower crust similar to that reached previously on petrological grounds.

An essentially consistent but probably more realistic interpretation of the heat-flow data has been suggested by Lachenbruch (1970). Instead of a two-layer crust, he proposed a model in which the abundance of heat-producing elements decreased exponentially with depth in the crust. Such an exponential decrease could readily be explained in terms of repeated partial-melting processes in the crust accompanied by vertical redistribution of U, Th, and K. This model was demonstrated to produce a linear relation between surface heat-flow and heat-production, as shown in Fig. 2-4.

The low abundances of U, Th, and K which must be assigned to the lower crust throughout extensive continental provinces in order to explain the relationship between surface heat-flow and the radioactivity of the upper crust further restricts the role which amphibolites are likely to play as compared with high-grade granulite- and eclogite-facies rocks. It was noted previously that a lower crust composed mainly of amphibolite with smaller volumes of associated acid and metamorphic rocks in the amphibolite facies would provide a satisfactory seismic velocity in the 6.6 to 7.2 km/sec range and would constitute a stable mineral assemblage in the P-T field of the lower crust in the presence of an appreciable partial pressure of water. However, the abundances of radioactive elements in a heterogeneous lower crust of this composition would probably be too high to be

[1]Roy, Blackwell, and Birch (1968).

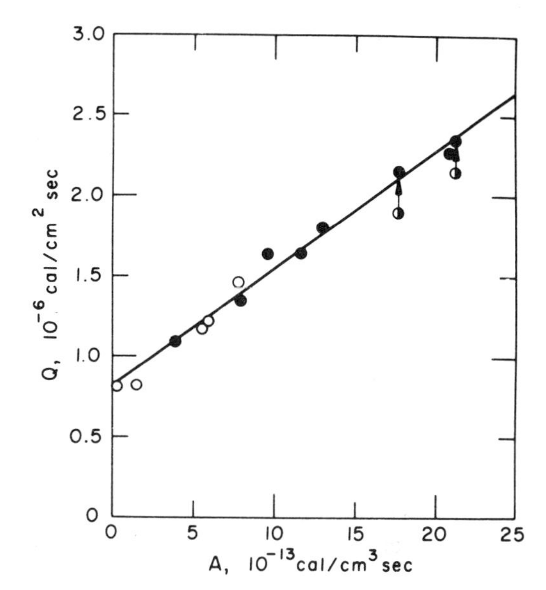

FIGURE 2-4
Heat-flow versus near-surface heat-production data for plutons in the New England area (solid circles) and the Central Stable Region of the USA (open circles). (*From Roy, Blackwell, and Birch, 1968, with permission.*)

compatible with the inferred heat-production in the stable lower continental crust in many regions.

2-2 THE CRUST IN CONTINENTAL OROGENIC REGIONS

Emphasis in the previous section has been directed towards the nature of the stable continental crust because it is possible to arrive at a number of constraints. The composition of the crust in unstable orogenic regions presents greater problems. The principal seismic features which differentiate the crust in some orogenic regions from the crust in stable regions are the greater crustal thicknesses[1,2,3,4,5] of the former regions, the greater complexity of structure, the widespread presence of deep crustal layers characterized by seismic velocities in the range 7.2 to 7.8 km/sec,[5] and the absence in some (perhaps most) regions of a clearly defined M discontinuity owing to a more or less continuous transition in velocities between the lower crust and mantle.[1]

Crustal thicknesses in the Andes[3,6,7] and Appalachians[4] range up to 65 km. In the Alps,[1,2] crustal thickness (defined as the depth at which "mantle" P veloci-

[1]Fuchs, Müller, Peterschmitt, Rothe, Stein, and Strobach (1963).
[2]Giese (1968).
[3]Fisher and Raitt (1962).
[4]James, Smith, and Steinhart (1968).
[5]Drake and Nafe (1968).
[6]Woollard (1960).
[7]James (1971).

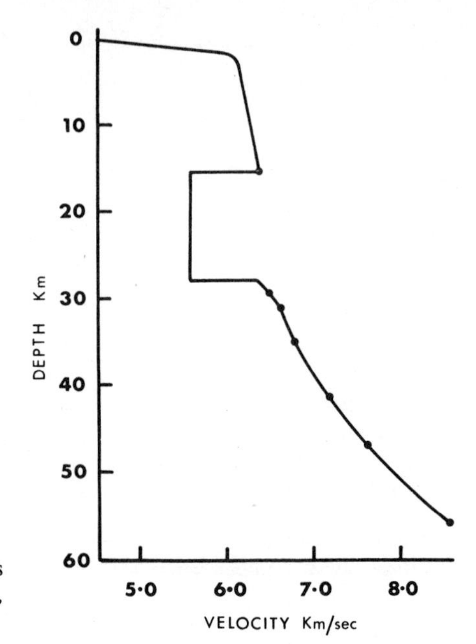

FIGURE 2-5
P velocity-depth distribution in the Alps near Lago-Lagorai. (*From Giese, 1968, p. 89, with permission.*)

ties of 8.2 km/sec are reached) ranges up to 55 km. An upper (sialic) crust of generally acid composition with velocities of 6.0 to 6.5 km/sec and thicknesses of 20 to 30 km is characteristic. Near the base of this layer a velocity reversal (low-velocity zone) is widespread.[1] Below about 30 km, a continuous increase in velocity from 6.5 to 8.2 km/sec commonly occurs (Fig. 2-5), thus constituting a transition zone. Giese's[1] seismic profiles below the Alps show that, in regions of normal crustal thickness, the transition zone is characterized by high velocity gradients, whereas in regions of greater crustal thickness, the velocity gradients between crust and mantle are much smaller. Drake and Nafe[2] have shown that layers possessing intermediate velocities of 7.2 to 7.8 km/sec are comparatively common beneath young orogenic belts.

There are at least three possible interpretations of the composition of the lower crust in such regions. The intermediate velocities could be provided by (1) amphibolites, (2) mafic rocks in the garnet-granulite facies, and (3) a physical mixture of ultramafic rocks and eclogites with acid-intermediate crustal rocks. Which of these alternatives may be applicable in any given region will depend upon the assessment of other kinds of evidence. The previous discussion (Sec. 2-1) regarding the factors controlling the occurrence of amphibolite and garnet granulite in the lower crust remains relevant and suggests that garnet granulites of mafic composition are likely to be of wider occurrence than amphibolites. Geothermal considerations—particularly the occurrence of relatively high heat-flow in the Alps[3,4]

[1]Giese (1968).
[2]Drake and Nafe (1968).
[3]Clark (1961).
[4]Lee and Uyeda (1965).

and a crustal low-velocity zone (probably due to high temperature gradients) —support this conclusion.[1] However, it also seems likely that the intermediate velocities in some regions may be caused by a physical mixture of mantle and crustal material,[2] as suggested by the occurrence at high levels in the alpine crust of the "Ivrea" body—presumably an extensive ultramafic intrusion accompanied by a large positive gravity anomaly.[1]

We have already noted[3] the relatively common occurrence of lower crustal layers beneath orogenic regions, with velocities in the range 7.2 to 7.8 km/sec, and the comparative rarity of these layers beneath stable continental regions—particularly shields. It should not be inferred, however, that intermediate-velocity layers are completely absent beneath shields. A lower layer with a velocity of 7.24 km/sec has been found beneath the southwestern portion of the West Australian shield.[4] There are also extensive continental regions which cannot be classified strictly as "orogenic" or "stable" but are generally intermediate in their degree of evolution. In these regions, lower crustal layers with velocities between 7.0 and 7.6 km/sec are of common occurrence. Two such profiles from southern United States[5] and from central western Germany[6] are shown in Figs. 2-6 and

[1]See also Giese (1968).
[2]Cook (1962).
[3]See also Drake and Nafe (1968) for detailed documentation.
[4]Everingham (1967).
[5]McCamy and Meyer (1966, p. 380).
[6]Fuchs and Landisman (1966, p. 450).

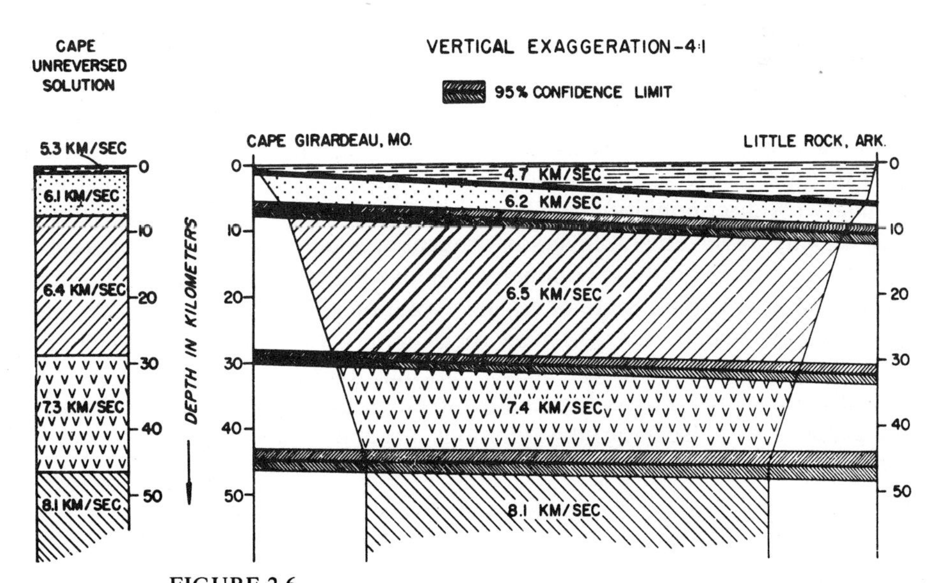

FIGURE 2-6
Crustal model for the Mississippi Embayment in the southern United States. (*From McCamy and Meyer, 1966, with permission. Copyright American Geophysical Union.*)

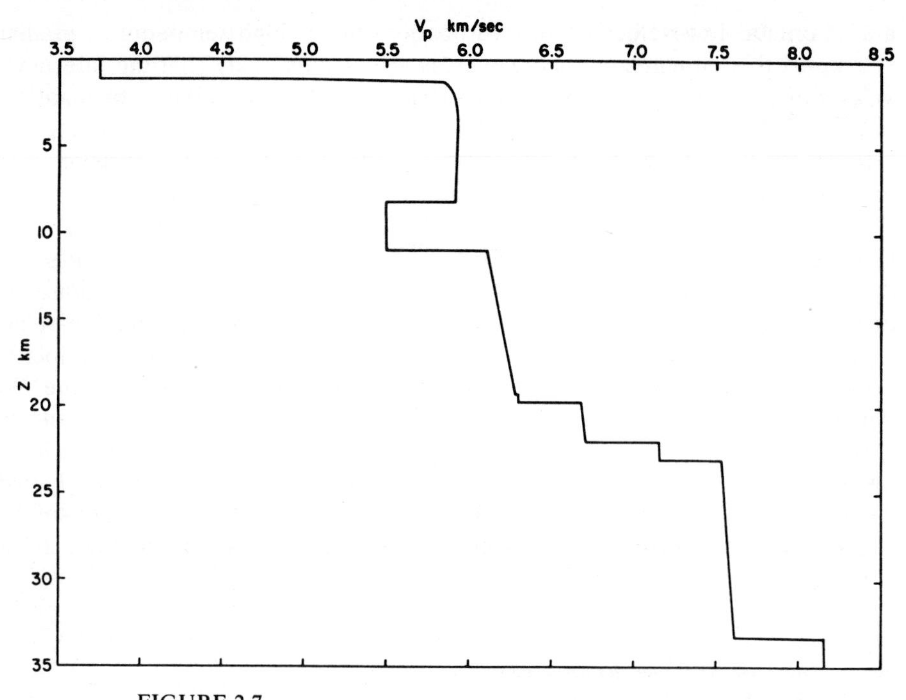

FIGURE 2-7
Velocity-depth relationship for central Western Germany. (*From Fuchs and Landisman, 1966, with permission. Copyright American Geophysical Union.*)

2-7. Note that the M discontinuity is clearly recognizable on these profiles, either as a discontinuity or as a narrow region of very high velocity gradient. However, there is a larger total increase in velocity associated with the transition from upper crust to lower crust (Conrad discontinuity) than across the lower crust-mantle boundary (M discontinuity). We have previously discussed the interpretation of high-velocity lower crustal layers in orogenic regions. Similar explanations are probably applicable in these continental regions which may be regarded as having reached an intermediate state of crustal evolution (Chap. 8); i.e., they are probably caused by the occurrence of mafic rocks in the garnet-granulite facies[1] or by intermediate rocks in the eclogite facies. The high average velocities in Fig. 2-6 are probably associated with a higher than average mean crustal density, which in turn explains, from isostatic considerations, the abnormal thickness of the crust in relation to its small elevation above sea level.[2]

[1]Ito and Kennedy (1971).
[2]Woollard (1966, 1968).

2-3 THE OCEANIC CRUST

Ocean Basins

The structure of the crust in deep oceanic basins (water depth > 4 km) has been reviewed by Hill (1957) and Raitt (1963). A distinctive uniformity of structure characteristic of all ocean basins has been established (Table 2-5).

Layer 1 is known to consist of unconsolidated sediments. The structure of this layer has been extensively investigated in recent years by reflection profiling[1,2] and more recently by direct drilling. The important results of these investigations are beyond the scope of this book.

Layer 2 is believed to consist mainly of unconsolidated basalt flows possessing substantial residual porosity. The lower part may contain swarms of mafic dykes. The evidence pointing in this direction is as follows: (1) the layer is thicker near some groups of Pacific Islands and seamounts, suggesting a volcanic origin;[3] (2) in regions where layer 2 is believed to outcrop, mafic rocks have been obtained in dredge hauls;[2,4] (3) the magnitudes and gradients of marine magnetic anomalies imply that the sources must lie in layer 2 and possess magnetizations in the range displayed by (oxidized) basaltic rocks;[5,6,7] (4) direct drilling through the sediments into the presumed surface of layer 2 during the Joides Project has repeatedly disclosed basaltic rocks.

Layer 3 is the main crustal layer. It is universally present in the deep oceanic basins and is characterized by comparatively uniform seismic velocities and thicknesses (Table 2-5). It has generally been concluded that layer 3 directly

[1]E.g., Ewing and Ewing (1967).
[2]E.g., Ewing, Ewing, Aitken, and Ludwig (1968).
[3]Raitt (1957).
[4]Talwani, Le Pichon, and Ewing (1965).
[5]Bullard and Mason (1963).
[6]Vine and Wilson (1965).
[7]Talwani, Windisch, and Langseth (1971).

Table 2-5 ***P*** **VELOCITIES AND THICKNESSES OF OCEANIC CRUST LAYERS WITH STANDARD DEVIATIONS.** (*After Hill,* 1957, *and Raitt,* 1963)

Layer	Thickness (km)	Velocity (km/sec)
Sea water	4.5	1.5
1	0.45	2
2	1.71 ± 0.75	5.07 ± 0.63
3	4.86 ± 1.42	6.69 ± 0.28
4	—	8.13 ± 0.24

overlies the mantle (mean velocity 8.1 km/sec). However, this conclusion may require modification.[1]

Irrespective of the possible occurrence of a fourth layer in the oceanic crust, the increase of seismic velocities from the lowermost crust to the uppermost mantle occurs in a very narrow depth range. Amplitude studies[2] imply that the transition occurs over a depth interval not exceeding 1 km. Hales and Nation (1966) recorded the occurrence of clearly defined reflected refractions from the oceanic M discontinuity. These require an even narrower transition interval between crust and mantle, probably on the order of 0.2 km. Ewing and Houtz (1969) have recently recorded direct reflections from the Moho, indicating that it may be even sharper than this.

Two principal hypotheses have been suggested to account for the nature of layer 3. Hess[3] has argued that layer 3 consists of serpentinized peridotite, whereas other workers[4] have regarded this layer as being of mafic composition.

The arguments advanced by Hess[3] in favour of the serpentine hypothesis are: (1) a mixture of 60 to 70% serpentine with 30 to 40% of peridotite would match the observed seismic velocity (6.7 km/sec) in layer 3; (2) the presumed occurrence of peridotite as the major constituent of the upper mantle and the dredging of serpentinized peridotite from fault scarps where layer 3 perhaps outcrops; (3) interpretation of magnitudes and gradients of marine magnetic anomalies suggests that the principal source is in layer 2 and that layer 3 is less strongly magnetized[5,6,7]—the relatively weak magnetization of some serpentines is consistent with this model;[6,7,8] and (4) Hess suggested that, in principle, a serpentinite crust could be readily generated from the mantle by hydration of rising mantle-peridotite beneath mid-oceanic ridges as the temperature falls below 500°C. The serpentine is then carried outward by the migrating lithosphere, and the M discontinuity represents a "fossil isotherm" for the peridotite-serpentine reaction. A serpentinite crust is "disposable" since, on reaching an oceanic trench and sinking into the mantle, it would become dehydrated to peridotite. According to Hess' model, layer 2 remains of basaltic composition.

Although the serpentine-crust hypothesis appears to possess some attractive aspects, it is confronted by serious difficulties:

1 Although the *P* velocity of layer 3 can be matched by a mixture of serpentine and peridotite, the corresponding *S*-wave velocity provided by the

[1]Recent detailed airgun-sonobuoy refraction profiling by Sutton, Maynard, and Hussong (1971) has revealed the presence of an intermediate layer between layer 3 and the mantle in deep oceanic basins. This layer possesses a mean velocity of 7.4 km/sec and an average thickness of 3 km. It is usually masked in normal refraction traverses.

[2]Helmberger and Morris (1969).

[3]Hess (1962, 1964, 1965).

[4]E.g., Hill (1957) and Oxburgh and Turcotte (1968).

[5]Vine and Wilson (1965).

[6]Pitman, Herron, and Heirtzler (1968).

[7]Vine (1966).

[8]However, this is contradicted by more recent measurements (Fox and Opdyke, 1973).

Table 2-6 MEAN VELOCITIES AT 10 KBARS FOR AGGREGATES OF OLIVINE, PYROXENE, AND SERPENTINE. (*After Christensen*, 1966)

	V_P	V_s	V_P/V_s	σ
Olivine	8.54	4.78	1.79	0.27
Pyroxene	7.93	4.65	1.70	0.24
Serpentine	5.10	2.35	2.17	0.38
Av. oceanic crust	6.7	3.72	1.80	0.28

serpentinized peridotite differs significantly from the value observed in the oceanic crust. This arises from the unusually high value of Poisson's ratio for serpentine as compared with olivines and pyroxenes (Table 2-6).

From Christensen's[1] detailed investigation, the average velocity of layer 3 (6.70 km/sec) would be provided by a mixture of serpentine 45% and peridotite 55% (the latter consisting of olivine 80% and pyroxene 20%). This mixture would possess a shear-wave velocity of 3.50 km/sec.

The velocities of *S* waves in layer 3 have been determined in several localities.[2,3,4] The average velocity is 3.7 km/sec, with a narrow range. The average ratio V_p/V_s is 1.80, again, with a narrow range. Taking V_p as 6.70 km/sec, the corresponding V_s is 3.72 km/sec. This is substantially higher than the value of 3.50 obtained above for a serpentinized peridotite, capable of yielding a 6.70 km/sec *P* velocity. Hess[5] has estimated that the degree of serpentinization in layer 3 may be as high as 70%. This would increase the discrepancy in *S*-wave velocity. Further difficulties with the serpentinite hypothesis have also been noted.[6,7]

2 The narrow range of seismic velocities characteristic of layer 3 and the sharpness of the oceanic M discontinuity have previously been noted. If layer 3 is to be regarded as consisting of *partially* serpentinized peridotite, it is extremely difficult to understand why the degree of serpentinization should be uniform throughout this layer.[6,7] The degree of serpentinization of natural peridotites characteristically varies widely within limited regions. It is even more difficult to understand the formation of a sharp M discontinuity.

3 According to the Hess hypothesis, water (20%) in the serpentinite is derived from the mantle. This implies a high (>1%) primary water content in the upper mantle. On the other hand, studies of basalt genesis (Chap. 4)

[1]Christensen (1966).
[2]Ewing and Ewing (1959, 1961).
[3]Francis and Shor (1966).
[4]Bunce and Fahlquist (1962).
[5]Hess (1962).
[6]Oxburgh and Turcotte (1968).
[7]Cann (1968).

imply that the primary water content of the upper mantle is much lower, probably in the neighbourhood of 0.1%.

4 Recent extensive measurements[1] have shown that oceanic serpentinized peridotites possess magnetic intensities and susceptibilities which are too high to permit them to be important components of layer 3.

Because of these difficulties, we turn to the alternative hypothesis that this layer consists of mafic rocks, displaying either an anhydrous basaltic mineral assemblage or consisting of low-grade metamorphic equivalents of basalts.

The seismic P velocity of 6.69 ± 0.28 km/sec for layer 3 is similar to the range observed in basaltic rocks.[2] Furthermore, the S-wave velocity of layer 3 (3.7 ± 0.2 km/sec) is closely matched by those observed in basalts. This evidence, when considered in the light of the widespread occurrence of basaltic volcanism on the floor of the ocean and the interpretation of mid-oceanic ridges as regions where extensive partial melting and basaltic volcanism are occurring, leading to the formation of new crust, is consistent with a mafic composition for layer 3.[3]

Direct observational evidence on the nature of layer 3 is provided by the rather convincing interpretations of some ophiolite complexes as consisting of complete sections of oceanic crust and mantle.[4] Some examples are described in Sec. 3-3. If these interpretations are correct, the dominant mafic nature of layer 3 is established beyond doubt.

The inferred low intensity of magnetization[5,6,7] of layer 3 no longer constitutes an objection to its presumed mafic nature. The slower cooling of mafic rocks in layer 3 has probably resulted in the formation of dolerites and gabbros with coarse grain sizes, which are commonly found to exhibit much less intense magnetizations than basalts.[5,1] Another possibility is that in layer 3, basaltic rocks have been converted to greenschist-facies rocks.[5,8,9] The metamorphism is observed to be accompanied by a strong decrease in magnetization.[1]

Assuming that layer 3 was initially formed by basaltic volcanism at mid-oceanic ridges, and recalling the discussion in Chap. 1, we would expect that, on cooling under dry conditions to normal oceanic crustal temperatures, the basaltic mineral assemblage, plagioclase + pyroxene, would not remain thermodynamically stable. It is possible, however, that, because of the relatively short time

[1]Fox and Opdyke (1973).
[2]Birch (1960, 1961).
[3]Oxburgh and Turcotte (1968).
[4]E.g., Papuan ultramafic belt—Davies (1968, 1969); Troodos and Vourinos complexes—Moores and Vine (1971).
[5]Vine and Wilson (1965).
[6]Pitman, Herron, and Heirtzler (1968).
[7]Vine (1966).
[8]Matthews, Vine, and Cann (1965).
[9]Luyendyk and Melson (1967).

scale for cooling ($\sim 10^7$ years) and the low temperatures ultimately reached, basaltic rocks would not transform on cooling but would remain preserved in a metastable state similar to many basaltic rocks in the upper continental crust (Sec. 2-1). Indeed, the retention of "basaltic" seismic velocities demonstrates that transformation to a denser anhydrous assemblage has not occurred.

On the other hand, a basaltic crust would be likely to transform extensively in the presence of a substantial partial pressure of water. This is readily supplied by the ocean. The problem is one of access, which is dependent upon the porosity and degree of fracturing of layer 3. If oceanic water is able to obtain widespread access to layer 3, basaltic rocks might be expected to recrystallize to greenschist facies or other hydrous metamorphic assemblages, depending upon the *P-T-* time path followed. The recovery of basaltic rocks in greenschist facies characterized by the mineral assemblage albite + epidote + chlorite + actinolite in several locations of the oceanic floor[1,2,3,4] provides firm evidence that the conditions for metamorphic recrystallization have been attained at least locally. Christensen (1970) has demonstrated that the *P* velocities of greenschist-facies metabasaltic rocks provide a close match to that of layer 3. The mean velocity of four such metabasalts (excluding an epidosite) was 6.81 km/sec.

The marine occurrences of greenschists have been interpreted as being formed by regional or burial metamorphism at substantial depths in the oceanic crust and later exposed at the surface by faulting.[2,3,5] However, other observations[4] have shown evidence of the development of greenschist metamorphism along veins and fractures in otherwise normal basaltic rock, suggesting that it is due to introduction of hydrothermal solutions and may be of a local and restricted nature, rather than representing a true regional metamorphism. Amphibolites have also been dredged from oceanic ridges and fracture zones,[5,6] and Cann[6] has suggested that amphibolite is a major constituent of layer 3. The high seismic velocity of amphibolites (Table 2-3) restricts the proportion of these rocks which might be present.[7]

Considering all available evidence, it appears likely that layer 3 is highly heterogeneous in composition, consisting mainly of a mixture of anhydrous dolerite and gabbro with mafic greenschists and amphibolites. The common occurrence of serpentinized peridotites in dredge hauls suggests that this rock may

[1]Matthews, Vine, and Cann (1965).
[2]Melson and Van Andel (1966).
[3]Cann and Vine (1967).
[4]Melson, Thomson, and Van Andel (1968).
[5]Miyashiro, Shido, and Ewing (1971).
[6]Cann (1968, 1970).
[7]However, the recently discovered 7.4 km/sec intermediate oceanic crust layer (see footnote on p. 52) may well be composed of amphibolite.

also be a significant and widely distributed constituent.[1,2] Although the serpentinite is not believed to be a major component of the crust, it is, nevertheless, of considerable importance as a carrier of water into the mantle in zones of crustal subduction—a topic which is discussed further in Chaps. 7 and 8.

Mid-Oceanic Ridges

Mid-oceanic ridges constitute a major oceanic geological province. The proposed crustal structure[3,4] across the Mid-Atlantic Ridge is shown in Figs. 2-8 and 2-9. The free-air anomaly shows the Ridge to be essentially in isostatic equilibrium, and the negative Bouguer anomaly reveals the presence of a substantial mass deficiency beneath the axial zone and somewhat smaller deficiencies on the flanks. The seismic section shows the absence of the sedimentary layer 1. Basaltic rocks with velocities appropriate for layer 2 outcrop at the surface. Velocities increase with depth in an irregular manner. Layer 2 is thicker than normal,

[1]Cann (1968).
[2]Melson and Thompson (1971).
[3]Ewing and Ewing (1959).
[4]Talwani, Le Pichon, and Ewing (1965).

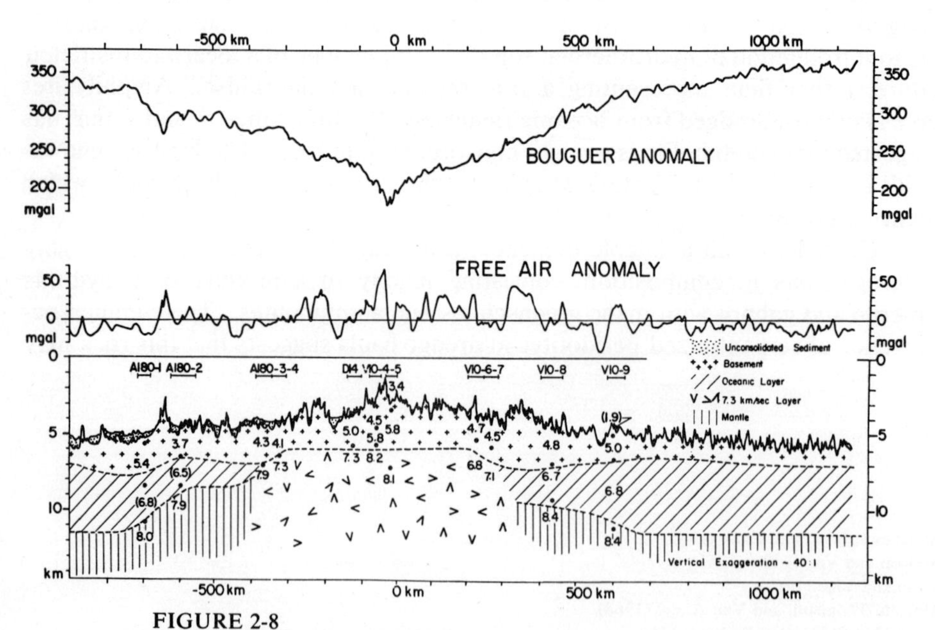

FIGURE 2-8
Gravity anomalies and seismically determined structure across the north Mid-Atlantic Ridge. (*From Talwani, Le Pichon, and Ewing, 1965, with permission. Copyright American Geophysical Union.*)

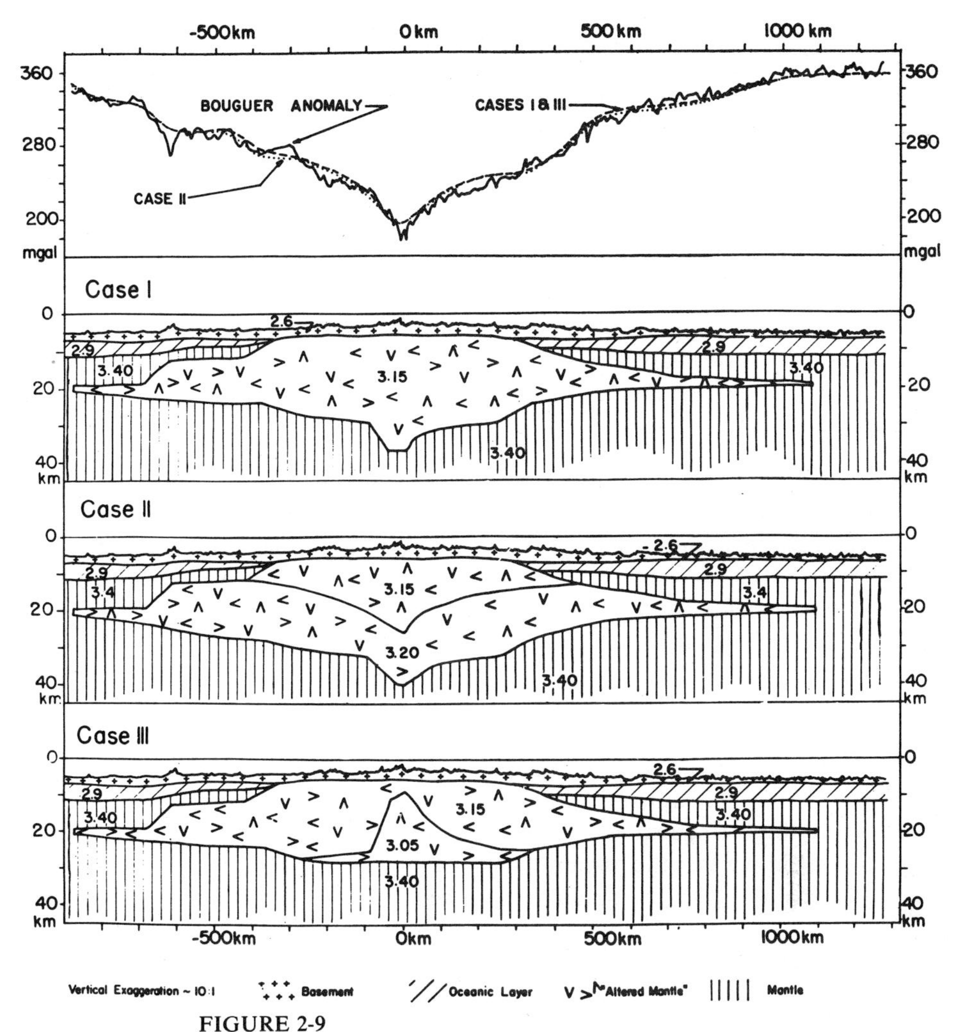

FIGURE 2-9
Three possible crustal models across the north Mid-Atlantic Ridge which satisfy gravity anomalies and are in accord with the seismic data of Fig. 2-8. (*From Talwani, Le Pichon, and Ewing, 1965, with permission. Copyright American Geophysical Union.*)

whereas layer 3 is thinner and appears to pass gradationally into the mantle which is characterized by abnormally low velocities. The M discontinuity is poorly defined and probably absent in many regions. The presence of "mantle" velocities (> 7.9 km/sec) at high levels in the crust in a few locations suggests an extremely heterogeneous and complex structure on a local scale.

The structure over the East Pacific Rise appears simpler,[1] with layers 2 and 3 clearly defined and of normal thickness. Beneath the axis of the ridge, velocities are subnormal (7.3 to 7.7 km/sec). The M discontinuity, in the sense of a narrow zone characterized by high velocity gradients, may not be present.

The structure and significance of mid-oceanic ridges have received a vast amount of attention in the recent literature in connection with the theory of sea-floor spreading.[2,3,4,5,6] The characteristic high heat-flow, low densities and seismic velocities, abundant basaltic volcanism, and heterogeneity of structure strongly indicate that the ridges are the loci of ascending convection currents in the mantle accompanied by partial melting, differentiation, and formation of new basaltic crust which migrates outward from the ridges.

Island Arcs

A third important province in the oceans is constituted by island-arc systems. Whereas the mid-oceanic ridges are believed to represent the sites of ascending convection currents and formation of new crust, the trenches in front of island arcs are believed to represent the sites of descending convection currents, whereby the oceanic crust is carried deep into the mantle.

The proposed structure of a typical island arc-trench system is shown in Fig. 2-10.[7] Morgan[8] and others have plausibly argued that the mass deficiency and negative gravity anomalies over the trench must be sustained by a dynamic process since they imply the presence of stress differences of the order of 500 bars at a shallow depth. Static stresses of this magnitude could not be supported by rocks for long periods. This dynamic process is now believed to arise from the sinking of the lithosphere (tectosphere) into the mantle because of its higher density. Morgan[8] has shown that the higher density of the sinking lithosphere may contribute to the positive gravity anomalies and mass excess which occur over island arcs.[9]

The sinking lithosphere is accompanied by the generation of earthquakes in the so-called Benioff zone, extending down to depths of about 700 km. Oliver and Isacks (1967) have shown that this zone is characterized by unusually low seismic attenuation, suggesting that it is much cooler than surrounding mantle. There is also evidence that seismic velocities along this zone are higher than normal.[10,11,12,13]

[1]Talwani, Le Pichon, and Ewing (1965).
[2]Le Pichon (1968).
[3]Morgan (1968).
[4]Isacks, Oliver, and Sykes (1968).
[5]Oxburgh and Turcotte (1968).
[6]Talwani, Windisch, and Langseth (1971).
[7]Talwani, Sutton, and Worzel (1959).
[8]Morgan (1965).
[9]See also Hatherton (1969), Elsasser (1967).
[10]Cleary (1967).
[11]Oliver and Isacks (1967).
[12]Mitronovas and Isacks (1971).
[13]The structure of island arc-trench systems are further considered in Chaps. 7 and 8.

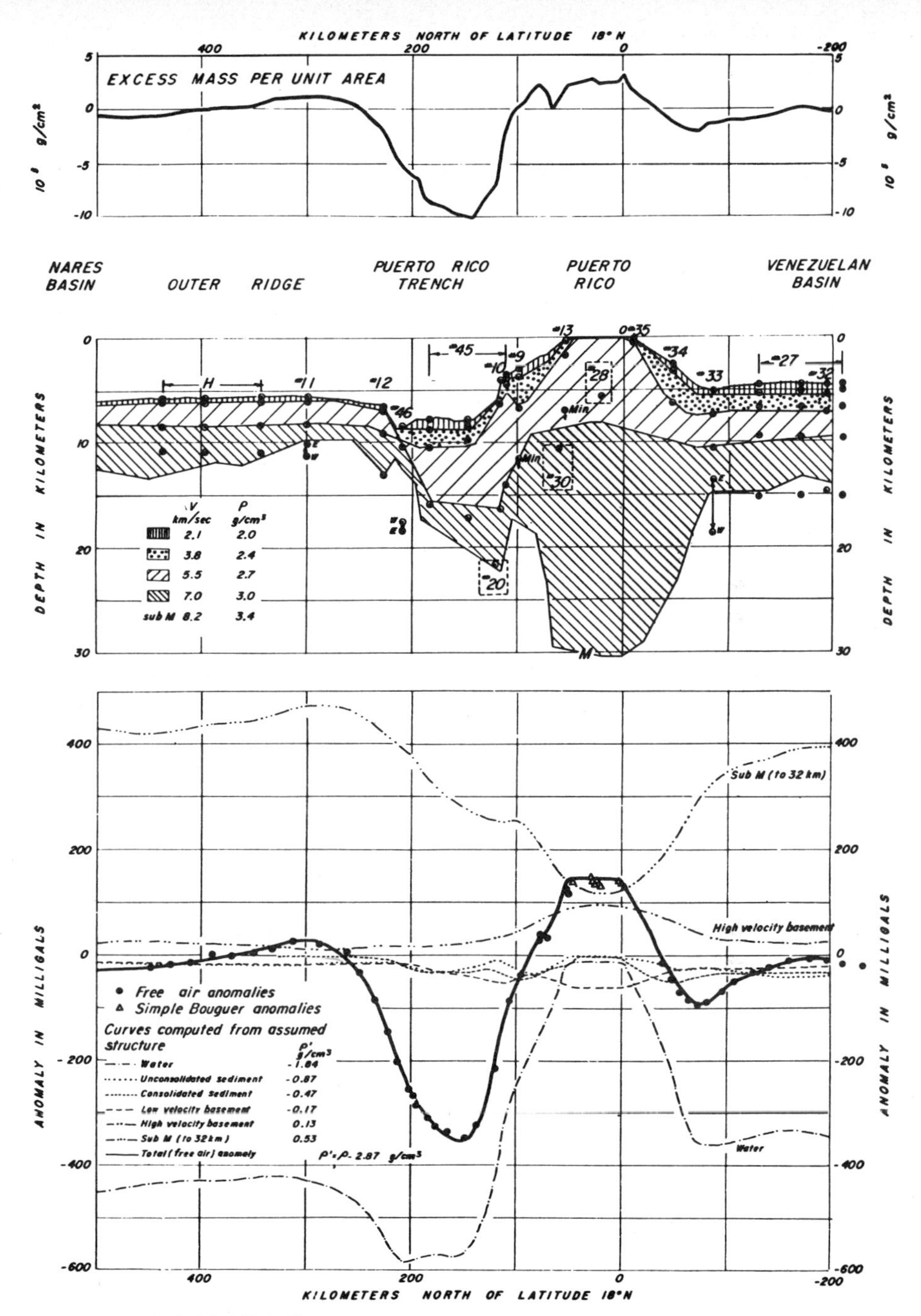

FIGURE 2-10
Crustal section across the Puerto Rico Trench from seismic refraction and gravity data. The points in the middle figure are from seismic refraction data, and the position of the M discontinuity has been obtained from gravity data in such a way that the computed free-air anomaly matches the observed values (below). The mass-anomaly curve (top) has been obtained from the crustal section. (*From Talwani, Sutton, and Worzel, 1959, with permission. Copyright American Geophysical Union.*)

The structure beneath island arcs may be more complex than indicated in Fig. 2-10. In some regions the M discontinuity may not be present. Heat-flow on the island-arc side of trenches is characteristically higher than normal,[1] and the attenuation of seismic waves, particularly *S* waves, from earthquakes below island arcs is much higher than average.[2,3,4] The presence of magma chambers was inferred from Fedotov's investigations;[3] this is consistent with the widespread volcanism observed on many island arcs.

The composition of the crust in island-arc regions is probably very heterogeneous. Volcanic rocks of the orogenic suite (Chap. 7) are the most abundant class represented in surface exposures, and they may well be predominant throughout the crust. Markhinin (1968) has shown that volcanic rocks, mainly andesitic pyroclastics, in the Kurile Islands were erupted at the annual rate of 0.08 km^3 over the period 1930–1963. If this rate has been maintained since the Upper Cretaceous, it would be sufficient to account for the formation of the entire crust now represented by the Kurile Islands.

In addition to orogenic-type igneous rocks, it appears probable that deep-sea sediments, scraped off at the edge of the trench as the oceanic crust descends into the mantle, may form a significant component of the island-arc crust.[5] The presence of significant thicknesses of low-velocity material between island-arc ridges and the axes of trenches, as disclosed by seismic refraction, may perhaps be due to scraped-off distorted oceanic sediments, although alternative explanations are possible.[6] Ultramafic rocks constitute another significant component of some island arcs. Their emplacement is clearly a consequence of the specialized tectonic conditions existing in these regions (Chap. 3).

2-4 THE NATURE OF THE MOHOROVICIC DISCONTINUITY

The hypothesis that the Mohorovicic (M) discontinuity is caused by an isochemical phase transformation from a gabbroic lower crust into an eclogite upper mantle was proposed by Fermor[7] and Holmes[8] and has received a great deal of discussion during recent years.[9] Much of the attraction of the hypothesis may be attributed to its possible tectonic comsequences—changes in temperature at

[1]Uyeda and Vacquier (1968).
[2]Oliver and Isacks (1967).
[3]Fedotov (1968).
[4]See also Fig. 8-3.
[5]Oxburgh and Turcotte (1970).
[6]Ewing, Ewing, Aitken, and Ludwig (1968).
[7]Fermor (1913, 1914).
[8]Holmes (1926a,b, 1927).
[9]E.g., Birch (1952), Sumner (1954), Kennedy (1956, 1959), Lovering (1958), Harris and Rowell (1960), McDonald and Ness (1960), Bullard and Griggs (1961), Yoder and Tilley (1962), Broecker (1962), Wyllie (1963), Stishov (1963), O'Connell and Wasserburg (1967), Ringwood and Green (1964, 1966a,b), T. Green (1967), Cohen, Ito, and Kennedy (1967), Ito and Kennedy (1968, 1970, 1971), Wyllie (1971).

the crust-mantle boundary might cause transformation of gabbro to eclogite, or vice versa, accompanied by large volume changes, resulting in crustal uplift or depression.

The phase-change hypothesis was examined in the light of detailed experiments on the gabbro-eclogite transformation by Ringwood and Green (1966a), who demonstrated that the Moho in stable continental regions and beneath deep oceanic basins could not be caused by the gabbro-eclogite transformation. It was concluded, therefore, that the Moho must generally be caused by a chemical change from intermediate-mafic lower crustal rocks into an ultramafic (peridotitic) mantle. They pointed out, however, that, in some tectonically active regions where the Moho often is not clearly defined, the transformation may indeed occur, and transitional garnet granulites may occur extensively in the lower crust.

The hypothesis that eclogite may be a major component of the upper mantle and that the Moho may yet represent a phase change has recently been reopened by developments in two fields. Firstly, Press[1] has concluded from inversion of free-oscillation data that the density of large regions of the upper mantle is closer to that of eclogite, rather than peridotite. Secondly, Ito and Kennedy (1970, 1971) have produced new experimental data on the gabbro-eclogite transformation which, they conclude, proceeds discontinuously in two separate stages. They claimed that the Moho is caused by the second subdiscontinuity from garnet granulite to eclogite. The subject has thus become surrounded by a measure of confusion, and in view of its importance, further discussion appears desirable.

A wide range of evidence on the composition of the uppermost mantle is reviewed in Chap. 3. It is concluded that this region is indeed dominantly composed of peridotite and that the free-oscillation data which were held by Press to require a high abundance of eclogite are equivocal. The experimental results of Ito and Kennedy were discussed in Chap. 1. The earlier work by these authors was admitted by them to be error and their claim of a density "discontinuity" between the garnet granulite, and eclogite fields in basalt NM5 has been shown to be erroneous (Fig. 1-8, Sec. 1-4). The later results of Ito and Kennedy (1971) on NM5 are indeed completely concordant with Ringwood and Green's (1966a) experimental study of the gabbro-eclogite transformation. Accordingly, the arguments on the nature of the Moho employed by Ringwood and Green are still applicable. In view of the interest and controversy raised by the phase-change hypothesis during recent years, these are restated below. They provide a good example of the way in which data from experimental petrology (reviewed in Chap. 1) can be applied to some fundamental geological-geophysical problems.

Seismic Velocity Distribution in the Crust

A critical requirement of the phase-change hypothesis is its ability to explain the seismic velocity distributions in the crust. As discussed previously, these vary

[1]Section 3-2.

widely on a regional basis. It is useful to restrict discussion initially to stable continental regions, where the M discontinuity seems to be defined most clearly. Even in such regions, the nature of the transition from lower crust to mantle varies widely from place to place. In some regions, there is evidence for vertical reflections from the M discontinuity[1] which imply the presence of a sharp velocity increase within a thickness of about 0.3 km. In other regions the observations of wide-angle reflections[2] and critical refraction[3] phenomena accompanied by large amplitudes imply the presence of high velocity gradients at the M discontinuity. Similar conclusions have been reached by other workers[4,5,6] using different techniques. Even in regions where such specialized studies as those above have not been carried out, it is widely believed by seismologists that a large part of the velocity increase from ≈7 km/sec in the lower crust to ≈8 km/sec in the upper mantle occurs in a restricted depth interval, probably smaller than 5 km[e.g., 5] and sometimes smaller than 1 km.[e.g., 6] The permissible range of velocity-depth relationships obtained by Tuve et al. (1954)[5] for the Maryland region of the United States (Fig. 2-11) is probably applicable to many other stable continental regions.

Accordingly, it is important to see whether the experimental observations on the transformation are consistent with inferred crust-mantle velocity profiles. The

[1]Dix (1965), Liebscher (1964), Clowes, Kanasewich, and Cumming (1968).
[2]Hill and Pakiser (1966), O'Brien (1968).
[3]Tatel and Tuve (1955).
[4]Nakamura and Howell (1964), Phinney (1964).
[5]Tuve, Tatel, and Hart (1954).
[6]Clowes and Kanasewich (1970).

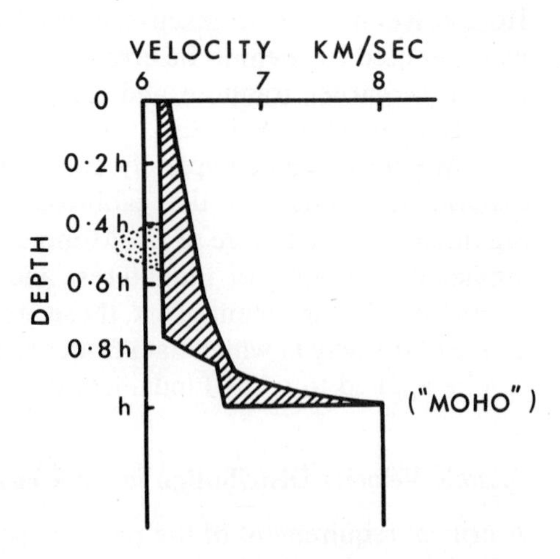

FIGURE 2-11
Permissible range (shaded) of velocity-depth relationships for the Maryland region. (*After Tuve, Tatel, and Hart, 1954.*)

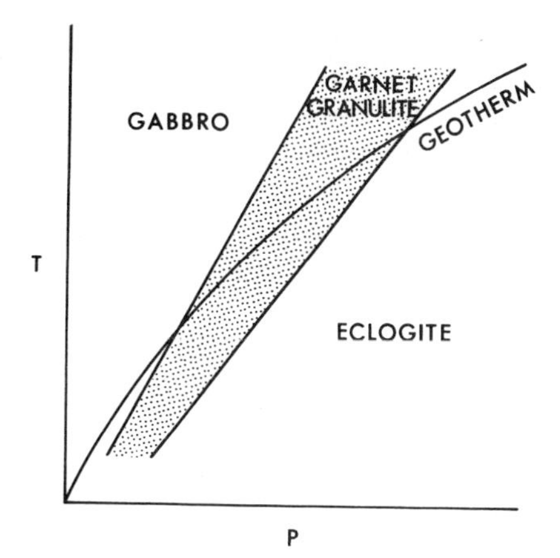

FIGURE 2-12
Diagrammatic representation of the intersection of a geotherm with the gabbro-garnet granulite-eclogite stability boundaries, illustrating the increase in effective transition width caused by oblique intersection of the geotherm with phase boundaries.

experimental results at 1100°C (Fig. 1-1) indicated that, in most basaltic compositions, the transformation between gabbro and eclogite proceeded via a transition zone of garnet granulite varying in width between 4 and 15 kbars, equivalent to depth intervals of about 15 to 50 km.[1] It was observed that the principal mineralogical changes (i.e., incoming of garnet and disappearance of plagioclase) were spread approximately uniformly throughout the transition zone. Accordingly, the corresponding changes in physical properties would also occur approximately uniformly across this zone. This is demonstrated in Figs. 1-7 and 1-8, which show the densities observed in alkali basalt and in NM5 as a function of pressure.[2]

Most of the experimental data were obtained near 1100°C, whereas the temperatures at the continental Moho are probably largely in the range 400 to 700°C. The width of the transition zone is somewhat smaller at these lower temperatures (Fig. 1-6). However, this is more than offset by the fact that the transition field boundaries and the temperature gradients in the earth are generally similar, causing the geotherms to intersect the garnet granulite field boundaries at low angles, thereby greatly increasing the effective width of the transition zone (Fig. 2-12). The situation has been discussed in greater detail by Ringwood and Green (1966a), who show that the gabbro-eclogite transformation is quite incapable of providing velocity gradients of the magnitude inferred by seismologists to be present at the M discontinuity in many continental regions.

Relationship between Temperature at the M Discontinuity and Thickness of the Crust

The geothermal data discussed in Sec. 2-1 implied the existence of temperature differences exceeding 200°C at depths of 40 km in different heat-flow provinces.

[1]Highly undersaturated rocks poor in alkalis provide a partial exception to this generalization (e.g., No. 6, Fig. 1-1) and are discussed separately below.

[2]An inflection is observed in the P-ρ relation of NM5 but is not sufficiently pronounced to alter the basic conclusion stated above (Fig. 1-8).

These large temperature differences are unaccompanied by any systematic effect upon the depth to the M discontinuity.[1,2] If the M discontinuity is caused by a phase change, these relationships imply that the pressure at which the phase change occurs must be relatively insensitive to temperature, with a gradient smaller than about 5 bars/°C. However, the experimental evidence (Sec. 1-2) indicates a gradient in the vicinity of 20 bars/°C. This value, even allowing for a generous uncertainty on the basis of experimental error, cannot be reconciled with the requirements of the phase-change hypothesis. It would result in large variations in crustal thickness in different heat-flow provinces—much greater than those which are observed. Indeed, as Pakiser (1963, 1965) has shown, to explain the M discontinuity beneath the Basin and Range province and Great Plains province in the United States by a phase transformation, it would be necessary to assume a high negative dP/dT because of the thin crust combined with high subsurface temperatures in the former province as compared to thick crust and low temperatures of the latter province. Such a gradient would be directly opposite to that observed for the gabbro-eclogite transformation.

Effect of Change of Chemical Composition upon Transition Parameters

The pressures required to stabilize eclogite in several basaltic rocks varying significantly in chemical composition (Table 1-1) were found to vary by about 12 kbars (Fig. 1-1). Although the corresponding pressure range at the average temperature at the M discontinuity might be somewhat smaller than this, a major reduction is not likely since the gradients of the transformation in individual basalts are probably rather similar.

If the lower crust and upper mantle is generally of mafic composition, it is reasonable to expect that all the common classes of basaltic compositions would be represented. The large effect on transition pressures caused by rather minor changes in chemical compositions would lead to wide fluctuations in depth to the M discontinuity or to further "smearing out" of the velocity distribution in the transformation zone, depending upon the scale of occurrence of particular basaltic compositions. This would make it difficult to interpret the comparative uniformity of crustal thicknesses in stable continental regions and the nature of the seismic velocity distribution in the crust (Fig. 2-1).

Stability of Gabbro in the Lower Crust

Most statements of the hypothesis that the Moho is caused by the gabbro-eclogite transformation have presumed that the lower crust is composed of gabbro. However, it was demonstrated in Sec. 2-1 that rocks of gabbroic composition are not

[1]Bullard and Griggs (1961).
[2]Ringwood and Green (1966a).

stable under the *P-T* conditions existing in the lower crust. The stable mineral assemblage for such rocks would be eclogite or garnet granulite so that this version of the phase-transition hypothesis can be rejected outright.

Ito and Kennedy (1970) have attempted to restate the phase-change hypothesis by proposing that the lower crust consists of garnet granulite and that the Moho is caused by a sharp transition from garnet granulite to eclogite. Unfortunately, experimental data show that the transition between garnet granulite and eclogite is also smeared out over a broad pressure interval of several kilobars and that the final disappearance of plagioclase is not marked in any sense by a discontinuity (Figs. 1-1, 1-3, 1-7, and 1-8). Moreover, the seismic velocities of high-grade mafic garnet granulites ($V_p \sim 7.5$ to 8.0 km/sec) are not usually characteristic of the lower crust in most continental regions.

Stability of Low-Alkali Olivine Tholeiite

The garnet granulite field in rocks of this composition is relatively narrow (3 kbars), and transition to a plagioclase-free eclogitic assemblage occurs at a comparatively low pressure (Fig. 1-1). It is not possible, however, to explain the Moho in terms of a phase change in rocks of this composition since, because of the low pressure needed to achieve the eclogite assemblage, these compositions would occur as eclogites throughout the normal continental crust (Sec. 1-3).

Conclusion

The above evidence, when considered overall, leads to the firm conclusion that the Mohorovicic discontinuity beneath stable continental regions is not generally caused by an isochemical phase transformation between gabbro and eclogite nor between garnet granulite and eclogite. It is therefore necessary to accept the alternative explanation that the M discontinuity is caused by a change in chemical composition. The nature and origin of the Moho are discussed further in Chaps. 3 and 8.

These conclusions are not necessarily applicable to some orogenic regions, however, where the upper crust is separated from the mantle by thick layers possessing *P* velocities between 7 and 8 km/sec. In some such regions, e.g., parts of the Alps,[1] the existence of a clearly defined M discontinuity is in doubt, and the seismic data are consistent with a continuous transition of velocities with no evidence of a discontinuity. In such regions, the gabbro-eclogite transformation may play a significant role.[2,3,4]

The previous discussion has been devoted to the nature of the continental

[1]Giese (1968).
[2]Pakiser (1965).
[3]Ringwood and Green (1966a).
[4]Ito and Kennedy (1971).

rather than the oceanic M discontinuity because of the smaller extrapolation required from the experimental results. Most advocates of the hypothesis that the M discontinuity is caused by a gabbro-eclogite transformation have believed that this hypothesis is more applicable to the subcontinental than to the suboceanic M discontinuity.[1] In view of the above arguments that the continental M discontinuity is not generally caused by a phase change, the proposition that the oceanic discontinuity is caused by an isochemical phase change does not appear enticing. The specific objections to the phase-change hypotheses for the oceanic M discontinuity are similar to those made in the subcontinental case. The problem of a transition interval is the most severe since the increase in seismic velocity between the lower crust and upper mantle occurs within about 1 km, equivalent to a pressure of 0.25 kbar. In some regions, the transition interval is probably smaller than 0.2 km (0.05 kbar, Sec. 2-3). Even allowing for the likelihood that the mineral assemblages involved in a gabbro-"eclogite" transformation under the *P-T* conditions at the oceanic M discontinuity differ substantially from those which were investigated experimentally (Sec. 1-2), it appears extremely improbable that it would be possible to explain the sharpness of the oceanic M discontinuity in terms of an isochemical phase transformation.

REFERENCES

BIRCH, F. (1952). Elasticity and constitution of the earth's interior. *J. Geophys. Res.* **57**, 227–286.

——— (1958). Interpretation of the seismic structure of the crust in the light of experimental studies of wave velocities in rocks. In: H. Benioff, M. Ewing, B. F. Howell, and F. Press (eds.), "*Contributions in Geophysics in Honour of B. Gutenberg*," pp. 158–170. Pergamon, London.

——— (1960). The velocity of compressional waves in rocks to 10 kilobars, 1. *J. Geophys. Res.* **65**, 1083–1102.

——— (1961). The velocity of compressional waves in rocks to 10 kilobars, 2. *J. Geophys. Res.* **66**, 2199–2224.

BROECKER, W. (1962). The contribution of pressure-induced phase changes to glacial rebound. *J. Geophys. Res.* **67**, 4837–4842.

BULLARD, E. C., and D. T. GRIGGS (1961). The nature of the Mohorovicic discontinuity. *Geophys. J.* **6**, 118–123.

——— and R. G. MASON (1963). The magnetic field over the oceans. In: M. N. Hill (ed.), "*The Sea*," vol. 3, chap. 9. Interscience, a division of Wiley, New York.

BUNCE, E. A., and D. A. FAHLQUIST (1962). Geophysical investigation of the Puerto Rico Trench and outer ridge. *J. Geophys. Res.* **67**, 3955–3972.

CANN, J. R. (1968). Geological processing at the mid-oceanic ridge crests. *Geophys. J.* **15**, 331–341.

[1]E.g., Yoder and Tilley (1962), Wyllie (1963), Stishov (1963), and Van Bemmelen (1964).

——— (1970). A new model for the structure of the ocean crust. *Nature* **226,** 928–930.
——— and F. J. VINE (1967). An area on the crest of the Carlsberg Ridge: petrology and magnetic survey. *Phil. Trans. Roy. Soc. London* **A259,** 198–217.
CHRISTENSEN, N. I. (1965). Compressional wave velocities in metamorphic rocks at pressures up to 10 kilobars. *J. Geophys. Res.* **70,** 6147–6164.
——— (1966). Elasticity of ultrabasic rocks. *J. Geophys. Res.* **71,** 5921–5931.
——— (1970). Possible greenschist metamorphism of the oceanic crust. *Bull. Geol. Soc. Am.* **81,** 905–908.
CLARK, S. P. (1961). Heat flow in the Austrian Alps. *Geophys. J.* **6,** 55–63.
CLEARY, J. (1967). Azimuthal variation of the Longshot source term. *Earth Planet. Sci. Letters* **3,** 29–37.
CLOWES, R. M., and E. R. KANASEWICH (1970). Seismic attenuation and the nature of reflecting horizons within the crust. *J. Geophys. Res.* **75,** 6693–6705.
———, ———, and G. L. CUMMING (1968). Deep crustal seismic reflections at near vertical incidence. *Geophysics* **33,** 441–451.
COATS, R. R. (1962). Magma type and crustal structure in the Aleutian arc. In: "*Crust of the Pacific Basin,*" pp. 92–109. *Am. Geophys. Union, Geophys. Monograph* **6.**
COHEN, L. H., K. ITO, and G. C. KENNEDY (1967). Melting and phase relations in an anhydrous basalt to 40 kilobars. *Am. J. Sci.* **265,** 475–518.
COOK, K. L. (1962). The problem of the crust-mantle mix: lateral inhomogeneity in the uppermost part of the earth's mantle. *Advan. Geophys.* **9,** 295–360.
DAVIES, H. L. (1968). Papuan ultramafic belt. *Inter. Geol. Congr., 23d, Prague, Rept.* **1,** 209–220.
——— (1969). Peridotite-gabbro-basalt complex in eastern Papua: An overthrust plate of oceanic mantle and crust. Ph. D. Thesis, Stanford Univ. pp. 1–89.
DEN TEX, E. (1965). Metamorphic lineages of orogenic plutonism. *Geol. Mijnbouw* **44e,** 105–132.
DIX, C. H. (1965). Reflection seismic crustal studies. *Geophysics* **30**; 1068–1084.
DRAKE, C. L. and J. E. NAFE (1968). The transition from ocean to continent from seismic refraction data. In: "*The Crust and Upper Mantle of the Pacific Area,*" pp. 174–186. *Am. Geophys. Union, Geophys. Monograph* **12.**
ELSASSER, W. W. (1967). Convection and stress propagation in the uppermantle. (Preprint.)
ENGEL, A. E. J., and C. G. ENGEL (1962). Progressive metamorphism and amphibolite, Northwest Adirondack Mountains, New York. In: A. E. J. Engel, H. L. James, and B. F. Leonard (eds.), "*Petrologic Studies,*" Buddington Volume, pp. 37–82, Geological Society of America, New York.
ENGEL, C. G., and A. E. J. ENGEL (1961). Composition of basalt cored in Mohole project. *Bull. Am. Assoc. Petrol. Geologists* **45,** 1799.
EVERINGHAM, I. B. (1967). The crustal structure of the south-west of Western Australia. Comm. of Australia, Bureau of Mineral Resources Record No. 1965/97.
EWING, J., and M. EWING (1959). Seismic refraction measurements in the Atlantic Ocean basins, in the Mediterranean Sea, on the mid-Atlantic Ridge and in the Norwegian Sea. *Bull. Geol. Soc. Am.* **70,** 291–318.
——— and ——— (1961). A telemetering ocean-bottom seismograph. *J. Geophys. Res.* **66,** 3863–3878.

——— and ——— (1967). Sediment distribution on the mid-ocean ridges with respect to spreading of the sea floor. *Science* **156**, 1590–1592.

———, ———, T. AITKEN, and W. J. LUDWIG (1968). North Pacific sediment layers measured by seismic profiling. In: "*The Crust and Upper Mantle of the Pacific Area*," pp. 147–173. *Am. Geophys. Union, Geophys. Monograph* **12**.

——— and R. HOUTZ (1969). Mantle reflections in airgun-sonobuoy profiles. *J. Geophys. Res.* **74**, 6706–6709.

FEDOTOV, S. A. (1968). On deep structure, properties of the upper mantle and volcanism of the Kuril-Kamchatka Island Arc according to seismic data. In: "*The Crust and Mantle of the Pacific Area*," pp. 131–139. *Am. Geophys. Union, Geophys. Monograph* **12**.

FERMOR, L. L. (1913). Preliminary note on garnet as a geological barometer and on an infra-plutonic zone in the earth's crust. *Records Geol. Surv. India* **43**.

——— (1914). The relationship of isostasy, earthquakes and vulcanicity to the earth's infra-plutonic shell. *Geol. Mag.* **51**, 65–67.

FISHER, R. L., and R. W. RAITT (1962). Topography and structure of the Peru-Chile Trench. *Deep Sea Res.* **9**, 423–443.

FOX, P. J., and N.D. OPDYKE (1973). Geology of the oceanic crust: magnetic properties of oceanic rocks. *J. Geophys. Res.* **78**, 5139–5154.

FRANCIS, T. J., and G. G. SHOR (1966). Seismic refraction measurements in the northwest Indian Ocean. *J. Geophys. Res.* **71**, 427–450.

FUCHS, K., and M. LANDISMAN (1966). Detailed crustal investigation along a north-south section through the central part of Western Germany. In: "*The Earth Beneath theContinents*," pp. 433–452. *Am. Geophys. Union, Geophys. Monograph* **10**.

———, ST. MULLER, E. PETERSCHMITT, J. ROTHE, A. STEIN, and K. STROBACH (1963). Krustenstructur der Westalpen nach refractionsseismischen Messungen. *Gerlands Beitr. Geophys.* **72**, 149–169.

GIESE, P. (1968). Versuch einer Gliederung der Erdkruste in nördlichen Alpenvorland, in den Ostalpen und in Teilen der Westalpen mit Hilfe characteristicher Refractions-Laufzeit-Kurven sowie eine geologische Deutung. "*Geophysickalische Abhand lungen*" vol. 1 (2) pp. 1–201. Inst. Meteor. Geophysik, Frei. Univ. Berlin.

GREEN, D. H. , and I. B. LAMBERT (1965). Experimental crystallization of anhydrous granite at high pressures and temperatures. *J. Geophys. Res.* **70**, 5259–5268.

——— and A. E. RINGWOOD (1967). An experimental investigation of the gabbro to eclogite transformation and its petrological applications. *Geochim. Cosmochim. Acta* **31**, 767–833.

GREEN, T. H. (1967). An experimental investigation of sub-solidus assemblages formed at high pressure in high-alumina basalt, kyanite eclogite and grosspydite compositions. *Contr. Mineral. Petrol.* **16**, 84–114.

——— (1970). High pressure experimental studies on the mineralogical composition of the lower crust. *Phys. Earth Planet. Interiors* **3**, 441–450.

——— and A. E. RINGWOOD (1968). Genesis of the calc-alkaline igneous rock suite. *Contr. Mineral. Petrol.* **18**, 105–162.

GROUT, F. F. (1938). Petrographic and chemical data on the Canadian shield. *J. Geol.* **46**, 486–504.

GUTENBERG, B. (1955). Wave velocities in the earth's crust. *Geol. Soc. Am. Spec. Paper* **62**, 19–34.

——— (1959a). "*Physics of the Earth's Interior.*" [International Geophysics Series, vol. 1, J. V. Mieghem (ed.).] Academic, New York. 240 pp.

——— (1959b). The asthenosphere low-velocity layer. *Ann. Geofis. Rome* **12**, 439–460.

HALES, A. L., and J. B. NATION (1966). Reflections at the M discontinuity and the origin of microseisms. In: "*The Earth Beneath the Continents*," pp. 529–537. *Am. Geophys. Union, Geophys. Monograph* **10**.

HARRIS, P. G., and J. A. ROWELL (1960). Some geochemical aspects of the Mohorovicic discontinuity. *J. Geophys. Res.* **65**, 2443–2459.

HATHERTON, T. (1969). Gravity and seismicity of asymmetric active regions. *Nature* **221**, 353–355.

HEIER, K. S. (1965a). Metamorphism and the chemical differentiation of the crust. *Geol. Fören. Stóckh. Förhandl.* **87**, 249–256.

——— (1965b). Radioactive elements in the continental crust. *Nature* **208**, 479–480.

——— and J. ADAM (1965). Concentration of radioactive elements in deep crustal material. *Geochim. Cosmochim. Acta* **29**, 53–61.

HELMBERGER, D. V., and G. B. MORRIS (1969). A travel time and amplitude interpretation of a marine refraction profile: Primary waves. *J. Geophys. Res.* **74**, 483–494.

HESS, H. H. (1962). History of ocean basins. In: A. E. J. Engel, H. L. James, and B. F. Leonard (eds.), "*Petrologic Studies*," Buddington Volume, pp. 599–620. Geological Society of America, New York.

——— (1964). The oceanic crust, the upper mantle and the Mayaguez serpentinised peridotite. In: C. A. Burk (ed.) "*A Study of Serpentine*," pp. 169–175. *Nat. Acad. Sci–Nat. Res. Coun. Publ.*

——— (1965). Mid-oceanic ridges and tectonics of sea floor. In: W. F. Whittard and R. Bradshaw (eds.), "*Submarine Geology and Geophysics*," pp. 317–332. Butterworth, London. 464 pp.

HILL, D. P., and L. C. PAKISER (1966). Crustal structure beneath the Nevada Test Site, and Boise, Idaho, from seismic refraction measurements. In: "*The Earth Beneath the Continents*," pp. 391–419. *Am. Geophys. Union, Geophys. Monograph* **10**.

HILL, M. N. (1957). Recent exploration of the ocean floor. In: L. Ahrens, F. Press, K. Rankama, and S. K. Runcorn (eds.). "*Physics and Chemistry of the Earth*," vol. 2, pp. 129–163. Pergamon, London.

HOLMES, A. (1926a). Contributions to the theory of magmatic cycles. *Geol. Mag.* **63**, 306–329.

——— (1926b). Structure of the continents. *Nature* **118**, 586–587.

——— (1927). Some problems of physical geology and the earth's thermal history. *Geol. Mag.* **64**, 263–278.

HYNDMAN, R. D., I. B. LAMBERT, K. S. HEIER, J. C. JAEGER, and A. E. RINGWOOD (1968). Heat flow and surface radioactivity measurements in the Precambrian shield of Western Australia. *Phys. Earth Planet. Interiors* **1**, 129–135.

ISACKS, B., J. OLIVER, and L. R. SYKES (1968). Seismology and the new global tectonics. *J. Geophys. Res.* **73**, 5855–5899.

ITO, K., and G. C. KENNEDY (1968). Melting and phase relations in the plane, tholeiite-lherzolite-nepheline basanite to 40 kilobars with petrological implications. *Contr. Mineral. Petrol.* **19**, 177–211.

——— and ——— (1970). The fine structure of the basalt-eclogite transition. *Mineral. Soc. Am. Spec. Paper* **3**, 77–83.

——— and ——— (1971). An experimental study of the basalt-garnet granulite-eclogite transition. In: "*The Structure and Physical Properties of the Earth's Crust*," pp. 303–314. *Am. Geophys. Union, Geophys. Monograph* **14.**

JAMES, D. E. (1971). A plate tectonic model for the evolution of the Central Andes, *Bull. Geol. Soc. Am.* **82,** 3325–3346.

———, T. J. SMITH, and J.S. STEINHART (1968). Crustal structure of the middle Atlantic states. *J. Geophys. Res.* **73,** 1983–2007.

KANAMORI, H., and H. MIZUTANI (1965). Ultrasonic measurement of elastic constants of rocks under high pressures. *Bull. Earthquake Res. Inst.* **43,** 173–194.

KENNEDY, G. C. (1956). Polymorphism in the felspars at high temperatures and pressures. *Bull. Geol. Soc. Am.* **67,** 1711–1712 (abstract).

——— (1959). The origin of continents, mountain ranges, and ocean basins. *Am. Sci.* **47,** 491–504.

KOSMINSKAYA, I. P. (1962). Comparisons between detailed and less-detailed observations. (In Russian.) In: "*Glubinnoye Seysmicheskoye Zondirovaniye Zemnoy Kory V SSSR,*" pp. 479–481. Akad. Nauk SSSR, Moscow.

LACHENBRUCH, A. H. (1970). Crustal temperatures and heat production: Implications of the linear heat-flow relation. *J. Geophys. Res.* **75,** 3291–3300.

LAMBERT, I. B. and K. S. HEIER (1967). The vertical distribution of uranium, thorium and potassium in the continental crust. *Geochim. Cosmochim. Acta* **31,** 377–390.

——— and ——— (1968). Geochemical investigations of deep-seated rocks in the Australian shield. *Lithos* **1,** 30–53.

LANGSETH, M. C., X. LE PICHON, and M. EWING (1966). Crustal structure of the mid-oceanic ridges (5), Heat flow through the Atlantic Ocean floor and convection currents. *J. Geophys. Res.* **71,** 5331–5335.

LEE, W., and S. UYEDA (1965). Review of heat flow data. In: "*Terrestrial Heat Flow,*" pp. 87–190. *Am. Geophys. Union, Geophys. Monograph* **8.**

LE PICHON, X. (1968). Sea-floor spreading and continental drift. *J. Geophys. Res.* **73,** 3661–3697.

LIEBSCHER, H. J. (1964). Deutungsversuchen fur die Structur der tieferen Erdruste nach reflexionsseimischen and gravimetrischen Messungen im deutschen Alpenvorland. *Z. Geophys.* **30,** 51–96.

LOVERING, J. F. (1958). The nature of the Mohorovicic discontinuity. *Trans. Am. Geophys. Union* **39,** 947–955.

LUYENDYK, B. P., and W. G. MELSON (1967). Magnetic properties and petrology of rocks near the crest of the mid-Atlantic Ridge. *Nature* **215,** 147–149.

MARKHININ, E. K. (1968). Volcanism as an agent of formation of the earth's crust. In: "*The Crust and Upper Mantle of the Pacific Area,*" pp. 413–422. *Am. Geophys. Union, Geophys. Monograph* **12.**

MATTHEWS, D. H., F. J. VINE, and J. R. CANN (1965). Geology of an area of the Carlsberg Ridge, Indian Ocean. *Bull. Geol. Soc. Am.* **76,** 675–682.

MCCAMY, K., and R. P. MEYER (1966). Crustal results of fixed multiple shots in the Mississippi Embayment. In: "*The Earth Beneath the Continents,*" pp. 370–381. *Am. Geophys. Union, Geophys. Monograph* **10.**

MCDONALD, G. J. F. , and N. F. NESS (1960). Stability of phase transitions in the earth. *J. Geophys. Res.* **65,** 2173–2190.

MCKENZIE, D. (1967). Some remarks on heat flow and gravity anomalies. *J. Geophys. Res.* **72,** 6261–6273.

MELSON, W. G., and T. H. VAN ANDEL (1966). Metamorphism in the mid-Atlantic Ridge, 22° N latitude. *Marine Geol.* **4**, 165–186.

——— and G. THOMPSON (1971). Petrology of a transform fault zone and adjacent ridge segments. *Phil. Trans. Roy. Soc. London* **A268**, 423–442.

———, ———, and T. H. VAN ANDEL (1968). Volcanism and metamorphism in the mid-Atlantic Ridge, 22° N latitude. *J. Geophys. Res.* **73**, 5925–5941.

MITRONOVAS, W., and B. ISACKS (1971). Seismic velocity anomalies in the upper mantle beneath the Tonga-Kermadec Island arc. *J. Geophys. Res.* **76**, 7154–7180.

MIYASHIRO, A., F. SHIDO, and M. EWING (1971). Metamorphism in the mid-Atlantic Ridge near 24° and 30° N. *Phil. Trans. Roy. Soc. London* **A268**, 589–603.

MOORES, E. M., and F. J. VINE (1971). The Troodos Massif, Cyprus and other ophiolites as oceanic crust: evaluation and implications. *Phil. Trans. Roy. Soc. London* **A268**, 443–446.

MORGAN, W. J. (1965). Gravity anomalies and convection currents, 2. The Puerto Rico Trench and the Mid-Atlantic Rise. *J. Geophys. Res.* **70**, 6189–6204.

——— (1968). Rises, trenches and crustal blocks. *J. Geophys. Res.* **73**, 1959–1982.

NAKAMURA, Y., and B. F. HOWELL (1964). Maine seismic experiment: frequency spectra of refraction arrivals and the nature of the Mohorovicic discontinuity. *Bull. Seismol. Soc. Am.* **54**, 9–18.

NOCKOLDS, S. R. (1954). Average chemical compositions of some igneous rocks. *Bull. Geol. Soc. Am.* **65**, 1007–1032.

O'BRIEN, P. N. S. (1968). Lake Superior crustal structure—a reinterpretation of the 1963 seismic experiment. *J. Geophys. Res.* **73**, 2669–2689.

O'CONNELL, R. J., and G. J. WASSERBURG (1967). Dynamics of the motion of a phase change boundary to changes in pressure. *Rev. Geophys.* **5**, 329–410.

OLIVER, J., and B. ISACKS (1967). Deep earthquake zones, anomalous structures in the upper mantle and the lithosphere. *J. Geophys. Res.* **72**, 4259–4275.

OXBURGH, E. R., and D. L. TURCOTTE (1968). Mid-ocean ridges and geotherm distribution during mantle convection. *J. Geophys. Res.* **73**, 2643–2661.

——— and ——— (1970). Thermal structure of island arcs. *Bull. Geol. Soc. Am.* **81**,1665–1668.

PAKISER, L. C. (1963). Structure of the crust and upper mantle in the western United States. *J. Geophys. Res.* **68**, 5747–5756.

——— (1965). The basalt-eclogite transformation and crustal structure in the western United States. *U.S. Geol. Surv. Prof. Paper* **525-B**, 1–8.

——— and R. ROBINSON (1966). Composition of the continental crust as estimated from seismic observations. In: "*The Earth Beneath the Continents*," *Am. Geophys. Union, Geophys. Monograph* **10**.

PHINNEY, R. A. (1964). Structure of the earth's crust from spectral behaviour of long-period body waves. *J. Geophys. Res.* **69**, 2997–3018.

PITMAN, W. C., E. M. HERRON, and J. R. HEIRTZLER (1968). Magnetic anomalies in the Pacific and sea-floor spreading. *J. Geophys. Res.* **73**, 2069–2085.

POLDERVAART, A. (1955). Chemistry of the earth's crust. *Geol. Soc. Am. Spec. Paper* **62**, 119–144.

RAITT, R. (1957). Seismic refraction studies of Einiwetok Atoll. *U.S. Geol. Surv. Prof. Paper* **260-S**, 685–698.

———(1963). The crustal rocks. In: M. N. Hill (ed.), "*The Sea*," vol. 3, chap. 6. Interscience, a division of Wiley, New York. 963 pp.

RICHARDS, T. C., and D. J. WALKER (1959). Measurement of the earth's crustal thickness in Alberta. *Geophysics* **24**, 262–284.

RICHTER, D.H., and J. G. MOORE (1966). Petrology of the Kilauea Ikilava lake, Hawaii. *U.S. Geol. Surv. Prof. Paper* **537-B**, 1–26.

RINGWOOD, A. E. (1969). Composition and evolution of the upper mantle. In: "*The Earth's Crust and Upper Mantle*," pp. 1–17. *Am. Geophys. Union, Geophys. Monograph* **13.**

——— and D. H. GREEN (1964). Experimental investigations bearing on the nature of the Mohorovicic discontinuity. *Nature* **201**, 566–567.

——— and ——— (1966a). An experimental investigation of the gabbro to eclogite transformation and some geophysical implications. *Tectonophysics* **3**, 383–427.

——— and ——— (1966b). Petrological nature of the stable continental crust. In: "*The Earth Beneath the Continents*," pp. 611–619. *Am. Geophys. Union, Geophys. Monograph* **10.**

ROY, R. F., D. D. BLACKWELL, and F. BIRCH (1968). Heat generation and continental heat flow provinces. *Earth Planet. Sci. Letters* **5**, 1–12.

SEDERHOLM, J. J. (1925). The average composition of the earth's crust in Finland. *Bull. Comm. Geol. Finlande* **12**(70), 1–20.

STEINHART, J. S., and R. P. MEYER (1961). Explosion studies of continental structure. *Carnegie Inst. Washington Publ.* **622**, 409 pp.

STISHOV, S. M. (1963). The nature of the Mohorovicic discontinuity. *Akad. Nauk SSSR, Ser. Geofiz.* **1**, 42–48.

SUMNER, J. S. (1954). Consequences of a polymorphic transition at the Mohorovicic discontinuity. *Trans. Am. Geophys. Union* **35**, 385.

SUTTON, G. H., G. MAYNARD, and D. HUSSONG (1971). Widespread occurrence of a high-velocity basal layer in the Pacific crust found with repetitive sources and sonobuoys. In: "*The Structure and Physical Properties of the Earth's Crust*," pp. 193–209. *Am. Geophys. Union, Geophys. Monograph* **14.**

TALWANI, M., X. LE PICHON, and M. EWING (1965). Crustal structure of the mid-ocean ridges. *J. Geophys. Res.* **70**, 341–352.

———, G. H. SUTTON, and J. L. WORZEL (1959). A crustal section across the Puerto Rico Trench. *J. Geophys. Res.* **64**, 1545–1555.

———, C. WINDISCH, and M. LANGSETH (1971). Reykjanes ridge crust: A detailed geophysical study. *J. Geophys. Res.* **76**, 473–517.

TATEL, H. A., and M. A. TUVE (1955). Seismic exploration of a continental crust. In: A. Poldervaart (ed.), "*Crust of the Earth*," pp. 35–50. *Geol. Soc. Am. Spec. Paper* **62.**

TAYLOR, S. R., and A. J. WHITE (1965). Geochemistry of andesites and the growth of continents. *Nature* **208**, 271–273.

TUVE, M. A., H. E. TATEL, and P. J. HART (1954). Crustal structure from a seismic exploration. *J. Geophys. Res.* **59**, 415–422.

UYEDA, S., and V. VACQUIER (1968). Geothermal and geomagnetic data in and around the island arc of Japan. In: "*The Crust and Upper Mantle of the Pacific Area*," pp. 349–366. *Am. Geophys. Union, Geophys. Monograph* **12.**

VAN BEMMELEN, R. W. (1964). The evolution of the Atlantic Mega-Undation. *Tectonophysics* **1**, (5), 385–430.

VINE, F. J. (1966). Spreading of the ocean floor; New evidence. *Science* **154**, 1405–1415.

——— and J. T. WILSON (1965). Magnetic anomalies over a young oceanic ridge off Vancouver Island. *Science* **150**, 485–489.

VOGT, J. H. L. (1931). On the average composition of the earth's crust. *Skrifter Norske Videnskaps Akad. Oslo, (1), Mat-Naturv.* Kl. **7**, 1–48.

WETHERILL, G. W. (1961). Steady-state calculations bearing on geological implications of a phase-transition Mohorovicic discontinuity. *J. Geophys. Res.* **66**, 2983–2993.

WILSON, J. T. (1954). The development and structure of the crust. In: G. P. Kuiper (ed.), "*The Earth as A Planet*," pp. 138–214. Univ. Chicago Press, Chicago.

WOOLLARD, G. P. (1960). Seismic crustal studies during the I.G.Y. *Trans. Am. Geophys. Union* **41**, 351–355.

——— (1962). The relation of gravity anomalies to surface elevation, crustal structure and geology. *Dept. Geol., Univ. Wisconsin Res. Rept.* **62-9** **(II)**, 57–59.

——— (1966). Regional isostatic relations in the United States. In: "*The Earth Beneath the Continents*," pp. 557–594. *Am. Geophys. Union, Geophys. Monograph* **10**.

——— (1968). The interrelationship of the crust, the upper mantle and isostatic gravity anomalies in the United States. In: "*The Crust and Upper Mantle in the Pacific Area*," pp. 312–341. *Am. Geophysical Union, Geophys. Monograph* **12**.

WYLLIE, P. J. (1963). The nature of the Mohorovicic discontinuity. A compromise. *J. Geophys. Res.* **68**, 4611–4619.

——— (1971). "*The Dynamic Earth*," Wiley, New York. 416 pp.

YODER, H. S., and C. E. TILLEY (1962). Origin of basalt magmas: an experimental study of natural and synthetic rock systems. *J. Petrol.* **3**, 342–532.

3
COMPOSITION OF THE UPPER MANTLE

3-1 INTRODUCTION

The seismic *P*-wave velocity in the mantle immediately beneath stable continental regions and deep oceanic basins is usually within the range 8.2 ± 0.2 km/sec. This property, together with certain broad petrological and chemical limitations,[1] effectively restricts the dominant mineralogical composition of the upper mantle to some combination of olivine, pyroxene, garnet, and perhaps, in restricted regions, amphibole. The two principal rock types carrying these minerals are peridotite (olivine-pyroxene) and eclogite (pyroxene-garnet). *P*-wave velocities of fresh[2] peridotites and eclogites mostly fall in the range 8.0 to 8.5 km/sec.[3,4,5] Complete mineralogical transitions between the two major rock types are uncommon and usually of only local significance. These relationships have given rise to the alternative hypotheses that the upper mantle is of peridotitic or eclogitic composition.

[1]Sections 5-1 and 6-2.

[2]Velocities lower than 8 km/sec for eclogites and peridotites have frequently been recorded, but petrographic investigations revealed the presence of extensive secondary alteration in these rocks. The velocity range cited refers to measurements at pressures of 6 kbars, which largely eliminates the effects of initial porosity.

[3]Birch (1960, 1961).

[4]Verma (1960).

[5]Kanamori and Mizutani (1965).

The eclogitic hypothesis is often coupled with the hypothesis that the Mohorovicic discontinuity is caused by an isochemical phase change from a gabbroic or granulitic lower crust into an eclogitic upper mantle. It was demonstrated in Chap. 2 that this hypothesis is no longer tenable. Nevertheless, the possibility that the Moho is caused by a chemical change from an acid-intermediate lower crust into an eclogitic upper mantle might still be considered.[1]

During the last 50 years, most earth scientists have favoured the view that the upper mantle is generally of peridotitic composition rather than eclogitic. Nevertheless, there have been periods when the latter hypothesis was strongly supported, e. g., by Goldschmidt and Holmes during the 1920s. A second revival occurred during the 1950s but succumbed to advances in experimental petrology some 10 years later.[2] More recently, there have been fresh attempts to revive this hypothesis jointly upon the basis of some further developments in experimental petrology[1,3] and upon studies of density distribution in the upper mantle obtained primarily by the inversion of free-oscillation data. Although the peridotitic hypothesis is probably still favoured by a majority of earth scientists (including the author), the eclogitic hypothesis has received a substantial amount of favourable consideration in some quarters. In view of this revival, a review of evidence bearing upon the roles of peridotite and eclogite in the upper mantle appears desirable.

3-2 DIAGNOSTIC PHYSICAL CRITERIA

Poisson's Ratio

In this section we will consider some physical properties of the upper mantle which might aid in discriminating between peridotitic and eclogitic hypotheses. Although the *P*- and *S*-wave velocities found for fresh peridotites and eclogites overlap extensively and cannot be used as discriminants, there is a possibility that the ratio of these velocities (V_p/V_s) or Poisson's ratio σ given by

$$\sigma = \frac{1}{2}\left(\frac{R^2 - 2}{R^2 - 1}\right) \qquad \text{where } R = \frac{V_p}{V_s}$$

may be a more sensitive indicator.[4] This ratio is only slightly affected by minor secondary alteration and porosity. Values of Poisson's ratio have been determined for the mantle beneath Precambrian shields and stable continental regions, where *P* and *S* arrivals have been clearly recorded and accurately timed from the

[1]Ito and Kennedy (1970, 1971).
[2]Section 2-4.
[3]Critically discussed in Secs. 2-4 and 3-2.
[4]Kanamori and Mizutani (1965).

same explosion.[e.g.,1,2,3] The values found are in the range 0.245 to 0.260. Poisson's ratios for Mg-rich olivines (Fo ~ 90%) have been accurately determined and fall in the range 0.245 to 0.255.[4,5,6] Peridotite is clearly a satisfactory candidate for this region of the upper mantle. On the other hand, Poisson's ratios for two fresh eclogites were found to lie between 0.30 and 0.32.[7] On these grounds, it was concluded[7] that eclogite was unlikely to be the major constituent of the upper mantle. However, some caution is necessary before accepting this conclusion, and many more σ determinations on fresh eclogites[8] and on the individual minerals of eclogites are necessary. If Poisson's ratios exceeding 0.27 are found to be characteristic of fresh eclogites of deep-seated origin, an important limitation to the composition of the mantle underlying stable continental regions will have been established.

Anisotropy

Hess (1964a) pointed out that upper mantle *P*-wave velocities in the northeastern Pacific between California and Hawaii were correlated with the direction of the seismic profiles. He suggested that nearly all the variations could be explained by anisotropy of velocity, with a maximum value in the eastward direction and a minimum in the northward direction. This suggestion was tested in detailed surveys of two locations near California by Raitt and colleagues (1969). At each location, anisotropy of about 0.3 km/sec was observed, agreeing in direction with the apparent anisotropy of the entire region. A further detailed survey of a region near Hawaii[9] revealed an even stronger anisotropy, amounting to 0.6 km/sec (Fig. 3-1) and agreeing in orientation with that suggested by Hess. Several further occurrences of anisotropy in the upper mantle have since been found[10,11,12] and are shown in Fig. 3-1.

The existence of anisotropy may also be a useful discriminant among petrological models of the upper mantle.[13,14,15] Hess suggested that it could be

[1]Bolt, Doyle, and Sutton (1958).
[2]Doyle and Everingham (1964).
[3]Denham et al. (1972).
[4]Verma (1960).
[5]Kumazawa and Anderson (1969).
[6]Graham and Barsch (1969).
[7]Kanamori and Mizutani (1965).
[8]Poisson's ratio for the Healdsburg ecolgite (California) was 0.26. In this case, *P* and *S* velocities were determined by different authors, and it is not certain whether the specimen used was the same. Furthermore, the Healdsburg "eclogite" is a crustal type of unusual composition with a relatively low *P* velocity (7.94 km/sec at 6 kbars). It contains only 24% garnet and, petrologically, might be classed as a garnet pyroxenite. (Simmons, 1964.)
[9]Morris, Raitt, and Shor (1969).
[10]Keen and Tramontini (1970).
[11]Keen and Barrett (1971).
[12]Raitt et al. (1972).
[13]Avé Lallemant and Carter (1970).
[14]Kumazawa et al. (1971).
[15]Carter et al. (1972).

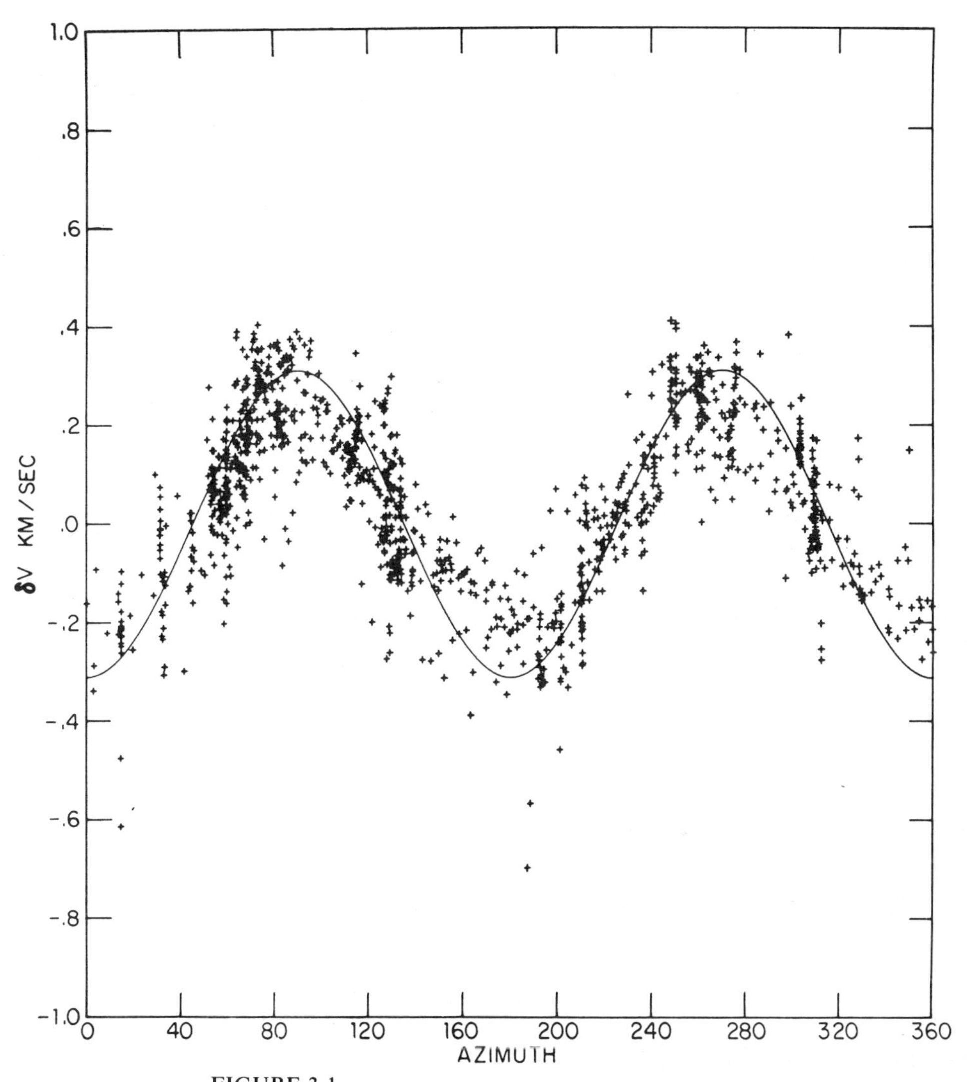

FIGURE 3-1
Combined data from Hawaii region showing velocity anisotropy given as deviations from the mean velocity of 8.159 km/sec plotted as a function of azimuth. (*From Morris, Raitt, and Shor*, 1969, *with permission. Copyright American Geophysical Union.*)

explained by preferred orientation of olivine crystals, which individually possess extremely high anisotropy.[1] It is seen from Fig. 3-2 that anisotropy is a common feature of peridotites and that the observed anisotropy of the upper mantle could be readily explained by a peridotitic upper mantle. On the other hand, eclogites containing about 50% garnet which is isotropic, shows much smaller anisotropies,

[1]Verma (1960).

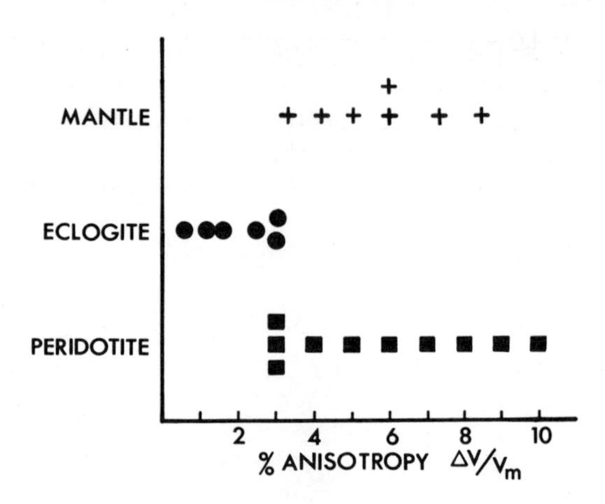

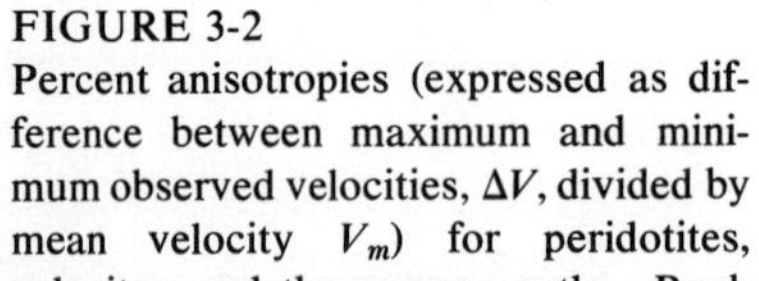

FIGURE 3-2
Percent anisotropies (expressed as difference between maximum and minimum observed velocities, ΔV, divided by mean velocity V_m) for peridotites, eclogites, and the upper mantle. Rock data from Birch (1960), Simmons (1964), and Christensen (1966). For mantle data, see text.

mostly falling well below the observed mantle values. The anisotropy observations to date thus indicate a peridotitic rather than eclogitic suboceanic mantle. Nevertheless, it should be noted that the evidence so far available is rather limited, and further determinations of the anisotropy of mantle-type eclogite from diamond pipes[1] and of anisotropy in the upper mantle are desirable.

Density from Isostatic Considerations

The densities of peridotitic rocks which might qualify on petrological grounds[2] as major upper mantle constituents range between 3.25 and 3.40 g/cm^3, with an average close to 3.32 g/cm^3. The densities of fresh, unaltered eclogites, on the other hand, range between 3.4 and 3.6 g/cm^3, with an average of about 3.5 g/cm^3.[3] The types of eclogites which best qualify as upper mantle constituents (quartz-free varieties such as are found in diamond pipes) fall in the higher part of this range. The density distributions appropriate for peridotitic and eclogitic upper mantles are shown in Fig. 3-3. It is clear that independent knowledge of the density distribution in the upper mantle could discriminate between peridotitic and eclogitic models.[3,4]

The interpretation of gravity observations combined with the theory of isostasy provide important limitations to the density of the uppermost mantle. Subject to certain simplifying assumptions, these are capable of providing the mean density difference between crust and mantle. If the density of the crustal sections can be estimated by independent methods, the density of the mantle is thereby obtained. Detailed gravity and seismic investigations of normal[5] oceanic and conti-

[1]Eclogites and garnet pyroxenites of metamorphic crustal origin are less relevant to this problem.
[2]Sections 3-3, 3-4, and 4-5.
[3]Ringwood and Green (1966a).
[4]Clark and Ringwood (1964).
[5]I.e., excluding tectonically active regions such as trenches, island arc, mid-ocean ridges, and young mountain belts.

nental sections yielded a density contrast between crust and mantle of 0.43 g/cm.[1,2] Limitations on the density of the crust are obtained from direct observations of the occurrence and densities of crustal rocks combined with geologic inferences concerning their abundances. Another method of obtaining the mean crustal density is from the seismic velocity distribution in the crust, combined with knowledge of the empirical relationship between seismic velocity and density for common rock types.[3,4] Arguments based upon the above methods have led to the widely accepted view that the mean density of the normal continental crust is between 2.8 and 2.9 g/cm^3. From these values, together with the density contrast between crust and mantle as deduced from gravity and seismic data, it can be concluded[5,6] that the most probable mean density of the upper mantle is between 3.3 and 3.4 g/cm^3, thus implying a peridotitic uppermost mantle, rather than an eclogitic.[5] Woollard (1970) has been responsible for the most comprehensive development of these arguments. He concluded that the mean crustal density was in the range 2.87 to 2.89 g/cm^3 and that the crust-mantle density contrast was in the range 0.38 to 0.42 g/cm^3, yielding mantle densities of 3.24 to 3.32 g/cm^3. Woollard pointed out that this density range was consistent with a peridotitic upper mantle but could not be reconciled with an upper mantle of eclogite, which would imply a crust-mantle density contrast exceeding 0.5 g/cm^3.

The principal assumption underlying this conclusion is that isostatic compensation is dominantly achieved at about the depth of the continental Mohorovicic discontinuity. Although small regional differences in mantle density undoubtedly persist below this depth,[7] Woollard's investigation[8] led him to conclude that "no firm case can therefore be made in general for mass distributions deeper than the [continental] M-discontinuity being important in maintaining isostatic equilibrium."

A detailed seismic and gravity investigation of the crustal structure of the Middle Atlantic states (United States) by James, Smith, and Steinhart (1968) provided further information on the density of the upper mantle beneath a localized region where crustal thickness varied from 30 to 60 km. From the determined seismic velocity-depth structure, three-dimensional computations of Bouguer gravity anomalies were made, using a range of possible linear velocity-density relationships. It was found that the predicted Bouguer anomalies were in close agreement with the observed anomalies for a range of velocity-density relationships where $d\rho/dV$ varied between 0.20 and 0.23. A value of "$d\rho/dV$ of 0.23 yielded gravity anomalies that are as large as can be tolerated." Birch's velocity-density relationship shows that, for rocks of constant mean atomic weight $\bar{M}$,

[1]Worzel and Shurbet (1955).
[2]Drake, Ewing, and Sutton (1959).
[3]Birch (1961).
[4]Woollard (1962, 1970).
[5]Ringwood and Green (1966a).
[6]Ringwood (1969).
[7]Clark and Ringwood (1964).
[8]Woollard (1970).

$d\rho/dV$ is 0.33, which is far outside the permissible range. The observed $d\rho/dV$ thus implied a substantial *decrease* in mean atomic weight in the region of the lower crust and upper mantle. This could be explained most readily by a transition from a mafic-intermediate (amphibolitic) lower crust ($\bar{M} = 21.5$ to 22.0) into a peridotitic upper mantle ($\bar{M} = 21$). The authors concluded that there is no reasonable way in which a decrease in mean atomic weight could be provided if the upper mantle were composed of eclogite ($\bar{M} = 22.0$).

In an analogous investigation of the structure of the Lake Superior region, where crustal thicknesses varying from 25 to 55 km were encountered, Smith, Steinhart, and Aldrich (1966) demonstrated that a self-consistent solution to the gravity and seismic observations in the light of Birch's empirical velocity-density relationship also required a decrease of mean atomic weight with depth by about 0.3 to 0.6. Ringwood and Green (1966b) pointed out that this was consistent with the presence of a peridotitic upper mantle but could not be explained if the upper mantle were composed of eclogite.

Crust-Mantle Density Ratio Obtained by Inversion of Surface-Wave Data

A further indication of the density of the upper mantle beneath the Eastern United States was obtained by Dorman and Ewing (1962) from the inversion of surface-wave data. In this case, the crustal structure and S-wave velocity distribution had previously been obtained by refraction methods so that the density solution was in principle determinate. The method yielded a ratio of 0.866 for the ratio of crust-to-mantle density. This ratio, obtained without reference to gravity observations, represented the best agreement between refraction results and the surface-wave data. Taking a mean crustal density of 2,88 g/cm^3, the corresponding mantle density is 3.33 g/cm^3, again implying a peridotitic uppermost mantle rather than an eclogitic. Uncertainties in this estimate should not be ignored—it is not clear to what extent possible errors in the refraction data might alter the crust-mantle density ratio derived from surface-wave data. Nevertheless, the fact that the result represents the best solution to extensive data from a well-studied area and that the solution is independent of, but nevertheless agrees closely with, inferences from gravity data in the same area[1] is encouraging.

Density Distributions Obtained by Monte Carlo Inversion

In an important series of papers, which have been widely quoted, Press[2] has attempted to derive density and shear velocity distributions throughout the earth by means of systematic searches by Monte Carlo methods for models which satisfy, with specified precisions, the periods (97) of free oscillation of the earth, surface-wave phase velocities, travel times of compressional and shear waves, and mass and moment of inertia of the earth. Several millions of models were inves-

[1] James, Smith, and Steinhart (1968).
[2] Press (1968, 1969, 1970, 1972).

tigated, only a few tens of which were found to satisfy the boundary conditions. All these successful models were characterized by the occurrence of densities in the vicinity of 3.5 g/cm^3 between depths of 80 and 150 km. From these high densities Press concluded that this region is composed largely of eclogite (Fig. 3-3).

This conclusion, however, is at variance with other evidence discussed elsewhere in this chapter, which implies that the upper mantle is composed dominantly of peridotitic rocks, with eclogite a widely distributed but relatively minor component, except perhaps in certain localized regions. In view of these differing conclusions, and because Press' models have had a considerable influence upon recent geophysical and petrological thought, further discussion appears desirable.

One doubtful feature of all the successful models[1] is that the ratios of bulk modulus K to shear modulus μ decrease sharply from about 2.5 to 2.0 (average) between depths of about 200 and 350 km. This implies that the ratio V_p/V_s decreases by 10% in this depth interval. Such a decrease is inconsistent with recent body-wave P and S distributions which show that V_p/V_s is either constant or, more probably, increases with depth in this region.[e.g.,2,3,4] Furthermore, a decrease of K/μ with depth between 200 to 350 km is not easily reconciled with the elastic properties of minerals believed to be present in this region unless the existence of unreasonably high temperature gradients are assumed (Table 3-1).

[1]Press (1970, fig. 19).
[2]Johnson (1967).
[3]Julian and Anderson (1968).
[4]Anderson and Julian (1969).

Table 3-1 PRESSURE AND TEMPERATURE DERIVATIVES OF BULK (K) AND SHEAR (μ) MODULI OF OLIVINE AND GARNET. (NOTE LARGE DIFFERENCES BETWEEN PRESSURE DERIVATIVES OF K AND μ COMPARED TO SMALL DIFFERENCES BETWEEN CORRESPONDING TEMPERATURE DERIVATIVES)

		Olivine*	Garnet†
$\left(\frac{\partial K_s}{\partial P}\right)_T$	Mbar^{-1}	5.1	5.4
$\left(\frac{\partial \mu}{\partial P}\right)_T$	Mbar^{-1}	1.8	1.4
$\left(\frac{\partial K_s}{\partial T}\right)_P$	10^{-4}Mbar/°C	−1.6	−0.2
$\left(\frac{\partial \mu}{\partial T}\right)_P$	10^{-4}Mbar/°C	−1.3	−0.1

*Kumazawa and Anderson (1969).
†Anderson, Schreiber, Liebermann, and Soga (1968).

A second characteristic is that all except one of the successful models[1] possess density minima between 200 and 400 km which are substantially smaller than the density of pyrolite at the corresponding depth. This feature is particularly evident in the family of successful tectonic models,[2] *all* of which are closely grouped at 300 km around a density of 3.3 g/cm^3.

The average density distribution for the successful tectonic models is shown in Fig. 3-3. The pyrolite model provides the lowest mantle densities which can be reconciled with geochemical and petrological evidence. The densities indicated by Press' models at depths of 250 to 320 km (Fig. 3-3) imply that the silicates in this region contain negligible amounts of FeO. In view of the considerable degree of convective "stirring" in the upper mantle required by modern theories of plate tectonics and the evidence supplied by considerations of basalt genesis[3] and mantle inclusions in diamond pipes of deep-seated origin,[4] this implication must be regarded as being improbable. The investigation by Graham (1970) of the elasticity and constitution of the mantle between depths of 150 to 350 km should also be mentioned. Using recent measurements of the elastic properties of olivine, combined with seismic data, Graham concluded that a widely self-consistent explanation of the elastic properties and density of the mantle could be provided if the region were dominantly composed of olivine with Mg/Mg + Fe ratios in the range 0.88 ± 0.03. This provides independent evidence of the applicability of the pyrolite model in this depth interval.

It is instructive to examine the range of *average* densities obtained for the upper 400 km of the mantle according to different models. If inversion methods are unable to resolve the mean density of the upper 400 km of the mantle with adequate precision, there are grounds for scepticism regarding the details of fine structure which can be resolved within this region. The mean upper mantle densities for a number of models are given in Table 3-2. It is seen that Press' (1970)

[1]Press (1970, fig. 15).
[2]Press (1972, fig. 12).
[3]Section 4-6.
[4]Section 3-4.

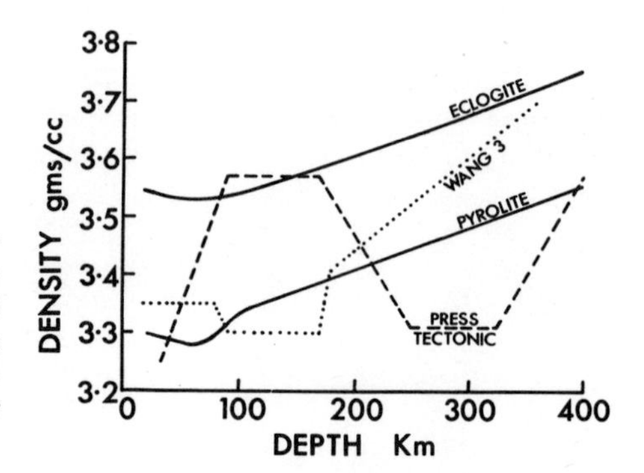

FIGURE 3-3
Upper mantle density distributions corresponding to pyrolite and eclogite compositions (Clark and Ringwood, 1964) compared to models given by Wang (1970) and Press (1972, fig. 12). The latter represents the mean of solutions for the successful tectonic models.

Table 3-2 MEAN UPPER MANTLE DENSITIES OF SOME RECENT EARTH MODELS WHICH SATISFY FREE-OSCILLATION DATA COMPARED WITH MEAN DENSITIES APPROPRIATE TO PYROLITIC AND ECLOGITIC UPPER MANTLES

Model	Depth range (km)	Mean density (g/cm³)	Ref.
Haddon-Bullen I	15–350	3.379	1
Dziewonski O_1	20–400	3.395	2
Wang SEM	20–360	3.414	3
Wang III	20–360	3.446	4
Derr	30–400	3.501	5
Press tectonic	35–400	3.438	6
Press average	10–400	3.481	7
Press maximum	10–371	3.554	8
Press minimum	10–371	3.420	9
Pyrolite	20–400	3.414–3.430	10
Eclogite	20–400	3.630	10

References and notes:

1 Haddon and Bullen (1969).
2 Dziewonski (1971).
3 Wang (1972).
4 Wang (1970).
5 Derr (1969).
6 Press (1972, fig. 12).
7 Press (1970). Average of successful models shown in fig. 15.
8 Press (1970, figs. 9–13). Maximum upper mantle density of successful model.
9 Press (1970, figs. 9–13). Minimum upper mantle density of successful model.
10 Clark and Ringwood (1964). Range due to uncertainty in thermal expansion of olivine. Clark and Ringwood's model has been modified slightly to allow for stability of garnet pyrolite at lower pressures as demonstrated by later experimental data.

successful models yield average densities for the upper mantle ranging from 3.420 to 3.554. This wide range suggests that the inversion methods employed possess a rather limited resolving power.

Haddon and Bullen,[1] Dziewonski,[2] and Wang[3] have employed alternative methods independent of free-oscillation data to formulate approximate density models for the earth based upon the use of *P* and *S* body-wave velocities via the Williamson-Adams equation, semi-empirical equations of state, and constrained by mass and moment of inertia considerations. These models, based upon a limited number of parameters, have then been perturbed by successive approximations until close agreement with the earth's free-oscillation periods was obtained. Resultant earth models were very different from those of Press, although

[1]Haddon and Bullen (1969).
[2]Dziewonski (1970, 1971).
[3]Wang (1970, 1972).

they satisfied essentially similar constraints. Wang's model for the upper mantle (Fig. 3-3) is almost a mirror image of the average Press model. In a second investigation, Wang[1] chose a density distribution appropriate to a pyrolite upper mantle and demonstrated that this model was capable of matching the free periods within experimental uncertainties. Dziewonski[2] found that a constant upper mantle density of 3.395 g/cm^3 extending down to 400 km was consistent with observational data.

The mean densities of the upper mantle to depths of 350 to 400 km obtained by a number of authors and claimed to be consistent with free-oscillation data are shown in Table 3-2. These results cast serious doubts upon the uniqueness of Press' results. Press suggests that his own models fit the data slightly better than some of the alternatives, but this appears to be debatable when adequate account is taken of lateral heterogeneities and anisotropy in the upper mantle, of possible observational errors in free-oscillation periods and body-wave travel times,[3,4] and the possibility of imperfections in elasticity influencing the free periods and shear velocity distributions.[5,6] Numerical experiments by Dziewonski[3] and Wang[4] demonstrated that minor changes in shear-wave velocity distribution which were much smaller than the experimental uncertainties were capable of causing substantial changes in derived density in the upper mantle. From these experiments, Wang[4] concluded that the density distribution within the upper mantle cannot yet be adequately fixed by inversion methods.

Several workers have investigated the reasons for the differing results. Dziewonski (1971) has pointed out that the thickness of the high-velocity lithosphere (lid) has an important effect upon upper mantle density distribution. Furthermore, the latter is especially sensitive to small changes in *S*-wave velocity distribution between depths of 400 and 600 km. This aspect has been further studied by Worthington et al. (1972), who conclude that the principal reason for the difference between the results of Press and Wang is that the *S*-velocity envelope chosen by Press is overconstrained near 400 km. Wang's *S*-velocity distribution is consistent with travel-time data but falls slightly below Press' envelope near this depth. This small difference nevertheless has a drastic effect on densities in the 70 to 200 km depth interval.

Conclusion

The hypothesis that the uppermost mantle to a depth of about 70 km is composed dominantly of peridotitic rocks rather than of eclogite is supported by several independent lines of physical evidence—(1) Poisson's ratio, (2) anisotropy, (3) den-

[1]Wang (1972, Model SEM, Table 3-2).
[2]Dziewonski (1971).
[3]Dziewonski (1970, 1971).
[4]Wang (1970, 1972).
[5]Haddon and Bullen (1969).
[6]Solomon (1972).

sity derived from gravity observations, and (4) density derived by inversion of surface-wave data in a region where the seismic crust-mantle structure had been previously determined by refraction methods. These lines of evidence have varying degrees of uncertainty attached to them. Nevertheless, the fact that they all point in the same direction must be regarded as significant. Inferences on the composition of the uppermost mantle provided by density limitations obtained from combined gravity-seismic investigations and by anisotropy observations are probably the best founded of those discussed.

Evidence on the composition of the mantle between depths of 70 and 200 km obtained by Monte Carlo inversion of free-oscillation and other data is conflicting, and it appears that more accurate resolution of shear-wave velocity distributions will be necessary before this conflict can be resolved. Perhaps the most significant result obtained by inversion methods to date is that the average density of the upper mantle to a depth of 400 km, obtained by most investigators, corresponds more closely to a peridotite or pyrolite composition with $Mg/Mg + Fe \sim 0.88$ than to the composition of an eclogite of the type found in diamond pipes. This conclusion is supported by petrological evidence discussed in the following sections.

3-3 SIGNIFICANCE OF ALPINE PERIDOTITES AND OPHIOLITE COMPLEXES

An important question is whether information on the composition of the upper mantle may be obtained directly by the study of rock associations which have been derived from the mantle and emplaced into the crust by tectonic processes. It is widely, but not universally, believed that certain classes of ultramafic rocks—particularly, alpine-type peridotites—are, indeed, derived from the upper mantle. This hypothesis has been strongly supported by Benson[1] and Hess[2] among others. Hess pointed out that alpine-type peridotites characteristically have been intruded close to the axes of maximum deformation along orogenic belts and island arcs. Intrusion has frequently been controlled by major faults with strike lengths up to several hundred miles and which almost certainly extend into the mantle. Ultramafic bodies may occur in the form of innumerable separated bodies along these fault zones, as in the Appalachians and the Great Serpentine belt of New South Wales. Elsewhere, as in New Caledonia, Cuba, the Philippines, and New Guinea, ultramafics may occur in the form of large individual intrusions covering hundreds of square miles and closely connected with major tectonic features. Hess and many others concluded that alpine peridotites of this type have been derived directly from the mantle by tectonic processes and are representative of the rocks occurring in this part of the earth. Unaltered alpine

[1] Benson (1926).
[2] Hess (1939, 1955a,b, 1964b).

peridotites possess the required uppermost mantle seismic velocities and density (Sec. 3-2).

A closely related family of rocks which is also widely believed to be ultimately of mantle origin is the ophiolites.[1] These are characteristic of the Tethyan belt, extending from Western Europe to the Himalayas. Ophiolite complexes consist characteristically of basal ultramafic layers from several hundred metres to several kilometres thick overlain by gabbroic and doleritic rocks which in turn grade upward into mafic pillow lavas and breccias, frequently associated with radiolarian cherts.

Hess tended to emphasize the distinctions between alpine ultramafics and other types of ultramafic bodies, such as the ophiolites, which are more closely associated with basaltic and gabbroic rocks. The trend of more recent and detailed studies has been to emphasize the close genetic relations between alpine ultramafic and ophiolite complexes. According to these views, alpine peridotites and serpentinites are regarded as representing the basal zones of ophiolite complexes which have sometimes been remobilized or selectively exposed by erosion. Nevertheless, certain kinds of ultramafic intrusions, specifically the recently recognized class of high-temperature peridotites,[2] have been regarded as being directly derived from the upper mantle by diapiric intrusion and genetically unrelated to mafic rocks which may occur in the same neighbourhood. Excellent reviews of the petrogenesis of ultramafic rocks have recently appeared, and the reader is referred to these for an account of current viewpoints.[3,4,5] Brief descriptions of some individual complexes follow.

The Lizard High-Temperature Peridotite (Green, 1964)

The Lizard peridotite occurs in Cornwall and covers about 80 km^2. The primary mineral assemblage consists of olivine (Fo_{89}), aluminous enstatite, aluminous diopside, and aluminous spinel, possessing a coarse grain size and anhedral texture. The pyroxenes contain up to 7% Al_2O_3 in solid solution. The primary assemblage is surrounded by a recrystallized anhydrous assemblage consisting of olivine, enstatite, diopside (low in Al), and plagioclase and locally by a recrystallized hydrous assemblage consisting mainly of olivine and amphibole. The chemical compositions of all three assemblages are similar, and they clearly represent equilibration under different P-T-P_{H_2O} conditions. The recrystallized assemblages are much finer grained than the primary assemblage and are characterized by cataclastic and finely foliated textures.

A notable feature is the presence of a high-grade metamorphic aureole surrounding the intrusion. Green concludes that the peridotite was emplaced at a temperature exceeding 800°C, probably in the vicinity of 1000°C. At these tem-

[1] Steinmann (1926).
[2] Green (1967).
[3] Wyllie (1967).
[4] Symposium on deep-seated foundations of geological phenomena, Pt. II, *Tectonophysics* **7**, 1969.
[5] Wyllie (1970).

peratures, the mineralogy of the primary assemblage, characterized by the incompatibility of olivine and plagioclase, indicates minimum pressures of about 10 kbars,[1,2] indicating an origin in the mantle. The peridotite is believed to have risen from the mantle into the crust as a hot crystalline diapir under conditions of tectonic deformation. The recrystallized marginal plagioclase and amphibole assemblages represent re-equilibration under conditions of lower pressure in the crust and in the presence of water. Recrystallization was facilitated by intense deformation of the peripheral regions of the intrusion which occurred during emplacement.

Similar high-temperature peridotites characterized by contact metamorphic aureoles have been recognized elsewhere; e.g., the Tinaquillo,[3,4] Mt. Albert,[5] and Ronda[6,7] peridotites are also believed to have been emplaced as hot crystalline diapirs. The Tinaquillo peridotite has a bulk composition similar to the Lizard (Table 3-3) and is also characterized by the occurrence of aluminous pyroxenes in a primary assemblage of high-pressure origin. The similarity in composition between the Lizard and Tinaquillo intrusions[3] and the bulk composition of a differentiated peridotite-gabbro ophiolite complex (Vourinos—Table 3-3) is notable.

[1]Green (1964, 1967).
[2]Green and Hibberson (1970a).
[3]MacKenzie (1960).
[4]Green (1963).
[5]Smith and MacGregor (1960).
[6]Dickey (1970).
[7]Loomis (1972).

Table 3-3 CALCULATED VOURINOS BULK COMPOSITION COMPARED WITH COMPOSITIONS OF LIZARD, TINAQUILLO, AND ST. PAUL ROCKS HIGH-TEMPERATURE PERIDOTITES

	Vourinos* (A)	(B)	Lizard†	Tinaquillo‡	St. Paul Rocks§
SiO_2	46.3	44.9	44.8	44.9	43.6
TiO_2	0.2	0.1	0.2	0.1	0.3
Al_2O_3	4.7	3.0	4.2	3.2	3.7
FeO¶	5.2	5.1	8.2	7.6	8.0
MnO	0.1	0.1	0.1	0.1	0.1
MgO	35.5	39.9	39.2	40.0	38.5
CaO	3.5	2.5	2.4	3.0	2.6
Na_2O	0.6	0.4	0.2	0.2	0.3
K_2O	0.1	0.1	0.05	0.02	0.1
Cr_2O_3	0.2	0.3	0.4	0.5	0.5
NiO	0.2	0.3	0.2	0.3	0.3
P_2O_5	0.04	0.02	0.01	—	0.1

*Moores (1970) A and B analyses represent use of different methods in arriving at bulk composition.
† Green (1964).
‡ Green (1963).
§ Melson et al. (1970).
¶ All iron is calculated as FeO.

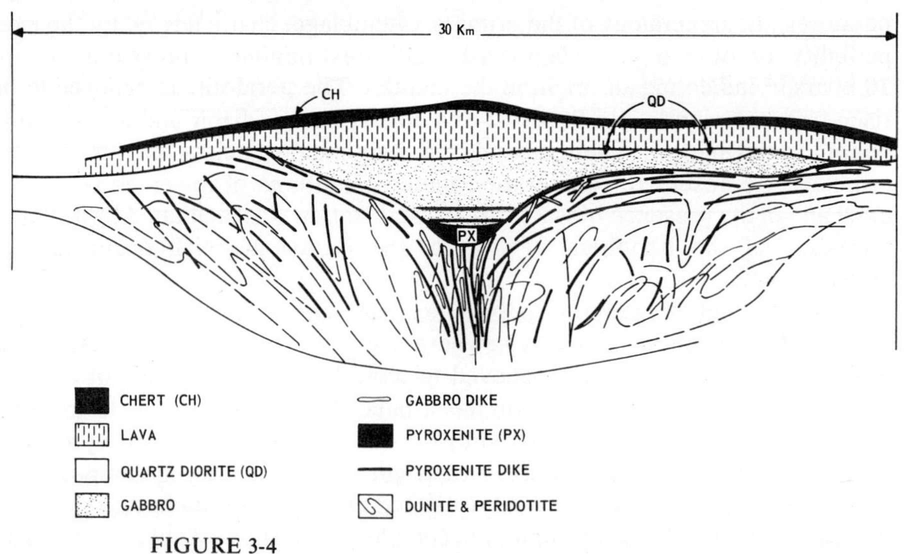

FIGURE 3-4
Idealized reconstruction of Vourinos ophiolite complex showing complex structure of ultramafic rocks plunging towards the center, pyroxenite and gabbro dykes concentrated in the center, and overlying mafic rocks. (*From Moores, 1970, with permission.*)

The mylonitic spinel peridotites of St. Peter and St. Paul Rocks in the mid-Atlantic also have mean compositions and mineralogies similar to the Lizard and Tinaquillo peridotites and have also been interpreted as high-temperature diapiric intrusions derived directly from the mantle.[1] Amphibole- and plagioclase-bearing facies are also present in these oceanic peridotites.[2]

Vourinos Complex, Greece

A detailed description of this well-exposed ultramafic complex which extends over 200 km^2 has been provided by Moores (1970). An idealized section is shown in Fig. 3-4.

The Vourinos complex is composed of 75 to 85% ultramafics, mainly dunite and harzburgite with uniform mineral composition (Fo_{90-94}). The ultramafics occur at the base of the structure and range up to 7 km in thickness. The layered ultramafic zone is characterized by complex isoclinal folding and deformation resulting from nearly solid penetrative flow. Overlying the ultramafics are some 4.5 km of gabbro, diorite, and basaltic lavas, characterized by much simpler layering of magmatic origin, accompanied by evidence of fractionation during crystallization. Compositional similarity and complex gradational contacts between intrusive and extrusive mafic rocks imply common parentage. A transitional zone

[1]Melson, Jarosevich, Bowen, and Thompson (1967).
[2]The peridotites of Galicia, Spain, provide further examples of this class (Maaskant, 1970).

Table 3-4 **THICKNESSES OF ZONES IN VOURINOS COMPLEX.** (*After Moores, 1970*)

Zone	Thickness (km)
Ultramafic	7
Dunite 32%	
Harzburgite 63%	
Pyroxenite 23%	
Transition	1.9
Pyroxenite 23%	
Gabbro 14%	
Dunite 63%	
Gabbro	0.8
Diorite	0.7
Basaltic lavas	1.0
Quartz diorite and dacite	0.2
Total	11.6

of pyroxenite, dunite, and gabbro separates the ultramafics from the mafic zone. The structural discordance between the ultramafic zone and the overlying mafic zones (Fig. 3-4) is a most significant feature, indicating a difference in the modes of origin of the two members. The estimated bulk composition of the complex is given in Table 3-3 and the section and average thicknesses of principal rock types in Table 3-4.

Moores' preferred hypothesis of origin is that the complex formed by partial fusion of mantle material followed by emplacement as a deforming and differentiating solid-liquid mass on the ocean floor. The ultramafic zone represents the refractory, largely unmelted component of parental mantle material which suffered complex deformation mainly by solid penetrative flow. The complementary mafic liquid crystallized above the residual ultramafic floor to form a layered complex of much simpler structure. As an alternative hypothesis, Moores considers it possible that the complex may consist of an overthrust slice of oceanic crust and upper mantle, with the latter representing the ultramafic zone and the former the mafic and transition zones. He points out that the thicknesses of the latter zones agree well with that of the oceanic crust (Table 3-4). As noted by Moores and discussed later, these alternative hypotheses are not necessarily mutually exclusive. The strong similarities between the Vourinos complex and the Papuan ultramafic belt (discussed next) and between their proposed modes of origin are to be remarked.

Papuan Ultramafic Belt[1]

The Papuan ultramafic belt is exposed over a length of 400 km and a width of up to 40 km (Fig. 3-5). It consists of a complex of peridotite, gabbro, and basalt, the

[1]Davies (1968, 1969).

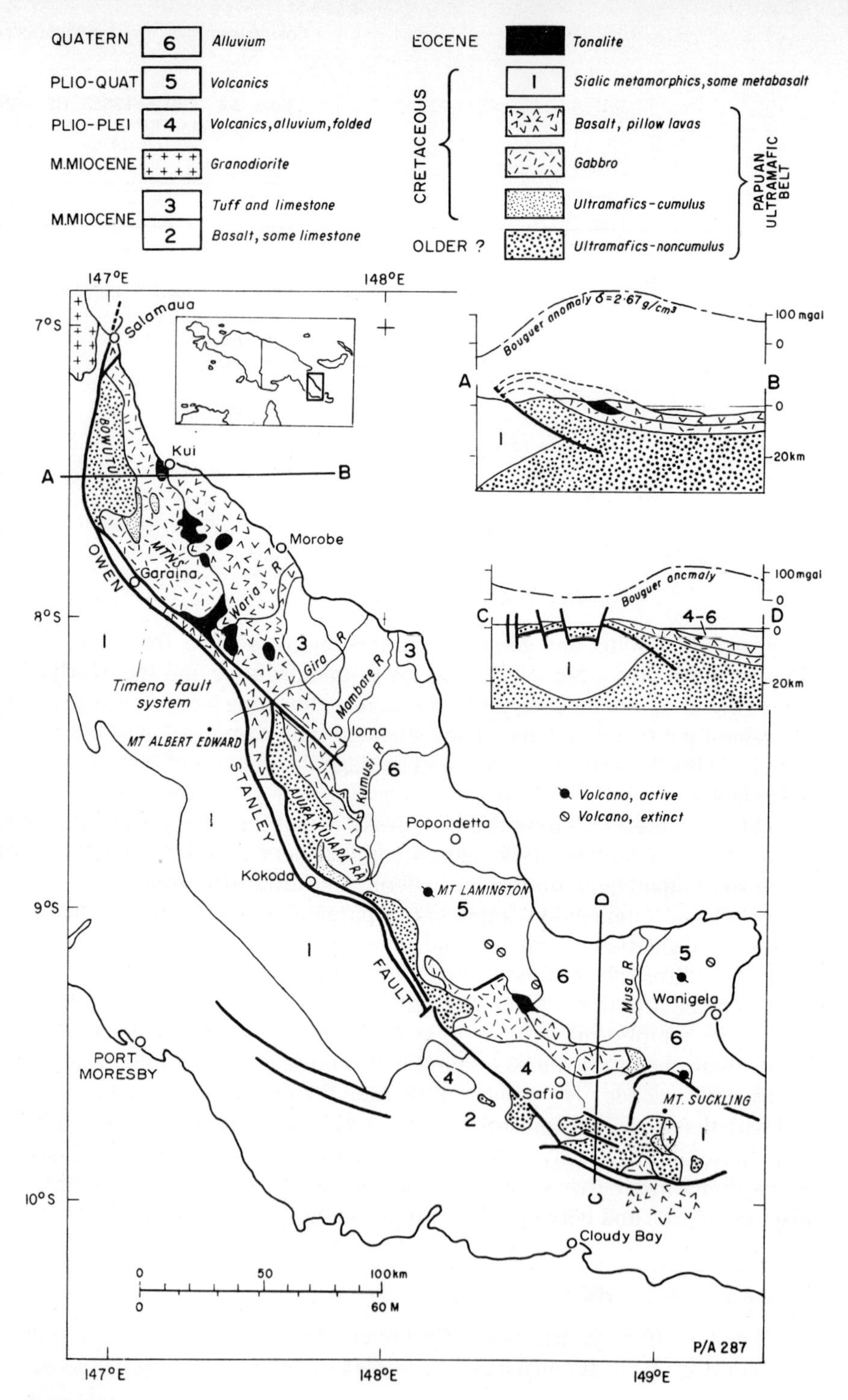

FIGURE 3-5
Papuan ultramafic belt. (*From Davies, 1969, with permission.*)

Table 3-5 ROCK UNITS OF THE PAPUAN ULTRAMAFIC BELT. *(After Davies, 1969)*

Zone	Petrology	Thickness (km)
Basalt zone	Basalt, spilite, lavas, pillow lavas, dacite in one area	4–6
Gabbro zone	High-level gabbro: ophitic, may grade upward into basalt zone; up to 1-km thick granular gabbro, includes accumulates, grain size 1–2 mm	4
Ultramafic zone	Cumulus ultramafics, probably grade upward into cumulus gabbro	0.2–0.5
	Noncumulus ultramafics: harzburgite, dunite, pyroxenite; deformed metamorphic textures; grains 4–20 mm	4–8

distribution of which suggests a crude layering, with peridotite inland and near the base, overlain successively by gabbro at the center, and basalt oceanward. A section through the belt is summarized in Table 3-5. The dip of the complex is towards the ocean, and Davies has interpreted it as actually being continuous with the oceanic mantle and crust to the northeast. The thickness of the mafic zone is observed to be similar to that of the neighbouring oceanic crust. The Papuan ultramafic belt is thus regarded as part of a faulted plate of oceanic crust and mantle which was thrust over the sialic margin of southern Papua, the compression arising from interaction between the north-moving Australian plate and the west-moving Pacific plate.

More than 90% of the ultramafics are believed to represent primary mantle material which formed a floor for the pile of gabbroic accumulates. The ultramafics are composed dominantly of harzburgite (60 to 80% olivine, Fo_{93}) and completely lack cumulus-type textures and plagioclase. These rocks are similar to typical alpine ultramafics and display extensive deformation textures. Between this noncumulus ultramafic member and the mafic zone there is a thin layer (100 to 500 m thick) of finer grained ultramafics possessing typical cumulus textures and showing lesser evidence of deformation. Of considerable significance is the existence of a break in chemical composition between the olivines and pyroxenes of noncumulus and cumulus members.[1]

The cumulus ultramafic member is in turn overlain by a great thickness of highly magnesian alkali-poor gabbro, much of which also appears to be accumulative in nature, with well-developed layering and lacking in deformation textures.

[1]England and Davies (1970).

The gabbro zone is overlain by a zone of massive basalt and spilitic submarine lavas some 4 to 6 km thick. Contacts between the zones are gradational, and it was inferred that the ultramafic cumulates, gabbros, and basalts developed by crystallization differentiation of a single parent magma. In contrast, the deformed noncumulus ultramafics were believed to have a separate origin, representing a floor of refractory, unmelted upper mantle material.

Troodos Plutonic Complex, Cyprus (Wilson, 1959, Gass, 1967, 1968)

The Troodos plutonic complex of gabbroic and peridotitic rocks outcrops over several hundred square kilometres and is associated with two other major units—the Sheeted intrusive complex and the Troodos pillow lava series (Fig. 3-6). The pillow lava series consist dominantly of submarine basaltic (tholeiitic) rocks with a thickness of 1 km. These overlie and are younger than the Sheeted intrusive complex, which is from 2 to 4 km thick. The latter complex consists of a remarkably dense swarm of parallel, N-S striking, steeply dipping mafic dykes. In a 120-km section across the strike, about 48,000 dykes up to 3 m wide, and lacking chilled margins, occur. More than 90% of the section is composed of tholeiitic dykes which are separated by thin screens of mafic rocks of similar composition. The swarm resembles in some respects the dyke swarm occurring in Iceland[1] on the Mid-Atlantic Ridge, and Gass has indeed interpreted the Cyprus structure as forming at the axis of a former mid-oceanic ridge subject to E-W tension.[2]

Rocks of the Troodos plutonic complex range from dunite and peridotite, through pyroxenite and olivine gabbro, to gabbro and granophyre. The ultramafic rocks (olivine, Fo_{92}) are surrounded by a larger annulus of gabbroic rocks. Contact relationships are not clear, but it is believed that the proportion of ultramafic to mafic rocks increases markedly with depth and that the gabbros "although abundant at the surface, are in fact, minor differentiates of a vast mass of ultrabasic material underlying Troodos." The latter conclusion is based upon an interpretation of gravity data (Fig. 3-6), which show the presence of a major positive gravity anomaly increasing in amplitude up to 250 mgal northwesterly from the outcrop of ultramafics. This is one of the largest positive gravity anomalies known and implies the presence of a vast mass of dense ultramafic rocks beneath this region of Cyprus.

According to Gass' interpretation of the gravity data, the high-density mass consists of a near-surface, rectangular horizontal slab, 200 km east-west, 120 km north-south, with a thickness of at least 12 km and a density of at least 3.3 g/cm^3. Gass suggests that the rocks of the Troodos igneous suite originally represented a volcanic-plutonic edifice formed from the mantle at the crest of a mid-oceanic ridge between Eurasia and Africa. During the Alpine orogeny, the Eurasian and

[1]Bodvarsson and Walker (1964).
[2]See also Moores and Vine (1971).

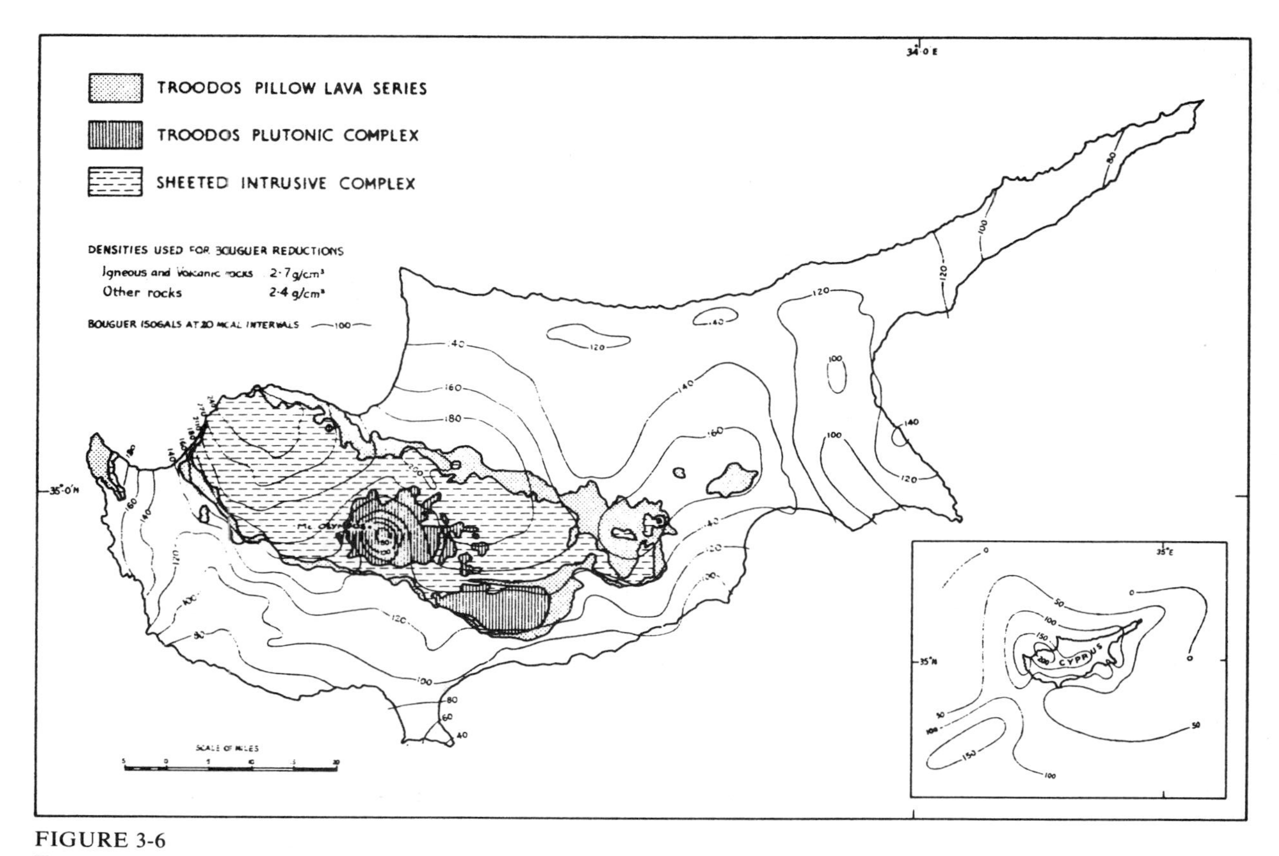

FIGURE 3-6
Troodos complex, Cyprus, in relation to Bouguer gravity anomalies. (*From Gass, 1967, with permission.*)

African plates converged, and the ultramafic rocks of Troodos were underthrust by the African sialic plate, resulting in the detachment of a slab of the upper mantle and its association with the underlying sialic crust responsible for the elevation of Cyprus. The Troodos plutonic complex is thus regarded as upper mantle material, partly fused and differentiated in situ to provide the volcanic rocks of the massif.

In a subsequent study of the Troodos massif, Moores and Vine (1971) arrived at an interpretation broadly concordant with that of Gass. They concluded that

> The Troodos Massif as a whole represents a slice of oceanic crust and uppermost mantle, that the harzburgite and dunite of the ultramafic rocks represent depleted mantle, the olivine pyroxenite and gabbro represent intrusive or cumulate magmatic material, that the gabbro and Lower Sheeted Complex represent seismic layer three of the oceanic crust and that the Upper Sheeted Complex, Lower Pillow Lavas and Upper Pillow Lavas represent layer two.

Ultramafic Rocks of the Ocean Basins

Investigations of mid-ocean ridges have shown that they contain major occurrences of ultramafic rocks. Wilson (1968) has compiled a list of such occurrences.[1] They have been found at widely separated localities along the Mid-Atlantic Ridge, the Indian Ocean Ridge, and the axis of the Red Sea. In view of the limited amount of exploration so far conducted and the frequency with which ultramafic rocks have been discovered, it appears possible that the mid-oceanic ridges contain the largest exposures of ultramafic rocks which are present on the earth. This is particularly significant in view of the shallow depth—as little as 3 km below the surface of the crust—at which mantle velocities in the vicinity of 8 km/sec are found in some regions of the Mid-Atlantic and Pacific Ridges.[2,3,4] The shallow occurrence of mantle material as revealed by seismology, the occurrence of pronounced mantle seismic anisotropy (Sec. 3-2), the intensely fractured and deformed structural nature of the mid-oceanic ridges, and the widespread occurrence of ultramafic rocks combine to strengthen the widely held presumption that the latter represent actual samples of the suboceanic upper mantle.

Of particular significance are the dredgings carried out in deep trenches along the Chain, Romanche, and St. Paul fracture zones where they intersect the Mid-Atlantic Ridge.[5,6] Peridotites were usually dredged from the deepest levels and basalts from the upper slopes, with gabbro and greenschists from intermediate levels. At one locality on the Romanche fracture, a minimum thickness of 3.5 km

[1]See also Udintsev and Dmitriev (1971).
[2]Le Pichon et al. (1965).
[3]Keen and Tramontini (1970).
[4]Menard (1960).
[5]Bonatti (1968).
[6]Bonatti, Honnorez, and Ferrara (1971).

of ultramafics (mainly lherzolites) with compositions similar to those of the Lizard and Tinaquillo was found. Bonatti concluded that these occurrences support the view that mantle-derived ultramafics are the prevalent rock types in the interior of mid-oceanic ridges. Also in the Romanche fracture zone, dredging revealed the occurrence of a differentiated, layered gabbroic complex in the oceanic crust.[1] The petrology of the rocks encountered was comparable with the gabbroic rocks occurring at Troodos, Bowutu Mts. (Papuan ultramafic belt), and the Camaguey complex (Cuba).

Oceanic trenches (e.g., Tonga and Puerto Rico trenches) constitute another environment in which ultramafic rocks have been found. The resemblances between ultramafics found at mid-oceanic ridges, trenches, and island-arc and orogenic environments take on a new dimension in the light of theories of sea-floor spreading and plate tectonics. In the first place, the dynamic behaviour and high mobility displayed by the mantle provide new and plausible mechanisms of intruding dense ultramafic rocks into orogenic sialic environments. Secondly, it has been suggested that the alpine-type ultramafics and ophiolites presently occurring in island arcs and in the Alpine-Himalayan orogenic belt may actually have originated on mid-oceanic ridges (cf Troodos) and may have been carried with the migrating sea floor into their present environments.[2,3,4,5]

This theme has been successfully developed by Dewey and Bird (1971), Moores and Vine (1971), and Coleman (1971), who emphasize the close correspondence both in structure, dimensions, and geophysical and petrological properties between ophiolite complexes and sections of the oceanic crust and uppermost mantle. They have proposed that ophiolite complexes generally represent slices of oceanic crust and mantle which have been *obducted* (thrust) into their present sialic environments during collisions between plates and have applied this hypothesis to provide an integrated explanation of many geological and petrological characteristics of the ophiolite association.

Petrogenesis

We have discussed above a group of peridotites which, after careful studies, have been concluded by their investigators to be of mantle origin. Many more examples, e.g., the large peridotitic complexes in Cuba, the Phillippines, and New Caledonia, could be added to this category. It appears likely, therefore, that such bodies provide us with relatively direct evidence concerning the nature of the upper mantle. Of course, not all peridotites are derived from the upper mantle—a large and important class are formed as olivine-rich cumulates from stratiform mafic intrusions within the crust, e.g., the Bushveldt and Stillwater complexes.

[1]Melson and Thompson (1970).
[2]Thayer (1969a).
[3]Bonatti (1968).
[4]Green (1970a).
[5]Bonatti, Honnorez, and Ferrara (1971).

Remobilization of the ultramafic zones of such complexes may occur during orogenies, and some alpine peridotites may have formed in this manner.[1] It is evident that each peridotitic occurrence must be studied individually and the genetic evidence considered on its own merit.

Although the petrogenetic interpretations of the authors who described the Vourinos complex, the Papuan ultramafic belt, and the Troodos complex differ in significant details, there is an overall unity about fundamentals, and it is likely that these and other comparable bodies have been formed by generally similar processes. Moores and Gass have drawn particular attention to the close genetic relationships existing between ultramafics, gabbros, and basalts, whilst Thayer[2] has emphasized the association between ultramafics and nearby gabbros. The overall relationship of this group of complexes may be explained in terms of partial fusion of primary mantle material followed by separation of the mafic magma from residual dunite and peridotite. The initial formation of such complexes in the uppermost mantle has been strongly advocated by Thayer.[2] Subsequent upward emplacement of the complexes into the crust may be accompanied by deformation and folding of the mainly solid residual dunites and peridotites, whereas the complementary mafic liquids crystallize at higher levels to form layered stratiform complexes with much simpler structures and cumulus textures. In this manner, structural and mineralogical discordances are produced between the high-level, mainly mafic cumulates and the floors of residual ultramafics, although ultimately, these units possess related and complementary origins.

The similarity between the compositions of high-temperature peridotites and the bulk compositions of some peridotite-gabbro complexes—e.g., Vourinos, Table 3-3—has already been noted. The relationships between these bodies are most readily interpreted in terms of a hypothesis involving varying degrees of partial melting during upward movement of diapirs of parental mantle material from the low-velocity zone.[3,4,5] High-temperature peridotites represent hot mantle

[1]Smith (1958).

[2]Thayer (1960, 1967, 1969a, b, c) has been a prime advocate of the existence of a genetic relationship between many large alpine ultramafic and associated gabbros. He has presented an impressive case favouring an ultimate origin of these complexes in the upper mantle. Somewhat surprisingly, he maintains that basalts are not genetically associated with this lineage, although the evidence from the Troodos, Papuan, and Vourinos complexes is clearly contrary to this opinion. Thayer stresses the Mg-rich and alkali-poor nature of the gabbros commonly found associated with alpine ultramafics and the difference in composition between these gabbros and common types of basalt. From the investigations of Davies, Moores, Gass, and others, it appears more likely that these characteristics of the gabbros are at least partly connected with their accumulative nature. It is also probable that partial melting of parental material with the composition of high-temperature peridotites (e.g., the Lizard, Table 3-3) which appear to have suffered previous fractionation episodes characterized by loss of a small proportion of highly alkalic magmas (Sec. 8-5) would result in the formation of Mg-rich and alkali-poor mafic differentiates. Green (1970a) has reconciled the essential elements of Thayer's interpretation with those of other authors.

Layered structures often found in the ultramafic zones of large alpine-type peridotites have sometimes been thought to require an origin by accumulation from primary ultramafic magmas (e.g., O'Hara and Mercy, 1963), but this rather extreme hypothesis appears unnecessary. It is entirely conceivable that such layering could develop during *partial* melting of primary mantle material accompanied by differential elutriation of suspended crystals in a mobile crystal-liquid system.

[3]Green and Ringwood (1967a).

[4]Green (1970a).

[5]Section 4-5.

diapirs which have undergone only a very small degree of partial melting and magma segregation and hence approach the composition of the primitive mantle (pyrolite). Diapirs originating from greater depth, however, will intersect the solidus before reaching the crust (Fig. 4-7), thereby undergoing extensive partial melting and magma segregation within the uppermost mantle, leading to the development of peridotite-gabbro complexes possibly accompanied by extensive basaltic volcanism. These processes are believed to be particularly important beneath mid-oceanic ridges.

Finally, we consider an alternative hypothesis of origin of some alpine-type ultramafics and possible means of discriminating between the differing modes of origin. The alternative hypothesis regards the ultramafic zones of some ophiolite complexes as representing cumulates from a parental basaltic magma. Such cumulate zones are found in large mafic stratiform intrusions, e.g., the Bushveldt and Stillwater complexes. Remobilization of the ultramafic zones during orogenesis may result in the formation of detached intrusions interpreted as alpine ultramafics. A related explanation has been applied by Challis (1965) to the origin of a group of alpine-type ultramafic intrusions in New Zealand. She regards these as representing precipitates from deep-level magma chambers along a line of former volcanoes which produced large volumes of olivine-poor basaltic rocks. The New Zealand intrusions are compared with the ultramafic zones of the Stillwater and Great Dyke intrusions.

It appears likely that alpine-type ultramafics are indeed polygenetic and that the origin of each occurrence has to be considered separately in the light of evidence available. In the cases of examples discussed earlier, the large volumes and structural discordances of ultramafic zones in relation to associated gabbros provides key evidence. In contrast, ultramafic, plagioclase-free cumulate zones of large mafic stratiform bodies, e.g., the Bushveldt and Stillwater complexes,[1] amount to only 10 to 20% of the (original) volume of the entire complex. Furthermore, the layering of ultramafic and overlying mafic zones is generally parallel and concordant.

Mineralogical studies also provide significant evidence. Hess (1939) noted the uniformity in mineral compositions of alpine ultramafics, which typically are characterized by olivines in the range Fo_{90} to Fo_{95}, and nickel contents of $0.3 \pm 0.1\%$. On the other hand, ultramafic zones known to have accumulated from basaltic magmas are usually characterized by a substantial range of compositions and are generally more iron-rich and nickel-poor. For example, the olivine in the ultramafic zone of the Stillwater complex ranges in composition from Fo_{80} to Fo_{90},[1] and in the dunite zones of the Muskox intrusion the range is from Fo_{70} to Fo_{85}.[2] Although this criterion is very useful it is not infallible, since a few cases are known where cumulates believed to have settled from mafic magmas have forsterite contents overlapping the alpine ultramafic range.[3]

A more decisive criterion is the *range* of compositions displayed by individ-

[1]Jackson (1967).
[2]Irvine and Smith (1967).
[3]E.g., the Great Dyke—Fo_{86} to Fo_{94} (Jackson, 1967).

ual minerals in the ultramafic zones of a given intrusion. The relative constancy in composition of the olivine of most alpine-type peridotites is an important feature which is to be expected for bodies that represent refractory residues from partial melting processes. On the other hand, closed-system differentiation of large volumes of mafic magmas almost inevitably leads to fractionation of the liquid and parallel changes in composition of the crystals which separate.[1] The earliest crop of liquidus olivine may well possess forsterite contents in the range Fo_{89} to Fo_{94} and similar to those in many alpine peridotites. However, with increasing degrees of crystallization, the crystals become more iron-rich,[2,1] passing outside the range exhibited by most alpine peridotites. The behaviour of nickel during fractionation also provides an important clue. The partition coefficient for nickel between olivine and liquid is very high, in the range 10 to 15.[1,3] This implies that the concentration of nickel in the olivines crystallizing from a mafic (or picritic) magma drops by a factor of 2 after only 5 to 8% crystallization.

Recent studies[4,5,6] of the isotopic composition of strontium in alpine ultramafics have shown that it is frequently much more radiogenic (Sr^{87}/Sr^{86} ratios up to 0.725 and averaging 0.711) than is characteristic of basalt of mantle origin (average 0.703). This discordance renders it extremely improbable that alpine ultramafics with this characteristic can represent *precipitates* from parental mafic (or picritic) magmas. The discordance, however, is not incompatible with the hypothesis that the ultramafics represent residues from partial melting processes in the mantle (see next section).

It appears that the assignation of any given ultramafic to a direct origin in the mantle or as a cumulate from a crystallizing basaltic magma requires assessment of several lines of evidence. These include the sizes and tectonic relationships of the intrusion, the relative volumes of ultramafic and mafic components, and the structural relationships of these components. Detailed mineralogical studies on the fractionation of Fe, Mg, Ni, and Cr,[7] together with measurements of strontium isotopic composition, may also provide decisive criteria. The combination of these sources of evidence should usually suffice to establish the mode of origin of a given intrusion, although some occurrences will remain which cannot be unambiguously interpreted.

Discussion

The evidence considered in this section strongly suggests that certain kinds of large ultramafic bodies represent samples of the upper mantle which have been

[1]Irvine and Smith (1967).
[2]Jackson (1967).
[3]Wager and Mitchell (1951).
[4]Hurley (1967).
[5]Stueber and Murthy (1966)
[6]Bonatti, Honnorez, and Ferrara (1971).
[7]Thayer (1969c) has drawn attention to the importance of chromite mineral chemistry and the scales of local accumulation of chromite in distinguishing between the origins of ultramafic cumulates from parent basaltic magmas and ultramafics of mantle origin.

subsequently emplaced into the crust. The characteristic occurrences of alpine- and ophiolite-type ultramafic rocks along mid-oceanic ridges, in oceanic trenches and island arcs, and in orogenic belts take on added significance in the light of theories of sea-floor spreading and plate tectonics which are beginning to provide an integrated interpretation of their origin and mode of emplacement into the crust.[1,2,3] It should be acknowledged that many detailed problems of the petrogenesis and petrochemistry of these rocks remain to be solved. Such problems, e.g., the highly radiogenic strontium found in many alpine ultramafics,[4,5,6,7] are of considerable importance, and their solution will require extensive further investigations. Nevertheless, the first-order result is that studies of many crustal ultramafic bodies provide significant evidence strongly suggesting that extensive regions of the uppermost mantle are also of ultramafic composition. This is supported by the interpretation of physical properties of the uppermost mantle as discussed in Sec. 3-2.

In contrast, studies of eclogites which occur in the crust (apart from those found as xenoliths in diamond pipes) provide little support to the hypothesis that eclogite is the major component of the uppermost mantle. Compared to peridotites, eclogites are extremely rare. The size of individual eclogite occurrences is much smaller than that of peridotites—dimensions of individual bodies rarely exceed 1 km^2—more often the maximum dimension is of the order of a few tens of metres. In the past, many geologists suggested that eclogites occurring in the crust were emplaced from deeper levels, perhaps from the mantle, by tectonic action. In most cases, the field evidence supporting this hypothesis was tenuous, to say the least. The hypothesis was based primarily upon the preconception that eclogites were unstable under the *P-T* conditions obtained in crustal environments and must therefore have been formed under the much higher pressure conditions existing in the mantle. This preconception is now known to be false—recent ex-

[1]Coleman (1971).
[2]Moores and Vine (1971).
[3]Dewey and Bird (1971).
[4]Hurley (1967).
[5]Stueber and Murthy (1966).
[6]Bonatti, Honnorez, and Ferrara (1971).
[7]Strontium from alpine ultramafics has a considerable range of isotopic compositions with an average 87/86 ratio of 0.711[4,5,6] compared to 0.703 for oceanic basalts. This has been considered to indicate that alpine ultramafics are genetically unrelated to associated gabbros and basalts and instead represent portions of the upper mantle which were fractionated with respect to rubidium at a very early stage in the history of the earth. There are some problems with this explanation in view of the dynamic behaviour now attributed to the mantle and the probability of rather efficient mixing (Chap. 8). It is also difficult to ignore much petrological evidence which often supports a genetic relationship between ultramafics and associated mafics.

The anomalous nature of the strontium may be connected with processes of separation of magmas from parental mantle material (pyrolite) under conditions during which bulk isotopic equilibrium between magma and residual ultramafic phases was not maintained. Olivines and enstatites from pyrolite are likely to be characterized by high Rb/Sr ratios and highly radiogenic strontium, whereas most of the "common" strontium resides in diopside and accessory interstitial phases, e.g., apatite. Partial melting to liberate basalt results in decomposition of these phases and incorporation of their strontium and rubidium into the liquid, but the radiogenic strontium locked in residual olivine and enstatite may not be affected. Subsequent deformation and recrystallization of the residual ultramafics during emplacement into the crust may be accompanied by selective loss of Rb from the olivines and orthopyroxenes and redistribution of radiogenic strontium, giving rise to the patterns now observed. This interpretation is further discussed by Graham and Ringwood (1971).

perimental data (reviewed in Chaps. 1 and 2) have demonstrated that, under dry conditions, many eclogites are stable under the range of *P-T* conditions existing in crustal environments. Concordant with the experimental data, detailed studies of the occurrences of many eclogites in regional metamorphic terrains have demonstrated that they have indeed been transformed from mafic material in situ (Chap. 2) and do not represent tectonically transported mantle material. The only eclogites for which a mantle origin can be firmly established are those occurring as inclusions in diamond pipes and in some types of alkali basalts which are discussed in the next section.

3-4 XENOLITHS OF MANTLE ORIGIN

Inclusions in Kimberlite Pipes

A classic paper dealing with the interpretation and significance of kimberlite inclusions was published by Wagner (1928). Kimberlite pipes carrying diamonds are of frequent occurrence throughout 1 million square miles of southern Africa and throughout a comparable region in Siberia. They are also known in India, Brazil, the United States, and Australia. Where pipes have been well exposed, they have been found to carry numerous xenoliths of crustal rocks which they are known to have intruded on their journey upward.[1] They contain, also, large numbers of xenoliths of rocks which are not known to occur in the vicinity, particularly, peridotites, pyroxenites, and eclogites. Wagner concluded that these xenoliths had been derived from deeper levels of the earth and represented a random sample of deep-lying rock types cut by the pipes. The occurrence of diamonds both in the pipes and in the inclusions implied that the pipes were derived from substantial depths in the upper mantle. Accordingly, he concluded that the peridotite-eclogite suite of xenoliths in kimberlite pipes provides a random sample of upper mantle rocks over a vast, subcontinental area.

Wagner's hypothesis has been supported by many subsequent investigations. The temperature of the kimberlite magma (prior to explosive eruption and adiabatic cooling) was probably not much less than 1000°C,[2,3,4] and at this temperature, a minimum pressure of about 45 kbars[5] is required for the stable synthesis of diamond from graphite. There seems little doubt that the well-crystallized diamonds found in kimberlite grew under thermodynamically stable conditions, which would imply the derivation of diamondiferous kimberlites and diamond-bearing xenoliths from a minimum depth of 140 km. An experimental investiga-

[1] E.g., Williams (1932).
[2] Harris and Middlemost (1970).
[3] Kushiro, Syono, and Akimoto (1968).
[4] McGetchin (1968).
[5] Bundy (1963) (revised pressure scale).

tion of the conditions of formation of the remarkable diopside-ilmenite intergrowths found as nodules in some pipes provided strong evidence that they were derived from much greater depths—possibly as great as 300 km.[1]

The temperature of equilibration of garnet lherzolites may be estimated from experimental data on the solid solubility of orthopyroxene in clinopyroxene.[2,3] Temperatures in the range 800 to 1300°C are found, with an average of about 1000°C. When combined with other high *P-T* data on the solid solubility of Al_2O_3 in pyroxenes,[4,5,6,7] they imply that most garnet lherzolite xenoliths from kimberlites have equilibrated at pressures of 20 to 60 kbars or corresponding depths of about 60 to 200 km and are thus of mantle origin. Investigations of Fe-Mg-Ca tie lines between coexisting garnet and pyroxenes of eclogites from kimberlites demonstrated that they were unlike the corresponding tie lines for eclogites of known crustal origin, and it was also concluded that the eclogites were of mantle origin.[8] Occurrence of diamonds in some eclogites[9] supports this conclusion. Further studies of the mechanism of intrusion of kimberlites (discussed later) have shown how it was possible for the kimberlite magmas to have transported dense xenoliths from such great depths to the surface.

It is of great significance that, in the African and Siberian kimberlites, where sampling has been the most thorough, *peridotitic inclusions are found to be much more common than eclogitic inclusions.*[10,11,12,13,14] From a study of South African pipes, MacGregor and Carter (1970) estimated that about 95% of xenoliths from all but one pipe were of ultramafic composition, and 5% were eclogitic. (The exception was the Roberts Victor pipe which contained about 80% eclogite xenoliths and 20% ultramafic xenoliths.) The overall predominance of peridotitic xenoliths strongly suggests that the upper mantle sampled by kimberlite pipes is dominantly of peridotitic composition, with eclogite a minor but widely distributed constituent.[10,15]

Comprehensive studies of the petrology and mineralogy of kimberlite xenoliths have been published by Williams[11] and Sobolev (1959).[16] References to more recent literature are given by Wyllie (1967) and Dawson (1968). The peridotitic nodules are dominantly composed of varying proportions of four prin-

[1] Ringwood and Lovering (1970).
[2] Boyd (1967).
[3] Sobolev (1970).
[4] Boyd and MacGregor (1964).
[5] Green and Ringwood (1967b, 1970).
[6] Boyd (1970).
[7] O'Hara (1967a,b).
[8] Kushiro and Aoki (1968).
[9] Sobolev and Kuznetsova (1966).
[10] Wagner (1928).
[11] Williams (1932).
[12] Dawson (1962).
[13] Nixon et al. (1963).
[14] Sobolev (personal communication).
[15] Ringwood (1958).
[16] See also N.V. Sobolev (1973).

Table 3-6 PROPORTIONS OF ROCK TYPES FOUND IN LARGE SUITES OF ULTRAMAFIC AND MAFIC XENOLITHS FROM SOUTH AFRICAN KIMBERLITE PIPES. (*After Mathias, Siebert, and Rickwood,* 1970, *and MacGregor, personal communication,* 1969)

Rock	Mineralogy	MSR %* (1970)	McGregor %† (1969)
A: Peridotite-pyroxenite association			
Dunite	Ol,	0.3	0.5
Harzburgite	Ol, Opx ± Sp	16	26
Lherzolites	Ol, Opx, Cpx, ± Sp	14	11
Garnet harzburgite	Ol, Opx, Ga	18	21
Garnet lherzolite	Ol, Opx, Cpx, Ga	43	39
Pyroxenites	Ol, Cpx ± Ga	6	2.5
Others		3	
B: Eclogitic association‡			
Eclogite	Ga, Cpx	63	
2-px eclogite	Ga, Cpx, Opx	2	
Kyanite eclogite	Ga, Cpx, Ky	8	
Corundum eclogite	Ga, Cor, Cpx	6	
Quartz eclogite	Ga, Cpx, Qz	0.6	
Plagioclase eclogite§	Ga, Cpx, Plag	8	
Garnet granulite§	Ga, Cpx, Plag, Qz	3	
Others		9	

*Population of 295 xenoliths.
†Population of about 200 xenoliths.
‡Population of 171 xenoliths.
§Possibly granulites of crustal origin.
Abbreviations:
Ol = olivine, Opx = orthopyroxene, Cpx = Ca-rich clinopyroxene, Ga = pyrope-rich garnet, Ky = kyanite, Cor = corundum, Qz = quartz, Plag = plagioclase, Sp = spinel.

cipal minerals—olivine, orthopyroxene, clinopyroxene, and garnet. The proportions of different rock types are given in Table 3-6. Among the ultramafic xenoliths, olivine (Fo_{88-94}) is the dominant mineral, and the average olivine/orthopyroxene ratio is 2/1. The most abundant mineral assemblage is garnet lherzolite—average composition olivine 64%, orthopyroxene 27%, clinopyroxene 3%, pyrope-rich garnet 6%. The wide variation in proportions of individual minerals within the different rock groups of the peridotitic association should be emphasized. Essentially, there is a continuous mineralogical gradation between harzburgite, lherzolite, garnet harzburgite, and garnet lherzolite.[1] The chemical compositions of peridotitic xenoliths have recently been investigated by Carswell and Dawson (1970). Their results are summarized in Table 3-7. The general similarity between the compositions of the garnet peridotites from diamond pipes and the analyses of ultramafic rocks from high-temperature peridotites and ophiolites (Table 3-3) is to be remarked.

An important discontinuity in modal mineralogy between the peridotitic and eclogitic associations (Table 3-6) occurs. Garnet peridotites characteristically

[1]Mathias et al. (1970).

Table 3-7 COMPOSITIONS OF 15 GARNET PERIDOTITE XENOLITHS FROM SOUTH AFRICAN DIAMOND PIPES. (*After Carswell and Dawson*, 1970)

	(1)	(2)	(3)
SiO_2	46.5	45.5	44.5 –47.9
TiO_2	0.3	0.2	0.02– 2.3
Al_2O_3	1.8	2.7	1.1 – 3.3
Cr_2O_3	0.4	0.3	0.2 – 0.5
FeO	6.7	7.0	5.9 – 8.4
MnO	0.1	0.1	0.1 – 0.2
NiO	0.3	0.3	0.25– 0.4
MgO	42.0	41.9	37.7 –45.7
CaO	1.5	1.9	0.9 – 3.5
Na_2O	0.2	0.2	0.06– 0.4
K_2O	0.2	0.1	0.0 – 0.4
P_2O_5	0.02	0.03	0 – 0.05

Explanation:
(1) Mean of 9 analyses by Carswell and Dawson.
(2) Mean of 6 analyses collected from previous literature.
(3) Composition range in all 15 analyzed garnet peridotites.

have less than 15% garnet, whereas eclogites characteristically have more than 30% garnet.[1] Exceptions are very rare. Moreover, no eclogites containing olivine were found in the extensive search conducted by Mathias et al. (1970). This discontinuity is accompanied also by characteristic differences in mineral chemistry between garnet peridotites and eclogites, which have been discussed by many authors.[e.g.,2,3,4,5,6,7,8,9]

Eclogites from diamond pipes have been interpreted most widely as being derived from basaltic or picritic magmas generated at considerable depths in the mantle. It appears likely that a significant proportion of such magmas may not reach the surface and, accordingly, would crystallize at depth as eclogite.[10,11,4,7]

[1] Rickwood, Mathias, and Siebert (1968).
[2] O'Hara and Mercy (1963).
[3] Nixon, von Knorring, and Rooke (1963).
[4] Kushiro and Aoki (1968).
[5] Carswell and Dawson (1970).
[6] Mathias, Siebert, and Rickwood (1970).
[7] MacGregor and Carter (1970).
[8] Dawson (1962).
[9] V. S. Sobolev, N. V. Sobolev, and coworkers—numerous detailed studies. See especially V. S. Sobelev (1959) and N. V. Sobelev (1970, 1973).
[10] Ringwood (1958).
[11] O'Hara and Yoder (1967).

This explanation, however, fails to provide an explanation of the absence of olivine eclogite from diamond-pipe assemblages.[1] At pressures greater than about 30 kbars, liquids forming by partial melting in the mantle will fall on the olivine side of the eclogite plane, which constitutes a thermal divide[2,3] (Fig. 4-5). Unmodified liquids separating and crystallizing at depths greater than 90 km would thus crystallize as olivine eclogites. Moreover, as such olivine-saturated liquids ascend to higher levels, the primary field of crystallization of olivine expands[2,4] so that, in general, the precipitates would consist of mixtures of olivine + pyroxene(s) ± garnet throughout a wide depth interval (Chap. 4). Although certain paths of crystallization may lead to olivine-free eclogitic accumulates, the conditions required are somewhat restrictive[4,3] and unlikely to be generally operative, although they are probably applicable to a limited proportion of xenoliths.

It appears likely that the dominant bimineralic nature of diamond-pipe eclogites (Table 3-6) is frequently a direct consequence of the behaviour of eclogite as a thermal divide at high pressures in the system olivine-garnet-diopside-quartz (Fig. 4-5). This suggests the alternative hypothesis[5,6,7] that many mantle eclogites are derived ultimately from the oceanic crust which subsides into the mantle beneath deep oceanic trenches as implied by recent developments in mantle dynamics. The oceanic crust probably has a mean chemical composition similar to that of an oceanic tholeiite. Such a composition transforms to quartz eclogite at high pressure, and this transformation will occur at quite shallow depths in the mantle. In view of the vast amounts of oceanic crust believed to enter the mantle, the extreme rarity of quartz eclogite xenoliths in diamond pipes may appear surprising (Table 3-6). The explanation proposed[6,7] is that the quartz eclogite undergoes partial melting during subsidence, leading to the formation of calc-alkaline magmas, together with a refractory residuum of quartz-free eclogite, poor in alkalis.[7]

Eclogites may also be formed as high-pressure precipitates from water-rich highly undersaturated magmas, including the parental kimberlites,[8] and from water-rich tholeiitic magmas generated in the mantle beneath seismic zones, as discussed in Sec. 7-8. It appears that eclogites found in diamond pipes may possess a multiplicity of origins and, when correctly interpreted, may provide guides to major fractionation processes which have occurred within the mantle. Mineralogic evidence of the presence of at least two distinct classes of bimineralic

[1] Mathias et al. (1970).
[2] O'Hara (1963).
[3] Davis (1964).
[4] O'Hara and Yoder (1967).
[5] Ringwood and Green (1966a).
[6] Green and Ringwood (1968).
[7] Section 7-8.
[8] E.g., Williams (1932).

eclogites in diamond pipes has been produced,[1] and further classes of kyanite eclogites, corundum eclogites, and grosspydites have also been recognized.[2,3]

Apart from the included xenoliths, kimberlites consist of two principal components which vary widely in their relative abundances. The first is an assemblage of individual mineral fragments, often rounded, which are compositionally similar to the minerals occurring in peridotitic and eclogitic xenoliths. Olivine ($Fo_{88\text{-}95}$) is by far the dominant mineral and may constitute more than 70% of the entire kimberlite. Accompanying the olivines are orthopyroxenes, diopsidic pyroxenes, and garnets. The assemblage is believed[4,5] to have been derived largely by the disintegration of mantle xenoliths and by attrition of mantle wall rocks of the conduits during emplacement of the kimberlite. Some of these minerals, together with large crystals of Mg-rich ilmenite and phlogopite, may also have crystallized from the parent kimberlite magma at depth.[6]

The second component represents the primary kimberlite magma—this is highly variable in composition but is generally an ultramafic type, extremely rich in MgO, H_2O, and CO_2 and often alkali-rich.[6] This family of magmas is apparently related to some of the end-members of the alkali basalt suite—specifically, the olivine-melilite-nephelinites—and appears to have been derived by generally similar fractionation processes operating to a more extreme degree.[6,7,8,9] As a group, kimberlites carry high concentrations of incompatible elements—e.g., U, Th, Ba, Sr, Ta, P, and rare earths—and the rare earths are strongly fractionated in a manner suggesting participation of garnet in the relevant equilibria.[7,8]

The processes of formation of kimberlite magmas at great depths in the upper mantle have been discussed by Harris and Middlemost (1970). The kimberlite magma is believed to form by a small degree of partial melting accompanied by extensive zone refining and wall-rock reaction processes, leading to enrichment of incompatible elements in the magma by factors of 200 and more over the original mantle concentration. These processes also result in strong enrichment of water and carbon dioxide in the magma—up to 40% of these components may be present.

In a favourable structural environment, the volatile-rich kimberlite magma may rise upward into lower pressure regions within the uppermost mantle or crust, where the load pressure is unable to contain the carbon dioxide and water in solution. When this point is reached, exsolution of a gas phase occurs, and further

[1]MacGregor and Carter (1970).
[2]Sobolev, Kuznetsova, and Zyuzin (1968).
[3]Table 3-6.
[4]Nixon et al. (1963).
[5]McGetchin (1968).
[6]Williams (1932).
[7]Dawson (1962).
[8]Gast (1968).
[9]Chapter 4.

emplacement is dominated by gas fluidization. A detailed quantitative study of this phase of evolution has been made by McGetchin (1968). A crack propagates upward until the surface has been reached. At this point a sharp reduction in fluid pressure occurs and rapidly propagates downward. Exsolution of gas from liquid correspondingly propagates downward, accompanied by a drastic increase in upward flow velocity of the fluidized system. Expansion is accompanied by adiabatic cooling and solidification of the liquid so that temperatures of final emplacement are low. McGetchin demonstrated that flow velocities at high levels during the catastrophic stage may exceed 400 m/sec, thus providing the capacity to transport large xenoliths of mantle material from great depths. The violence of the eruption also causes attrition and reaming of the conduit walls and disintegration of xenoliths, accompanied by introduction of individual fragments of mantle minerals into the kimberlite. The process is intrinsically nonselective, and accordingly, there are strong grounds for supposing that the peridotitic and eclogitic xenolith and xenocryst assemblage found in kimberlites represent an average sample of the mantle path traversed.

To summarize, studies of the relative abundances of ultramafic and mafic xenoliths, and of corresponding xenocrystal mineral components in kimberlite pipes, provide some of the strongest evidence that the upper mantle beneath continental regions is dominantly composed of ultramafic rocks, with eclogite a widely distributed but relatively minor component except perhaps in some limited regions.

Peridotitic Xenoliths in the Alkali Basalt Suite

More than 200 continental and oceanic localities[1] are known where peridotitic xenoliths ("olivine nodules") are found in rocks of the alkali basalt-basanite-nephelinite suite. The nodules occur most commonly associated with the more vesicular and tuffaceous members of this suite and also in diatremes, strongly indicating that the host magmas were originally rich in volatiles. The inclusions are mineralogically and chemically heterogeneous, usually ranging between dunite and lherzolite.[2,3,4] In most localities spinel lherzolite is the dominant rock type. The lherzolites are dominantly composed of olivine (Fo_{88-92}), and the proportions of accompanying orthopyroxene and clinopyroxene vary widely (Fig. 3-7). The pyroxenes characteristically contain substantial amounts of alumina—usually between 3 and 7%.

A detailed study by Ross, Foster, and Myers (1954) demonstrated that the petrology and mineral chemistry of the xenoliths were very similar to those of alpine peridotites, and they concluded that these two classes of rocks were genetically related and derived from the upper mantle. A particularly close rela-

[1]Forbes and Kuno (1965).
[2]Vilminot (1965).
[3]White (1966).
[4]Jackson (1968).

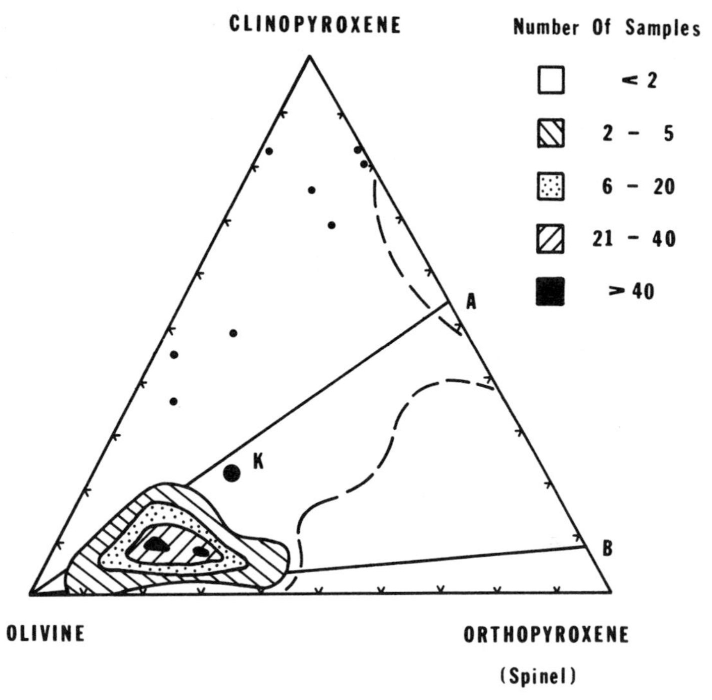

FIGURE 3-7
Contoured triangular plot of major phase mineralogy of nodules from Kilbourne Hole, New Mexico. (*From Carter, 1970, with permission.*) Similar results have been obtained by Vilminot (1965) and White (1966) for other collections of nodules.

tionship, especially in the alumina contents of the pyroxenes, was demonstrated to exist between the mineralogies of lherzolite nodules and high-temperature peridotites.[1] The mineral fabrics of alpine peridotites and olivine nodules were also demonstrated to be very similar, being produced by complex metamorphic processes involving extensive solid state deformation.[2,3,4] The deformational mechanisms and fabrics were studied and reproduced experimentally by Avé Lallemant and Carter (1970), who showed that they were also displayed by an ultramafic xenolith from a diamond pipe. These authors concluded that the deformation had probably occurred in the mantle.[5]

Chemical compositions of olivine nodules have been recently discussed by several authors.[6,7,8] The range of compositions found corresponds to expectations

[1]Green (1964).
[2]Collee (1963).
[3]Avé Lallemant (1967).
[4]Den Tex (1969).
[5]Avé Lallemant and Carter (1970).
[6]Harris, Reay, and White (1967).
[7]Kuno and Aoki (1970).
[8]Hutchison, Paul, and Harris (1970).

Table 3-8 COMPOSITIONS OF COLLECTIONS OF SPINEL PERIDOTITE NODULES OCCURRING IN THE ALKALI BASALT SUITE

	(1)	(2)	(3)	(4)
SiO_2	41.10	44.4	44.5	45.0
TiO_2	0.08	0.04	0.07	0.07
Al_2O_3	0.56	1.66	2.69	3.01
Cr_2O_3	0.35	0.47	0.43	0.41
Fe_2O_3	1.24	1.41	1.46	1.28
FeO	9.31	7.50	6.71	6.70
NiO	0.44	0.27	0.26	0.25
MnO	0.15	0.13	0.11	0.11
MgO	46.33	42.3	40.9	39.7
CaO	0.17	1.64	2.61	3.15
Na_2O	–	0.11	0.22	0.24
K_2O	–	0.04	0.01	0.04

Explanation:

(1) Mean of three analyses of nodules with Al_2O_3 contents of less than 1% (Harris, Reay, and White, 1967).

(2) Mean of analyses of 27 nodules from Puy Beanite, west of Riom, Puy-de-Dome (Hutchison, 1970).

(3) Mean of analyses of 42 nodules from Le Puy, Haute-Loire (Hutchison, 1970).

(4) Mean of analyses of 20 nodules from the basalte de Rocher du Lion, (Vilminot, 1965).

based upon observed variations in mineralogy (Fig. 3-7). Most compositions fall between those of dunites ($< 0.5\%$ CaO and Al_2O_3) and lherzolites, containing up to 4% each of Al_2O_3 and CaO. Only a small proportion of lherzolite nodules contain more than 3% each of Al_2O_3 and CaO, and it is probable that these are over-represented in collections of analyses owing to a tendency by some of the earlier investigators to choose "interesting" nodules for analyses, richer than average in deep-green chrome diopside and dark spinel. A collection of more recent representative analyses is given in Table 3-8. The chemical compositions correspond closely to those displayed by alpine ultramafic rocks ranging from dunite to high-temperature peridotite (Table 3-3).

The compositional ranges of olivine nodules (Table 3-8) extensively overlap those of the garnet peridotite xenoliths from diamond pipes (Table 3-7). Green and Ringwood (1963)[1] pointed out that the two classes represent equilibrium by material of similar composition in different *P*-*T* environments, the denser garnet peridotite assemblage being indicative of higher pressures. This conclusion has been fully verified by subsequent experimental work discussed in Sec. 6-2.

The spinel lherzolites in turn represent high-pressure assemblages compared to isochemical plagioclase peridotites.[1,2] Olivine and plagioclase in the latter

[1]See also Ringwood (1962).

[2]Green and Ringwood (1963).

rocks react to form aluminous pyroxenes and spinel as found in the spinel lherzolites. Experimental data on the partition of Mg, Fe, and Ca between olivine, orthopyroxene, and clinopyroxene[1] compared with observed partitions in nodules[2,3] suggest that most lherzolite nodules have equilibrated at temperatures in the range 800 to 1100°C. Corresponding investigations of the conditions required for the reaction of olivine plus plagioclase to form aluminous pyroxenes plus spinel within this temperature range indicate minimum pressures of equilibration in the range 8 to 15 kbars.[4,5,6] This implies that lherzolite nodules found in oceanic regions (e.g., Hawaii) are derived from the mantle and that a large proportion of nodules found in continental regions are also probably of mantle origin.

Studies on the genetic relationships between different classes of alkaline mafic and ultramafic magmas[7,8] lead to the conclusion that kimberlites and associated melilite nephelinites, melilite basalts, alnoites, and carbonatites which carry garnet lherzolite nodules in diamondiferous provinces are genetically related to the more abundant alkali basalt-basanite-olivine nephelinite suite in which spinel lherzolite nodules predominate. The kimberlitic magmas appear to have formed at greater depth and have evolved via more extensive fractionation processes, leading to higher abundances of volatile components and incompatible elements as compared with the other members of the alkali basalt suite.[7,9] Nevertheless, the fundamental mechanisms of eruption are probably similar, and the occurrence of olivine nodules in the more vesicular and tuffaceous members of the alkali basalt probably indicates that the nodules were also transported by fluidized gas-liquid systems which ascended rapidly to the surface.

The dominant occurrence of garnet lherzolites as mantle-derived xenoliths in kimberlite pipes as compared to spinel lherzolites in the alkali basalt suite is probably due to a combination of causes.[10,11,12,13,14] Spinel lherzolite transforms to garnet lherzolite at a depth of about 70 km in regions of normal heat-flow.[15] However, alkali basalts tend to occur more frequently in unstable regions—including rifts and continental margins characterized by higher-than-normal heat-flow—and the transition from spinel to garnet lherzolite may not occur until depths of 90 to 100 km. Since the more abundant members of the alkali basalt suite are probably generated at depths between 40 and 100 km,[12,4] the material which they penetrate

[1]Davis and Boyd (1966).
[2]Ross, Foster, and Myers (1954).
[3]White (1966).
[4]Green (1970b).
[5]Green and Hibberson (1970a).
[6]The "pyroxene geobarometer" of O'Hara (1967a) is inapplicable to the spinel lherzolite assemblage.
[7]O'Hara and Yoder (1967).
[8]Section 4-5.
[9]Harris and Middlemost (1970).
[10]Section 6-2.
[11]Boyd and MacGregor (1964).
[12]Green and Ringwood (1967a).
[13]MacGregor (1968).
[14]O'Hara (1968).
[15]Green and Ringwood (1967b, 1970).

on their journey to the surface lies mainly in the spinel lherzolite field. On the other hand, kimberlites are believed to originate at much greater depth and penetrate large thicknesses of mantle material in the garnet lherzolite field. The contrast is enhanced by the circumstance that most kimberlites occur in stable continental platforms or shields characterized by low heat-flow and lower than average subcrustal temperatures.[1] Under these conditions the transition of spinel to garnet lherzolite occurs at lower pressures,[2,3,4] and garnet lherzolite may be stable relative to spinel lherzolite at all depths below the Mohorovicic discontinuity.

A notable feature is the essential restriction of spinel lherzolite xenoliths to members of the alkali basalt suite. They are extremely rare in tholeiites. This appears to be caused by two factors. Alkali basalts are richer in volatile components and, accordingly, are capable of providing a mechanism (gas-liquid fluidization) for transport of xenoliths from the mantle, whereas tholeiites, being poorer in volatiles, are rarely associated with violent eruptive activity and lack the capacity to travel at the required high velocities from the mantle into the crust. Secondly, experimental studies show that the mineral assemblage of lherzolite nodules is not stable in tholeiitic magmas. If accidentally incorporated, such nodules would undergo extensive partial melting accompanied by complete disintegration until their former presence was revealed only by the occurrence of scattered olivine and (perhaps) orthopyroxene xenocrysts. In contrast, Green and Ringwood[5] demonstrated that lherzolite nodules incorporated in alkali basalts at pressures of 10 to 20 kbars would remain intact as refractory, subsolidus assemblages.

Most workers[e.g.,6,7,8,9,10,11,12,13,14,15,16] who have studied lherzolite nodules during recent years have concluded that they represent samples of mantle material transported by their alkali basalt hosts to the surface. According to this interpretation, their origin is analogous to the garnet lherzolite nodules of diamond pipes. Nevertheless, a minority[e.g.,17,18,19] have maintained that the nodules represent cognate precipitates at depth from the host basaltic magmas. One of the

[1]Clark and Ringwood (1964).
[2]Boyd and MacGregor (1964).
[3]Green and Ringwood (1967b, 1970).
[4]Section 6-2.
[5]Green and Ringwood (1964, 1967a).
[6]Ross, Foster, and Myers (1954).
[7]Wilshire and Binns (1961).
[8]Forbes and Kuno (1965).
[9]White (1966).
[10]Harris, Reay, and White (1967).
[11]Jackson (1968).
[12]Carter (1970).
[13]Hutchison, Paul, and Harris (1970).
[14]Kuno and Aoki (1970).
[15]Green, Morgan, and Heier (1968).
[16]Cooper and Green (1969).
[17]Brothers (1960).
[18]O'Hara (1967b).
[19]O'Hara (1968).

grounds for this view has been the observed relationship between nodule types and the nature of the magmas with which the nodules are associated. However, we have seen above that this relationship may be caused by a combination of factors, and does not necessarily imply consanguinity.

Intensive mineralogical and chemical studies of large xenolith populations in restricted localities have done much to clarify the situation during recent years.[1,2,3] White demonstrated the occurrence at Hawaii of two distinct classes of xenoliths in the alkali basalt suite. One class, which occurred mainly in olivine nephelinites and basanites, consisted dominantly of typical spinel lherzolites characterized by uniform mineral compositions, high Mg/Mg + Fe ratios, and often, metamorphic and deformation textures. The other class, occurring mainly in alkali olivine basalts and hawaiites, consisted of dunite, wehrlite, feldspathic peridotite, pyroxenite, and gabbro. The minerals of this latter class were richer in iron and covered much wider compositional ranges. Textures of this class were more typically igneous, although some members displayed evidence of deformation.[3] White concluded that the lherzolites represented genuine mantle xenoliths, whereas the other class represented accumulates of crystals precipitated at depth. Corresponding dichotomies in xenolith populations have since been recorded in several regions and interpreted similarly.[4,5,6] In all cases, a suite of characteristic metamorphic lherzolite nodules (Fo_{88-94}) has been found associated with a mineralogically and chemically heterogeneous class of xenoliths displaying typical igneous textures and higher Fe/Fe + Mg ratios. The latter ratios are consistent with the minerals of these xenoliths having been precipitated from the host basalts, whereas the Fe/Fe + Mg ratios of the lherzolite suite are usually too low for this to have been possible. Experimental investigations on a host basalt demonstrated that aluminous pyroxenes which occurred on the liquidus at elevated pressures, corresponding to mantle conditions, possessed closely similar chemical compositions to the large unstrained glassy xenocrysts occurring in the basalt.[8] Almost certainly these assemblages and others resembling them[4,7,8,9] represent cognate precipitates from the host magmas.

The recognition of two distinct classes of xenoliths and xenocrysts goes far towards reconciling earlier differences concerning their origins. We have seen that a strong case can be made for a cognate origin of the second class (discussed above). The contrast in properties between the latter class and the spinel lherzolite class renders it unlikely that the two have a common origin. The follow-

[1] White (1966).
[2] Jackson (1968).
[3] Jackson and Wright (1970).
[4] Aoki and Kushiro (1968).
[5] Carter (1970).
[6] Green (1970b).
[7] Binns, Duggan, and Wilkinson (1970).
[8] Green and Hibberson (1970b).
[9] Kuno (1964).

ing evidence shows rather clearly that the spinel lherzolite suite is not generally of cognate origin, in the sense of having crystallized at depth from the host magmas.

1 Spinel lherzolite nodules characteristically are metamorphic rocks displaying textural evidence of deformation and recrystallization.[1,2,3,4] Furthermore, the low Ca contents of the orthopyroxenes and the low degree of solid solubility of orthopyroxene in diopsidic clinopyroxene[1] demonstrate that the nodules have equilibrated at temperatures which are well below those of the magmas in which they are included.[3,5] Metamorphism and recrystallization must also have occurred under subsolidus conditions. These characteristics are not readily explained if the nodules are interpreted as loose aggregates of crystals which have been precipitated from their host magmas at depth along the walls of magmatic conduits and have then been elutriated to the surface.[6,7]

2 The comparative uniformity of compositions (particularly $Fe/Fe + Mg$ ratios and Cr and Ni contents) of individual minerals from spinel lherzolite nodules has already been noted. Even if the parent magmas were more magnesian and richer in normative olivine than the derived magmas in which the nodules are now found,[8] closed-system crystallization of such parent magmas necessarily involves Rayleigh-type fractionation which would lead to a wide range of $Fe/Fe + Mg$ ratios and Ni and Cr contents. The very high partition coefficients of Ni in olivine and Cr in pyroxene and spinels should cause particularly wide variations in the abundances of these elements.[9]

3 The similarities in mineralogy and compositions between nodules and alpine peridotites have already been discussed. These are not readily explained by the cognate accumulation hypothesis.

4 Strontium isotope ratios found in lherzolite nodules display a wide range, with Sr^{87}/Sr^{86} varying from 0.703 to 0.710,[10,11,12,13] compared to 0.703 to 0.705 for most basalts. Whereas the isotopic data do not exclude a cognate origin for some xenoliths,[10,12] the substantial differences in strontium isotope ratios

[1]White (1966).
[2]Den Tex (1969).
[3]Green and Ringwood (1967a).
[4]Jackson (1968).
[5]Green (1970b).
[6]O'Hara (1967b).
[7]O'Hara (1968) maintained that "spinel lherzolite nodules in nepheline-normative magmas are cognate accumulates whose genesis is an integral part of the process by which some hypersthene-normative magmas become nepheline-normative. . . ."
[8]As hypothesized by O'Hara (1968).
[9]Section 4-6.
[10]Stueber and Murthy (1966).
[11]Leggo and Hutchison (1968).
[12]O'Neil, Hedge, and Jackson (1970).
[13]Peterman, Carmichael, and Smith (1970).

between other xenoliths and their host rocks[1,2,3] clearly demonstrate that the xenoliths cannot represent cognate precipitates from these magmas. An investigation of Sr^{87}/Sr^{86} isotopic ratios in the individual minerals of a lherzolite xenolith enclosed in a basanite gave the following results:[2] olivine 0.7087, diopside 0.7016, orthopyroxene 0.708, host basanite 0.7031. Not only were there substantial differences between the minerals and basanite, but the coexisting minerals in the xenolith were in isotopic disequilibrium. Peterson et al. concluded that the lherzolite was of accidental origin and did not represent a crystal residuum related to differentiation of the basanite.

5 Cooper and Green (1969) determined the lead isotopic compositions of a group of lherzolite inclusions and their host basalts. The lead from the xenoliths was shown to be isotopically different from that in the basanites, precluding the hypothesis that the inclusions were cognate.

6 Green, Morgan, and Heier (1968) determined the U and K abundances in a suite of peridotite xenoliths and their host basanites. The K/U ratios of the xenoliths were found on the average to be about one-third of those in the basanites. This is not readily explicable if the xenoliths represent cumulates contaminated by varying small proportions of trapped (intercumulus) basanite magma. On the other hand, the authors point out that the difference is readily explicable if the xenoliths represent residual material remaining behind after extraction of a basalt magma.

Garnet Pyroxenite and "Eclogite" Xenoliths in Nephelinites

In contrast to the abundance of lherzolite nodules in alkali basalts, xenoliths of eclogitic rocks are comparatively rare. One of the best described occurrences is at Salt Lake Crater, Hawaii.[4,5,6,7] These detailed studies have shown that the dominant rock type is a garnet pyroxenite rather than a true eclogite. In addition to the predominant mineralogy of two pyroxenes and garnet, varying amounts of olivine and spinel may be present. The pyroxenes contain large amounts of alumina in solid solution as Tschermak's molecules[8] and differ in this respect from the pyroxenes of eclogites from diamond pipes.[9] The textures are characterized by evidence of extensive exsolution of clinopyroxenes and orthopyroxenes, of garnet from aluminous pyroxenes, and by complex reactions involving spinel, pyroxenes, garnet, and olivine. Some workers have interpreted these as high-pressure accumulates from the magmas in which they are found or from closely

[1] Leggo and Hutchison (1968).
[2] Peterman, Carmichael, and Smith (1970).
[3] Hutchison and Dawson (1970).
[4] Yoder and Tilley (1962).
[5] White (1966).
[6] Green (1966).
[7] Jackson and Wright (1970).
[8] Section 4-4.
[9] White (1964).

related magmas.[1,2] The nodules do not display evidence of the deformation and recrystallization which are so common in the lherzolites.[3] However, Jackson and Wright (1970) believe that they may represent accidental inclusions of inhomogeneous mantle material.

Green[2] has shown that the bulk chemical compositions of many of the garnet pyroxenite xenoliths are similar to those of the aluminous subcalcic clinopyroxenes which are observed on the liquidi of alkali basalts and basanites at pressures of 13 to 18 kbars. He concluded that the xenoliths represent former pyroxenite precipitates from such magmas which cooled at constant pressure to about 1000°C, causing exsolution of garnet from the original highly aluminous clinopyroxene and reaction of spinel and pyroxenes to yield garnet and olivine.

Related garnet pyroxenite nodules have been found in a nephelenite diatreme at Delegate, New South Wales,[4,5] and in a volcanic breccia at Kakanui, New Zealand.[6] Both assemblages contain some xenoliths with subequal proportions of garnet and pyroxenes, but the pyroxenes contain very high proportions of Tschermak's molecules, in contrast to the pyroxenes in genuine eclogites from diamond pipes. Irving and Green[5] demonstrated experimentally that the conditions of final equilibration of the minerals from some of the Delegate xenoliths were in the range 14 to 16 kbars at 1050 to 1100°C. They interpreted the minerals as precipitates from a parent basaltic magma within the mantle, which had subsequently been recrystallized at lower temperatures before being incorporated in the nephelinite magma.

REFERENCES

ANDERSON, O. L., and B. R. JULIAN (1969). Shear velocities and elastic parameters of the mantle. *J. Geophys. Res.* **74**, 3281–3286.

———, E. SCHREIBER, R. C. LIEBERMANN, and N. SOGA (1968). Some elastic constant data on minerals relevant to geophysics. *Rev. Geophys.* **6**, 491–524.

AOKI, K., and I. KUSHIRO (1968). Some clinopyroxenes from ultramafic inclusions in Dreiser Weiher, Eifel. *Contr. Mineral. Petrol.* **18,** 326–337.

AUBOUIN, J. (1965). "*Geosynclines*." (Developments in Geotectonics, vol. 1.) Elsevier, Amsterdam. 335 pp.

AVÉ LALLEMANT, H. G. (1967). Structural and petrofabric analysis of an alpine-type peridotite: the lherzolite of the French Pyrenees. *Leidse Geol. Mededel.* **42**, 1–57.

——— and N. L. CARTER (1970). Syntectonic recrystallization of olivine and modes of flow in the upper mantle. *Bull. Geol. Soc. Am.* **81**, 2203–2220.

BENSON, W. N. (1926). The tectonic conditions accompanying the intrusion of ultrabasic and basic igneous rocks. *Nat. Acad. Sci. Mem.* **19**, mem. **1**, 1–90.

[1] Kuno (1964).
[2] Green (1966).
[3] White (1966).
[4] Lovering and White (1969).
[5] Irving and Green (1970).
[6] Mason (1968).

BINNS, R. A., M. B. DUGGAN, and J. F. G. WILKINSON (1970). High pressure megacrysts in alkaline lavas from northeastern New South Wales. *Am. J. Sci.* **269**, 132–168.

BIRCH, F. (1960). The velocity of compressional waves in rocks to 10 kilobars, 1. *J. Geophys. Res.* **65**, 1083–1102.

———(1961). The velocity of compressional waves in rocks to 10 kilobars, 2. *J. Geophys. Res.* **66**, 2199–2224.

BODVARSSON, G., and G. P. WALKER (1964). Crustal drift in Iceland. *Geophys. J.* **8**, 285-300.

BOLT, B. A., H. A. DOYLE, and D. J. SUTTON (1958). Seismic observations from the 1956 atomic explosions in Australia. *Geophys. J. Roy. Astr. Soc.* **1**, 135–145.

BONATTI, E. (1968). Ultramafic rocks from the mid-Atlantic Ridge. *Nature* **219**, 363.

———, J. HONNOREZ, and G. FERRARA (1971). Peridotite-gabbro-basalt complex from the equatorial mid-Atlantic Ridge. *Phil. Trans. Roy. Soc. London* **A 268**, 385–402.

BOYD, F. R. (1967). Electroprobe study of diopside inclusions from kimberlite. *Am. J. Sci.* **267-A**, 50–69.

——— (1970). Garnet peridotites and the system $CaSiO_3$-$MgSiO_3$-Al_2O_3. *Mineral. Soc. Am. Spec. Paper* **3**, 65–75.

——— and J. L. ENGLAND (1964). The system enstatite-pyrope. *Carnegie Inst. Washington Yearbook* **63**, 157–161.

——— and I. D. MCGREGOR (1964). Ultramafic rocks. *Carnegie Inst. Washington Yearbook* **63**, 152–156.

BROTHERS, R. N. (1960). Olivine nodules from New Zealand, *Intern. Geol. Congr. 21st, Copenhagen, Rept.* **13**, 68–81.

BUNDY, F. P. (1963). Direct conversion of graphite to diamond in static high pressure apparatus. *J. Chem. Phys.* **38**, 631–643.

CARSWELL, D. A., and J. B. DAWSON (1970). Garnet peridotite xenoliths in South African kimberlite pipes and their petrogenesis. *Contr. Mineral. Petrol.* **25**, 163–184.

CARTER, J. L. (1970). Mineralogy and chemistry of the earth's upper mantle based on the partial fusion–partial crystallization model. *Bull. Geol. Soc. Am.* **81**, 2021–2034.

CARTER, N. L., D. W. BAKER, and R. P. GEORGE (1972). Seismic anisotropy, flow and constitution of the upper mantle. (Preprint—Griggs Volume.)

CHALLIS, G. A. (1965). The origin of New Zealand ultramafic intrusions. *J. Petrol.* **6**, 322–364.

CHRISTENSEN, N. I. (1966). Elasticity of ultrabasic rocks. *J. Geophys. Res.* **71**, 5921–5931.

CLARK, S. P. and A. E. RINGWOOD (1964). Density distribution and constitution of the mantle. *Rev. Geophys.* **2**, 35–88.

COLEMAN, R. G. (1971). Plate tectonic emplacement of upper mantle peridotites along continental edges. *J. Geophys. Res.* **76**, 1212–1222.

COLLEE, A. L. (1963). A fabric study of lherzolites, with special reference to ultrabasic nodular inclusions in the lavas of Auvergne, France. *Leidse Geol. Meded.* **28**, 1–102.

COOPER, J. A., and D. H. GREEN (1969). Lead isotope measurements on lherzolite inclusions and host basanites from Western Victoria, Australia. *Earth Planet. Sci. Letters* **6**, 69–76.

DAVIES, H. L. (1968). Papuan ultramafic belt. *Intern. Geol. Congr., 23d, Prague, Rept.* **1**, 209–220.

——— (1969). Peridotite-gabbro-basalt complex in eastern Papua: An overthrust plate of oceanic mantle and crust. Ph.D. Thesis, Stanford Univ., pp. 1–89.

DAVIS, B. T. (1964). The system diopside-forsterite-pyrope at 40 kilobars. *Carnegie Inst. Washington Yearbook* **63**, 165–171.

——— and F. R. BOYD (1966). The join Mg_2SiO_2-$CaMgSi_2O_6$ at 30 kilobars pressure and its application to pyroxenes from kimberlites. *J. Geophys. Res.* **71**, 3567–3576.

DAWSON, J. B. (1962). Basutoland kimberlites. *Bull. Geol. Soc. Am.* **73**, 545–560.

——— (1968). Recent researches on kimberlite and diamond geology. *Econ. Geol.* **63**, 504–511.

DENHAM, D., D. SIMPSON, D. SUTTON, and P. GREGSON (1972). Travel times from the Ord River explosions in Northern Australia. *Geophys. J.* **28**, 225–235.

DEN TEX, E. (1969). Origin of ultramafic rocks, their tectonic setting and history. *Tectonophysics* **7**, 457–488.

DERR, J. S. (1969). Internal structure of the earth inferred from free oscillations. *J. Geophys. Res.* **74**, 5202–5220.

DEWEY, J. F., and J. M. BIRD (1971). Origin and emplacement of the ophiolite suite. *J. Geophys. Res.* **76**, 3179–3206.

DICKEY, J. S. (1970). Partial fusion products in alpine-type peridotites: Serrania de la Ronda and other examples. *Mineral. Soc. Am. Spec. Paper* **3**, 33–49.

DORMAN, J., and M. EWING (1962). Numerical inversion of seismic surface wave dispersion data and crust-mantle structure in the New York–Pennsylvania area. *J. Geophys. Res.* **67**, 5227–5241.

DOYLE, H. A., and I. EVERINGHAM (1964). Seismic velocities and crustal structure in southern Australia. *J. Geol. Soc. Aust.* **11**, 141–150.

DRAKE, C. L., M. EWING, and G. H. SUTTON (1959). Continental margins and geosynclines. *Phys. Chem. Earth* **3**, 110–198.

DZIEWONSKI, A. M. (1970). Correlation properties of free period partial derivatives and their relation to resolution of gross earth data. *Bull. Seism. Soc. Am.* **60**, 741–768.

——— (1971). Upper mantle models from "pure path" dispersion data. *J. Geophys. Res.* **76**, 2587–2601.

ENGLAND, R. N., and H. L. DAVIES (1970). Mineralogy of cumulus and noncumulus ultramafic rocks from eastern Papua. Bureau Min. Resources Australia, Record 1970/66.

FORBES, R., and H. KUNO (1965). The regional petrology of peridotite inclusions and basaltic host rocks. In: "*Upper Mantle Symposium*," pp. 161–179. New Delhi.

GASS, I. G. (1967). The ultrabasic volcanic assemblage of the Troodos massif, Cyprus. In: P. J. Wyllie (ed.), "*Ultramafic and Related Rocks*," pp. 121–134. Wiley, New York.

——— (1968). Is the Troodos Massif of Cyprus a fragment of the Mesozoic ocean floor? *Nature* **220**, 39–42.

GAST, P. W. (1968). Trace element fractionation and the origin of tholeiitic and alkaline magma types. *Geochim. Cosmochim. Acta* **32**, 1057–1086.

GRAHAM, A. L., and A. E. RINGWOOD (1971). Lunar basalt genesis: The origin of the europium anomaly. *Earth Planet. Sci. Letters* **13**, 105–115.

GRAHAM, E. K. (1970). Elasticity and composition of the upper mantle. *Geophys. J.* **20**, 285–302.

——— and G. R. BARSCH (1969). Elastic constants of single crystal forsterite as a function of temperature and pressure. *J. Geophys. Res.* **74**, 5949–5960.

GREEN, D. H. (1963). Alumina content of enstatite in a Venezuelan high-temperature peridotite. *Bull. Geol. Soc. Am.* **74**, 1397–1402.

——— (1964). The petrogenesis of the high-temperature peridotite intrusion in the Lizard area. Cornwall. *J. Petrol.* **5**, 134–188.

——— (1966). The origin of the "eclogites" from Salt Lake Crater, Hawaii. *Earth Planet. Sci. Letters* **1**, 414–420.

——— (1967). High temperature peridotite intrusions. In: P. J. Wyllie (ed.), "*Ultramafic and Related Rocks*," pp. 212–222. Wiley, New York.

——— (1970a). Peridotite-gabbro complexes as keys to petrology of mid-oceanic ridges: A discussion. *Bull. Geol. Soc. Am.* **81**, 2161–2166.

——— (1970b). The origin of basaltic and nephelinitic magmas. *Trans. Leicester Literary Philos. Soc.* **64**, 28–54.

——— and W. HIBBERSON (1970a). The instability of plagioclase in peridotite at high pressure. *Lithos* **3**, 209–221.

——— and ——— (1970b). Experimental duplication of conditions of precipitation of high-pressure phenocrysts in a basaltic magma. *Phys. Earth Planet. Interiors* **3**, 247–254.

———, J. W. MORGAN, and K. S. HEIER (1968). Thorium, uranium and potassium abundances in peridotite inclusions and their host basalts. *Earth Planet. Sci. Letters* **4**, 155–166.

——— and A. E. RINGWOOD (1963). Mineral assemblages in a model mantle composition. *J. Geophys. Res.* **68**, 937–945.

——— and ——— (1964). Fractionation of basalt magmas at high pressures. *Nature* **201**, 1276–1279.

——— and ——— (1967a). The genesis of basaltic magmas. *Contr. Mineral. Petrol.* **15**, 103–190.

——— and ——— (1967b). The stability fields of aluminous pyroxene peridotite and garnet peridotite and their relevance in upper mantle structure. *Earth Planet. Sci. Letters* **3**, 151–160.

——— and ——— (1970). Mineralogy of peridotitic compositions under upper mantle conditions. *Phys. Earth Planet. Interiors* **3**, 359–371.

GREEN, T. H., and A. E. RINGWOOD (1968). Genesis of the calc-alkaline igneous rock suite. *Contr. Mineral. Petrol.* **18**, 105–162.

HADDON, R. A., and K. E. BULLEN (1969). An earth model incorporating free earth oscillation data. *Phys. Earth Planet. Interiors* **2**, 35–49.

HARRIS, P. G., and E. A. MIDDLEMOST (1970). The evolution of kimberlites. *Lithos* **3**, 79–90.

———, A. REAY, and I. G. WHITE (1967). Chemical composition of the upper mantle. *J. Geophys. Res.* **72**, 6359–6369.

HESS, H. H. (1939). Island arcs, gravity anomalies and serpentinite intrusions. *Intern. Geol. Congr., 17th, Moscow*, pt. 2, 263–283.

——— (1955a). Serpentines, orogeny and epeirogeny. In: A. Poldervaart (ed.), "*Crust of the Earth*," pp. 391–408. *Geol. Soc. Am. Spec. Paper* **62**.

——— (1955b). The oceanic crust. *J. Marine Res.* **14**, 423–439.

——— (1964a). Seismic anisotropy of the uppermost mantle under oceans. *Nature* **203**, 629–631.

——— (1964b). The oceanic crust, the upper mantle, and the Mayaguez serpentinized peridotite. In: C. A. Burk (ed.), "*A Study of Serpentinite*," pp. 169–175. *Nat. Acad. Sci.–Nat. Res. Coun. Publ.* **1188**.

HURLEY, P. M. (1967). Rb^{87}-Sr^{87} relationships in the differentiation of the mantle. In: P. J. Wyllie (ed.), "*Ultramafic and Related Rocks*," pp. 372–375. Wiley, New York.

HUTCHISON, R., and J. B. DAWSON (1970). Rb, Sr, and $Sr^{87/86}$ in ultrabasic xenoliths and host rocks, Lashaine volcano, Tanzania. *Earth Planet. Sci. Letters* **9**, 87–92.

———, D. K. PAUL, and P. G. HARRIS (1970). Chemical composition of the upper mantle. *Min. Mag.* **37**, 726–729.

IRVINE, T. N., and C. H. SMITH (1967). The ultramafic rocks of the Muskox Intrusion, Northwest Territories, Canada. In: P. J. Wyllie (ed.), "*Ultramafic and Related Rocks*," pp. 38–49. Wiley, New York.

IRVING, A., and D. H. GREEN (1970). Experimental duplication of mineral assemblages in basic inclusions of the Delegate breccia pipes. *Phys. Earth Planet. Interiors* **3**, 385–389.

ITO, K., and G. C. KENNEDY (1970). The fine structure of the basalt-eclogite transition. In: B. A. Morgan (ed.), "*Fiftieth Anniversary Symposia.*" *Mineral. Soc. Am. Spec. Paper* **3.**

——— and ——— (1971). An experimental study of the basalt-garnet granulite-eclogite transition. In: "*The Structure and Physical Properties of the Earth's Crust,*" pp. 303–314. *Am. Geophys. Union, Geophys. Monograph* **14.**

JACKSON, E. D. (1967). Ultramafic cumulates in the Stillwater, Great Dyke and Bushveldt intrusions. In: P. J. Wyllie (ed.), "*Ultramafic and Related Rocks,*" pp. 20–38. Wiley, New York.

——— (1968). The character of the lower crust and upper mantle beneath the Hawaiian Islands. *Intern. Geol. Congr., 23d Prague, Rept.* **1**, 48–55.

——— and T. L. WRIGHT (1970). Xenoliths in the Honolulu Volcanic Series, Hawaii. *J. Petrol.* **11**, 405–430.

JAMES, D. E., T. J. SMITH, and J. S. STEINHART (1968). Crustal structure of the middle Atlantic states. *J. Geophys. Res.* **73**, 1983–2007.

JOHNSON, L. R. (1967). Array measurements of P velocities in the upper mantle. *J. Geophys. Res.* **72**, 6309–6324.

JULIAN, B. R., and D. L. ANDERSON (1968). Travel times, apparent velocities and amplitudes of body waves. *Bull. Seism. Soc. Am.* **58**, 339–366.

KANAMORI, H., and H. MIZUTANI (1965). Ultrasonic measurement of elastic constants of rocks under high pressures. *Bull. Earthquake Res. Inst.* **43**, 173–194.

KEEN, C. E., and D. L. BARRETT (1971). A measurement of seismic anisotropy in the northeast Pacific. *Can. J. Earth Sci.* **8**, 1056–1064.

——— and C. TRAMONTINI (1970). Seismic refraction survey on mid-Atlantic ridge. *Geophys. J.* **20**, 473–491.

KUMAZAWA, M., and O. L. ANDERSON (1969). Elastic moduli, pressure derivatives, and temperature derivatives of single crystal olivine and single crystal forsterite. *J. Geophys. Res.* **74**, 5961–5972.

———, H. HELMSTAEDT, and K. MASAKI (1971). Elastic properties of eclogite xenoliths from diatremes of the East Colorado Plateau and their implication to the upper mantle structure. *J. Geophys. Res.* **76**, 1231–1247.

KUNO, H. (1964). Aluminium augite and bronzite in alkali olivine basalt from Taka-Sima, north Kyusyu, Japan. In: "*Advancing Frontiers in Geology and Geophysics.*" Volume dedicated to Dr. Krishnan (India), pp. 205-220. Osmania Univ. Press, Hyderabad.

——— and K. AOKI (1970). Chemistry of ultramafic nodules and their bearing on the origin of basaltic magmas. *Phys. Earth Planet. Interiors* **3**, 273–301.

KUSHIRO, I., and K. AOKI (1968). Origin of some eclogite inclusions in kimberlite. *Am. Min.* **53**, 1347–1367.

———, Y. SYONO, and S. AKIMOTO (1968). Melting of a peridotite nodule at high pressures and at high water pressures. *J. Geophys. Res.* **73**, 6023–6029.

LEGGO, P. J., and R. HUTCHISON (1968). A Rb-Sr isotope study of ultrabasic xenoliths and their basaltic host rocks from The Massif Central, France. *Earth Planet. Sci. Letters* **5**, 71–75.

LE PICHON, X., R. E. HORITZ, C. L. DRAKE, and J. E. NAFE (1965). Crustal structure of the mid-oceanic ridges (1). Seismic refraction measurements. *J. Geophys. Res.* **70**, 319–339.

LOOMIS, P. L. (1972). Contact metamorphism of pelitic rock by the Ronda ultramafic intrusion, southern Spain. *Bull. Geol. Soc. Am.* **83**, 2449–2474.

LOVERING, J. F., and A. J. WHITE (1969). Granulitic and eclogitic inclusions from basic pipes at Delegate, Australia. *Contr. Mineral. Petrol.* **21**, 9–52.

MAASKANT, P. (1970). Chemical petrology of polymetamorphic ultramafic rocks from Galicia, New Spain. *Leidse Geol. Mededel.* **45**, 237–325.

MACGREGOR, I. D. (1968). Mafic and ultramafic inclusions as indicators of the depth of origin of basaltic magmas. *J. Geophys. Res.* **73**, 3737–3745.

——— and J. L. CARTER (1970). The chemistry of clinopyroxenes and garnets of eclogite and peridotite xenoliths from the Roberts Victor Mine, South Africa. *Phys. Earth Planet. Interiors* **3**, 391–397.

MACKENZIE, D. B. (1960). High temperature alpine-type peridotite from Venezuela. *Bull. Geol. Soc. Am.* **71**, 303–318.

MASON, B. (1968). Eclogitic xenoliths from volcanic breccia at Kakanui, New Zealand. *Contr. Mineral. Petrol.* **19**, 316–327.

MATHIAS, M., J. C. SIEBERT, and P. C. RICKWOOD (1970). Some aspects of the mineralogy and petrology of ultramafic xenoliths in kimberlite. *Contr. Mineral. Petrol.* **26**, 75–123.

MCGETCHIN, T. R. (1968). The Moses Rock dike: geology, petrology and mode of emplacement of a kimberlite-bearing breccia-dyke, San Juan County, Utah. Ph.D. Thesis, California Institute of Technology.

MELSON, W. G., F. JAROSEVICH, V. T. BOWEN, and G. THOMPSON (1967). St. Peter and St. Paul Rocks: A high temperature mantle-derived intrusion. *Science* **155**, 1532–1535.

——— and G. THOMPSON (1970). Layered basic complex in oceanic crust, Romanche Fracture, equatorial Atlantic Ocean. *Science* **168**, 817–820.

MENARD, H. W. (1960). The east Pacific rise. *Science* **132**, 1737–1746.

MOORES, E. M. (1970). Petrology and structure of the Vourinos ophiolitic complex of northern Greece. *Geol. Soc. Am. Spec. Paper* **118**, 1–74.

——— and F. J. VINE (1971). The Troodos Massif, Cyprus and other ophiolites as oceanic crust: Evaluation and implications. *Phil. Trans. Roy. Soc.* **A 268**, 443–466.

MORRIS, G. B., R. W. RAITT, and G. G. SHOR (1969). Velocity anisotropy and delay-time maps of the mantle near Hawaii. *J. Geophys. Res.* **74**, 4300–4316.

MUIRHEAD, K. J., and J. R. CLEARY (1969). The D'' layer and the free oscillations of the earth. *Nature* **223**, 1146.

NIXON, P. H., O. VON KNORRING, and J. M. ROOKE (1963). Kimberlites and associated inclusions of Basutoland: a mineralogical and geochemical study. *Am. Min.* **48**, 1090–1131.

O'HARA, M. J. (1963). Melting of garnet peridotite at 30 kilobars. *Carnegie Inst. Washington Yearbook* **62**, 71–76.

——— (1967a). Mineral parageneses in ultrabasic rocks. In: P. J. Wyllie (ed.), "*Ultramafic and Related Rocks*," pp. 393–403. Wiley, New York.

——— (1967b). Crystal-liquid equilibria and the origins of ultramafic nodules in basic igneous rocks. *Ibid.*, pp. 346–349.

——— (1968). The bearing of phase equilibria studies on synthetic and natural systems on the origin and evolution of basic and ultrabasic rocks. *Earth Sci. Rev.* **4**, 69–133.

——— and E. L. P. MERCY (1963). Petrology and petrogenesis of some garnetiferous peridotites. *Trans. Roy. Soc. Edinburgh* **65**, 251–314.

——— and H. S. YODER (1967). Formation and fractionation of basic magmas at high pressures. *Scottish J. Geol.* **3**, 67–117.

O'NEIL, J. R., C. E. HEDGE, and E. D. JACKSON (1970). Isotopic investigations of xenoliths and host basalts from the Honolulu Volcanic Series. *Earth Planet. Sci. Letters* **8**, 253–257.

PETERMAN, Z. E., I. S. CARMICHAEL, and A. L. SMITH (1970). Strontium isotopes in quaternary basalts of southeastern California. *Earth Planet. Sci. Letters* **7**, 381–384.

PRESS, F. (1968). Earth models obtained by Monte Carlo inversion. *J. Geophys. Res.* **73**, 5223–5234.

——— (1969). The suboceanic mantle. *Science* **165**, 174–178.

——— (1970). Earth models consistent with geophysical data. *Phys. Earth Planet. Interiors* **3**, 3–22.

——— (1972). The earth's interior as inferred from a family of models. In: E. C. Robertson (ed.), "*The Nature of the Solid Earth*," chap. 7, pp. 147–171. McGraw-Hill, New York.

RAITT, R. W., G. G. SHOR, T. J. FRANCIS, and G. B. MORRIS (1969). Anisotropy of the Pacific upper mantle. *J. Geophys. Res.* **74**, 3095–3109.

———, ———, H. K. KILK, and M. HENRY (1972). Anisotropy of the oceanic upper mantle. *Geol. Soc. Am. 68*th Ann. Meeting, Cordilleran Sect. Abstracts, p. 222.

RICKWOOD, P. C., M. MATHIAS, and J. C. SIEBERT (1968). A study of garnets from eclogite and peridotite xenoliths found in kimberlite. *Contr. Mineral. Petrol.* **19**, 271–301.

RINGWOOD, A. E. (1958). Constitution of the mantle (3); Consequences of the olivine-spinel transition. *Geochim. Cosmochim. Acta* **15**, 195–212.

——— (1962). A model for the upper mantle, 2. *J. Geophys. Res.* **67**, 4473–4477.

——— (1969). Composition and evolution of the upper mantle. In: "*The Earth's Crust and Upper Mantle*," pp. 1–17. *Am. Geophys. Union, Geophys. Monograph* **13**.

——— (1970). Petrogenesis of Apollo 11 basalts and implications for lunar origin. *J. Geophys. Res.* **75**, 6465.

——— and D. H. GREEN (1966a). An experimental investigation of the gabbro-eclogite transformation and some geophysical consequences. *Tectonophysics* **3**, 383–427.

——— and ——— (1966b). Petrological nature of the stable continental crust. In: "*The Earth Beneath the Continents*," pp. 611–619. *Am. Geophys. Union, Geophys. Monograph* **10**.

——— and J. F. LOVERING (1970). Significance of pyroxene-ilmenite intergrowths among kimberlite xenoliths. *Earth Planet. Sci. Letters* **7**, 371–375.

ROSS, C. J., M. D. FOSTER, and A. T. MYERS (1954). Origin of dunites and olivine-rich inclusions in basaltic rocks. *Am. Min.* **39**, 693–737.

SIMMONS, G. (1964). Velocity of shear waves in rocks to 10 kilobars, 1. *J. Geophys. Res.* **69**, 1123–1130.

SMITH, C. H. (1958). Bay of Islands igneous complex, Western Newfoundland. *Geol. Surv. Canada Mem.* **290**, 1–132.

——— and J. D. MACGREGOR (1960). Ultrabasic intrusive conditions illustrated by the Mount Albert ultrabasic pluton, Gaspé, Quebec. *Bull. Geol. Soc. Am.* **71**, 1978 (Abstr.).

SMITH, T. J., J. S. STEINHART, and L. T. ALDRICH (1966). Lake Superior crustal structure. *J. Geophys. Res.* **71**, 1141–1172.

SOBOLEV, N. V. (1970). Eclogites and pyrope peridotites from the kimberlites of Yakutia. *Phys. Earth Planet. Interiors* **3**, 398–404.

——— (1973). "*Deep Seated Inclusions in Kimberlites and the Problem of the Composition of the Earth's Mantle.*" Nauka, Moscow. (In press.)

——— and I. K. KUZNETSOVA (1966). Mineralogy of diamond bearing eclogites. *Dokl. Akad. Nauk SSSR* **167**, No. 6 (in Russian).

———, ———, and N. I. ZYUZIN (1968). The petrology of grosspydite xenoliths from the Zagadochnaya kimberlite pipe in Yakutia. *J. Petrol.* **9**, 253–280.

SOBOLEV, V. S. (ed.) (1959). "*Diamond Deposits of Yakutia.*" Moscow.

SOLOMON, S. C. (1972). Seismic wave attenuation and partial melting in the upper mantle of North America. *J. Geophys. Res.* **77**, 1483–1502.

STEINMANN, G. (1926). Die ophiolitischen Zonen in dem Mediterranean Kettengebirgen, *Intern. Geol. Congr. 14th, Madrid, Compt. Rend.*, pt. 2, pp. 638–667.

STUEBER, A. M., and V. R. MURTHY (1966). Strontium isotope and alkali element abundances in ultramafic rocks, *Geochim. Cosmochim. Acta* **30**, 1243–1259.

THAYER, T. P. (1960). Some critical differences between alpine-type and stratiform peridotite-gabbro complexes. *Intern. Geol. Congr., 21st, Copenhagen, Rept.*, pt. 13, 247–259.

——— (1967). Chemical and structural relations of ultramafic and feldspathic rocks in alpine intrusive complexes. In: P. J. Wyllie (ed.), "*Ultramafic and Related Rocks,*" pp. 222–239. Wiley, New York.

——— (1969a). Peridotite-gabbro complexes as keys to petrology of mid-oceanic ridges. *Bull. Geol. Soc. Am.* **80**, 1515–1522.

——— (1969b). Alpine-type sensu structu (ophiolitic) peridotites: Refractory residues from partial melting or igneous sediments? *Tectonophysics* **7**, 511–516.

——— (1969c). Gravity differentiation and magmatic re-emplacement of podiform chromite deposits. *Econ. Geol. Monograph* **4**, 132–146.

UDINTSEV, G. B., and L. V. DMITRIEV (1971). Ultrabasic rocks. In: J. Maxwell (ed.), "*The Sea,*" vol. 4, pt. 1, chap. 14, pp. 521–573. Wiley, New York.

VERMA, R. K. (1960). Elasticity of some high density crystals. *J. Geophys. Res.* **65**, 757–766.

VILMINOT, J. C. (1965). Les enclaves de peridotite et de pyroxenolite a spinelle dans le basalte du Rocher du Lion. *Bull. Soc. Franc. Mineral. Crist.* **88**, 109–118.

VINOGRADOV, A. P., G. B. UDINTSEV, L. V. DMITRIEV, V. F. KANAEV, Y. P. NEPROCHNOV, G. N. PETROVA, and L. N. RIKUNOV (1969). The structure of the mid-oceanic rift zone of the Indian Ocean and its place in the world rift system. *Tectonophysics* **8**, 377–401.

WAGER, L. R., and R. L. MITCHELL (1951). The distribution of trace elements during strong fractionation of basic magma—a further study of the Skaergaard Intrusion, East Greenland. *Geochim. Cosmochim. Acta* **1**, 129–208.

WAGNER, P. A. (1928). The evidence of the kimberlite pipes on the constitution of the outer part of the earth. *S. African J. Sci.* **25**, 127–148.

WANG, C. Y. (1970). Density and constitution of the mantle. *J. Geophys. Res.* **75**, 3264–3284.

——— (1972). A simple earth model. *J. Geophys. Res.* **77**, 4318–4329.

WHITE, A. J. (1964). Clinopyroxenes from eclogites and basic granulites. *Am. Min.* **49**, 883–888.

WHITE, R. W. (1966). Ultramafic inclusions in basaltic rocks from Hawaii. *Contr. Mineral. Petrol.* **12**, 245–314.

WILLIAMS, A. F. (1932). "*The Genesis of the Diamond*" (2 vols.). Ernest Benn, London. 636 pp.

WILSHIRE, H. G., and R. A. BINNS (1961). Basic and ultrabasic xenoliths from volcanic rocks of northeastern New South Wales. *J. Petrol.* **2**, 185–208.

WILSON, J. T. (1968). Report of Upper Mantle Committee: Working group on tectonics. Presented at 23d International Geological Congress, Prague.

WILSON, R. A. (1959). The geology of the Xeros-Troodos area. *Cyprus Geol. Surv. Dep. Mem.* **1**, 1–135.

WOOLLARD, G. P. (1962). The relation of gravity anomalies to surface elevation, crustal structure and geology. *Dept. Geol. Univ. Wisconsin Res. Rept.* 62–9.

——— (1970). Evaluation of the isostatic mechanism and role of mineralogic transformations from seismic and gravity data. *Phys. Earth Planet. Interiors* **3**, 484–498.

WORTHINGTON, M. H., J. R. CLEARY, and R. S. ANDERSSEN (1972). Density modelling by Monte Carlo Inversion. II. Comparison of recent earth models. *Geophys. J.* **29**, 445–457.

WORZEL, J. L. and G. L. SHURBERT (1955). Gravity interpretations from standard oceanic and continental sections. *Geol. Soc. Am. Spec. Paper* **62**, 87–100.

WYLLIE, P. J. (ed.) (1967). "*Ultramafic and Related Rocks*," pp. 1–464. Wiley, New York.

——— (1970). Ultramafic rocks and the upper mantle. *Mineral. Soc. Am. Spec. Paper* **3**, 3–32.

YODER, H. S, and C. E. TILLEY (1962). Origin of basalt magmas; an experimental study of natural and synthetic rock systems. *J. Petrol.* **3**, 342–532.

4

ORIGIN OF BASALTIC MAGMAS

4-1 INTRODUCTION

Basaltic magmas have been erupted on a vast scale throughout geological time, both in continental and oceanic regions. From the days of Daly and Bowen to the present, problems relating to the genesis and fractionation of basaltic magmas have occupied a unique position in igneous petrogeny. Several lines of evidence lead to the conclusion that basalts are derived from the mantle. Their compositions accordingly constitute a reservoir of information, the correct interpretation of which should throw a great deal of light on the chemical and mineralogical nature of this region. In the present chapter, we will be concerned primarily with exploring the relationships of basaltic magmas, both among themselves and with their mantle source regions, in the light of recent advances in experimental petrology.

Petrologists have long recognized the existence of several classes of basalts, characterized by different mineralogical and chemical compositions (Table 4-1).

Table 4-1 CLASSIFICATION OF BASALTS ACCORDING TO CIPW NORMATIVE MINERALOGICAL COMPOSITIONS

Suite	Composition	
Tholeiitic suite	Quartz tholeiite	Contain quartz + hypersthene
Characterized by normative hypersthene	Olivine tholeiite	Contain olivine + hypersthene
Transitional suite	Olivine basalt	Arbitrary term for tholeiites falling very close to critical plane of undersaturation, Fig. 4-1 (< 3% hypersthene)
Alkaline basaltic suite	Alkali olivine basalt	Contain olivine; nepheline 0–5%
Characterized by normative nepheline	Basanite	Contain olivine; nepheline > 5%
	Olivine nephelinite	Contain olivine, diopside, and nepheline as major minerals, but not albite
High-alumina suite	Basalts containing 15–18% Al_2O_3 and which have been erupted as *liquids*. Normative compositions mostly fall in the olivine tholeiite volume of Fig. 4-1, but range from alkali olivine basalt through olivine basalt to olivine tholeiite.	

A convenient classification in terms of the CIPW normative mineralogy[1] has been proposed by Yoder and Tilley[2] (Fig. 4-1). The classification has a degree of artificiality, in that natural basaltic rocks display a complete continuum of compositions between the different classes. Nevertheless, it is very useful in practice.

Extensive studies of simplified model systems and of the crystallization of natural basaltic rocks provided an understanding of the differentiation trend from olivine tholeiite to quartz tholeiite in terms of Bowen's olivine reaction relationship.[3] The existence of a continuum of basalt compositions between *B* and *A* (Fig. 4-1) was thus readily explicable. However, the existence of a continuum of basalt compositions between *B* and *D* (Fig. 4-1) remained a major problem for many years. Experimental investigations on several simplified systems combined with observations of the crystallization behaviour of natural tholeiites and alkali basalts demonstrated that the clinopyroxene-plagioclase-olivine plane of Fig. 4-1 constituted a thermal barrier[2] and that, under normal low-pressure conditions,

[1]The CIPW norm refers to a low-pressure mineralogical composition which is calculated from the bulk chemical composition of the rock. It is particularly useful in the classification of fine-grained and/or glassy rocks (e.g., many basalts) in which the actual proportions of minerals cannot be readily determined.

[2]Yoder and Tilley (1962).

[3]Bowen (1928).

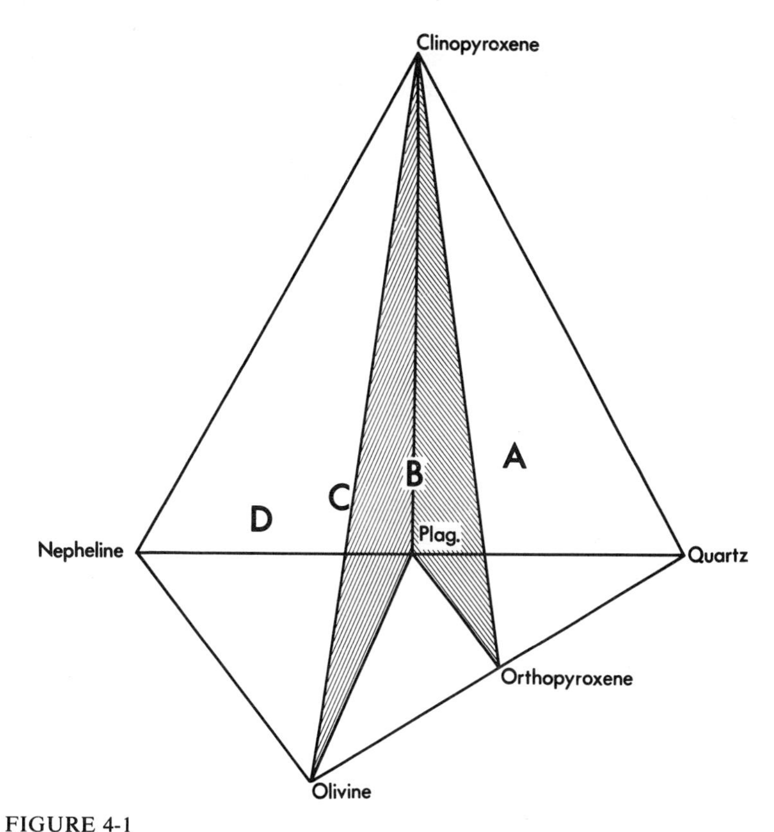

FIGURE 4-1
Diagrammatic representation of the major mineralogy of basalts using Yoder and Tilley's "basalt tetrahedron." The plane olivine-clinopyroxene-plagioclase is referred to as the "critical plane of undersaturation." *A*—field of quartz tholeiites; *B*—field of olivine tholeiites; *C*—field of alkali olivine basalts; *D*—field of olivine basanites. High-alumina basalts generally fall within the olivine tholeiite field with compositions relatively displaced towards the plagioclase subapex.

fractionation trends in basaltic magmas move *away* from this plane on either side. Accordingly, it did not appear possible for a parental olivine tholeiite magma to penetrate the critical plane of undersaturation and thus to generate an alkali olivine basalt by crystallization differentiation, or vice versa. On the other hand, petrologists studying the field, mineralogical, and petrological relationships between alkali basalts and tholeiites had established that a continuous spectrum of compositions existed, and many were convinced that these magmas were related to each other by some kind of crystallization differentiation process.

The high-alumina basalts constituted another group whose genesis remained in doubt. These occurred widely in continental, island-arc, and mid-oceanic-ridge environments, being particularly abundant in the last mentioned. A continuum of transitional types between high-alumina basalts, on the one hand, and olivine tholeiites, olivine basalts, and alkali olivine basalts, on the other, was recognized. Textural evidence showed that they had been erupted as liquids, yet melting experiments at low pressure showed a wide field of primary crystallization of plagioclase ± olivine, indicating that the high-alumina liquids were not the products of low-pressure fractionation processes.

Several petrologists[e.g.,1] had speculated that crystallization equilibria in basalts might be greatly altered by high pressures and that the inferred pathways between normal tholeiites, alkali basalts, and high-alumina basalts might be the results of expansions in the primary fields of crystallization of orthopyroxene, clinopyroxene, and garnet at high pressures. Support for these speculations was provided by the pioneering experiments of Yoder and Tilley (1962) on the crystallization of basaltic compositions at high pressures. These experiments, although few in number and of a reconnaissance nature, demonstrated that the thermal divide of Fig. 4-1 did not exist at high pressure and that differentiation at pressures greater than 20 kbars would be controlled by the crystallization of garnets and pyroxenes. They suggested that crystallization of garnet from a tholeiitic liquid might produce an alkali basalt, whilst crystallization of omphacite from alkali basalt would cause the residual liquid to differentiate towards a tholeiite. These specific suggestions have been found to be inadequate.[2,3] Nevertheless, Yoder and Tilley's application of experimental high pressure-temperature methods to study the crystallization of natural basalts marked an important milestone.

For several decades prior to 1960, most petrologists had neglected natural rock systems and had concentrated upon rigorous and comprehensive studies of analogue systems containing up to four components. It was hoped that with the accumulation of sufficient data on simple systems it would be possible ultimately to obtain a detailed and quantitative understanding of the petrogenesis of igneous rocks. These "simple" systems were often found to exhibit highly complex phase relationships and it was widely assumed that natural rock systems containing eight or more components would be so complex as to be experimentally intractable, especially when the effects of pressure were considered in addition to the other variables.

It appears now that this latter view was unduly pessimistic. The order which is observed in the petrology of igneous rocks and the fact that clearly defined magma types and rock associations are recognizable on a worldwide basis demonstrate that fundamental petrogenetic processes must possess a high degree of universality and must have operated in magmas possessing broad compositional ranges. The differentiation processes responsible for the different classes of basalts clearly fall within this category. It follows that it should be possible to discover these fundamental differentiation processes by an inductive approach—i.e.,by choosing a number of typical and significant basalt compositions

[1]Holmes and Harwood (1932).
[2]Green and Ringwood (1967a).
[3]Sections 4-6 and 4-7.

and studying their crystallization behaviour over a wide range of pressure and temperatures.

The introduction of electronprobe microanalysis of phases and liquids formed during high-pressure runs[1] greatly facilitated the realization of this objective. By analyzing the compositions of multicomponent phases and residual liquids in experimental runs over a wide range of pressure and temperature conditions, it was possible to follow quantitatively and in detail the crystallization differentiation of natural rock systems as a function of pressure and temperature. The marriage of the electronprobe to high *P-T* investigations of natural rock systems provides the capacity to solve many of the fundamental problems of petrology. This chapter will be devoted mainly to developments in this field during the period 1964–1972, specifically applied to the petrogenesis of basalts. During this period, most of the combined electronprobe-high pressure work on natural basaltic systems was carried out at the Australian National University, and the present chapter will be primarily concerned with a review of this work.

Green and Ringwood (1964, 1967a) described an extensive investigation of crystallization equilibria in an olivine tholeiite, an olivine basalt, an alkali olivine basalt, and a picrite. Compositions of these magmas are given in Table 4-2. Before describing the results, some discussion of the rationale for choosing these particular compositions appears desirable.

The composition of an uncontaminated basalt magma at the earth's surface is principally the result of the interplay of two processes. The first is the partial melting process in the mantle by which the *primary magma* was initially generated. The second is the partial crystallization of the magma during its journey to the surface, which modifies the primary composition. For experiments of the type under discussion, we are interested in the compositions of the primary magmas which were in equilibrium with source minerals in the mantle. The two most important source minerals are found to be olivine and orthopyroxene.

O'Hara (1965) has argued vigorously that nearly all observed basaltic magmas have suffered extensive fractionation during their journeys to the surface and hence do not represent primary liquids. The evidence supporting this view is the observation that many basaltic liquids are close to an olivine-plagioclase-pyroxene cotectic when erupted, whereas experimental studies show that liquids formed at depth in the mantle and in equilibrium with olivine (and orthopyroxene) depart significantly from the low-pressure cotectic. The departure is in the sense that liquids formed at depth contain more normative olivine.

There is little doubt that this view is correct in many instances, particularly in the cases of continental tholeiites.[2] The role of high-level olivine fractionation in modifying the compositions of Hawaiian tholeiites[3] is well documented. Nevertheless, there are many cases where good evidence exists that fractionation during

[1]Green and Ringwood (1964, 1967a).
[2]This was also recognized by Hess (1960).
[3]MacDonald and Katsura (1961).

Table 4-2 COMPOSITIONS AND CIPW NORMS AND "ECLOGITE" NORMS OF BASALTS AND PICRITE STUDIED EXPERIMENTALLY BY GREEN AND RINGWOOD (1967a)

	Olivine tholeiite	Olivine basalt	Alkali olivine basalt	Picrite
SiO_2	47.0	47.1	45.4	45.5
TiO_2	2.0	2.3	2.5	1.9
Al_2O_3	13.1	14.2	14.7	12.4
Fe_2O_3	1.0	0.4	1.9	0.9
FeO	10.1	10.6	12.4	8.7
MnO	0.2	0.2	0.2	0.2
MgO	14.6	12.7	10.4	18.8
CaO	10.2	9.9	9.1	9.7
Na_2O	1.7	2.2	2.6	1.6
K_2O	0.1	0.4	0.8	0.1
P_2O_5	0.2	—	0.0	0.2
Total	100.0	100.0	100.0	100.0
$\frac{Mg}{Mg + Fe^{++}}$ (atomic ratio)	0.72	0.68	0.60	0.79
CIPW norms:				
Or	0.6	2.7	4.5	0.5
Ab	14.7	18.9	18.0	13.9
Ne	—	—	2.2	—
An	27.6	27.3	26.2	26.3
Di	17.0	17.6	15.7	16.5
Hy	12.3	1.3	—	2.8
Ol	21.9	27.2	25.8	34.6
Ilm	3.8	4.4	4.8	3.7
Mt	1.4	0.6	2.9	1.3
Ap	0.5	—	—	0.4
"Eclogite" norm: assuming TiO_2 in ilmenite, small "K_2O" solubility in acmite-omphacite. Garnet is $Ca_{0.5}(Mg, Fe)_{2.5}Al_2Si_3O_{12}$ in all compositions				
Pyroxene	50	52	47	40
Garnet	46	44	46	43
Olivine	—	—	2	13
Quartz	Trace	Trace	—	—
Ilmenite	4	4	5	4

ascent had not been a major influence—e.g., in alkali basalts, basanites, and nephelinites which erupted very rapidly from their source regions with sufficient velocity to carry large mantle xenoliths and high-pressure xenocrysts to the surface[1] (Secs. 4-4 and 4-6). Clearly, magmas which have transported such large objects are unlikely to have left small crystals of olivine behind. Another case where

[1]Bultitude and Green (1971).

the degree of prior fractionation can be demonstrated to have been small is with many oceanic tholeiites (Sec. 4-6). Even where fractionation during ascent is believed to have occurred, there are ways of estimating the primary liquid composition by using the results of high-pressure experimentation which define the nature of the fractionation that might have occurred[1] and by using detailed local petrologic studies.[2] At Hawaii the most undersaturated basaltic glass from Kilauea Iki (1959 eruption) contained only 6.5% normative olivine, whereas from study of the role of olivine settling, it was estimated[2] that the average magma composition of the entire lava lake contained 18% normative olivine. Other criteria which are useful in selecting primary magma compositions and in evaluating the role of fractionation during ascent are the nickel and chromium contents of the magmas and the Mg/Mg + Fe ratios. The application of these criteria is discussed in Sec. 4-6. The compositions used for experimental investigation (Table 4-2) were selected with the above criteria in mind and were believed to represent possible primary magmas—i.e., magmas which could have been in equilibrium with a peridotitic residuum at depths of 10 to 80 km in the mantle. This inference, together with some minor but necessary modifications, is discussed further in Sec. 4-5.

The olivine tholeiite, olivine basalt, and alkali-olivine basalt compositions straddled the critical plane of undersaturation (Fig. 4-1). The initial objective of the experiments by Green and Ringwood was to study the crystallization behaviour of each of the chosen basaltic compositions at a series of pressures in an attempt to find a direct differentiation pathway from the olivine tholeiite to the alkali basalt. These experiments would thus explore one widely discussed petrologic hypothesis which maintained that alkali basalts may be formed by some process of crystallization differentiation from parental olivine tholeiite magma. A further objective was to use the compositions of phases found experimentally to be on or near the liquidi of the chosen basalts to constrain the composition of the source regions in the mantle and to elucidate the processes by which alkali basalts and tholeiites might be formed by partial melting. The third objective was to gain an understanding of the genesis of high-alumina basalts.

Petrologists have recognized that another important genetically related series of magmas is represented by the alkali basalt-basanite-olivine nephelinite association (Fig. 4-1). It seems very likely that these are related by a common fractionation process. The origin of this group of magmas was investigated by Green and coworkers during 1968–1972. It is convenient to discuss the experimental work on petrogenesis of the olivine tholeiite-high alumina basalt-alkali basalt magmas separately from that on the alkali basalt-basanite-nephelinite association.

[1]Green and Ringwood (1967a).
[2]MacDonald and Katsura (1961).

4-2 CRYSTALLIZATION EQUILIBRIA IN THOLEIITES AND ALKALI OLIVINE BASALT AT HIGH PRESSURES AND TEMPERATURES

Experimental results on the crystallization of the olivine tholeiite and alkali olivine basalt over the pressure range 0 to 30 kbars and temperatures of 1100 to 1500°C are shown in Figs. 4-2 and 4-3.[1,2]

Low-Pressure Regime (0 to 5 kbars)

At atmospheric pressure, a wide field of primary crystallization of olivine occurs in the olivine tholeiite before entry of plagioclase and, at lower temperature, clinopyroxene. Comprehensive melting experiments by other workers on basalts similar to the olivine tholeiite[3,4] showed that the primary field of crystallization of olivine continued until residual liquids had become quartz-normative. Thus, following Bowen's earlier conclusion,[5] crystallization of olivine tholeiite by separation of olivine at low pressure is capable of explaining the continuous observed spectrum of compositions between olivine tholeiite and quartz tholeiite.

Moderate-Pressure Regime (5 to 10 kbars)

Olivine remains the liquidus phase in this interval, but the crystallization relations of pyroxenes and plagioclase show a sharp change from the low-pressure regime. Whereas plagioclase was the second phase to crystallize at low pressure, its place is taken by orthopyroxene in the olivine tholeiite, orthopyroxene + subcalcic clinopyroxene in the olivine basalt, and clinopyroxene in the alkali olivine basalt. Thus the primary field of crystallization of plagioclase is markedly depressed, whilst the fields of pyroxenes are considerably expanded. Electronprobe analyses of the latter show them to be notably aluminous (4 to 9% Al_2O_3).

Using the compositions of the phases observed to crystallize as determined by electronprobe microanalysis, and their observed proportions, the fractionation trends of the residual liquid can be calculated (Table 4-3). The derivative liquids are characterized by high contents of alumina and normative plagioclase, resembling in composition the high-alumina olivine tholeiites which are so abundant on the deep oceanic floor and along the mid-oceanic-ridge system.[6] Thus, in Fig. 4-1, the residual liquid fractionates not towards the quartz apex, as in the low-

[1]Green and Ringwood (1967a)

[2]Oxidation states of samples were checked by chemical analyses and showed no significant changes during runs. The first batch (170) of runs was carried out in platinum containers, and the extent of iron-loss to the platinum during runs was carefully monitored so that corrections could be applied for this effect in calculations of differentiation trends. Average loss of iron during runs was 2.0% FeO. Subsequently, all key runs were repeated using graphite crucibles in which no iron was lost. These runs verified the corrections made in the case of runs in platinum.

[3]Tilley, Yoder, and Schairer (1963, 1964, 1965).

[4]Yoder and Tilley (1962).

[5]Bowen (1928).

[6]Engel, Engel, and Havens (1965).

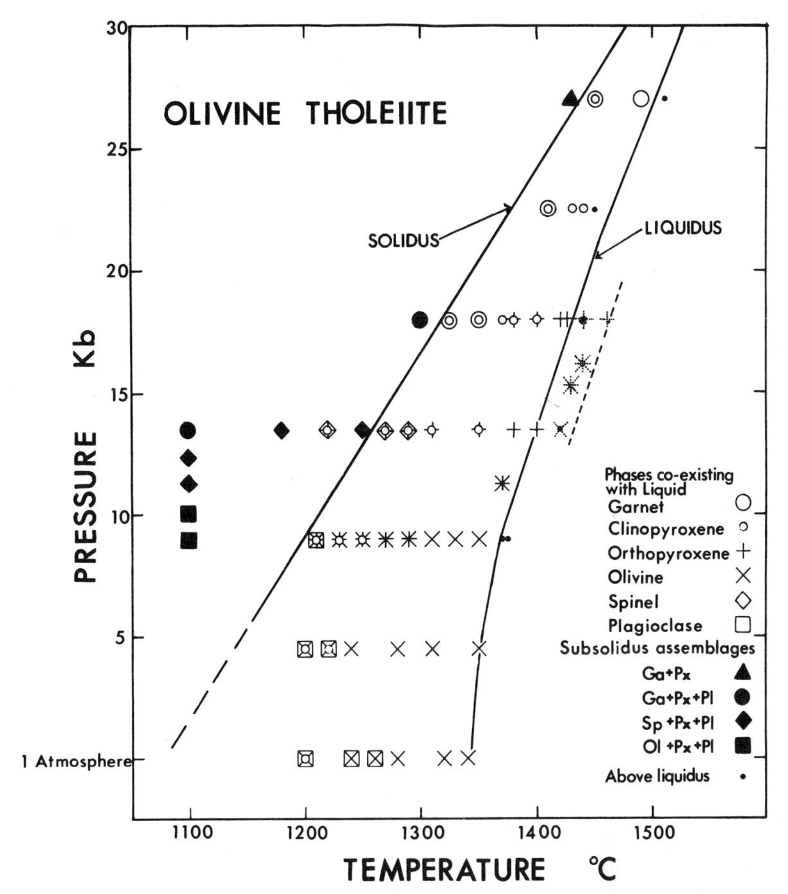

FIGURE 4-2
Detail of crystallization experiments carried out on the olivine tholeiite composition at high pressures and temperatures. (*From Green and Ringwood, 1967a, with permission.*) Experiments on modified olivine-rich composition from Green (1970a): Five runs indicated by broken symbols.

pressure case, but towards the plagioclase subapex. A similar trend towards high-alumina characteristics was demonstrated in the case of the alkali basalt composition. Further detailed experimental studies of the formation of high-alumina basalts by differentiation in the moderate-pressure regime have been described.[1]

Intermediate-Pressure Regime (10 to 20 kbars)

With increasing pressure in this regime, further important differences in crystallization behaviour become apparent. Referring to the olivine tholeiite (Fig.

[1]T. Green, D. Green, and Ringwood (1967).

Table 4-3 FRACTIONATION OF THE OLIVINE THOLEIITE AT 9 kbars. (*After Green and Ringwood, 1967a*)

***P-T* conditions (kbars)**		**9**	**9**	**9**
(°C)		**1290**	**1250**	**1230**
Nature and estimated percentage of crystals		**11% Ol**	**12% Ol 3% Opx**	**12% Ol 4% Opx 14% Cpx**
Composition of crystal extract:				
SiO_2		40.3	42.5	45.9
Al_2O_3		—	1.4	6.2
FeO		12.3	13.0	11.3
MgO		47.1	42.3	31.6
CaO		0.3	0.8	5.0
Composition of liquid phase:				
	(Initial liquid)			
SiO_2	47.0	47.6	47.6	47.3
TiO_2	2.0	2.3	2.4	2.9
Al_2O_3	13.1	14.8	15.2	16.1
Fe_2O_3	1.0	1.1	1.2	1.5
FeO	10.1	9.8	9.6	9.5
MnO	0.2	0.2	0.2	0.2
MgO	14.6	10.5	9.5	7.2
CaO	10.2	11.4	11.8	12.4
Na_2O	1.7	2.0	2.0	2.5
K_2O	0.1	0.1	0.1	0.1
P_2O_5	0.2	0.2	0.2	0.3
CIPW norm of liquid phase:				
Or	0.7 } 43.0	0.7 } 48.3	0.7 } 49.6	0.8 } 54.2
Ab	14.7	16.5	16.8	21.0
An	27.6	31.1	32.1	32.4
Di	17.0	19.2	20.2	22.5
Hy	12.3	13.8	12.9	5.2
Ol	21.9	12.2	10.5	10.1
Ilm	3.8	4.3	4.6	5.5
Mt	1.4	1.6	1.8	2.1
Ap	0.5	0.5	0.5	0.6
$\frac{Mg}{Mg + Fe^{++}}$	0.72	0.66	0.64	0.57

4-2), the most significant difference is the large contraction in the primary field of olivine and a corresponding expansion in the field of crystallization of aluminous orthopyroxene (4 to 10% Al_2O_3) which joins olivine on the liquidus at 11 kbars. At 13.5 kbars, olivine no longer crystallizes and orthopyroxene becomes the liquidus phase, being joined by subcalcic aluminous clinopyroxene about 100°C lower. This trend is maintained at 18 kbars, although the gap between the appearance of orthopyroxene and clinopyroxene has decreased. A further important feature is the disappearance of plagioclase in this pressure interval. Fractionation of the olivine tholeiite in the intermediate-pressure regime is thus governed by crystallization of aluminous orthopyroxene and, to a lesser extent, of aluminous clinopyroxene.

Referring to the basalt tetrahedron (Fig. 4-1), it is seen that separation of orthopyroxene has the effect of driving residual liquids directly towards the plane of critical undersaturation. The fractionation can be calculated directly using the proportions and compositions of phases observed to crystallize (Table 4-4). It is found that the residual liquid resulting from separation of 15% pyroxene has a composition similar to the olivine basalt (Table 4-2) and lying on the plane of criti-

Table 4-4 FRACTIONATION OF THE OLIVINE THOLEIITE AT 13.5 kbars AND 18 kbars. (*After Green and Ringwood, 1967a*)

P-T conditions (kbars) (°C)		13.5 1350	18 1400
Nature and estimated percentage of crystals		15% Opx	10% Opx 5% Cpx
Composition of crystal extract:			
SiO_2		53.8	53.0
Al_2O_3		6.0	7.3
FeO		7.2	6.3
MgO		30.2	28.6
CaO		2.8	4.8
Composition of liquid phase:	(Initial liquid)		
SiO_2	47.0	45.7	45.9
TiO_2	2.0	2.4	2.4
Al_2O_3	13.1	14.3	14.1
Fe_2O_3	1.0	1.2	1.2
FeO	10.1	10.6	10.7
MnO	0.2	0.2	0.2
MgO	14.6	11.9	12.1
CaO	10.2	11.3	11.1
Na_2O	1.7	2.0	2.0
K_2O	0.1	0.1	0.1
P_2O_5	0.2	0.2	0.2
CIPW norm of liquid phase:			
Or	0.6 } 42.9	0.6 } 47.6	0.6 } 46.7
Ab	14.7	17.3	17.3
An	27.6	29.7	28.8
Di	17.0	20.5	20.3
Hy	12.3	0.2	2.0
Ol	21.9	24.9	24.2
Ilm	3.8	4.5	4.6
Mt	1.4	1.8	1.8
Ap	0.5	0.5	0.5
$\frac{Mg}{Mg + Fe^{++}}$	0.72	0.67	0.67

Table 4-5 FRACTIONATION OF THE OLIVINE BASALT AT 13.5 kbars AND 18 kbars. (*After Green and Ringwwod, 1967a*)

P-T conditions (kbars)		13.5	18
(°C)		1310	1335
Nature and estimated percentage of minerals		10% Opx (as at 1320°C, 13.5 kbars)	5% Opx 10% Cpx
Composition of crystal extract:			
SiO_2		53.8	51.9
Al_2O_3		6.4	8.6
FeO		6.6	5.8
MgO		30.3	25.6
CaO		2.9	8.1
Composition of residual liquid:			
	(Initial liquid)		
SiO_2	47.1	46.4	46.3
TiO_2	2.3	2.6	2.7
Al_2O_3	14.2	15.0	15.1
Fe_2O_3	0.4	0.5	0.5
FeO	10.6	11.1	11.5
MnO	0.2	0.2	0.2
MgO	12.7	10.8	10.4
CaO	9.9	10.6	10.2
Na_2O	2.2	2.5	2.6
K_2O	0.4	0.5	0.5
CIPW norm of residual liquid:			
Or	2.7	3.0	3.3
Ab	18.9	16.8	17.8
Ne	—	2.4	2.3
An	27.3	28.1	27.8
Di	17.6	19.9	18.6
Hy	1.3	—	—
Ol	27.2	24.2	24.3
Ilm	4.4	5.0	5.1
Mt	0.6	0.7	0.8
$\frac{Mg}{Mg + Fe^{++}}$	0.68	0.63	0.62

cal undersaturation. Further studies of fractionation are thus best conducted by investigating the crystallization of the olivine basalt. In this composition, aluminous orthopyroxene remains on the liquidus[1] but is joined after a relatively smaller amount of crystallization by subcalcic clinopyroxene. It is evident from inspection of Fig. 4-1 that continued orthopyroxene crystallization will drive the residual liquid composition across the critically undersaturated plane into the alkali basalt field. This is confirmed quantitatively using proportions and compositions of phases observed to crystallize from the olivine basalt (Table 4-5). Re-

[1]Green and Ringwood (1967a, fig. 5).

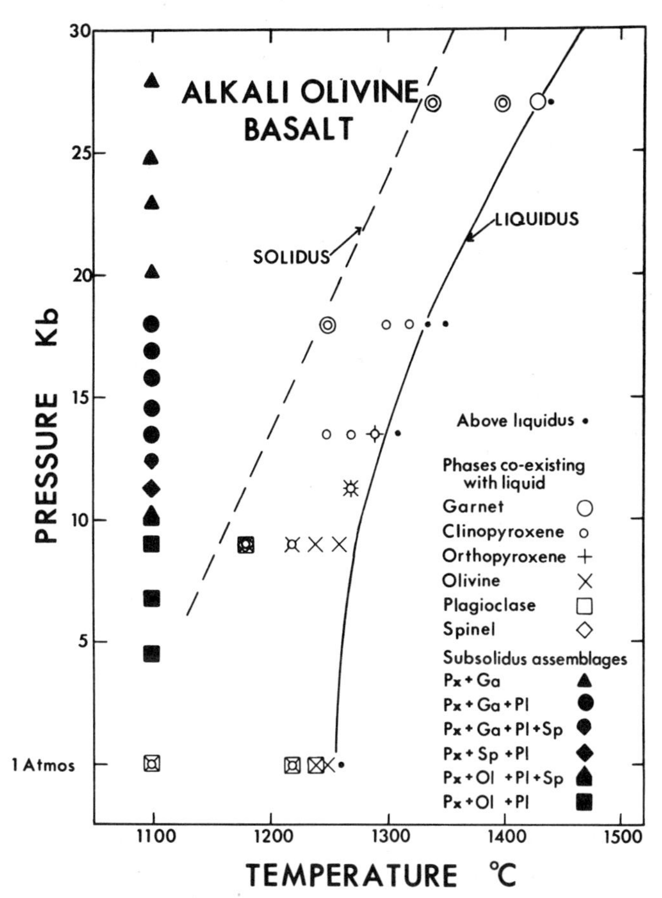

FIGURE 4-3
Detail of crystallization experiments on alkali olivine basalt composition at high pressures and temperatures. (*From Green and Ringwood, 1967a, with permission.*)

sidual liquids of olivine-rich, alkali basalt composition containing over 2% normative nepheline are thereby derived. These results thus demonstrate that, in the intermediate-pressure regime, the low-pressure thermal barrier is broken, and crystallization dominated by the separation of aluminous orthopyroxenes drives residual liquids directly into the alkali basalt field. Earlier petrologic speculations[1,2,3] concerning the possible role of orthopyroxene fractionation in producing alkalic magmas are thus confirmed.

Experiments on the crystallization of the alkali olivine basalt are shown in Fig. 4-3. It is seen that aluminous orthopyroxene remains on the liquidus between

[1]Holmes and Harwood (1932).
[2]Powers (1935).
[3]Larsen (1940).

about 11 and 15 kbars, where it is accompanied by subcalcic aluminous clinopyroxene. Crystallization of these phases causes the residual liquid composition to move deeper into the alkali basalt field. At 13.5 kbars, crystallization of 2.5% aluminous orthopyroxene and 7.5% subcalcic clinopyroxene yield a residual liquid containing 5 to 6% normative nepheline and 23% normative olivine—i.e., an olivine-rich basanite.[1] Similar results are produced by crystallization of 20% subcalcic clinopyroxene at 18 kbars.[1]

The picrite composition investigated (Table 4-2) is similar normatively to the olivine basalt, except that it contains 9% additional olivine. Accordingly, it lies almost on the critically undersaturated plane (Fig. 4-1). Between 13 to 18 kbars the liquidus phase is found to be olivine (Fo_{92}), closely followed by orthopyroxene (En_{90}). Crystallization of these phases would drive the residual liquid into the critically undersaturated volume.[1,2]

The quantitative experimental results quoted above show that, in the intermediate-pressure regime, a parental olivine tholeiite or tholeiitic picrite magma would fractionate directly towards and through the critical plane of undersaturation (Fig. 4-1), producing a continuous spectrum of basaltic compositions between the parental tholeiite extending to alkali basalts and basanites containing up to 6% normative nepheline. These results clearly have a crucial bearing on the petrogenesis of the alkali basalt suite.

High-Pressure Regime (20 to 30+ kbars)

As pressure exceeds 20 kbars, a further major shift of the crystallization behaviour of the olivine tholeiite and alkali basalt (Figs. 4-2 and 4-3) takes place. At 22.5 kbars, the liquidus phase is clinopyroxene, joined at lower temperatures by garnet. The clinopyroxene is believed to contain more calcium than at lower pressures and also has a significant sodium content. At 27 kbars, the liquidus phase in the basalts is garnet, closely followed by clinopyroxene. This regime is thus dominated by eclogite fractionation, as found earlier by Yoder and Tilley.[3] For a wide range of basaltic compositions in the high-pressure regime, either garnet or clinopyroxene may be the liquidus phase. However, only a small degree of crystallization of one of these phases is permitted before it is joined by the other, and an extensive degree of garnet-clinopyroxene cotectic crystallization occurs over a relatively small decrease in temperature (~50°C).[3,4] The small intervals over which either garnet or clinopyroxene crystallize alone before being joined by the others are unfavourable to Yoder and Tilley's hypothesis[3] that alkali basalts could be formed by extensive crystallization of garnet from an olivine tholeiite or that tholeiites could form by extensive crystallization of omphacitic clinopyrox-

[1]Green and Ringwood (1967a, table 18). Key runs demonstrating the appearance of orthopyroxene on the liquidus in the olivine basalt and alkali olivine basalt were confirmed in subsequent work using graphite capsules in which the problem of iron-loss was avoided (Green, 1970a).

[2]Green and Ringwood (1964).

[3]Yoder and Tilley (1962).

[4]Green and Ringwood (1967a).

ene from alkali basalt. Even if this difficulty is disregarded, detailed calculations of the effects of extensive separation of garnet *of the composition actually observed on the liquidus of tholeiites* show that the residual liquids generated are unlike alkali basalts.[1] Similarly, residual liquids generated by separation of observed clinopyroxene from alkali basalts do not resemble natural tholeiites.[1]

Since the major element composition of the garnet + clinopyroxene assemblage crystallizing from olivine-rich basaltic compositions at high pressures is roughly similar to those of the parental compositions, a large degree of eclogite crystallization is possible without driving the residual liquid into nonbasaltic compositions[2] (major element). However, extensive degrees of eclogite fractionation have drastic effects upon the abundances of some minor elements and upon Mg/Mg + Fe ratios of residual liquids. These effects are discussed in Secs. 4-6 and 4-7.

Summary of Fractionation Trends

The experimental results considered above demonstrate that an olivine-rich tholeiite magma may fractionate to yield three distinct derivative magmas in clearly defined pressure regimes (Fig. 4-4).

[1]Green and Ringwood (1967a).
[2]O'Hara (1968a).

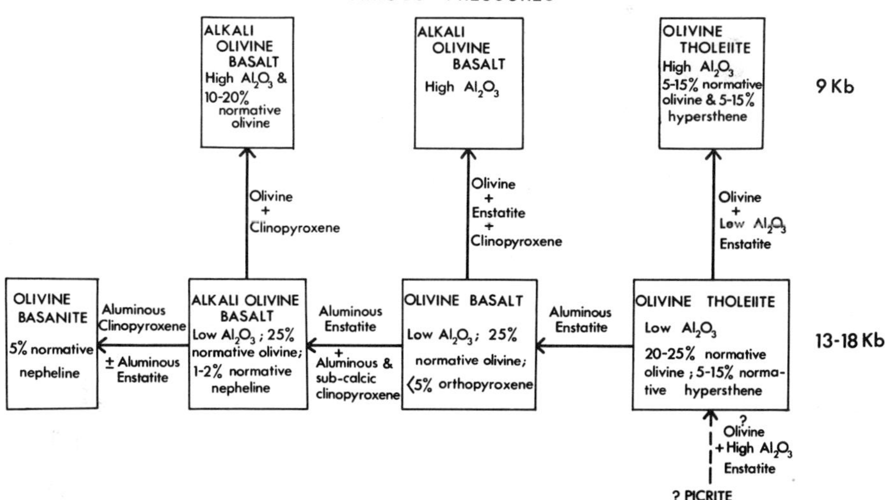

FIGURE 4-4
Diagrammatic summary of the effects and direction of fractionation of basaltic magmas at moderate to intermediate pressures. (*From Green and Ringwood, 1967a, with permission.*)

At low pressure (0 to 5 kbars), corresponding to depths less than 15 km, crystallization is dominated by olivine, with plagioclase or clinopyroxene as the second phase appearing at lower temperatures. Residual liquids are driven in the direction of the SiO_2 apex of the basalt tetrahedron (Fig. 4-1). The primary field of crystallization extends across the plane of silica saturation, accompanied by a reaction relationship so that residual liquids are of quartz tholeiite composition.

At moderate pressures (5 to 10 kbars), corresponding to depths of 15 to 35 km, early stages of fractionation are dominated by olivine, as at low pressure, but the olivine is joined by orthopyroxene in tholeiitic compositions. The field of crystallization of plagioclase is relatively depressed in all basaltic compositions, and residual liquids fractionate directly towards the plagioclase apex of the basalt tetrahedron (Fig. 4-1), giving rise to high-alumina olivine basalts.

At intermediate pressures (10 to 20 kbars), corresponding to depths of 35 to 70 km, the olivine field is largely suppressed, and crystallization is dominated by the separation of aluminous orthopyroxene $\pm$ subcalcic aluminous clinopyroxene. The residual liquids fractionate directly towards the critical plane of undersaturation and across into the alkali basalt fields, yielding alkali olivine basalts and basanites containing up to 6% normative nepheline.

The existence of these three radically different fractionation trends within such a small pressure or depth interval is most remarkable. Petrologists have long recognized the existence of distinctive olivine tholeiite, quartz tholeiite, high-alumina basalt, and alkali magma types and the transitional relationships between these magma types. The experimental results described in this and other sections provide a firm and quantitative foundation for interpreting these relationships.

4-3 EXPERIMENTAL DATA BEARING ON THE RELATIONSHIPS OF ALKALI OLIVINE BASALTS, BASANITES, AND NEPHELINITES

Olivine basanites containing up to 15% nepheline and olivine nephelinites containing up to 30% nepheline and lacking in plagioclase occur widely in alkali basalt provinces. Compositions and norms of two typical examples are given in Table 4-6. A continuous spectrum of chemical compositions appears to exist in natural rocks between alkali olivine basalts, olivine basanites, and olivine nephelinites, suggesting the existence of a common fractionation mechanism. The experimental results on basalt fractionation described in Sec. 4-2 revealed a pathway extending from tholeiites through alkali olivine basalts to the edge of the basanite field (6% nepheline). However, the mechanisms studied did not appear capable of yielding the series of basanites (10 to 15% nepheline) and nephelinites (15 to 30% nepheline).

Detailed studies of the genesis of this series of rocks have been conducted by D. H. Green and colleagues.[1,2,3] Under dry conditions, the dominant phases

[1]Bultitude and Green (1968, 1971).
[2]Green and Hibberson (1970).
[3]Green (1969, 1970a, 1973a).

Table 4-6 COMPOSITION AND NORMS OF A TYPICAL BASANITE AND OLIVINE NEPHELINITE. (*Green, 1970a, table 1*)

	Olivine basanite, Mt. Leura, Victoria	Olivine nephelinite, Scotsdale, Tasmania
SiO_2	44.6	39.3
TiO_2	2.9	3.9
Al_2O_3	11.7	9.5
Fe_2O_3	3.0	5.1
FeO	9.4	10.7
MnO	0.2	0.2
MgO	13.9	13.9
CaO	7.7	11.2
Na_2O	3.7	3.0
K_2O	2.0	1.5
P_2O_5	1.0	2.3
$\frac{Mg}{Mg + Fe^{++}}$	0.73	0.70
CIPW norms:		
Lc	—	0.3
Or	11.7	8.3
Ab	12.0	—
Ne	10.2	13.4
An	9.8	8.4
Di	18.7	29.1
Hy	—	—
Ol	25.8	22.3
Ilm	5.6	6.4
Mt	4.3	7.4
Ap	2.0	4.5

crystallizing at pressures up to 30 kbars are olivine and clinopyroxene. Extraction of these phases does not, however, lead to substantial increase in nepheline content.[1,2] It is seen from Fig. 4-1 that further separation of orthopyroxene (±subcalcic clinopyroxene) is necessary if basanites and nephelinites are to be produced.

Bultitude and Green[1] made the important discovery that the primary field of crystallization of orthopyroxene from alkaline basaltic magmas at high pressures is greatly extended by the addition of water to the melts. Since highly alkaline magmas frequently display explosive eruptions indicative of high water contents, the addition of this component is petrologically plausible. Reconnaissance experiments were carried out upon alkali olivine basalt, olivine nephelinite, and picritic nephelinite at pressures of 18 to 27 kbars and temperatures of 1150 to 1250°C. The melts were estimated to contain about 5 to 7% water. Aluminous orthopyroxene was found to be an important phase in each of these compositions, indicating a large extension of the fractionation pathway (Fig. 4-4) discussed in Sec. 4-2. Under anhydrous conditions, the corresponding liquidus and near-

[1]Bultitude and Green (1968, 1971).
[2]Green and Hibberson (1970).

liquidus phases were olivine, clinopyroxene, and garnet, which crystallized at temperatures of 1530 to 1400°C.

Some scepticism was expressed concerning the significance of these experiments because of the possibility of contamination of the charge with material carried in solution by the water which entered from the pressure medium.[1,2] Subsequently, however, a detailed study of the crystallization of an olivine basanite (Table 4-6) containing 11% normative nepheline in sealed platinum and silver-palladium tubes was carried out.[3] This demonstrated unequivocally the existence of a limited window (22 to 27 kbars, 1230 to 1320°C, 2 to 5% H_2O) in which aluminous orthopyroxene appeared on the liquidus of the basanite, accompanied in some runs by olivine and aluminous clinopyroxene. The occurrence of an enlarged field of orthopyroxene on the liquidus of alkali olivine basalt at 18 kbars in the presence of water was also confirmed using sealed platinum tubes.[4,5]

The data demonstrate[3,4,5,6] that, provided water (and CO_2) are available in the source regions of basaltic magmas, fractional crystallization by the separation of orthopyroxene as the major crystallizing phase extends over the entire compositional range from olivine tholeiite through alkali olivine basalt and olivine basanite to olivine nephelinite. In olivine-rich compositions, olivine is a coprecipitating phase, and either clinopyroxene or garnet is the third phase to appear, depending on *P-T* conditions and bulk magma composition. The effect of fractional crystallization by separation of major orthopyroxene is to deplete the residual liquid in SiO_2 rather rapidly whilst enriching it in CaO and alkalis. The chemical variation accords well with the trend observed in mantle-derived basalts ranging in composition from alkali olivine basalt to olivine nephelinite.[7]

A significant feature is that the fractionation pathways leading to formation of basanites and nephelinites exist only at pressures (20 to 35 kbars) which are higher than required for the olivine tholeiite-alkali basalt pathway (10 to 20 kbars) discussed in Sec. 4-2. This confirms views expressed by other workers[8,9,10] that the highly undersaturated nephelinites may be derived from generally greater depths than is alkali olivine basalt.

[1]Kushiro (1969).
[2]O'Hara (1968a).
[3]Green (1973a).
[4]Green (1970a).
[5]Eggler (1973) has demonstrated in a study of crystal-liquid equilibria in the system MgO-SiO_2-H_2O-CO_2 that dissolved CO_2 has the effect of expanding the primary field of crystallization of orthopyroxene relative to olivine. A similar effect of expansion of orthopyroxene crystallization into an olivine nephelinite liquid containing dissolved CO_2 was noted by Green (1973a). Brey and Green (1974) have recently studied the crystallization of olivine nephelinite and olivine melilitite (38% SiO_2) with added volatile species (50 mol%CO_2–50 mol%H_2O) in sealed Ag-Pd capsules at 1100 to 1200° and 30 to 35 kbars. Orthopyroxene was found to be a liquidus phase. These results strongly suggest that dissolved CO_2 (as CO_3^{--}) plays an important role in the genesis of magmas more alkalic than olivine basanite.
[6]Green and Hibberson (1970).
[7]Bultitude and Green (1968).
[8]Kuno (1959).
[9]Kushiro and Kuno (1963).
[10]Kushiro (1968).

4-4 DIRECT EVIDENCE OF THE ROLE OF HIGH-PRESSURE ALUMINOUS PYROXENE FRACTIONATION IN THE PETROGENESIS OF ALKALI BASALTS

The widespread occurrence of lherzolite and peridotite xenoliths in rocks of the alkali basalt suite has long been known and was discussed in Sec. 3-4. Much more recently, the occurrence of another distinctive class of inclusions in these basalts has become recognized. In 1964, Kuno described the occurrence of large megacrysts or xenocrysts of orthopyroxene and clinopyroxene up to 3 cm across in an alkali olivine basalt. The clinopyroxene was found to contain 10% Al_2O_3, whilst the orthopyroxene contained over 4% Al_2O_3. The compositions of the xenocrysts differed markedly from those of the pyroxenes in accompanying lherzolite nodules. Kuno suggested that the pyroxenes had crystallized from the enclosing magma at high pressure in the upper mantle. This interpretation was strongly supported by Green and Ringwood's demonstration in the same year that orthopyroxene crystallizing at high pressure from an olivine tholeiite magma was distinctive in the large amounts of alumina contained. Kuno also drew attention to several other analogous occurrences of pyroxene xenocrysts in alkali basalts.

Many new occurrences of such xenocrysts have since been found, particularly among the alkali basalts of eastern Australia. Binns and coworkers[1,2] have provided a detailed description of a xenocryst suite found in basanites in northeastern New South Wales. Clinopyroxene xenocrysts are up to 10 cm in length and are characteristically undeformed and glassy, exhibiting conchoidal fracture. Their compositions are uniform throughout and are characterized by the presence of large amounts of Al_2O_3 (up to 9.5%). They also contain significant amounts of the jadeite component and are subcalcic. They are accompanied rarely by orthopyroxenes containing up to 8.0% Al_2O_3 and about 2% CaO. The compositions of these xenocrysts are very similar to the compositions of pyroxenes experimentally observed to crystallize near the liquidus of olivine basalt and alkali basalts at about 10 to 20 kbars.[3] The pyroxenes were accompanied by other kinds of large xenocrysts, including olivine, spinel, kaersutite, and anorthoclase. Apart from the olivine, the roles of these other phases in fractionation processes have not yet been established. Binns and coworkers concluded that the xenocrysts are of cognate origin and crystallized from the associated magmas at pressures of 10 to 15 kbars.

Green and Hibberson (1970) described the occurrence of corresponding aluminous orthopyroxene and clinopyroxene xenocrysts in an olivine basalt from the Auckland Islands (Table 4-7). In an experimental study of the host basalt, they found that aluminous orthopyroxene and clinopyroxene were near-liquidus phases at 11 to 18 kbars, but the degree of solid solution between the pyroxenes crystallizing from the dry magma was much greater than in the natural pyroxene

[1]Binns (1969).
[2]Binns, Duggan, and Wilkinson (1970).
[3]Green and Ringwood (1967a).

Table 4-7 COMPOSITIONS OF ORTHOPYROXENE AND CLINOPYROXENE XENOCRYSTS OBSERVED IN AUCKLAND ISLAND OLIVINE BASALT COMPARED WITH COMPOSITIONS OF ORTHOPYROXENE AND CLINOPYROXENE OBSERVED TO CRYSTALLIZE ON OR NEAR THE LIQUIDUS OF THE BASALT AT HIGH PRESSURE. (*After Green and Hibberson, 1970*)

		Orthopyroxene (xenocryst)	Orthopyroxene (18 kbars, 1200°C)	Clinopyroxene (xenocryst)	Clinopyroxene (18 kbars, 1200°C)
SiO_2		56.0	54.9	52.9	52.3
TiO_2		0.3	0.4	0.7	0.7
Al_2O_3		3.3	3.1	4.3	5.9
FeO		8.5	9.1	6.5	6.4
MgO		31.3	30.5	19.1	17.5
CaO		2.1	2.0	15.6	16.2
Na_2O		<0.1	—	0.7	1.0
$\frac{Mg}{Mg + Fe}$		0.87	0.86	0.84	0.83
Molecular proportions	Ca	4	4	33	36
	Mg	83	82	56	53
	Fe	13	14	11	11

xenocrysts. Accordingly, they crystallized the basalt in this pressure interval in the presence of 2% water, which lowered the liquidus to 1200°C. Aluminous clinopyroxene and orthopyroxene continued to be found on and near the liquidus; however, because of the lower temperature, the degree of solid solubility of the pyroxenes was decreased. In runs carried out at 1130 to 1230°C, a close correspondence between compositions of the naturally occurring xenocrysts and the experimental near-liquidus phases of the basalt was found. These experiments leave little room for doubt that the natural xenocrysts indeed crystallized from the host magma under the stated conditions and that, furthermore, the host magma contained about 2% water.

Green (1966) demonstrated that several of the garnet pyroxenite xenoliths occurring in olivine nephelinite tuff at Salt Lake Crater, Oahu, have bulk compositions similar to those of subcalcic clinopyroxene ± olivine found experimentally to occur on the liquidus of alkali basalt at 10 to 20 kbars. If the clinopyroxene ± olivine assemblage were cooled to lower temperatures, exsolution into orthopyroxene and Ca-rich clinopyroxene would occur. This would be accompanied by a decrease in the solid solubility of Al_2O_3 in the pyroxenes, leading to exsolution of garnet and/or reaction of Al_2O_3 with olivine to form spinel. The naturally occurring garnet pyroxenite xenoliths from Hawaii display abundant evidence of the operation of these exsolution and reaction effects, thus strongly supporting Green's interpretation.

The widespread natural occurrence of pyroxene xenocrysts, similar in composition to the pyroxenes which are observed experimentally on the liquidus of their host basalts at high pressures, adds a new dimension to the experimental investigations. The xenocrysts provide a direct link between natural occurrences and the high-pressure experiments, strongly indicating that the compositions of the host basalts have been influenced by fractional crystallization at depth in the mantle along the lines inferred from the experimental investigations.

4-5 GENERATION OF BASALTIC MAGMAS BY PARTIAL MELTING IN THE MANTLE

Application of experimental methods described in Secs. 4-2 and 4-3 has emphasized the role of fractional crystallization. The experimental results, however, are equally applicable to the formation of basaltic magmas by partial melting processes. It so happens that, owing to experimental constraints, it is more convenient to approach the problem by examining the nature of near-liquidus phases in a series of related basaltic compositions than by performing the partial melting experiments directly. Thus, the choice of this approach is without implications as to whether fractional crystallization or partial melting is considered to be the dominant influence in causing the diversity of basaltic rocks observed at the surface. This question must be settled with the aid of additional kinds of evidence.

The rationale behind application of the experimental results to partial melting processes is simple. The compositions of phases which occur on and near the liquidus of a series of basalts have been determined over a wide range of pressures and temperatures. These phases are in equilibrium with their respective basaltic liquids. However, chemical equilibrium is independent of the relative proportions of crystalline phases and liquid which may be present. Although a given basaltic composition might contain 5% crystals in equilibrium with 95% liquid at a given pressure and temperature, the compositions of the phases in equilibrium would be unchanged if the proportions were reversed to 5% liquid and 95% solid at the same pressure and temperature. The latter situation corresponds to the case of partial (batch) melting,[1] and it is seen that the compositions of the residual, refractory phases remaining after partial melting are given directly by the compositions of the phases on the liquidus of the basalt.

We have previously (Sec. 3-3) identified the residual refractory phases with the alpine peridotites and with peridotitic and lherzolitic xenoliths in alkali basalts and kimberlite pipes. It must be remembered that most of these rocks have

[1]Two kinds of partial melting processes can be distinguished:

In *fractional melting*, an increment of equilibrium liquid is formed from one or more solid phases. Once formed, the liquid is removed from the solid phase. The liquids from all increments are continuously mixed together to form a single liquid which is not in thermodynamic equilibrium with the solid that gave rise to the last increment which joined the liquid.

In contrast to fractional melting, *batch melting* involves an equilibrium of residual crystals with the total amount of liquid produced. A fraction F of liquid is equilibrated with $(1-F)$ residual crystals in a closed system and is then withdrawn from the crystals without alteration in composition of either phase. In the mantle, the partial melting process probably corresponds more closely to batch melting than to fractional melting. The experimental method based upon determination of compositions of near-liquidus phases is directly applicable to batch melting as described above, and the use of the term *partial melting* in the text is in the sense of batch melting.

recrystallized at temperatures in the vicinity of 1000°C or less and that this has an important effect on their mineralogies. On the other hand, we are interested in the mineral assemblages which these ultramafic compositions would display at the liquidus temperatures of basaltic magmas at appropriate pressures. These are well established from the experimental investigations described in Sec. 6-2. The principal effects are increased degrees of mutual solid solution between orthopyroxenes and clinopyroxenes and increases in the solubility of alumina in pyroxenes. As a result of this latter effect, the spinels or garnets which are present in the majority of alpine ultramafics and ultramafic xenoliths from alkali basalts (mostly spinel peridotites) and in the majority of ultramafic xenoliths from kimberlite pipes (mostly garnet lherzolites and garnet harzburgites) would react to form aluminous pyroxenes at the liquidus temperatures of basaltic magmas in the pressure range 5 to 25 kbars. The mineral assemblage displayed by most of these rocks would then consist of olivine + aluminous orthopyroxene ± aluminous subcalcic clinopyroxene (Mg/Mg + Fe ~ 0.90 to 0.94). It is these latter phases which represent the refractory residual mineral assemblages complementary to many basaltic magmas. The existing lower temperature mineral assemblages of naturally occurring ultramafic intrusions and xenoliths are not directly relevant in this context.

If residual peridotites and basaltic magmas are regarded as complementary differentiates of a common source material, what is the composition of this source material? Clearly, it will be intermediate between those of average basalt and peridotite. The question of the detailed composition of primitive mantle source material (termed *pyrolite*) is discussed in Sec. 5-2. It is concluded that the composition of pyrolite is approximately similar to that of a mixture of 1 part of olivine tholeiite to 3 parts peridotite. The subsolidus phase relations of pyrolite are discussed in Sec. 6-2, Fig. 6-4. The pyrolite composition crystallizes in three anhydrous mineral assemblages under near-solidus mantle conditions:[1]

1 *0 to 10 kbars (approx.)*: Olivine + orthopyroxene + clinopyroxene + plagioclase + spinel
2 *10 to 27 kbars (approx.)*: Olivine + aluminous orthopyroxene + aluminous clinopyroxene ± minor spinel
3 27 *kbars*: Olivine + orthopyroxene + clinopyroxene + garnet

In melting anhydrous pyrolite, the first drop of liquid formed at a given pressure will be in equilibrium with one of the mineral assemblages listed above, and with increasing degree of melting, the residual phases will change in composition and be eliminated one by one.[2] The composition of the partial melt will thus change with the degree of melting as the solid phases buffering it also change. Because of the olivine-rich nature of the source composition, all partial melts will be saturated with respect to olivine. Experimental studies also show that enstatite

[1]Green and Ringwood (1967b, 1970).
[2]Green (1971).

will be the next to last phase to disappear during progressive partial melting. Thus, primary basaltic liquids produced by partial melting of pyrolite will have olivine and orthopyroxene as liquidus phases and may also have clinopyroxene, plagioclase, spinel, or garnet as liquidus phases under appropriate *P-T* conditions. In the pyrolite composition, plagioclase, spinel, and garnet tend to be eliminated after relatively small degrees of melting at appropriate pressures so that the predominant residual phases remaining after extraction of the more common types of basalt magmas are olivine + orthopyroxene ± clinopyroxene.

Physical Processes of Magma Generation

The formation of a magma in the mantle requires the supply of a large amount of thermal energy, in excess of 100 cal/g of magma, to a localized region. Physical processes (e.g., radioactive heat generation or pressure release during convection) which might be responsible for the supply of this energy operate on a comparatively long time scale. In contrast, the time scale for separation of crystals from liquid within the mantle, directly by gravity or indirectly by deformational processes ultimately of gravitational origin, is probably smaller by orders of magnitude. Because of these conditions, and also because of the large temperature interval between the solidus and liquidus of mantle material,[1] the formation of magmas in the mantle will almost always be the result of *partial* melting, rather than of complete melting. Where a substantial degree of partial melting occurs through a large volume, the magma will tend to segregate from residual crystals into a self-contained magma body which thereafter evolves independently of the refractory residuum with which it was formerly associated. The degree of partial melting which is required before the magma separates from residual crystals doubtless varies according to physical conditions—particularly the degree of local deformation which may be operative. Estimates are rather subjective but perhaps mostly range between 2 to 30% (by volume). Under some circumstances the degree of melting prior to magma segregation may be smaller than 2%.[2]

The low-velocity zone between about 70 and 150 km appears to provide the most likely ultimate source of basaltic magmas. It is a region possessing a high degree of potential convective instability in which the actual temperature gradient greatly exceeds the adiabatic gradient. Indeed, the density actually decreases with depth in this region.[3] As discussed in Sec. 6-3, the low-velocity zone in most regions is in a state of incipient melting and hence lacks strength and possesses great mobility.

The generation of magmas in the mantle may be visualized as follows.[4]

[1]E.g., Ito and Kennedy (1967).

[2]Trace element data (Kay, 1971) suggest that some lamprophyres, kimberlites, and high-potassium basalts may segregate from pyrolite after smaller degrees of melting than 1%. Alternatively, processes of wall-rock reaction (Green and Ringwood 1967a) might be responsible for their high abundances of incompatible elements.

[3]Clark and Ringwood (1964).

[4]Green and Ringwood (1967a).

Gravitational instability causes a source mass of incipiently (~1%) molten pyrolite to rise diapirically (in the manner of a salt dome) from the low-velocity zone. The rising diapir is sufficiently large, and hence possesses sufficient thermal inertia in relation to its velocity, so that it cools adiabatically and does not interact by thermal conduction with the surrounding mantle. The adiabatic gradient, of the order of 0.3°C/km,[1] is much smaller than the melting-point gradient of pyrolite. Accordingly, the degree of partial melting of the diapir will increase as it rises to higher levels so that the diapir actually consists of a mush of crystals and interstitial liquid. It is probable that, at this stage, the velocity of upward movement is sufficiently slow to permit the liquid component of the mush to remain in chemical equilibrium with the residual unmelted crystals. Eventually, the degree of partial melting becomes sufficiently extensive[2] (perhaps 2 to 30%) so that the liquid segregates from residual refractory crystals and forms an independent homogeneous magma body. This may be termed the stage of *magma segregation*. From this stage onward, the magma is no longer in equilibrium with the residual crystals with which it was originally associated. Instead, it may fractionate independently by cooling and crystal settling as it rises towards the surface.

The above outline suggests that the processes of magma generation in the mantle are more complex than sometimes assumed. As we shall see, the composition of the interstitial liquid depends sensitively upon depth, and the magma which finally segregates at a much shallower level will have a composition differing greatly from that of the first-formed liquid in the source region of the low-velocity zone. As long as crystals and liquid in the rising crystal mush remain in chemical equilibrium, as appears likely, the nature of the liquid will change continuously with pressure and will retain no "memory" of its earlier, deeper origin. The overall chemistry of the magma is not determined until it segregates from residual refractory crystals. Thus the depth of *magma segregation* is decisive in determining the nature of the magma.

Partial Melting of Pyrolite—Anhydrous Conditions

The principles involved in applying the experimental results to the problem of partial melting in the mantle were discussed earlier. These will now be applied to a consideration of the nature of magmas which may be produced in different depth intervals. Later, it is shown that the water content of the mantle has an important role in the generation of some magmas. It is convenient, however, to commence with a discussion of magma genesis under anhydrous conditions and then to extend the discussion to cover the role of water.

(i) Depth of magma segregation 0 to 15 km At pressures of about 0 to 5 kbars (0 to 15 km), the primary field of olivine extends into quartz-normative basaltic

[1]Birch (1952).

[2]Green and Ringwood (1967a) suggested that 20 to 40% melting might be required before magmas segregated. This estimate now appears to be too high. Because of continuous deformation of the diapirs during ascent, liquids may separate from residual crystals after a smaller degree of partial melting than was contemplated previously.

compositions (Sec. 4-2). The olivine may be accompanied by clinopyroxene and/or plagioclase. Accordingly, in this depth interval, a quartz tholeiite magma could be in equilibrium with residual dunite or with assemblages of olivine ± clinopyroxene ± plagioclase. The composition of pyrolite is such that a small degree of partial melting (e.g., 5%) would yield a quartz tholeiite in equilibrium with residual olivine + clinopyroxene ± plagioclase, whilst a larger degree of partial melting (e.g., 20%) would yield a quartz tholeiite in equilibrium with residual dunite.[1] The two kinds of quartz tholeiite could be distinguished by their abundance patterns of incompatible trace elements (Sec. 4-6). The magma produced by the smaller degree of partial melting would possess higher abundances of incompatible trace elements and would be characterized by a strongly fractionated rare earth pattern, owing to the presence of clinopyroxene as a residual phase. On the other hand, with the larger degree of partial melting in which the residual phase is solely olivine, little or no fractionation of rare earths would be anticipated.

With an even larger degree of partial melting (>20%), still more olivine would enter the liquid, changing its composition to that of an olivine tholeiite or a picrite.

(ii) Depth of magma segregation 15 to 35 km (5 to 10 kbars) Experimental crystallization of the olivine tholeiite (Fig. 4-2, Table 4-3) showed that, at 1250°C, 9 kbars, the separation of about 12% olivine (Fo_{91-86}) and 3% orthopyroxene (En_{87}, 5% Al_2O_3) produced residual liquid similar to high-alumina tholeiite. Since the liquid is saturated with these phases under the stated conditions, it would be possible to increase these proportions to any desired amount without affecting liquid composition; e.g., 56% olivine and 27% enstatite (5% Al_2O_3) could be added, giving a bulk composition essentially identical to pyrolite (Sec. 5-4). It follows that partial melting of pyrolite to the extent of about 15 to 20%, followed by magma segregation under these *P-T* conditions, would yield a high-alumina olivine tholeiitic liquid containing about 13% normative hypersthene.

The experimental results on the olivine tholeiite (Fig. 4-2, Table 4-3) showed that, at 1230°C, 9 kbars, the separation of about 12% olivine, 4% orthopyroxene, and 14% clinopyroxene also produced a high-alumina basaltic liquid which, however, contained only 5% normative hypersthene. These results indicate that such a basalt might be produced by a smaller degree (10 to 15%) of partial melting of pyrolite under similar *P-T* conditions, leaving behind a residuum of olivine and aluminous (5 to 6%Al_2O_3) orthopyroxene and subcalcic clinopyroxene.[2]

Recrystallization at lower temperatures of the residual assemblage (olivine + aluminous orthopyroxene) in the first example would yield a typical spinel harzburgite. Recrystallization at lower temperatures of the residual assemblage (olivine + aluminous orthopyroxene + aluminous subcalcic clinopyroxene) in the second case would yield a typical spinel lherzolite.

[1]Reay and Harris (1964).
[2]T. Green, D. Green, and Ringwood (1967).

(iii) Depth of magma segregation 35 to 70 km The compositions of olivines and orthopyroxene (Mg/Mg + Fe $\sim$ 0.92) on the liquidus of the picrite at 13.5 kbars closely match the compositions of olivines and orthopyroxene found in many natural peridotites.[1] Also, the compositions of orthopyroxenes on the liquidi of the olivine tholeiite and olivine basalt are close to those of orthopyroxenes from natural peridotites.[2] Olivine did not occur on the liquidus of the olivine tholeiite and olivine basalt between 12 and 18 kbars. However, Green and Ringwood pointed out from comparison with the picrite that these basalts were very close to being saturated with olivine and that an olivine tholeiite containing 3 to 5% more normative olivine, but otherwise having the same composition as the olivine tholeiite studied, would have both olivine and orthopyroxene on its liquidus in this pressure interval. Furthermore, the composition of the olivine and orthopyroxene would be similar to those in many peridotites.[3] This was confirmed in subsequent experiments.[4] It follows that a fairly large degree of partial melting of pyrolite under these conditions would yield an olivine tholeiite magma closely resembling the chosen composition (Table 4-1) but containing 3 to 5% more normative olivine, together with a harzburgitic (olivine + orthopyroxene) residuum.[5]

The experiments on the olivine tholeiite and olivine basalt at 12 to 20 kbars showed that these magmas would fractionate by means of the crystallization of substantial amounts of aluminous orthopyroxene $\pm$ aluminous clinopyroxene to the composition of alkali basalts. The picrite would fractionate similarly, except that the pyroxenes would be accompanied by olivine. The experiments thus show that an alkali basalt (containing 3 to 5% more normative olivine than the experimental composition, Table 4-1) can be in equilibrium with an assemblage of olivine + aluminous orthopyroxene $\pm$ aluminous clinopyroxene at 12 to 20 kbars. This assemblage would be identical on cooling at pressure to the assemblage found in natural lherzolites which occur both as intrusions and as xenoliths in alkali basalts. These relationships show that olivine-rich alkali basalt may be formed by direct partial melting of pyrolite.

Whether an olivine tholeiite or an alkali olivine basalt is formed by partial melting of pyrolite under these conditions depends simply upon the extent of partial melting. With a relatively small degree (5 to 10%) of partial melting of pyrolite, the residual crystals consist of olivine, abundant aluminous orthopyroxene $\pm$ aluminous clinopyroxene, and the liquid has the composition of an olivine-rich alkali basalt. With further increase of temperature accompanied by an increased degree of partial melting, the clinopyroxene and then a large amount of aluminous orthopyroxene enter the liquid, changing its composition through olivine basalt to olivine tholeiite at about 25% of partial melting.[6]

[1] Green and Ringwood (1964, table 2).
[2] Green and Ringwood (1967a, table 9).
[3] Green and Ringwood (1967a, p. 166).
[4] Green (1970, table 2).
[5] O'Hara (1968a) drew attention to the absence of olivine on the liquidus of the olivine tholeiite at 13 to 18 kbars but failed to notice that this point had been fully considered by Green and Ringwood (1967a, p. 166).
[6] Green and Ringwood (1964, 1967a).

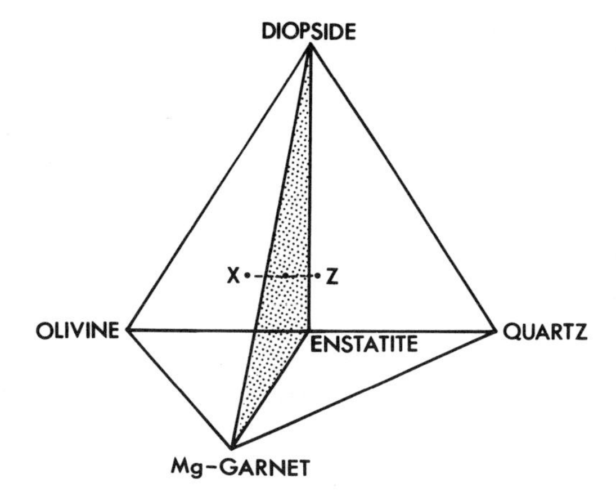

FIGURE 4-5
Diagram illustrating melting behaviour in anhydrous mantle at pressures greater than 25 kbars. The shaded plane is the eclogite plane, which forms a thermal divide. The composition of the first liquid to form by partial melting, *X*, is always on the olivine side of the eclogite plane.

(iv) Depth of magma segregation greater than 70 km At pressures higher than 30 kbars (100 km), garnet becomes stable in the pyrolite composition at the solidus (Sec. 6-2, Fig. 6-4). Liquids formed by relatively small degrees of partial melting of anhydrous garnet pyrolite are of necessity saturated with olivine, orthopyroxene, clinopyroxene, and garnet. O'Hara (1963) first showed that, under these conditions, the liquids are picritic in nature, containing over 30% normative olivine. The melting relationships are shown in Fig. 4-5. At pressures above 30 kbars, the eclogite plane Ga-Cpx-Opx behaves as a thermal divide,[1] and in the presence of olivine as in the mantle, the first liquid to form lies on the olivine side of the eclogite plane, near the point *X*, corresponding to about 30% olivine in the low-pressure norm.[2] With increasing pressure, the point *X* moves towards the olivine apex, and the liquids become even richer in olivine. These relationships appear to be of general occurrence under anhydrous conditions and have been confirmed in several independent studies of different systems.[3,4,5,6,7] The assumed picritic nature of primary magmas plays a central role in O'Hara's scheme of basalt petrogenesis,[3] which is discussed in Sec. 4-7.

At pressures between 20 and 30 kbars (70 to 100 km), melting relationships are intermediate between those observed in the 10 to 20 kbar interval and those above 30 kbars. As shown in Fig. 6-4, garnet does not occur at the solidus of anhydrous pyrolite until a pressure close to 30 kbars so the liquids formed between 20 and 30 kbars are saturated with olivine and aluminous pyroxenes. Liq-

[1]O'Hara (1963).
[2]When the compositions of typical bimineralic eclogites (omphacite + MgFe garnet) are recalculated into low-pressure norms, the latter usually contain 20 to 30% olivine, in addition to plagioclase and pyroxene. Examples are given in Table 4-2.
[3]O'Hara (1963, 1968).
[4]Davis (1964).
[5]Green and Ringwood (1967a).
[6]Ito and Kennedy (1967, 1968).
[7]Note, however, that water causes the olivine field to expand and that, at high water pressures, the eclogite plane is no longer a thermal divide (ref. Sec. 7-6).

uids may be alkalic or tholeiitic, according to the degree of partial melting, as discussed in the previous section. However, they are richer in normative olivine and are more properly classed as alkali picrites and tholeiitic picrites.

Partial Melting of Pyrolite—Hydrous Conditions

In the previous section, discussion was restricted to the partial melting of anhydrous pyrolite. However, there are strong grounds for believing that pyrolite in the mantle contains a small amount of water, and the influence of this additional component upon melting equilibria must be considered. From a comparison with the amount of nitrogen which has been degassed from the earth's interior, Ringwood (1966) estimated that the mantle has retained at least 3 times as much water as has been degassed into the oceans; this would suggest an average content of about 0.1% water. This estimate is supported by analyses of rapidly chilled nonvesicular tholeiitic glasses from the ocean floor, which usually contain from 0.2 to 0.5% water.[1,2] The frequent presence of excess argon combined with isotopic evidence suggests that this water is primordial and has not exchanged with ocean water.[1,2,3] If most of the water in pyrolite had entered these liquids, which represent 15 to 20% partial melts, the water content of primary pyrolite would be in the vicinity of 0.05 to 0.1%. The study of crystallization of the Auckland Island basalt[4] previously discussed provided strong evidence that the magma contained about 2% water. If this magma represented 5 to 10% of its original source material, water content of the latter would be between 0.1 and 0.2%. We have seen that the presence of water appears essential in the development of basanites and nephelinites.[5] A widely self-consistent explanation of the origin of these magmas emerges if the mantle is assumed to contain about 0.1% water. In subsequent discussion we will assume a water content of 0.1% in the mantle[6] (exclusive of regions overlying Benioff zones), recognizing that this is an approximation which may well be wrong by a factor of 2 or more and that substantial local variations probably occur.

The presence of water, even in such small amounts, nevertheless has some important effects upon the properties and behaviour of the mantle. Some of these effects are discussed below; others are considered in greater detail in Sec. 6-3.

The solidi of anhydrous pyrolite and of pyrolite containing excess water[7,8] ($P_{H_2O} = P_{total}$) are shown in Fig. 4-6. It is seen that high water pressures have a rather drastic effect in lowering the solidus. The solidus of pyrolite containing

[1]Funkhouser, Fisher, and Bonatti (1968).
[2]Moore (1965).
[3]Shepard and Epstein (1970).
[4]Green and Hibberson (1970).
[5]See also the footnote on page 140 concerning the probable role of CO_2 in the genesis of nephelinitic and melilitic magmas.
[6]This figure is probably much higher in regions of mantle subduction.
[7]Kushiro, Syono, and Akimoto (1968).
[8]Green (1970a, 1973b).

0.1% water[1] (Fig. 4-6) is approximately equally intermediate between these two extremes at depths up to 75 km. In this interval, the water is held in amphibole, which is stable to quite high temperatures (well above the $P_{H_2O} = P_{total}$ solidus) under these water-deficient conditions. As a result, the pressure of water vapour in pyrolite (0.1% H_2O) is much smaller than the total pressure in the region where amphibole remains stable. This is responsible for the observation that the pyrolite solidus (0.1% H_2O) lies at much higher temperatures than the corresponding solidus for $P_{H_2O} = P_{total}$.

At depths greater than 75 to 90 km, amphibole is no longer stable, owing to a pressure-induced transformation (Sec. 6-2). The water is liberated as a free phase, with P_{H_2O} nearly [2] as high as P_{total}. The solidus is accordingly greatly depressed, approaching the solidus for $P_{H_2O} = P_{total}$ conditions. At depths greater than about 150 km, the pressure of water vapour is again lowered owing to the stability of various hydroxylated silicates (Sec. 6-3). As a result, the solidus rises steeply[3] around 150 km, as in Figs. 4-6 and 4-7.

The overall configuration of the pyrolite solidus (0.1% H_2O) is thus characterized by a deep depression between 75 to 150 km. This depression is intersected by oceanic geotherms, implying the existence of a small degree of partial melting in this interval (Figs. 4-6 and 4-7). Implications with respect to the low-velocity zone are considered in Sec. 6-3.

At depths up to 75 km, the influence of water present in pyrolite upon melting relationships involving basalt genesis is not very marked. The principal effect is a moderate expansion and depression of the temperature interval over which partial melting occurs. However, the melting equilibria, particularly at higher degrees of partial melting, are not substantially altered. In this depth interval, the dry-melting experiments on basalts and their interpretation in terms of fractional crystallization or partial melting as discussed earlier remain applicable.

Below 75 km, however, the reversal in slope of the solidus causes a remarkable broadening of the temperature interval over which partial melting occurs (Fig. 4-7). Moreover, a large increase in temperature above the solidus (200 to 300°C) causes only a small increase in the degree of partial melting. Thus a liquid formed by a very small degree of partial melting of anhydrous pyrolite at 25 kbars, 1400°C, would be in equilibrium with olivine, aluminous enstatite (~6% Al_2O_3, 2.2% CaO), and aluminous subcalcic clinopyroxene (~8% Al_2O_3, 11% CaO), whereas the liquid produced with a similarly small degree of melting at 25 kbars, 1200°C would be in equilibrium with olivine, enstatite (~3%, Al_2O_3, <1.5% CaO) clinopyroxene (~5% Al_2O_3, >20% CaO), and garnet (or spinel).[4] The differences in mineralogy are largely a consequence of the marked decreases in mutual solid solubility between orthopyroxene and clinopyroxene, which occur on cooling

[1]Green (1970a, 1973b).

[2]P_{H_2O} is not quite as high as P_{total} because the activity of water is lowered by solution of other components and also by surface effects along grain boundaries and in small pores.

[3]More recent investigations (Fig, 13-3) do not support this interpretation.

[4]Green (1970b).

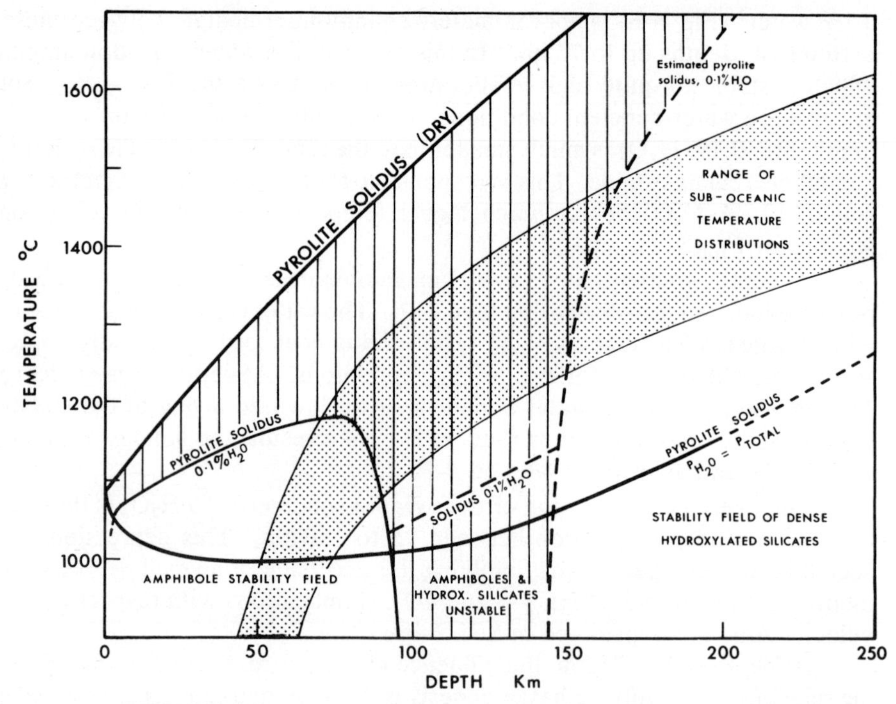

FIGURE 4-6
Solidus curves for pyrolite containing varying amounts of water in relation to possible range of suboceanic temperature distributions (stippled region) centered on the Clark and Ringwood (1964) oceanic geotherm. Anhydrous pyrolite solidus after Green and Ringwood (1967b). Hydrous solidus ($P_{H_2O} = P_{total}$) after Kushiro, Syono, and Akimoto (1968) and Green (1973b). Solidus of pyrolite containing 0.1% H_2O is based upon experimental work of Green (1971, 1973b) at pressures up to 40 kbars. At higher pressures, this solidus is based upon estimates which have a wide margin of uncertainty, particularly above 50 kbars. The sharp depression of the pyrolite (0.1% H_2O) solidus between 80 and 150 km is caused by the instability of amphibole and hydroxylated silicates, resulting in high P_{H_2O} in this region. Intersection of the temperature distributions with the depressed solidus in this interval implies a small degree of partial melting in the low-velocity zone.

from 1400 to 1200°C, and the corresponding decrease in the solubility of Al_2O_3 in the pyroxenes, which reacts out to form garnet or spinel. Thus, the effect of water makes it possible for liquid to coexist with crystals in a *P-T* field in which solid solubility and mineral stability relationships are considerably different from those which pertain under anhydrous conditions. It is not surprising therefore, that the liquids formed by small degrees of partial melting of hydrous pyrolite at relatively low temperatures differ from those formed at higher temperatures by similar degrees of partial melting of anhydrous pyrolite.

The key experiments which established the nature of the liquid phase formed by small degrees of partial melting of pyrolite under hydrous conditions were carried out by Green and colleagues.[1,2,3] These demonstrated that, in a *P-T* window around 27 kbars, 1250°C, an olivine basanite magma having 10% normative nepheline, 26% olivine, and 2 to 4% water crystallized with olivine (Fo_{86}), orthopyroxene (En_{81}, 6% Al_2O_3), clinopyroxene ($En_{56}Wo_{34}Fs_{10}$,7% Al_2O_3), and garnet ($Py_{69}Al_{17}Gr_{14}$) as liquidus or near-liquidus phases. These experiments demonstrated unambiguously that, in this *P-T* window, such a basanite could be formed by a small degree (3 to 6%)[4] of partial melting of pyrolite, leaving behind a residual garnet peridotite. Reconnaissance experiments upon an olivine nephelinite[1,5,6] indicated the existence of a similar window at slightly lower temperatures, higher pressures, and higher water (and CO_2) contents within which the nephelinite magma was saturated with olivine, orthopyroxene, garnet, and clinopyroxene. It is estimated that such an olivine nephelinite magma could form 1 or 2% partial melting of pyrolite.

It appears likely that highly undersaturated magmas such as olivine melilite nephelinites, olivine melilitites, and kimberlites may also be formed by extremely small degress ($<$ 1%) of partial melting under hydrous conditions and at higher pressures, perhaps combined with the operation of extensive wall-rock-reaction processes.[7,8,9,10] However, the experimental methods necessary to determine crystallization sequences in these water-rich undersaturated compositions are difficult, owing to quenching and other problems, and progress towards a quantitative understanding of their genesis is likely to be rather slow. An important effect of water in lowering the temperatures of partial melting is to reduce the pressure at which garnet is stable on the pyrolite solidus (Fig. 6-4), and it is probable that these extremely undersaturated magmas have formed in equilibrium with residual olivine, orthopyroxene, clinopyroxene, and garnet—i.e., garnet peridotite.[7,8,9,10] Under some conditions, a significant amount of residual garnet may be left in the refractory residue remaining after extraction of olivine nephelinite or basanite magmas. Further work is required in order to clarify the role of garnet in influencing the generation of highly undersaturated magmas. Studies of rare-earth fractionation are likely to be helpful in solving this problem.[11]

[1]Bultitude and Green (1968).

[2]Green (1968, 1970a,b, 1971, 1973a,b).

[3]Green and Hibberson (1970).

[4]Green (1971) had earlier estimated the degrees of partial melting of pyrolite needed to produce olivine basanite and olivine nephelinite as 10% and 5%, respectively. Studies of trace element abundances in basanites and nephelinites by Gast (1968), Kay, Hubbard, and Gast (1970), and Kay (1971) suggest that these estimates are too high. The matter is further discussed in Sec. 4-6. See also Green (1973a).

[5]Green (1970a).

[6]See also the footnote on page 140 concerning the probable role of CO_2 in the genesis of nephelinitic and melilitic magmas.

[7]Harris (1957, 1969).

[8]O'Hara (1968a).

[9]Green (1971, 1973a).

[10]Harris and Middlemost (1970).

[11]Kay (1971).

Summary

A petrogenetic scheme based upon diapiric uprise of pyrolite from the incipiently molten low-velocity zone (Sec. 6-3) is shown in Fig. 4-7. This demonstrates the critical role of water in widening the *P-T* field over which relatively small degrees of partial melting occur. Pyrolite diapirs rising from points *A*, *B*, *C*, and *D* follow the *P-T* courses given by the corresponding straight lines. As the diapirs rise, the degrees of partial melting increase, and following previous discussion, the composition of the magma changes according to percent melting, pressure, and temperature.

A diapir ascending from point *A* (83 km) would freeze as soon as it left the low-velocity zone. Such a diapir might ultimately ascend into a near-surface environment as an amphibole-bearing peridotite. Diapirs rising from slightly greater depths may lose a small proportion of an alkalic liquid before encountering the shoulder of the solidus at 75 to 80 km, where they would freeze to lherzolitic assemblages.[1] Such diapirs may be recognizable as high-temperature peridotites—e.g., St. Paul Rocks and the Lizard (Sec. 3-3).

Diapirs arising from a slightly greater depth (e.g., point *B*, 100 km) avoid the shoulder on the solidus and hence display a continually increasing degree of partial melting as they rise upward. If a magma segregates at *B*1 (1% partial melting, 90 km), it would have the composition of an olivine nephelinite, whilst the residual assemblage is that of a garnet lherzolite. If, however, the diapir ascended to *B*2 (50 km) before the magma segregated, the composition of the liquid would be that of an olivine basanite. At a shallower depth of magma segregation, the degree of partial melting increases as more clinopyroxene enters the liquid, and high-alumina basalt (alkalic affinities) *B*3 is produced at a depth of 25 km. Finally, near the surface, the liquid which segregates (*B*4) has the composition of a quartz tholeiite.

The course followed by diapirs rising from *C* is slightly different. The first magma formed with a small degree (1%) of partial melting is an olivine nephelinite (*C*1, 100 km) in equilibrium with a residual garnet lherzolite assemblage. With increasingly shallower depths of magma segregation, the liquid composition changes through olivine basanite (*C*2, 75 km) to alkali olivine basalt (*C*3, 50 km) in equilibrium with a residual lherzolitic assemblage of olivine + aluminous orthopyroxene and aluminous subcalcic clinopyroxene. With further uprise of the diapir, accompanied by an increased degree of partial melting (*C*4, 25 km), more of the aluminous clinopyroxene enters the liquid, which is made over into a high-alumina basalt (alkalic affinities), with a residual assemblage of olivine, orthopyroxene ($\sim$5%, Al_2O_3) + subcalcic clinopyroxene ($\sim$6% Al_2O_3). Near the surface (*C*5), a quartz tholeiite is produced.

Diapirs which rise from the base of the low-velocity zone (*D*) evolve analogously but, because of their excess superheat, are able to attain greater degrees of partial melting at shallower depths. The first liquid to segregate near 140 km (*D*1) is probably kimberlitic. This evolves through an olivine nephelin-

[1]Green (1970c).

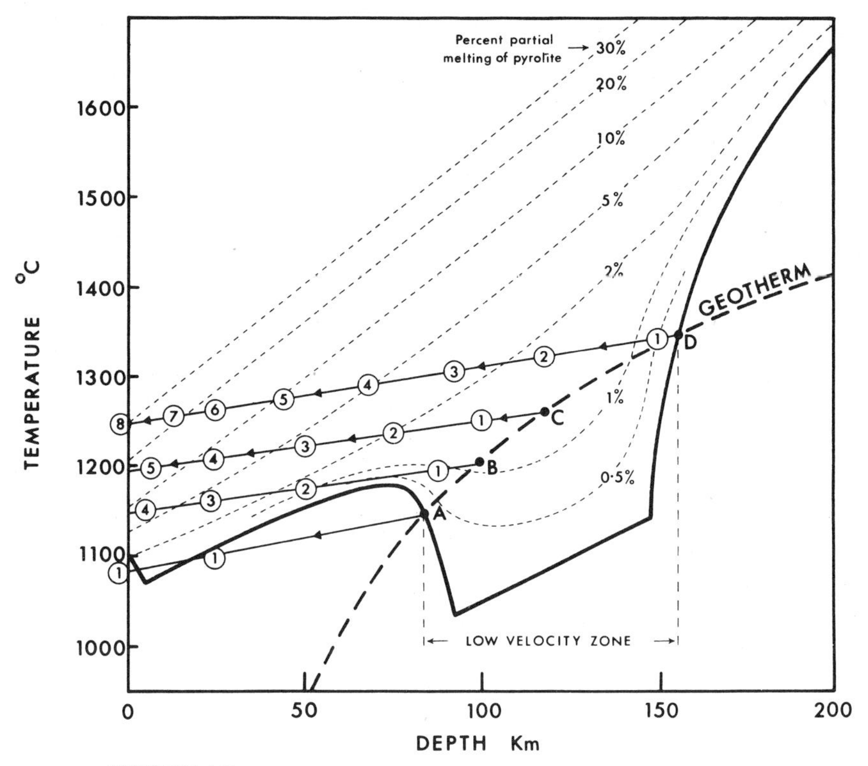

FIGURE 4-7
Possible relationships between mantle solidus (0.1% H_2O), mantle temperature distribution, degrees of partial melting of pyrolite, and nature of magmas produced. Letters *A*, *B*, *C*, and *D* indicate possible source regions in the low-velocity zone for diapiric uprise of pyrolite. Straight lines emanating from these points indicate *P-T* paths followed by diapirs as they rise towards surface. Numbers on lines denote magma types which are formed by magma segregation at different stages of partial melting (light broken lines). *Key* (see text for further detail):

*A*1 High-temperature peridotite

*B*1 Olivine nephelinite
*B*2 Olivine basanite
*B*3 High-Al alkali basalt
*B*4 Quartz tholeiite

*C*1 Olivine nephelinite
*C*2 Olivine basanite
*C*3 Alkali olivine basalt
*C*4 High-Al basalt
*C*5 Quartz tholeiite

*D*1 Kimberlite?
*D*2 Olivine nephelinite
*D*3 Olivine basanite
*D*4 Alkali olivine basalt
*D*5 Olivine basalt (alkalic affinities)
*D*6 High-alumina olivine tholeiite
*D*7 Olivine tholeiite
*D*8 Tholeiitic picrite

ite (*D*2, 115 km) to an olivine basanite composition for a liquid segregated at *D*3, 90 km, with 3% partial melting. Magma segregation at *D*4 (65 km, 5% melted) would produce an alkali olivine basalt, whereas at *D*6 (25 km, 18% melted) a typical high-alumina oceanic tholeiite would segregate, leaving residual harzburgite.

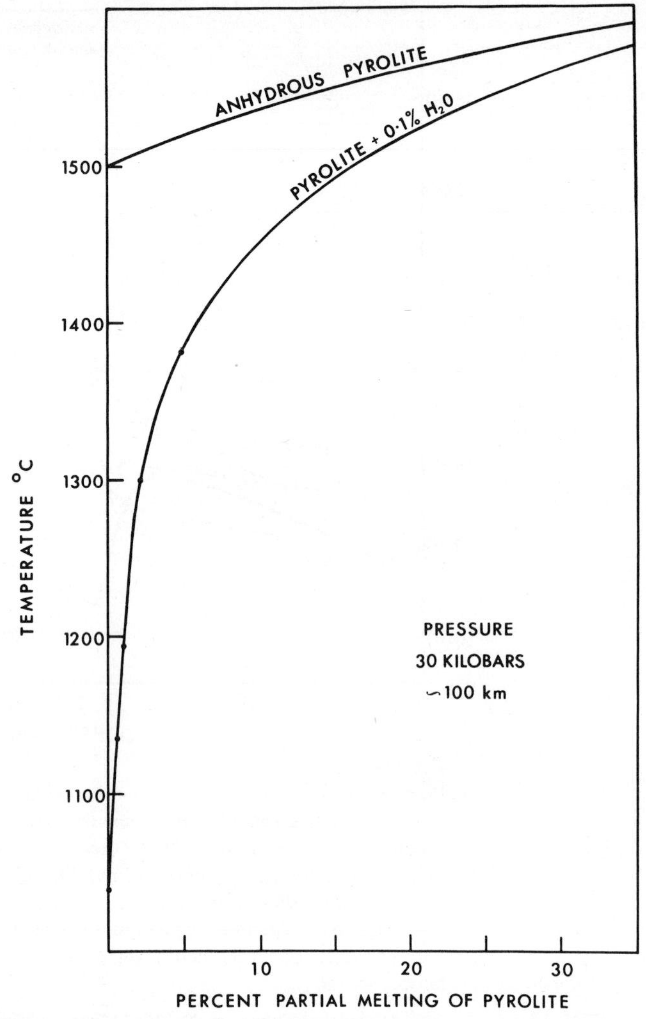

FIGURE 4-8
Estimated degree of melting versus temperature in pyrolite under anhydrous conditions (Green and Ringwood, 1967b) and in the presence of 0.1% water at 30 kbars. (*Modified from Ringwood, 1969, using data of Green, 1971, 1973b.*)

At higher levels (*D*7, 10 km, 25% melted), an olivine tholeiite is produced, whereas near the surface (*D*8, 30% melted), the magma which segregates is a tholeiite picrite, leaving a residuum of dunite.

In Fig. 4-8, the degree of melting of anhydrous pyrolite and of pyrolite containing 0.1% H_2O is plotted against temperature for a pressure of 30 kbars. This diagram emphasizes the very large increase in the melting interval caused by the presence of such a small amount of water. As discussed in Sec. 6-3, this effect appears to play an important role in causing the formation of a low-velocity zone.

4-6 PARTIAL MELTING AND FRACTIONAL CRYSTALLIZATION IN BASALT PETROGENESIS

The experimental investigations described earlier revealed the existence of characteristic pathways through which parental olivine tholeiite magmas could differentiate by fractional crystallization to form quartz tholeiites, high-alumina basalts, alkali olivine basalts, olivine basanites, and olivine nephelinites. Each of these differentiation trends was possible only within well-defined P-T-P_{H_2O} "windows." Likewise, the same experimental data implied that, within these windows, varying degrees of partial melting of pyrolite could produce the above magmas.

A fundamental problem lies in assessing the relative roles of fractional crystallization and partial melting in producing the spectrum of basaltic compositions observed at the surface. Clearly, both processes have been important. The initial production of a primary basaltic magma is the result of a partial melting process; this magma may undergo further modification by crystallization differentiation en route to the surface.

Innumerable detailed petrologic studies have shown that, when mafic magmas are held for extended periods of time in reservoirs, they are likely to differentiate to form a wide range of derivative magmas. The density of basaltic magmas (2.7 to 2.8)[1] is similar to the mean density of the crust but much less than the density of mantle rocks. It follows that large, basaltic reservoirs are likely to be more stable within the crust than within the mantle. Because of the density contrast, basaltic magmas within the mantle will have a strong tendency to rise. When the conduit to the surface or to a crustal reservoir is ultimately opened, flow of magma from the mantle would be expected to occur rather rapidly. Such considerations suggest that extensive differentiation by fractional crystallization is likely to be of more general occurrence under low-pressure crustal conditions than within the mantle. Nevertheless, the occurrence of high-pressure xenocrysts (Sec. 4-4) shows that some fractional crystallization also occurs within the mantle.

The nature of the differentiation processes displayed by basaltic magmas under shallow conditions is reasonably well understood, both from observational studies and from experimental investigations. The critically undersaturated plane of Fig. 4-1 forms a thermal divide, and in the nepheline-normative volume, alkaline magmas differentiate to form a wide range of trachytic and phonolitic residual liquids. On the other hand, tholeiitic magmas differentiate principally by crystallization of olivine, plagioclase, and pyroxenes to form a series of quartz-normative residual tholeiitic liquids in which the Fe/Fe + Mg and Na/Na + Ca ratios rise steadily as the extent of fractionation increases. Magmas of this kind

[1]Clark (1966).

have been developed on a vast scale in continental regions, e.g., Deccan traps and other flood basalts,[1] and also occur in oceanic regions, e.g., Hawaii[2,3,4] and mid-oceanic ridges.[5]

The origins and relationships of the above classes of magmas which have evolved via high-level crystallization differentiation processes principally in crustal environments will not be further pursued. Discussion will instead be focused on the petrogenesis of magmas which have been derived directly from the mantle unmodified or only slightly modified by subsequent differentiation in the crustal environment. Such magmas include those alkali basalts, basanites, nephelinites, melililites, and kimberlites which contain xenoliths and xenocrysts demonstrably of mantle origin (Sec. 4-4). Clearly, if a magma is able to ascend sufficiently rapidly to the surface to transport large dense xenoliths for several tens of kilometres, the amount of near-surface differentiation caused by settling of small cognate crystals will be relatively minor.[6] Other classes of magmas for which a strong case can be made for mantle origins unmodified by extensive crustal differentiation include high-alumina oceanic tholeiites, olivine tholeiites with Mg/Mg + Fe ratios greater than 0.70,[7] and some picrites.

Trace Element-Major Element Relationships

Traditionally, the recognition of different types of basaltic rocks and their classification has rested upon major element chemical compositions, which in turn govern the mineralogies displayed by the rocks. The primary objective of petrologists has always been to attain an understanding of the genetic relationships between different classes of basalts in terms of their major element compositions. The experimental investigations discussed earlier have gone far towards achieving this objective.

The accumulation of accurate trace element data on basalts, particularly during the last 10 years, has drawn attention to the geochemical relations between different magma types. Gast and colleagues have demonstrated that important constraints on the roles of fractional crystallization and partial melting in mantle-derived magmas arise from considerations of trace element abundances.[8,9,10,11]

The petrological approach towards basalt genesis, which is based primarily

[1]Turner and Verhoogen (1951).
[2]Tilley (1950).
[3]MacDonald and Katsura (1961).
[4]Powers (1955).
[5]Miyashiro, Shido, and Ewing (1969) and Kay, Hubbard, and Gast (1970) have provided detailed studies of differentiation of oceanic olivine tholeiite by separation of olivine and plagioclase to produce a range of derivative tholeiites enriched in silica and iron and depleted in magnesia and alumina.
[6]Bultitude and Green (1971).
[7]Excluding basalts and picrites containing accumulative olivine.
[8]Gast (1968).
[9]Kay, Hubbard, and Gast (1970).
[10]Kay (1971).
[11]Hubbard (1969).

upon major element relationships, and the geochemical approach, based upon trace element abundances, must ultimately be compatible. Although some differences existed initially,[1] this objective now appears close to being realized.

Green and Ringwood (1967a) showed, on the basis of major element equilibria, that, in the 10 to 20 kbar interval, some 30 to 50% of crystallization of a parental olivine tholeiite would lead to the formation of a residual alkali basalt liquid. Likewise, in the same interval, an alkali basalt would be produced by a relatively small degree of partial melting (~ 15%) of pyrolite, whereas an olivine tholeiite would be produced by a greater degree of partial melting (25 to 30%). However, the abundances of incompatible elements[2] such as K, U, Zr, Ba, and La are frequently from 2 to 5 times higher in alkali basalts than in olivine tholeiites. This difference was too large to be produced solely by the fractional crystallization-partial melting processes envisaged, and accordingly, Green and Ringwood invoked the occurrence of "wall-rock-reaction" processes in order to selectively concentrate the incompatible elements during fractional crystallization or partial melting.[3] It was envisaged that, under some conditions, a body of magma could cool by reaction with and solution of the lowest melting fraction of wall-rock material with which it was in contact, particularly where the temperature contrast between magma and wall rock was small, as would be the case near the source region in the mantle. Under these conditions, a considerable enrichment of incompatible trace elements was possible, whereas major element compositions did not change substantially, being buffered by the principal mantle phases.

Gast (1968) suggested a simple alternative to the wall-rock-reaction process. He proposed that the alkali basalt suite was produced dominantly by partial melting rather than by fractional crystallization and that the degree of partial melting involved in the formation of the alkali basalt suite was much smaller (3 to 7%) than had earlier been suggested by Green and Ringwood (15 to 20%).[4] With a very small degree of partial melting, the incompatible elements would be strongly concentrated in the liquid and their high abundances thereby explained.

Subsequent studies of more undersaturated magmas such as basanites and nephelinites have left little doubt that their compositions were established dominantly by direct partial melting of the mantle rather than by fractional crystallization of olivine tholeiite (Sec. 4-5). For these magmas in which the abundances of incompatible elements are frequently 10 or more times higher than in tholeiites, Gast's proposal of a very small degree of partial melting now appears to

[1]Gast (1968).

[2]Incompatible elements (Ringwood, 1966) possess ionic radii and charges which prevent them from readily entering the major mantle minerals. Hence, they possess very low crystal-liquid distribution coefficients and are strongly fractionated during crystal-liquid differentiation processes.

[3]The process suggested was similar in some respects to the "zone-refining" process suggested by Harris (1957).

[4]This estimate was based upon subjective estimates of the degree of melting which would be required before liquid would separate from crystals within the mantle, and the estimate may well have been in error. It seems likely that deformation processes accompanying the upward movement of partly molten pyrolite diapirs may lead to magma segregation at a much lower degree of partial melting than was envisaged earlier. Figure 4-7 represents a revision of these earlier views in the light of these considerations.

provide a satisfactory explanation both of major element and minor element chemistry.

On the other hand, it still appears likely that, in some cases, fractional crystallization of parental olivine tholeiite in the mantle has produced the transitional series of basalts between olivine tholeiite, olivine basalt, and alkali olivine basalt ($< 6\%$ normative nepheline), although these magmas could also be produced by varying degrees of partial melting as discussed earlier. The occurrence in the Auckland Island basalt (Sec. 4-4) of aluminous pyroxene xenocrysts possessing lower $Mg/(Mg + Fe)$ ratios than in pyrolite provides direct evidence of the operation of this process. Although wall-rock reaction does not now appear to be the single decisive process in causing the very high abundances of incompatible elements in basanites and nephelinites, there is increasing evidence[1] of its actual occurrence. Indeed, it is difficult to see how a very small amount of liquid might be extracted from a relatively large volume of source material without the concomitant occurrence of wall-rock-reaction processes. Wall-rock reaction may also play a significant role in the genesis of the olivine tholeiite-olivine basalt-alkali olivine basalt series in which the degree of enrichment of incompatible elements is much lower. Further detailed trace element data on this transitional series is needed in order to establish the relative roles of partial melting and fractional crystallization.

Fractional Crystallization and the Behaviour of Ni, Cr, Mg, and Fe

In a magma subjected to fractional crystallization under conditions where crystals are removed from contact with the residual liquid as soon as they are formed (e.g., by sinking), the concentration of an element distributed between crystals and liquid is given by the Rayleigh fractionation law. For a given element, the enrichment or depletion in the residual liquid is given by[2]

$$\frac{C^L}{C^0} = F^{K-1} \tag{1}$$

where F = weight fraction of liquid that remains, i.e., $1 > F > 0$
C^L = weight fraction of element in residual liquid during fractionation
C^0 = initial weight fraction of element in liquid (prior to fractionation)
K = distribution coefficient for the given element defined as the ratio of the equilibrium concentration of the element in the solid phase (S) to its equilibrium concentration in the liquid (L) from which the solid is crystallizing, i.e.,

$$K = \frac{C^S}{C^L} \tag{2}$$

If more than one phase is crystallizing, K represents an average of individual distribution coefficients weighted according to the propor-

[1]Green (1971) has extended the concept and discussed its operation in greater detail.
[2]E.g., Gast (1968).

tion in which different phases crystallize. It is assumed that the distribution coefficient K remains constant during the process.

The concentration in the crystals separating from the liquid in which crystallization has proceeded to fraction F is given by

$$\frac{C^S}{C^0} = KF^{K-1} \tag{3}$$

The *mean* concentration, $\bar{C}^S$, of the element in crystals which have formed throughout the entire fractionation process (e.g., in ideally zoned crystals) is

$$\frac{\bar{C}^S}{C^0} = \frac{1 - F^K}{1 - F} \tag{4}$$

Equation (4) also gives the mean concentration of the element in a liquid produced by fractional melting,[1] where K is independent of the composition of the solid being melted, with appropriate changes in definition of symbols, i.e.,

$$\frac{C^L}{C^{\text{initial solid}}} = \frac{1 - F^{1/K}}{1 - F} \tag{5}$$

Curves showing the relation between the trace element contents of residual liquids as compared to the concentration in the initial liquid (i.e., C^L/C^0) as a function of the degree of crystallization F for different partition coefficients K are shown in Fig. 4-9. It is seen that, when an element is enriched in accumulating crystals by a factor of 5 or more relative to the liquid, it is rapidly depleted in the liquid. This behaviour is particularly important in the case of the crystallization of nickel and chromium from mafic magmas.

A list of distribution coefficients for nickel and chromium is given in Table 4-8. It is seen that the distribution coefficient for nickel in olivine is very high, in the vicinity of 10 to 15. Thus, the crystallization of only 5 to 8% olivine will cause the concentration of nickel in the residual liquid to fall by a factor of 2 (Fig. 4-9). Clearly, the abundance of nickel provides a sensitive monitor of the degree to which a given magma has undergone fractionation by separation of olivine. Where the amount of olivine which has separated is as high as 40%, the concentration of nickel in the residual liquid will have decreased by a factor of about 100 (Fig. 4-9). The behaviour of nickel during crystallization of the Muskox and Skaergaard intrusions illustrates and confirms these principles.[2,3] Likewise, the distribution coefficients of chromium in garnet and clinopyroxene are also high (Table 4-8). These place tight restrictions upon the amount of fractionation which a given magma has undergone by the separation of eclogite, a topic which is discussed in the next section.

[1] Fractional melting was defined in Sec. 4-5.

[2] Irvine and Smith (1967).

[3] Wager and Mitchell (1951).

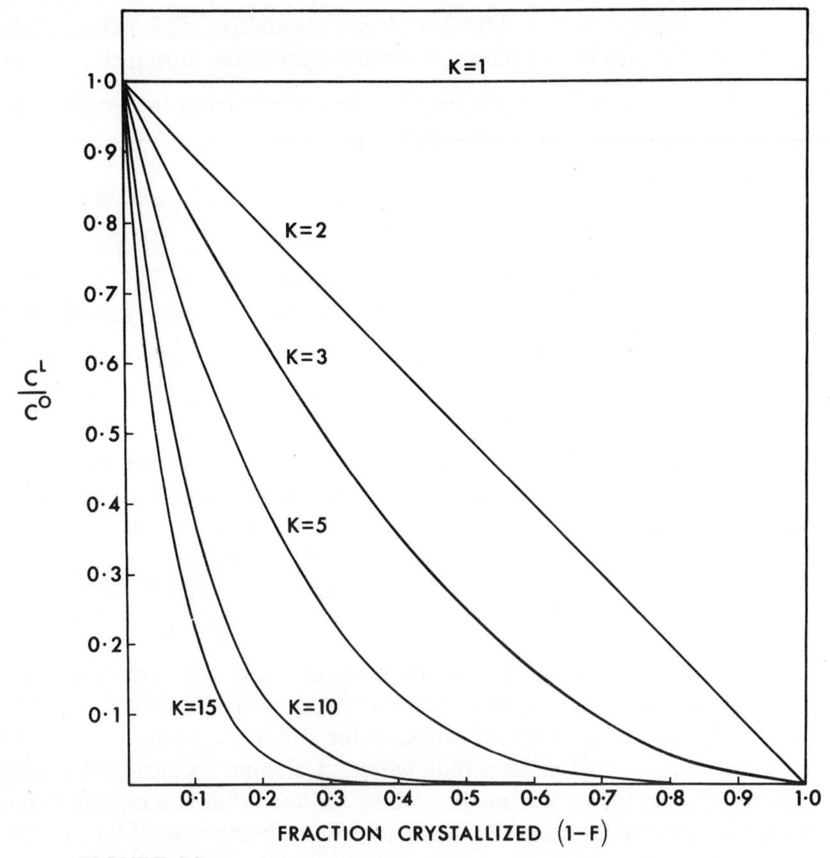

FIGURE 4-9
Rayleigh fractionation curves showing *relative* concentration of element in residual liquid (C^L/C^0) plotted against degree of crystallization ($1-F$) for various values of the distribution coefficient K.

Application of nickel and chromium distribution coefficient data to problems of basalt petrogenesis is hindered by the paucity of good analytical data for Ni and Cr on basalts for which major element compositions are also known. Whilst there is a considerable amount of trace element data available upon basaltic rocks generally termed *tholeiitic* or *alkalic*, this is of limited use because these rocks have frequently suffered extensive near-surface fractionation, causing drastic decreases in Ni and Cr contents, which accordingly bear little relation to those of the parental mantle-derived magmas. Gast (1968) collected available data and noted that "the Cr and Ni content of abyssal (tholeiitic) basalts is somewhat greater than that observed in alkaline rocks; however, the Cr and Ni content of alkaline basalt varies so widely that it is difficult to characterize the magma type by a mean value."

Table 4-8 DISTRIBUTION COEFFICIENTS K FOR NICKEL AND CHROMIUM IN PHASES CRYSTALLIZING FROM BASALTIC MAGMAS

System	Ni_K	Cr_K	Ref.
Olivine-liquid	12	—	1
	10-15	—	2
	16	—	3
Clinopyroxene-liquid	4	18	1
	3.5	—	3
Subcalcic clinopyroxene-liquid	—	6	4
Orthopyroxene-clinopyroxene	2.5	0.2	5
Garnet-clinopyroxene	—	2	6
	—	3	7
Garnet-clinopyroxene	0.15	0.7	5,8,9
Eclogite on liquidus of basaltic magma (estimated)	2	4-10	10

References:
1 Wager and Mitchell (1951).
2 Irvine and Smith (1967).
3 Hakli and Wright (1967).
4 Ringwood (unpublished experimental data).
5 Taylor et al. (1969). Data on eclogites.
6 Green (personal comm.) Data on garnet peridotites.
7 Nicholls (personal comm.). Experimental measurements on garnets and pyroxenes on liquidus of natural eclogite at high pressures.
8 Turekian (1963). Data on eclogites.
9 Taylor et al. (1971).
10 Partition coefficients of Cr in pyroxenes and garnets vary over a wide range with composition. Data for high-temperature Mg-rich garnets and pyroxenes similar to eclogite liquidus phases (6,7,9) are more relevant than data for eclogites with higher Fe/Mg ratios (5,8).

Following Green (1971), the significance of alkali basaltic magmas containing high-pressure xenocrysts and xenoliths of mantle origin has previously been emphasized. The average nickel content of 15 alkali basaltic magmas which contained high-pressure xenocrysts and/or xenoliths was 290 parts per million,[1] whilst the average chromium content of these rocks was 380 ppm. By comparison, the most primitive and least fractionated oceanic tholeiites contain 200 ppm Ni[2] and 300 to 400 ppm Cr.[3] These figures demonstrate that the nickel and chromium contents of the most primitive and least fractionated alkali basalts are not significantly smaller than those of corresponding primitive tholeiites. A similar conclusion follows from a plot of Ni versus MgO for a collection of tholeiites and alkali basalts.[4] It is seen that there is no significant difference between the Ni-Mg relationship for tholeiites and alkali basalts (Fig. 4-10).

[1]Data from Binns et al. (1970), Jackson and Wright (1970), and Green and Kiss (unpublished analyses).
[2]Kay, Hubbard, and Gast (1970, fig. 12).
[3]Engel, Engel, and Havens (1965).
[4]Hedge (1971).

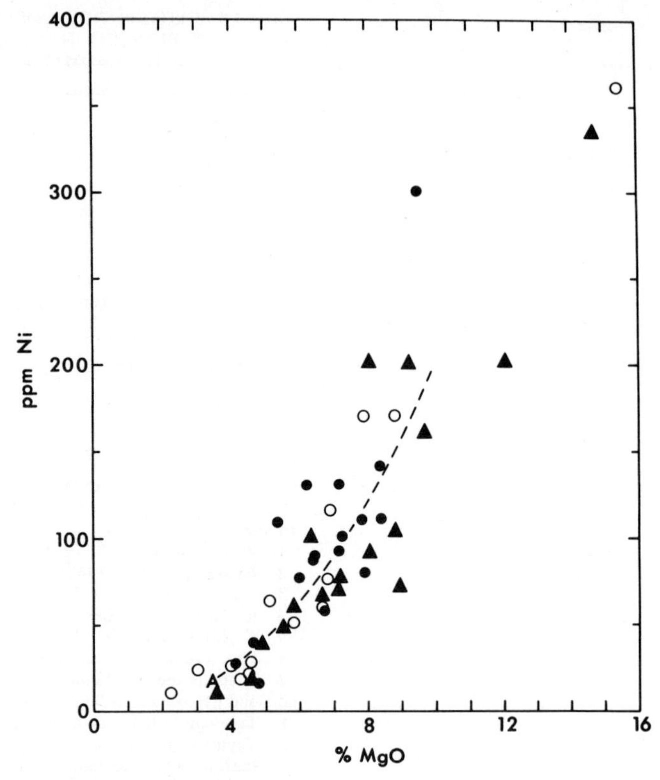

FIGURE 4-10
The variation of Ni and MgO in some representative basalts. Triangles are high-alumina basalts from Japan, the Lesser Antilles, and the Cascades. Closed circles are tholeiites from Japan, oceanic ridges, and Hawaii. Open circles are alkali basalts from St. Helena and various Pacific islands. The dashed line is the path of successive liquids formed by removal of olivine. (*From Hedge*, 1971, *with permission.*)

In order to generate a basanite or nephelinite by fractional crystallization of an olivine tholeiite, by the processes discussed in Secs. 4-2 and 4-3, more than 50% of the parent magma would need to crystallize as orthopyroxene, subcalcic clinopyroxene, and olivine. Application of the relevant distribution coefficients (Table 4-8) in terms of Fig. 4-9 shows that the derived basanites and nephelinites would contain less than one-tenth of the Ni abundances of the parental tholeiites. Chromium would also be substantially depleted. The observation that no significant difference in Ni and Cr abundances exists between corresponding tholeiitic and alkalic basalts is in conflict with the fractional crystallization model and demostrates rather clearly that these highly undersaturated alkalic magmas were not formed in this manner. On the other hand, the Ni and Cr abundances are generally consistent with formation of inclusion-bearing alkaline basaltic magmas by partial (batch) melting[1] of pyrolite in the mantle, leaving behind residual lherzolite (Ni 2000 to 4000 ppm, Cr 2000 to 4000 ppm).[2]

A high-alumina olivine basalt (0.2% nepheline) from the Juan de Fuca Ridge was found to contain 152 ppm Ni.[3] Such a rock could have formed by the fractional crystallization of about 12% aluminous orthopyroxene and 3% olivine from a parental high-alumina olivine tholeiite containing 250 ppm Ni. The indications are, therefore, that rocks transitional between olivine tholeiites and alkali olivine basalts containing up to about 2% nepheline may form either by fractional crystallization of parental olivine tholeiite or by direct partial melting in the mantle. However, the more undersaturated basanites and nephelinites of mantle origin probably form directly by small degrees of partial melting of pyrolite and not by fractional crystallization from olivine tholeiite.

The Mg/Mg + Fe ratios of ferromagnesian silicates and residual liquids provide another important petrogenetic indicator. This ratio does not change as rapidly as the Cr and Ni abundances during fractionation. However, this is compensated by the existence of accurate measurements of Mg/Mg + Fe ratios on coexisting mafic crystals and liquids, both in natural and in experimental high-pressure systems.[4,5] The crystallization of olivines, pyroxenes, and garnets from mafic magmas results in a marked increase of iron relative to magnesium in the residual liquids. When this effect is used in conjunction with experimentally measured Mg and Fe crystal-liquid distribution coefficients, strong constraints can be placed upon the extent and nature of fractionation undergone by a given magma.[6,7]

These data show that basaltic liquids in equilibrium with pyrolite (Mg/Mg + Fe = 0.89) or residual peridotites with Mg/Mg + Fe ~ 0.92 are relatively magnesian, with Mg/Mg + Fe ratios in the range 0.68 to 0.72.[6] On the other hand, basaltic liquids which segregate from their complementary peridotite and undergo closed-system fractionation by separation of olivine, pyroxenes, and garnet will display lower Mg/Mg + Fe ratios, the decrease in these ratios being a measure of the degree and nature of the subsequent closed-system fractionation.

From a study of 94 analyses of alkaline basaltic magmas containing dense high-pressure xenocrysts and xenoliths, Green (1971) found a strong concentration of Mg/Mg + Fe ratios in the range 0.63 to 0.73. Assuming the Mg/Mg + Fe ratio of the mantle to be around 0.9 (Chaps. 3 and 5), and using the measured partition coefficient,[4,5] this implied that many magmas with Mg/Mg + Fe ratios in the higher part of this range had been derived rather directly by partial melting of

[1]Section 4-5.
[2]Tables 3-7 and 3-8.
[3]Kay, Hubbard, and Gast (1970); rock C10D3 containing 18% Al_2O_3.
[4]Green and Ringwood (1967a).
[5]Roeder and Emslie (1970).
[6]Bultitude and Green (1971).
[7]Green (1971).

the mantle and had ascended to the surface without significant fractionation. Other magmas with somewhat lower Mg/Mg + Fe ratios probably suffered limited fractionation en route to the surface, e.g., the liquidus phases of Auckland Island xenocryst-bearing basalt (Sec. 4-4) had Mg/Mg + Fe ratios of 0.86 to 0.87. The liquid could not have been a direct partial melt from pyrolite (Mg/Mg + Fe = 0.89) but must have fractionated some way along the path towards alkali olivine basalt by separation of about 5 to 10% orthopyroxene, clinopyroxene, and minor olivine at depths of 50 to 55 km.

Kay, Hubbard, and Gast (1970) have made extensive use of trace element partition data combined with Mg/Mg + Fe ratios to elucidate the petrogenesis of basaltic rocks occurring along oceanic ridges. They demonstrated that much of the observed chemical variations had been caused by varying degrees of crystallization of olivine and plagioclase at shallow levels. Basalts with the highest Mg and Ni contents and the lowest Fe contents had clearly been subject to the least high-level olivine fractionation. Sharp cutoffs were observed at 9.5% MgO and 200 ppm Ni (upper limit) associated with 8% FeO. These criteria defined a high-alumina (15 to 18% Al_2O_3) olivine tholeiite which is extremely broadly distributed over the ocean floors.[1] Olivine on the liquidus of this tholeiite would have a Mg/Mg + Fe ratio of 0.90 and a nickel content of 2000 to 3500 ppm. These values are very close to those occurring in olivines from residual alpine peridotites and peridotite inclusions in basalts. These considerations demonstrate that the parental oceanic tholeiites are very close to being primary magmas directly derived from the mantle. The nickel contents of these magmas imply that not more than 10% olivine could have crystallized after they had segregated from residual peridotite.[2] Similar limitations follow from their Mg/Mg + Fe ratios (the latter restricting the amount of permissible pyroxene fractionation).

The high-pressure experimental investigations reviewed in Secs. 4-2 and 4-5 demonstrated that oceanic high-alumina tholeiites could be formed by partial melting and magma segregation within a narrow window between 5 and 10 kbars (depths of 15 to 35 km) and that a small amount (5 to 10%) of fractionation involving the separation of olivine and clinopyroxene may have occurred after segregation and during ascent of the magmas to the surface. Data on the distribution of Ni, Cr, Mg, and Fe in these magmas are fully concordant with this interpretation.

4-7 PETROGENETIC HYPOTHESIS OF M. J. O'HARA

O'Hara (1968a)[3] has proposed a different hypothesis of basalt petrogenesis than that considered above. He suggests that partial melting and magma segregation occur at higher pressures ($\geq$ 30 kbars) than those envisaged in the Green-

[1] Engel, Engel, and Havens (1965).
[2] Kay, Hubbard, and Gast (1970).
[3] See also O'Hara and Yoder (1967).

Ringwood models (mostly 5 to 25 kbars). The source rock is believed to be a garnet peridotite of the kind found in diamond pipes (rather similar in composition to pyrolite). Under the postulated higher pressure conditions, O'Hara points out that partial melts (5 to 30%) of the parental garnet peridotite will be picritic in nature, containing 30 to 40% normative olivine. Moreover, at pressures over 30 kbars, garnet will, in many cases, be an important phase remaining behind in the residual unmelted portion. Garnet harzburgites found as inclusions in diamond pipes are assumed to represent this material.

O'Hara emphasizes the role of closed-system fractionation controlled by the crystallization of olivine during ascent of the magma so that, by the crystallization of 20 to 40% olivine at lower pressures, parental picrites produce a spectrum of tholeiitic magmas ranging from olivine tholeiite to quartz tholeiite.

The origin of alkaline basaltic magmas is believed to be more complex according to O'Hara's model. The primary tholeiitic picrite magma fractionates towards nepheline-normative magmas by the precipitation of large proportions (40%) of olivine and aluminous pyroxenes at lower pressures (~15 kbars). These olivine and aluminous pyroxene precipitates are identified as the lherzolite xenoliths which are found so commonly in alkali basalts (Sec. 3-4). However, O'Hara recognizes that such a fractionation scheme will not produce the high concentration of incompatible elements (La, Ba, Zr, K, Ti, P, Sr, and Ta) which are characteristic of the alkaline basalt suite. To overcome this problem, he proposes that the parental picritic magma was subjected to extensive fractionation by the crystallization of garnet and clinopyroxene (eclogite) near the depth of origin (≥ 100 km) before ascending to higher levels where olivine fractionation predominates. Because eclogite has a roughly similar major element composition to that of the parental picrite, he argues that extraction of large quantities (~ 40%) of eclogite will not alter the major element composition substantially but will effectively result in an enrichment of incompatible elements in the residual magma.

According to Green and Ringwood (1967a), the depth of magma segregation for most mafic magmas occurs at depths less than 100 km, whereas O'Hara (1968a) appeals to depths of segregation greater than 100 km so that the parental magmas are picritic. Both of these positions are defensible since there is little direct evidence relating to the depth of magma segregation. Primary picritic magmas do indeed occur,[1,2,3] but they are relatively rare and there is no direct evidence that they are parental to most normal basaltic magmas. It seems likely that many mafic magmas have suffered significant amounts of fractionation in the mantle en route to the surface and there are innumerable well-documented examples of the occurrence of extensive fractionation of basaltic magmas within the crust by crystallization differentiation. What is debated is not the occurrence of these processes but the extent to which they can be considered to be the dom-

[1]Drever and Johnston (1958).
[2]Jamieson (1966).
[3]Viljoen and Viljoen (1969).

inant influence in producing the observed spectrum of basaltic magmas which have risen from the mantle.

O'Hara[1] illustrates his hypothesis with the aid of highly simplified projections into pseudoquaternary diagrams which attempt to reduce the ten-component complex natural system to an imaginary four-component system. At no stage does he attempt to explore the consequences of his hypothesis *quantitatively* by calculating the compositions of residual liquids, component by component, on the basis of his assumed initial composition, combined with the proportions of phases which are required to crystallize according to his model. When these exercises are carried out, and when the predictions of O'Hara's hypothesis with regard to relations between basaltic magmas and inferred refractory residues and cognate accumulates are compared with observational data (below), a number of serious deficiences is revealed.

1 In order to generate alkaline basaltic magmas, O'Hara[2] postulates a stage of partial melting (5 to 30%) of an assumed parental garnet peridotite to produce a primary magma of tholeiitic picrite composition which undergoes crystallization at depth, resulting in the separation of 40% garnet and clinopyroxene (eclogite). The residual liquid then rises to the surface accompanied by the crystallization of 40% olivine, generating an assumed alkali basaltic liquid representing about 20% of the original picritic magma.

We have already noted that the Ni, Cr, and Mg/Mg + Fe ratios of the most primitive alkali basalts (inclusion bearing) are similar to those which would be formed by batch melting of pyrolite using the observed distribution coefficients (Table 4-8). The Ni and Cr contents and Mg/Mg + Fe ratio of O'Hara's assumed parental garnet peridotite are similar to those of pyrolite. The picritic magma formed by partial melting of the parental garnet peridotite would contain about 300 to 400 ppm Ni and 400 to 500 ppm Cr (Tables 3-7 and 4-8). Fractional crystallization of 40% eclogite would reduce Ni to 200 ppm and Cr to less than 50 ppm (Fig. 4-9). Fractional crystallization of a further 40% of olivine would reduce Ni in the residual liquid to 2 ppm and would increase Cr to an upper limit of 80 ppm. This nickel abundance is more than 100 times smaller than the observed nickel content of inclusion-bearing alkali basalts, whilst the chromium content is deficient by about a factor of 4 or more. These large discrepancies demonstrate rather conclusively that inclusion-bearing alkali basalts have not formed in the manner suggested by O'Hara. There are other discrepancies, moreover, involving major elements such as Fe, Mg, Ca, and Al. Fractional crystallization on the scale envisaged causes a sharp decrease in the Mg/Mg + Fe ratio of the residual liquid, and the relevant distribution coefficients have been ac-

[1]O'Hara (1968a, p. 87).
[2]O'Hara (1968a, p. 118).

curately determined. Green (1971)[1] showed that crystallization of 40% olivine and 40% eclogite from a liquid formed by partial melting of pyrolite or garnet peridotite would produce a residual liquid having Mg/Mg + Fe ratios smaller than 0.55, compared to ratios of 0.63 to 0.73 which are commonly observed in alkali basaltic magmas of direct mantle derivation. Analogous calculations[1,2] show that extensive eclogite fractionation also leads to anomalies in residual liquids involving other major elements, particularly the Ca/Al ratio.

2 O'Hara has also suggested that oceanic tholeiites are derived from tholeiitic picrite magmas formed by partial melting at considerable depth.[3] The picrites are envisaged to have fractionated by crystallization of 30% or more of olivine en route to the surface. This would cause a 25-fold reduction in nickel content (Fig. 4-9), and the residual olivine tholeiites would contain <8 ppm Ni compared to the 200 ppm observed in the least fractionated oceanic high-alumina olivine tholeiites (Sec. 4-6, Fig. 4-10). The observed nickel content of these rocks shows that they have not lost more than 10% olivine after separating from residual peridotite in the mantle.[4]

3 O'Hara places great emphasis on the role of olivine crystallization during the ascent of picritic primary magma from depths greather than 100 km in producing the spectrum of fractionated olivine and quartz tholeiitic basalts which are observed in the crust.[5,6] This fractionation is believed to occur dominantly in the mantle rather than in the crust. The distribution of chromium and nickel in basalts argues against this view, however. Turekian (1963) demonstrated the strong covariance of Cr and Ni in a suite of over 100 basalts (mostly tholeiites) from all over the world (Fig. 4-11) in which absolute Cr and Ni abundances varied by a factor of 50. He pointed out that the fractional crystallization processes which had been responsible for the diversity of compositions were such that the Cr/Ni ratio was not markedly affected.

We have already noted that the crystallization of olivine alone causes a drastic decrease in nickel content of the residual magma but does not strongly affect the chromium abundance. It is not feasible to avoid this difficulty by postulating crystallization of spinel in addition to olivine since spinel does not occur on the liquidus of most basaltic magmas at pressures above 5 kbars. The observed covariance of nickel and chromium over a

[1]See also Bultitude and Green (1971).

[2]Green and Ringwood (1967a).

[3]O'Hara (1968a, 1968b).

[4]Kay, Hubbard, and Gast (1970).

[5]"Quartz-normative tholeiites have probably attained their present compositions by continuous olivine fractionation unaccompanied by extensive fractionation of other phases."—O'Hara (1968a, p. 96).

[6]"There is no evidence in favour of the assumption that these liquids [Hawaiian tholeiites] have fractionated in any other equilibrium than olivine-liquid prior to their arrival high in the volcanic superstructure."—O'Hara (1968a, p. 119).

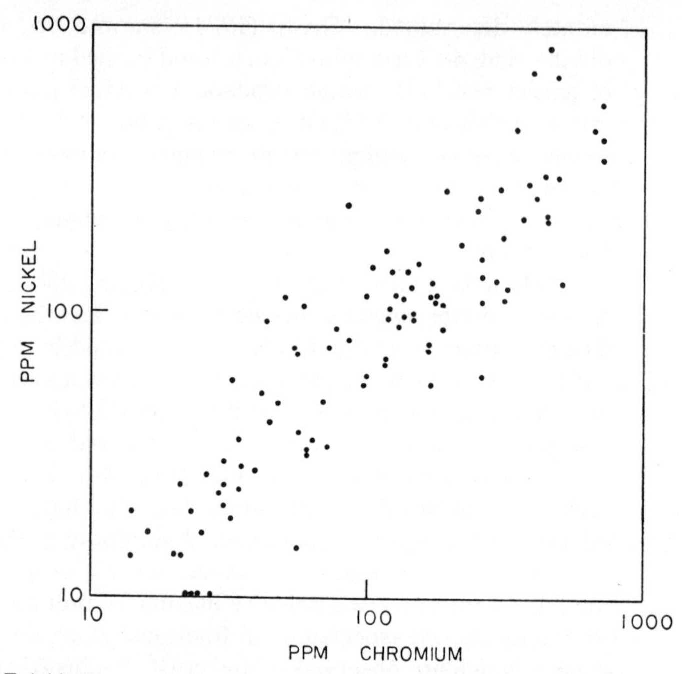

FIGURE 4-11
Covariance of chromium and nickel in a worldwide sampling of basaltic rocks. (*From Turekian, 1963, with permission.*)

wide concentration range is thus not readily explained by O'Hara's hypothesis of olivine fractionation dominantly within the mantle.

On the other hand, Turekian demonstrated that, during fractional crystallization of basaltic magmas in the crust, dominated by crystallization of pyroxenes and plagioclase, the Cr/Ni ratio was not significantly altered. This suggests that the large range of Ni and Cr concentrations observed in basaltic rocks, combined with constant Cr/Ni ratios, has been caused dominantly by high-level crustal fractionation processes rather than by olivine fractionation in the mantle, although the initial Cr/Ni ratio of the parental magmas was probably established directly by partial melting.

These arguments are not aimed at minimizing the role of olivine fractionation under low-pressure *crustal* conditions. This has been well documented in numerous instances—e.g., Hawaii (Sec. 4-1). However, under such conditions, olivine may be accompanied by minor Cr spinel and Cr-rich clinopyroxene, which tend to maintain the constant Cr/Ni ratio. On the other hand, in the mantle, the phases crystallizing from tholeiitic picrite magmas over a wide range of pressures are olivine and Cr-poor orthopyroxene without spinel. Crystallization of these phases on the large scale envisaged by O'Hara can lead only to great increases in Cr/Ni ratios in residual magmas with increased degrees of fractionation.

4 O'Hara (1968a) regards the lherzolite nodules occurring in alkali basalt as cognate accumulates precipitated during ascent to the surface and as being responsible for the nepheline-normative nature of the residual liquid. However, isotopic, compositional, and mineralogical data discussed in Sec. 3-4 conclusively demonstrated that most lherzolite nodules could *not* have crystallized from their associated alkaline magmas. Moreover, following from the discussion in Sec. 3-4 and 4-6, the high and relatively uniform Ni and Cr contents and Mg/Mg + Fe ratios of these nodules show that they represent residual refractory material remaining behind after partial melting processes. If they represented accumulates from basaltic magmas, the laws of fractional crystallization in conjunction with observed distribution coefficients dictate that they would display a wide range of Ni and Cr contents, and Mg/Mg + Fe ratios, and that on the average these would be much lower than observed [equation (4), page 161.]. Indeed, the high-pressure xenocrysts of aluminous pyroxenes have previously been demonstrated to represent genuine cognate precipitates, and as expected, they display a range of compositions with substantially lower Mg/Mg + Fe ratios than lherzolite nodules. Moreover, among the typical unstrained glassy cognate xenocrysts, aluminous pyroxenes are much more common than olivines, which is contrary to expectations from O'Hara's hypothesis.[1]

Finally, although eclogite fractionation plays a comparable role to olivine fractionation in O'Hara's scheme, it is notable that the proportion of eclogite xenoliths to peridotite xenoliths in alkali basalts is extremely small, and indeed, most of the "eclogites" appear to be garnet pyroxenites derived by cooling of high-temperature olivine + aluminous pyroxene assemblages.[2] Moreover, the extreme rarity (Table 4-3) of olivine eclogites among diamond pipe xenoliths is difficult to explain.

It must be concluded in the light of the preceding discussion that O'Hara's scheme of basalt petrogenesis is inadequate, failing to meet the obvious quantitative tests which can be applied and inconsistent with an array of qualitative observational data.

[1]Binns, Duggan, and Wilkinson (1970).
[2]Green (1966).

REFERENCES

BINNS, R. A. (1969). High pressure megacrysts in basanitic lavas near Armidale, New South Wales. *Am. J. Sci.* **267A,** 33–49.

———, M. DUGGAN, and J. F. G. WILKINSON (1970). High pressure megacrysts in alkaline lavas from north-eastern New South Wales. *Am. J. Sci.* **269**, 132–168.

BIRCH, F. (1952). Elasticity and constitution of the earth's interior. *J. Geophys. Res.* **57,** 227–286.

BOWEN, N. L. (1928). "*The Evolution of the Igneous Rocks.*" Princeton Univ. Press. 332 pp.

BREY, G., and D. H. GREEN (1974). The role of carbon dioxide in the genesis of olivine melililite. (In press.)

BULTITUDE, R. J., and D. H. GREEN (1968). Experimental study at high pressures on the origin of olivine nephelinite and olivine melilite nephelinite magmas. *Earth Planet. Sci. Letters* **3,** 325–337.

——— and ——— (1971). Experimental study of crystal liquid relationships at high pressures in olivine nephelinite and basanite compositions. *J. Petrol.* **12,** 121–147.

CLARK, S. P. (ed.) (1966). "*Handbook of Physical Constants.*" *Geol. Soc. Am. Mem.* **97.**

———and A. E. RINGWOOD (1964). Density distribution and constitution of the mantle. *Rev. Geophys.* **2,** 35–88.

DAVIS, B. T. C. (1964). The system diopside-forsterite-pyrope at 40 kilobars. *Carnegie Inst. Washington Yearbook* **63,** 165–171

DREVER, H. I., and R. JOHNSTON (1958). The petrology of picritic rocks in minor intrusions—a Hebridaen group. *Trans. Roy. Soc. Edinburgh* **63,** 459–499.

EGGLER, D. H. (1973). Role of CO_2 in melting processes in the mantle. *Carnegie Inst. Washington Yearbook* **72,** 457–467.

ENGEL, A. E., C. G. ENGEL, and R. G. HAVENS (1965). Chemical characteristics of oceanic basalts and the upper mantle. *Bull. Geol. Soc. Am.* **76,** 719–734.

FUNKHOUSER, J. G., D. E. FISHER, and E. BONATTI (1968). Excess argon in deep sea rocks. *Earth planet. Sci. Letters* **5,** 95–100.

GAST, P. W. (1968). Trace element fractionation and the origin of tholeiitic and alkaline magma types. *Geochim. Cosmochim. Acta* **32,** 1057–1086.

GREEN, D. H. (1966). The origin of the "eclogites" from Salt Lake Crater, Hawaii. *Earth Planet. Sci. Letters* **1,** 414–420.

——— (1969). The origin of basaltic and nephelinitic magmas in the earth's mantle. *Tectonophysics* **7,** 409–422.

——— (1970a). A review of experimental evidence on the origin of basaltic and nephelinitic magmas. *Phys. Earth Planet. Interiors* **3,** 221–235.

——— (1970b). The origin of basaltic and nephelinitic magmas. *Trans. Leicester Lit. Philos. Soc.* **64**, 28–54.

——— (1970c). Peridotite-gabbro complexes as keys to petrology of mid-oceanic ridges: A discussion. *Bull. Geol. Soc. Am.* **81,** 2161–2166.

——— (1971). Composition of basaltic magmas as indicators of conditions of origin: application to oceanic volcanism. *Phil. Trans. Roy. Soc. London* **A268,** 707–725.

——— (1973a). Conditions of melting of basanite magma from garnet peridotite. *Earth Planet. Sci. Letters* **17,** 456–465.

——— (1973b). Experimental melting studies on model upper mantle compositions at high pressure under both water-saturated and water-unsaturated conditions. *Earth Planet. Sci. Letters* **19,** 37–53.

——— and W. HIBBERSON (1970). Experimental duplication of conditions of precipitation of high pressure phenocrysts in a basaltic magma. *Phys. Earth Planet. Interiors* **3,** 247–254.
——— and A. E. RINGWOOD (1964). Fractionation of basalt magmas at high pressures. *Nature* **201,** 1276–1279.
——— and ——— (1967a). The genesis of basaltic magmas. *Contr. Mineral. Petrol.* **15,** 103–190.
——— and ——— (1967b). The stability fields of aluminous pyroxene peridotite and garnet peridotite and their relevance in upper mantle structure. *Earth Planet. Sci. Letters* **3,** 151–160.
——— and ——— (1970). Mineralogy of peridotitic compositions under upper mantle conditions. *Phys. Earth Planet. Interiors* **3,** 359–371.
GREEN, T. H., D. H. GREEN, and A. E. RINGWOOD (1967). The origin of high-alumina basalts and their relationships to quartz tholeiites and alkali basalts. *Earth Planet. Sci. Letters* **2,** 41–52.
HAKLI, T., and T. L. WRIGHT (1967). The fractionation of nickel between olivine and augite as a geothermometer. *Geochim. Cosmochim. Acta* **31,** 877–884.
HARRIS, P. G. (1957). Zone refining and the origin of potassic basalts. *Geochim. Cosmochim. Acta* **12,** 195–208.
——— (1969). Basalt type and rift valley tectonism. *Tectonophysics* **8,** 427–436.
——— and A. K. MIDDLEMOST (1970). The evolution of kimberlites. *Lithos* **3,** 79–90.
HEDGE, C. S. (1971). Nickel in high alumina basalts. *Geochim. Cosmochim. Acta* **35,** 522–524.
HESS, H. H. (1960). Stillwater igneous complex, Montana, a quantitative mineralogical study. *Geol. Soc. Am. Mem.* **80,** 230 pp.
HOLMES, A., and H. F. HARWOOD (1932). Petrology of the volcanic fields east and southeast of Ruwenzori, Uganda. *Quart. J. Geol. Soc. London* **88,** 370–342.
HUBBARD, N. J. (1969). A chemical comparison of oceanic ridge, Hawaiian tholeiitic and Hawaiian alkalic basalts. *Earth Planet. Sci. Letters* **5,** 346–352.
IRVINE, T. N., and C. H. SMITH (1967). The ultramafic rocks of the Muskox Intrusion, Northwest Territories, Canada. In: P. J. Wyllie (ed.), "*Ultramafic and Related Rocks*," pp. 38–49. Wiley, New York.
IRVING, A., and D. H. GREEN (1972). (In press.)
ITO, K., and G. C. KENNEDY (1967). Melting and phase relations in a natural peridotite to 40 kilobars. *Am. J. Sci.* **265,** 519–538.
——— and ——— (1968). Melting and phase relations in the plane tholeiite-lherzolite-nepheline basanite to 40 kilobars with geological implications. *Contr. Mineral. Petrol.* **19,** 177–211.
JACKSON, E. D., and T. L. WRIGHT (1970). Xenoliths in the Honolulu volcanic series, Hawaii. *J. Petrol.* **11,** 405–430.
JAMIESON, B. G. (1966). Evidence on the evolution of basaltic magma at elevated pressures. *Nature* **212,** 243–246.
KAY, R. W. (1971). The rare earth geochemistry of alkaline basaltic volcanics. Ph.D. Thesis, Columbia University.
———, N. J. HUBBARD, and P. W. GAST (1970). Chemical characteristics and origin of oceanic ridge volcanic rocks. *J. Geophys. Res.* **75,** 1585–1613.
KUNO, H., (1959). Origin of Cenozoic petrographic provinces of Japan and surrounding areas. *Bull Volc.* Ser 2, **20,** 37–76.
——— (1964). Aluminium augite and bronzite in alkali olivine basalt from Taka-sima,

north Kyusyu, Japan. In: "*Advancing Frontiers in Geology and Geophysics*," volume dedicated to Dr. Krishnan (India), pp. 205–220. Osmania Univ. Press, Hyderabad.

KUSHIRO, I. (1968). Compositions of magmas formed by partial zone melting of the earth's upper mantle. *J. Geophys. Res.* **73,** 619–634.

——— (1969). Discussion of paper: the origin of basaltic and nephelinitic magmas in the Earth's mantle. *Tectonophysics* **7,** 427–436.

——— and H. KUNO (1963). Origin of primary basalt magmas and classification of basaltic rocks. *J. Petrol.* **4,** 75–89.

———, Y. SYONO, and S. AKIMOTO (1968). Melting of a peridotite nodule at high pressures and high water pressures. *J. Geophys. Res.* **73,** 6023–6029.

LARSEN, E. S. (1940). Petrographic province of Central Montana. *Bull. Geol. Soc. Am.* **51,** 887–948.

MAC DONALD, G. A., and T. KATSURA (1961). Variations in the lava of the 1959 eruption in Kilauea Iki. *Pacific Sci.* **15,** 358–369.

MIYASHIRO, A., F. SHIDO, and M. EWING (1969). Diversity and origin of abyssal tholeiite from the Mid-Atlantic Ridge near 24° and 30° north latitude. *Contr. Mineral. Petrol.* **23,** 38–52.

MOORE, J. G. (1965). Petrology of deep sea basalt near Hawaii. *Am. J. Sci.* **263,** 40–52.

O'HARA, M. J. (1963). Melting of garnet peridotite and eclogite at 30 kilobars. *Carnegie Inst. Washington Yearbook* **62,** 71–77.

——— (1965). Primary magmas and the origin of basalts. *Scot. J. Geol.* **1,** 19–40.

——— (1968a). The bearing of phase equilibria studies on synthetic and natural systems on the origin and evolution of basic and ultrabasic rocks. *Earth Sci. Rev.* **4,** 69–133.

——— (1968b). Are ocean floor basalts primary magma? *Nature* **220,** 683–686.

——— (1970). Upper mantle composition inferred from laboratory experiments and observations of volcanic products. *Phys. Earth Planet. Interiors* **3,** 236–245.

——— and H. S. YODER (1967). Formation and fractionation of basic magmas at high pressures. *Scot. J. Geol.* **3,** 67–117.

POWERS, H. A. (1935). Differentiation of Hawaiian lavas. *Am. J. Sci.* **264,** 753–809.

——— (1955). Composition and origin of basaltc magma of the Hawaiian Islands. *Geochim. Cosmochim. Acta* **7,** 77–107.

REAY, A., and P. G. HARRIS (1964). The partial fusion of peridotite. *Bull. Volc.* **27,** 115–127.

RINGWOOD, A. E. (1966). The chemical composition and origin of the earth. In: P. M. Hurley (ed.), "*Advances in Earth Science*," pp. 287–356. M. I. T. Press, Cambridge, Mass.

——— (1969). Composition and evolution of the upper mantle. In: "*The Earth's Crust and Upper Mantle*," pp. 1–17. *Am. Geophys. Union, Geophys. Monograph* **13.**

ROEDER, P. L., and R. F. EMSLIE (1970). Olivine-liquid equilibrium. *Contrib. Mineral. Petrol.* **29,** 275–289.

SHEPARD, S. M., and S. EPSTEIN (1970). D/H and $^{18}O/^{16}O$ ratios of possible mantle or lower crustal origin. *Earth Planet. Sci. Letters* **9,** 232–246.

TAYLOR, S. R., M. KAYE, A. J. WHITE, A. R. DUNCAN, and A. EWART (1969). Genetic significance of Co, Cr, Ni, Sc and V content of andesites. *Geochim. Cosmochim. Acta* **33,** 275–286.

——— A. J. WHITE, A. EWART, and A. R. DUNCAN (1971). Nickel in high alumina basalts: A reply. *Geochim. Cosmochim. Acta* **35,** 525–528.

TILLEY, C. E. (1950). Some aspects of magmatic evolution. *Quart. J. Geol. Soc. London* **106,** 37–61.

———, H. S. YODER, and J. F. SCHAIRER (1963). Melting relations of basalts. *Carnegie Inst. Washington Yearbook* **62,** 77–84.

———, ———, and ——— (1964). New relations on melting of basalts. *Carnegie Inst. Washington Yearbook* **63,** 92–97.

———, ———, and ——— (1965). Melting relations of volcanic tholeiite and alkali rock series. *Carnegie Inst. Washington Yearbook* **64,** 69–82.

TUREKIAN, K. (1963). The chromium and nickel distribution in basaltic rocks and eclogite. *Geochim. Cosmochim. Acta* **27,** 835–846.

TURNER, F. J., and J. VERHOOGEN (1951). "*Igneous and Metamorphic Petrology.*" McGraw-Hill, New York. 602 pp.

VILJOEN, M. J., and R. P. VILJOEN (1969). Evidence for the existence of a mobile extrusive peridotitic magma from the Komati Formation of the Onverwacht Group. *Geol. Soc. S. Africa Spec. Pub.* **2,** 87–112.

WAGER, L. R., and R. L. MITCHELL (1951). The distribution of trace elements during strong fractionations of basic magma—a further study of Skaergaard Intrusion, East Greenland. *Geochim. Cosmochim. Acta* **1,** 129–208.

WILSHIRE, H. G., and R. A. BINNS (1961). Basic and ultrabasic xenoliths from volcanic rocks of northeastern New South Wales. *J. Petrol.* **2,** 185–208.

YODER, H. S., and C. E. TILLEY (1962). Origin of basalt magmas: an experimental study of natural and synthetic rock systems. *J. Petrol.* **3,** 342–532.

5

THE PYROLITE MODEL

5-1 INTRODUCTION

In Chap. 3 we discussed evidence leading to the conclusions that the upper mantle is dominantly of ultramafic composition and that many alpine peridotites, together with certain classes of xenoliths in kimberlites and alkali basalts, represent actual samples of the upper mantle. On the other hand, in Chap. 4 we emphasized another key property—the upper mantle must be capable of yielding the range of basaltic magmas which are erupted at the earth's surface.

For some time, it appeared that the composition of the upper mantle as inferred from observed ultramafic rocks did not provide a ready explanation of the occurrence of basaltic magmas. The relative abundances of many minor and trace elements in most alpine ultramafics were found to be too low and variable for these rocks to serve as sources of basaltic magmas.[e.g.,1,2,3] Some petrologists[4] sought to avoid this difficulty by appealing to selected and rarer classes of natural

[1]Lovering (1958).
[2]Tilton and Reed (1963).
[3]Hamilton and Mountjoy (1965).
[4]E.g., Kushiro and Kuno (1963).

peridotites containing relatively high abundances of Na, Al, and Ca as representing the parental mantle. Although a step in the right direction, this postulate did not prove successful. Although some of these natural peridotites appeared capable of providing the major element compositions of basalts by suitable partial melting processes, serious discrepencies arose in the cases of many minor and trace elements and often isotopic compositions. For example, high-temperature peridotites and lherzolite xenoliths containing 3 to 4% Al_2O_3 and CaO were sometimes found to contain extremely low abundances of potassium and related incompatible elements and to be characterized by highly fractionated rare earth abundance patterns[1,2,3,4,5,6] (Fig. 5-2). Some workers[7,8] favoured selected garnet lherzolite nodules from kimberlites as representing primitive mantle material. It has not been possible to test this proposal adequately since extensive contamination from the enclosing kimberlitic magma has been shown to have greatly modified the minor element compositions of kimberlite xenoliths, which are rarely fresh.[9,10,11,12] It was shown in Sec. 3-4 that the range of major element compositions displayed by peridotite xenoliths in alkali basalts and kimberlites overlap extensively and perhaps completely. The two classes of xenoliths essentially represent similar ranges of mantle compositions which have crystallized in different *P-T* environments. It appears likely that the minor element characteristics of primary garnet peridotite xenoliths, prior to contamination, were generally similar to those in peridotite xenolith from alkali basalts.

The accumulation of geochemical and petrological evidence during the last 10 years has led to widespread realization that most alpine ultramafic rocks and ultramafic xenoliths do not possess a parental relationship to basaltic magmas. Their properties and occurrences are consistent instead with the hypothesis that they represent differentiated refractory material which has remained after a basaltic component has been extracted—i.e., they mostly possess a *complementary relationship* to basaltic magmas. This hypothesis is supported by a broad array of evidence—specifically, the observed genetic relationships between alpine ultramafics and gabbroic rocks in ophiolite complexes, as discussed in Chap. 3, and the experimental phase equilibria and geochemical relationships considered in Chap. 4.

The widespread sampling of the uppermost mantle by kimberlite pipes, alkali

[1]E.g., J. Morgan (unpublished) found by neutron activation methods that the Tinquillo peridotite contained only 6 ppm potassium.
[2]Green, Morgan, and Heier (1968).
[3]Stueber and Murthy (1966).
[4]Frey (1969, 1970a).
[5]Frey, Haskin, and Haskin (1971).
[6]White (1966).
[7]Ito and Kennedy (1967).
[8]O'Hara (1968).
[9]Heier (1963).
[10]Berg (1968).
[11]Allsop, Nicolaysen, and Hahn-Weinheimer (1968).
[12]Manton and Tatsumoto (1971).

basalts, and direct tectonic emplacement of upper mantle rocks into the crustal environment has led to the conclusion that the uppermost mantle is dominantly composed of refractory, residual peridotites, similar in composition to the observed alpine ultramafics and ultramafic xenoliths. The primitive mantle from which the basaltic component has not yet been extracted must lie generally below this differentiated residual material—thus, a chemically zoned model of the upper mantle is indicated, as shown in Fig. 5-1. Estimates of "upper mantle composition" based directly upon observed compositions of alpine ultramafics and xenoliths [e.g.,1,2,3] thus refer to the layer of residual refractory mantle and not to the deeper parental mantle.

A zonal structure of this kind for the upper mantle was implicit in the early work of Bowen (1928, pp. 315–320) on basalt genesis and of Rubey (1951) on mantle differentiation. These workers envisaged that basalts and crustal rocks were produced by partial melting of primitive (meteoritic) mantle material, leaving residual or "barren" peridotite. A more formal statement of this hypothesis was developed by Ringwood (1958) in which a mantle structure similar to Fig. 5-1 was hypothesized. According to this model, the primitive meteoritic source material would crystallize to a garnet peridotite assemblage under the relevant *P-T* conditions.

Without further detailed justification, identification of the primitive undepleted upper mantle composition with a composition derived from chondritic meteorites is arbitrary. Clearly, it would be desirable to obtain estimates of this primitive composition directly from terrestrial evidence. Recognition of the essential complementary relationship between basaltic magmas and many ultramafics combined with the inference that the latter represent residual refractory residues provide a basis for estimates of the primitive composition. This must lie somewhere between those of basalts on the one hand and peridotites on the other. Studies of the physical processes of basalt genesis and magma segregation (Chap. 4) imply that basalts represent partial melts of this primitive material and that the proportion of basalt was smaller, often much smaller, than that of peridotite.

Considerations of this nature led to the proposal[4,5,6,7] that the primitive composition lay somewhere between alpine peridotite/basalt ratios of 4:1 and 1:1. The most probable ratio was considered to be in the vicinity of 3:1. A series of specific models was calculated as additional constraints became apparent, although the differences between the models were rather small. The latest was based upon a mixture of 3 parts alpine peridotite (79% Ol, 20% Opx, 1% Sp) and 1

[1] Hess (1964).
[2] Harris, Reay, and White (1967).
[3] White (1966).
[4] Ringwood (1962a,b).
[5] Green and Ringwood (1963).
[6] Clark and Ringwood (1964).
[7] Ringwood (1966a).

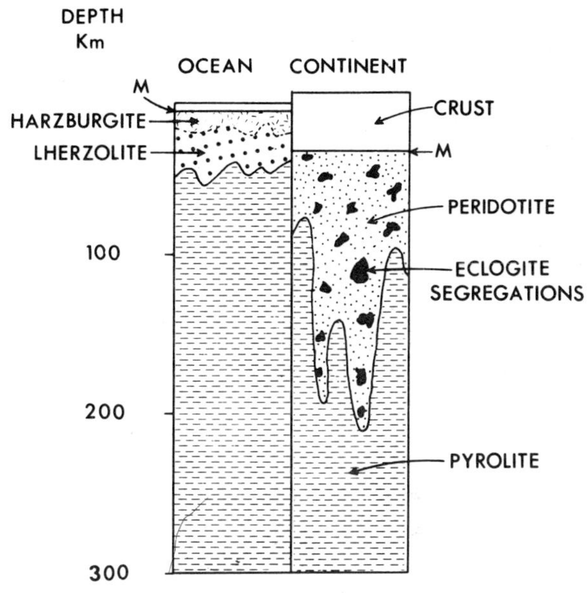

FIGURE 5-1
Chemically zoned model for the upper mantle. (*After Ringwood, 1966a.*)

part Hawaiian tholeiite (Table 5-2). There is little doubt that this model could be further improved—e.g., the Hawaiian tholeiite chosen was abnormally rich in TiO_2, and the abundance of this component in the model composition is probably too high.[1] It would probably have been preferable to base the model upon an average oceanic tholeiite composition.

This discussion simply emphasizes the flexibility of the model and the view that the most appropriate parental composition must be approached via a series of successive approximations as additional information becomes available. It is also probable that the parental mantle is not uniform in composition so that the quest is essentially for an acceptable composition range, which provides the most broadly self-consistent explanation of the petrological, mineralogical, and geochemical relationships between basalts and ultramafics. Ringwood (1962a) coined the term *pyrolite* (pyroxene + olivine ± pyrope garnet rock) for this inferred parental composition.

Some petrologists[2] have questioned the need for a new name for this parental mantle composition, pointing out that, since it contains more than 50% olivine, it falls within the petrological classification of the peridotite family and can be defined as a particular variety of peridotite according to its specific mineralogy. I believe to the contrary that this exclusively petrological approach is altogether too

[1]Kuno and Aoki (1970).
[2]E.g., Wyllie (1970).

narrow and misses the fundamental point that the definition of pyrolite is based upon *chemical composition* and not upon mineralogy. The composition of pyrolite is *defined* by the property that it is required to produce a basaltic magma upon partial melting, leaving behind a residual refractory peridotite. As we have already observed, naturally occurring ultramafic rocks appear to be generally residual and fractionated in nature and rarely, if ever, possess the capacity to produce basaltic magmas of observed compositions on partial melting.

The widespread occurrences of basaltic rocks which have been erupted throughout geological time both in continental and oceanic regions imply that pyrolite must be a major component of the upper mantle. The extreme rarity or absence of pyrolite at the earth's surface is apparently a consequence of the mechanisms by which ultramafic rocks are transported to the surface. These mechanisms are commonly associated with some form of magmatism accompanied by intense tectonic deformation. It is not surprising that these conditions should result in varying degrees of differentiation, particularly of incompatible elements such as K, U, Ba, Ti, and P which appear to become fractionated very readily—perhaps owing to the formation of a low-melting liquid.[1]

5-2 COMPOSITION OF PYROLITE

Early estimates[2,3] of the pyrolite model composition were based upon somewhat subjective judgements of the degree of partial melting involved when tholeiites separated from refractory residua. The 3:1 peridotite/basalt ratio was based primarily on this consideration; however, the flexibility of the model and the uncertainty involved in this estimate were emphasized.[1] An upper limit of 4:1 appeared likely from cosmochemical and geochemical considerations, which suggested that the abundances of Ca and Al relative to Si and Mg were unlikely to be smaller in the earth's mantle than in the sun or in chondritic meteorites. A lower limit for the ratio of 1:1 appeared necessary if basalts were to be produced by physically reasonable processes of *partial* melting of pyrolite. Since these early crude statements of the model, additional sources of evidence have become available and new approaches to the problem have been made. These are showing encouraging signs of convergence and are summarized below.

Derivation of Pyrolite Composition from Alpine Ultramafics

In Chap. 4 we considered how the observed range of basaltic magmas, olivine tholeiite-alkali basalt-basanite-nephelinite, might form by varying degrees of partial melting of pyrolite—ranging from about 25 down to 1% or less. Accordingly, the complementary ultramafic residues should display a range of compositions

[1]Ringwood (1966a).
[2]Ringwood (1962a,b).
[3]Green and Ringwood (1963).

corresponding to the differing degrees of depletion of low-melting components. Clearly, ultramafics which are the residues from systems characterized by the smallest degrees of partial melting would approach most closely the pyrolite composition.

Alpine ultramafics are observed to display a continuum of compositions ranging from pure dunite, containing virtually no CaO, Al_2O_3, and Na_2O, through to lherzolite, containing about 4% Al_2O_3, 3% CaO, and 0.4% Na_2O.[1,2,3] Alpine-type ultramafics containing higher bulk concentrations of these low-melting components are rare, and it appears that rocks approaching the above composition represent residues from the smallest degree of partial melting of pyrolite. Green[3] has emphasized the significance of a particular class of high-temperature peridotites in this respect. These contain 3 to 4% Al_2O_3 and CaO (Table 3-3) and appear to represent the closest approach to unfractionated mantle material. Nevertheless, these rocks display strong depletions and internal fractionation among incompatible elements—e.g., the light rare earths, potassium, uranium—and have clearly lost a low-melting-point component.[4] This component was presumably a basanite or nephelinite produced by a very small degree of partial melting.[5]

Rare earth abundances (normalized to chondritic) exhibited by a typical high-temperature peridotite are shown in Fig. 5-2. The extreme internal fractionation and depletion of light rare earths are to be remarked. Also shown are the rare earth abundances of a typical nephelinite. The complementary nature of these patterns is immediately obvious. Cosmochemical considerations combined with the observed near-chondritic rare earth patterns of oceanic tholeiites strongly indicate that pyrolite possesses a near-chondritic rare earth relative-abundance pattern. Production of this pattern from the complementary patterns of the peridotite and nephelinite (Fig. 5-2) requires the addition of about 1% nephelinite to the peridotite. This is consistent with the experimental evidence that nephelinites are produced by a very small degree of partial melting of pyrolite.[5]

These considerations make it possible to calculate an ideal pyrolite composition by combining 99% of the peridotite with 1% of the nephelinite. The calculated composition is given in Table 5-2, column 2.

Derivation from Ultramafic Xenoliths

Studies of the compositions, mineralogy, and petrology of peridotitic xenoliths in alkali basalts demonstrated the occurrence of two distinct populations, one interpreted as refractory material remaining from partial melting processes involved in basalt production and the other interpreted as accumulates from basaltic

[1]Green and Ringwood (1963).
[2]Ringwood (1966a).
[3]Green (1964, 1967, 1970).
[4]Ref. Sec. 5-1.
[5]Chapter 4.

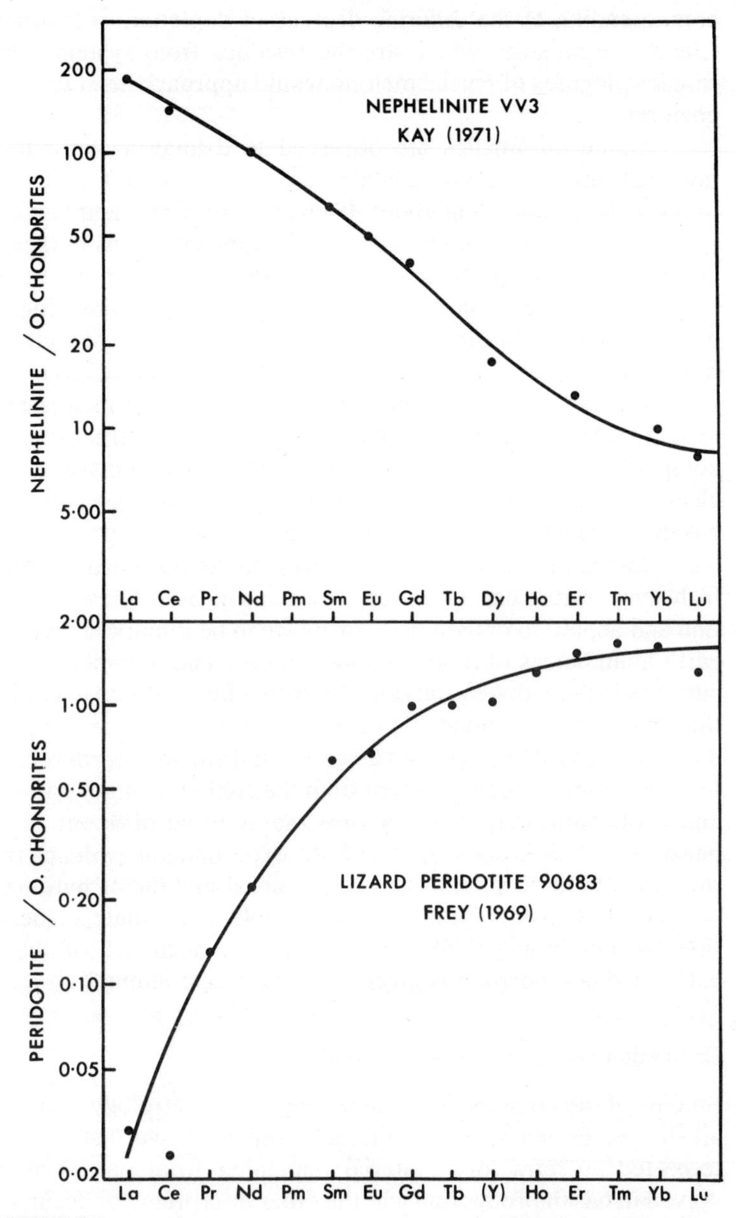

FIGURE 5-2
Rare earth abundances in the Lizard high-temperature peridotite (*after Frey, 1970a*) and in a Cape Verdes nephelinite (*VV3-1, after Kay, 1971*). Rare earth abundances have been divided, element by element, by the corresponding abundances in *ordinary* chondrites.

magmas.[1] Compositions of the first group reveal a continuum from pure high-Mg dunite to lherzolites containing up to 3 to 4% Al_2O_3 and CaO and 0.4% Na_2O.[e.g.,2,3] This series forms a close parallel to the alpine ultramafic series discussed previously and is interpreted similarly, the nodules with the highest contents of low-melting components being regarded as representing the closest approach to pyrolite.[4]

Trace element abundances in the nodules reveal complex patterns which are not interpreted as simply as those of some of the high-temperature peridotites (Fig. 5-2). A serious problem is the possibility of contamination of xenoliths by components derived from the enclosing basanitic and nephelinitic magmas which may contain up to 1000 times higher concentrations of some incompatible elements (e.g., light rare earths, K, Ba, and U) than the xenoliths. Clearly, a minute amount of contamination would be sufficient to distort the trace element patterns. It is difficult to see how small xenoliths surrounded by mobile and water-rich alkali basaltic magma could escape some degree of contamination. For this reason, the trace element abundances in relatively unserpentinized, massive high-temperature peridotites, not obviously associated with alkali basalt magmas, are likely to be more significant since they would be less liable to contamination.[5] Studies of the compositions of peridotite xenoliths (mostly garnetiferous) from kimberlitic diamond pipes reveal a closely analogous situation to that discussed above. A continuum of compositions ranging from pure dunite to garnet peridotites containing 3.5% CaO and Al_2O_3 and 0.4% Na_2O is found,[6] but garnet peridotites with higher levels of these components rarely, if ever, occur. This is confirmed by extensive mineralogic and petrographic investigations.[7,8]

Detailed studies[9,10,11] have shown that the xenoliths from kimberlites have usually been contaminated by trace components introduced from the surrounding kimberlites so that the proportion and composition of the complementary differentiate needed for calculation of the model pyrolite composition are not readily estimated. Nevertheless, we can conclude that the nodules containing the maximum amounts of CaO, Al_2O_3, and Na_2O have suffered the smallest degree of partial melting and are closest to the ideal pyrolite composition. It is most significant that this least fractionated composition is essentially identical in populations of garnet

[1]Reviewed in Secs. 3-4 and 4-4.
[2]Harris, Reay, and White (1967).
[3]Carter (1970).
[4]Green and Ringwood (1963).
[5]Nevertheless, contamination of massive peridotites appears to have occurred in some instances—e.g., St. Paul Rocks (Frey, 1970b). In this case, contamination might have been related to the occurrence of associated alkali basaltic magma now represented by hornblende mylonites.
[6]Table 3-7.
[7]Rickwood, Mathias, and Siebert (1968).
[8]Mathias, Siebert, and Rickwood (1970).
[9]Heier (1963).
[10]Berg (1968).
[11]Allsop, Nicolaysen, and Hahn-Weinheimer (1968).

peridotites from diamond pipes, lherzolites from alkali basalts, and high-temperature peridotites.[1,2]

An alternative approach towards estimating the composition of pyrolite from peridotite nodules from alkali basalt has been made by Carter (1970). This is based upon recognition of two populations of nodules, each with distinctive overall petrologic features. The population with olivine compositions between Fa_7 and Fa_{11} has the petrologic characteristics of residua, whilst that with olivine compositions between Fa_{14} and Fa_{30} has the characteristics of accumulates and solidified liquids. Nodules with olivine compositions in the range Fa_{12} and Fa_{14} were found to be very rare. Carter considers that this minimum represents the olivine composition of the primary mantle which was partially melted to form the Mg-rich residual population on the one hand and the Fe-rich cumulate and liquid-derived population on the other. Extensive mineralogic and chemical data on the residual population of nodules were obtained as a function of fayalite content of olivine. By extrapolating these data into the Fa_{12} to Fa_{14} gap, the composition of primary undepleted mantle was estimated (Table 5-2). This composition is seen to be somewhat higher in Al_2O_3 than other estimates of the pyrolite composition and is also depleted in K_2O. The latter is presumably because the occurrence of accessory phases (e.g., phlogopite) in the nodules was not considered. Otherwise the overall agreement is quite good.

Bulk Compositions of Ophiolite Complexes

Ophiolite complexes provide some of the most direct evidence of the complementary relationship between basaltic magmas and residual unmelted peridotites. Moores (1970) has estimated the bulk composition of the unusually intact and well-exposed Vourinos ophiolite complex which was described in Sec. 3-3. This composition is given in Table 5-2 and is seen to agree well with other estimates of the pyrolite composition.

Naturally Occurring Ultramafic Liquids

An alternative approach to seeking the least fractionated ultramafic residual material as an approximation to the pyrolite composition is to search for natural mafic and ultramafic liquids which represent the greatest degree of partial melting of pyrolite source material. It is hoped that these two approaches should converge.

As discussed in Chap. 4, the vast majority of primary basaltic magmas formed by partial melting of pyrolite contained up to about 30% normative olivine. However, primary picritic magmas containing more than 30% olivine have been recorded. Some of the best examples of these are the picritic sills and dykes described by Drever and Johnston (1958, 1967).[3] The border facies of some of

[1] Green and Ringwood (1963).
[2] Tables 3-3, 3-7, and 3-8.
[3] See also Gass (1958).

these sills contain up to 40% small skeletal olivines (Fo_{90}), and textural evid showed that the olivine had crystallized very rapidly from a completely mo picritic magma. The overall compositions of these sills would correspond approximately to a 1:1 mixture of basalt and residual peridotite, representing an upper limit to the basalt/peridotite ratio of pyrolite.

Important evidence of the existence of natural ultramafic magmas approaching the pyrolite composition has been presented by Viljoen and Viljoen (1969a, b). These authors described the occurrence of a series of thin peridotitic lava flows from the Archaean Komati formation of the Barberton region of South Africa. Field and petrologic evidence (e.g., occurrence of pillow structures, fine crystal size and habit, and stratigraphic relationships) indicated that these "peridotitic komatiites" had been erupted under subaqueous conditions and were completely molten when erupted. Convincing evidence establishing the latter characteristic is supplied by the widespread occurrence of "quench textures" (Fig. 5-3) within the flows. Experimental investigations[1] have shown that these textures can be reproduced only by rapid quenching of ultramafic liquids representative of the compositions of komatiites. Similar rocks have been described from Archaean regions in Australia[2,3] and Canada.[4] In Australia, the peridotites with typical quench textures are known as *spinifex rocks.* Nesbitt[3] showed that these represent the marginal facies of associated ultramafic bodies.

Compositions of some peridotites which texture, mineralogy, and field occurrence prove to have been entirely liquid are listed in Table 5-1.[5] Experimental studies[1] showed that olivine (Fo_{93}) is on the liquidus of the peridotitic komatiite composition up to very high pressures ($\sim$40 kbars), and the principal differences (in Mg/Mg + Fe and pyroxene/olivine ratios) are consistent with derivation of the komatiite by very high degrees of melting (60 to 80%) of pyrolite, leaving only olivine (Fo_{93}) as a residual phase, thus confirming the conclusions of Viljoen and Viljoen (1969a, b).

The peridotitic komatiite flows are overlain and interbedded with olivine and pyroxene-rich rocks which have been referred to as *basaltic komatiites*[6] or *high-magnesian basalts.*[3] The chemical compostions of these basalt types are indicative of genesis from more olivine-rich parents at low pressure. Their genesis may reflect olivine settling and extraction from parental peridotitic komatiite magma at or near the earth's surface, or alternatively, they may be partial melts of a pyrolite source rock with magma segregation occurring at very low pressures. For the latter mechanism, the degree of melting would be lower than that required to develop peridotitic komatiite magmas.[7]

[1]Green, Nicholls, Viljoen, and Viljoen (1972).
[2]Glikson (1970).
[3]Nesbitt (1972).
[4]Naldrett and Mason (1968).
[5]Green (1972b).
[6]Viljoen and Viljoen (1969a,b).
[7]Green (1972a, b).

FIGURE 5-3
"Quench" peridotite which originally crystallized rapidly from the molten state. The original intergrowth of olivine and pyroxenes has since been replaced by alternating layers of tremolite, magnetite-chlorite, and serpentine. Magnification ×20. (*From Viljoen and Viljoen, 1969a, with permission.*) These textures have since been reproduced experimentally by rapid quenching of melts of komatiite composition (*Green, Nicholls, Viljoen, and Viljoen, 1972*).

Table 5-1 COMPOSITIONS OF PERIDOTITIC KOMATIITES REPRESENTING FORMER ULTRAMAFIC LIQUIDS

	(1)	(2)	(3)	(4)	(5)
SiO_2	45.2	46.6	46.5	44.8	47.1
TiO_2	0.7	0.3	0.2	0.2	0.3
Al_2O_3	3.5	3.0	3.6	5.3	6.6
Cr_2O_3	0.4	—	0.4	—	—
Fe_2O_3	0.5	1.0	1.0	1.0	1.0
FeO	8.0	9.6	9.4	9.5	7.7
MnO	0.1	0.2	0.2	0.2	0.2
NiO	0.2	—	—	—	—
MgO	37.5	34.2	33.0	34.3	30.2
CaO	3.1	4.8	5.1	4.4	6.8
Na_2O	0.6	0.15	0.5	0.35	0.2
K_2O	0.1	0.03	0.2	0.03	.02
P_2O_5	0.06	—	0.01	—	—

Explanation:
Columns 2 to 5 recalculated from original data to 100% anhydrous, with Fe_2O_3 arbitrarily made 1.0% (after Green, 1972a).
(1) Pyrolite model composition (Ringwood, 1966a).
(2) Average peridotitic komatiite, Barberton area, South Africa (Viljoen and Viljoen, 1969a).
(3) Freshest sample of quenched peridotitic komatiite. Demonstrated ultramafic liquid from Barberton area, South Africa (Green, Nicholls, Viljoen, and Viljoen, 1972).
(4) Peridotite with quench texture, Mt. Ida, Western Australia (Nesbitt, 1972).
(5) Peridotite with quench texture, Scotia, Western Australia (Nesbitt, 1972).

Thus far, the occurrences of the peridotitic komatiite rock association appear to be restricted to the most ancient greenstone belts of the Archaean shields. The production of these magmas, requiring much higher temperatures and degrees of melting than for basaltic magmas, suggests that processes of magma generation may have differed in important respects in the early Archaean from those which predominated subsequently.[1,2] Possible reasons for this difference have been discussed by Green.[1]

Derivation of Pyrolite Composition from Oceanic Tholeiites

Another approach is to use the observed composition of primitive oceanic tholeiites as the liquid phase combined with a residual refractory component (harzburgite), the composition of which is obtained from the experimental partial melting relationships described in Chap. 4. The unknown quantity is the propor-

[1]Green (1972a, b).
[2]Viljoen and Viljoen (1969a, b).

tion in which basalt and harzburgite are to be combined to obtain the parental composition. This question is discussed in greater detail in Secs. 5-3 and 5-4. To be consistent with an earth model containing chondritic abundances of *involatile* elements, as indicated by cosmochemical considerations, the required proportions are about 17% basalt to 83% harzburgite. The pyrolite composition resulting from this model is given in Table 5-2.

Summary

The compositions of some model pyrolites derived in different manners are shown in Table 5-2, together with some key limiting compositions.

The first column shows the pyrolite composition derived from a somewhat arbitrary 3:1 peridotite-basalt mix by Ringwood (1966a). This composition is probably too high in TiO_2 and, perhaps, K, Na, and P owing to choice of the particular basaltic component.[1] Nevertheless, it compares well with the other model compositions of Table 5-2.

[1] Kuno and Aoki (1970).

Table 5-2 PYROLITE COMPOSITIONS

Model compositions						Limiting compositions		Average mantle pyrolite
	(1)	(2)	(3)	(4)	(5)	(6)	(7)	(8)
SiO_2	45.2	44.9	46.1	45.6	42.9	44.9	46.5	45.1
TiO_2	0.7	0.24	0.2	0.2	0.2	0.1	0.2	0.2
Al_2O_3	3.5	4.3	4.3	3.9	5.8	3.2	3.6	4.6
Cr_2O_3	0.4	0.4	—	0.4	0.2	0.5	0.4	0.3
Fe_2O_3	0.5	—	—	—	0.3	—	1.0	0.3
FeO	8.0	8.2	8.2	5.2	8.9	7.6	9.4	7.6
MnO	0.14	0.1	—	0.1	0.14	0.1	0.2	0.1
NiO	0.2	0.2	—	0.3	0.2	0.3	—	0.2
MgO	37.5	38.9	37.6	37.7	37.2	40.0	33.0	38.1
CaO	3.1	2.5	3.1	3.0	3.7	3.0	5.1	3.1
Na_2O	0.6	0.23	0.4	0.5	0.4	0.2	0.5	0.4
K_2O	0.13	0.02	0.03	0.1	0.003	0.0006	0.2	0.02
P_2O_5	0.06	0.02	—	0.03	—	—	0.01	0.02
								100.0

Explanation:

(1) Pyrolite model composition (Ringwood, 1966a).

(2) Pyrolite model composition: 99% Lizard peridotite (Green, 1964) + 1% nephelinite (VV3-1 Kay, 1971).

(3) Pyrolite model composition: 83% residual harzburgite + 17% primitive oceanic tholeiite (Table 5-5).

(4) Bulk composition of Vourinos ophiolite complex (Table 3-3, average of A and B).

(5) Upper mantle model composition from Carter (1970, table 4) corresponding to a parental mantle olivine composition of Fo_{88}.

(6) Least fractionated alpine ultramafic, representing smallest degree of partial melting of pyrolite (Tinaquillo peridotite, Table 3-3).

(7) Ultramafic liquid representing highest degree of partial melting of pyrolite (Table 5-2).

(8) Mean of columns 2,3,4, and 5. Alumina content is probably too high because of column 5 (see text).

Column 2 shows the pyrolite model composition obtained by combining those of one of the least fractionated alpine ultramafics with the complementary nephelinite, the proportions chosen being dictated by considerations of experimental basalt petrogenesis and cosmochemistry. The slight deficiency in CaO could have been remedied by choice of another high-temperature peridotite (e.g., Tinaquillo, column 6). The pyrolite composition derived from oceanic tholeiite + harzburgite is shown in column 3. It agrees quite well with the previous estimates. Both these compositions agree closely with the overall composition of an ophiolite complex (column 4).

Carter's (1970) preferred composition (column 5) based upon the partial melting-partial crystallization model agrees fairly well with these models, although the CaO and, more particularly, the Al_2O_3 abundances are significantly higher. There are grounds for questioning the Al_2O_3 value, which gives a substantially higher Al/Ca ratio than is present in all classes of meteorites.[1] These components are not readily fractionated in the cosmochemical processes believed to have been involved in the formation of the earth,[1,2,3] and it appears likely that the mantle may possess the meteoritic Al/Ca ratio. Better agreement can be secured within the assumptions and uncertainties of Carter's model by selecting the model composition corresponding to a slightly lower Fe/(Fe + Mg) ratio than was chosen by Carter.

Columns 6 and 7 show key limiting compositions which should bracket the pyrolite model composition. Column 6 gives the composition of a peridotite which is interpreted to represent the residuum remaining after pyrolite has been subjected to a very small degree of partial melting. Similar residual compositions are obtained from studies of lherzolite and garnet peridotite xenoliths from alkali basalts and kimberlites (Sec. 3-4). Column 7 shows the composition of an ultramafic liquid believed to represent the largest degree of partial melting of pyrolite. The similarity between these two limiting compositions provides strong constraints for the ideal model pyrolite composition.

It appears therefore that estimates of the primitive mantle composition obtained by several methods are in close agreement and that the composition of pyrolite is defined within quite narrow bounds. In all probability there is no unique composition because of a limited degree of primary compositional heterogeneity within the mantle. The range of compositions in Table 5-2 gives an idea of the possible extent of this heterogeneity. An average mantle pyrolite composition is given in column 8.

5-3 PYROLITE MANTLE AND THE CHONDRITIC EARTH MODEL

During the 1950s and early 1960s it was widely assumed that the earth had formed from material resembling the ordinary chondritic meteorites and that the

[1] Ahrens (1970).
[2] Ringwood (1966a).
[3] Larimer and Anders (1967, 1970).

composition of these objects was therefore similar to that of the entire earth. This assumption gave rise to the so-called chondritic earth model which was to form the basis of a vast number of geochemical and geothermal papers during this period. The demonstrations[1,2] that the total surface heat-flow of the earth divided by its mass was similar to the rate of heat production per unit mass of chondritic meteorites was generally taken to imply that the earth contained, on the average, the same abundances of uranium, thorium, and potassium as chondrites. This was widely regarded as providing convincing support to the chondritic earth model.

Birch (1958) developed this model to demonstrate that it implied a very efficient differentiation of the earth, with most (~ 60%) of the uranium and thorium being segregated in the crust. Curiously, however, potassium, which was usually covariant with uranium during magmatic fractionation processes, did not appear to be strongly concentrated in the crust—indeed, Birch demonstrated that the chondritic earth model implied that about 80% of the earth's potassium must be buried within the mantle. This was not held to be an objection against the model.

The first serious questioning of the chondritic earth model began with an important paper by Gast (1960), who demonstrated, from a study of the isotopic development of terrestrial strontium, that the Rb/Sr ratio of the upper mantle-crust system was at least 4 times lower than the corresponding ratio in chondritic meteorites. Moreover, he produced supporting geochemical evidence to show that the abundances of K, Rb, and Cs compared to Ba, Sr, and U in the upper mantle-crust system were systematically smaller than in chondritic meteorites. Gast suggested two alternative explanations of these observations:

1 The alkali metals K, Rb, and Cs may have become depleted relative to chondrites by a process of selective volatilization which occurred prior to, or during, the earth's formation, thus implying that the earth is not presently of chondritic composition.

2 The heavy alkali metals may have been selectively retained in the deep mantle by a specialized process of differentiation during which U, Ba, and Sr were strongly concentrated in the crust. This explanation permitted retention of the chondritic earth model. Gast did not express a preference between the two alternatives.

Ringwood (1962c) reconsidered these alternatives and rejected the second as being improbable on geochemical and crystal chemical grounds. Previously, he had proposed a model for the formation of the earth which provided a mechanism for volatilizing Rb, K, and related volatile elements from the material which was accreting on the earth.[3] The volatile components were believed to have entered a massive primitive atmosphere which was subsequently lost. Accordingly, he supported the first of Gast's alternative hypotheses and generalized it by suggesting

[1]Urey (1956).
[2]Birch (1958).
[3]Ringwood (1960, p. 255).

that the earth may also be depleted relative to primordial abundances in a wide range of elements which are volatile under high-temperature reducing conditions, e.g., Pb, Hg, and Tl.[1] Subsequently, detailed studies of the geochemistry of a range of volatile elements, including Na, Zn, Cd, Hg, Pb, In, Bi, Ge, Cl, S, and F, provided strong evidence that they were strongly depleted in the earth relative to primordial abundances.[2]

During this period, it had also become clear that ordinary chondrites were not representative of primordial abundances and were strongly depleted in many volatile elements relative to carbonaceous and enstatite chondrites.[3] Moreover, further discrepancies between abundances in ordinary chondrites and in the earth emerged[2] in addition to those noted by Gast. The entire rationale of the "chondritic earth model" appeared to have collapsed.

At the same time, mounting evidence showed that one particular group of chondrites—the Type 1 carbonaceous chondrites—had had a simpler thermal and chemical history than other chondrite groups and appeared to have retained the primordial or solar abundances of most elements, except for some extremely volatile gases.[4,5,6,7,8] The accumulation of increasingly accurate data on the chemical composition of the solar nebula from which the earth, planets, and meteorites were ultimately derived focused attention on theories of origin of these bodies and the accompanying chemical fractionations.

Two classes of theories emerged. Curiously, both had rather similar consequences with regard to the predicted compositions of the earth and meteorites.

Ringwood[9] argued that the earth, other terrestrial planets, and noncarbonaceous chondrites had formed directly by accretion from parental material similar to Type 1 carbonaceous chondrites in an initially cool solar nebula. During accretion, relatively volatile elements were boiled off into primitive atmospheres which were lost. Anders and coworkers[10,11] argued for selective condensation of planetesimals from a hot solar nebula followed by accretion of planetisimals into planets and meteorites.

Both theories predict that selective volatility is the principal property responsible for differences between the present composition of the earth (and chondrites) and the primordial abundances. Both theories imply that resemblances between the present composition of the earth and the primordial abun-

[1]Ringwood (1962c).
[2]Ringwood (1966a,b).
[3]Reed, Kigoshi, and Turkevich (1960).
[4]Mason (1960).
[5]Ringwood (1961).
[6]Anders (1964).
[7]Ringwood (1966c).
[8]Anders (1971a,b).
[9]Ringwood (1959, 1960, 1961, 1966a,b,c).
[10]Anders (1964, 1971a,b).
[11]Larimer and Anders (1967, 1970).

Table 5-3 COMPOSITION OF EARTH AS DERIVED BY REDUCTION FROM COMPOSITION OF TYPE 1 CARBONACEOUS CHONDRITES. (*After Ringwood, 1966a*)

	(1)	(2)	(3)	(4)
SiO_2	33.3	35.9	29.8	43.2
MgO	23.5	25.2	26.3	38.1
FeO	35.5	6.1	6.4	9.3
Al_2O_3	2.4	2.6	2.7	3.9
CaO	2.3	2.5	2.6	3.7
Na_2O	1.1	1.2	1.2	1.8
NiO	1.9			
Total	100.0	73.5	69.0	100.0
Fe		24.9	25.8	
Ni		1.6	1.7	
Si		—	3.5	
Total		26.5	31.0	

Explanation:

(1) Average composition of principal components of Type 1 carbonaceous chondrites (Orgueil and Ivuna) on a C-, S-, and H_2O-free basis (analyses by Wiik, 1956).

(2) Analysis from column 1 with FeO/MgO + FeO reduced to be consistent with probable value for earth's mantle (0.12).

(3) Analysis from column 2, with sufficient SiO_2 reduced to elemental silicon to yield a total silicate to metal ratio of 69/31 as in earth.

(4) Model mantle composition: silicate phase from column 3 recalculated to 100%.

dances should be greatest for relatively involatile elements, whereas the earth is expected to be depleted relative to primordial abundances in comparatively volatile elements. In the case of the earth, the involatile group may be considered to include elements less volatile than sodium (under appropriate redox conditions).

These theories provided the basis for a restatement of the chondritic earth model. Since it is widely accepted that the most reliable estimates of primordial solar nebula abundances are based upon compositions of Type 1 carbonaceous chondrites, the revised statement[1,2] of the chondritic earth model simply hypothesizes that the relative abundances of involatile elements (defined above) in the earth are similar to those in Type 1 carbonaceous chondrites. No such similarity is required for abundances of volatile elements (including oxygen) so that the present overall composition of the earth is not required to be similar to that of any group or mixture of groups of meteorites.

Ringwood (1959, 1966a) explored the above hypothesis and showed that a self-consistent earth model could be constructed from the chondritic abundances

[1]Ringwood (1962c).
[2]Ringwood (1966a,b,c).

of major involatile elements—Fe, Ni, Si, Mg, Ca, and Al (Table 5-3). This model required the presence of some silicon in the core in order to produce the correct core-to-mantle ratio.[1] The composition of the mantle derived in this manner (Table 5-3) was similar to that of pyrolite (Table 5-2), except for sodium, which has since been reclassified as a "volatile" element.

On the other hand, it did not appear at first as if the abundances of some involatile incompatible trace elements (e.g., U, Th, Ba, Sr, and rare earths) in pyrolite could be simply explained in terms of chondritic abundances. The abundances of these elements in the upper mantle appeared to be substantially higher than could be provided by chondritic abundances.[2,3,4]

One possible explanation[3,5,6] was that the mantle had been subjected to a rather specialized differentiation process characterized by thorough convective mixing of a crystal mush containing a small amount of residual liquid. This would have the effect of chemically homogenizing the major rock-forming minerals, together with the trace elements which were able to enter these minerals freely. However, the incompatible elements which were unable to enter the major minerals freely would be strongly partitioned in the small amount of residual liquid, and it was suggested that this liquid had been squeezed upward into the uppermost mantle, thereby accounting for high concentrations of incompatible elements in this region. A second explanation[2,4] simply maintained that the abundance of *highly* involatile elements (e.g., Ca, Al, U, Th, Ba, Sr, and rare earths) throughout the earth was 3 to 4 times higher than in chondrites and that this represented a primary accretional feature of the earth. This view, in essence, amounted to a rejection of the chondritic earth model in the revised sense as stated above.

Re-examination of Earlier Hypotheses on Mantle Composition

These alternative hypotheses have important implications for theories of the earth's origin and early evolution, and a reappraisal appears desirable in the light of more recent developments. The case for enrichment of incompatible elements in the upper mantle relative to Mg, Si, Ca, and Al was based partly upon geothermal arguments which appeared to require the presence of relatively high concentrations of U, Th, and K in this region.[3] However, these arguments assumed that heat transport in the oceanic mantle was dominated by thermal conduction, whereas it is now known that convection plays a major role (Chap. 8). The case for postulating a relative concentration of U, Th, and K in the upper mantle on geothermal grounds has accordingly disappeared.

Gast[4,7] has subsequently shown from strontium isotope studies that the

[1]See also Ringwood (1958), MacDonald and Knopoff (1958), and MacDonald (1959).
[2]Wasserburg, MacDonald, Hoyle, and Fowler (1964).
[3]Clark and Ringwood (1964).
[4]Gast (1968a).
[5]Wager (1958).
[6]Ringwood (1966a,b).
[7]Gast (1972).

Rb/Sr ratio of the entire earth is probably smaller than that of chondrites by a factor of about 8. If the earth has the chondritic abundance of strontium, this would require that about 70% of the total rubidium originally present in the earth is now in the crust. A similar situation exists in the case of potassium. Likewise, if the earth is assumed to contain the chondritic abundances of U and Ba, Gast estimated that 46 and 30%, respectively, of the earth's original endowment of these elements now resides in the crust (Sec. 8-7). Gast considers this degree of concentration of incompatible elements in the crust to be inherently improbable and points out that the difficulty would be greatly lessened if the earth was assumed to contain higher absolute abundances of involatile elements U, Th, Ba, and Sr (and also Ca and Al but *not* Mg and Si) than chondrites. He suggests specifically a threefold enrichment of these elements within the earth relative to chondrites. A similar proposal was made largely on geothermal grounds in the cases of U and Th and is examined in Sec. 5-5.[1]

Kay (1971) has carried out a detailed investigation of the partition of rare earths between a number of highly alkaline basalts formed by very small degrees (0.5 to 2%) of partial melting of pyrolite and their refractory ultramafic complements. He finds that the original abundances of rare earths required in the pyrolite source regions range from 1.4 to 3.6 times the chondritic abundance, with twice chondritic abundance being the most frequent value.

Abundances of heavy rare earths in the Lizard and Tinaquillo high-temperature peridotites are about 1.5 to 2 times the chondritic abundances.[2,3] These bodies are believed to approach closely the pyrolite composition and to have suffered only very small degrees of partial melting (Sec. 5-2). Their rare earth abundances, combined with complementary relationships to derived magmas (Fig. 5-2), indicate that pyrolite contains 1.5 to 2 times the chondritic abundances of rare earths.[4]

Discussion

The geochemical evidence cited above suggests that pyrolite contains about twice the chondritic abundances of rare earths. The arguments used by Gast[5] also support the view that the abundances of U, Ba, and Sr in the mantle are significantly higher than chondritic. He estimates the enrichment to be about threefold, but the uncertainties are such that his arguments would be compatible with a twofold enrichment over chondrites.

At first sight, the inferred overabundance in pyrolite of rare earths, U, Ba, Sr, and related elements compared to Mg and Si might appear to make it difficult to derive the pyrolite composition from chondritic abundances of involatile ele-

[1]Wasserburg, MacDonald, Hoyle, and Fowler (1964).
[2]Frey (1969, 1970a).
[3]Frey, Haskin, and Haskin (1971).
[4]Kay (1971).
[5]Gast (1968a, 1972).

ments as in Table 5-3. This, indeed, is the conclusion of Gast, Kay, Frey, and others. There are grounds for questioning this conclusion, however. Confusion has been introduced into the discussion of these workers by their use of abundances by weight in *ordinary chondrites* for purposes of normalization. There are two problems here. The pyrolite model refers only to the earth's mantle, whereas the chondrite model relates to the entire earth (mantle + core + crust). In order to apply the chondrite model to the mantle, it is necessary to remove metallic iron into the core and also to reduce some FeO from the chondrite silicate phase to metallic iron so that the chondritic MgO/MgO + FeO ratio matches that of the mantle (Table 5-3). These exercises have the effect of increasing the abundances of oxyphile elements in the mantle composition by a factor of about 1.5. Secondly, it is known that significant fractionations of highly involatile elements (e.g., Ca, Al, U, Ti, rare earths, etc.) have occurred between different groups of chondrites.[1] Available data indicate that enstatite chondrites are depleted in these elements by about 50% compared to ordinary chondrites, whereas carbonaceous chondrites are enriched in these involatile elements by about a factor of 1.4 (compared to ordinary chondrites).[1]

The combined effect of segregation of the core, together with the model requiring the earth to be derived from *carbonaceous chondrite-like* material rather than from *ordinary chondrites*, results in abundances of Ca, Al, U, Th, Ba, Sr, rare earths, and other involatile elements in the earth's mantle which are about twice as high as their net abundances in *ordinary* chondrites.[2] Thus the inferred abundances of these elements in the earth's mantle, i. e., in pyrolite, are compatible with an earth model derived from carbonaceous chondrites.[2] It is not necessary to assume that the earth is characterized by a marked absolute enrichment in these involatile elements relative to Mg + Si + Fe or that incompatible elements have been selectively concentrated in the upper mantle by a specialized differentiation process. This is not to deny the possibility of some relative enrichment of Ca, Al, and related elements relative to Mg, Si, and Fe, when normalized to primordial carbonaceous chondritic abundances. The situation is, rather, that existing data do not require a significant enrichment of this kind. If such an enrichment exists, it is likely to be small, probably not exceeding a factor of 1.5.

5-4 OCEANIC THOLEIITES AND THE PYROLITE MODEL

Having concluded that the pyrolite composition is consistent with an earth model derived from material resembling carbonaceous chondrites, we must next enquire whether this model is capable of explaining the combined major and trace element compositions of oceanic or abyssal tholeiites which represent the most extensive class of erupted volcanic magmas.

[1]Larimer and Anders (1970).
[2]Larimer (1971).

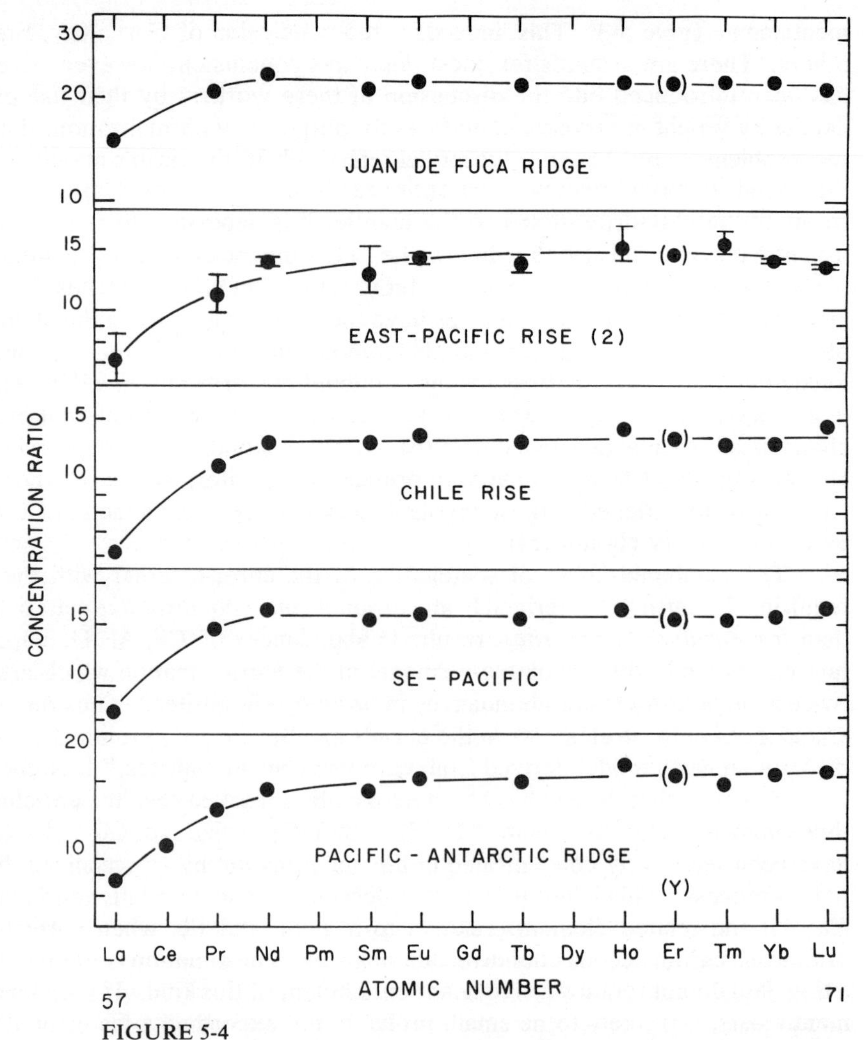

FIGURE 5-4
Abundance of rare earths relative to ordinary chondrites in submarine basalts of the Pacific Ocean. (*a*) Juan de Fuca Ridge; (*b*) East Pacific Rise; (*c*) Chile Rise; (*d*) S.E. Pacific; (*e*) Pacific-Antarctic Ridge. (*From Schilling, 1971, with permission.*)

Rare earth abundance patterns for several oceanic tholeiites are shown in Fig. 5-4. Ignoring the slight depletion of the lightest rare earths (discussed later), the most important feature is the uniform degree of enrichment of rare earths from Nd to Lu relative to chondrites. The average enrichment factors for the vast majority of oceanic tholeiites relative to *ordinary* chondrites fall between 10 to 20, with an average of 15.[1,2,3,4] However, in an important study of the petrochemistry of a wide range of major and minor elements in oceanic tholeiites, Kay et al. (1970) demonstrated that the compositions of many of these rocks had been affected by varying degrees of high-level crystal fractionation involving the separation of plagioclase and, to a lesser extent, olivine. The effect of these differentiations was to cause a substantial enrichment of rare earths (except Eu) and Fe combined with depletions of Mg, Ni, Sr, Eu, Ca, and Al. The least differentiated class of primary magmas could be recognized by their higher abundances of these latter elements combined with absence of europium anomalies. They contain 16 to 18% Al_2O_3, 11 to 12% CaO, 7.5 to 8.5% MgO, 7 to 9% FeO,[5] 150 to 200 ppm Ni, and 120 to 180 ppm Sr. The rare earth abundances in this class of basalts mostly fall within the range of 10 to 15 times those in ordinary chondrites.[6] We will assume in subsequent discussion that the average *primary* oceanic tholeiite magma contains 12 times the ordinary chondritic abundances of rare earths. This is equivalent to 6 times the rare earth abundances in pyrolite composition derived from the carbonaceous chondrite earth model as discussed in Sec. 5-3, Table 5-3. Assuming total partition of rare earths in the magma during partial melting of pyrolite, this implies that primitive oceanic tholeiites represent 15 to 20% partial melts of pyrolite, with an average of about 17%.

The chemical and mineralogical equilibria involved in the formation of high-alumina oceanic tholeiite magmas were discussed in Secs. 4-2 and 4-5. It was demonstrated that they were formed in the pressure interval 4 to 10 kbars and that the residual mineral assemblage for the most primitive type of oceanic tholeiite consisted[7,8] of olivine (Fo_{90}, 0.4% CaO) and enstatite (En_{90}, 2.8% CaO, 5.4% Al_2O_3). A possible residual mineral assemblage would be that of a harzburgite containing 67% olivine and 33% orthopyroxene. Combining 83% of this harzburgite with 17% primitive oceanic tholeiite,[9] we obtain a model pyrolite composition given in column 3 of Table 5-4.

The composition is seen to agree well with the other pyrolite models, although the CaO content may be slightly low. This could be remedied by including 5% subcalcic clinopyroxene (18% CaO, 6.2% Al_2O_3)[8] in the residual refractory assemblage—an assumption which would be consistent with experimental petrological considerations (Secs. 4-2 to 4-5)[8,7] Reversing the argument, we

[1]Gast (1968b).
[2]Kay, Hubbard, and Gast (1970).
[3]Schilling (1971).
[4]Frey, Haskin, Poetz, and Haskin (1968).
[5]All iron as FeO.
[6]Some abundant continental tholeiite magmas—e.g., the chilled zones of the Bushveldt and Stillwater complexes—contain only 2 to 5 times the chondritic abundances of rare earths (Frey et al., 1968).
[7]Green and Ringwood (1967).
[8]Green, Green, and Ringwood (1967 and unpublished measurements).
[9]Rock KD11, selected as one of the most primitive types of oceanic tholeiites by Kay, Hubbard, and Gast (1970).

Table 5-4 PYROLITE MODEL COMPOSITION DERIVED FROM PRIMITIVE OCEANIC THOLEIITE AND COMPLEMENTARY RESIDUAL HARZBURGITE

	Oceanic tholeiite* KD11	Harzburgite† 67% Ol 33% Opx	Pyrolite 17% KD11 83% Hzb
SiO_2	50.3	45.3	46.1
TiO_2	1.2	—	0.2
Al_2O_3	16.5	1.8	4.3
FeO‡	8.5	8.1	8.2
MgO	8.3	43.6	37.6
CaO	12.3	1.2	3.1
Na_2O	2.6	—	0.4
K_2O	0.2	—	0.03

*KD11 selected as primitive oceanic tholeiite by Kay, Hubbard, and Gast (1970).

†Compositions of olivine and orthopyroxene from experimental near-liquidus phase compositions observed in high-Al olivine tholeiites at 5 to 9 kbars (see text).

‡All iron calculated as FeO.

conclude that a pyrolite model derived from carbonaceous chondrite abundances appears capable of providing a self-consistent explanation both of the major element chemistry and the rare earth abundances of oceanic tholeiites.[1]

Gast[2] and Schilling[3] have emphasized the significance of the depletion of light rare earths (La, Ce, and Pr) which are characteristic of most, but not all,[3] oceanic tholeiites (Fig. 5-4). Gast[2] demonstrated that, when appropriate normalization procedures are applied, oceanic tholeiites are found to be correspondingly depleted in other large cations, e.g., Ba^{++}, Th^{4+}, U^{4+}, Rb^{+}, and Cs^{+}. These are just the elements which are strongly enriched in highly alkaline basalts,[4,5] and hence an apparent complementary geochemical relationship exists for these elements between oceanic tholeiites and alkaline basalts. Gast argued that the pyrolite source material of oceanic tholeiites had previously been subjected to partial melting episodes during which a very small amount of highly alkaline liquid had been removed, thereby resulting in an average twofold depletion of these large cations. Studies of the isotopic evolution of lead in basaltic rocks and in their source regions indicate that the inferred fractionations were not isochronous with formation of oceanic tholeiites—i.e., the depletions were caused not by contemporaneous alkali basaltic magmatism but by a series of earlier minor differentiation episodes involving the extraction of small amounts of highly alkaline magmas

[1]See also Larimer (1971).
[2]Gast (1968a,b).
[3]Schilling (1971).
[4]Kay, Hubbard, and Gast (1970).
[5]Kay (1971).

from the mantle. These episodes may have occurred periodically during the last 2.5 billion years.[1]

A possible explanation of these earlier differentiation episodes is considered in Sec. 8-4.

5-5 THERMAL HISTORY OF THE EARTH

We concluded in the previous section that the abundances of involatile elements in pyrolite are consistent with an earth model derived from carbonaceous chondrite-like material.[2] It has been proposed that, during the formation of the earth from parent material of this nature, reduction of oxidized iron occurred to form a metal phase which segregated into the core, whilst elements more volatile than sodium were strongly depleted by volatilization processes.[3,4,5]

Ringwood[3,4] drew attention to the geothermal consequences of selective loss of potassium. Whereas the earlier versions of the chondritic earth model[6,7] implied that the total amount of radiogenic heat produced within the earth by U, Th, and K was similar to the total heat lost at the surface by conduction (i.e., a state of thermal equilibrium prevailed), loss of most of the earth's potassium implied that the radiogenic heat produced in the earth was smaller than the heat loss and that therefore the earth is cooling.

In 1963, Tilton and Reed pointed out that the K/U ratio of a wide range of terrestrial rocks was close to 10,000, which is much smaller than the chondritic ratio of 80,000, and pointed out that this circumstance was of considerable importance to considerations of the earth's thermal history. This point was developed in greater detail in 1964 in two papers which were published simultaneously.[8,9] Clark and Ringwood[8] examined available data on K/U ratios of terrestrial rocks and concluded that the average value of this ratio for the upper mantle-crust system was between 10^4 and 2×10^4, i.e., that potassium was depleted in this region of the earth by a factor of 4 to 8 as compared with chondrites. Detailed calculations of the thermal structure of the upper mantle were carried out using the empirical terrestrial K/U ratio as a key boundary condition. Wasserburg et al.[9] also concluded from essentially similar considerations that the K/U ratio of the earth was close to 10^4 and drew attention to the geothermal consequences. However,

[1] Gast (1968a,b).
[2] See also Larimer (1971).
[3] Ringwood (1960, p. 255).
[4] Ringwood (1962c, p. 212).
[5] Ringwood (1966a,b).
[6] Urey (1956).
[7] Birch (1958).
[8] Clark and Ringwood (1964).
[9] Wasserburg, MacDonald, Hoyle, and Fowler (1964).

further development of these ideas by the two groups of workers differed in important respects.

Clark and Ringwood favoured the view that earth was derived from chondritic material but had been selectively depleted in potassium whilst retaining the chondritic abundances of uranium, and thorium. This model implied that the earth is now cooling and has been cooling for most of its existence. The difference between the earth's total heat-flow and the proportion due to radioactivity was about 0.5 μcal/cm^{2}sec. This deficiency must have been supplied by the earth's "original heat," implying that the earth formed originally in a high-temperature condition.[1]

On the other hand, Wasserburg et al. preferred to retain the assumption that the net production of radiogenic heat within the earth is similar to the integrated heat-flux. This requires that the mean uranium and thorium contents of the earth are about 3 times higher than in chondrites, providing that the K/U ratio is 10^4. Once the earlier version[2,3] of the chondritic earth model is rejected (as now agreed by all), retention of the assumption that the total production of radiogenic heat within the earth is equal to that lost by conduction at the surface must be regarded as rather arbitrary. Nevertheless, some support is afforded by nucleosynthetic considerations, which suggest that the primordial uranium and thorium abundances should be 2 to 4 times higher than in carbonaceous chondrites.[4] These estimates, however, are likely to possess rather large uncertainties. Moreover, they are based on the assumption that the abundance of other involatile elements (e.g., the rare earths) in chondrites are similar to the primordial abundances. It is difficult to propose plausible cosmochemical processes which might explain why U and Th have been fractionated to this extent relative to the rare earths both in chondrites and in the earth.

The Wasserburg et al. equilibrium heat-production model implies that the contribution of "original heat" of nonradiogenic derivation to the present terrestrial heat-flux is negligible.[5] In contrast to the Clark-Ringwood model, this requires, in effect, a "cold" origin for the earth.

An alternative approach is to consider sources of evidence bearing on the initial temperature of the earth. A demonstrated "hot" origin would favour the chondrite-derived model since an appreciable proportion of the present heat-flux would consist of original heat transported by mantle convection. On the other hand, demonstration of a "cool" origin would imply a model with higher than chondritic abundances of U and Th. During the 1950s, a cool origin was widely favoured on geochemical and astrophysical grounds.[6,7,8,9] More recent develop-

[1]Ringwood (1960, 1962b).
[2]Urey (1956).
[3]Birch (1965).
[4]Wasserburg, MacDonald, Hoyle, and Fowler (1964).
[5]MacDonald (1964, 1965).
[6]Urey (1952).
[7]Urey (1957).
[8]Urey (1962).
[9]Schmidt (1958).

ments, however, have challenged this view.[1,2,3,4,5] The geochemical arguments in favour of an initially cool earth no longer appear compelling.[4] Strong arguments have been advanced that core formation occurred early in the earth's history, most probably within 10^8 years of its formation.[1,2,3,4,6] The process was highly exothermic owing to liberation of gravitational potential energy, which caused a mean rise in temperature of about 2000°C,[3] resulting in widespread early partial melting within the earth.[7]

REFERENCES

AHRENS, L. H. (1970). The composition of stony meteorites (IX). *Earth Planet. Sci. Letters* **10,** 1–6.

ALLSOP, H. L., L. O. NICOLAYSEN, and P. HAHN-WEINHEIMER (1968). Rb/K ratios and Sr isotopic compositions of minerals in eclogitic and peridotitic rocks. *Earth Planet. Sci. Letters* **5,** 231–244.

ANDERS, E. (1964). Origin, age and compositon of meteorites. *Space Sci. Rev.* **3,** 583-714.

——— (1971a). Meteorites and the early solar system. *Ann. Rev. Astron. Astrophys.* **9,**1–34.

——— (1971b). How well do we know "cosmic" abundances? *Geochim. Cosmochim. Acta* **35,** 516–522.

BERG, G.W. (1968). Secondary alteration in eclogites from kimberlite pipes. *Am. Min.* **53,** 1336–1346.

BIRCH, F. (1958). Differentiation of the mantle. *Bull. Geol. Soc. Am.* **69,** 483–486.

——— (1965). Speculations on the earth's thermal history. *Bull. Geol. Soc. Am.* **76,** 133–154.

BOWEN, M. (1928). "*The Evolution of the Igneous Rocks.*" Princeton Univ. Press, Princeton, N. J. 332 pp.

CARTER, J. L. (1970). Mineralogy and chemistry of the earth's upper mantle based on the partial fusion–partial crystallization model. *Bull. Geol. Soc. Am.* **81,** 2021–2034.

CLARK, S. P., and A. E. RINGWOOD (1964). Density distribution and constitution of the mantle. *Rev. Geophys.* **2,** 35–88.

DREVER, H. I., and R. JOHNSTON (1958). The petrology of picritic rocks in minor intrusions—a Hebredaen group. *Trans. Roy. Soc. Edinburgh* **63,** 459–499.

——— and ——— (1967). Picritic minor intrusions. In: P. J. Wyllie (ed.), "*Ultramafic and Related Rocks,*" pp. 71–82. Wiley, New York.

ELSASSER, W. M. (1963). Early history of the earth. In: J. Geiss and E. Goldberg (eds.), "*Earth Science and Meteoritics,*" pp. 1–30. North-Holland, Amsterdam.

FREY, F. (1969). Rare earth abundances in a high-temperature peridotite intrusion. *Geochim. Cosmochim. Acta* **33,** 1429–1447.

[1]Ringwood (1960).
[2]Elsasser (1963).
[3]Birch (1965).
[4]Ringwood (1966a,b).
[5]Hanks and Anderson (1969).
[6]Oversby and Ringwood (1971).
[7]See also Chap. 16.

——— (1970a). Rare earth abundances in Alpine ultramafic rocks. *Phys. Earth Planet. Interiors* **3,** 323–330.

——— (1970b). Rare earth and potassium abundances in St. Paul's Rocks. *Earth Planet. Sci. Letters* **7,** 351–360.

———, L. A. HASKIN, and M. HASKIN (1971). Rare earth abundances in some ultramafic rocks. *J. Geophys. Res.* **76,** 2057–2070.

———, M. HASKIN, J. POETZ, and L. A. HASKIN (1968). Rare earth abundances in some basic rocks. *J. Geophys. Res.* **73,** 6085–6098.

GASS, I. G. (1958). Ultrabasic pillow lavas from Cyprus. *Geol. Mag.* **95,** 241–251.

GAST, P. W. (1960). Limitations on the composition of the upper mantle. *J. Gephys. Res.* **65,** 1287–1297.

——— (1968a). Trace element fractionation and the origin of tholeiitic and alkaline magma types. *Geochim. Cosmochim. Acta* **32,** 1057–1086.

——— (1968b). Upper mantle chemistry and evolution of the earth's crust. In: R. Phinney (ed.), *"History of the Earth's Crust,"* pp. 15–27. Princeton Univ. Press, Princeton, N. J.

——— (1972). The chemical composition of the earth, the moon and chondritic meteorites. In: E. C. Robertson (ed.), "*The Nature of the Solid Earth,*" pp. 19–40. McGraw-Hill, New York.

GLIKSON, A. Y. (1970). Geosynclinal evolution and geochemical affinities of early Precambrian systems. *Tectonophysics* **9,** 397–433.

GREEN, D. H. (1964). The petrogenesis of the high-temperature peridotite intrusion in the Lizard area, Cornwall. *J. Petrol.* **5,** 134–188.

——— (1967). High temperature peridotite intrusions. In: P. J. Wyllie (ed.), "*Ultramafic and Related Rocks,*" pp. 212–222. Wiley, New York.

——— (1970). The origin of basaltic and nephelinitic magmas. *Trans. Leicester Literary Phil. Soc.* **64,** 26–54.

——— (1972a). Magmatic activity as the major process in chemical evolution of the earth's crust and mantle. *Tectonophysics* **13,** 47–71.

——— (1972b). Archaen greenstone terrains: possible terrestrial equivalents of lunar maria. *Earth Planet. Sci. Letters* **15,** 263–270.

———, J. MORGAN, and K. HEIER (1968). Thorium, uranium and potassium abundances in peridotite inclusions and their host basalts. *Earth Planet. Sci. Letters* **4,** 155–166.

———, I. A. NICHOLLS, M. H. VILJOEN, and R. P. VILJOEN (1972). Experimental study of extremely high temperature ultramafic extrusions. (In press.)

——— and A. E. RINGWOOD (1963). Mineral assemblages in a model mantle composition. *J. Geophys. Res.* **68,** 937–945.

——— and ——— (1967). The genesis of basaltic magmas. *Contr. Mineral. Petrol.* **15,** 103–190.

GREEN, T. H., D. H. GREEN, and A. E. RINGWOOD (1967). The origin of high-alumina basalts and their relationships to quartz tholeiites and alkali basalts. *Earth Planet. Sci. Letters* **2,** 41–51.

HAMILTON, W., and W. MOUNTJOY (1965). Alkali content of alpine ultramafic rocks. *Geochim. Cosmochim. Acta* **29,** 661–671

HANKS, T. C., and D. L. ANDERSON (1969). The early thermal history of the earth. *Phys. Earth Planet. Interiors* **2,** 19–29.

HARRIS, P. G., A. REAY, and I. G. WHITE (1967). Chemical composition of the upper mantle. *J. Geophys. Res.* **72,** 6359–6369.

HEIER, K. S. (1963). Uranium, thorium and potassium in eclogitic rocks. *Geochim. Cosmochim. Acta* **27,** 849–860.

HESS, H. H. (1964). The oceanic crust, the upper mantle, and the Mayaguez serpentinized peridotite. In: C. A. Burk (ed.), "*A Study of Serpentinite*," pp. 169–175. *Nat. Acad. Sci.—Nat. Res. Coun. Publ.* 1188.

ITO, K., and G. C. KENNEDY (1967). Melting and phase relations in a natural peridotite to 40 kilobars. *Am. J. Sci.* **265,** 519–538.

KAY, R. W. (1971). The rare earth geochemistry of alkaline basaltic volcanics. Ph. D. Thesis, Columbia University.

———, N. J. HUBBARD, and P. W. GAST (1970). Chemical characteristics and origin of oceanic ridge volcanic rocks. *J. Geophys. Res.* **75,** 1585–1613.

KUNO, H., and K. AOKI (1970). Chemistry of ultramafic nodules and their bearing on the origin of basaltic magmas. *Phys. Earth Planet. Interiors* **3,** 273–301.

KUSHIRO, I., and H. KUNO (1963). Origin of primary basalt magmas and classification of basaltic rocks. *J. Petrol.* **4,** 75–89.

LARIMER, J. (1971). Composition of the earth: chondritic or achondritic? *Geochim. Cosmochim. Acta* **35,** 769–786.

——— and E. ANDERS (1967). Chemical fractionations in meteorites **II**. *Geochim. Cosmochim. Acta* **31,** 1239–1270.

——— and ——— (1970). Chemical fractionations in meteorites III. *Geochim. Cosmochim. Acta* **34,** 367–388.

LOVERING, J. F. (1958). The nature of the Mohorovicic Discontinuity. *Trans. Am. Geophys. Union* **39,** 947–955.

MACDONALD, G. J. F. (1959). Chondrites and the chemical composition of the earth. In: P. Abelson (ed.), "*Researches in Geochemistry*," pp. 476–494. Wiley, New York.

——— (1964). Dependence of the surface heat flow on the radioactivity of the earth. *J. Geophys. Res.* **69,** 2933–2946.

——— (1965). Geophysical deductions from observations of surface heat flow. In: *Am. Geophys. Union, Geophys. Monograph* **8,** pp. 191–210.

——— and L. KNOPOFF (1958). The chemical composition of the outer core. *Geophys. J.* **1,** 284–297.

MANTON, W. I., and M. TATSUMOTO (1971). Some Pb and Sr isotopic measurements on eclogites from the Roberts Victor mine, South Africa. *Earth Planet. Sci. Letters* **10,** 217–226.

MASON, B. (1960). The origin of meteorites. *J. Geophys. Res.* **65,** 2965–2970.

MATHIAS, M., J. C. SIEBERT, and P. C. RICKWOOD (1970). Some aspects of the mineralogy and petrology of ultramafic xenoliths in kimberlite. *Contr. Mineral. Petrol.* **26,** 75–123.

MOORES, E. M. (1970). Petrology and structure of the Vourinos ophiolitic complex of northern Greece. *Geol. Soc. Am. Spec. Paper* **118,** 1–74.

NALDRETT, A. J., and G. D. MASON (1968). Contrasting Archaean ultramafic igneous bodies in Dundonald and Clergue Townships, Ontario. *Can. J. Earth Sci.* **5,** 111–143.

NESBITT, R. W. (1972). Skeletal crystal forms in the ultramafic rocks of the Yilgarn block, Western Australia; evidence for an Archaean ultramafic liquid. *Geol. Soc. Australia Spec. Pub.* **3,** pp. 331–350.

O'HARA, M. J. (1968). The bearing of phase equilibria on synthetic and natural systems on the origin and evolution of basic and ultrabasic rocks. *Earth Sci. Rev.* **4,** 69–133.

OVERSBY, V. M., and A. E. RINGWOOD (1971). Time of formation of the earth's core. *Nature* **234,** 463–465.

REED, G. W., K. KIGOSHI, and A. TURKEVICH (1960). Determinations of concentrations of heavy elements in meteorites by activation analysis. *Geochim. Cosmochim. Acta* **20,** 122–140.

RICKWOOD, P. C., M. MATHIAS, and J. C. SIEBERT (1968). A study of garnets from eclogite and peridotite xenoliths found in kimberlite. *Contr. Mineral. Petrol.* **19,** 271–301.

RINGWOOD, A. E. (1958). Constitution of the mantle (3); Consequences of the olivine-spinel transition. *Geochim. Cosmochim. Acta* **15,** 195–212.

——— (1959). On the chemical evolution and densities of the planets. *Geochim. Cosmochim. Acta* **15,** 257–283.

——— (1960). Some aspects of the thermal evolution of the earth. *Geochim. Cosmochim. Acta* **20,** 241–259.

——— (1961). Chemical and genetic relationships among meteorites. *Geochim. Cosmochim. Acta* **24,** 159–197.

——— (1962a). A model for the upper mantle. *J. Geophys. Res.* **67,** 857–866.

——— (1962b). A model for the upper mantle, 2. *J. Geophys. Res.* **67,** 4473–4477.

——— (1962c). Present status of the chondritic earth model. In: C. B. Moore (ed.), "*Researches in Meteorites,*" pp. 198–216. Wiley, New York.

——— (1966a) The chemical composition and origin of the earth. In: P. Hurley (ed.), "*Advances in Earth Science,*" pp. 287–356. M. I. T. Press, Cambridge, Mass.

——— (1966b). Chemical evolution of the terrestrial planets. *Geochim. Cosmochim. Acta* **30,** 41–104.

——— (1966c) Genesis of chondritic meteorites. *Rev. Geophys.* **4,** 113–175.

RUBEY, W. W. (1951). Geologic history of sea water. *Bull. Geol. Soc. Am.* **62,** 1111–1147.

SAFRONOV, V. S. (1959). On the primeval temperature of the earth. *Bull. Acad. Sci. USSR., Geophys. Ser.* No. 1, 85–89.

SCHILLING, J. G. (1971). Sea-floor evolution: rare earth evidence. *Phil. Trans. Roy. Soc. London* **A268,** 663–706.

SCHMIDT, O. Y. (1958). "*A Theory of the Origin of the Earth: Four Lectures,*" pp. 1–138. Foreign Languages Publishing House, Moscow; Lawrence and Wishart, London (1959).

STUEBER, A. M., and V. R. MURTHY (1966). Strontium isotope and alkali element abundances in ultramafic rocks. *Geochim. Cosmochim. Acta* **30,** 1243–1259.

TILTON, G. R., and G. W. REED (1963). Radioactive heat production in eclogite and some ultramafic rocks. In: J. Geiss and E. Goldberg (eds.), "*Earth Science and Meteorites,*" pp. 31–43. North-Holland, Amsterdam.

UREY, H. C. (1952). "*The Planets.*" Yale Univ. Press, New Haven, Conn. 245 pp.

——— (1956). The cosmic abundances of potassium, uranium, and thorium, and the heat balances of the earth, the moon and Mars. *Proc. Nat. Acad. Sci. U.S.* **42,** 889–891.

——— (1957). Boundary conditions for the origin of the solar system. In: L. Ahrens, F. Press, K. Rankama, and S. K. Runcorn (eds.), "*Physics and Chemistry of the Earth,*" vol. 2, pp. 46–76. Pergamon, London.

——— (1962). Evidence regarding the origin of the earth. *Geochim. Cosmochim. Acta* **26,** 1–13.

VILJOEN, R. P., and M. J. VILJOEN (1969a). Evidence for the existence of a mobile extrusive peridotitic magma from the Komati Formation of the Onverwacht Group. *Geol. Soc. S. Africa Spec. Pub.* **2,** 87–112.

——— and ——— (1969b). Evidence for the composition of the primitive mantle and its products of partial melting, from a study of the rocks of the Barberton Mountain Land. *Geol. Soc. S. Africa Spec. Pub.* **2,** 275–295.

WAGER, I. R. (1958). Beneath the earth's crust. *Advanc. Sci.* **58,** 1–14. (Presidential address to Section C, British Association for the Advancement of Science.)

WASSERBURG, G. J., G. J. F. MACDONALD, F. HOYLE, and W. A. FOWLER (1964). Relative contributions of uranium, thorium and potassium to heat production in the earth. *Science* **143**, 465–467.

WHITE, R. (1966). Ultramafic inclusions in basaltic rocks from Hawaii. *Contr. Mineral. Petrol.* **12**, 245–314.

WIIK, H. B. (1956). The chemical composition of some stony meteorites. *Geochim. Cosmochim. Acta* **2**, 91–117.

WYLLIE, P. J. (1970). Ultramafic rocks and the upper mantle. *Mineral. Soc. Am. Spec. Paper* **3**, 3–32.

6

CONSTITUTION OF THE UPPER MANTLE

6-1 SEISMIC STRUCTURE

Prior to about 1960, seismic velocity distributions in the upper mantle were obtained from travel times of body waves, and two distributions, those of Jeffreys and Gutenberg, were widely used (Fig. 0-1). In Gutenberg's distribution, a low-velocity (LV) zone was present in the upper mantle, whereas this feature did not appear in Jeffrey's solution. Since 1960, new techniques based upon surface-wave dispersion and free-oscillation data have greatly advanced our understanding of this region. Parallel advances have been made in body-wave seismology, owing to the use of accurately timed and located explosive sources combined with large numbers of improved seismometers. These new methods have largely explained the differences between Jeffreys' and Gutenberg's interpretations.

The principal results have been the recognition of widespread regional differences in upper mantle velocity profiles between continents and oceans and also within individual continents and oceans. These differences extend to depths of at least 400 km. Gutenberg's velocity distributions were obtained primarily from data gathered in western United States, and it has since been confirmed that well-defined low-velocity zones for both *P* and *S* waves exist in this region. On the other hand, a low-velocity zone for *P* waves does not appear to exist in stable continental regions, e.g., eastern United States and the Canadian Shield, in agreement with Jeffreys' interpretation. An important result has been the confirmation of a low-velocity zone for S waves as an *average* feature of the upper mantle. However, the velocity structure and depth of this channel vary widely in different regions, and it is much less marked beneath shield areas. A useful summary of the earlier evidence for the existence of a low-velocity zone beneath western United States is given by Gutenberg (1959).

It is convenient to discuss seismic velocity profiles on the basis of a regional classification consisting of oceanic, stable continental (shield), and tectonically active regions. This classification is necessarily simplified, and additional regional subdivisions of these major provinces might well be considered.[1] In particular, the tectonically active regions embrace a range of different geophysical-geological provinces—e.g., island arcs, Pacific-type continental margins, and alpine mountain belts. For the present purposes, mid-oceanic ridges are not included in this category—their structure was discussed in Sec. 2-3.

Toksöz and Anderson (1966) used phase velocities of mantle Love waves which had traversed a range of great-circle paths comprising different proportions of oceanic, shield, and tectonic regions to estimate the phase velocities corresponding to "pure" provinces as defined above (Fig. 6-1). This figure illustrates

[1]Brune (1969).

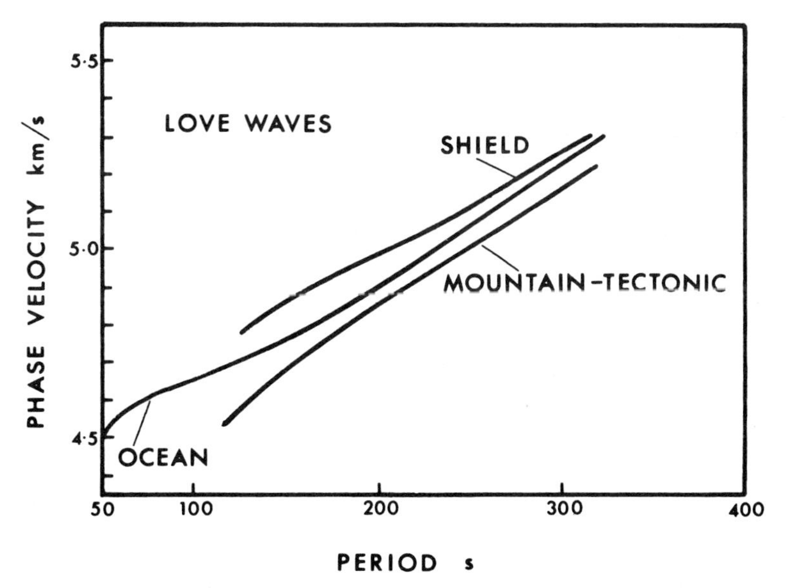

FIGURE 6-1
Pure-path Love-wave dispersion curves for oceanic, Precambrian shield, and tectonically active regions. The "pure paths" were extracted from observed composite-paths data. (*After Toksöz and Anderson, 1966.*)

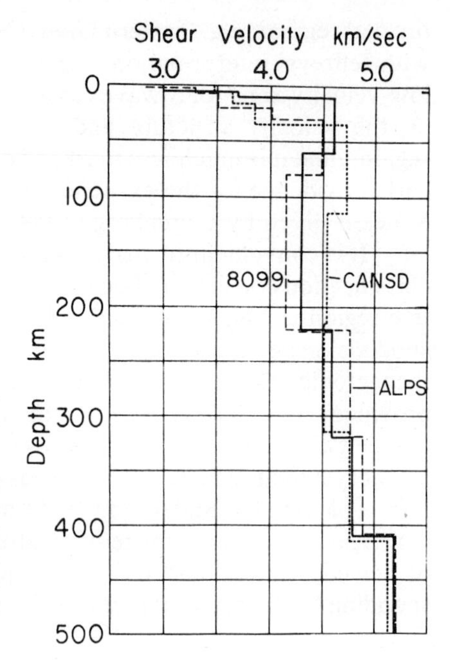

FIGURE 6-2
Three shear velocity models for the upper mantle derived from Rayleigh-wave dispersion. (*From Dorman, 1969, with permission.*) CANDSD is the Canadian Shield structure of Brune and Dorman (1963), 8099 is the oceanic structure of Dorman et al. (1960), and ALPS represents the European Alps structure of Seidl et al. (1966). (*Copyright American Geophysical Union.*)

well the occurrence of regional differences in velocity structures. Shear velocity profiles for some individual regions are shown in Fig. 6-2.

Stable Continental Regions

The highest average seismic velocities in the upper mantle occur beneath Precambrian shields (Figs. 6-1 and 6-2). Beneath the Canadian Shield[1] there is a thick "lid" with approximately constant *S*-wave velocity, below which a slight decrease in velocity occurs. *P*-wave velocities beneath shields[1,2] and other stable continental areas[3,4] do not generally display a minimum. Velocities fall in the range 8.1 to 8.5 km/sec, and some evidence exists for a small increase of velocity with depth.[1,2,5] Attenuation of S_N and P_N body waves is much smaller on shields as compared to other provinces. *P* and *S* waves characteristically arrive early[6,7] at shield stations owing to the relatively high average upper mantle velocities.

[1]Brune and Dorman (1963).
[2]Doyle (1957)
[3]Roller and Jackson (1966).
[4]Hales, Cleary, Doyle, Green, and Roberts (1968).
[5]Green and Hales (1968).
[6]Hales and Doyle (1967).
[7]Cleary and Hales (1966).

Oceanic Regions (Exclusive of Mid-Oceanic Ridges and Island Arc-Trench Systems)

Surface waves reveal the presence of a well-defined low-velocity layer[1] between depths of about 70 and 200 km, with a velocity in the channel of about 4.3 km/sec. The detailed structure of the low-velocity zone is not well resolved, and variations in the minimum velocity within the channel may be compensated by corresponding changes in thickness of the channel. S_N body waves from explosions are strongly attenuated, and few recordings exist.[2] P-wave travel-time data are limited. Amplitude studies along a traverse in the Gulf of Mexico indicated the presence of a low-velocity zone between depths of about 100 and 200 km.[3]

The oceanic low-velocity zone is overlain by a high-velocity "lid" about 70 km thick[4] in which P-wave velocities are usually in the range 7.9 to 8.5 km/sec, with average S-wave velocities of 4.6 km/sec. P-wave velocities show a slight increase with depth[3,5,6] in the lid. The bottom of the lid appears to be marked by a relatively abrupt decrease of S velocity, reaching a minimum of 4.3 km/sec in the channel.

Active Tectonic Regions

Western United States represents the most extensively studied region of this kind. Body-wave studies show the presence of low-velocity zones both for P and for S waves.[7] A detailed study of the P-wave-velocity distribution has been carried out by Archambeau et al. (1969), using extensive spectral amplitude and travel-time data (including later arrivals and travel-time delays). They conclude that a lid with a thickness of 10 to 50 km and a velocity close to 8.0 km/sec exists in most regions. Below the lid, velocity probably decreases sharply to an average minimum of 7.7 km/sec (Fig. 6-3). The bottom of the low-velocity zone at about 150 km is also probably characterized by high positive velocity gradients.[8] In the Basin and Range province, the lid is absent, and the low-velocity zone extends to the M discontinuity.

Nuttli and Bolt (1969) have analyzed the P-wave travel-time delays as a function of azimuth of source and have obtained striking evidence of large local variations in the width and depth of the low-velocity zone. They obtain a minimum P-wave velocity in the channel of only 7.2 km/sec. The low-velocity channel is overlain by a 45 to 95 km thick lid with a V_P of 7.9 km/sec.

[1]E.g., Dorman, Ewing, and Oliver (1960).
[2]Doyle and Webb (1963).
[3]Hales, Helsley, and Nation (1970).
[4]Kanamori and Press (1970).
[5]Shurbet (1964).
[6]Helmberger and Morris (1969).
[7]Gutenberg (1959).
[8]See also Johnson (1967).

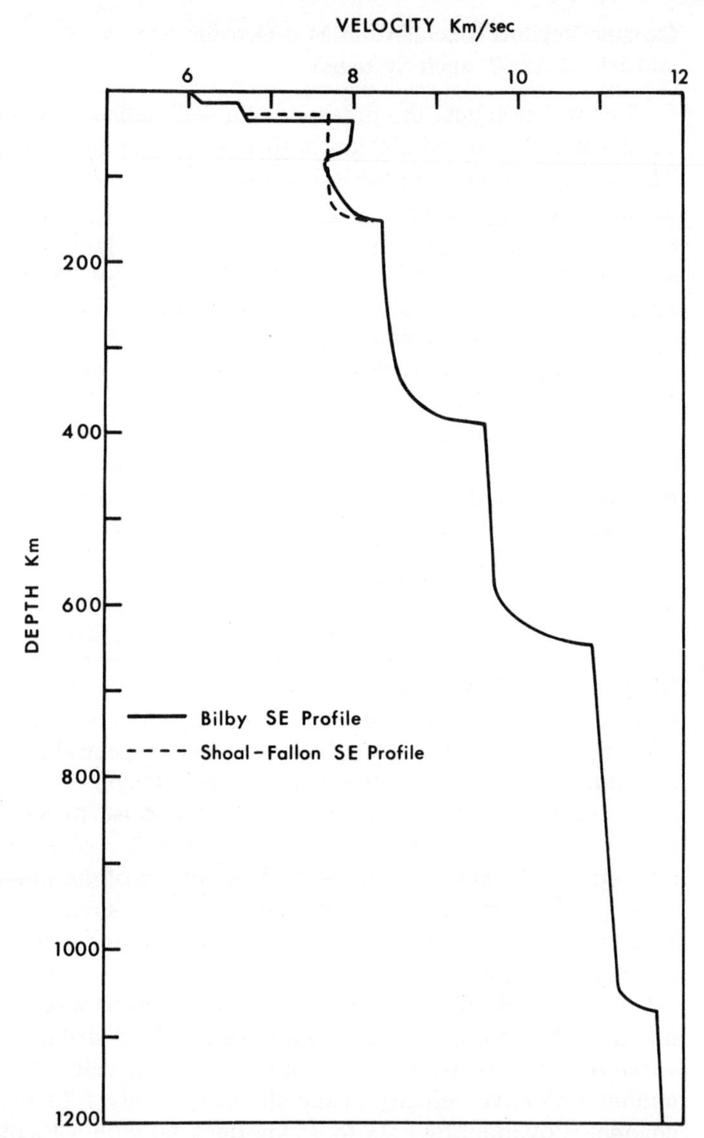

FIGURE 6-3
Upper mantle compressional velocity models for tectonic provinces of the continental United States (*After Archambeau, Flinn, and Lambert, 1969, with permission. Copyright American Geophysical Union.*)

- - - Basin and Range province (Shoal-Fallon SE profile).
——— Colorado Plateau–Rocky Mountain province (Bilby SE profile).

Brooks (1962) has demonstrated that a low-velocity zone for both *P* and *S* waves exists in a second tectonically active zone—the New Guinea–Solomon Island region. The channel is thicker and has lower minimum velocities than found by Gutenberg for western United States. The *P* velocity reaches a minimum of about 7.5 km/sec at a depth of 120 km, whilst the *S*-wave velocity indicated is about 4.2 km/sec at the same depth.

The subalpine profile (Fig. 6-2) indicates an average *S*-velocity minimum smaller than occurs beneath oceanic regions. Further detailed studies[1] suggest considerable local variations of velocity profiles in tectonically active regions. In the western Mediterranean and probably extending under the Alps, a channel with a minimum *S*-wave velocity of 4.10 km/sec was found.[1] This was overlain by a lid with variable thickness and with *P*- and *S*-wave velocities of 8.15 and 4.8 km/sec, respectively.

6-2 PETROLOGICAL ZONING

Chemical zoning of the upper mantle was discussed in Sec. 5-1, where it was concluded that the fundamental structure consisted of a layer of refractory peridotite of varying thickness passing downward into pyrolite (Fig. 5-1). An important property of rocks approaching the pyrolite composition is their capacity to crystallize in four distinct mineral assemblages[2,3] as follows:

Olivine + amphibole:	ampholite
Olivine + Al-poor pyroxenes + plagioclase:	plagioclase pyrolite
Olivine + aluminous pyroxenes ± spinel:	pyroxene pyrolite
Olivine + Al-poor pyroxenes + pyrope-rich garnet:	garnet pyrolite

These mineral assemblages possess distinctly different physical properties (Table 6-1) clearly indicative of different P-T-P_{H_2O} conditions of crystallization and equilibration. This raises the possibility that large-scale mineralogical zoning controlled by the stability fields of these mineral assemblages may be present in the upper mantle.

Ampholite

At low total pressures, the stabilities of amphiboles are controlled principally by temperature and by water pressure. For $P_{H_2O} = P_{load} = 2$ to 10 kbars, most common amphiboles of the types found in mafic and ultramafic rocks decompose to anhydrous mineral assemblages at temperatures in the vicinity of 1000°C.[4] At

[1] Berry and Knopoff (1967).
[2] Ringwood (1962a,b).
[3] Green and Ringwood (1963).
[4] Ernst (1968).

pressures higher than 20 kbars and at 1100°C, Green and Ringwood (1967a) obtained evidence that amphibole was unstable in olivine-rich basaltic compositions under P_{H_2O} conditions similar to those which caused formation of amphibole between 10 and 15 kbars. It was inferred that amphibole would decompose to a denser eclogitic mineral assemblage plus water under these conditions and that the breakdown curves would have negative slopes at pressures above 15 to 20 kbars because the large decrease in the specific volume of water at high pressure would change the Δv of the reaction

$$\text{Amphibole} \rightleftharpoons \text{eclogitic assemblage} + H_2O$$

from positive to negative. Green and Ringwood also pointed out that these relationships necessarily imply a limit to the occurrence of amphibole in the upper mantle. The inferred instability of amphibole at high pressures was confirmed by Lambert and Wyllie (1968), who showed that an amphibole separated from a granodiorite transformed to garnet and pyroxene at 25 kbars, 750°C in the presence of excess water. Gilbert (1969) also obtained analogous results upon a series of synthetic end-member amphibole components. Several studies of the stability field of amphibole in mafic[1,2,3] and ultramafic[4,5,6] rocks have since been reported. The boundaries of the stability fields in the subsolidus region are not accurately established because the equilibria are sluggish and difficult to reverse.

The stability field of amphibole in pyrolite containing 0.1% water extends to the solidus (0.1% H_2O) shown in Figs. 4-6 and 6-4. As discussed above, amphibole is unstable at pressures above about 25 to 30 kbars subsolidus conditions over a wide range of temperatures. The stability limits define a broad region, extending to depths of about 90 km at temperatures below about 1100°C, in which the ampholite assemblage could be stable within the upper mantle. The stability field of ampholite thus overlaps extensively the stability fields of the anhydrous plagioclase and pyroxene pyrolite assemblages (Fig. 6-4).

The actual occurrence of the ampholite assemblage in the upper mantle will be stongly controlled by the amount of water present. In order to crystallize completely to an assemblage of (olivine + amphibole), a water content of about 0.3 to 0.4% would be required in pyrolite. With the preferred mean water content of 0.1% in primary pyrolite (Sec. 4.5), there is insufficient water to convert all the pyroxenes to amphibole so that transitional (olivine + amphibole + pyroxene) assemblages are formed.

The ideal ampholite assemblage is likely to occur in regions of the mantle where water has been somewhat concentrated—e.g., in the wedges overlying Benioff zones and perhaps in the uppermost mantle immediately beneath the

[1]Lambert and Wyllie (1970a,b).
[2]Essene, Hensen, and Green (1970).
[3]Hill and Boettcher (1969).
[4]Green and Ringwood (1970).
[5]Kushiro (1970).
[6]Green (1973).

Mohorovicic discontinuity. Since water is probably a relatively mobile component within the mantle, ampholite may be expected to occur wherever differentiation processes have caused its local enrichment. It is probable that the transitional assemblage (olivine + pyroxene + amphibole), corresponding to pyrolite containing about 0.1% water is more widely distributed in the outer 90 km of the mantle than the ideal ampholite assemblage (0.3% H_2O).

The upper 90 km of the mantle beneath oceans, and to a greater depth beneath Precambrian shields, is believed to have been extensively fractionated owing to extraction of varying amounts of basaltic magmas (Chap. 4). Since water is strongly partitioned into the liquid phase, the extraction of even very small amounts of alkalic basalt magmas would cause strong depletion of water (and other incompatible elements) in the mantle, whilst not greatly affecting the major element composition (cf compositions of high-temperature peridotites). Magmatic processes of this kind, leading to effective dehydration of large regions of the upper mantle will greatly limit the occurrence of amphibole in this region.

The transition from the ampholite to the pyroxene pyrolite assemblage is accompanied by a significant increase of seismic velocity (Table 6-1), and the occurrence of these mineral assemblages within the upper mantle, controlled by small variations in water content, is likely to influence the seismic structure of this region. Temperature gradients in the uppermost 70 km of the oceanic upper mantle are sufficiently high to cause seismic velocities to decrease with depth, providing this layer is homogeneous.[1,2] However, seismic velocities are observed to increase with depth in this layer,[3,4] implying the existence of mineralogical inhomogeneity. This may well be due to a decrease in the abundance of

[1]Birch (1952).
[2]Gutenberg (1959).
[3]Shurbet (1964).
[4]Helmberger and Morris (1969).

Table 6-1 CALCULATED DENSITIES AND *P*-WAVE VELOCITIES AT ATMOSPHERIC TEMPERATURE AND PRESSURE FOR PYROLITE MINERAL ASSEMBLAGES AND FOR PERIDOTITE AND DUNITE. (*After Ringwood, 1966*)

Rock	Density (g/cm^3)	V_p (km/sec)
Ampholite*	3.27	7.98
Plagioclase pyrolite	3.26	8.01
Pyroxene pyrolite	3.33	8.18
Garnet pyrolite	3.38	8.38
Peridotite†	3.31	8.32
Dunite	3.32	8.48

*35% amphibole.
†20% orthopyroxene.

amphibole with depth caused by a corresponding decrease in water content.[1] Within this region, ampholite ($V_P = 8.0$ km/sec) would transform gradually to pyroxene pyrolite ($V_P \sim 8.2$ km/sec) with increasing depth. Alternatively, if sufficient water (0.3%) were present to maintain the stability of ampholite to 60 to 70 km, and the water content below this depth decreased to, say, 0.1%, then ampholite would transform directly to garnet pyrolite ($V_P \sim 8.4$ km/sec). Hales et al. (1970) have observed a rapid increase in P velocity from 8.1 to 8.6 km/sec at a depth of about 60 km beneath the Gulf of Mexico. This increase of velocity and the position of the discontinuity would be consistent with the ampholite to garnet pyrolite transformation. The absolute velocities are somewhat higher than anticipated for ampholite and garnet pyrolite of the model composition, but considering the various sources of uncertainty, including possible mantle anisotropy and slope of the discontinuity, the discrepancy is not serious.

Plagioclase Pyrolite

The plagioclase pyrolite field in Fig. 6-4 consists of the minerals olivine, diopsidic pyroxene and orthopyroxene (low in aluminum), and plagioclase. With increasing pressure, plagioclase becomes unstable and breaks down according to the following simplified equilibria:

$$\underset{\text{albite}}{NaAlSi_3O_8} + \underset{\text{forsterite}}{Mg_2SiO_4} \rightarrow \underset{\text{jadeite}}{NaAlSi_2O_6} + \underset{\text{enstatite}}{2MgSiO_3} \qquad \text{(I)}$$

$$\underset{\text{anorthite}}{CaAl_2Si_2O_8} + \underset{\text{forsterite}}{Mg_2SiO_4} \rightarrow \underset{\text{Ca-Tschermak's molecule}}{CaAl_2SiO_6} + \underset{\text{enstatite}}{2MgSiO_3} \qquad \text{(II)}$$

The jadeite and Tschermak's molecule enter into solid solution in pyroxenes. Thus the net effect is the formation of omphacite and aluminous pyroxene at the expense of plagioclase.

Reactions between olivine and plagioclase to yield aluminous pyroxenes ± spinel have been studied for the system Fo-An[2] and for the systems Fo-An, olivine (Fo_{92})-labradorite (Ab_{41}-An_{59}), and for pyrolite composition.[3] The boundary shown in Fig. 6-4 is based on the latter results, combined with a gradient obtained from the Fo-An system.[2] These investigations show that plagioclase pyrolite will not occur as the stable mantle assemblage along typical geothermal gradients in either continental or oceanic regions, but it may be present in areas possessing particularly high heat-flows and thin crusts where near-solidus temperatures are reached at depths of 25 to 35 km. Regions where plagioclase pyrolite may play a significant role include mid-oceanic ridges, the Basin and Range province of western United States, and perhaps, beneath evolving island-arc structures such as Japan. The low mantle velocities observed in

[1]Ringwood (1962b, 1966).
[2]Kushiro and Yoder (1966).
[3]Green and Hibberson (1970).

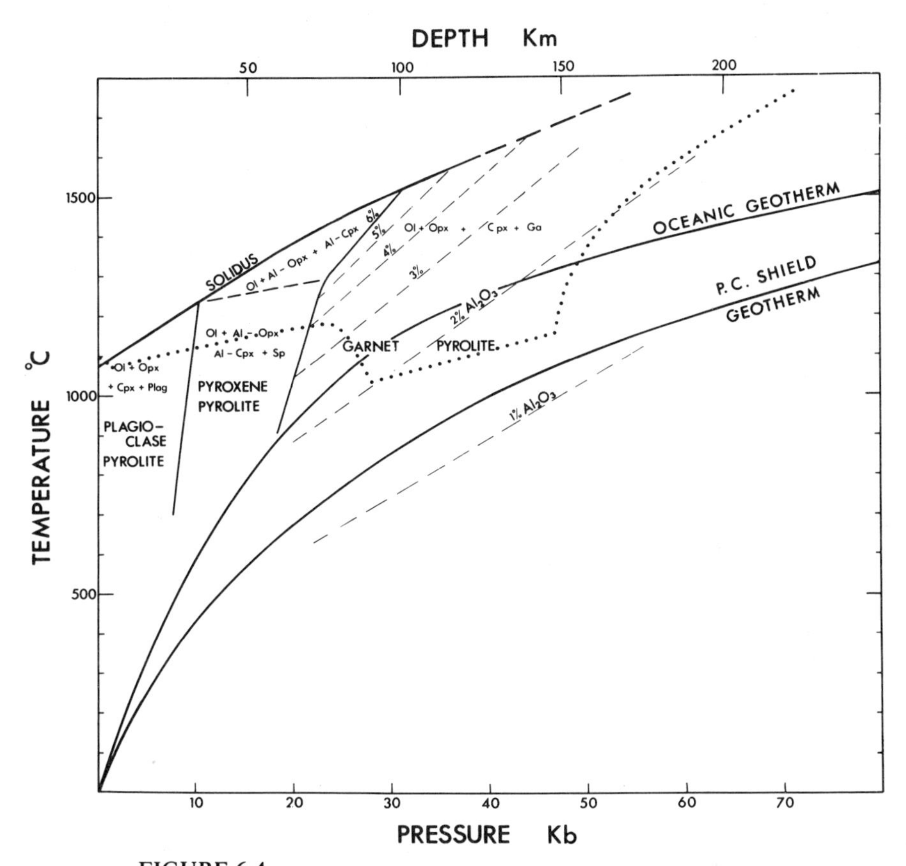

FIGURE 6-4
Diagram illustrating the P-T fields of pyrolite mineral assemblages. (*After Green and Ringwood, 1967b.*) The figures 1% Al_2O_3, 2% Al_2O_3, etc., refer to the Al_2O_3 content of orthopyroxene in equilibrium with garnet in the garnet pyrolite field. The oceanic and Precambrian shield geotherms are those given by Clark and Ringwood (1964). The dotted line represents the solidus for pyrolite containing 0.1% water (from Fig. 4-6).

these regions may be partly due to the presence of plagioclase, although other factors—high temperatures and partial melting—are perhaps of comparable importance (Sec. 6-3).

The relationship between thermal structure and elevation of mid-oceanic ridges for a zoned pyrolite model based on Figs. 5-1 and 6-4 was investigated by Sclater and Francheteau (1970). When allowance was made for phase changes in pyrolite, particularly the occurrence of a zone of plagioclase pyrolite extending to a depth of 35 km beneath the ridges,[1] excellent agreement between calculated and observed elevations was obtained. Transformations between pyrolite mineral assemblages accounted for about 20% of the total elevation of the ridge.

[1]See also Miyashiro, Shido, and Ewing (1970).

Pyroxene Pyrolite

The pyroxene pyrolite field consists of the assemblage olivine + aluminous enstatite + aluminous diopside ± spinel (Fig. 6-4). The boundary of this assemblage with the higher pressure garnet pyrolite assemblage has been the subject of extensive investigations, both in simplified model systems[1,2,3,4,5,6,7] and in "natural" pyrolite[8,9] and peridotite[10] systems. An early estimate of the high-pressure boundary of the pyroxene pyrolite field[11,12] based mainly upon a consideration of simplified equilibria was subsequently found to be about 5 kbars too high.[8,9,10] This is indicative of uncertainties attached to extrapolations of phase boundaries determined in simplified systems to complex natural systems. Nevertheless, studies of simplified model systems have provided invaluable insight into the chemistry of the more complex systems, particularly in understanding the role of specific compositional variables in controlling individual equilibria. Noteworthy are the studies of MacGregor (1964, 1970) on the equilibrium

$$\underset{\text{pyroxene}}{4MgSiO_3} + \underset{\text{spinel}}{MgAl_2O_4} \rightleftharpoons \underset{\text{forsterite}}{Mg_2SiO_4} + \underset{\text{pyrope}}{Mg_3Al_2Si_3O_{12}} \qquad \text{(III)}$$

and on the influence of chromium and calcium upon this equilibrium. These demonstrate in particular that increasing the Cr_2O_3/Al_2O_3 ratio of the system causes a substantial increase in the pressure required to transform spinel peridotite to garnet peridotite. Other significant series of experiments upon relevant simplified systems are those of Boyd and England (1964) on enstatite-pyrope solubility relations and Boyd (1970) on the system $CaSiO_3$-$MgSiO_3$-Al_2O_3.

The complexity of the equilibria indicated the desirability of determining the stability fields of mineral assemblages directly upon natural compositions. This was undertaken on the pyrolite composition by Green and Ringwood (1967b, 1970) and on a natural garnet peridotite by Ito and Kennedy (1967). Results of experimental runs on the pyrolite composition are shown in Fig. 6-5. Equilibrium was demonstrated by reversing the reactions along key boundaries. Also, the compositions of the pyroxenes in the garnet pyrolite field were determined by electronprobe microanalysis, permitting the Al_2O_3 isopleths to be constructed as in Fig. 6-4.

Between *AB* and *ELF* (Fig. 6-5), both garnet and plagioclase are absent, and the mineral assemblage is that of pyroxene pyrolite. This field is divided into two

[1] MacGregor (1964).
[2] MacGregor and Boyd (1964).
[3] MacGregor and Ringwood (1964).
[4] MacGregor (1968).
[5] MacGregor (1970).
[6] Boyd and England (1964).
[7] Boyd (1970).
[8] Green and Ringwood (1967b).
[9] Green and Ringwood (1970).
[10] Ito and Kennedy (1967).
[11] Ringwood, MacGregor, and Boyd (1964).
[12] Ringwood (1966).

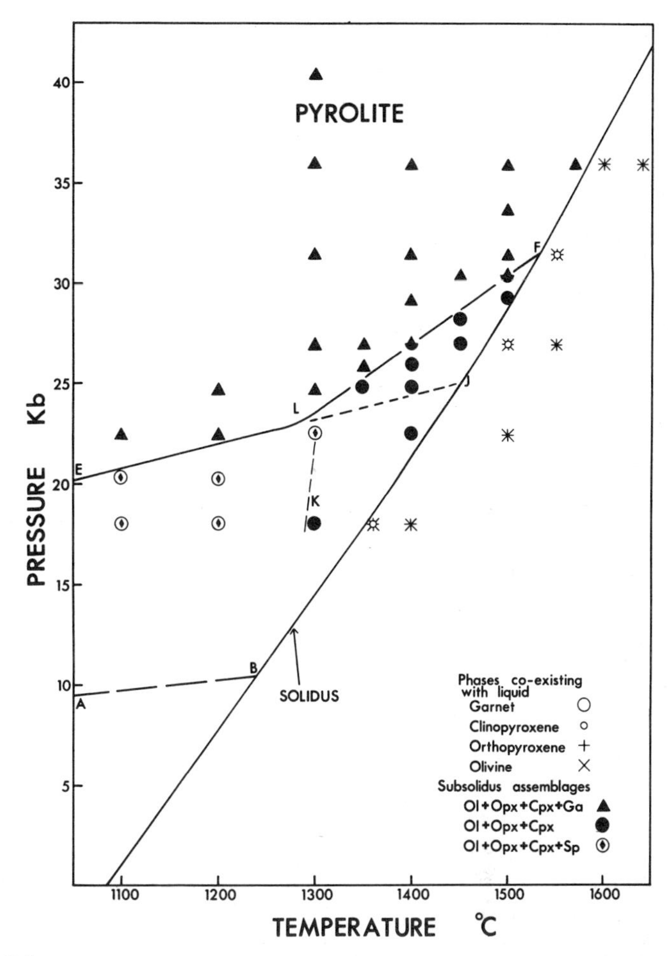

FIGURE 6-5
Results of experimental runs on the pyrolite composition illustrating detailed relationships between the garnet and pyroxene pyrolite fields. (*From Green and Ringwood, 1967b, with permission.*)

subfields by the line *K*. To the left of *K*, pyroxenes are unable to hold in solid solution the entire amount of R_2O_3 ($Al_2O_3 + Cr_2O_3 + Fe_2O_3$) present in pyrolite, and the excess R_2O_3 occurs in the form of (MgFe) (Al, Cr, Fe)$_2$O$_4$ spinel solid solution. As temperature increases, the solubility of R_2O_3 in pyroxene in equilibrium with spinel increases, and the amount of spinel therefore decreases. Along *K*, spinel is finally consumed according to the simplified reaction

$$\underset{\text{spinel}}{MgAl_2O_4} + \underset{\text{enstatite}}{mMgSiO_3} \rightleftharpoons \underset{\text{aluminous enstatite}}{(m-2)MgSiO_3 \cdot MgAl_2SiO_6} + \underset{\text{forsterite}}{MgSiO_4} \qquad \text{(IV)}$$

To the right of *K*, the phases present are olivine + aluminous pyroxenes. The extent of this latter field is sensitively dependent upon the R_2O_3/pyroxene ratio of the bulk rock and also upon the Cr/Al ratio of the R_2O_3 component.[1] Experiments upon modified pyrolite compositions possessing substantially higher R_2O_3/pyroxene ratios showed the (spinel + pyroxenes) field extending along *ELJ* (Fig. 6-5), and the spinel-free field *LJF* was not encountered.[2] In these compositions, garnet was encountered on the high-pressure side of *LJ*.

At higher pressures, pyroxene pyrolite becomes unstable and transforms to the denser garnet pyrolite assemblage consisting of olivine, orthopyroxene, diopsidic pyroxene (both lower in Al), and pyrope-rich garnet. The transformation involves two distinct equilibria, and there is a sharp change of gradient where they intersect. Along the boundary *LE*, spinel and pyroxene react to produce garnet and olivine as described earlier. However, along boundary *LF*, pyrope garnet is formed by the breakdown of aluminous pyroxene according to a different equilibrium (simplified):

$$\underset{\text{aluminous enstatite}}{m\mathrm{MgSiO_3} \cdot n\mathrm{MgAl_2SiO_6}} \rightleftharpoons$$

$$\underset{\text{pyrope garnet}}{\mathrm{Mg_3Al_2Si_3O_{12}}} + \underset{\text{aluminous enstatite}}{(m-2)\mathrm{MgSiO_3} \cdot (n-1)\mathrm{MgAl_2SiO_6}} \qquad \text{(V)}$$

An analogous reaction can be written for aluminous diopsidic pyroxene. Since garnet and pyroxene have similar basic formulae of the R_2O_3 type (e.g., pyrope can be written as $3MgSiO_3 \cdot Al_2O_3$ and aluminous enstatite as $MgSiO_3 \cdot xAl_2O_3$), we are dealing essentially with a *P-T*-controlled solid solubility of garnet in pyroxene. As the pressure increases, and as temperature decreases, the solid solubility of garnet in pyroxene falls.

Ito and Kennedy (1967) presented data on subsolidus phase relations in a garnet lherzolite nodule from kimberlite. Their determinations of *P-T* conditions for transformation of spinel to garnet peridotite (22 kbars, 1300°C) agrees with the results of Green and Ringwood (23 kbars, 1300°C). However, they failed to locate a spinel-free field of aluminous pyroxenes, despite the fact that the R_2O_3/pyroxene ratio in their starting composition was smaller than that of the pyrolite of Fig. 6-5. This was apparently caused primarily by the fact that their solidus lay some 100°C lower than the pyrolite solidus. Entry of water from the pressure medium into their sample is perhaps responsible for the discrepancy.

The experimentally determined boundaries of the pyrolite stability fields are shown in relation to possible suboceanic and subcontinental temperature distributions[3] in Fig. 6-4. It is seen that, in "dry" ($H_2O < 0.1\%$) regions of the oceanic upper mantle, the pyroxene pyrolite (+ spinel) assemblage is stable to a depth of about 70 km, beyond which the garnet pyrolite field is entered. This is consistent with the occurrence and mineralogy of many alpine peridotites[4,5] which contain

[1]MacGregor (1970).
[2]Green and Ringwood (1967b, 1970).
[3]Clark and Ringwood (1964).
[4]MacGregor and Boyd (1964).
[5]Ito and Kennedy (1967).

spinel rather than garnet and occur in orogenic environments in which mantle temperature distributions were probably higher at the time of peridotite intrusion than the oceanic geotherm of Fig. 6-4.

With increasing depth below the Moho along the oceanic geotherm, the amount of spinel decreases and the Al_2O_3 contents of the pyroxenes increase. Prior to the incoming of garnet at about 70 km and 1000°C, aluminous spinel would coexist with orthopyroxene containing about 3% Al_2O_3. An intersection of the geotherm with the phase boundary at a higher temperature than that illustrated in Fig. 6-4 would yield assemblages with less spinel and with orthopyroxene of higher Al_2O_3 content (4 to 5% Al_2O_3). An extremely steep geothermal gradient would be required to enter the olivine + aluminous pyroxene (6% Al_2O_3) field. Such gradients must be attained in regions of partial melting and basaltic magma generation. It may be noted that garnet does not appear on the anhydrous pyrolite solidus until depths of over 100 km.

At depths of 60 to 70 km on the oceanic geotherm, garnet appears from reaction (III) and is in equilibrium with orthopyroxene containing about 3% Al_2O_3. It is estimated[1] that about 6% garnet would appear at 60 to 70 km on the geothermal gradient illustrated. If the geothermal gradient intersected the boundary at a higher temperature, the amount of garnet appearing would be correspondingly less. The incoming of garnet due to reaction (III) at about 60 to 70 km in the oceanic mantle probably occurs over a relatively small depth interval (5 to 15 km) and would be accompanied by small ($\sim$ 1%) increases in density and seismic velocities. The latter may contribute to the fine structure of the seismic velocity distribution in this region.

Along the Precambrian shield geotherm, there is only a very small region immediately below the crust which lies in the pyroxene pyrolite field, and the geotherm passes at about 45 km into the garnet pyrolite field. This is consistent with the dominant occurrence of garnet peridotite xenoliths in kimberlite pipes which penetrate shields. The presence of chemical zoning (Fig. 5-1) in the mantle beneath Precambrian shields also influences the mineral assemblages present. Reduction in the R_2O_3/pyroxene ratio of residual peridotite due to earlier extraction of basaltic magmas may cause the pressure required for the transition to garnet peridotite to be increased, as discussed earlier. Preceding partial melting episodes also tend to cause an increase in the Cr_2O_3/R_2O_3 ratio of the residual peridotite. MacGregor (1970) has shown that increases in this ratio cause a correspondingly marked increase in the pressure required for the transition of spinel peridotite to garnet peridotite (Fig. 6-6). Thus, the stability fields of residual spinel peridotites and of garnet peridotites and garnet pyrolite may overlap substantially. This explains the occurrence of some spinel peridotites associated with garnet peridotites in diamond pipes.[2] (MacGregor, 1970, also pointed out that the garnet peridotite intrusions which occur in some crustal metamorphic environ-

[1]Green and Ringwood (1967b, 1970).
[2]Williams (1932).

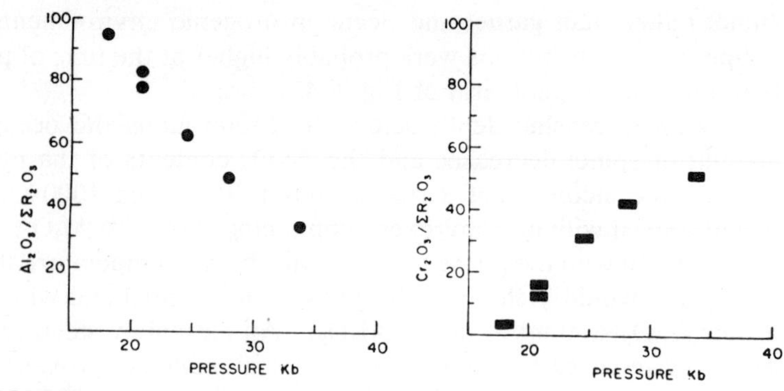

FIGURE 6-6
Pressure-composition plots at 1200°C showing pressure at which spinel peridotite transforms to garnet peridotite for particular $Al_2O_3/\Sigma R_2O_3$ and $Cr_2O_3/\Sigma R_2O_3$ ratios, where $\Sigma R_2O_3 = Al_2O_3 + Cr_2O_3 + Fe_2O_3$. (*From MacGregor, 1970, with permission.*)

ments have low Cr_2O_3/Al_2O_3 ratios and that this factor is capable of explaining their occurrence.)

Chemical zoning (Fig. 5-1) caused by extraction of the low-melting basaltic component from the uppermost mantle may also influence velocity profiles. The amount of pyroxene remaining in the residual peridotite after extraction of basalt may vary from zero to about 40%, depending upon the degree and depth of partial melting and the nature of the basalt extracted (Sec. 4-5). The *P*-wave velocities of common pyroxenes (jadeite excepted) are smaller than those of associated olivines, and significant velocity variations (8.1 to 8.5 km/sec) would occur among residual peridotites according to the proportion of pyroxene present. This factor may contribute towards a complex fine structure, particularly in the mantle beneath shields and evolved continental regions. For example, the small increase in *P* velocity observed at a depth of about 90 km beneath Central United States[1,2] could be caused by a transition from fractionated residual peridotite containing 30 to 40% orthopyroxene into unfractionated garnet pyrolite.

Garnet Pyrolite

The characteristic occurrence of this assemblage in the mantle beneath Precambrian shields has been mentioned above. The garnet pyrolite assemblage appears to be stable over a very wide *P-T* field, extending to depths of about 350 km.[3] The determination of detailed element partition relationships between minerals of this assemblage holds the promise of being able to estimate the *P-T* conditions of for-

[1]Roller and Jackson (1966).
[2]Hales (1969).
[3]Chapter 14.

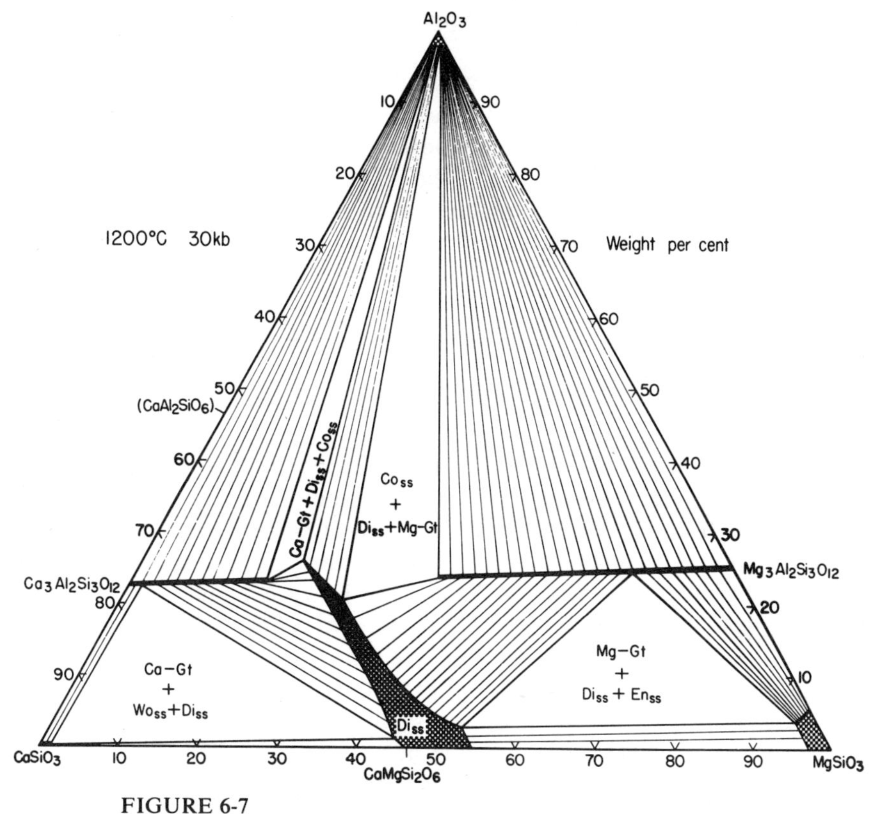

FIGURE 6-7
Phase relationships in the system $CaSiO_3$-$MgSiO_3$-Al_2O_3 at 30 kbars and 1200°C. (*From Boyd, 1970, with permission.*)

Wo = wollastonite	Di = diopside
Co = corundum	En = enstatite
Gt = garnet	ss = solid solution

mation and equilibration of natural garnet peridotites. The alumina contents of orthopyroxenes in equilibrium with garnet (Fig. 6-4) vary systematically with pressure and temperature. When these data[1] are combined with solid-solubility relationships between coexisting orthopyroxenes and clinopyroxenes,[2,3,1] an estimate of temperature and pressure of equilibration becomes possible. O'Hara (1967) has constructed a petrogenetic grid using these and other data. Although of limited value in the pyroxene pyrolite field because of uncertainties introduced by changes in Cr/Al ratios and other compositional variables, the application of the grid to determining *P-T* conditions in the garnet pyrolite field is promising, al-

[1]Green and Ringwood (1967b, 1970).
[2]Davis and Boyd (1966).
[3]Boyd and Schairer (1964).

though here, also, the effects of compositional variables such as the Cr/Al ratio remain to be established.

Boyd (1970) has carried out a detailed investigation of subsolidus equilibria in the system $CaSiO_3$-$MgSiO_3$-Al_2O_3 at 30 kbars, in the course of which the compositions of coexisting phases were determined by electronprobe microanalysis. Phase relationships and tie lines are shown in Fig. 6-7. The further accumulation of data of this type[1] should ultimately lead to a detailed understanding of the pressure-temperature conditions of equilibration among naturally occurring garnetiferous ultramafic rocks.

6-3 THE LOW-VELOCITY ZONE

The occurrence and seismic structure of the low-velocity zone was described in Sec. 6-1. In a homogeneous medium, seismic velocity generally increases with rising pressure and decreases with rising temperature. Consequently, a critical temperature gradient exists within the earth at which these effects cancel each other and velocity remains constant with depth. A list of such critical gradients for different minerals is given in Table 6-2. Early explanations[2,3,4] of the occurrence of the low-velocity zone were based on this effect. The critical temperature gradients, particularly for *S* waves (Table 6-2), are probably exceeded within the outer 150 km so that there is little doubt that this explanation is at least partly correct. Nevertheless, other factors must also be involved. If high-temperature gradients were solely responsible, the decrease of velocity with depth would be most rapid immediately beneath the Moho, where the temperature gradients are highest. On the contrary, velocities are observed to increase initially with depth in the lid beneath the oceanic Moho, showing that chemical and mineralogical heterogeneity also play a role.[5]

Beneath Precambrian shields there is evidence for a small decrease of *S* velocity occurring at a depth of about 120 km, whereas *P*-wave velocity does not display a minimum.[6] Clark and Ringwood (1964) pointed out that this might be explained by the fact that the critical gradients for *P* waves were generally much higher than for *S* waves. It appears likely that the seismic velocity distributions beneath shields can be satisfactorily explained in terms of critical temperature gradients combined with the presence of mineralogical and chemical heterogeneity.

However, these factors do not appear to provide a satisfactory explanation of the nature of the low-velocity zone beneath tectonic and oceanic regions. Birch

[1] See also Kushiro, Syono, and Akimoto (1967).
[2] Birch (1952).
[3] Valle (1956).
[4] MacDonald and Ness (1961).
[5] Section 6-2.
[6] Section 6-1.

Table 6-2 CRITICAL THERMAL GRADIENTS FOR A LOW-VELOCITY LAYER IN SOME MINERALS OF GEOPHYSICAL INTEREST.* (*After Liebermann and Schreiber, 1969*)

	$(\partial T/\partial r)_{v_p}$ (°C/km)	$(\partial T/\partial r)_{v_s}$ (°C/km)
Olivine	7.0	3.0
Pyroxene†	9.5	3.9
Garnet	8.2	4.1
Spinel	5.7	0.7
Corundum	5.5	2.8
Periclase	5.9	3.2
Rutile	3.8	1.3

*These critical gradients apply to conditions of low pressure and temperature. Birch (1969) has estimated that the critical gradients for olivine, periclase, and corundum at 1200°C are about 20 to 30% smaller than in this table.
†Birch (1960).

(1969) has carried out a detailed review of the elastic properties of Mg-rich olivine as a function of pressure and temperature and has calculated the temperature distribution required to produce Gutenberg's and Anderson's velocity-depth distributions. Birch found that the temperatures required were several hundred degrees higher than the dry pyrolite solidus (Fig. 6-4). On the other hand, assuming a reasonable temperature distribution for the upper mantle, he found that the temperature-induced decrease in seismic velocity in the outer 150-km zone composed of olivine was unlikely to exceed about 2%. This may be compared with seismically inferred velocity decreases ranging from 3 to 10% in the low-velocity zone. Anderson and Sammis (1970) carried out an analogous study, reaching a similar conclusion. They also cited evidence indicating that the low-velocity channel was bounded above[1] and below by regions possessing high negative[1] and positive velocity gradients, respectively. Anderson and Sammis pointed out that it was not possible to explain the presence of these high velocity gradients in terms of corresponding high temperature gradients. They also noted that the temperature gradients required to cause a low-velocity zone implied the occurrence of an unacceptably high heat-flux at depth in the mantle.[2]

The possibility that the low-velocity zone is caused by transformations between pyrolite stability fields has also been explored.[3] Preliminary results suggested that these might provide a satisfactory explanation; however, subsequent, more precise determinations of the P-T boundary between pyroxene

[1]This conclusion was also reached by Kanamori and Press (1970).
[2]This point has been previously made by Ringwood (1962a).
[3]Ringwood (1962a,b, 1966).

pyrolite and garnet pyrolite[1,2] showed that this transition occurred at too shallow a depth to be responsible for the termination of the low-velocity zone (150 to 200 km). Furthermore, velocity changes caused by this transition amounted to about 2% (Table 6-1), which is insufficient to account for the magnitude of the velocity-decrease in the channel[3] (3 to 10%). Nevertheless, the small changes in velocity caused by mineralogical transformations may result in subtle effects upon velocity gradients in the region of the low-velocity zone which contribute to the seismic fine structure in this region. These are discussed in greater detail elsewhere.[1]

The above considerations suggest that some other factor is at least partially responsible for the occurrence of a low-velocity zone beneath tectonically active and oceanic regions. The factor which has been given the widest consideration in this respect is partial melting.[4,5,6,7,8,9,10]

Data on the elastic properties of partially molten systems are limited. Spetzler and Anderson (1968) studied the effects of grain-boundary melting in the NaCl-ice system. As the sample was warmed, both the compressional and shear velocities dropped abruptly at the eutectic temperature, i.e., at the onset of partial melting. For a dilute solution containing 3.3% melt at the eutectic temperature, the compressional and shear velocities were 9.5% and 13.5% less, respectively, than in the unmelted solid. The attenuation increased at the same time by about 45%. Thus, a small amount of melt can have significant effects upon the elastic and anelastic properties of a material. These effects would be enhanced if the liquid wets the crystals, forming thin intergranular films. Experimental observations of partial melting in relevant silicate systems[11] show that this is likely to be the case in the mantle.[12] It has been estimated[13] that about 1% of *incipient melting*[14] would be sufficient to explain the occurrence of the low-velocity zone beneath tectonic and oceanic regions. On the other hand, the small total decrease (~ 2%) in S velocity in the channel beneath Precambrian shields and stable continental regions, com-

[1]Green and Ringwood (1967b).
[2]Ito and Kennedy (1967).
[3]Anderson and Sammis (1970).
[4]Shimozuru (1963).
[5]Anderson (1962, 1967).
[6]Oxburgh and Turcotte (1968).
[7]Aki (1968).
[8]Hales and Doyle (1967).
[9]Kushiro, Syono, and Akimoto (1968).
[10]Press (1959) also came close to this explanation when he attributed the low-velocity zone to a "state near the melting point."
[11]Author's unpublished observations.
[12]Birch (1969) considered the case of partial melting leading to the formation of small spherical liquid inclusions rather than intergranular films. He showed that the decrease of velocity for the same degree of partial melting would be smaller than that observed by Spetzler and Anderson.
[13]Spetzler and Anderson (1968).
[14]It is conceivable that the amount of intergranular liquid film may be even smaller than 1%. The term *incipient melting* is used to describe this state in distinction to the much larger degrees of melting usually considered to be involved when magmas form by partial melting followed by segregation.

bined with the absence of a corresponding minimum for *P* waves, suggests that partial melting, at least to this degree, may not be involved.

The conditions which cause large volumes of the upper mantle to remain in a quasi-stable state of incipient melting deserve some comment. It does not appear likely that this would be possible if the mantle were completely anhydrous. At pressures of 20 to 30 kbars, the melting interval between solidus and liquidus of olivine tholeiite and picrite is smaller than 100°C.[1] It is probable that dry pyrolite would become 30% molten within this interval. Temperature fluctuations as small as 20°C in the low-velocity zone would produce substantial (~ 5%) partial melting. Such a state would not remain stable for long; liquid would soon segregate into magma bodies which would rise to the surface, producing volcanism. Clearly, if the upper mantle were in this condition, widespread volcanism should occur on a much larger scale than is now observed. To account for a state of incipient melting throughout the low-velocity zone, some kind of stabilization mechanism must be present. Moreover, it is difficult to devise plausible stable suboceanic temperature distributions which intersect the anhydrous pyrolite solidus regularly between depths of about 70 to 150 km.

It is probable that the presence of a small amount of water in pyrolite may account for both the stabilization mechanism and the required intersection of the pyrolite solidus by mantle geotherms. We concluded previously (Sec. 4-5) that primary pyrolite probably contained about 0.1% H_2O and that this was mainly held in amphiboles[2] at depths of less than 75 km. The key role played by the instability of amphibole at high pressure and its transformation into an eclogitic mineral assemblage plus H_2O[3,4] in controlling the mantle solidus and the formation of a low-velocity zone were noted independently by Lambert and Wyllie[3] and by Ringwood.[5] The instability of amphibole above about 25 kbars caused a strong increase in the water-vapour pressure, resulting in a sharp depression of the mantle solidus, as shown in Figs. 4-6, 4-7, and 6-4. The consequence of this transformation was to make it possible for reasonable mantle temperature-distribution curves to intersect the solidus at depths of 70 to 100 km. Below this intersection, a small amount of interstitial liquid, probably less than 1%, would be present (Fig. 4-7), thereby accounting for the physical characteristics of the low-velocity zone. Because of the deep depression in the solidus between 75 and 150 km, a situation is produced where a comparatively large increase in temperature results in the formation of a relatively small amount of liquid (Fig. 4-8). Thus, temperature fluctuations up to about 200°C are possible without causing a sufficiently extensive degree of partial melting (possibly $\geq$ 1%) which would result in segregation of magma.[5]

[1]Green and Ringwood (1967c).

[2]Although phlogopite is stable to much higher pressures than amphibole, the probable abundance of potassium in pyrolite limits the amount of water held in phlogopite to about 0.02%.

[3]Green and Ringwood (1967a).

[4]Lambert and Wyllie (1968).

[5]Ringwood (1969).

Some seismic body-wave studies[1] suggest that the downward termination of the low-velocity zone may be relatively abrupt. This could indicate the sudden disappearance of the incipient melting condition responsible for the low-velocity zone. It has been suggested[1,2] that disappearance of melt below 150 to 200 km might be caused by a corresponding decrease in the fugacity of water, caused in turn by the occurrence of the dense hydrated magnesium silicate (DHMS) phase[3] which was believed to become stable below these depths. This hypothesis is analogous to the amphibole-instability mechanism invoked to explain the upper boundary of the low-velocity zone. The pyrolite solidus (0.1% H_2O) in Figs. 4-6 and 6-4 below 150 km was constructed with this hypothesis in mind. Unfortunately, however, recent experimental determinations[4] of the stability field of the DHMS phase cannot be reconciled with this hypothesis.

A detailed study of the attenuation of seismic waves in the upper mantle by Solomon (1972) indicates that the low-velocity zone may possess a more complex structure than was formerly believed. Very small amounts of intergranular liquid were inferred to occur at depths up to 300 to 400 km. Solomon concluded that both the shear modulus μ and quality factor Q were frequency-dependent. Most of the delays and attenuation of short-period body waves occurred in the "upper asthenosphere" and were probably caused by a small amount (say, about 1%) of intergranular liquid. The extent of the upper asthenosphere ($\sim$ 60 to 160 km) was essentially similar to that of the low-velocity zone as delineated also by short-period body waves. On the other hand, most of the attenuation of long-period teleseismic S and P waves was found to occur in the "lower asthenosphere," extending from about 160 to 300–400 km. This layer was characterized by an abnormally low shear modulus for long-period waves (but not for short-period waves). The elastic and anelastic properties of this region also appeared to require the presence of a very small amount of intergranular liquid—perhaps amounting to 1 to 10% of the proportion of liquid which was present in the upper asthenosphere. The lower asthenosphere appeared to extend both beneath western North America (a tectonically active area) and beneath eastern North America (a stable region). However, temperatures in this layer beneath the latter region would be up to 100 to 200°C cooler than beneath the former.

According to Solomon's model, the properties of the upper and lower asthenosphere layers are controlled by the proportions and distribution of intergranular liquid believed to be present. In the upper asthenosphere, about 1% of a highly alkalic magma containing perhaps 20% water may occur along grain boundaries. With increasing pressures above about 50 kbars ($\sim$ 150 km), the silicate components of the melt may be induced to crystallize out so that, between 150 and 300–400 km, the fluid phase consists principally of supercritical dense

[1]Anderson and Sammis (1970).
[2]Lambert and Wyllie (1968).
[3]Sections 8-3 and 13-2.
[4]Discussed in Sec. 13-2.

water containing some dissolved components. The amount and pressure of this phase will be determined by the solid solubility of OH^- ions in normal mantle minerals—olivines, pyroxenes, and garnets. A small proportion of OH^- ions is expected to replace O^- ions at normal lattice sites under mantle conditions.[1,2]

We assume that, between 150 and 300 to 400 km, these minerals are saturated with OH^- so that a very minute amount (0.1 to 0.01%) of a highly mobile fluid phase is present along grain boundaries, giving rise to the observed anelastic properties of this layer. Below 300 to 400 km, pyroxenes and olivines transform to a denser assemblage, consisting of a spinel-like phase and a complex garnet solid solution.[3] These minerals have the capacity for taking a wide range of cations and anions possessing differing charges and radii into solid solution. It is likely that the solubility of OH^- ions in these phases will be much higher than in olivines and pyroxenes so that the free H_2O fluid phase would become totally dissolved in the dense phases present below 300 to 400 km. With the disappearance of the free-fluid phase, the large decrease in attenuation below 300 to 400 km becomes explicable.

Several authors[4] have pointed out the profound tectonic implications of an incipiently molten low-velocity zone. It may explain the extreme mobility of the oceanic crust, which appears to slide on the upper part of the low-velocity zone (upper asthenosphere) with very low friction. Furthermore, lateral flow in the upper asthenosphere may play an important role in the mass transfer process involved in plate movements. On the other hand, the thickness of the rigid lithosphere plates beneath Precambrian shields is probably about 150 to 200 km. Motions of these thick plates apparently depend upon a high mobility for the lower asthenosphere as suggested by Solomon's observations.

REFERENCES

AKI, K. (1968). Seismological evidence for the existence of soft, thin layers in the upper mantle under Japan. *J. Geophys. Res.* **73,** 585–594.

ANDERSON, D. L. (1962). The plastic layer of the earth's mantle. *Sci. Am.* **205** (July) 2–9.

——— (1967). Latest information from seismic advances. In: T. Gaskell (ed.), "*The Earth's Mantle,*" chap. 12, pp. 355–420. Academic, London.

——— and C. SAMMIS (1970). Partial melting in the upper mantle. *Phys. Earth Planet. Interiors* **3,** 41–50.

ANDERSON, O. L., E. SCHREIBER, R. C. LIEBERMANN, and N. SOGA (1968). Some elastic constant data on minerals relevant to geophysics. *Rev. Geophys.* **6,** 491–524.

ARCHAMBEAU, C. B., E. A. FLINN, and D. G. LAMBERT (1969). Fine structure of the upper mantle. *J. Geophys. Res.* **74,** 5825–5865.

BERRY, M. J., and L. KNOPOFF (1967). Structure of the upper mantle beneath the western Mediterranean basin. *J. Geophys. Res.* **72,** 3613–3626.

[1]Fyfe (1970).

[2]Martin and Donnay (1972).

[3]Cf Chap. 14.

[4]E.g., Anderson and Sammis (1970).

BIRCH, F. (1952). Elasticity and constitution of the earth's interior. *J. Geophys. Res.* **57,** 227–286.

——— (1960). The velocity of compressional waves in rocks to 10 kilobars, 1. *J. Geophys. Res.* **65,** 1083–1102.

——— (1969). Density and composition of the upper mantle: First approximation as an olivine layer. In: "*The Earth's Crust and Upper Mantle,*" pp. 18–36. *Am. Geophys. Union, Geophys. Monograph* **13.**

BOYD, F. R. (1970). Garnet peridotites and the system $CaSiO_3$-$MgSiO_3$-Al_2O_3. *Mineral. Soc. Am. Spec. Paper* **3,** 63–75.

——— and J. L. ENGLAND (1964). The system enstatite-pyrope. *Carnegie Inst. Washington Yearbook* **63,** 157–161.

——— and J. F. SCHAIRER (1964). The system $MgSiO_3$-$CaMgSi_2O_6$. *J. Petrol.* **5,** 275–309.

BROOKS, J. P. (1962). Seismic wave velocities in the New Guinea–Solomon Islands region. In: "*The Crust of the Pacific Basin,*" pp. 2–10. *Am. Geophys. Union, Geophys. Monograph* **6.**

BRUNE, J. N. (1969). Surface waves and crustal structure. In: "*The Earth's Crust and Upper Mantle,*" pp. 230–242. *Am. Geophys. Union, Geophys. Monograph* **13.**

——— and J. DORMAN (1963). Seismic waves and earth structure in the Canadian Shield. *Bull. Seism. Soc. Am.* **53,** 167–209.

BULTITUDE, R. J., and D. H. GREEN (1968). Experimental study at high pressures on the origin of olivine nephelinite and olivine melilite nephelinite magmas. *Earth Planet. Sci. Letters* **3,** 325–337.

CARDER, D. S. (1964). Travel times from central Pacific nuclear explosions and inferred mantle structure. *Bull. Seism. Soc. Am.* **54,** 2271–2294.

CLARK, S. P., and A. E. RINGWOOD (1964). Density distribution and constitution of the mantle. *Rev. Geophys.* **2,** 35–88.

CLEARY, J. R., and A. L. HALES (1966). An analysis of the travel times of P waves to North American stations in the distance range 32 to 100°. *Bull. Seism. Soc. Am.* **56,** 467–489.

DAVIS, B. T. C., and F. R. BOYD (1966). The join $Mg_2Si_2O_6$-$CaMgSi_2O_6$ at 30 kilobars pressure and its application to pyroxenes from kimberlites. *J. Geophys. Res.* **71,** 3567–3576.

DORMAN, J. (1969). Seismic surface wave data on the upper mantle. In: "*The Earth's Crust and Upper Mantle,*" pp. 257–265. *Am. Geophys. Union, Geophys. Monograph* **13.**

———, M. EWING, and J. OLIVER (1960). Study of shear velocity distribution in the upper mantle by mantle Rayleigh waves. *Bull. Seism. Soc. Am.* **50,** 87–115.

DOYLE, H. A. (1957). Seismic recordings of atomic explosions in Australia. *Nature* **180,** 132–134.

——— and J. P. WEBB (1963). Travel times to Australian stations from Pacific nuclear explosions in 1958. *J. Geophys. Res.* **68,** 1115–1120.

ERNST, W. G. (1968). "*Amphiboles.*" Springer-Verlag, New York. 125 pp.

ESSENE, E. J., B. J. HENSEN, and D. H. GREEN (1970). Experimental study of amphibolite and eclogite stability. *Phys. Earth Planet. Interiors* **3,** 378–384.

FYFE, W. S. (1970). Lattice energies, phase transformations and volatiles in the mantle. *Phys. Earth Planet. Interiors* **3,** 196–200.

GILBERT, M. C. (1969). Reconnaissance study of the stability of amphiboles at high pressure. *Carnegie Inst. Washington Yearbook* **67,** 167–170.

GREEN, D. H. (1973). Experimental melting studies on model upper mantle compositions under both water-saturated and water-unsaturated conditions. *Earth Planet. Sci. Letters* **19,** 37–53.

——— and W. HIBBERSON (1970). The instability of plagioclase in peridotite at high pressure. *Lithos* **3,** 209–222.

——— and A. E. RINGWOOD (1963). Mineral assemblages in a model mantle composition. *J. Geophys. Res.* **68,** 937–945.

——— and ——— (1967a). An experimental investigation of the gabbro to eclogite transformation and its petrological applications. *Geochim. Cosmochim. Acta* **31,** 767–833.

——— and ——— (1967b). The stability fields of aluminous pyroxene peridotite and garnet peridotite and their relevance in upper mantle structure. *Earth Planet. Sci. Letter* **3,** 151–160.

——— and ——— (1967c). Genesis of basaltic magmas. *Contr. Mineral. Petrol.* **15,** 103–190.

——— and ——— (1970). Mineralogy of peridotitic compositions under upper mantle conditions. *Phys. Earth Planet. Interiors* **3,** 359–371.

GREEN, R. W. E., and A. L. HALES (1968). The travel times of P waves to 30 degrees in the Central United States and upper mantle structure. *Bull. Seism. Soc. Am.* **58,** 267–289.

GUTENBERG, B. (1959). "*Physics of the Earth's Interior.*" 240 pp. Academic, New York. [International Geophysics Series, vol. 1, J. V. Mieghem (ed.).]

HALES, A. L. (1969). A seismic discontinuity in the lithosphere. *Earth Planet. Sci. Letters* **7,** 44–46.

———, J. R. CLEARY, H. A. DOYLE, R. GREEN, and J. ROBERTS (1968). P-wave station anomalies and the structure of the Upper Mantle. *J. Geophys. Res.* **73,** 3885–3896.

——— and H. A. DOYLE (1967). P and S travel time anomalies and their interpretation. *Geophys. J. Roy. Astron. Soc.* **13,** 403–415.

———, C. E. HELSLEY, and J. B. NATION (1970). P travel times for an oceanic path. *J. Geophys. Res.* **75,** 7362–7381.

HELMBERGER, D. V., and G. B. MORRIS (1969). A travel time and amplitude interpretation of a marine reflection profile: Primary waves. *J. Geophys. Res.* **74,** 483–494.

HILL, R. E. T., and A. L. BOETTCHER (1969). Water in the earth's mantle: Melting curves of basalt-water and basalt-carbon dioxide. *Science* **167,** 980–982.

ITO, K., and G. C. KENNEDY (1967). Melting and phase relations in a natural peridotite to 40 kilobars. *Am. J. Sci.* **265,** 519–538.

JOHNSON, L. R. (1967). Array measurements of P velocities in the upper mantle. *J. Geophys. Res.* **72,** 6309–6325.

KANAMORI, H., and F. PRESS (1970). How thick is the lithosphere? *Nature* **226,** 330–331.

KUSHIRO, I. (1970). Formation of amphibole in a peridotite composition. *Carnegie Inst. Washington Yearbook* **68,** 245–247.

——— and H. S. YODER (1966). Anorthite-forsterite and anorthite-enstatite reactions and their bearing on the basalt-eclogite transformation. *J. Petrol.* **7,** 337–362.

———, Y. SYONO, and S. AKIMOTO (1967). Effect of pressure on garnet-pyroxene equilibrium in the system $MgSiO_3$-$CaSiO_3Al_2O_3$. *Earth Planet. Sci. Letters* **2,** 460–464.

———, ———, and ——— (1968). Melting of a peridotite nodule at high pressures and high water pressures. *J. Geophys. Res.* **73,** 6023–6029.

LAMBERT, I. B., and P. J. WYLLIE (1968). Stability of hornblende and a model for the low velocity zone. *Nature* **219,** 1240–1241.

——— and ——— (1970a). Melting in the deep crust and upper mantle and the nature of the low-velocity layer. *Phys. Earth Planet. Interiors* **3,** 316–322.
——— and ——— (1970b). Low-velocity zone of the earth's mantle: Incipient melting caused by water. *Science* **169,** 764–766.
LIEBERMANN, R. C., and E. SCHREIBER (1969). Critical geothermal gradients in the mantle. *Earth Planet. Sci. Letters* **7,** 77–81.
MACDONALD, G. J. F., and N. NESS (1961). A study of the free oscillations of the earth. *J. Geophys. Res.* **66,** 1865–1911.
MACGREGOR, I. D. (1964). The reaction 4 enstatite + spinel = forsterite + pyrope. *Carnegie Inst. Washington Yearbook* **63,** 156–157.
——— (1968). Mafic and ultramafic inclusions as indicators of the depth of origin of basaltic magmas. *J. Geophys. Res.* **73,** 3737–3745.
——— (1970). The effect of CaO, Cr_2O_3, Fe_2O_3 and Al_2O_3 on the stability of spinel and garnet peridotites. *Phys. Earth Planet. Interiors* **3,** 372–377.
——— and F. R. BOYD (1964). Ultramafic rocks. *Carnegie Inst. Washington Yearbook* **63,** 152–156.
——— and A. E. RINGWOOD (1964). The natural system enstatite-pyrope. *Carnegie Inst. Washington Yearbook* **63,** 161–163.
MARTIN, R. F., and G. DONNAY (1972). Hydroxyl in the mantle. *Am. Min.* **57,** 554–570.
MIYASHIRO, A., F. SHIDO, and M. EWING (1970). Petrologic models for the mid-Atlantic Ridge. *Deep Sea Res.* **17,** 109–123.
MOORE, J. G. (1965). Petrology of deep sea basalt near Hawaii. *Am. J. Sci.* **263,** 40–52.
NUTTLI, O. W., and B. A. BOLT (1969). P wave residuals as a function of azimuth, 2. *J. Geophys. Res.* **74,** 6594–6602.
O'HARA, M. J. (1967). Mineral parageneses in ultrabasic rocks. In: P. J. Wyllie (ed.), "*Ultramafic and Related Rocks,*" pp. 393–403. Wiley, New York.
——— and E. L. P. MERCY (1963). Petrology and petrogenesis of some garnetiferous peridotites. *Trans. Roy. Soc. Edinburgh* **65,** 251–314.
OXBURGH, E. R., and D. L. TURCOTTE (1968). Mid-ocean ridges and geotherm distributions during mantle convection. *J. Geophys. Res.* **73,** 2643–2661.
PRESS, F. (1959). Some implications on mantle and crustal structure from G waves and Love waves. *J. Geophys. Res.* **64,** 565–568.
RINGWOOD, A. E. (1962a). A model for the upper mantle. *J. Geophys. Res.* **67,** 857–867.
——— (1962b). A model for the upper mantle, 2. *J. Geophys. Res.* **67,** 4473–4477.
——— (1966). Mineralogy of the mantle. In: P. M. Hurley (ed.), "*Advances in Earth Science,*" pp. 357–399. M.I.T. Press, Cambridge, Mass.
——— (1969). Composition and evolution of the upper mantle. In: "*The Earth's Crust and Upper Mantle,*" pp. 1–17. *Am. Geophys. Union, Geophys. Monograph* **13.**
———, I. D. MACGREGOR, and F. R. BOYD (1964). Petrological constitution of the upper mantle, *Carnegie Inst. Washington Yearbook* **63,** 147–152.
ROLLER, J. C., and W. H. JACKSON (1966). Seismic wave propagation in the upper mantle: Lake Superior, Wisconsin to Central Arizona. *J. Geophys. Res.* **71,** 5933–5941.
SCLATER, J. G., and J. FRANCHETEAU (1970). The implications of terrestrial heat flow observations on current tectonic and geochemical models of the crust and upper mantle of the earth. *Geophys. J.* **20,** 509–542.
SEIDL, D., ST. MÜLLER, and L. KNOPOFF (1966). Dispersion von Rayleigh-wellen in Sudwestdeutschland und in den Alpen. *Geophys.* **32,** 472–481.
SHIMOZURU, D. (1963). On the possibility of the existence of the molten portion in the upper mantle of the earth. *J. Phys. Earth* **11,** 49–55.

SHURBET, D. H. (1964). The high frequency S phase and the structure of the upper mantle. *J. Geophys. Res.* **69,** 2065–2070.

SOLOMON, S. C. (1972). Seismic-wave attenuation and partial melting in the upper mantle of North America. *J. Geophys. Res.* **77,** 1483–1502.

SPETZLER, H., and D. L. ANDERSON (1968). The effect of temperature and partial melting on velocity and attenuation in a simple binary system. *J. Geophys. Res.* **73,** 6051–6060.

TOKSÖZ, M. N., and D. L. ANDERSON (1966). Phase velocities of long period surface waves and structure of the upper mantle. *J. Geophys. Res.* **71,** 1649–1658.

WILLIAMS, A. F. (1932). "*The Genesis of the Diamond*" (2 vol.). Ernest Benn, London. 636 pp.

VALLE, P. E. (1956). On the temperature gradient necessary for formation of a low-velocity layer. *Ann. Geofis. Roma* **9,** 371–377.

7
THE OROGENIC IGNEOUS ROCK ASSOCIATION

7-1 INTRODUCTION

The orogenic igneous rock association is composed dominantly of the basalt-andesite-dacite-rhyolite volcanic series, together with the gabbro-diorite-granodiorite-granite plutonic series. Representatives of this association, particularly intermediate and acid members, are the most abundant igneous rocks in orogenic regions. The characteristic occurrence of these rock types in present and past tectonically active zones is suggestive of a close relationship between the genesis of this rock association, the fundamental mechanism of orogenesis, and the evolution of continental regions. Active orogenic-type volcanism is currently observed to be closely associated with zones of lithosphere subduction beneath island arcs and "Pacific-type" continental margins. Volcanoes of this class are characteristically located in regions which are 80 to 150 km above the upper boundaries of sinking lithosphere plates, as revealed by the inclined planes of seismic activity extending several hundred kilometres downward into the mantle (Benioff zones). This relationship has suggested to many that the origins of the orogenic volcanic series are connected with processes occurring in and near the Benioff zones.

The principle of uniformitarianism suggests that the vast volumes of the basalt-andesite-dacite-rhyolite suite which occur in older orogenic provinces were generated similarly and were once associated with processes of mantle subduction which have since ceased. The deeper levels of erosion of older provinces have exposed the plutonic gabbro-diorite-granodiorite-granite members of the series, often in the form of huge batholiths extending over tens of thousands of square kilometres. In many cases, the plutonic members of the orogenic igneous series are derived from the same magmas which were responsible for the volcanic series. These plutons simply represent orogenic-type magmas which have crystallized slowly at depth to form coarser grained rocks. In other cases, however, members of the plutonic orogenic series appear to have undergone a more complex petrological evolution. The corresponding volcanic and plutonic members (e.g., dacite and granodiorite) of the orogenic igneous association are not always isochemical.[1,2] Moreover, there is a much greater preponderance of more-acid members (granodiorites and granites) among the plutonic series than among the volcanic series in which andesites predominate. This chapter will be mainly concerned with the petrogenesis of the volcanic orogenic series and with those members of the plutonic series which are essentially cogenetic with them. The origins of the plutonic series are discussed in more general terms in Sec. 7-9.

7-2 MINERALOGY AND CHEMISTRY OF OROGENIC VOLCANIC SERIES

The members of this series are usually porphyritic, the phenocrysts consisting mainly of plagioclase, orthopyroxene, clinopyroxene, and quartz, with rarer biotite, amphibole, olivine, and garnet. In andesite and basaltic andesite, plagioclase is by far the most abundant phase, comprising 50 to 70% of the rock. Phenocrysts typically display zoning, both normal and oscillatory, often with resorbed margins to some of the zones which may range from anorthite to oligoclase. The principal ferromagnesian phenocrysts are hypersthene and augite. Minor olivine phenocrysts occur in some basaltic andesites and less frequently in andesites where they are sometimes surrounded by hypersthene coronas. Opaque mineral phenocrysts (magnetite and ilmenite) are not common and, if they occur, make up less than 5% of the rock. Hornblende is the characteristic ferromagnesian phenocryst in some provinces, but it is uncommon in others. Biotite and rare almandine garnet, together with abundant quartz (frequently resorbed), occur as phenocrysts in the silica-rich members of the series. The groundmass in basaltic andesites and andesites consists dominantly of plagioclase, clinopyroxene, and hypersthene, whereas in the more acid members of the series, plagioclase, silica minerals, and glass predominate.

It appears likely that the characteristic porphyritic textures and nonequilib-

[1]Taylor (1969).
[2]Gill (1970).

rium features of the mineralogy (e.g., oscillatory zoning and resorption phenomena) are connected with the former presence of water in the parental magmas. Escape of water at temperatures below the anhydrous solidus may cause forced crystallization and other complex effects. The role of water in the genesis of the orogenic volcanic series is discussed later.

Major Element Chemistry

Chemical analyses of groundmasses of members of the basalt-basaltic andesite-andesite-dacite-rhyodacite-rhyolite series[1] have demonstrated the existence of magmatic liquids corresponding in composition to each of these rock types.[2] This is an important property because it indicates the existence of a true line of liquid descent, which must be explained by an acceptable theory of petrogenesis. The existence of a liquid line of descent for the volcanic series is incompatible with petrogenetic hypotheses involving hybridism which have been successfully invoked to explain the origin of some plutonic members of the association.[3] When the compositions of members of the orogenic series occurring in a given region are plotted on variation diagrams, they tend to fall on smooth and regular curves, indicating continuous genetic relationships between members (Fig. 7-1). These diagrams are suggestive of the operation of crystal-liquid fractionation processes. In principle they could also be caused by contamination or mixing of mafic magmas with acidic rocks; however, the occurrence of the series in island arcs which are not underlain by normal sialic crust is difficult to reconcile with these hypotheses.[4]

Chemical analyses of a representative suite of orogenic volcanic rocks are given in Table 7-1, and a typical variation diagram is given in Fig. 7-1. A notable feature is the high alumina content of rocks falling between basalt and andesite in composition. The basalts occurring in orogenic igneous provinces form an important class, which, as Tilley (1950) has pointed out, should be distinguished from the principal classes of basalts discussed in Chap. 4. Although tholeiitic and characteristically high in alumina, they are usually distinct from the high-alumina oceanic tholeiites by virtue of higher K_2O and SiO_2 and lower MgO and TiO_2 (Table 7-1, columns 2 and 3). Corresponding differences occur in minor and trace elements. Basalts of the orogenic igneous association typically contain much lower abundances of Ni, Cr, and Co than oceanic tholeiites.[5,6,7] Detailed petrochemical studies of this class of rocks have been carried out by Jakeš and Gill (1970), who propose that they be termed *island-arc tholeiites*. Their characteristic high

[1] Or of groundmass compositions calculated from known bulk compositions and known phenocryst mineralogies and compositions.

[2] Reviewed by T. Green and Ringwood (1968a). See also Wilkinson (1971).

[3] E.g., Wilkinson, Vernon, and Shaw (1964).

[4] Gorshkov (1962, 1969).

[5] Taylor (1969).

[6] Taylor and White (1966).

[7] Baker (1968a).

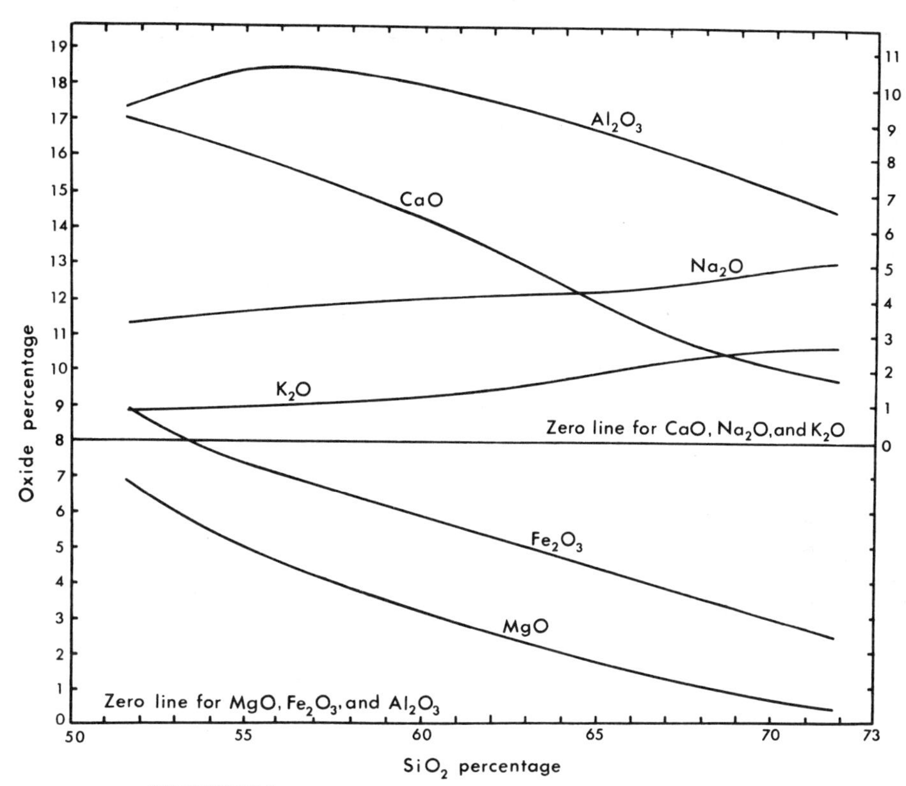

FIGURE 7-1
Variation diagram for typical orogenic volcanic series—Crater Lake, northwestern United States. (*After Williams, 1942.*)

FeO/MgO ratios, combined with low abundances of Cr and Ni, indicate that they are not primary magmas derived directly by partial melting of pyrolite but have probably suffered fractionation en route to the surface, most probably by separation of olivine.

Basalts transitional in composition between oceanic tholeiites and island-arc tholeiites are found,[1,2] and in some dominantly orogenic igneous provinces, the typical oceanic series of oceanic tholeiite, high-alumina basalt, and alkali basalt also occurs.[1] In other provinces, the most primitive basalts may approach oceanic tholeiites in composition.[2,3]

Important characteristics of the differentiation of igneous rock series are brought out in the FMA diagram.[4,5] This is a triangular plot of total iron (as FeO),

[1]Kuno (1966).
[2]E.g., South Sandwich Islands (Baker, 1968b).
[3]Le Masurier (1968).
[4]Poldervaart (1949).
[5]Tilley (1950).

Table 7-1 CHEMICAL ANALYSES OF A REPRESENTATIVE SUITE OF OROGENIC VOLCANIC ROCKS, CASCADE PROVINCE, NORTHWESTERN UNITED STATES. *(After Turner and Verhoogen, 1960)*

Constituent	(1)	(2)	(3)	(4)	(5)	(6)	(7)	(8)	(9)	(10)
SiO_2	47.1	50.7	51.5	55.8	60.1	63.2	67.7	68.6	72.4	73.6
TiO_2	0.9	1.3	1.1	0.8	0.5	0.5	0.3	0.5	0.3	0.3
Al_2O_3	18.5	18.1	17.5	18.0	17.9	18.2	16.3	15.6	14.0	14.0
Fe_2O_3	Trace	1.6	1.5	2.6	2.0	1.4	0.3	1.3	0.6	0.4
FeO	7.9	7.0	6.7	4.1	3.5	3.3	3.2	2.2	1.8	1.4
MnO	Trace	0.4	0.2	0.1	Trace	Trace	Trace	Trace	—	0.0
MgO	10.9	7.6	6.7	5.1	3.5	2.3	1.3	1.1	0.3	0.4
CaO	12.0	9.7	8.8	7.4	6.3	5.2	3.4	3.1	1.3	1.4
Na_2O	2.3	2.7	3.3	3.6	4.2	4.1	3.9	4.5	5.0	4.0
K_2O	Trace	0.7	0.8	1.2	1.3	1.2	3.2	2.2	3.9	4.3
H_2O	0.3	—	0.7	0.9	0.4	0.5	0.3	0.8	0.5	0.2
P_2O_5	0.1	0.2	0.3	0.1	0.2	0.1	0.1	0.1	Trace	0.1
Total	100.0	100.0	99.1	99.7	99.9	100.0	100.0	100.0	100.1	100.1

Explanation:
(1,2,3) High-alumina olivine basalt.
(4) Basaltic andesite.
(5) Hypersthene andesite.
(6) Pyroxene andesite.
(7, 8) Dacites.
(9, 10) Rhyolitic obsidians.

magnesia, and alkali contents of series of petrologically related rocks. The tholeiitic differentiation trends displayed during the crystallization of basaltic or gabbroic rocks in crustal environments are particularly well documented. The tholeiitic trend shows marked to strong iron enrichment in the early and middle stages of fractionation, as in the Skaergaard[1,2] intrusion and the differentiated lavas of Thingmuli volcano in Iceland[3] (Fig. 7-2). This is caused by crystallization of olivines and pyroxenes possessing much higher Mg/Fe ratios than the parental magma. At a relatively late stage of differentiation, the differentiation trend swings around and proceeds towards the alkalic apex (Fig. 7-2), often culminating in silica-rich granophyric rocks. The amounts of acid and intermediate differentiates produced during this kind of tholeiitic fractionation are usually relatively small compared to the volumes of the parental mafic magmas.

Whereas the tholeiitic differentiation trend is usually followed by mafic magmas crystallizing under relatively stable crustal conditions, the magmas generated in regions of lithosphere subduction (island arcs and continental margins) often follow a distinctly different trend termed *calc-alkaline*.[4] On the

[1] E.g., Wager and Deer (1939).
[2] Wager (1960).
[3] Carmichael (1964).
[4] The term *calc-alkaline* derives from a rock classification index suggested by Peacock (1931), based on crossover points on variation diagrams of alkalis and lime versus silica for petrogenetically related rock series. Although this index has fallen into disuse, the term *calc-alkali* has survived to denote certain major rock series characteristic of orogenic regions.

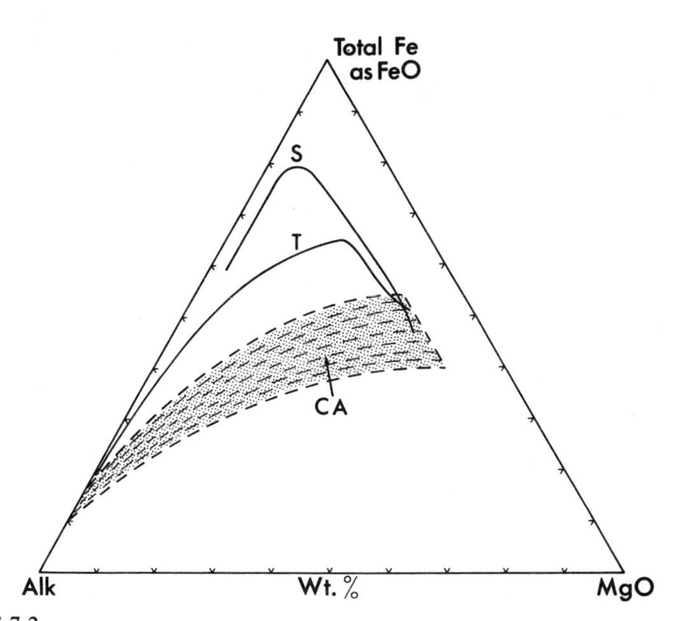

FIGURE 7-2
FMA diagram showing characteristic tholeiitic and calc-alkaline trends. The tholeiitic region is confined by differentiation trends observed for the Skaergaard Intrusion (Wager and Deer, 1939) and Thingmuli volcano (Carmichael, 1964). The calc-alkaline band (shaded) embraces the differentiation trends displayed by magmas from the Cascade, Aleutian, and New Zealand calc-alkaline provinces. Curve S represents the Skaergaard differentiation trend and curve T the Thingmuli.

FMA diagram, the rocks of the calc-alkaline series fall on an almost straight band extending from the Fe-Mg side to the alkali apex (Fig. 7-2). There is only a relatively small amount of iron enrichment relative to magnesium with increasing total alkali (and silica) content. In order to generate the calc-alkaline trend from basaltic starting material, it is necessary to crystallize phases with higher average Fe/Mg ratios than the olivines and pyroxenes which are observed to crystallize from crustal mafic magmas.[1] A further important difference from tholeiitic provinces is that the calc-alkaline provinces are characterized by the occurrence of vast amounts of magmas possessing intermediate silica contents—andesites containing about 60% SiO_2 are the most voluminous magma type, and large proportions of dacites, rhyodacites, and rholites are also present.

The trends in Fig. 7-2 represent the extremes of tholeiitic and calc-alkaline lines of descent. In nature, a continuum of intermediate trends is observed in different petrologic suites, filling the gap between the ideal tholeiite and ideal calc-

[1]Alternatively, the primary partial melting process may produce parental magmas possessing higher Mg/Fe ratios in relation to silica contents than are produced during the generation of normal basalt magmas by partial melting of pyrolite under "dry" conditions—see Sec. 7-6 and page 272.

alkaline trends. Such intermediate trends are commonly observed during the early stages of development of island arcs. Although they show certain tholeiitic characteristics, including distinct iron enrichment during the early stages of differentiation, this trend is rarely as strong[1] as is shown by true tholeiitic magmas, and moreover, much larger relative volumes of intermediate differentiates are produced. Clearly, these intermediate trends occurring together with the strictly calc-alkaline trends in orogenic regions are indicative of different processes of magmatic fractionation than those which are observed to be operative during the normal tholeiitic fractionation of mafic magmas in the crust.

In the subsequent discussion we regard the entire spectrum of magmas generated in regions of lithosphere subduction as belonging to the orogenic igneous association, whether or not their differentiation trends may be classed strictly as calc-alkaline or as possessing variable degrees of tholeiitic characteristics (with reference to the FMA diagram).

Geochemical Characteristics

Dickinson and Hatherton (1967)[2] drew attention to a significant chemical feature of the orogenic volcanic series. They constructed K_2O versus SiO_2 variation diagrams for a large number of rock suites from different orogenic volcanic provinces. These trends were approximately linear for given provinces (Fig. 7-3). However, the slopes and intercepts varied substantially among the different volcanic provinces. From the K_2O versus SiO_2 diagrams, they determined the mean

[1]However, the volcanic rocks of Tonga provide an example of a strongly tholeiitic trend (fig. 7-11, Ewart, Bryan, and Gill, 1973).

[2]See also Kuno (1966) and Dickinson (1968, 1969).

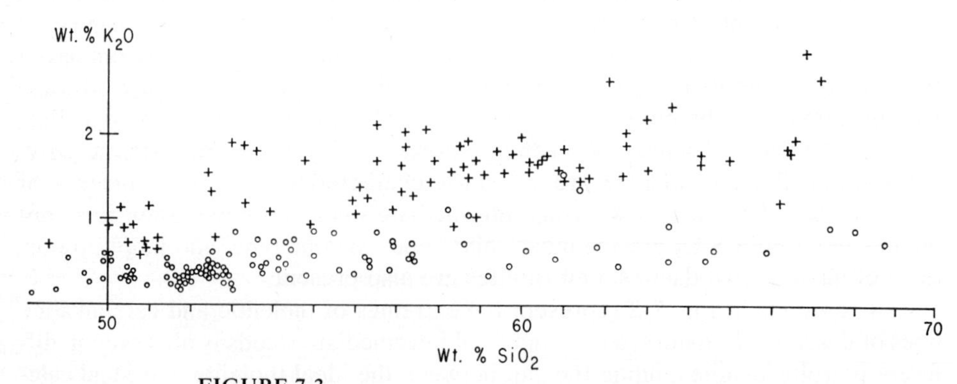

FIGURE 7-3
Potash versus silica variation for the Ryukyu-Kyushu arc (crosses) and the Izu arc (circles) in Japan. (*From Dickinson, 1968, with permission. Copyright American Geophysical Union.*)

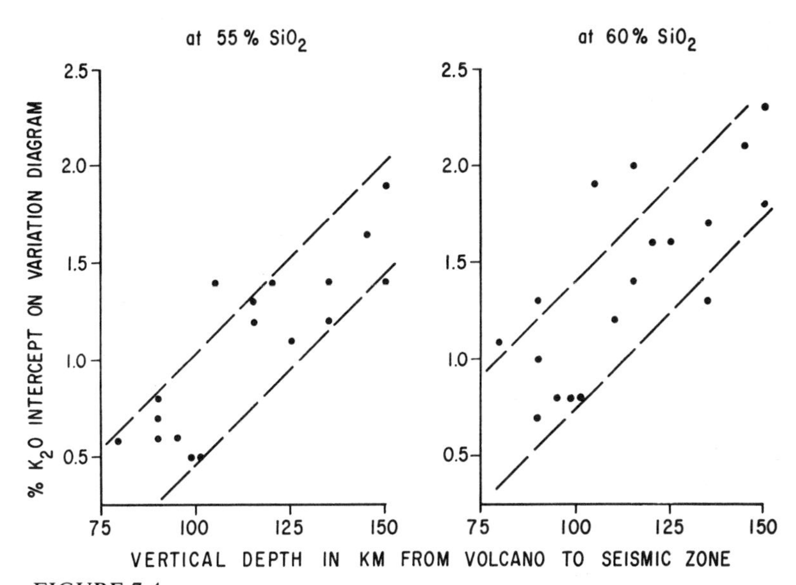

FIGURE 7-4
Plot of potash levels in selected arc lavas versus the depths from corresponding volcanoes to the Benioff zone beneath the arcs. (*From Dickinson, 1968, with permission. Copyright American Geophysical Union.*)

potash contents at a given level of silica enrichment, e. g., 60% SiO_2, for individual volcanic provinces. These were used as characteristic parameters for particular provinces. When the characteristic potash contents were plotted against depths of the Benioff zones, a distinctive trend emerged (Fig. 7-4). The degree of potassium enrichment increased as the depth of the Benioff zone beneath the volcanic province increased. Dickinson and Hatherton concluded that a genetic relationship existed between the physical mechanisms responsible for earthquakes along the Benioff zone and the processes involved in the generation of the basalt-andesite-dacite-rhyolite series. They suggested that the relevant crystal-liquid partition equilibria varied substantially with pressure and controlled the degree of enrichment of potash in the magmas. Their model also envisaged the generation of orogenic-type magmas at considerable depth in the mantle.

Systematic studies of the trace element chemistry of andesites and related rocks have been carried out by Taylor and others[1,2,3,4,5] and provide many clues to the origins of these rocks. An important characteristic is the wide range of abun-

[1]Taylor (1967, 1969).
[2]Taylor and White (1966).
[3]Gill (1970, 1974).
[4]Jakeš and Gill (1970).
[5]Jakeš and White (1969, 1971).

dances (fivefold or more) displayed by potassium and related incompatible elements (U, Th, Rb, Cs, Ba, Sr, Pb, rare earths, Zr, and Hf) in andesites of approximately similar major element compositions. The abundance patterns are related to differentiation trends on the FMA diagram. The overall abundance levels of incompatible elements tend to be relatively low in rock series following the tholeiitic trend and high in rock series following the calc-alkaline trend. Rare earth patterns are particularly significant. In rocks of orogenic volcanic series following the tholeiitic trend, the rare earths frequently display chondritic relative abundances, whilst the absolute concentrations are often fairly low—in the range 5 to 20 times chondritic. On the other hand, the rare earths in typically calc-alkaline volcanic rocks are usually strongly fractionated. The light REE (La, Ce, etc.) are often present at levels 30 to 50 times chondritic in andesitic calc-alkaline rocks, whereas the heavy REE (Yb, Lu, etc.) in the same rocks may be present at levels only 10 times chondritic.

Strontium 87/strontium 86 isotopic ratios in young orogenic volcanic rocks[1,2,3,4,5,6,7] are usually found to fall in the range 0.703 to 0.707, with values close to 0.704 being most commonly observed. The strontium is thus similar to the strontium found in oceanic tholeiites, which falls most frequently in the range 0.7025 to 0.7030. This demonstrates that the source materials of orogenic volcanic rocks have evolved for most of their history in an environment possessing a Rb/Sr ratio generally similar to that of the mantle and differing strongly from that in the average crust. Nevertheless, the strontium found in most orogenic volcanics is slightly but significantly more radiogenic than the average strontium from oceanic tholeiites, and this is a factor which is relevant to petrogenetic considerations.

A number of lead isotopic investigations have been carried out on orogenic volcanic rocks.[6,7,8,9,10,11,12] These data have an important bearing on the degree of possible contamination of orogenic volcanic rocks by oceanic sediments which may have occurred (Sec. 7-5).

7-3 SPACE-TIME RELATIONSHIPS

The proportions of the various members of the basalt-andesite-dacite-rhyolite series vary widely in different provinces. In the overall Pacific province, andesite

[1]Hedge (1966).
[2]Tilley, Yoder, and Schairer (1967).
[3]Ewart and Stipp (1968).
[4]Doe et al. (1969).
[5]Gill (1970, 1974).
[6]Armstrong (1968, 1971).
[7]Ewart, Bryan, and Gill (1973).
[8]Church and Tilton (1973).
[9]Doe (1967).
[10]Tatsumoto (1969).
[11]Armstrong and Cooper (1971).
[12]Oversby and Ewart (1972).

is by far the most abundant member of the series.[1] However, in some localities, andesite may be subordinate to basalt[2] or to rhyolite and rhyodacite.[3] Most orogenic volcanic provinces appear to contain representatives of all major magma types. However, this does not apply in all provinces; thus, in the Chilean Andes, basalt appears to be absent.[3,4]

Recent studies suggest that the nature and relative volumes of members of the orogenic volcanic series may vary systematically with time and with the stage of magmatic evolution of the province. Baker (1968b) has carried out an instructive comparison of the ages, structures, and petrology of two recent volcanic arcs—the Lesser Antilles and the South Sandwich Islands. Judged by several criteria—age relationships, size, volcanic activity, structural relationships—the South Sandwich Islands appear to represent an island arc at an earlier stage of evolution than the Lesser Antilles. Significant petrologic differences exist between these arcs. Basalt is by far the predominant rock type in the South Sandwich Islands, whereas andesite predominates in the Lesser Antilles (Fig. 7-5). More-

[1]E.g., Kuno (1969).
[2]E.g., South Sandwich Islands (Baker, 1968b).
[3]E.g., the Chilean Andes (Pichler and Zeil, 1969).
[4]Basalts are also rare or absent in the Northern Cascades (Hopson et al. 1965).

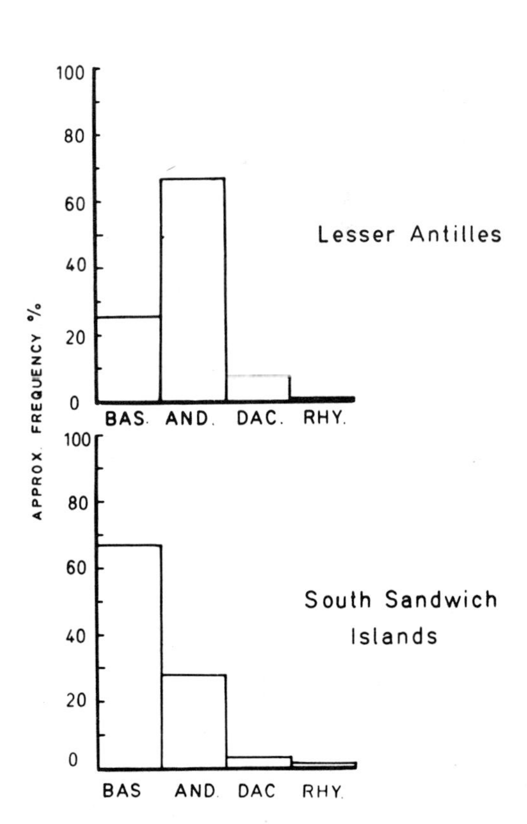

FIGURE 7-5
Estimates of proportions of basalt, andesite, dacite, and rhyolite occurring in the Lesser Antilles and South Sandwich Islands. (*After Baker, 1968b.*)

over, the differentiation trend on FMA diagrams of the South Sandwich Island rocks is intermediate between the tholeiitic and calc-alkaline trends, whereas the Lesser Antilles trend is typically calc-alkaline. Compared with the basalts of the Lesser Antilles, those of the South Sandwich Islands are distinguished by lower concentrations of K, Sr, and Ba and higher Cr, Co, and Ni. These latter basalts show certain similarities to oceanic tholeiites, whereas the Lesser Antilles basalts differ markedly from oceanic tholeiites.

Baker concludes that these two island groups represent two largely distinct stages in volcanic island-arc evolution. The South Sandwich Islands represent an early stage in which the lavas emitted are mainly tholeiites, and the more acid differentiates tend to follow a tholeiitic trend on the FMA diagram. The Lesser Antilles, on the other hand, may represent a later and more evolved stage of evolution in which the bulk of the lava emitted is of andesitic composition and the differentiation trend is typically calc-alkaline (Fig. 7-2).

Several detailed petrochemical studies of island-arc systems in the Melanesian region of the Pacific have shown that this pattern is of widespread occurrence and appears to represent a fundamental feature of the evolution of island-arc systems.[1,2,3,4] As noted earlier, the basalts in these environments are usually chemically distinct from normal oceanic tholeiites and have been termed *island-arc tholeiites*.[1,2,3] Rocks of tholeiitic affinities, showing marked iron enrichment with increase of SiO_2, appear to have predominated during the early stages of formation of several island arcs, including Fiji,[2,3] Tonga,[4] the Izu and Mariana Islands,[1] and New Guinea-New Britain.[1] They differ from normal calc-alkaline rocks by having a lower silica mode, more iron enrichment, higher Na/K ratios, less K and associated trace elements, K/Rb $\sim$ 1000, Th/U $\sim$ 1 to 2, and near-chondritic rare earth patterns.[2] There is frequently a continuous series of differentiates, ranging from basalt through basaltic andesite to andesite, dacite, and rhyolite, although the entire series shares the chemical characteristics outlined above. The average composition of the entire series appears to be more mafic (SiO_2 $\sim$ 55%) than the average composition of calc-alkaline suites (SiO_2 $\sim$ 60%).

Rocks of the calc-alkaline suite appear to have been developed in many regions at a later and more mature stage of island-arc evolution in accordance with Baker's suggestion. Examples are to be found in Fiji, New Zealand, Indonesia, the West Indies, western North America, the Aleutians, and Kamchatka. In addition to the characteristic trend on an FMA diagram, and larger proportions of more silicic members, the calc-alkaline series usually contain larger amounts of potassium and related incompatible elements and are often characterized by strongly fractionated rare earth patterns.[5,3]

[1] Jakeš and White (1969, 1971).
[2] Jakeš and Gill (1970).
[3] Gill (1970).
[4] Ewart, Bryan, and Gill (1973).
[5] Taylor (1969).

The occurrences of orogenic volcanic rocks in the Superior and Slave provinces of the Canadian Shield have been studied by Baragar and Goodwin (1969). These rocks occur widely in enormous volumes, some sequences exceeding 40,000 feet in thickness. The early compositions of these sequences from different regions are very similar to each other and appear to differ significantly from average Pacific andesites in their lower abundances of K, Ti, Ba, and Sr. An evolutionary sequence is also apparent. Among the older members of geologic sections, basaltic rocks generally predominate, whereas andesitic rocks are more abundant in the younger sections. Basalt is estimated to be relatively more abundant on the average than in the Pacific province. Proportions of major rock types are estimated as follows: basalt 60%, andesites 28%, and dacites and rhyolites 12%. These proportions should be regarded with some caution. Because they are frequently pyroclastic, andesites and dacites may be much more readily eroded and weathered than more massive basalts. A large proportion of some of the geosynclinal belts of western North America are composed of weathered andesitic debris.[1] In estimating the abundance of andesitic volcanism in Precambrian shield sequences, allowance should be made for the andesitic component of associated geosynclinal sediments.

Although the evolutionary trends in orogenic volcanism described above appear to be widespread, they are not of universal occurrence. In Japan, Kuno[2] has long recognized two orogenic volcanic lineages—the pigeonitic and hypersthenic series, which are closely associated in space and time. The former correspond to the less evolved series, showing trends towards tholeiitic differentiation. In the south Aegean arc, the usual evolutionary trend appears to be reversed.[3]

7-4 CRYSTALLIZATION BEHAVIOUR AT ATMOSPHERIC PRESSURE

The melting relations of suites of differentiated tholeiitic or alkalic rocks usually display regular trends. Liquidus temperatures of rocks decrease more or less regularly as the ratio of ($FeO + Fe_2O_3$) to ($MgO + FeO + Fe_2O_3$) increases (Fig. 7-6).[4] The melting behaviour of some typical orogenic volcanic rocks is summarized in Table 7-2.[5] The liquidus temperatures of these rocks are plotted on Fig. 7-6. It is seen that no systematic relation exists between liquidus temperatures and degree of fractionation, as indicated by $FeO + Fe_2O_3/MgO + FeO + Fe_2O_3$ ratios. The behaviour of the orogenic series is fundamentally different from that of tholeiitic and alkalic series. The outstanding feature is the high temperature of crystallization of plagioclase and the broad temperature interval through which plagioclase crystallizes alone. This is particularly marked in the dacite where plagioclase appears at 1275°C and crystallizes over an interval of 100°C before being joined by pyroxene.

[1]Dickinson (1962).
[2]Kuno (1950, 1966, 1968, 1969).
[3]Nicholls (1971).
[4]Tilley, Yoder, and Schairer (1967).
[5]Brown and Schairer (1968).

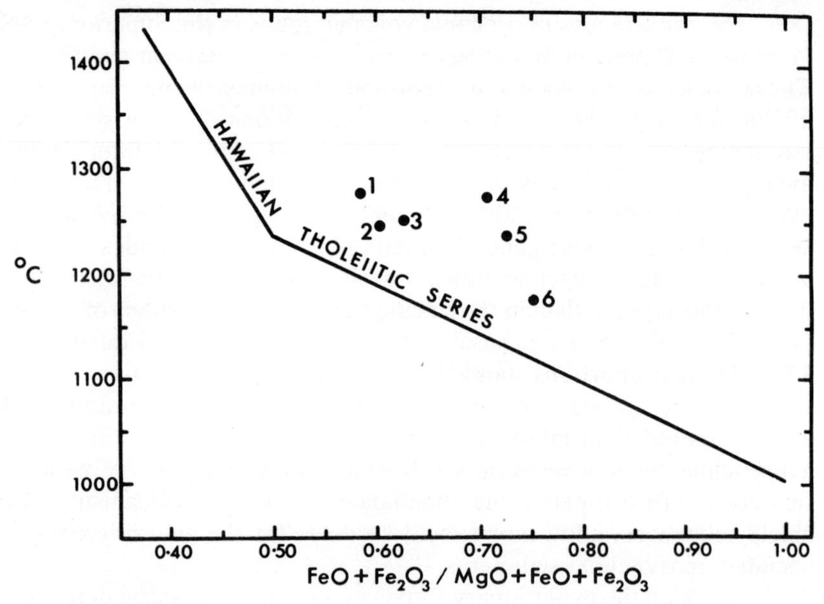

FIGURE 7-6
Plots correlating liquidus temperature with iron enrichment for Hawaiian tholeiite series (full line) compared with corresponding relations for a group of orogenic-type rocks (Table 7-2). (*After Tilley et al., 1967, and Brown and Schairer 1968.*) Numbers identify rocks referred to in Table 7-2.

Table 7-2 CRYSTALLIZATION BEHAVIOUR OF A SERIES OF WEST INDIES CALC-ALKALINE ROCKS*

7-6†	BS‡	Rock	SiO_2 content (%)	Highest temperatures of crystallization of major phases
1	16K	Olivine basalt	47.9	P1 (1280°)
2	20L	Olivine basalt	50.5	P1 (1245°)
	27V	Olivine basalt	50.5	P1 (1215°), Ol (1185°), Px (1175°)
5	19K	Hypersthene andesite	59.7	Pl (1240°), Px (1180°)
3	21L	Hypersthene andesite	60.7	Pl (1255°), Px (1180°)
4	23L	Biotite dacite	64.9	Pl (1275°), Px (1180°)

*Data from Brown and Schairer (1968).
†Column identifies rocks plotted in Fig. 7-6. Point No. 6 (Fig. 7-6) = rock 19K-A of Brown and Schairer.
‡Column gives rock identification numbers used by Brown and Schairer.

These relationships make it highly unlikely that the orogenic volcanic series have developed from parental basaltic magmas by crystallization differentiation at 1 atm. Rocks with high normative plagioclase, such as andesite and dacite, effectively constitute a thermal barrier between basalt on the one hand and rhyolites and rhyodacites, which probably have much lower liquidus and solidus temperatures,[1] on the other. Yet the chemical and textural evidence cited earlier showed clearly that andesites and dacites were on the liquid line of descent between basalt and rhyolite.

These considerations suggest that a fundamental requirement of any hypothesis which seeks to explain the petrogenesis of the orogenic volcanic series by crystal-liquid fractionation processes is a mechanism to depress the crystallization field of plagioclase relative to ferromagnesian minerals. This can be accomplished by the effects of high water pressure,[2,3] high load pressure,[4,5] or some combination of these factors. A second requirement is an explanation of the characteristic differentiation trends on the FMA diagram (Fig. 7-2), which imply that the ferromagnesian minerals involved in the fractionation of orogenic-type magmas possess higher Fe/Mg ratios, on the average, than the olivines and pyroxenes which control fractionation of normal tholeiitic types.

Compositional data on iron-titanium oxide phases crystallizing from andesitic and dacitic magmas have indicated crystallization temperatures in the vicinity of 900 to 1050°C.[6,7,8] These are much smaller than the expected crystallization temperatures of anhydrous andesitic and dacitic magmas (Table 7-2), strongly suggesting that substantial pressures of water vapour were present during crystallization of these rocks in the natural environment.

7-5 EARLY THEORIES OF ORIGIN OF OROGENIC VOLCANIC SERIES

Melting of Pre-existing Sial, Hybridism, and Contamination of Basaltic Magma by Sialic Crust

A wide range of theories of these types has been proposed and favourably discussed in the earlier literature.[9] However, a number of serious objections exists. All theories requiring the participation of pre-existing continental sialic rocks in

[1]Tuttle and Bowen (1958).
[2]Yoder (1969).
[3]Hamilton (1964).
[4]Green and Ringwood (1967).
[5]T. Green and Ringwood (1968a).
[6]Carmichael (1967).
[7]Wilkinson (1971).
[8]Nicholls (1971).
[9]Reviewed by T. Green and Ringwood (1968a).

their petrogenesis fail to explain the origin of orogenic volcanic rocks in oceanic regions where pre-existing sialic rocks did not exist.[1,2] Nor do they account for the similarities between suites of orogenic volcanic rocks erupted from intraoceanic island-arc environments with suites which have been erupted through the sialic crust near continental margins.[1,2] These theories cannot account for the evolution of sialic continents from the mantle as discussed in Chap. 8. Initial strontium isotope ratios of andesites are usually low and similar to those of related basalts and show that they cannot have formed from pre-existing old crustal material enriched in radiogenic strontium.[3,4,5] Similarly, the observed trace element abundances in many andesites preclude large-scale contributions from sialic material.[6] These objections effectively dispose of this class of hypotheses as providing a *general* explanation of the petrogenesis of *volcanic* members of the orogenic association. The mechanisms may, however, be of localized importance in some individual regions and may sometimes be involved in the genesis of the rhyolite-rhyodacite members of the series. We will also see that this class of hypotheses is probably of importance in connection with the origin of the *plutonic* members of the orogenic series.

Contamination of Basaltic Magma by Oceanic Sediments

A related theory holds that deep oceanic sediments dragged into the mantle along subduction zones play a key role in the petrogenesis of the orogenic volcanic series.[2,7] However, studies of trace element abundances in oceanic sediments and andesites demonstrated that the latter could contain only a few percent, at the most, of oceanic sediments.[8] In the case of orogenic volcanic rocks from Tonga, the lead isotopic compositions demonstrated that less than 1% of the lead present could have been derived from oceanic sediments.[9] Armstrong[7] has argued that the proportion of sediments incorporated in some orogenic volcanic series—e.g., New Zealand andesites—may be as high as 10%, but this conclusion has been criticized.[9] In the light of available data, it seems clear that the overall major element characteristics of orogenic volcanic rocks are essentially of primary origin and do not reflect substantial degrees of contamination by sediments. Nevertheless, the incorporation of small proportions of sediments cannot be excluded in some cases, and indeed, this may help explain certain element and isotopic character-

[1]Gorshkov (1962, 1969).
[2]Coats (1962).
[3]Hedge (1966).
[4]Ewart and Stipp (1968).
[5]Doe et al. (1969).
[6]Taylor and White (1966).
[7]Armstrong (1968, 1971).
[8]Taylor (1969).
[9]Oversby and Ewart (1972).

istics such as the abundance of barium and the isotopic composition of strontium.[1]

Fractional Crystallization of Basaltic Magma

Bowen (1928) considered that the basalt-andesite-dacite-rhyolite series was derived by simple fractional crystallization of anhydrous basaltic magmas. However, detailed studies of the fractionation of basaltic magmas in the crust showed that they followed the tholeiitic rather than the calc-alkaline trend.[2] Furthermore, fractional crystallization as envisaged does not appear capable of generating liquids of andesitic or dacite composition characterized by the presence of calcic plagioclase on the liquidus at temperatures above the anhydrous basalt liquidus (Table 7-2). The hypothesis is not supported by the observation that, in some regions, huge volumes of andesites have been erupted either unaccompanied by basalt or associated with minor amounts of basalt. A number of specific objections to Bowen's hypothesis have been listed by Poldervaart and Elston (1954).

Kennedy (1955) and Osborn[3] have extended Bowen's hypothesis by considering the effect of varying oxygen pressures on the courses of crystallization of model systems and basalt magmas. They proposed that the normal tholeiitic trend is a consequence of fractionation of basaltic magma at constant total composition (closed system). If, however, basalt is crystallized in an environment which is capable of introducing excess oxygen into the magma during crystallization, a large amount of magnetite may be precipitated at an early stage of crystallization. This would prevent the tholeiitic fractionation trend towards iron enrichment. In the simplified model system studied (MgO-FeO-Fe_2O_3-SiO_2, both alone and saturated with anorthite), the phases which crystallize early are rich in both MgO and FeO and poor in SiO_2. Residual liquids should thus follow the calc-alkaline trend on the FMA diagram.

Applying this to natural rocks, Osborn suggests that olivine basaltic magmas are intruded at high levels into wet piles of geosynclinal sediments, from which they absorb a few percent of water. This becomes dissociated, producing oxygen which causes the precipitation of magnetite, whilst hydrogen escapes back into the sediments. Osborn (1969a) argues that extensive fractionation under these conditions leads to the development of andesitic magmas and that the complementary crystalline differentiates represent the alpine peridotites.

This hypothesis has been widely discussed in the literature and has won a measure of acceptance.[4] Nevertheless, it is open to several serious objections. Observational evidence favouring *extensive* precipitation of magnetite on or near the liquidi of basalt, basaltic andesite, and andesite magmas is weak. Careful stud-

[1]Armstrong (1968, 1971).
[2]Wager and Deer (1939).
[3]Osborn (1959, 1962, 1969a, 1969b).
[4]E.g., Kuno (1968).

ies of some basalt-andesite sequences have shown that early magnetite separation was *not*[1,2] involved as a differentiation mechanism. Orogenic-type rocks are not invariably strongly oxidized[3]—many of the high Fe^{3+}/Fe^{++} ratios observed may be attributed to subaerial oxidation during eruption. Studies of the mineralogy of orogenic-type volcanic rocks have revealed oxidation states comparable to tholeiitic rocks.[2,4,5,6] Osborn's hypothesis implies fractionation at high crustal levels in an environment of hydrated sediments. Vast quantities of complementary precipitates consisting of magnetite + olivine + plagioclase should abound in regions of orogenic volcanism which have been moderately dissected by erosion. These have yet to be found. Osborn argues that alpine peridotites represent the complementary differentiates of andesitic magmas. Few petrologists would accept this claim. The mineral assemblage and the composition of the individual minerals of alpine peridotites (olivines and pyroxenes with $Fe/Fe + Mg \sim 0.1$ and chrome-rich spinels) bear no resemblance to the minerals, which, according to Osborn's hypothesis, should be on or near the liquidus of liquids fractionating from basalt to andesite (magnetite, plagioclase, olivines, and pyroxenes with $Fe/Fe + Mg \sim 0.3$ to 0.4). This is far from completing the list of objections. The experiments on which the hypothesis is based were carried out in simplified systems in which either plagioclase was not a component or the liquids were saturated with plagioclase. The experiments did not disclose any way of penetrating the "plagioclase barrier" discussed earlier (Table 7-2) unless the presence of excessive water pressure is assumed, in which case magnetite may no longer be a liquidus phase in natural basaltic systems. The magnetite crystallizing from orogenic magmas is very rich in vanadium (usually 0.5 to 1.4%),[7,8,2] the partition coefficient for vanadium being about 35. In passing from basalt to andesite in orogenic suites, the vanadium content may be approximately constant[7,8] or may show a twofold decrease.[9] These observations place an upper limit of 2 to 3% on the amount of magnetite which could have crystallized in differentiating from basalt to andesite.[7,8]

It appears unlikely that Osborn's ingenious hypothesis can withstand the combined weight of these objections. This is not to argue that oxidation state is a negligible factor in the petrogenesis of orogenic magmas. In order to obtain the required oxygen pressures, Osborn was obliged to introduce water into the magma. This, however, tends to negate his hypothesis since, with a sufficient water pressure to break the "plagioclase barrier," the fractionation trend of the

[1] Carmichael and Nicholls (1967).
[2] Smith and Carmichael (1968).
[3] Green and Ringwood (1968a).
[4] Wilkinson (1966).
[5] Best and Mercy (1967).
[6] Carmichael (1967).
[7] Taylor and White (1966).
[8] Duncan and Taylor (1968).
[9] Baker (1968a).

liquid usually becomes governed by separation of amphibole, and magnetite is only occasionally involved (when oxygen pressures are abnormally high).[1] This introduces an entirely new hypothesis which is discussed later. The amount of iron which is removed from the liquid by amphibole crystallization will increase with oxygen pressure[2,3] since a larger amount of ferric iron will enter amphiboles under these circumstances. Thus the oxidation state may play a significant but subsidiary role in influencing fractionation trends of residual liquids, and to this extent, an element of the Kennedy-Osborn hypothesis may well be of lasting importance.

7-6 FORMATION OF OROGENIC VOLCANIC SERIES BY PARTIAL MELTING OF MANTLE UNDER HIGH WATER PRESSURE

Poldervaart (1955) suggested that the product of partial melting of an ultramafic upper mantle under conditions of high water-vapour pressure might be andesitic rather than basaltic. A similar proposal was made by O'Hara (1965) on the basis of an analogy with crystallization equilibria in a simplified model system. This hypothesis has since been strongly supported by Kushiro and others, again mainly on the basis of experiments upon simple model systems,[4,5,6,7,8] particularly MgO-SiO_2-H_2O. Phase relations along the join Mg_2SiO_4-SiO_2 at 20 kbars and in the presence of excess water are shown in Fig. 7-7, which indicates that, above 1375°C, enstatite melts incongruently to yield a small amount of forsterite plus a liquid which contains slightly more silica than the pure $MgSiO_3$ composition. Analogous experiments[8] have been carried out in systems in which diopside, anorthite, and nepheline have been used as components (separately) in addition to MgO, SiO_2, and H_2O. The pressures used covered the range 10 to 30 kbars, and most or all of the experiments were carried out in the presence of excess water vapour. In all these hydrous systems, the field of crystallization of olivine showed a substantial expansion when compared with its field in the corresponding dry system, and in two of the systems (but not the third) extended into silica-saturated liquids.

In the systems investigated by Kushiro and Yoder, the maximum degree of oversaturation of liquids from which olivine was inferred to have crystallized was 4% SiO_2 (in the simple MgO-SiO_2-H_2O system). In other systems, addition of further components reduced this figure. Thus, if the experimental results on simple systems are to be regarded as analogous to natural systems, it would

[1]E.g., Nicholls (1971).
[2]T. Green and Ringwood (1968a).
[3]Ref. Sec. 7-7.
[4]Sclar, Carrison, and Stewart (1968).
[5]Kushiro, Yoder, and Nishikawa (1968).
[6]Kushiro and Yoder (1969).
[7]Yoder (1969).
[8]Kushiro (1969, 1970).

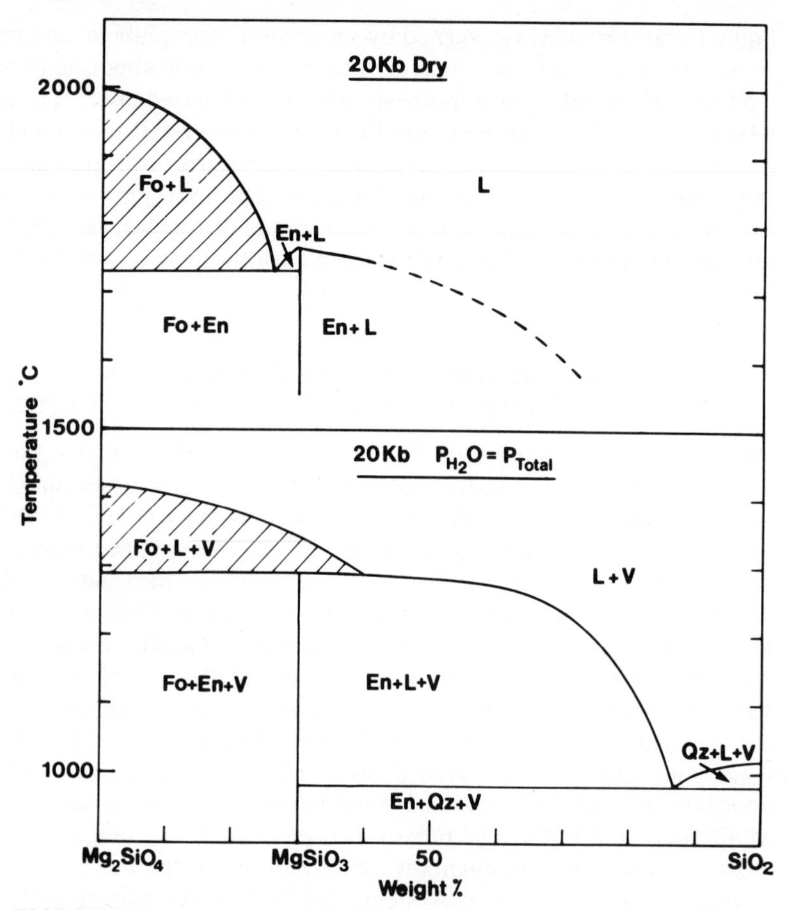

FIGURE 7-7
The system Mg_2SiO_4-SiO_2 at 20 kbars (anhydrous) and at 20 kbars ($P_{load} = P_{H_2O}$). Based on Kushiro and Yoder, 1969.

be reasonable to suggest that the first liquid to form in partial melting of a peridotitic mantle under high water pressures (20 to 30 kbars) would be an oversaturated tholeiite carrying a few percent normative quartz. This, indeed, was the initial conclusion which Kushiro and Yoder drew from their results.[1,2] However, the subsequent inference[3,4] that the first liquid to form would be of andesitic or dacitic composition, i.e., containing 10 to 25% normative quartz, is hardly warranted on the basis of analogies with the simple systems discussed above.

[1]Kushiro, Yoder, and Nishikawa (1968).
[2]Kushiro and Yoder (1969).
[3]Kushiro (1969, 1970).
[4]Yoder (1969).

Some preliminary melting experiments on a natural spinel lherzolite at 26 kbars and 1190°C in the presence of water were held to support this latter hypothesis.[1] Approximately 20% of the charge melted to a glass coexisting with residual olivine, orthopyroxene, and clinopyroxene. After correction for water content (13%), the anhydrous glass was found to contain 68% SiO_2, 10.2% CaO, 0.6% MgO, and 1.1% FeO. Although the silica content is remarkably high and similar to a dacite, the abundances of other components do not resemble dacitic compositions (Table 7-1). A subsequent experimental study[2] of the melting relationships of a similar synthetic "dacite" under identical P-T-P_{H_2O} conditions to those employed by Kushiro et al.[1] showed that a liquid of this composition could not have been in equilibrium with the observed residual mineral assemblage and that the "dacite" could not therefore have formed by partial melting of lherzolite under equilibrium conditions.[2] Reasons for the discrepancy were explored in a further study[3] of the melting of pyrolite under conditions where $P_{H_2O} = P_{total}$. In runs at 20 kbars, it was demonstrated that the high silica content of the liquid was the result of rapid metastable precipitation of olivine and amphibole during the quenching process.

An alternative approach towards evaluating this hypothesis was employed by Nicholls and Ringwood (1973), who studied the crystallization equilibria of an olivine tholeiite (Table 1-1, column 6) and a silica-saturated tholeiite[4] (Table 7-3) at high pressures, both under anhydrous conditions and in the presence of controlled amounts of water. In the olivine tholeiite, under anhydrous conditions the liquidus phase was found to be olivine up to a pressure of 14 kbars, above which olivine was replaced by clinopyroxene and, above 18 kbars, garnet. However, the maximum pressure of crystallization of olivine was found to increase markedly and regularly with addition of water to the basalt. With 10% water, olivine persisted on the liquidus to 18 kbars, whilst addition of 20% water ($P_{H_2O} < P_{total}$) caused olivine to remain a liquidus phase to 25 kbars. Finally, under conditions of water saturation ($P_{H_2O} = P_{total}$, 1090°C), olivine was found on the liquidus at 27 kbars.

These results confirm that, in the natural basaltic system, addition of water at high pressures greatly expands the primary field of crystallization of olivine. Indeed, at pressures between 18 and 27 kbars, where garnet and clinopyroxene (eclogite) are liquidus phases under dry conditions and constitute a thermal divide in the system Ol-Ga-Cpx-Opx-Qz (Fig. 7-8), addition of water causes the olivine field to expand through the eclogite plane into the quartz eclogite volume. The significance of the piercing of the eclogite thermal divide by olivine under these conditions is discussed further in Secs. 7-6 and 7-8.

The extent to which the primary field of olivine extends into more silica-rich

[1]Kushiro, Shimizu, Nakamura, and Akimoto (1972).
[2]Nicholls and Ringwood (1973).
[3]Green (1973).
[4]I.e., no normative quartz or olivine.

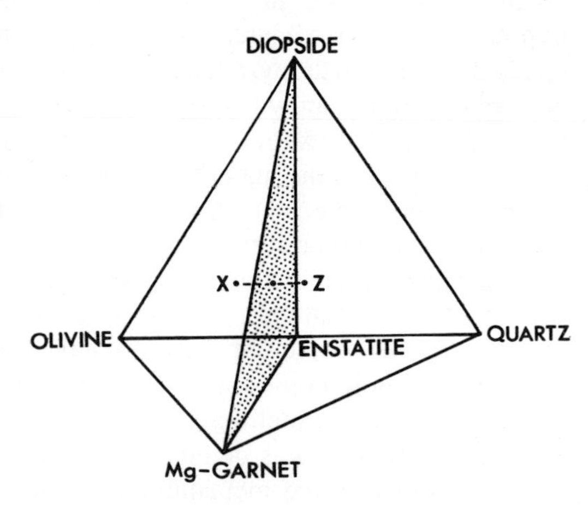

FIGURE 7-8
Diagram illustrating melting behaviour in mantle at pressures greater than about 20 kbars. The shaded plane is the eclogite plane, which forms a thermal divide under dry conditions but which is pierced by the primary field of olivine at high water pressures. Under dry conditions, the composition of the first liquid to form by partial melting, X, is always on the olivine side of the eclogite plane. At high water-vapour pressures, the composition of the first liquid is at Z, on the silica side of the eclogite plane. Further closed-system fractionation by separation of eclogite minerals would drive residual liquids to compositions rich in SiO_2.

systems under high water pressure was investigated[1] by studying the crystallization of the SiO_2-saturated tholeiite (Table 7-3). Results are shown in Fig. 7-9. Under dry conditions, olivine was on the liquidus at atmospheric pressure, but disappeared below 4.5 kbars.[2] However, with the addition of 10% water, olivine remained the liquidus phase to 14 kbars and, under water-saturated conditions ($P_{H_2O} = P_{total}$), olivine was on the liquidus at 20 kbars, 1030°C (20% H_2O).

Corresponding experiments were carried out[1] on a basaltic andesite (56% SiO_2, 8% normative quartz and, Mg/Mg + Fe = 0.63). Under conditions of water saturation ($P_{H_2O} = P_{total}$), olivine was present on the liquidus to a maximum pressure of 6 kbars.

The bearing of these results on the formation of tholeiitic and andesitic magmas in the mantle was discussed in detail.[1] They demonstrate that, where sufficient water can be introduced into the mantle, olivine tholeiite magmas could be formed by partial melting at depths of up to 100 km, whilst silica-saturated tholeiitic magmas could be formed at depths up to 70 km. These are much greater than the corresponding depths at which these magmas may be formed under dry conditions, which are 70 km and 15 km, respectively.[3] Moreover, a basaltic andesite magma of the kind studied could form by partial melting of an olivine-rich mantle at depths up to 20 km under water-saturated conditions.[1,4] Subsequent investigations by Nicholls (1974) on a wider range of compositions have extended this limit significantly. Nicholls found that andesitic liquids ($\geq$ 10% normative quartz) con-

[1]Nicholls and Ringwood (1973).
[2]T. Green, Green, and Ringwood (1967).
[3]D. Green and Ringwood (1967).
[4]See also Eggler (1972a,b) and Green (1973).

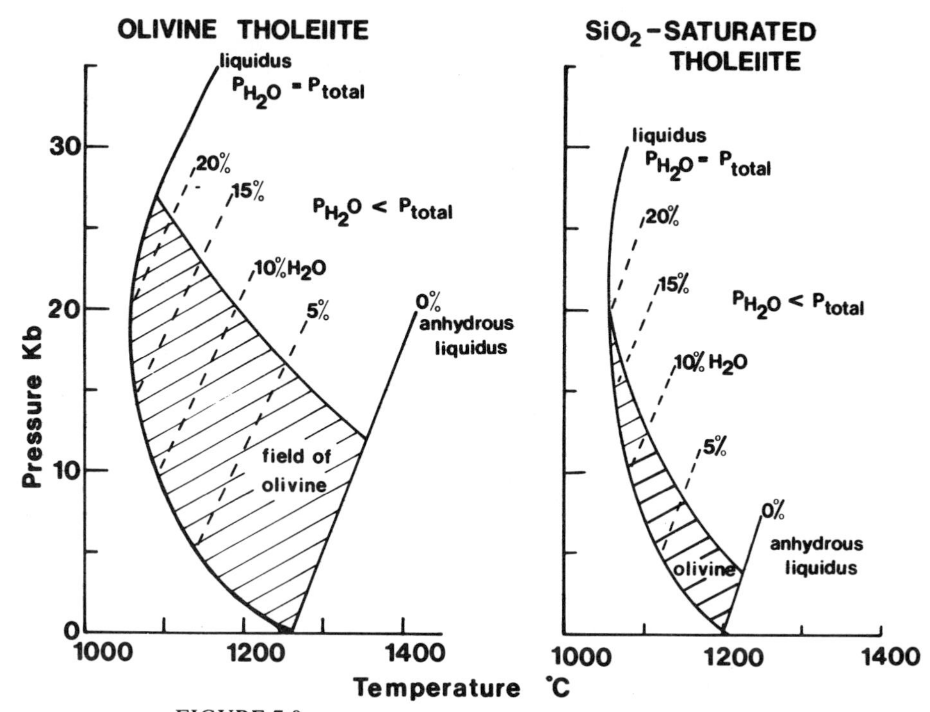

FIGURE 7-9
Liquidi of olivine tholeiite and silica-saturated tholeiite under conditions of $P_{H_2O} \leq P_{total}$ for a range of water contents. Shaded regions indicate fields of olivine crystallization on and near the liquidi. Based on results of Nicholls and Ringwood (1973).

taining up to 60% SiO_2 can be in equilibrium with olivine-bearing residual mineral assemblages at *maximum pressures* up to 10 kbars. At higher pressures, the silica and normative quartz contents of hydrous liquids in equilibrium with peridotitic mineral assemblages fall sharply. These results clearly provide severe limitations to the possibility of producing andesitic magmas by direct partial melting at considerable depths (70 to 100 km) as advocated by Kushiro.

There are, moreover, additional constraints on the possibility of producing primary quartz tholeiite and basaltic andesite magmas by hydrous partial melting. The liquidus temperatures of water-saturated tholeiites at high pressures are greatly depressed and are smaller than the 1-atm liquidus temperatures (Fig. 7-9). Accordingly, water-saturated magmas formed at considerable depths in the mantle are unable to reach the surface without undergoing forced crystallization as confining pressure is reduced and water escapes. In order to avoid this situation, the basalts produced by partial melting in the mantle must be undersaturated in water, but this in turn requires that the depth interval under which such tholeiites can form in equilibrium with mantle peridotite is considerably restricted.

Table 7-3 COMPARISON WITH NATURAL BASALTIC ANDESITE OF A COMPOSITION PRODUCED BY REMOVAL OF OLIVINE FROM AN EXPERIMENTAL THOLEIITE. (*After Nicholls and Ringwood, 1973*)

	(1) SiO_2-saturated tholeiite	(2) Column 1 minus 13% Ol (Fo_{86})	(3) Basaltic andesite (Tonga)
SiO_2	51.5	53.9	53.7
TiO_2	1.8	2.0	0.6
Al_2O_3	13.8	16.0	17.7
Fe_2O_3	2.2	2.3	2.6
FeO	8.9	7.3	7.6
MnO	0.2	0.1	0.2
MgO	9.4	4.1	4.3
CaO	8.9	10.3	11.1
Na_2O	2.5	2.9	1.8
K_2O	0.7	0.8	0.4
P_2O_5	0.2	0.2	0.1
$\frac{Mg}{Mg + Fe}$ (mol)	0.65	0.5	0.5

Explanation:
(1) Experimental composition studied by Nicholls and Ringwood (1973).
(2) Composition produced from experimental tholeiite by subtraction of 13% equilibrium olivine.
(3) Basaltic andesite L1, Late Island, Tonga group (Ewart et al., 1973).

An analysis of the relevant phase relationships[1] reveals that water-saturated magmas of olivine tholeiite or SiO_2-saturated type, originating in the mantle at depths of 60 to 100 km, are obliged to crystallize and fractionate as they rise to shallower levels where the confining pressures are smaller. Moreover, the (adiabatic) cooling paths of these magmas pass *only through the field of olivine crystallization.* Crystallization of 10 to 15% olivine from the silica-saturated tholeiite magma produces a basaltic andesite composition (Table 7-3). The water-saturated liquidus of basaltic andesite at about 5 kbars lies about 100°C below those of tholeiitic basalts and olivine is the liquidus phase.[1] These relationships show that a continuum of fractionated magma compositions ranging between olivine tholeiite-quartz tholeiite-basaltic andesite will be generated during the rise of water-rich basalt magmas formed initially at depths of 60 to 100 km. The basaltic andesites may differentiate by olivine segregation ultimately to andesitic compositions ($SiO_2 \leq 60\%$) in the crust or uppermost mantle (depths of 20 to 40 km).

Kushiro's hypothesis that andesites and dacites are formed as primary partial melts of mantle pyrolite at water pressures of 20 to 30 kbars implies that these magmas could be produced along the Benioff zone, where water is introduced by subducted lithosphere. The results of Nicholls and Ringwood are inconsistent

[1]Nicholls and Ringwood (1973). See also Fig. 7-9.

with this hypothesis and imply, on the other hand, that the magmas formed by partial melting under high P_{H_2O} at the Benioff zone will be basaltic. Such water-rich magmas will rise and may differentiate by olivine separation towards andesitic compositions only at relatively shallow depths. Regardless of the details, there is no doubt that water plays a key role in the genesis of magmas of the orogenic volcanic series, a subject which is further developed in the following section.

7-7 ROLE OF WATER IN FORMATION OF OROGENIC MAGMAS: AMPHIBOLE FRACTIONATION

Many authors[1,2,3,4,5] have suggested that the orogenic volcanic series may have formed by crystallization-differentiation of basalt magmas under high water-vapour pressures or by partial melting of basaltic material in the lower crust or upper mantle in the presence of water. Although this hypothesis has been the subject of much speculative discussion for many years, direct supporting evidence was rather weak. The situation has since been changed by parallel advances in experimental petrology and in careful petrological and mineralogical studies of orogenic magma suites. We will consider the experimental results first.

Pioneering investigations on the crystallization of several natural basalts under controlled water-vapour pressures (0 to 10 kbars, $P_{H_2O} = P_{load}$) were carried out by Yoder and Tilley (1962) using a gas apparatus.[6] These studies (Fig. 7-10) demonstrated the major effect of water pressure in depressing the crystallization field of plagioclase[7] relative to the ferromagnesian minerals and in extending the crystallization field of amphibole. Yoder and Tilley pointed out that crystallization of basalt under hydrous conditions would lead to an increase in the alumina content of the magma and suggested that high-alumina basalts and anorthosites may form in this manner. Hamilton (1964) called attention to the depression of the plagioclase field by water and suggested that fractionation of basalt under high water-vapour pressures would lead to the formation of andesite and dacite magmas. Although this appears to be a feasible mechanism of breaking the "plagioclase barrier" (Sec. 7-4), it did not offer an explanation of another major characteristic displayed by many members of the series—the calc-alkaline differentiation trend on the FMA diagram.

Experiments by Nicholls and Ringwood (1973) on the near-liquidus crystallization relationships of basaltic magmas were discussed in the previous section. These showed that the presence of water under high pressures also has the

[1]Daly (1933).
[2]Coats (1962).
[3]Hamilton (1964).
[4]Lidiak (1965).
[5]Branch (1967).
[6]See also Hamilton, Burnham, and Osborn (1964).
[7]This was also demonstrated by Yoder's (1965) investigation of the system anorthite-diopside at high water pressure.

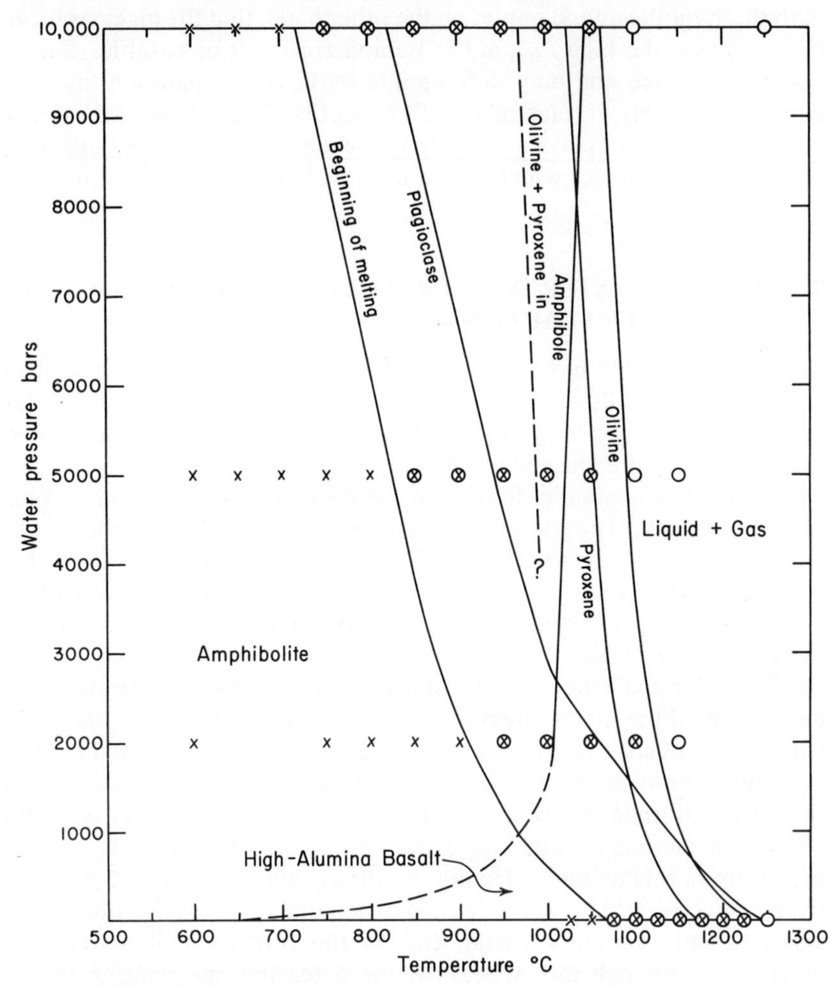

FIGURE 7-10
Pressure-temperature projection of a natural high-alumina basalt-water system. (*From Yoder and Tilley, 1962. Copyright 1962 Oxford University Press. By permission of the Clarendon Press, Oxford.*)

effect of expanding the primary crystallization field of olivine relative to pyroxene. It was demonstrated that water-rich basaltic magmas formed at depths of 60 to 100 km would fractionate by olivine crystallization on rising to higher levels to form a continuum of magma compositions ranging from olivine tholeiite-quartz tholeiite-basaltic andesite-andesite. This differentiation trend would be strictly tholeiitic in terms of the FMA diagram (Fig. 7-2). In order to produce the more acid andesite-dacite-rhyolite magmas of the orogenic volcanic series, further fractionation processes controlled by the crystallization of phases other than olivine would by necessary. Three possibilities representing crystallization with increasing pressures are apparent: (1) pyroxene-plagioclase fractionation, (2) amphibole fractionation, and (3) eclogite fractionation. A continuum between these processes probably occurs in nature.

Starting with a parental basaltic andesite magma generated as discussed previously, further differentiation may proceed by means of the crystallization of pyroxenes and plagioclase. This is the nearest approach to Bowen's original basalt fractionation hypothesis (Sec. 7-5). However, for this kind of differentiation to occur, the "plagioclase barrier" (Sec. 7-4) must be broken in order that pyroxenes and plagioclase may occur together on the liquidi of basaltic andesite and andesite compositions. The required suppression of the crystallization field of plagioclase relative to pyroxene may be accomplished by the presence of high water pressures,[1,2,3] high load pressures,[3,4] or a combination of these effects. Experimental investigations[3,5,6,7,8] on relevant compositions show that these conditions can be realized approximately at temperatures of 1000 to 1100°C, water pressures of 1 to 3 kbars, combined with total pressures up to about 6 kbars. This kind of differentiation is effectively restricted, therefore, to magma chambers at shallow depths (< 20 km).

Because the pyroxenes separating under these conditions have much higher Mg/Fe ratios than the parent magmas, the liquid differentiates produced follow a strongly tholeiitic trend on the FMA diagram.

Detailed investigations of the geochemistry and petrogenesis of basaltic andesites, andesites, and dacites of the Tonga group indicated that they had formed by fractionation processes of this kind.[9,10] Nevertheless, the extreme tholeiitic trend (Fig. 7-11) displayed by this series appears to be relatively rare, and the very restricted P-T-P_{H_2O} conditions required for the operation of this kind of fractionation process suggest that it plays a limited role in the genesis of the orogenic volcanic association, most members of which display smaller degrees of iron enrichment during differentiation.

Gabbroic xenoliths consisting of pyroxenes and highly calcic plagioclase are commonly found in rocks of the orogenic volcanic association.[11] Experimental investigations revealed that the plagioclase crystallizing from andesitic magma under high water pressure was more calcic than that crystallizing under anhydrous conditions.[3] It seems likely that these gabbros have crystallized in shallow reser-

[1]Hamilton (1964).
[2]Yoder and Tilley (1962).
[3]T. Green and Ringwood (1968a).
[4]D. Green and Ringwood (1967).
[5]T. Green (1972).
[6]Hill and Boettcher (1970).
[7]Holloway and Burnham (1972).
[8]Eggler (1972a,b).
[9]Ewart, Bryan, and Gill (1973).
[10]Oversby and Ewart (1972).
[11]E.g., Nicholls (1971).

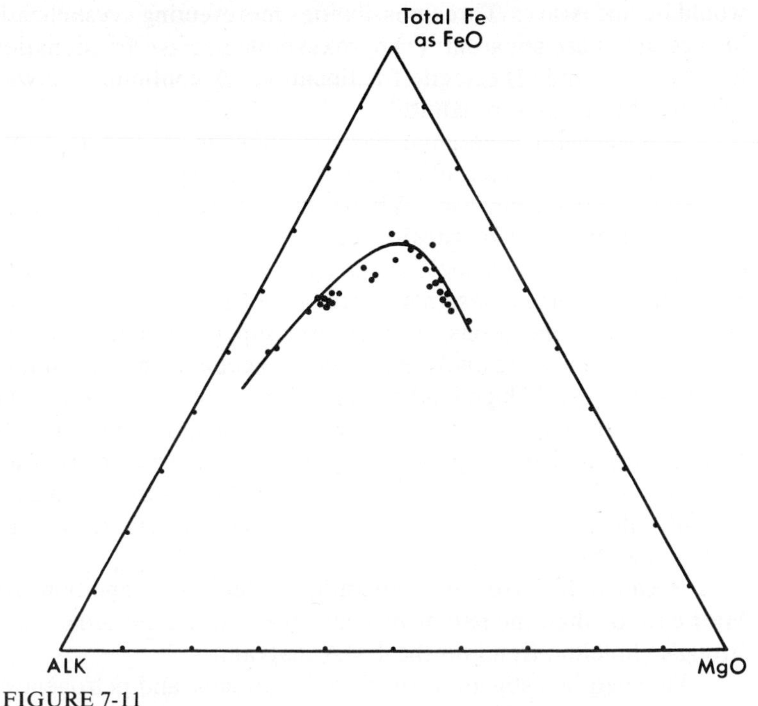

FIGURE 7-11
FMA differentiation trend displayed by basalt-andesite-dacite series from Tonga. (*After Ewart, Bryan, and Gill, 1973.*)

voirs from basalt, basaltic andesite, and andesite magmas[1,2,3] under the P-T-P_{H_2O} conditions discussed above. The fact that such "tholeiitic" xenoliths are frequently found in orogenic volcanic rocks possessing typical *calc-alkaline* trends suggests that high-level gabbroic crystallization has not been predominant in generating this differentiation trend. Most probably, these xenoliths represent relatively small volumes of precipitates formed when hydrous magmas generated at deeper levels ascended to shallow reservoirs and were subjected to forced crystallization owing to reduction of water pressure.[4]

In order to generate magma series following the calc-alkaline trend on the FMA diagram, or magma series following intermediate paths between the extreme tholeiitic and calc-alkaline trends (Fig. 7-2), it is necessary to crystallize phases having higher Fe/Mg ratios than the olivines and pyroxenes responsible for the tholeiitic trend. It seemed possible that fractionation processes dominated by the crystallization of amphibole might satisfy this requirement. To test this

[1]Nicholls (1971).
[2]Gill (1974).
[3]Ewart, Bryan, and Gill (1973).
[4]Plagioclase-amphibole xenoliths found in many calc-alkaline magmas may have a similar origin.

Table 7-4 COMPOSITIONS OF SYNTHETIC GLASSES USED BY T. GREEN AND RINGWOOD (1968a) IN THEIR EXPERIMENTAL INVESTIGATION OF CALC-ALKALINE CRYSTALLIZATION EQUILIBRIA

	(1) High-alumina olivine tholeiite	(2) High-alumina quartz tholeiite	(3) Basaltic andesite	(4) Andesite	(5) Dacite	(6) Rhyodacite
SiO_2	50.3	52.9	56.4	62.2	65.0	69.6
TiO_2	1.7	1.5	1.4	1.1	0.7	0.6
Al_2O_3	17.0	16.9	16.6	17.3	16.1	14.7
Fe_2O_3	1.5	0.3	3.0	0.3	1.4	1.7
FeO	7.6	7.9	5.7	5.9	3.5	1.8
MnO	0.2	0.2	0.1	0.1	0.1	0.1
MgO	7.8	7.0	4.3	2.4	1.8	1.0
CaO	11.4	10.0	8.5	5.2	5.0	2.5
Na_2O	2.8	2.7	3.0	3.3	3.6	3.4
K_2O	0.2	0.6	1.0	2.3	2.1	4.6
Total	100.4	100.0	100.0	100.1	99.3	100.0
Mol. prop.: $\frac{100\ MgO}{MgO + FeO_{total}}$	60.7	60.4	47.7	41.0	40.3	36.0

hypothesis, Green and Ringwood[1] carried out a reconnaissance experimental investigation of the crystallization of a series of orogenic volcanic rocks at 10 kbars and under a substantial water pressure. The rocks studied were basalt, basaltic andesite, andesite (10 kbars), and dacite and rhyodacite at 10 to 30 kbars (Table 7-4). Water vapour in these experiments was supplied by dehydration of the talc pressure medium and was not well controlled. Most experiments were carried out under conditions where the liquidus was lowered by about 200°C. This required pressures estimated to be in the 2 to 5 kbar range. The objective was to have the water pressure smaller than load pressure so as to simulate more closely the probable geologic situation. In addition to establishing the crystallization sequence and proportions of phases present, the compositions of major phases were determined with an electronprobe microanalyzer, thus permitting a quantitative investigation of fractionation trends under these conditions.

Results on the basalt and basaltic andesite compositions showed a broad field of crystallization of amphibole, clinopyroxene, and minor orthopyroxene at 10 kbars combined with relative suppression of the plagioclase field. In the basalt, clinopyroxene was the liquidus phase at 1100°C, joined by orthopyroxene and amphibole at lower temperatures (1040°C). Amphibole was the dominant phase below 960°C, and plagioclase appeared well below the liquidus at 920°C. A similar sequence of crystallization occurred in the basaltic andesite composition.

[1]T.H. Green and Ringwood (1967, 1968a,b).

Table 7-5 TYPICAL COMPOSITIONS OF PHASES CRYSTALLIZING FROM HIGH-ALUMINA QUARTZ THOLEIITE (TABLE 7-4) AT HIGH PRESSURES AND TEMPERATURES AND UNDER HYDROUS CONDITIONS. COMPOSITIONS WERE DETERMINED ON CRYSTALS FROM EXPERIMENTAL RUNS USING AN ELECTRONPROBE MICROANALYZER. (*After T. Green and Ringwood 1968a*)

Phase	Orthopyroxene	Clinopyroxene	Amphibole	Amphibole	Plagioclase
Condition of runs	9 kbars 1040°C	9 kbars 1040°C	9 kbars 1040°C	10 kbars 920°C	10 kbars 920°C
Coexisting phases	Amph Cpx	Amph Cpx	Cpx Opx	Cpx Opx Plag	Cpx Opx Amph
SiO_2	46.3	46.8	39.3	39.9	52.9
TiO_2	0.8	1.7	3.8	2.9	—
Al_2O_3	7.3	8.5	15.6	15.6	30.0
FeO	15.4	7.6	9.8	12.6	—
MgO	24.5	13.9	13.7	11.6	—
CaO	1.7	19.9	11.8	12.2	13.9
Na_2O	—	0.6	2.9	2.5	3.9
K_2O	—	—	0.2	0.4	0.1
Total	96.0	99.0	97.1	97.7	100.8
Mol. prop: $\frac{100\ Mg}{Mg + Fe}$	74	77	71	62	

Abbreviations:
Amph = amphibole, Cpx = clinopyroxene, Opx = orthopyroxene, Plag = plagioclase.

However, in the andesite composition at 10 kbars, clinopyroxene *and* plagioclase were near-liquidus phases at 940°C and were joined by garnet, amphibole, and orthopyroxene at 900°C.

Analyses of clinopyroxene, orthopyroxene, amphibole, and plagioclase crystallizing from the basaltic composition were obtained (Table 7-5). The pyroxenes showed high alumina contents (7.5 to 10.6%) and low silica contents (47.2 to 48.0%). The amphiboles were also extremely subsilicic, containing 39.8 to 40.5% SiO_2 and were high in alumina (14.8 to 15.8%). The Fe/Mg ratio in amphibole was always higher than the same ratio in the coexisting pyroxenes. This is an important feature in connection with generating the calc-alkaline FMA trend. In the andesite and rhyodacite compositions, plagioclase occurred near the liquidus and was joined by garnet and amphibole at lower temperatures. With increasing pressure, garnet became more abundant as a near-liquidus phase until, at 18 kbars in the rhyodacite and at 27 kbars in the dacite (this composition was not studied at 18 kbars), garnet was the liquidus phase.

In the basalt and the basaltic andesite, the broad field of crystallization of subsilicic amphibole, together with subordinate clinopyroxene, provides an efficient mechanism for causing silica and alkali enrichment in residual liquids. Marked iron enrichment is prevented by extensive crystallization of amphibole with a relatively high Fe/Mg ratio. It was suggested that this ratio would have been even higher had the experiments been carried out under more oxidizing conditions which would have led to the incorporation of a substantial amount of Fe^{3+} in the amphiboles. [1,2] Calculations based upon the proportions and compositions of the phases which were observed to crystallize showed that fractionation of basaltic magmas under these conditions follows the calc-alkaline trend and provides an efficient mechanism of generating andesitic and dacitic magmas.

These reconnaissance results have been followed by more rigorous experimental investigations of the crystallization of andesite,[3,4] basalt,[5,6,7] and rhyodacite[8] under conditions where water pressure was controlled independently of load pressure and where oxidation state was also controlled by a buffer.[4,5] These experiments have broadly confirmed the earlier results and provided important new data. Green's results on the crystallization behaviour of an andesite are shown in Figs. 7-12 and 7-13.

Holloway and Burnham (1972) studied the melting behaviour of a basalt between 2 and 8 kbars in a system with P_{O_2} controlled by the quartz-fayalite-magnetite buffer and with $P_{H_2O} = 0.6P_{total}$. They found amphibole stable up to 975°C, 2 kbars; 1025°C, 5 kbars; and 1060°C, 8 kbars. The Al_2O_3 contents of the amphibole increased with pressure, whilst the silica content decreased, analogous to the behaviour observed in pyroxenes. [9,1] Amphibole crystallized at 1000°C, 5 kbars contained 43% SiO_2 and 11.5% Al_2O_3, whilst amphibole crystallized at 1050°C, 8 kbars contained 41% SiO_2 and 14.5% Al_2O_3. The kaersutite composition of the latter amphibole closely approached that found by Green and Ringwood at 10 kbars (Table 7-5). Holloway and Burnham showed by electron-microprobe analysis that the liquid coexisting with amphibole, clinopyroxene, plagioclase, and magnetite was dacitic near the solidus (800 to 900°C), whilst the liquid coexisting with amphibole, clinopyroxene, and magnetite approached a silica-rich andesite near 1000 to 1050°C and 5 to 8 kbars.

The Fe/Mg ratios of coexisting amphibole, clinopyroxene, olivine, and the initial liquid found by Holloway and Burnham are shown in Table 7-6. Note that the Fe/Mg ratio of amphibole is much higher than coexisting olivine and pyroxene and is similar to that of the parental liquid. This is necessary if differentiated liquids are to follow the calc-alkaline trend.

Experimental investigations[10,5,3] indicate that high water pressures

[1]T. Green and Ringwood (1968a).
[2]Confirmed by Holloway and Burnham (1972). See also Table 7-6.
[3]T. Green (1972).
[4]Eggler (1972a,b).
[5]Holloway and Burnham (1972).
[6]Nicholls and Ringwood (1973).
[7]Hill and Boettcher (1970).
[8]T. Green and Ringwood (1972).
[9]D. Green and Ringwood (1967).
[10]Yoder and Tilley (1962).

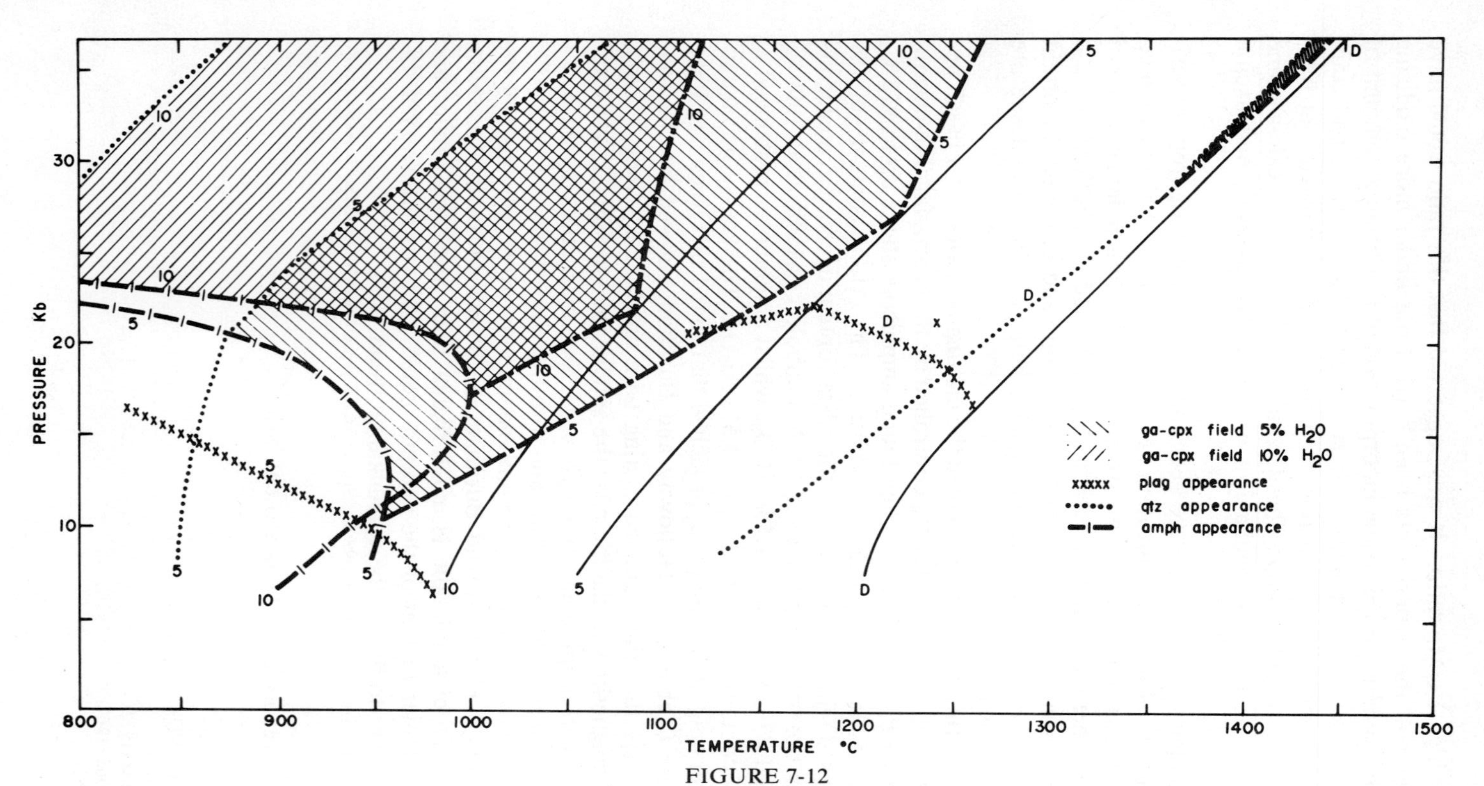

FIGURE 7-12
Crystallization of andesite under controlled high-pressure hydrous conditions showing effect of water on (*a*) enlargement of garnet-clinopyroxene crystallization field, (*b*) suppression of quartz and plagioclase fields, (*c*) stability of amphibole. Numbers beside curves refer to water content applying to that particular stability curve. Solid lines represent liquidi at indicated water contents, "D" represents anhydrous conditions. (*From T. Green, 1972, with permission.*)

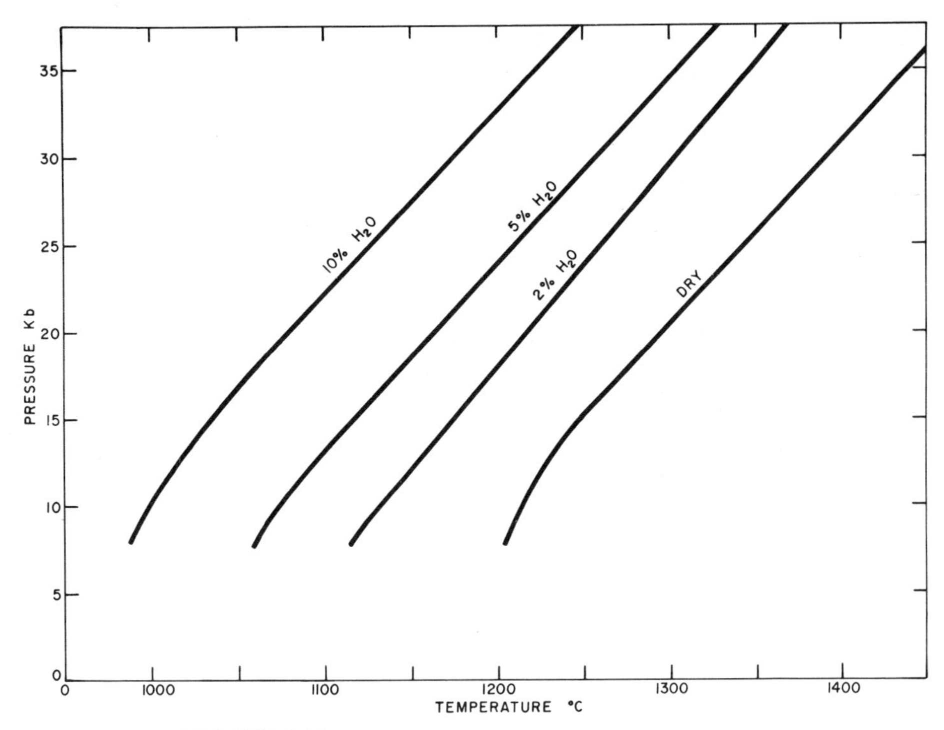

FIGURE 7-13
P-T plot showing effect of addition of water on lowering of the andesite liquidus. (*From T. Green, 1972, with permission.*)

Table 7-6 Fe/Mg RATIOS OF OLIVINE, CLINOPYROXENE, AND AMPHIBOLE OBSERVED TO CRYSTALLIZE FROM THOLEIITIC BASALT AT $P_{H_2O} = 0.6\ P_{total}$, $P_{total} \sim 8$ kbars, AND $T \sim 1000°C$, WITH f_{O_2} CONTROLLED BY THE QUARTZ-FAYALITE-MAGNETITE BUFFER. IRON CONTENTS REPRESENT TOTALS OF $Fe^{++} + Fe^{3+}$. (*After Holloway and Burnham, 1972*)

Phase	Fe/Mg (atom ratio)
Parental basalt	0.64
Olivine	0.21
Clinopyroxene	0.23
Amphibole	0.69

($P_{H_2O} > 0.5P_{total}$) are necessary in order for amphibole to control the fractionation of basaltic magmas as discussed earlier. At lower water pressures, and at load pressures between 5 and 15 kbars, there is a very wide field in which fractionation is controlled by simultaneous separation of amphibole and pyroxene or by pyroxene followed by amphibole. All gradations between pyroxene-controlled and amphibole-controlled fractionation trends are possible,[1,2,3] whilst plagioclase crystallization is suppressed. These mechanisms will produce a corresponding continuum of fractionation trends on the FMA diagram between tholeiitic (pyroxene ± plagioclase crystallization, low P_{H_2O}), intermediate (amphibole + pyroxene crystallization, moderate P_{H_2O}), and calc-alkaline (amphibole crystallization, high P_{H_2O}).

In the interval between 5 and 15 kbars, the amphiboles crystallizing from basaltic to andesitic magmas possess broadly similar K/Na ratios to the parental magmas, whilst the pyroxenes incorporate only small amounts of Na and negligible amounts of K.[1,3,2] Thus fractionation in this pressure regime will not cause drastic changes in the K/Na ratios of residual liquids. Although the relevant partition coefficients for rare earths are poorly known, it does not appear that moderate degrees of pyroxene-amphibole fractionation will lead to strong relative fractionations within this group. However, separation of amphibole and pyroxenes will lead to strong depletion of Ni and Cr in residual liquids and marked depletion of Ti.

These characteristics suggest that amphibole-pyroxene-controlled fractionation may contribute to the early phase of island-arc volcanism when magmas show tholeiitic to intermediate trends on the FMA diagram (Sec. 7-3). A related application of these fractionation mechanisms is to the formation of intermediate and acid rocks occurring in the ophiolite association. Thayer (1967) has emphasized the widespread occurrence of small amounts (usually < 10%) of diorites, dacites, and related rocks which occur in many ophiolite complexes and also that these rocks frequently follow intermediate to tholeiitic trends on the FMA diagram. Differentiation of mafic magmas in the presence of water to produce acid-intermediate magmas has also occurred occasionally along oceanic ridges.[4,5]

Two further examples of this kind of differentiation may be mentioned. Best and Mercy (1967)[6] carried out a detailed study of the petrology and mineral chemistry of the Guadalupe igneous complex, consisting of a plutonic series of gabbro, diorite, grandiorite, and granite, displaying a generally calc-alkaline

[1]T. Green and Ringwood (1968a).
[2]Holloway and Burnham (1972).
[3]T. Green (1972).
[4]Hald, Noe-Nygaard, and Pedersen (1971).
[5]Aumento (1969).
[6]See also Best (1963).

FMA trend. They demonstrated that the parental gabbroic magma had crystallized under a high water-vapour pressure and that crystallization was dominated by amphibole which appeared early and precipitated throughout most of the crystallization sequence. Best and Mercy pointed out that the crystallization of low-silica amphibole (45% SiO_2) provided an efficient means of enriching residual liquids in alkalis and silica, whilst at the same time, the relatively high Fe/Mg ratio of amphibole (as compared to pyroxenes) prevented marked iron enrichment at intermediate stages of differentiation and was responsible for the calc-alkaline FMA trend.

A second example is furnished by Watterson's (1968) study of the crystallization of a series of mafic dykes emplaced during the closing stages of a period of plutonic activity. Mafic minerals, dominantly hornblende, had an extended period of crystallization before plagioclase. It was estimated that over half of the magma had crystallized before plagioclase appeared and that the residual liquid contained 60 to 80% normative plagioclase. Watterson estimates that a water pressure of 3 to 4 kbars was necessary to achieve this crystallization sequence and that the temperature of crystallization was in the vicinity of 800 to 850°C based on Yoder and Tilley's (1962) results. The differentiates displayed a generally calc-alkaline trend (Fig. 7-14), attributed to fractionation dominated by amphibole crystallization.

Although we have been considering mainly the fractionation of basaltic magmas by crystallization differentiation under hydrous conditions, the equilibria are directly applicable to the partial melting of basaltic compositions, e.g., amphibolites, under hydrous conditions at pressures less than about 20 kbars.[1] At relatively low temperatures (e.g., ~ 900°C), where a small degree of partial melting has occurred, the liquid will be of a dacitic or rhyodacitic composition. With increasing temperature (e.g., ~ 1000°C) and a greater degree of partial melting, the liquid is of andesitic composition, and the residuum is of amphibole and pyroxene similar in composition to those found experimentally (Table 7-5). With still higher temperature, an increasing proportion of amphibole enters the liquid, which has the composition of basaltic andesite, and the residuum is dominantly of pyroxene.

Formation of orogenic volcanic rocks by partial melting of amphibolite might occur in some regions of the lower crust which are believed to consist mainly of amphibolite (Sec. 2-1). The required conditions would also be provided by the sinking of a plate of lithosphere beneath an oceanic trench. It appears probable that part of the oceanic crust is composed of amphibolite (Sec. 2-3) which could become partially melted by frictional dissipation[2] during sinking, leading to the formation of orogenic-type magmas above the sinking plate. These

[1]T. Green and Ringwood (1968a).
[2]Oxburgh and Turcotte (1968, 1970).

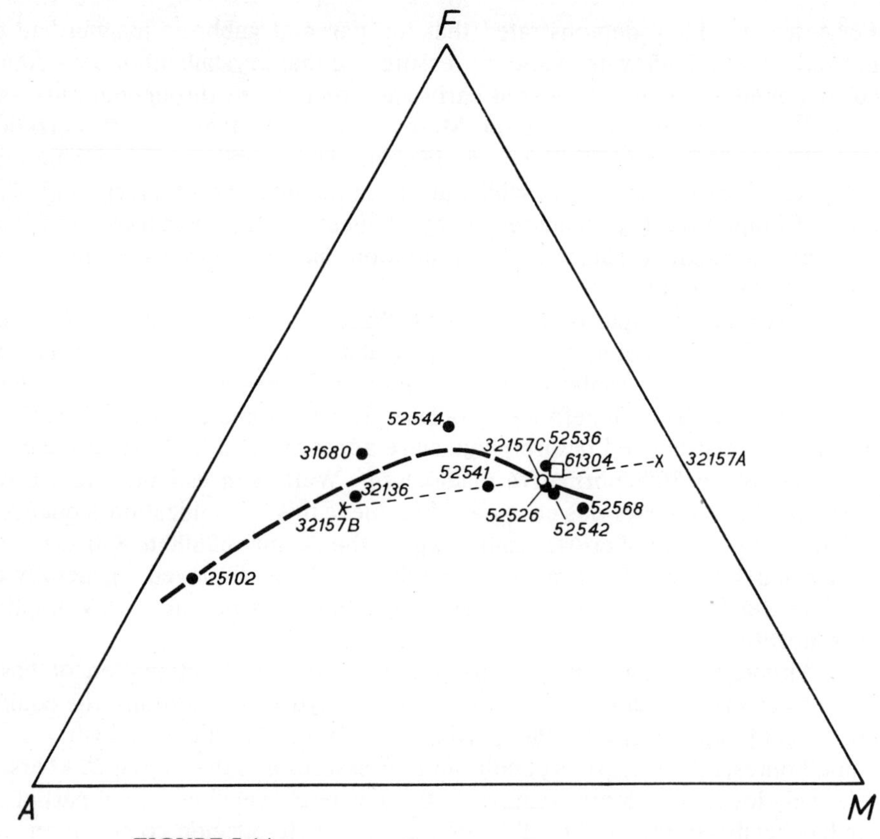

FIGURE 7-14
FMA compositional variation in Ilordleq dyke suite (solid circles) showing inferred differentiation trend. Separated components of a single strongly differentiated dyke (crosses) are shown, together with the calculated bulk composition of this dyke (open circle). (*From Watterson, 1968, with permission.*)

processes will be discussed in greater detail after we have considered another process for generating orogenic-type magmas.

7-8 ROLE OF ECLOGITE FRACTIONATION

Amphiboles are not stable at temperatures of 800 to 1000°C in basaltic compositions under pressures greater than about 27 kbars, equivalent to a depth of 90 km (Sec. 6-2).[1] On the other hand, many volcanoes of the orogenic series are situated

[1]See also Fig. 7-16.

some 100 to 150 km above the Benioff zones. If these magmas are indeed generated at or near the Benioff zones, their formation is unlikely to be controlled by equilibria involving amphiboles.

When mafic oceanic crust is introduced into the mantle by sinking lithosphere plates, it is transformed ultimately to quartz eclogite.[1] Regions of the crust which are anhydrous transform at very low pressures,[1] whereas hydrated regions containing amphibole require higher pressures, up to 27 kbars.[2] This suggested the possibility that orogenic volcanic magmas might be generated at depths greater than 100 km by partial melting of quartz eclogite[3,4] caused, for example, by frictional dissipation[5] in the boundary layer between sinking eclogite and mantle (Fig. 8-6). An alternative environment for this type of fractionation might be in the wedge-shaped region above the sinking lithosphere. Introduction of water from the sinking plate into the mantle above[4,6] may cause partial melting of pyrolite, giving rise to a near-SiO_2-saturated magma as discussed in Sec. 7-6. Subsequent closed-system fractionation of this magma by separation of garnet and clinopyroxene at depths greater than 80 km could also produce orogenic-type magmas.

To test these hypotheses, detailed experimental investigations of the crystallization behaviour between solidus and liquidus of a series of orogenic volcanic rocks under anhydrous conditions at pressures up to 40 kbars were undertaken.[4] Compositions are given in Table 7-4. Phases present and their proportions were obtained by microscopic examination and x-ray diffraction, and the compositions of near-liquidus phases were determined using an electronprobe microanalyzer, permitting the fractionation trends of the magma to be determined quantitatively as a function of pressure. The method used was similar to that employed in the studies of basalt fractionation described in Chap. 4.

In compositions from basalt to andesite, the near-liquidus phases were found to change from plagioclase-pyroxene dominated at 0 to 18 kbars to pyroxene-garnet dominated at 27 to 36 kbars. The same relationship held for the dacite and rhyodacite, except that the liquidus phase changed from garnet to quartz. Thus, as might have been anticipated, the plagioclase barrier which prevents the andesitic differentiation trend under low-pressure anhydrous conditions was not present at high pressures.

A notable feature was that andesite, rather than dacite or rhyodacite, occupies a marked thermal valley in the series (Fig. 7-15). The liquidus temperatures fall on passing from olivine tholeiite-quartz tholeiite-basaltic andesite-andesite, and fractionation is controlled by garnet and pyroxene. In compositions more acid than andesite, quartz becomes a major near-liquidus phase, and the

[1]Ringwood and Green (1966, fig. 10).
[2]See also Fig. 7-16.
[3]Ringwood and Green (1966).
[4]T. Green and Ringwood (1966, 1968a).
[5]Oxburgh and Turcotte (1968, 1970).
[6]Ringwood (1969).

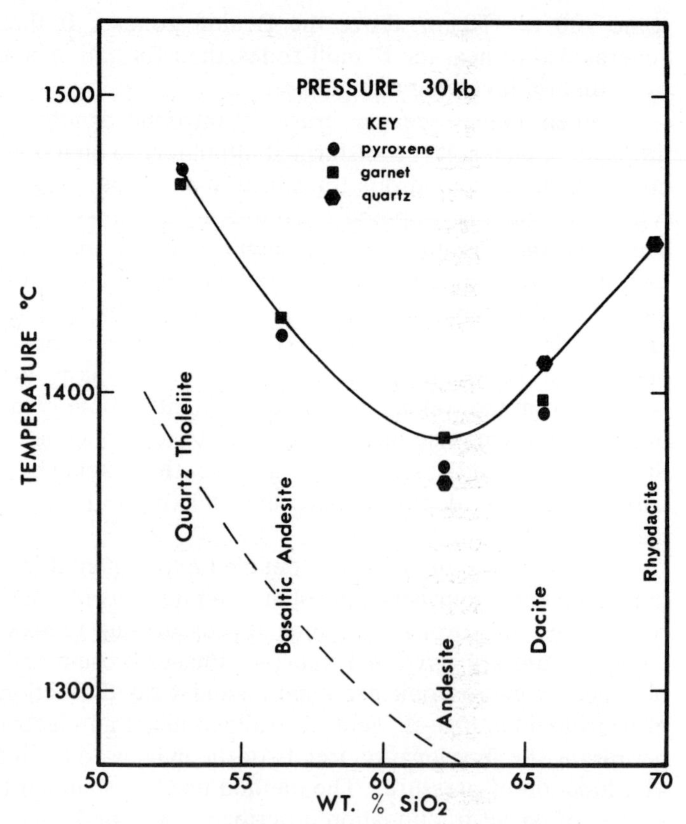

FIGURE 7-15
Extrapolated liquidus temperatures and sequence of crystallization at 30 kbars for the series of calc-alkaline rocks given in Table 7-4 (adamellite = rhyodacite). (*From T. Green and Ringwood, 1968a, with permission.*)

liquidus temperatures increase sharply. The higher liquidi of the dacite and rhyodacite, and the presence of quartz on the liquidus of these two compositions at high pressure, indicate that, at depth, andesite is a lower melting fraction than the more acid compositions.

Detailed calculation of fractionation trends based on proportions and compositions of phases observed in experiments confirmed the calc-alkaline nature of the fractionation at pressures above 25 kbars and demonstrated that the basalt-basaltic andesite-andesite sequence represents a true line of liquid descent under these conditions caused by separation of garnet (39% SiO_2) and clinopyroxene (49% SiO_2). Because of their subsilicic character, these phases provide a highly efficient means of enriching the fractionating liquid in silica and alkalis. The calc-alkaline FMA trend is caused by separation of garnet, which has a much higher Fe/Mg ratio than coexisting pyroxenes. Because of the increasing importance of garnet as the degree of fractionation increases, marked iron enrichment is prevented and the liquids were shown to follow the calc-alkaline trend.

The results have been discussed above in terms of fractional crystallization at high pressure. However, the same observations apply to the reverse case of partial melting of a quartz eclogite. With a moderate degree of partial melting, an andesitic magma could be formed, leaving behind residual garnet and pyroxenes (eclogite) with compositions similar to those observed during the fractional crystallization experiments. The magma will not move from the andesite minimum (Fig. 7-15) until all the quartz in the parental quartz eclogite has been incorporated into the liquid. Then, with a greater degree of melting, basaltic andesite might be formed.

Because of experimental limitations, most of the phase equilibrium data obtained were under anhydrous conditions. Nevertheless, reconnaissance experiments indicated that fractionation of basaltic magmas towards andesitic compositions at pressures above 25 kbars under hydrous conditions would not differ fundamentally from the dry fractionation trends, except that temperatures would be substantially reduced. Since amphiboles were not stable near the solidus above 25 kbars, it appeared that fractionation would continue to be controlled by garnet and pyroxene separation and that the calc-alkaline trend would be maintained.

However, an important effect may be expected in more acid compositions because of the drastic effect of water in suppressing the crystallization field of quartz relative to garnet and pyroxene. The presence of quartz on the liquidi of dacite and rhyodacite under high-pressure dry conditions has already been noted. Green and Ringwood (1968a,b) found that, in the presence of a substantial pressure of water vapour, quartz did not crystallize and was replaced by almandine-rich garnet as the liquidus phase. Thus, under these conditions, fractionation can continue from andesite into dacite, rhyodacite, and rhyolite compositions by further separation of almandine garnets and minor clinopyroxene. The widespread occurrence of almandine garnets in dacites and rhyodacites suggests the occurrence of this type of fractionation. Likewise, in the presence of substantial water-vapour pressure, partial melting of quartz eclogite at depths of 100 to 150 km in the earth would be expected to produce dacites and rhyodacites as the lowest melting fraction, rather than the andesitic liquid obtained under dry conditions.

The reconnaissance wet-melting experiments under conditions of uncontrolled P_{H_2O} described above have been followed by a series of experiments on basalt[1,2] (Fig. 7-16), andesite[3] (Fig. 7-12), and rhyodacite[4] compositions containing controlled amounts of water. These have confirmed and extended the earlier conclusions. There is no doubt that the liquid produced by partial melting of a quartz eclogite under high-pressure hydrous conditions would be calc-alkaline and highly silicic (probably $> 65\%$ SiO_2) in composition. The temperatures at which

[1]Hill and Boettcher (1970).
[2]Lambert and Wyllie (1970).
[3]T. Green (1972).
[4]T. Green and Ringwood (1972).

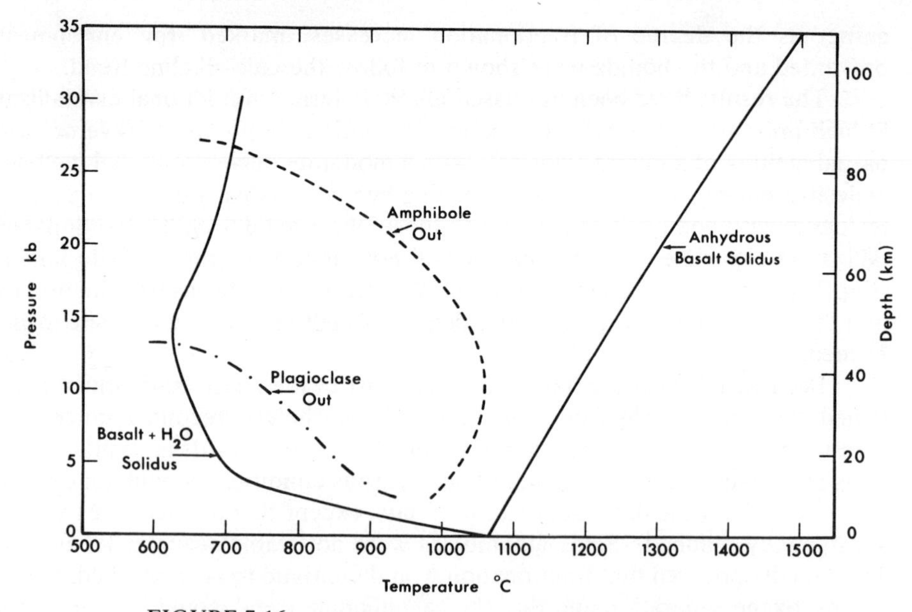

FIGURE 7-16
Solidus temperature for basalt at high pressure under conditions of $P_{H_2O} = P_{total}$. Also shown are the upper stability limits of amphibole and plagioclase, and the anhydrous solidus. (*After Hill and Boettcher*, 1970. *See also, Lambert and Wyllie, 1970.*)

such liquids would be generated depend upon the water pressures. For $P_{H_2O} = P_{total} = 20$ to 30 kbars, silica-rich liquids would be generated between 700 and 750°C (Fig. 7-16).[1] Under conditions of water undersaturation which may also be realized in the subducted lithosphere crust, temperatures of magma generation would be higher—possibly in the range 750 to 1000°C.[2,3]

The eclogite and amphibole fractionation mechanisms described in this and in the preceding section are complementary rather than exclusive and operate in different pressure regimes. At pressures up to about 20 kbars, amphibole may play an important role in the fractionation of hydrous basaltic magmas. At higher pressures above the amphibole stability limit, fractionation under both dry and wet conditions is governed by eclogite. There is a continuum between these regimes, and at intermediate depths, fractionation is governed by combinations of amphibole, pyroxenes, and garnet.[4,5]

The change from amphibole to eclogite-dominated fractionation with in-

[1]Hill and Boettcher (1970).
[2]Lambert and Wyllie (1970).
[3]T. Green (1972).
[4]T. Green and Ringwood (1969).
[5]Ringwood (1969).

creasing depth may be capable of explaining aspects of the minor element chemistry of orogenic-type rocks, in particular, the relationship between potassium enrichment and depth[1] (Fig. 7-4). Amphiboles crystallizing from orogenic-type magmas contain substantial amounts of both sodium and potassium (Sec. 7-7). On the other hand, pyroxenes crystallizing at high pressure from these magmas contain 2 to 3% sodium but almost no potassium.[2] Thus, on these models, the K/Na ratios of orogenic-type magmas would increase with depth as the residual, refractory assemblage changed from amphibolitic to eclogitic.[2] Analogous changes in abundances of many other trace elements might be expected. In particular, the liquids generated by partial melting of quartz eclogite would be expected to possess strongly fractionated rare earth abundance patterns enriched in light relative to heavy rare earths.

Further Development of Eclogite Fractionation Model

In the earlier models[3] for the generation of orogenic-type magmas by the partial melting of subducted oceanic crust, it was envisaged that andesitic-dacitic liquids formed by partial melting of quartz eclogite under either anhydrous or hydrous conditions ascended to the surface, undergoing variable degrees of crystallization by separation of garnet, pyroxene, amphibole, and, near the surface, plagioclase so that the effects of higher level fractionation were superimposed on the nature of the primary magmas produced at depth. However, it was assumed in these models that the magmas so produced did not interact substantially with the ultramafic mantle through which they penetrated. In the light of further considerations, this assumption now appears less plausible and requires modification, even at the expense of introducing more complex models.

One problem with the earlier model was that oceanic tholeiites in the subducted lithosphere appear to contain insufficient K, Rb, Cs, Ba, and radiogenic strontium to account for the abundances of these elements in many andesites.[4,5] In principle, this problem can be avoided by the introduction of additional assumptions—e.g., incorporation of *small* quantities of marine sediments,[4,5] prior chemical alteration of some of the subducted basalts by sea-water,[5,6] and incorporation of significant quantities of alkaline basalts in the oceanic crust.

Another aspect of the earlier model was that acid-intermediate magmas generated at depths of 100 to 150 km were required to ascend to the surface without reacting with the ultramafic mantle. This may well occur in some cases, where large bodies of acid-intermediate magmas are segregated at the Benioff

[1]Dickinson and Hatherton (1967).
[2]T. Green and Ringwood (1968a, 1969).
[3]T. Green and Ringwood (1966, 1968a).
[4]Armstrong (1971).
[5]Gill (1974).
[6]Hart (1971).

zone and ascend rapidly to the surface. It seems more likely, however, that, at any instant of time, the percentage of partial melting of subducted oceanic crust is rather small (< 5%) so that relatively *small* volumes of acidic liquids are intermittently generated and introduced into the overlying mantle.[1] Since the acid-intermediate magmas cannot be in equilibrium with olivine, it is likely that these magmas would rapidly react with the olivine in the ultramafic mantle immediately overlying the Benioff zone.

In the light of these considerations, it is now suggested that partial melting of the subducted oceanic crust frequently occurs under hydrous conditions at temperatures of 750 to 900°C and that the magmas produced are accordingly silica-rich, containing nearly all the incompatible elements formerly present in the oceanic crust. At depths greater than about 100 km, the residual refractory portion of the oceanic crust would be a bimineralic eclogite, lacking quartz and strongly depleted in low-melting components and incompatible elements. The silica-rich magmas rise and react extensively with the mantle immediately above the Benioff zone, transforming part of the olivine to pyroxene. In this manner, zones of olivine pyroxenite[2] are formed immediately overlying the Benioff zone at depths of 100 to 150 km. These zones also received all the incompatible elements formerly present in the oceanic crust. Because their density is slightly smaller than that of overlying pyrolite and because the presence of interstitial liquid endows them with high mobility, diapirs of pyroxenite rise episodically from the Benioff zone. Partial melting within the diapirs proceeds as they ascend, analogously to the formation of basalt magmas from rising pyrolite diapirs.[3,4] At depths greater than about 100 km, the pyroxenite is probably garnet-bearing. However, at shallower depths, the garnet would probably enter into solid solution in the pyroxene, as discussed in Sec. 6-2.

The nature of liquids formed by partial melting of the pyroxenite will be determined by the equilibria discussed in Sec. 7-6. For diapirs ascending from the Benioff zone at depths of 100 to 150 km and undergoing magma segregation at 60 to 100 km, magma compositions formed by, say, 20% of partial melting will range from magnesia-rich olivine tholeiite to quartz tholeiite.[5] Between about 40 and 60 km the segregating liquids will be generally of basaltic andesite composition (Mg-rich), grading into andesites for magma segregation at shallow depths[6,7] (20 to 40 km). Magmas formed by these processes are likely to be saturated or near-saturated in water. After segregation from their source diapirs, the magmas rise

[1] This represents a modification of the earlier models in which larger degrees of partial melting (e.g., 20 to 40%) of oceanic crust were assumed.

[2] There is probably a continuous gradation between pyroxenite and "modified" pyrolite.

[3] Section 4-5.

[4] See also Ringwood (1974).

[5] Nicholls and Ringwood (1973).

[6] Nicholls (1974).

[7] The high Mg/Fe ratios of these liquids in relation to their high silica contents (compared to normal olivine tholeiites formed by partial melting of pyrolite under "dry" conditions) probably contributes in large measure to the development of the calc-alkaline differentiation trend as discussed in Sec. 7-2.

towards the surface, undergoing fractionation under closed-system conditions, as discussed earlier. At depths of 70 to 100 km, the principal phases separating will be garnet and pyroxene; between 40 and 70 km, amphibole, pyroxene, and olivine; and at shallower depths, mainly pyroxene and plagioclase. Because of enforced crystallization during ascent, owing to loss of water, evidence of original high-pressure origin may be lost. A wide range of liquids from basaltic to rhyolitic may thereby be formed, with the mean (and most common) composition being that of a calc-alkaline andesite.

The essential property of this model is that silica-rich calc-alkaline components originally derived by partial melting of the oceanic crust under eclogitic conditions continue to constitute a major part of the orogenic volcanic magmas which ultimately reach the surface, although their evolutionary paths are now regarded as being more complex than in earlier models. The silica-rich magmas derived from the eclogitic zone of the subducted lithosphere will possess strongly fractionated rare earth patterns and other trace element characteristics which will imprint themselves on the pyroxenite diapirs formed by reaction with pyrolite mantle along the Benioff zone. These characteristics will ultimately be reproduced by the generation of magmas derived from partial melting of the pyroxenite diapirs during ascent. A sketch of this model is shown in Fig. 7-17.

There are two rather compelling reasons for believing that partial melting of the mafic oceanic crust indeed occurs along Benioff zones:

1 The oceanic crust is believed to contain a substantial amount of water, mainly held in amphiboles, zoisite, and serpentine (Sec. 2-4). As the crust becomes heated during its descent into the mantle, high water-vapour pressures are built up owing to dehydration of these minerals (Sec. 8-3). Investigations of the thermal structure of the sinking slab show that the low temperatures required for partial melting of quartz eclogite "crust" under hydrous conditions (700 to 900°C) are reached at quite shallow depths.[1,2,3,4] Indeed, provided 1 or 2% water is initially present in the oceanic crust, it is difficult to formulate models in which melting does *not* occur.[4]

2 If melting and extraction of siliceous liquid did *not* occur, the vast volumes of oceanic crust deposited in the mantle would be transformed to quartz eclogite (5 to 10% Qz). On the other hand, the mafic xenoliths of mantle origin found in diamond pipes consist dominantly of bimineralic garnet-clinopyroxene eclogites, and quartz eclogites are extremely rare.[5] The compositions of many of the eclogite xenoliths occurring in diamond pipes are consistent with their origin as refractory residua remaining after partial melting of oceanic crust.[5] The extreme rarity of quartz eclogite

[1]McKenzie (1969).
[2]Toksöz, Minear, and Julian (1971).
[3]Oxburgh and Turcotte (1970).
[4]Hill and Boettcher (1970).
[5]Section 3-4.

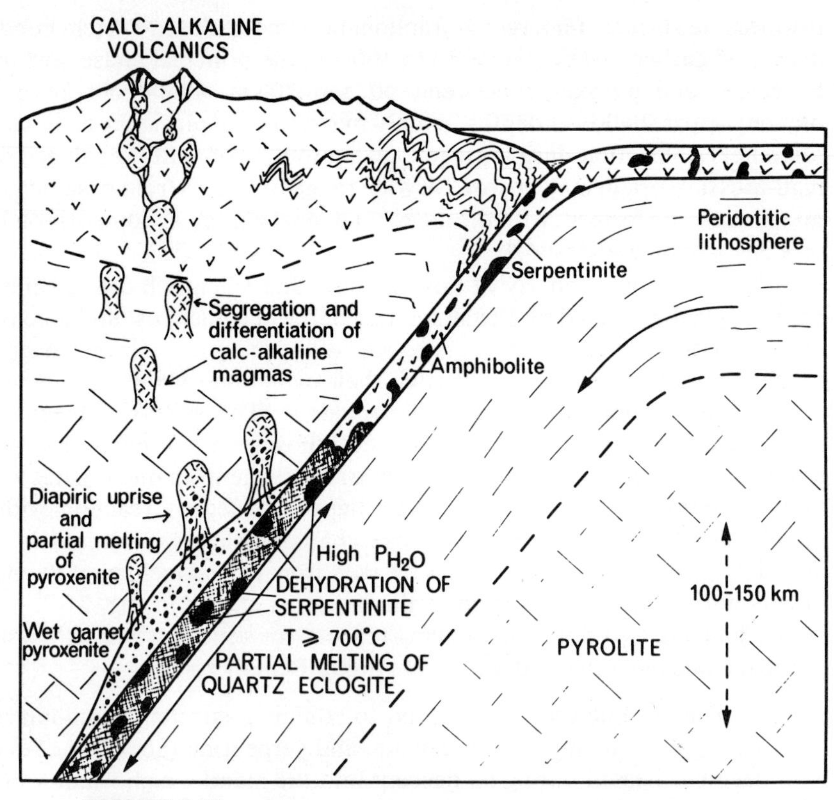

FIGURE 7-17
Mature phase of island-arc development involving partial melting of subducted oceanic crust and reaction of acidic liquids so produced with neighbouring mantle above Benioff zone to produce bodies of wet, mobile pyroxenite. These rise diapirically, leading to partial melting and formation of calc-alkaline-type magmas.

xenoliths is inexplicable unless the low-melting siliceous components are regularly extracted from the sinking oceanic crust.[1]

7-9 THE PLUTONIC OROGENIC SERIES

The members of the volcanic orogenic series, basalt-andesite-dacite-rhyolite, possess major element chemical compositions generally similar to the corresponding plutonic members of the series, gabbro-diorite-granodiorite-granite. Nevertheless, there are significant differences between the volcanic and plutonic series which make it desirable to treat them separately.[2] Whereas andesite is the

[1]Section 3-4.
[2]E.g., Turner and Verhoogen (1960).

most abundant volcanic member, the more silicic granodiorite predominates among the plutonic series. Detailed geochemical studies[1] suggest that significant differences in trace element abundances may exist between plutonic and volcanic members even when major element chemical compositions are similar.

It seems likely that the plutonic series are polygenetic to a much greater extent than the volcanic series. Direct evidence for a liquid line of descent is lacking in the plutonic series, and alternative hypotheses of hybridism and partial melting of sial must be seriously examined, in addition to the direct crystal-liquid fractionation mechanisms of forming orogenic volcanic magmas. Clearly, in view of the vast amount of the latter magmas which have been erupted, we must expect that, at deeper levels, below the volcanoes, large volumes must have crystallized to form plutons—the direct equivalents of the volcanic magmas. Numerous well-documented cases of such plutons connected with volcanic centers exist, particularly in western North America, South America, and Eastern Australia.[2,3,4,5,6] The problem is to establish the relative proportions of orogenic plutonic rocks which have direct magmatic origins of this kind and those which have been formed by other processes.

Numerous petrologic studies have concluded that many granitic magmas were formed by partial melting of sialic crustal material in the presence of substantial water-vapour pressures.[7] Basement complexes frequently display all grades of regional metamorphism, culminating in a stage at which relatively low-melting-point hydrous granitic liquids have been produced.[7] The P-T conditions under which crustal anatexis may occur are readily accessible in the laboratory. A large number of excellent experimental investigations[8,9,10,11,12,13,14,15,16,17,18] on the metamorphism and melting behaviour of igneous, metamorphic, and sedimentary rocks in the presence of water have provided detailed knowledge of the conditions under which anatexis may occur. A pleasing concordance between the conclusions of many field petrologists and experimentalists has been achieved.

Although it is established that some granitic magmas have formed by partial

[1]Taylor (1969).
[2]Wise (1969).
[3]Hamilton (1969).
[4]Smith (1960).
[5]Hills (1959).
[6]Branch (1967).
[7]Reviewed by Turner and Verhoogen (1960, pp. 329-388).
[8]Tuttle and Bowen (1958).
[9]Luth, Jahns, and Tuttle (1964).
[10]Brown (1963).
[11]Winkler (1957).
[12]Winkler and von Platen (1958, 1960, 1961a,b).
[13]Wyllie and Tuttle (1961).
[14]Von Platen (1965).
[15]Piwinskii and Wyllie (1968, 1970).
[16]Piwinskii (1968).
[17]Gibbon and Wyllie (1969).
[18]Brown and Fyfe (1970).

melting of sialic rocks, the low initial strontium-isotope ratios of most granitic rocks place limits on the origin of the sialic source rocks (Sec. 8-1). It can be demonstrated that ancient upper crustal rocks as are found on Precambrian shields have not contributed largely to the younger granites. It appears more likely that partial melting occurs deep within geosynclines consisting largely of relatively young orogenic-type volcanic rocks and sediments derived from the volcanics, which are probably of mantle origin with low initial strontium-isotope ratios, as discussed earlier. For example, Dickinson (1962) showed that andesitic debris derived from the weathering of andesitic volcanic rocks made up a major component of the Mesozoic geosyncline along the west coast of North America. He suggested that many of the plutonic orogenic rocks in this region were formed by the remelting of the sediments.

Whilst the experimental investigations discussed above provide an explanation of the formation of *granitic liquids*, during crustal anatexis, they do not suggest that other abundant plutonic members of the orogenic series—e.g., granodiorites, tonalites, and diorites—have been formed as liquids by crustal anatexis. Whereas a liquid of granitic composition may be formed at relatively low temperatures, much higher temperatures are necessary to produce the other orogenic-type magmas. For example,[1] melting of a granodiorite under a water-vapour pressure of 2 kbars resulted in the formation of about 50% granitic liquid within a narrow temperature interval between 700 and 730°C. The residual phases, mainly amphibole, biotite, and plagioclase, had a low solubility in the liquid so that there was a wide temperature interval, extending to about 900°C, over which the rock consisted of a suspension of these refractory phases in the quartzo-feldspathic liquid. Tonalites and monzonites showed generally similar melting behaviour displaced towards higher temperatures.[1,2,3,4] These results show that, whilst a granitic liquid may be produced rather readily by anatexis in the crust, the production of completely liquid granodioritic and tonalitic magmas by crustal anatexis would require much more extreme conditions which would rarely be attained. Piwinskii and Wyllie concluded that most of the large granodioritic-tonalitic intrusions believed to have formed by crustal anatexis have never existed as completely liquid magmas but, rather, were emplaced as crystal mushes consisting of varying proportions of plagioclase, amphibole, and biotite suspended in a granitic liquid.

A related process for the formation of intermediate orogenic plutonic rocks is that of hybridism.[5,6,7] Field studies of many smaller intermediate intrusions showed that pre-existing mafic rocks had been invaded and completely disin-

[1]Piwinskii and Wyllie (1968).
[2]Piwinskii and Wyllie (1970).
[3]Gibbon and Wyllie (1969).
[4]Piwinskii (1968).
[5]Nockolds (1934).
[6]Joplin (1959).
[7]Wilkinson, Vernon, and Shaw (1964).

tegrated by granitic liquids, producing intermediate "hybrid" rocks consisting of "phenocrysts" of relatively mafic minerals, e.g., amphiboles, plagioclase, and biotite, set in a granitic matrix. If cooling has been sufficiently rapid, the matrix may be distinguished because of finer crystal size, and nonequilibrium mineral assemblages may be preserved. However, with slower cooling, complete recrystallization to coarse equigranular rocks indistinguishable from normal plutonic rocks may occur. The range of rock types formed by the hybridism mechanism is thus due to mixing processes and not to magmatic differentiation. In many cases, it might not be possible to establish which of these processes has been operative.

The experiments of Wyllie and coworkers discussed earlier provide an adequate understanding of the naturally occurring process of hybridism. Relatively low-melting-point granitic liquids are apparently produced in abundance during the most intense stages of regional metamorphism. These liquids will rise to higher levels in the crust, sometimes encountering mafic rocks. Because of the insolubility of mafic minerals in the low-melting granitic liquid, the interaction is one of invasion of solid mafic rocks by granitic liquid, leading to varying degrees of mixing of the two components.

REFERENCES

ARMSTRONG, R. L. (1968). A model for the evolution of strontium and lead isotopes in a dynamic earth. *Rev. Geophys.* **6,** 175–199.

——— (1971). Isotopic and chemical constraints on models of magma genesis in island arcs. *Earth Planet. Sci. Letters* **12,** 137–142.

——— and J. COOPER (1971). Lead isotopes in island arcs. *Bull. Volc.* **35,** 27–63.

AUMENTO, F. (1969). Diorites from the Mid-Atlantic Ridge at 45°N. *Science* **165,** 1112–1113.

BAKER, P.E. (1968a). Petrology of Mt. Misery volcano, St. Kitts, West Indies. *Lithos* **1,** 124–150.

——— (1968b). Comparative volcanology and petrology of the Atlantic island arcs. *Bull. Volc.* **32**(1), 189–206.

BARAGAR, W. R. and A. M. GOODWIN (1969). Andesites and Archaean volcanism on the Canadian Shield. In: A. R. McBirney (ed.), "*Proceedings of the Andesite Conference*," pp. 121–142. Bull. 65, Dept. Geology and Mineral Resources, State of Oregon.

BEST, M. G. (1963). Petrology of the Guadalupe igneous complex, south-western Sierra Nevada foothills, California. *J. Petrol.* **4,** 223–259.

——— and E. L. P. MERCY (1967). Composition and crystallization of mafic minerals in the Guadalupe igneous complex, California. *Am. Min.* **52,** 436–474.

BOWEN, N. L. (1928). "*Evolution of the Igneous Rocks.*" Princeton Univ. Press, Princeton, N. J. 332 pp.

BRANCH, C. D. (1967). Genesis of magma for acid calcalkaline volcano plutonic formations. *Tectonophysics* **4,** 83–100.

BROWN, G. C., and W. S. FYFE (1970). The production of granitic melts during ultrametamorphism. *Contr. Mineral. Petrol.* **28,** 310–318.

BROWN, G. M. (1963). Melting relationship of Tertiary granitic rocks in Skye and Rhum. *Min. Mag.* **33,** 533–563.

——— and J. F. SCHAIRER (1968). Melting relations of some calcalkaline volcanic rocks. *Carnegie Inst. Washington Yearbook* **66,** 460–467.

CARMICHAEL, I. S. E. (1964). The petrology of Thingmuli, a Tertiary volcano in eastern Iceland. *J. Petrol.* **5,** 435–460.

——— (1967). The iron-titanium oxides of salic volcanic rocks and their associated ferromagnesian silicates. *Contr. Mineral. Petrol.* **14,** 36–64.

——— and J. NICHOLLS (1967). Iron-titanium oxides and oxygen fugacities in volcanic rocks. *J. Geophys. Res.* **72,** 4665–4687.

CHURCH, S. E., and G. R. TILTON (1973). Lead and strontium isotope geochemistryof the Cascade Mountains and their bearing on the genesis of andesitic magmas. *Bull. Geol. Soc. Am.* **84,** 431–454.

COATS, R. R. (1962). Magma type and crustal structure in the Aleutian arc. In: "*Crust of the Pacific Basin*," pp. 92–109. *Am. Geophys. Union, Geophys. Monograph* 6.

DALY, R. A. (1933). "*Igneous Rocks and the Depths of the Earth.*" McGraw-Hill, New York. 598 pp.

DICKINSON, W. R. (1962). Petrogenetic significance of geosynclinal andesitic volcanism along the Pacific margin of North America. *Bull. Geol. Soc. Am.* **73,** 1241–1256.

——— (1968). Circum-Pacific andesitic types. *J. Geophys. Res.* **73,** 2261–2269.

——— (1969). Evolution of calcalkaline rocks in the geosynclinal system of California and Oregon. A. R. McBirney (ed.), In: "*Proceedings of the Andesite Conference,*" pp. 151–156. Bull. 65, Dept. Geology and Mineral Resources, State of Oregon.

——— and T. HATHERTON (1967). Andesitic volcanism and seismicity around the Pacific. *Science* **157,** 801–803.

DOE, B. R. (1967). The bearing of lead isotopes on the source of granitic magma. *J. Petrol.* **8,** 51–83.

———, P. W. LIPMAN, and C. E. HEDGE (1969). Radiogenic tracers and the source of continental andesites: a beginning at the San Juan volcanic field, Colorado. In: A. R. McBirney (ed.), "*Proceedings of the Andesite Conference,*" pp. 143–149. Bull. 65, Dept. Geology and Mineral Resources, State of Oregon.

DUNCAN, A. R., and S. R. TAYLOR (1968). Trace element analyses of magnetites from andesitic and dacitic lavas from Bay of Plenty, New Zealand. *Contr. Mineral. Petrol.* **20,** 30–33.

EGGLER, D. H. (1972a). Water-saturated and undersaturated melting relationships in a Paricutin andesite and an estimate of water content in the natural magma. *Contr. Mineral. Petrol.* **34,** 261–271.

——— (1972b). Amphibole stability in H_2O-undersaturated calcalkaline melts. *Earth Planet. Sci. Letters* **15,** 28–34.

EWART, A., W. B. BRYAN, and J. GILL (1973). Mineralogy and geochemistry of the younger volcanic islands of Tonga, S. W. Pacific. *J. Petrol.* **14,** 429–465.

———, and J. J. STIPP (1968). Petrogenesis of the volcanic rocks of the Central North Island, New Zealand, as indicated by a study of Sr^{87}/Sr^{86} ratios and Sr, Rb, K, U, and Th abundances. *Geochim. Cosmochim. Acta* **32,** 699–736.

GIBBON, D. L., and P. J. WYLLIE (1969). Experimental studies of igneous rock series: the Farrington Complex, North Carolina and the Star Mountain rhyolite, Texas. *J. Geol.* **77,** 221–239.

GILL, J. B. (1970). Geochemistry of Vitu Levu, Fiji and its evolution as an island arc. *Contr. Mineral. Petrol.* **27,** 179–203.

——— (1974). Role of underthrust oceanic crust in the genesis of a Fijian calcalkaline suite. *Contr. Mineral. Petrol.* **43,** 29–45.

GORSHKOV, G. S. (1962). Petrochemical features of volcanism in relation to the types of the earth's crust. In: "*Crust of the Pacific Basin,*" pp. 110–115. *Am. Geophys. Union, Geophys. Monograph* 6.

——— (1969). Geophysics and petrochemistry of andesite volcanism of the circum-Pacific belt. In: A. R. McBirney (ed.), "*Proceedings of the Andesite Conference,*" pp. 91–98. Bull. 65, Dept. Geology and Mineral Industries, State of Oregon.

GREEN, D. H. (1973). Experimental melting studies on a model upper mantle composition under water saturated and water undersaturated condition. *Earth Planet Sci. Letters* **19,** 37–53.

——— and A. E. RINGWOOD (1967). The genesis of basaltic magmas. *Contr. Mineral. Petrol.* **15,** 103–190.

GREEN, T. H. (1972). Crystallization of calcalkaline andesite under controlled high pressure hydrous conditions. *Contr. Mineral. Petrol.* **34,** 150–166.

———, D. H. GREEN, and A. E. RINGWOOD (1967). The origin of high alumina basalts and their relationships to quartz tholeiites and alkali basalts. *Earth Planet. Sci. Letters* **2,** 41–52.

——— and A. E. RINGWOOD (1966). Origin of the calcalkaline igneous rock suite. *Earth Planet. Sci. Letters* **1,** 307–316.

——— and ——— (1967). Crystallization of basalt and andesite under high pressure hydrous conditions. *Earth Planet. Sci. Letters* **3,** 481–489.

——— and ——— (1968a). Genesis of the calcalkaline igneous rock suite. *Contr. Mineral. Petrol.* **18,** 105–162.

——— and ——— (1968b). Origin of garnet phenocrysts in calcalkaline rocks. *Contr. Mineral. Petrol.* **18,** 163–174.

——— and ——— (1969). High pressure experimental studies on the origin of andesites. In: A.R. McBirney (ed.), "*Proceedings of the Andesite Conference,*" pp. 21–32. Bull. 65, Dept. Geology and Mineral Industries, State of Oregon.

——— and ——— (1972). Crystallization of garnet-bearing rhyodacite under high pressure hydrous conditions. *J. Geol. Soc. Aust.* **19,** 203–212.

HALD. N., A. NOE-NYGAARD, and A. K. PEDERSEN (1971). The Kroksfjordur central volcano in northwest Iceland. *Acta Naturalia Islandica* **2** (10), 1–29.

HAMILTON, D. L., C. W. BURNHAM, and E. F. OSBORN (1964). The solubility of water and effects of oxygen fugacity and water content on crystallization in mafic magmas. *J. Petrol.* **15,** 21–39.

HAMILTON, W. (1964). Origin of high alumina basalt, andesite and dacite magmas. *Science* **146,** 635–637.

——— (1969). The volcanic central Andes—a modern model for the Cretaceous batholiths and tectonics of western North America. In: A. R. McBirney (ed.), "*Proceedings of the Andesite Conference,*" pp. 175–184. Bull. 65, Dept. of Geology and Mineral Resources, State of Oregon.

HART, S. R. (1971). K, Rb, Cs, Sr and Ba contents and Sr isotope ratios of ocean floor basalts. *Phil. Trans. Roy. Soc. London* **A268,** 573–587.

HEDGE, C. (1966). Variations in radiogenic strontium found in volcanic rocks. *J. Geophys. Res.* **71,** 6119–6126.

HILL, R. E. T., and A. L. BOETTCHER (1970). Water in the earth's mantle: melting curves of basalt-water and basalt-water-carbon dioxide. *Science* **167,** 980–982.

HILLS, E. S. (1959). Cauldron subsidences, granitic rocks and crustal fracturing in S.E. Australia. *Geol. Rundschau* **47,** 543–553.

HOLLOWAY, J. R., and C. W. BURNHAM (1972). Melting relations of basalt with equilibrium water pressure less than total pressure. *J. Petrol.* **13,** 1–29.

HOPSON, C. A., D. F. CROWDER, R. W. TABOR, F. W. CATER, and W. S. WISE (1965). Association of andesitic volcanoes in the Cascade Mountains with Late Tertiary epizonal plutons. *Geol. Soc. Am. Spec. Paper* 87.

JAKEŠ, P., and J. B. GILL (1970). Rare earth elements and the island arc tholeiite series. *Earth Planet. Sci. Letters* **9,** 17–28.

——— and A. J. WHITE (1969). Structure of the Melanesian Arcs and correlation with distribution of magma types. *Tectonophysics* **8,** 223–236.

——— and ——— (1971). Composition of island arcs and continental growth. *Earth Planet. Sci. Letters* **12,** 224–230.

JOPLIN, G. A. (1959). On the origin and occurrence of basic bodies associated with discordant bathyliths. *Geol. Mag.* **96,** 361–373.

KENNEDY, G. C. (1955). Some aspects of the role of water in rock melts. In: A. Poldervaart (ed.), "*Crust of the Earth,*" pp. 489–503. *Geol. Soc. Am. Spec. Paper* 62.

KUNO, H. (1950). Petrology of Hakone volcano and the adjacent areas, Japan. *Bull. Geol. Soc. Am.* **61,** 957–1014.

——— (1966). Lateral variation of basalt magma type across continental margins and island arcs. *Bull. Volc.* **29,** 195–222.

——— (1968). Origin of andesite and its bearing on the island arc structure. *Bull. Volc.* **32,** 141–176.

——— (1969). Andesite in time and space. In: A.R. McBirney (ed.), "*Proceedings of the Andesite Conference,*" pp. 13–20. Bull. 65, Dept. Geology and Mineral Industries, State of Oregon.

KUSHIRO, I. (1969). Liquidus relations in the system forsterite-diopside-silica-water at 20 kb. *Carnegie Inst. Washington Yearbook* **67,** 158–161.

——— (1970). Systems bearing on melting of the upper mantle under hydrous conditions. *Carnegie Inst. Washington Yearbook* **68,** 240–245.

———, N. SHIMIZU, Y. NAKAMURA, and S. AKIMOTO (1972). Compositions of coexisting liquid and solid phases formed upon melting of natural garnet and spinel lherzolites at high pressures: A preliminary report. *Earth Planet. Sci. Letters* **14,** 19–25.

——— and H. S. YODER (1969). Melting of forsterite and enstatite at high pressures under hydrous conditions. *Carnegie Inst. Washington Yearbook* **67,** 153–158.

———, ———, and M. NISHIKAWA (1968). Effect of water on the melting of enstatite. *Bull. Geol. Soc. Am.* **79,** 1685–1692.

LAMBERT, I. B., and P. J. WYLLIE (1970). Low-velocity zone of the earth's mantle: incipient melting caused by water. *Science* **169,** 764–766.

LE MASURIER, W. E. (1968). Crystallization behaviour of basalt magma, Santa Rosa Range, Nevada. *Bull. Geol. Soc. Am.* **79,** 949–972.

LIDIAK, E. G. (1965). Petrology of andesitic, spilitic and keratophyre flow rock, north-central Puerto Rico. *Bull. Geol. Soc. Am.* **76,** 57–88.

LUTH, W., R. JAHNS, and O. R. TUTTLE (1964). The granite system at pressures of 4 to 10 kilobars. *J. Geophys. Res.* **69,** 759–773.

MC KENZIE, D. P. (1969). Speculations on the consequences and causes of plate motions. *Geophys. J.* **18,** 1–32.

NICHOLLS, I. A. (1971). Petrology of Santorini Volcano, Cyclades, Greece. *J. Petrol.* **12,** 67–119.

——— (1974). Liquids in equilibrium with peridotitic mineral assemblages at high water pressures. *Contrib. Mineral. Petrol.* **45,** 289–316.

——— and A. E. RINGWOOD (1973). Effect of water on olivine stability in tholeiites and the production of SiO_2-saturated magmas in the island arc environment. *J. Geol.* **81,** 285–300.

NOCKOLDS, S. R. (1934). The production of normal rock types by contamination and their bearing on petrogenesis. *Geol. Mag.* **71,** 31–39.

O'HARA, M. J. (1963). Melting of garnet peridotite and eclogite at 30 kilobars. *Carnegie Inst. Washington Yearbook* **62,** 71–77.

——— (1965). Primary magmas and the origin of basalts. *Scottish J. Geol.* **1,** 19–40.

OSBORN, E. F. (1959). Role of oxygen pressure in the crystallization and differentiation of basaltic magma. *Am. J. Sci.* **257,** 609–647.

——— (1962). Reaction series for subalkaline igneous rocks based on different oxygen pressure conditions. *Am. Min.* **47,** 211–226.

——— (1969a). Experimental aspects of calcalkaline differentiation. In: A.R. McBirney (ed.), "*Proceedings of the Andesite Conference*," pp. 33–42. Bull. 65, Dept. Geol. and Mineral Industries, State of Oregon.

——— (1969b). The complementariness of orogenic andesite and alpine peridotite. *Geochim. Cosmochim. Acta* **33,** 307–325.

OVERSBY, V. M., and A. EWART (1972). Lead isotopic compositions of Tonga-Kermadec volcanics and their petrogenetic significance. *Contrib. Mineral. Petrol.* **37,** 181–210.

OXBURGH, E. R., and D. L. TURCOTTE (1968). Problems of high heat flow and volcanism associated with zones of descending mantle convective flow. *Nature* **218,** 1041.

——— and ——— (1970). The thermal structure of island arcs. *Bull. Geol. Soc. Am.* **81,** 1665–1688.

PEACOCK, M. A. (1931). Classifications of igneous rocks. *J. Geol.* **39,** 54–67.

PICHLER, H., and W. ZEIL (1969). Andesites of the Chilean Andes. In: A.R. McBirney (ed.), "*Proceedings of the Andesite Conference*," pp. 165–174. Bull. 65, Dept. Geology and Mineral Industries, State of Oregon.

PIWINSKII, A. J. (1968). Experimental studies of igneous rock series, Central Sierra Nevada Batholith, California. *J. Geol.* **76,** 548–570.

——— and P. J. WYLLIE (1968). Experimental studies of igneous rock series: a zoned pluton in the Wallowa batholith, Oregon. *J. Geol.* **76,** 205–234.

——— and ——— (1970). Experimental studies of igneous rock series: felsic body from the Needle Point pluton, Wallowa batholith, Oregon. *J. Geol.* **78,** 52–76.

POLDERVAART, A. (1949). Three methods of graphic representation of chemical analyses of igneous rocks. *Trans. Roy. Soc. S. Africa* **32,** 177–188.

——— (1955). Chemistry of the earth's crust. In: A. Poldervaart (ed.), "*Crust of the Earth*," pp. 119–144. *Geol. Soc. Am. Spec. Paper* 62.

——— and W. ELSTON (1954). The calcalkaline series and the trend of fractional crystallization of basaltic magma. *J. Geol.* **62,** 150–162.

RINGWOOD, A. E. (1969). Composition and evolution of the upper mantle. In: "*The Earth's Crust and Upper Mantle*," pp. 1–17. *Am. Geophys. Union Geophys. Monograph* 13.

——— (1974). The petrological evolution of island arc systems. *J. Geol. Soc. Lond.* **130,** 183–204.

——— and D. H. GREEN (1966). An experimental investigation of the gabbro-eclogite transformation and some geophysical implications. *Tectonophysics* **3,** 383–427.

SCLAR, C. B., L. C. CARRISON, and O. M. STEWART (1968). Effect of water vapour on the melting of forsterite and enstatite at 20 kilobars (abstract). *Trans. Am. Geophys. Union* **49,** 355–356.

SMITH, A. L., and I. S. E. CARMICHAEL (1968). Quaternary lavas from the southern Cascades, western U.S.A. *Contr. Mineral. Petrol.* **19,** 212–238.

SMITH, R. L. (1960). Ash flows. *Bull. Geol. Soc. Am.* **71,** 212–842.

TATSUMOTO, M. (1969). Lead isotopes in volcanıc rocks and possible ocean-floor thrusting beneath island arcs. *Earth Planet. Sci. Letters* **6,** 369–376.

TAYLOR, S. R. (1967). The origin and growth of continents. *Tectonophysics* **4,** 17–34.

——— (1969). Trace element chemistry of andesites and associated calcalkaline rocks. In: A.R. McBirney (ed.), "*Proceedings of the Andesite Conference*," pp. 43–63. Bull. 65, Dept. Geol. and Mineral Resources, State of Oregon.

——— and A. J. WHITE (1966). Trace element abundances in andesites. *Bull. Vol.* **29,** 177–194.

THAYER, T. P. (1967). Chemical and structural relations of ultramafic and feldspathic rocks in alpine intrusive complexes. In: P. J. Wyllie (ed.), "*Ultramafic and Related Rocks*," pp. 222–239. Wiley, New York.

TILLEY, C. E. (1950). Some aspects of magmatic evolution. *Quart. J. Geol. Soc. London* **106,** 37–61.

———, H. S. YODER, and J. F. SCHAIRER (1967). Melting relations of volcanic rock series. *Carnegie Inst. Washington Yearbook* **65,** 260–269.

———, ———, and ——— (1968). Melting relations of igneous rock series. *Carnegie Inst. Washington Yearbook* **66,** 452.

TOKSÖZ, M. N., J. W. MINEAR, and B. R. JULIAN (1971). Temperature field and geophysical effects of a downgoing slab. *J. Geophys. Res.* **76,** 1113–1138.

TURNER, F. J., and J. VERHOOGEN (1960). "*Igneous and Metamorphic Petrology*," 2d ed. McGraw-Hill, New York. 694 pp.

TUTTLE, O. F., and N. L. BOWEN (1958). Origin of granite in the light of experimental studies in the system $KAlSi_3O_8$-SiO_2-H_2O. *Geol. Soc. Am. Mem.* 74, 1–153.

VON PLATEN, H. (1965). Kristallisation granitischer schmelzen. *Contr. Mineral. Petrol.* **11,** 334–381.

WAGER, L. R. (1960). The major element variation of the layered series of the Skaergaard Intrusion and a re-estimation of the average composition of the hidden layered series and of the successive residual magmas. *J. Petrol.* **1,** 364–398.

——— and W. A. DEER (1939). Geological investigations in East Greenland, Part III. The petrology of the Skaergaard Intrusion, Kangerlugssaq, East Greenland. *Medd. Groenland* **105,** 1–352.

WATTERSON, J. (1968). Plutonic development of the Ilordleq area, South Greenland, Part II: Late-kinematic basic dykes. *Medd. Groenland* **185,** No. 3, 1–104.

WILKINSON, J. F. G. (1966). Some aspects of calcalkali rock genesis. *Proc. Roy. Soc. New South Wales* **99,** 69–77.

——— (1971). The petrology of some vitrophyric calcalkaline volcanics from the Carboniferous of New South Wales. *J. Petrol.* **12,** 587–619.

———, R. H. VERNON, and S. E. SHAW (1964). The petrology of an ademellite-porphyrite from the New England Bathylith (New South Wales). *J. Petrol.* **5,** 461–468.

WILLIAMS, H. (1942). The geology of Crater Lake National Park, Oregon. *Carnegie Inst. Washington Publ.* 540.

WINKLER, H. (1957). Experimentelle gesteinsmetamorphose, Pt. 1. *Geochim. Cosmochim. Acta* **13,** 42–69.

——— and H. VON PLATEN (1958). Experimentelle gesteinsmetamorphose, Pt. 2. *Geochim. Cosmochim. Acta* **15,** 91–112.

——— and ——— (1960). Experimentelle gesteinsmetamorphose, Pt. 3. *Geochim. Cosmochim. Acta* **18,** 294–316.

——— and ——— (1961a). Experimentelle gesteinsmetamorphose, Pt. 4. *Geochim, Cosmochim. Acta* **24,** 48–69.

——— and ——— (1961b). Experimentelle gesteirmetamorphase, Pt. 5. *Geochim. Coscochim. Acta* **24,** 250–259.

WISE, W. S. (1969). Geology and petrology of the Mt. Hood area: A study of High Cascade volcanism. *Bull. Geol. Soc. Am.* **80,** 969–1006.

WYLLIE, P. J., and O. F. TUTTLE (1961). Hydrothermal melting of shales. *Geol. Mag.* **98,** 56–66.

YODER, H.S. (1965). Diopside-anorthite-water at 5 and 10 kilobars and its bearing on explosive volcanism. *Carnegie Inst. Washington Yearbook* **64,** 82–89.

——— (1969). Calcalkalic andesites: experimental data bearing on the origin of their assumed characteristics. In: A.R. McBirney (ed.), "*Proceedings of the Andesite Conference,*" pp. 77–89. Bull. 65, Dept. Geol. and Mineral Industries, State of Oregon.

——— and C. E. TILLEY (1962). Origin of basalt magmas: an experimental study of natural and synthetic rock systems. *J. Petrol* **3,** 342–532.

8
PETROLOGIC EVOLUTION OF THE CRUST AND UPPER MANTLE

8-1 INTRODUCTION

The earlier geological literature contained two contrasted views regarding the origin of the earth's crust. The first held that the crust was formed very early in the earth's history, perhaps in the same episode of major differentiation which resulted in segregation of the core. According to the second view, the crust was derived ultimately from the mantle by igneous differentiation processes extending throughout geological time. Since 1950, the accumulation of evidence from several directions has generally favoured the latter evolutionary hypothesis, although whether the rate of evolution has been approximately constant with time is open to question.

The most decisive new line of evidence in relation to the oceanic crust has arisen from modern concepts of plate tectonics and sea-floor spreading which imply that, on the average, the crust in deep oceanic basins is less than 100 million years old. Evidence for an evolutionary model of continental crust is derived from several directions. One of the earlier arguments used in support is the zoned arrangement of younger geological provinces around Precambrian shield nuclei, with a general decrease of age outward from the shield, which is observed in some continents, e.g., Australia[1,2] and North America.[3] These configurations suggest that provinces of successively younger rocks have been welded on to the continental margins by orogenic activity. This version of continental growth by marginal accretion has been generally supported by recent geochronological investigations, although it is not universally applicable. Useful reviews of the evolution of continents from this viewpoint are given by Wilson (1954), Engel (1963), and Taylor (1967), who point to the importance of basaltic and andesitic volcanism in contributing new material from the mantle to the growing continental crust.

Geochemical evidence has become increasingly important in studying continental evolution. Rubey (1951, 1955) pointed to several problems which arise if it is assumed that all the earth's water and carbon dioxide were present in an early primitive atmosphere. To avoid these difficulties he appealed to a gradual evolution of hydrosphere and atmosphere by degassing of the earth associated with volcanic and plutonic igneous activity, during which crustal rocks were also formed. Although the situation is now known to be more complex than envisaged by Rubey because of the recycling of volatiles arising from subduction of crust in orogenic zones, some of his arguments still carry considerable weight.

The strongest geochemical evidence in favour of continental evolution derives from the isotopic development of strontium in the crust[4] as a result of radioactive decay of rubidium 87 to produce strontium 87. The average Rb/Sr ratio of rocks of the upper continental crust is estimated to be 0.25, with a corresponding average $^{87}Sr/^{86}Sr$ ratio of 0.720.[5] In contrast, rocks derived from the mantle, e.g., oceanic basalts, have strontium isotope ratios very close to 0.703, implying evolution in a reservoir with a Rb/Sr ratio of about 0.03. If continents were completely formed early in the earth's history, and if younger continental rocks are therefore regarded as being derived by recycling older continental material, e.g., by erosion of material from shields into younger geosynclines followed by metamorphism and partial melting, the strontium in younger continental rocks should contain a substantial radiogenic component; $^{87}Sr/^{86}Sr$ ratios in the range 0.710 to 0.720 might be expected.[4]

This expectation is rarely realized. The initial strontium-isotopic ratios of most granitic rocks and metamorphic complexes are usually found to be smaller than 0.710 and sometimes as low as 0.704.[6] The initial strontium-isotope ratios of orogenic volcanic rocks are frequently found to fall in the range 0.703 to 0.707.[7] This implies that rocks in most younger geological provinces have evolved for most of their history in an environment possessing a Rb/Sr ratio gen-

[1]Hills (1953).
[2]Richards and Evernden (1962).
[3]Hurley et al. (1962).
[4]E.g., Hurley et al. (1962).
[5]Hurley (1968).
[6]E.g., the large Precambrian granitic batholith in southwest Western Australia (Arriens, 1971).
[7]Section 7-2.

erally similar to the mantle and differing strongly from that in the upper crust. The low initial strontium ratios of continental igneous rocks are most readily interpreted if these rocks are dominantly or entirely composed of material which has been derived from a subsialic source region, most probably the mantle.[1] The observation that the initial strontium ratios of continental igneous rocks are often slightly higher than mantle strontium may imply small degrees of contamination by crustal strontium or else evolution in geosynclinal environments with higher Rb/Sr ratios than the mantle for a limited time followed by metamorphism and partial melting.

It was concluded on petrological grounds in Chap. 2 that the mean chemical composition of the continental crust resembled that of an intermediate igneous rock. Taylor (1967) has concluded from an investigation of trace element abundances that the average composition of the crust resembles that of an andesite, which is the most abundant volcanic member of the orogenic igneous rock association. Thus, the regions where active crustal growth is presently occurring are presumably those characterized by orogenic igneous activity generally and by andesitic volcanism specifically. Estimates of the current rate of extrusion of orogenic volcanic rocks, although admittedly imprecise, suggest that, if constant with time, it could account for the formation of the entire continental crust.[2] In a careful investigation of volcanic activity in the Kuriles from Upper Cretaceous to Recent, Markhinin (1968) showed that about 6.5×10^6 km^3 of mainly pyroclastic volcanic rocks with an average andesitic composition (58% SiO_2) had been erupted. This was sufficient to transform an original oceanic crust into a continental-type crust. Markhinin considered that, throughout geological history, the entire continental crust could have formed as products of volcanism in island arcs and continental margins.

The recent evolution of large volumes of continental crust is thus well documented. Several workers[2,3,4,5] have shown that extensive sequences of orogenic volcanic rocks and their characteristic sedimentary associates presently occur within the interior of continents and are widely found in provinces of all ages, extending back into the Precambrian. The uniformitarian concept thus leads to the view that additions to the crust via orogenic-type igneous activity have been a major and continuing feature of crustal evolution. The centres of ancient intracontinental orogenic volcanism were presumably located near the margins of then-existing continental blocks or in island-arc systems as in the contemporary cases. All these considerations strongly suggest that the origin and evolution of the continental crust is intimately connected with the origin of the orogenic igneous series.

The petrogenesis of orogenic igneous rocks was considered in the previous

[1]Hurley et al. (1965).
[2]Wilson (1954).
[3]Engel (1963).
[4]Dickinson (1969).
[5]Baragar and Goodwin (1969).

chapter, with primary emphasis on the physical and chemical differentiation processes which had caused the range of observed rock compositions. In the present chapter, the emphasis shifts to the relationship between the petrogenesis of orogenic igneous rocks and the processes occurring in regions of lithosphere subduction. This leads on to a consideration of the nature of the major differentiation processes which are responsible for the formation of the crust and the petrologic evolution of the upper mantle.

8-2 SOME GEOPHYSICAL CHARACTERISTICS OF SUBDUCTION ZONES

The distributions of oceanic trenches, island arcs, active volcanoes, and subduction zones in the western Pacific are shown in Fig. 8-1. The structure of the crust in an island arc-trench system was discussed in Sec. 2-3. A notable feature is the existence of the large negative free-air gravity anomaly (~300 mgal) over the trench and a corresponding positive anomaly of slightly smaller amplitude over the island arc (Fig. 2-10).

Most island arcs and active (Pacific-type) continental margins are characterized by intense seismic activity. Gutenberg and Richter (1941) showed that a large part of the seismicity is confined to an inclined plane dipping inward from the trench beneath the island arc or continental margin at an angle near 45°. Benioff (1955) suggested that these seismic planes represented major tectonic dislocations, and they have since been called *Benioff zones*.

By means of accurate locations of earthquake epicentres, Sykes (1966) demonstrated that the Benioff zone was 50 to 100 km thick in the Tonga-Fiji area and that seismic activity was continuous in places to depths of 650 km. Whilst the most common dip of Benioff zones is about 45°, a wide range of dips from 30 to 90° is found in different island arcs, and the dips may vary markedly even in single arcs, usually increasing with depth.

From an analysis of the Longshot nuclear explosion travel times, Cleary (1967) demonstrated that seismic *P* velocity parallel to the Benioff zone beneath the Aleutians was much higher than normal. A detailed analysis of the seismic structure of the Tonga-Kermadec arc revealed that *P*- and *S*-wave velocities along the Benioff zone were 6 to 7% higher than in comparable parts of aseismic normal mantle and that these differences appeared to persist down to the level of the deepest earthquakes.[1] The upper boundary of the Benioff zone appeared to be quite sharp, with a velocity discontinuity against normal mantle. These workers demonstrated that the zone of abnormally high velocities was slab-like in configuration. Oliver and Isacks (1967) had previously demonstrated that the Benioff zone beneath Tonga was characterized by abnormally low seismic attenuation (high *Q*), comparable to that exhibited by the uppermost 70 to 100 km of oceanic lithosphere. Since *Q* is strongly temperature-dependent, this suggested that the slab was much cooler than the surrounding mantle. A corresponding explanation

[1]Mitronovas and Isacks (1971).

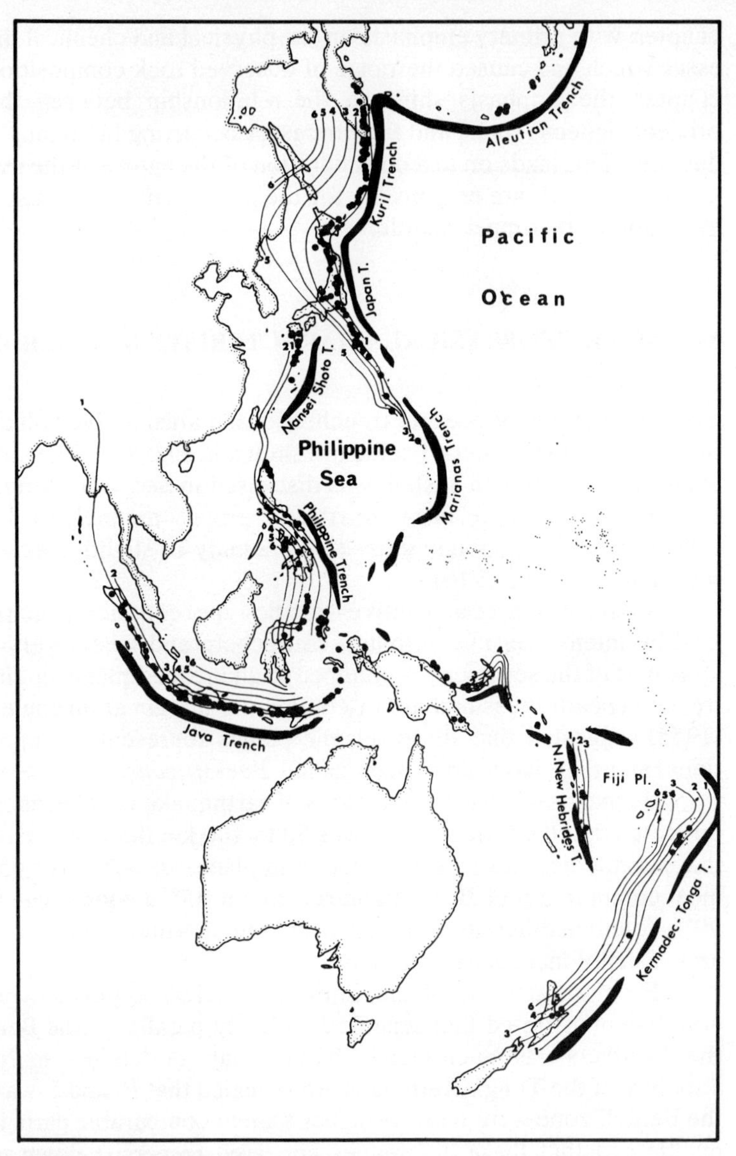

FIGURE 8-1
Trenches, Benioff zones, and active volcanoes of the western Pacific. Depth to the Benioff zone is indicated by 100-km contours. Volcanoes are shown as black dots. (*From Oxburgh and Turcotte, 1970, with permission.*)

is applicable, at least in part, for the anomalously high velocities, which could be caused if the slab were about 1000°C cooler than surrounding mantle.[1]

The above observations are readily interpreted in terms of plate tectonic models which regard the trenches and Benioff zones as the sites of lithosphere subduction. The distribution of stresses in the descending lithosphere has been studied by Isacks and Molnar (1969, 1971), who demonstrated that the earthquake-producing stresses are aligned parallel to the Benioff zone and within the high velocity-high *Q* slab. At depths greater than 300 km, the slab was found to be in a state of compression parallel to the dip. At smaller depths, however, tensional stresses predominated, although compressional stresses also occurred (Fig. 15-9).

The low temperatures inferred for the sinking slab from seismic observations are generally supported by thermal calculations.[2,3,4,5] Because of the large thermal inertia of the lithosphere in relation to its rate of sinking (typically 5 to 15 cm/year), temperatures within the interior of the lithosphere remain much smaller than the surrounding mantle to depths of several hundred kilometres (Fig. 15-4). Temperature differences may exceed 1000°C. The lithosphere is warmed principally by conduction from the surrounding mantle and by generation of heat by frictional dissipation at the interface between the sinking slab and overlying mantle. The latter source appears to be of considerable importance in causing partial melting of the oceanic crust.[3,4]

Heat-flow measured in oceanic trenches is generally low (Fig. 8-2) and consistent with the occurrence of a sinking cool slab; however, the heat-flow behind island-arc systems turned out to be unexpectedly high.[6,7] Heat-flow values as high as 5.6 hfu have been reported on the Fiji Plateau,[7] with a mean of 2.4 hfu. Heat-flows of this magnitude imply the existence of mass transport or convective processes occurring in the wedge between the Benioff zone and the surface. The observation that island arcs are composed dominantly of volcanic rocks is also indicative of the operation of these processes. Active volcanoes form a semicontinuous chain along the line of the island arcs parallel to the adjacent ocean trenches. The active or recently active volcanic belt is usually less than 300 km wide and may be considerably narrower; it occupies the oceanward side of the zone of high heat-flow. Active volcanoes are mostly situated along zones which are 80 to 150 km vertically above the Benioff zones. A relatively small number of volcanoes occur at greater vertical distances, up to 300 km.[8]

The wedge of mantle overlying the Benioff zone at depths smaller than

[1]Mitronovas and Isacks (1971).
[2]McKenzie (1969).
[3]Oxburgh and Turcotte (1970).
[4]Toksûz, Minear, and Julian (1971).
[5]Griggs (1972).
[6]Uyeda and Vacquier (1968).
[7]Sclater and Menard (1967).
[8]Fedotov (1963, 1968).

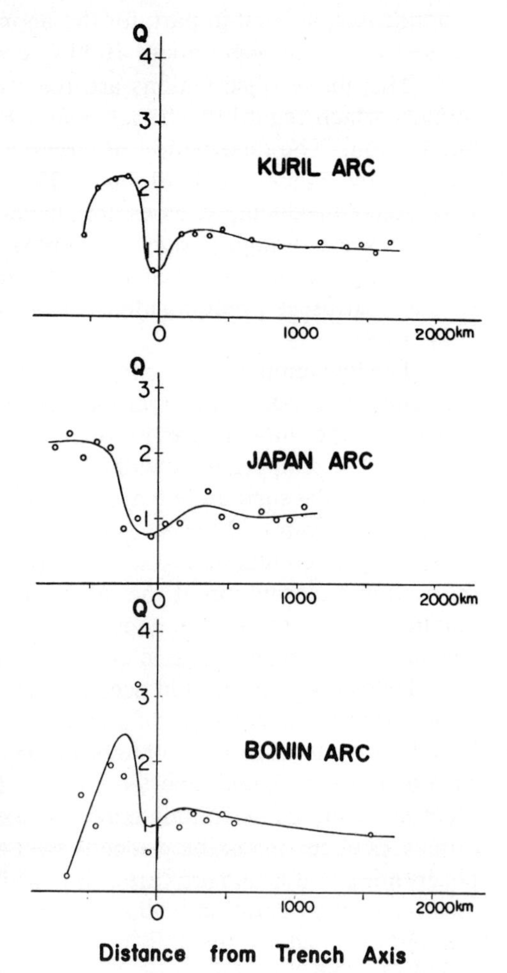

FIGURE 8-2
Profiles of heat-flow values across three island arcs averaged in 100-km intervals. The high heat-flow region is on the continental side of the arc in each case. (*From Uyeda and Vacquier, 1968, with permission. Copyright American Geophysical Union.*)

300 km sometimes displays unusual seismic characteristics. *P*- and *S*-wave velocities appear to be substantially smaller than normal and are also strongly attenuated.[1,2,3] Fedotov[1] has described seismic evidence for the existence of large magma chambers at depths up to 80 km beneath volcanoes on Kamchatka. A detailed investigation of *P*- and *S*-wave attenuation in the upper mantle in the vicinity of the Tonga island arc was carried out by Barazangi and Isacks (1971). Their results are shown in Fig. 8-3.

These results confirm the continuity of the high-Q seismically fast slab with the oceanic lithosphere.[4,5] The extremely low Q values found in the mantle be-

[1]Fedotov (1963, 1968).
[2]Barazangi and Isacks (1971).
[3]Molnar and Oliver (1969).
[4]Oliver and Isacks (1967).
[5]Barazangi, Isacks, and Oliver (1972).

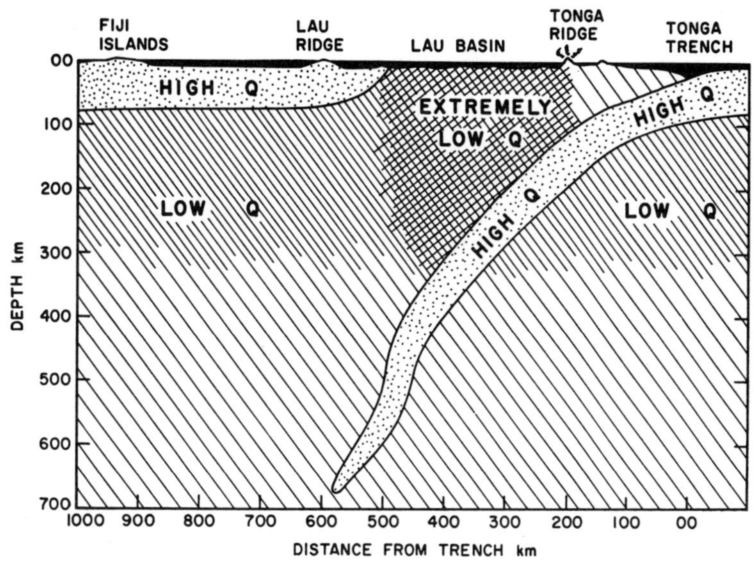

FIGURE 8-3
A schematic cross section perpendicular to the Tonga arc showing lithospheric plates (dotted) in relation to zones of high and low seismic-wave attenuation. (*From Barazangi and Isacks, 1971, with permission. Copyright American Geophysical Union.*)

neath the Lau basin, extending some 300 km to the west of the Tonga volcanic ridge, constitute an interesting and most significant feature. The entire prism of mantle below this region and extending to the Benioff zone at depths of 100 to 300 km is characterized by very high seismic attenuation for both *P* and *S* waves.

Karig (1971) assembled evidence on the bathymetry, sediment distribution, heat-flow, crustal structure, and geology of the Lau and Tonga islands and argued that these were best explained if the Lau basin had formed by rifting of an old frontal arc, with subsequent creation of new oceanic crust in the Lau basin by crustal spreading. The data on seismic attenuation, seismic velocities, and heat-flow described above strongly support this interpretation. It is difficult to avoid the conclusion that the mantle overlying the Benioff zone at depths from 100 to 300 km has been mobilized by some kind of interaction with the sinking lithosphere.[1]

8-3 PETROGENESIS OF MAGMAS IN THE ISLAND-ARC ENVIRONMENT

One of the earliest workers to link the origin of andesitic magmas with processes occurring at the Benioff zone and with continental growth was Wilson (1954), who

[1]Barazangi and Isacks (1971).

suggested that residual liquids and solutions rich in water, silica, and alkalis were generated at the Benioff zone and that "on the way up they could be expected to melt fractions of the mantle and of sediments, and thus to form the variety of acid to basic volcanics which are found along arcs." Although the origins of the residual liquids and solutions were not explained, these ideas, formulated in a period when most petrologists believed that andesites were produced by contamination of basalts by sediments or by remelting of sediments, were in advance of their time. Another notable early contribution was by Coats (1962), who suggested that eugeosynclinal sediments and basaltic volcanics were carried by underthrusting down the Benioff zone to a depth of at least 100 km:[1]

> Water and material of granitic composition were sweated out of these materials and were added to a molten fraction of basaltic composition that was interstitial to peridotite of the mantle. This magma rose in the block overlying the thrust zone, and was concentrated at moderate depth in magma chambers. There, differentiation of the water-rich magma under constant or increasing partial pressure of oxygen produced the observed variety of volcanic rocks.

Although Coats emphasized the roles of partial fusion of sediments and differentiation under high oxygen pressures in producing orogenic-type magmas—mechanisms which were rejected in Chap. 7—the general framework which he proposed for the petrogenesis of orogenic rocks was prescient. Mention should also be made of the work of Gorshkov (1962), who concluded on the basis of geophysical and petrological evidence that orogenic-type magmas were of primary origin, being generated in the mantle by unspecified processes at depths of 70 to 150 km beneath island arcs.

Kuno (1959) was one of the first to suggest the generation of magmas at the Benioff zone and to link the petrological nature of the magma with the depth of origin at the Benioff zone. Kuno applied this hypothesis (incorrectly) to the origin of tholeiites, high-alumina basalts, and alkali basalts and regarded different orogenic magma types as being formed by fractional crystallization of corresponding basaltic magmas. Dickinson and Hatherton (1967) applied Kuno's concept directly to orogenic magmas with considerable success, as discussed in Sec. 7-3. Ringwood (1966) suggested that processes occurring along the Benioff zone might trigger diapiric instability in the overlying wedge. Diapirs of mantle material rising upward from the Benioff zone might undergo partial melting at much shallower depths owing to release of pressure, thereby producing magmas.

The workers cited above were primarily concerned with the tectonic environment of formation of orogenic-type magmas, rather than with the detailed petrochemical processes by which they were generated. The latter problem was tackled experimentally by T. Green and Ringwood (1966, 1967, 1968). This work (reviewed in Chap. 7) demonstrated quantitatively that andesitic-dacitic magmas could be formed by partial melting of the mafic oceanic crust (either as amphibolite or quartz eclogite) along the Benioff zone where lithosphere was sub-

[1]See also Stille (1955).

ducted into the mantle. It was proposed,[1] however, that the tholeiitic magmas associated with andesites and dacites were formed not from the subducted oceanic crust but by partial melting of pyrolite in the wedge overlying the Benioff zone. Convective instability in the wedge, leading to partial melting and basalt generation, might be triggered by the uprise of orogenic magmas from the Benioff zone. Alternatively, water liberated by dehydration of subducted oceanic crust might enter the overlying wedge, causing partial melting and producing hydrous basaltic magmas which fractionate by amphibole separation to form a range of orogenic magmas associated with hydrous high-alumina basalts. At depths up to about 70 km, the oceanic crust subducted along the Benioff zone would consist largely of amphibolite; however, this would transform with increasing depth to eclogite.[2,3] Formation of orogenic magmas along the Benioff zone would occur by partial melting of amphibolite at shallower depths and of quartz eclogite at greater depths, the heat being supplied by viscous dissipation.[4] As a result, the K/Na ratios of derived magmas would increase with depth, in accordance with observation.[3]

From 1969 onward, a cascade of papers developing these themes more explicitly has appeared, and a detailed review is not practicable. The role of water liberated by dehydration of the oceanic crust and introduced into the overlying wedge has been repeatedly emphasized,[e.g.,2,5,6,7,8] as have the detailed effects of amphibole decomposition in the sinking lithosphere.[8,9,10,11] An attempt to arrive at a synthesis is described next.

Preferred Petrologic Model

According to the discussion in Sec. 2-3, the oceanic crust consists mainly of a heterogeneous mixture of anhydrous mafic rocks, mafic greenschists, amphibolites, and bodies of serpentine. Although the serpentine is believed to constitute a small proportion ($< 10\%$) of the total volume, it may, because of its high water content ($\sim 20\%$), account for a substantial proportion of the total water which is carried into the mantle. As the oceanic crust subsides into the mantle, greenschists will be converted to amphibolite at a relatively early stage, and the water evolved will react with surrounding anhydrous mafic rocks to produce more amphibolite. Much of the mafic component of the sinking crust is likely to be converted to amphibolite so that subsequent liberation of water is governed by the dehydration of amphibolite and serpentine.

[1]T. Green and Ringwood (1967, p. 486; 1968, pp. 113, 114, 158).
[2]Ringwood (1969).
[3]T. Green and Ringwood (1969).
[4]Oxburgh and Turcotte (1968a).
[5]Raleigh and Lee (1969).
[6]Hamilton (1969).
[7]McBirney (1969).
[8]Wylie (1971).
[9]Fitton (1971).
[10]D. Green (1972).
[11]T. Green (1972).

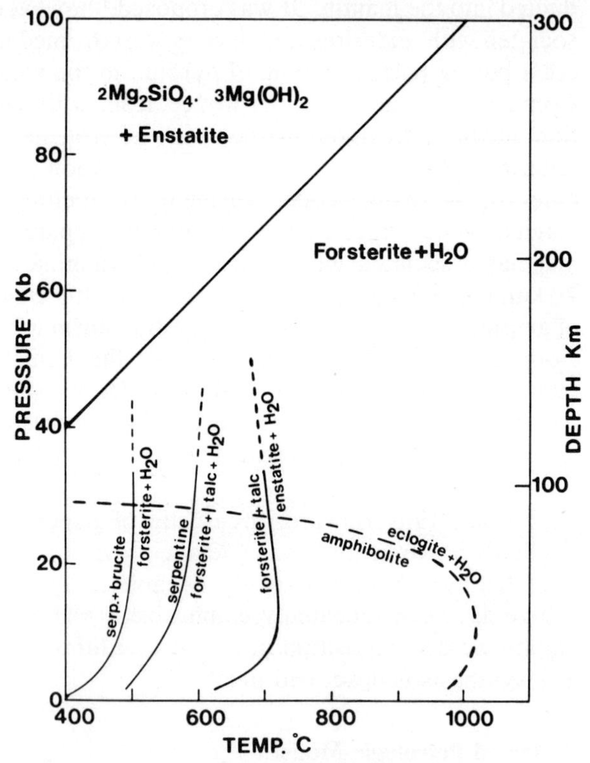

FIGURE 8-4
P-T equilibrium curves for dehydration reactions involving brucite, serpentine, talc, and $2Mg_2SiO_4 \cdot 3Mg(OH)_2$, together with an estimate of the amphibolite-eclogite transformation curve. All curves for conditions of $P_{H_2O} = P_{total}$.

The effects of pressure and temperature on the dehydration of these materials are radically different (Fig. 8-4). Whereas the transformation of amphibolite to eclogite plus water is approximately isobaric, occurring around 20 to 30 kbars over a wide range of temperatures, the dehydration of brucite, serpentine, and talc to form forsterite and enstatite plus water is approximately isothermal (between 5 and 30 kbars) and independent of total pressure.

Two groups of workers[1,2] independently discovered a new hydrated magnesium silicate phase which is stable over a very wide range of pressures and temperatures,[3] extending from 50 kbars, 500°C to over 130 kbars, 1300°C. Its composition[4] is $Mg_7Si_2O_8(OH)_6$ and its density[4] is 2.96 g/cm³. The structure of this phase has not yet been determined, and it is convenient to refer to it by the non-

[1]Ringwood and Major (1966, 1967).
[2]Sclar, Carrison, and Stewart (1967).
[3]The stability relations of this phase are further discussed in Sec. 13-2.
[4]Yamamoto and Akimoto (1974).

specific acronym "DHMS" (dense hydrated magnesium silicate). The stability relationships imply that serpentine will disproportionate into the DHMS phase plus enstatite at high pressures and, furthermore, that the DHMS phase constitutes an important host for water in the sinking lithosphere.[1,9]

As the oceanic crust sinks along the Benioff zone, heating of its upper margin by frictional dissipation occurs.[2,3,4] Estimates of the magnitude of this effect vary substantially. In the following discussion, we will not choose any of the specific models which have been advocated[2,3,4] but, rather, will assume a level of heating by this mechanism which is generally consistent with these models and that leads to a satisfactory petrologic model.

It is proposed specifically that the mafic oceanic crust follows a *P-T* path in which the isobaric transformation of amphibolite to eclogite + water vapour occurs under *subsolidus conditions*—i.e., the temperatures are sufficiently low so that an acid silicate melt is *not* generated during the transformation.[5] This requires that the temperature of the oceanic crust is generally lower than 700°C over the depth interval (70 to 100 km) at which the transition occurs (Sec. 7-8, Figs. 7-16 and 8-4). A large amount of water produced by the dehydration of amphibolite in the slab thereby rises into the mantle above, causing a drastic decrease in "viscosity" and initiating the uprise of diapirs of pyrolite from the Benioff zone.[6] Partial melting occurs in the rising diapirs in the presence of high water-vapour pressure, leading to the separation of hydrous tholeiitic magmas at shallower levels. As discussed by Nicholls and Ringwood (1973) and reviewed in Sec. 7-6, it is believed that these hydrous magmas fractionate by the separation of olivine, pyroxene, and amphibole, producing the early tholeiitic stage of island-arc development. The essential characteristic of this stage is that there is only a small amount of transfer of silicate components (carried in the vapour phase) from the subducted oceanic crust into the tholeiitic magmas which are ultimately developed in the wedge; moreover, the process occurs within the depth interval of 70 to 100 km, controlled by the isobaric subsolidus decomposition of amphibolite to eclogite + water.

At depths greater than 100 km, the mafic portion of the oceanic crust will have been converted to a quartz eclogite[7,8] (Fig. 8-6). The temperature near the top of the quartz eclogite layer, where heat is being generated by frictional dissipation, may reach about 700°C at 70 to 100 km; however, the deeper levels of the

[1]Nicholls and Ringwood (1973).
[2]Oxburgh and Turcotte (1968a, 1970).
[3]McKenzie (1969).
[4]Toksöz, Minear, and Julian (1971).
[5]The above proposal differe in this respect from those of Fitton (1971), Jakeš and Gill (1970), T. Green (1972), and D. Green (1972) but is in accord with the model of Nicholls and Ringwood (1973). See also Ringwood (1974).
[6]Wyllie (1971).
[7]Ringwood and Green (1966).
[8]Ringwood (1969).
[9]Recent experiments by Yamamoto and Akimoto (1975) suggest that hydroxyl chondrodite, $2Mg_2SiO_4 \cdot Mg(OH)_2$, may also be an important host for water in the sinking lithosphere.

quartz eclogite layer would be much cooler since conduction of heat from the surface is relatively slow. Consider now, the dehydration of discrete bodies of serpentine scattered throughout the quartz eclogite layer. There are several reactions which will occur over a wide range of temperatures (500 to 1200°C) (Fig. 8-4):

1 Serpentine + brucite → forsterite + water (~500°C)
2 Serpentine → forsterite + talc + water (~ 600°C)
3 Forsterite + talc → enstatite + water (~ 700°C)
4 Serpentine → DHMS + enstatite + water (~ 50 kbars, 500°C)
5 DHMS + enstatite → forsterite + water (~ 50 to 150 kbars, ~ 500 to 1500°C)

These dehydration reactions will occur gradually since the temperatures at different depths in the quartz eclogite layer are rising at different rates by thermal conduction from the upper boundary. Accordingly, there will be a very broad depth interval over which dehydration of serpentine blocks occurs. The water vapour produced must travel through surrounding quartz eclogite. Below 100 km, the temperature at the surface boundary of the quartz eclogite layer exceeds 700°C, and the 700°C isotherm penetrates deeper into this layer as the slab sinks. In the presence of high water pressures generated by sequential dehydration of serpentine bodies, the quartz eclogite will partially melt at temperatures above 700°C to form residual eclogite + water-rich rhyodacite magma, containing nearly all the incompatible elements formerly in the oceanic crust. As discussed in Sec. 7-8, this magma migrates upward into the wedge above the Benioff zone and reacts with pyrolite to form olivine pyroxenite.[1] Diapirs of wet pyroxenite thus formed rise and partially melt to form calc-alkaline-type orogenic magmas characterized by high K/Na ratios and strongly fractionated rare earth abundances.[2] These magmas may undergo further fractionation controlled mainly by crystallization of garnet, pyroxene, and amphibole as they rise to the surface. Rise of calc-alkaline magmas through the wedge may also trigger upward convective movements in the surrounding pyrolite, leading to the formation of basaltic magmas.[3] Acid hydrous magmas melted out of the sinking quartz eclogite layer may mix with these basaltic magmas, thereby altering their compositions and fractionation behaviour.[4]

There is probably a multiplicity of related processes occurring in the wedge which lead to the formation of calc-alkaline-type orogenic magmas. The depth interval over which these magmas are generated is mostly between 100 and 150 km, although smaller amounts may be generated as deep as 300 km.[5] The decrease in production of magmas below 150 km is probably because most of the low-melting

[1]See also Nicholls and Ringwood (1973).
[2]Section 7-8.
[3]T. Green and Ringwood (1968).
[4]T. Green and Ringwood (1972).
[5]Fedotov (1968).

acid component has been extracted from the sinking eclogitic layer by the time it reaches this depth. The fact that the region of high seismic attenuation extends to the Benioff zone at depths as great as 300 km (Fig. 8-3) suggests that water is still being introduced into the wedge from the sinking lithosphere and triggering convective instability even after magma extraction has ceased. The most likely source of this water is the DHMS phase, which is likely to be produced in former serpentine bodies trapped within the sinking lithosphere. At elevated temperatures, the DHMS phase would generate a high partial pressure of water vapour at depths of 150 to 400 km. Access of this vapour to unfractionated pyrolite in the overlying wedge would cause incipient (< 1%) melting of easily fusible components, resulting in high attenuation and mobility, as in the low-velocity zone. Upward flow of the resultant mobilized pyrolite, accompanied by partial melting and basalt formation, is presumably responsible for the creation of new oceanic lithosphere and crustal spreading behind some island arcs.[1,2]

To summarize, it is suggested that the early tholeiitic phase of orogenic magma genesis is caused by the introduction of water into the wedge overlying the Benioff zone at depths of 80 to 100 km, the water being produced by subsolidus dehydration of amphibolite in the sinking oceanic crust. The water causes partial melting of pyrolite to form bodies of hydrous tholeiite magmas which rise and fractionate as discussed in Secs. 7-6 and 7-7. An important point is that the orogenic magmas of this stage are derived ultimately from the pyrolite wedge and *not* from the sinking crust.

On the other hand, the later and more evolved calc-alkaline phase of petrogenesis is caused ultimately by the introduction of hydrous acid magma into the wedge above the Benioff zone at depths of 100 to 150 km (and greater), the acid magma being produced by partial melting of quartz eclogite in the sinking oceanic crust. The acid magmas transform regions of the overlying pyrolite to pyroxenite which rises diapirically and partially melts, forming calc-alkaline magmas. The calc-alkaline magmas thus represent a mixture of components derived from both the sinking oceanic crust and the overlying pyrolite wedge.

Below about 150 km, the former mafic oceanic crust has become converted to a highly refractory bimineralic eclogite occupying the thermal divide of Fig. 7-8 and strongly depleted in low-melting components and incompatible elements. Likewise, the underlying peridotite, which was originally formed near a mid-oceanic ridge and was complementary to the basaltic oceanic crust, is also residual and refractory in nature and strongly depleted in low-melting-point components and incompatible elements. Thus the lithosphere which sinks into the mantle has become differentiated into two highly refractory components—eclogite and peridotite. Both these components possess higher initial melting temperatures (solidi) than pyrolite (Fig. 7-8). These thermal characteristics, combined with the large size of each of the components and their depletion in incompatible elements,

[1]Karig (1971).
[2]Barazangi and Isacks (1971).

dictate that the sinking lithosphere can never again serve as a source of basaltic magma in any further cycle of melting—there appears to be no obvious way in which 5- to 50-km-sized blocks of fractionated refractory eclogite and peridotite can be intimately remixed at the centimetre to metre scale in the solid state in the deep mantle to reform a homogeneous pyrolite composition. Moreover, this would necessitate the reintroduction of the low-melting components and incompatible elements which had been extracted from the lithosphere and added to the continental crust. A plausible model for achieving this end is not in sight.

It follows that the lithosphere has become *irreversibly differentiated*, as emphasized by Ringwood (1969) and further discussed later in this chapter. The complementary products of this irreversible differentiation are continental crust rocks derived ultimately via orogenic-type magmatism and the depleted, refractory sinking slab of eclogite and peridotite.

8-4 EVOLUTION OF THE CRUST

Sinking of the lithosphere into the mantle occurs in two distinct environments —beneath trenches along continental margins (e.g., west coast of South America) and associated with island arcs and trenches within deep oceanic basins (e.g., Marianas, Tonga). Since trench and sinking lithosphere systems have finite lifetimes, perhaps on the order of 100 or 200 million years, it is interesting to speculate about the conditions responsible for causing a sinking lithosphere-trench system to develop in a region where such a system did not previously exist.

Lithosphere plates in oceanic regions are denser on the average than the underlying hotter mantle material of the low-velocity zone.[1] The system is thus potentially unstable with respect to convection. However, the mere existence of an unstable vertical density gradient is not a sufficient condition for convective overturn. The initiation of convection also requires the existence of a horizontal density gradient. Ringwood and Green (1966) proposed that initial sinking of the lithosphere might be caused by an instability generated by the transformation of basalt in the lower crust to eclogite.[2] This suggestion was based on experimental investigations (Chaps. 1 and 2) which showed that dry basaltic rocks were not stable under the *P-T* conditions existing in the normal crust and that, where kinetic conditions were favourable, transformation to eclogite possessing a density of 3.45 to 3.60 g/cm^3 would occur. If this transformation occurred throughout a sufficiently large volume of basalt, the mantle (density 3.30 g/cm^3) would be unable to support the load, and the eclogite would commence to sink into the mantle,

[1]Clark and Ringwood (1964).

[2]This paper was written whilst the concepts of sea-floor spreading and plate tectonics were at an early stage of development, and two models were considered—one showing the effects of the basalt-eclogite transformation on the vertical evolution of a continental margin without invoking sea-floor spreading, and one showing the same effects applied to sea-floor spreading. The present discussion directly combines these two models, using the vertical evolution model as the initial stage of sinking of the lithosphere implicit in sea-floor spreading. A detailed and illuminating discussion of a related model has been provided recently by Dewey (1969) and Dewey and Bird (1970).

deforming the crust and causing orogenesis. The series of sketches in Fig. 8-5 shows how a continental margin might evolve under these circumstances. A possible example chosen[1] was the eugeosyncline now believed to be forming off the east coast of the United States.[2]

The initial state (Fig. 8-5A) is taken to be a eugeosyncline overlying a continental slope composed mostly of anhydrous gabbroic material.[3] This is metastable relative to eclogite but, because of low temperatures, has not transformed. The deposition of sediments accompanied by isostatic subsidence pushes the gabbro further into the eclogite stability field (Fig. 8-5B). As temperatures increase to 300 to 400°C owing to thermal blanketing by sediments, gabbro commences to transform to eclogite at a substantial rate. A large region of thickened mafic crust transforms to eclogite (density $\sim$ 3.5 g/cm^3) which commences to sink into the less dense (3.3 g/cm^3) mantle, causing a bulge at the bottom of the crust and a marked deepening of the eugeosyncline (Fig. 8-5C).

Rupture of the crust then occurs, followed by initiation of sinking of lithosphere into mantle driven initially by eclogite "sinker" (Fig. 8-5D). This is accompanied by folding of geosynclinal sediments and intrusion of peridotites. The continental margin has now been transformed from "Atlantic type" to "Pacific type."[4,5] As the lithosphere sinks deeper, its own excess density becomes the dominant driving force (Chap. 15). Normal oceanic crust (basalt, gabbro, and amphibolite) transforms to eclogite. Dehydration of amphibolite produces water which rises into the overlying wedge to trigger convective instability and produce orogenic-type magmas with tholeiitic affinities.

A mature stage (E) of evolution along the Benioff zone is shown in Fig. 8-6. The lithosphere has now sunk below 300 km, and partial melting of the sinking quartz eclogite layer occurs, ultimately causing the generation of calc-alkaline-type orogenic magmas as discussed previously. The lithosphere continues to sink to depths of 700 km or more. Sinking ultimately ceases when the density differential between sinking lithosphere and surrounding mantle has been eliminated, either by thermal equilibration or by phase changes (Chap. 15).

By this stage (Fig. 8-5F), a large belt of newly formed crustal rocks has been differentiated from the mantle and welded on the continental margin. The new crust is probably a highly heterogeneous mixture of orogenic-type igneous rocks, metamorphic rocks, folded sediments, basalts and gabbros, and ultramafic intrusions (from stage D). Temperatures in the thickened crust become abnormally

[1]Ringwood and Green (1966).

[2]Drake et al. (1959).

[3]Ringwood and Green (1966) postulated an early stage of basaltic volcanism from source regions immediately below the eugeosyncline. However, Dewey (1969) pointed out that this postulate was unnecessary. The present crust beneath the continental slope and the eugeosyncline was the site of the initial rifting of the American-European landmass over an embryonic Atlantic mid-oceanic ridge. This would have been accompanied by extensive basaltic volcanism (plateau basalts) and intrusions of gabbro at depth. It is possible that this transitional region between continental and oceanic crust is dominantly composed of mafic material formed during the earliest stages of opening of the Atlantic. This view is accepted here.

[4]Dewey (1969).

[5]Dewey and Bird (1970).

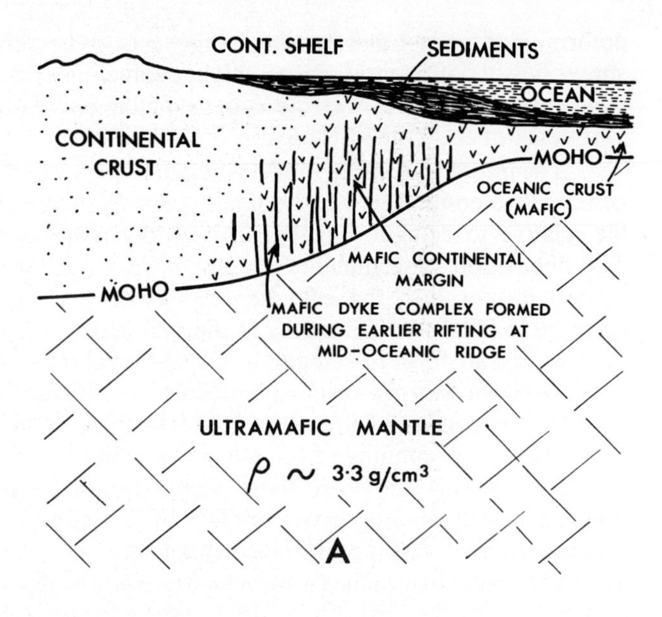

FIGURE 8-5
Model for the evolution of a continental margin based on Ringwood and Green (1966), figs. 8 and 9 (not to scale).

A Initial state—e.g. east coast of North America (Drake et al., 1959).
B Anhydrous mafic lower crust transforms to eclogite.
C Eclogitic lower crust slowly subsides into mantle causing eugeosyncline to deepen.
D Rupture of crust and commencement of subsidence of lithosphere driven by eclogite sinker. Folding of eugeosyncline and establishment of Benioff zone. Commencement of tholeiitic orogenic-type volcanism caused by migration of water from lithosphere into pyrolite wedge overlying the Benioff zone.
E See Fig. 8-6. Lithosphere sinks from 100 to below 300 km. Partial melting occurs in quartz eclogite layer along Benioff zone and in overlying pyrolite wedge.
F Sinking of lithosphere finally ceases after a depth of 700 km or more has been reached. A new belt of alpine-type continental crust has been differentiated from mantle and welded on to the edge of the normal continental crust. Relaxation of regional stress field permits isostatic uplift of alpine mountain chain.
G Erosion of mountains and major diapiric differentiation under gravity leading to re-establishment of Mohorovicic discontinuity and further cooling at depth, resulting finally in a stable shield.

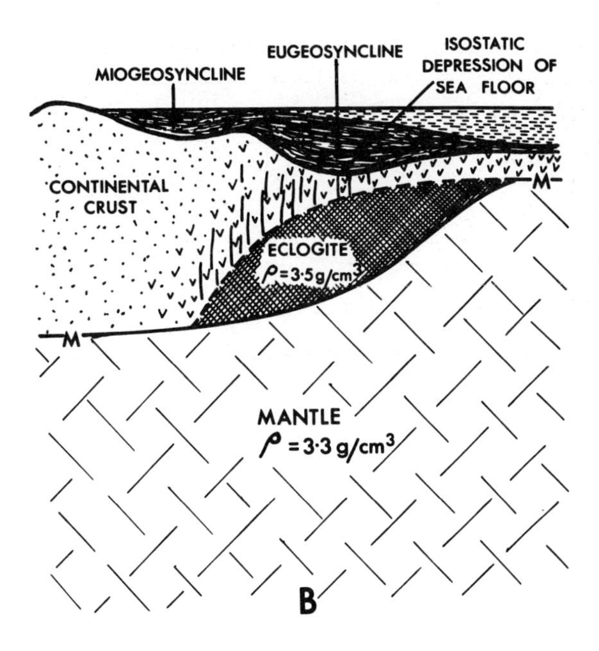

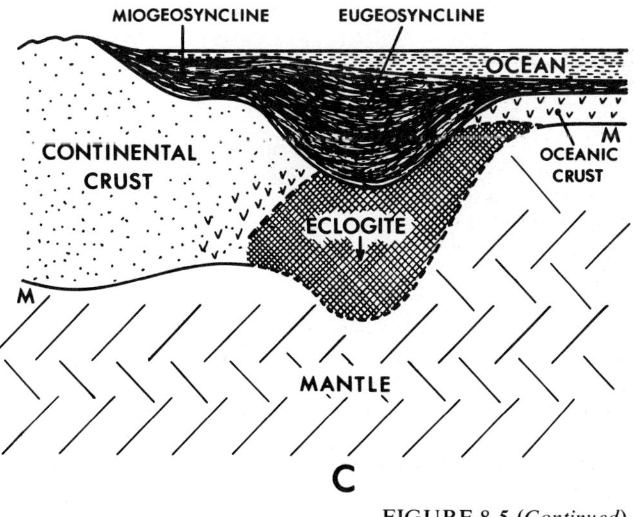

FIGURE 8-5 (*Continued*)

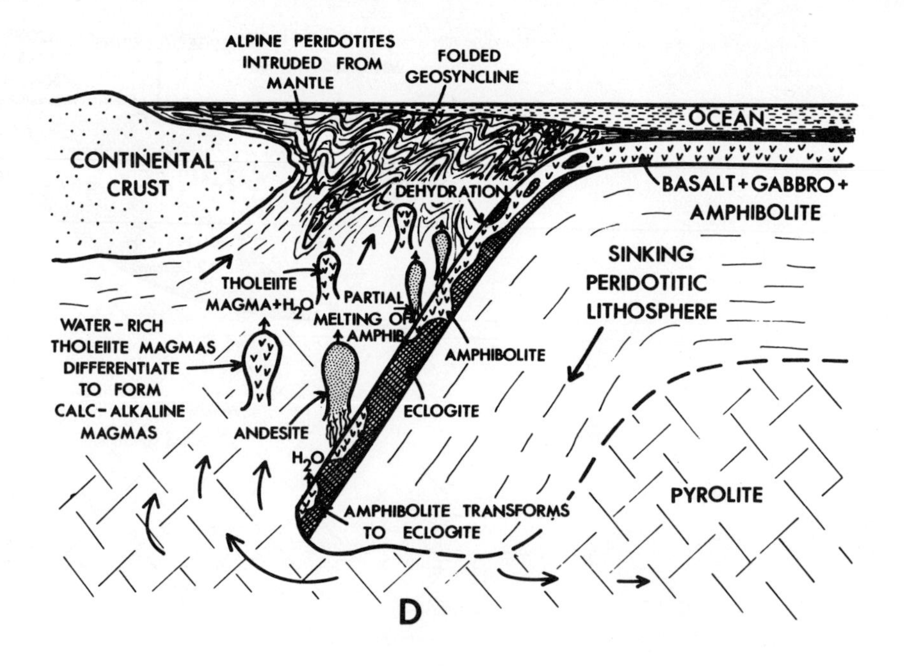

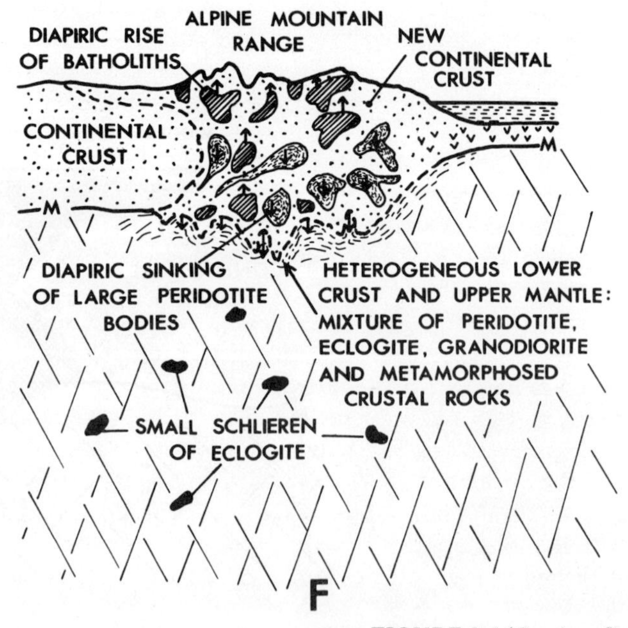

FIGURE 8-5 (*Continued*)

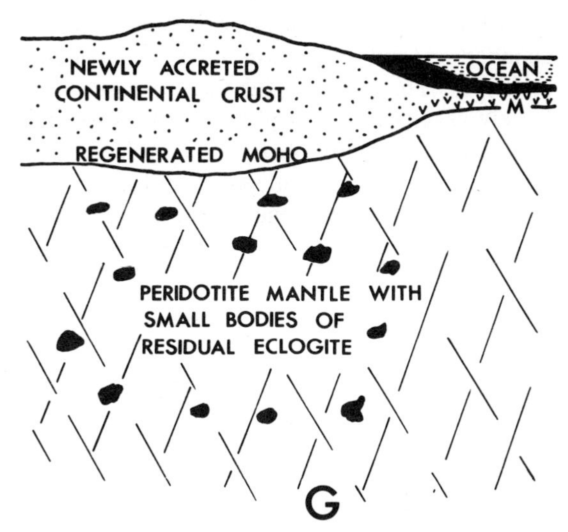

FIGURE 8-5 (*Continued*)

high due to the combined effects of radioactive heat generation in situ and conduction from large intrusions of primary magmas from the mantle. This leads to widespread regional metamorphism and to the generation of secondary granitic magmas by partial fusion of crustal rocks accompanied by hybridism and contamination. Because of the high crustal temperatures, basaltic rocks occur in the stability fields of pyroxene granulite and garnet granulite rather than eclogite. The physical heterogeneity of the rocks of the lower crust and upper mantle may cause the seismic velocity distribution between crust and upper mantle to become approximately continuous, and the Mohorovicic discontinuity may not be clearly recognized at this stage. With relaxation of the regional stress field owing to cessation of sinking, the folded geosynclinal wedge and its accompanying igneous intrusions rise isostatically to form an alpine mountain chain.

During the final and longest stage of crustal evolution (Fig. 8-5G), the mountains gradually become eroded and the crust develops a stable gravitational and rheological configuration. Large bodies of ultramafic rocks previously intruded into the crust sink diapirically back into the mantle, and large volumes of lighter acid rocks in the mantle rise diapirically into the crust. The end result of this separation of large volumes of rocks in the gravitational field according to their densities is the re-establishment of a distinct and recognizable boundary between generally intermediate but heterogeneous lower crustal rocks in the granulite facies and the ultramafic rocks of the upper mantle. Thus, the Mohorovicic discontinuity which was present beneath the oceanic crust at the beginning of the cycle, but which was obliterated during the active phase of orogenesis, becomes re-established in the stable continental crust.[1]

[1]Ringwood and Green (1966).

The occurrence of intense metamorphism and magmatism, accompanied by geochemical differentiation, leads to strong concentration of the radioactive elements, U, Th, and K in the upper crust. Further erosion of the upper crust ultimately results in net depletion of radioactive elements, leading to the low heat-flow values typically observed on Precambrian shields and to the occurrence of relatively low crustal and subcrustal temperatures. These, in turn, cause the crust and upper mantle to assume the state of relative stability and tectonic rigidity which is characteristic of Precambrian shields. Thus, the major geological cycle completes its course.

8-5 DIFFERENTIATION OF THE UPPER MANTLE

The theory of sea-floor spreading and plate tectonics is at the heart of current thinking regarding evolution of the crust and upper mantle. We are concerned here not so much with the structural and tectonic aspects of this theory as with the implications for geochemical and petrological differentiation of the mantle and segregation of the crust. Some of these are illustrated in Fig. 8-6.

The process of mantle differentiation may be regarded as commencing in the gravitationally unstable low-velocity zone beneath mid-oceanic ridges. Pyrolite diapirs rise upward from this region and undergo partial melting, leading to the generation of basaltic magma,[1] together with residual unmelted peridotite. It is envisaged that the immediate source region is the low-velocity zone and that replenishment in the first instance involves horizontal inward flow from the low-velocity zone[2] rather than upward vertical movement from source regions below the oceanic ridges and deeper than the low-velocity zone.[3] The suggested flow configuration is shown in Fig. 8-6. One advantage of this configuration with its shallow convective zone is the readier explanation provided of the horizontal mobility of mid-oceanic ridges with respect to one another.[4]

The axes of the ridges are characterized by high heat-flows, and the subsurface temperatures are high enough to maintain the stability of the basaltic mineral assemblages. The ridges thus develop as expanding features composed of heterogeneous mixtures of gabbro, dolerite, peridotite, and pyrolite with surficial basalts. The heterogeneity is indicated by seismic results (Fig. 2-8) which show the presence of localized mantle material ($V_P \sim 8$ km/sec) at high levels and the widespread occurrence of ultramafic rocks along fracture zones. The heterogeneous mixture of dense ultramafic rocks and less dense basaltic rocks is gravitationally unstable. At the high average temperatures present, the strength of the developing lithosphere is insufficient to prevent dense intrusions of peridotite from sinking back into the mantle. Thus, the Mohorovicic discontinuity is established by large-

[1]Ref. Chap. 4.

[2]Ringwood and Green (1966, p. 422), Isacks, Oliver, and Sykes (1968), and Ringwood (1969).

[3]As suggested, for example, by Oxburgh and Turcotte (1968b).

[4]E.g., the mid-oceanic ridges surrounding Africa.

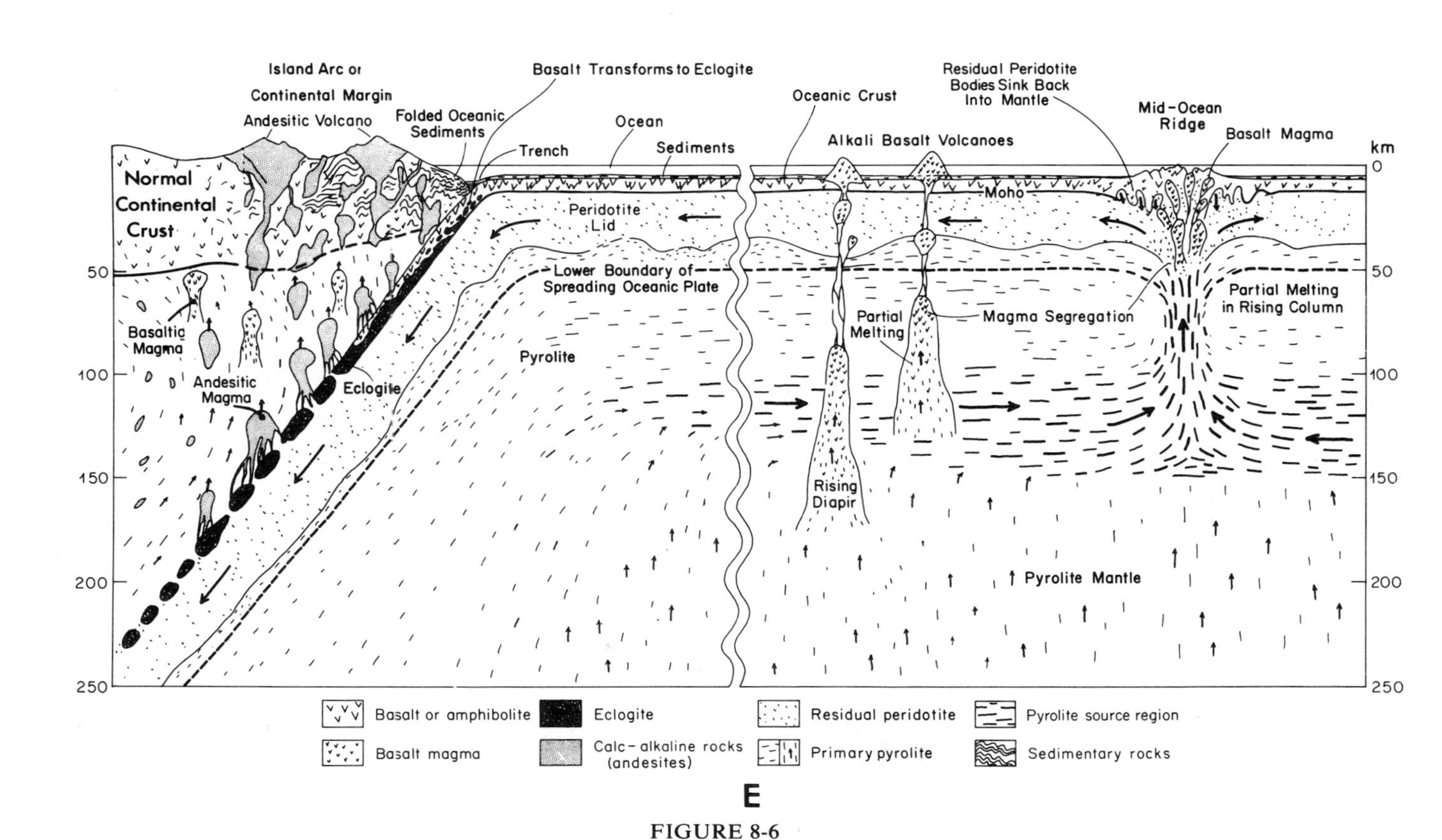

FIGURE 8-6
Petrologic model of plate generation and consumption. (*From Ringwood, 1969, with permission. Copyright American Geophysical Union.*)

scale solid-state differentiation of major rock types of contrasted density within the earth's gravitational field.[1]

In this way, a zoned oceanic lithosphere with a thickness of about 70 km is developed,[2] consisting of a layer of basaltic rocks overlying a layer of residual peridotite and, beneath this, a layer of pyrolite. Studies of temperature distribution in the rising plumes beneath mid-oceanic ridges in relation to the degree of partial melting strongly suggest that, between the depleted peridotite layer and the pyrolite layer, additional layers showing the effects of varying degrees of partial melting of primitive pyrolite might be present.[3,4] The upper layer of strongly depleted peridotite (harzburgite which was complementary to oceanic tholeiites) might be followed downward by a layer of lherzolite (complementary to alkali basalts) and then by a layer of pyrolite which has been subjected to only a very small degree of partial melting and has lost small amounts of highly alkaline mafic liquids strongly enriched in incompatible elements such as K, Rb, Ba, U, Th, and La. This layer would still be defined as pyrolite because, with further partial melting, it would be capable of yielding oceanic tholeiite magma. The presence of this slightly fractionated pyrolite layer is inferred from arguments developed by Gast (1968), who demonstrated that oceanic tholeiites are derived from a region of the mantle which has apparently undergone a 50% depletion of some incompatible elements (above) during earlier episodes of differentiation. A sketch of the proposed petrological-geochemical structure of the lithosphere plate is given in Fig. 5-1.

The plate then moves outward, sliding over the weak, incipiently melted low-velocity zone, ultimately reaching a trench and subsiding into the mantle as discussed earlier. Here, further differentiation occurs, leading to the formation of increasingly silica-rich orogenic-type magmas which rise upward and contribute to the growth of new continental crust. This may form either as a lateral addition to an existing continental margin or as an island-arc system which is subsequently accreted on to a continental margin.[5] Ultimately, the region beneath the growing continental crust becomes depleted in low-melting-point components and attains the chemical status of a refractory peridotite.

Generation of crust from mantle thus appears to have proceeded via two distinct types of differentiation processes dominated by vertical and lateral mass transport, respectively. Vertical differentiation involves partial melting of the mantle and transport of the low-melting components into the crust immediatley above. This process leads to the formation of a residual refractory ultramafic zone in the mantle immediately beneath the continental crust and has reached its maximum development in Precambrian shield regions. Lateral differentiation is an intrinsic part of the sea-floor-spreading process. Basalts which are formed by partial

[1]Ringwood and Green (1966).
[2]Kanamori and Press (1970).
[3]Oxburgh and Turcotte (1968b).
[4]Gast (1968).
[5]Dewey and Bird (1970).

melting of the mantle pyrolite beneath oceanic ridges are transported to trenches near continental margins and island areas, where they in turn are partially melted as the oceanic crust sinks into the mantle to form calc-alkaline magmas which also rise and enter the crust.

The cool descending plate of peridotite and eclogite continues to sink to depths of 700 km or greater. This highly differentiated residual material enters the deep mantle and does not return in a second cycle. Rather, it displaces upward relatively unfractionated pyrolite into the low-velocity zone which acts as the immediate source region for the mid-oceanic ridges (Fig. 8-6). The model implies that the fundamental mechanism of sea-floor spreading and plate tectonics is one of irreversible petrological and chemical differentiation of the mantle, leading to the formation of crustal rocks on the one hand and, on the other hand, to refractory, fractionated iron-magnesian silicates which enter the deep mantle.[1]

Returning to the model of the sinking lithosphere (Figs. 8-5 and 8-6), it was suggested that a layer of pyrolite which has been subjected to only a very small degree of partial melting may be present below the strongly differentiated basaltic (eclogitic) and peridotitic layers. Because of its greater depth, this layer is substantially warmer than the overlying peridotite layer, and its temperature will increase during sinking owing to conduction from underlying mantle and frictional dissipation.[2,3] It appears possible that this layer may become decoupled from the cool, sinking slab and be left behind in the upper mantle. Thus, the effective thickness of the sinking lithosphere plate would decrease with depth, and it is possible that, below a few hundred kilometres, all the pyrolite layer may have been left behind, the sinking slab consisting entirely of relatively cool residual peridotitic and eclogitic layers.

If this slightly fractionated pyrolite layer remains in the upper mantle, it will be available for a subsequent cycle of more extensive partial melting to form complementary peridotite and oceanic tholeiite. This may explain Gast's (1968) conclusion that, although the source regions of oceanic tholeiites contain remarkably uniform abundances of *most* major and trace elements, nevertheless, a few incompatible elements, notably those mentioned above, appear to be depleted by a factor of about 2, when compared to the abundances of other related elements. Gast interpreted this pattern to imply that the source region of present-day oceanic tholeiites has participated in earlier episodes of minor differentiation, resulting in the separation of a small amount of a highly alkaline mafic liquid strongly enriched in this group of elements.

The re-entry of slightly fractionated pyrolite from the lower lithosphere into the upper mantle to participate in a second stage of more complete partial melting does not conflict with the previous interpretation that sea-floor spreading is essentially an irreversible differentiation process. This interpretation applies to the gross differentiation of pyrolite into tholeiitic basalt and residual peridotite,

[1]Ringwood (1969).
[2]E.g., Oxburgh and Turcotte (1968a, 1970).
[3]Toksûz, Minear, and Julian (1971).

followed by ultimate disposal of the completely differentiated products in the deep mantle. Whether the differentiation occurs in one or in two stages is immaterial.

Green (1971) has proposed an alternative explanation of Gast's observations. He suggests that the low-velocity zone may be chemically zoned with respect to the distribution of incompatible trace elements. A continuous vertical upward migration of small amounts of the lowest melting temperature liquid fraction, enriched in Ba, La, U, Th, Rb, and Cs, may cause depletion of these elements throughout most of the low-velocity zone which constitutes the source region of oceanic tholeiites. These elements are reprecipitated at higher and cooler levels, causing corresponding enrichments. Partial melting of the overlying layer thus provides the characteristic abundance patterns found in Hawaiian-type tholeiite magmatism.

Hot Spots

One further topic which may be considered in terms of the model shown in Fig. 8-6 is the occurrence of basaltic volcanism within the deep oceanic basins that is unconnected with mid-oceanic ridges. The type examples are the Hawaiian chain of volcanoes and the Emperor seamounts. An explanation of these in terms of quasi-stationary mantle "hot-spots" over which lithosphere plates move has been proposed.[1,2] The hot spots are believed to be produced by upward convective motion of mantle material from the core-mantle boundary[2] or from the lower mantle.[1] There are some difficulties with a source as deep as this, however. In view of the rate of magma production observed, the cross-sectional area of the rising plumes can hardly exceed a few tens of square kilometres, whilst the depth is assumed to be 2000 to 3000 km. In the case of the Hawaiian-Emperor chain, the plume must have existed for some tens of millions of years. It is difficult to visualize how a plume with such an extreme ratio of length to cross-sectional area might be developed and remain stable for such a long period. Moreover, the rise of the plume requires the existence of a finite superadiabatic temperature gradient in the deep mantle. Unless the superadiabatic gradient were quite minute, lavas reaching the surface should be greatly superheated or, alternatively, should be of ultramafic composition (picrites and komatiites) since the superheat will lead to a greatly increased degree of partial melting of pyrolite.

The fact that the range of mafic magmas in the Hawaiian chain is generally similar to that found along mid-oceanic ridges and rift zones strongly suggests a general similarity of petrogenetic processes and source regions. It is suggested therefore, that these source regions also lie in the low-velocity zone, or asthenosphere, but at somewhat deeper levels (150 to 200 km) than the region (perhaps 70 to 150 km) at which maximum mobility occurs in the low-velocity zone. The suggested configuration is shown in Fig. 8-6, which illustrates the rise

[1]Wilson (1965).
[2]Morgan (1971).

of mantle diapirs from 150 to 200 km, leading to partial melting and generation of basaltic magmas.[1] The mantle at these depths and below is believed to be quasi-stagnant, and the diapirs rising from this region are believed to be decoupled from the shallower zone of rapid horizontal flow in the low-velocity zone which feeds the volcanism occurring beneath mid-oceanic ridges.[2]

Another effect of this kind of volcanism is to cause partial differentiation of the mantle on a different time scale from that which occurs beneath ridges. Some of the observations of Gast (1968) relating to episodes of differentiation which affected the source regions of oceanic tholeiites prior to eruption of the latter may be related to the occurrence of decoupled intraoceanic basaltic volcanism as illustrated in Fig. 8-6.

Yet another model of Hawaiian-type volcanism has been proposed by Green (1971). According to this model, which is aimed at explaining certain geochemical features of Hawaiian magmas, the source regions from which diapiric ascent of pyrolite commences are located in the upper levels of the low-velocity zone.

8-6 ORIGIN OF THE MOHOROVICIC DISCONTINUITY

The models discussed earlier imply that the Mohorovicic discontinuity both in oceanic and in continental regions has been formed by closely analogous processes. The crust which is generated at mid-oceanic ridges consists of a highly heterogeneous mixture of mafic and ultramafic rocks.[3] Likewise, the newly developed continental crust in zones of active orogenesis is also believed to consist of a heterogeneous mixture of folded sediments, granitic plutons, metamorphic rocks, and intrusions of dense peridotite, gabbro, and eclogite (Fig. 8-5F). In both provinces, crustal temperatures are relatively high and strength correspondingly low.

The diapiric rise of granitic plutons under the influence of gravity and the close analogy between the intrusion of many granitic bodies and salt domes has been well documented.[4] The density contrast between anhydrous ultramafic and average crustal rocks (+ 0.45 g/cm^3) is greater than between granites and mean crust (− 0.2 g/cm^3). If granites are able to rise diapirically within the crust, then *large* ultramafic intrusions must inevitably sink. It is no coincidence that most very large ultramafic intrusions are of Tertiary age. There can be little doubt that similar ultramafic bodies have been intruded into the crust at earlier periods. However, the stresses which they imposed on the crust apparently could not be sustained for periods of 10^8 to 10^9 years, and accordingly, they subsided into the mantle. Ramberg's[5] impressive model experiments and interpretation provide a

[1]Figure 8-6 portrays alkali basalt volcanoes formed in this manner. A more realistic model would also include tholeiitic volcanoes formed by strictly analogous processes.

[2]Figure 8-6.

[3]Section 2-3.

[4]E.g., Sorgenfrei (1971).

[5]Ramberg (1966, 1967, 1970, 1972).

foundation for understanding the role of diapiric differentiation in the earth's gravity field.

The establishment of a sharp chemical, density, and seismic boundary between crust and mantle in both oceanic and continental provinces is thus interpreted here as an evolutionary process, reaching maturity in the deep ocean basins and in continental shield regions.

8-7 PLATE TECTONICS AND THE DEEP MANTLE

The model depicted in Figs. 8-5 and 8-6 has implications relating to the differentiation of the entire mantle. An important question is the depth to which the refractory depleted plates of lithosphere descend—do they reach a maximum depth of only 700 km, or do they penetrate even deeper? In the latter case, the absence of earthquakes below 700 km would presumably be due to a change in deformation and failure mechanisms around that depth. This view is supported by Griggs (1972), who argues that the limiting depth of earthquakes is that at which the minimum temperature in the slab reaches a fixed proportion of the melting temperature. Above this critical temperature, the slab responds to stresses by flow rather than by seismic instability. According to this model, the temperature of the interior of the slab is substantially smaller than that of the surrounding mantle at depths exceeding 700 km.

Questions relating to the depth of sinking of lithosphere plates were discussed independently by Ringwood (1971, 1972) and Dickinson and Luth (1971).[1] Ringwood stated:

> The inference that slabs may descend below 700 km is consistent with some geochemical considerations. Mass balance calculations indicate that the continents contain 30 to 60 percent of the uranium and barium which would have been present originally in a chondritic earth.[2,3] This might suggest that a comparable proportion of the entire mantle had passed through the proposed irreversible geochemical-petrological cycle.[4] This estimate can be supported by other considerations. The oceanic floors appear to be renewed on a time-scale of about 10^8 years. If the oceanic lithosphere is about 75-100 km thick and the process has occurred continuously for 3.5×10^9 years, then about half of the mantle may have passed through the irreversible differentiation cycle. Although these estimates are crude and possess large uncertainties, they suggest that the deep mantle has participated to a considerable extent in the process and that plates must descend considerably deeper than 700 km. It also suggests that *the material now in the low-velocity zone was originally derived from very deep in the mantle and that the chemical compositions of basalts and peridotites erupted over geological time provide information about the composition of the entire mantle.*

Dickinson and Luth (1971) developed a similar line of thought independently and quantitatively. In addition, they added some entirely new concepts.

[1]Both interpretations were presented at a session of the Pan Pacific Science Congress, August 1971.
[2]Birch (1958).
[3]Gast (1960).
[4]As depicted in Fig. 8-6 and discussed in Sec. 8-5.

Table 8-1 DATA AND ASSUMPTIONS RELATING TO THE MANTLE EVOLUTION MODELS OF DICKINSON AND LUTH (1971)

Mass of asthenosphere (Moho–650 km)	117×10^{25}g
Mass of mesosphere (650–2900 km)	292×10^{25}g
Areal rate of generation and consumption of lithosphere over last 10^8 years	2 km²/year
Most probable thickness of depleted, refractory portion of sinking lithosphere slab	50 km
Present rate of production of depleted refractory lithosphere	32.5×10^{16} g/year
Total mass of refractory, residual lithosphere produced over 4.5×10^9 years, assuming production rate proportional to rate of generation of radioactive heat in earth	280×10^{25}g
Ratios of heat-producing elements assumed above	K/U = 10^4 Th/U = 3.7
Ratio: $\frac{\text{heat generation rate 3.5 b.y. ago}}{\text{heat generation at present}}$	2.6
Ratio: $\frac{\text{heat generation rate 4.5 b.y. ago}}{\text{heat generation at present}}$	4.3

They divide the mantle into two regions—the *asthenosphere*, extending to a depth of 650 km and marking the minimum depth to which plates are now known to descend, as indicated by seismic activity, and the *mesosphere*, extending between 650 km and the core at 2900 km. The principal boundary conditions in their model are given in Table 8-1. The areal rate of generation of new lithosphere is estimated from an analysis of dated oceanic marine anomaly profiles over the last 10^8 years and is found to be 2.0 km²/year, in close agreement with similar estimates by other authors. Their best estimate of the thickness of the residual, refractory portion of the lithosphere plate is 50 km. They make the key assumption that past rates of lithosphere production have varied in proportion to the rates of production of radiogenic heat in the earth. This is justified on the basis that plate-tectonic motions constitute a convective system powered ultimately by radioactive heat sources. On these assumptions they calculate the mass of depleted residual lithosphere produced during the last 4.5 billion years to be 280×10^{25} g, which is almost identical with the mass of the mesosphere (Table 8-1). Dickinson and Luth accordingly suggest that the mesosphere indeed consists of residual, refractory lithosphere which has been built outward from the core boundary by the accumulation of successive increments of lithosphere added progressively to its outer surface. This would provide a simple explanation for the cessation of deep-focus earthquake activity at the boundary of the mesosphere. Moreover, it implies that 71% of the entire mantle has passed through the irreversible differentiation cycle previously outlined by Ringwood.

Dickinson and Luth also link their model with the generation of new crust, which constitutes an integral part of the overall differentiation process. Their model thus implies a greater rate of growth of crust in the past. In view of the apparent lack of crustal rocks older than about 3.7 billion years, it would be necessary to postulate a rapid generation of crust in the period between 3.7 and 2.5 billion years ago—possibly the plate-tectonic mechanisms were radically different during this interval owing to the higher rates of radiogenic heat production.

Dickinson and Luth reach an analogous conclusion to that drawn by Ringwood regarding one of the major implications of these hypotheses of mantle evolution. They state:

> *The most obvious implication of the model in this context is the inference that the composition of the present lithosphere is a key to the composition of the inner mantle. A second logical consequence of the proposed model is that the material of the asthenosphere, rather than the mesosphere most closely approximates a primitive composition.*

Discussion

Three sources of evidence have a vital bearing on the models discussed above: (1) the rates of lithosphere production and consumption through time, (2) the rates of generation of continental crust through time, and (3) the proportions of the total amounts of the earth's incompatible elements (e.g., U, Ba, K, Rb, La, and Ta) which are now incorporated in the continental crust.

Whilst the rate of lithosphere production is quite well known during the past 10^8 years, there is less evidence relating to earlier periods. The fact that orogenic-type igneous rocks have been produced on a large scale over the earth throughout geological time, extending far back into the Precambrian, constitutes strong evidence that plate-tectonic activity has been a major and quasi-continuous process since then. Dickinson and Luth's proposal that, since the energy for plate motions is ultimately supplied by radioactivity, then the average rate of plate-tectonic evolution may have been correspondingly higher in the distant past, is plausible though not necessarily compelling.

Nevertheless, it seems hardly likely that the generation and consumption of plates in the past would have been very much smaller on the average than observed over the last 10^8 years. This inference is based on an analysis of global palaeomagnetic data relating to the rates of continental movements extending into the Precambrian.[1]

The thickness of the depleted refractory layer of lithosphere,[2] estimated as 50 km by Dickinson and Luth, may be evaluated in the light of the earlier studies of basalt genesis (Chap. 4, Sec. 5-4, Fig. 5-1). The abyssal tholeiite layer of the oceanic crust was believed to represent a 17% partial melt of pyrolite, leaving a complementary layer of harzburgite some 25 km thick. Below the harzburgite is a layer of lherzolite resulting from a small degree of partial melting (1 to 5%) of pyrolite to form members of the alkali basalt suite (Fig. 5-1). The proportion of these basalts in the oceanic crust unfortunately is not known, resulting in a corre-

[1]McElhinny (1973).

[2]Not to be confused with the total thickness of the oceanic lithosphere overlying the low-velocity zone, which is probably closer to 70 to 80 km. The lower region of this presumably consists of pyrolite which is reabsorbed by the mantle during descent of plates.

sponding uncertainty in the thickness of the lherzolite layer. A thickness of about 10 km would probably not be unreasonable. Thus the total thickness of depleted lithosphere (including former oceanic crust) in the sinking slab may be about 40 km, which is in reasonable agreement with Dickinson and Luth's estimate. It appears unlikely that this thickness would be less than 30 km.

Estimates of the rate of generation of orogenic-type volcanic magmas[1,2] indicate that, if constant with time, the presently observed rate could account for the formation of the entire continental crust. It must be conceded that the possible errors in such estimates are large—perhaps it would be more appropriate to state that the present rate of volcanism is *consistent* with the model. Nevertheless, it should be remembered that these estimates do not take into account the enormous volumes of plutonic igneous rocks which have been intruded into the crust, both at accessible levels where they may be exposed by erosion and at deeper, inaccessible levels. An appropriate allowance for the rate of addition of primary mantle-derived magmas in this category would strengthen the above argument substantially and would not be inconsistent with a greatly increased rate of crustal generation early in the Precambrian. Armstrong (1968) has pointed out that many Precambrian shield regions appear to have evolved very rapidly between 3.5 and 2.5 billion years ago and to have had a rather simple geologic history since then, apparently unaffected by major vertical movements and crustal deformation. Armstrong argues that the thickness of the crust (i.e., depth to Moho) in these regions has not changed greatly during the last 2 billion years. Accordingly, the relative elevations of these continental regions with respect to ocean basins also have not changed, and the oceans must have been at about their present size early in the Precambrian. If it is accepted that oceans were ultimately derived from the earth's interior by degassing processes connected with magmatism and differentiation, these considerations would suggest that, between 3.5 and 2.5 billion years ago, continental evolution may have proceeded at a much greater rate than, on the average, has occurred since then. Another source of evidence is provided by geochronological studies of the accretionary model of continental evolution for North America, which led to the conclusion that the areal rate of growth has been roughly uniform during the past 2.75 billion years.[3,4]

Finally, we consider the degree to which incompatible elements (e.g., U, Th, Ba, K, Rb, and La) have been concentrated in the crust. Because of large ionic radii, these elements are unable to substitute readily in the principal mantle minerals and hence are partitioned very strongly into the liquid phase during partial melting processes. Accepting the evidence that the crust is produced ultimately by magmatic processes of mantle origin, the total amount of these elements now in the crust is related to the amount of mantle whch has been differentiated and also to the concentration of these elements in the participating mantle source region.

[1]Wilson (1954).
[2]Markhinin (1968).
[3]Hurley et al. (1962).
[4]Engel (1963).

Gast (1972) has reviewed the crustal abundances of U, Ba, K, and Rb using a synthesis of petrologic, geochemical, and isotopic data. His most conservative model (C) leading to the lowest crustal abundances of these elements assumed a continental crust composed of 50% granodiorite and 50% diorite-granulite, together with an oceanic crust of abyssal (oceanic) tholeiite. If the original average abundances of *involatile* elements (U, Th, Ba, and Sr)[1] in the earth were those found in ordinary chondrites, Gast demonstrates that 46% of the total uranium, 30% of the barium, 63% of the potassium, and 67% of rubidium are now in the crust. The extreme degree of concentration of these elements in the crust (amounting to 0.44% of the earth's mass) implies the operation of an extremely efficient process of differentiation of the mantle. If these elements were originally uniformly distributed throughout the mantle, then some 30 to 70% of the mantle must have become irreversibly differentiated.

Actually, Gast's figures require some modification, but not enough to shake the qualitative conclusion given above. As was discussed in Sec. 5-3, there are grounds for preferring an earth model derived from the abundances of involatile elements in Type 1 carbonaceous chondrites, rather than in ordinary chondrites. This would have the effect of reducing Gast's estimates (above) by a factor of about 0.7. On the other hand, some increase in this factor would result from an allowance for the amounts of these elements which remain behind in the mantle during the partial melting processes which produced the crust and by the adoption of less conservative crustal abundance models.[2,3]

In Sec. 5-3 it was demonstrated that a mantle model derived from the abundances of involatile elements in Type 1 carbonaceous chondrites provided a satisfactory explanation of the geochemistry of basaltic magmas and their source regions. More especially, there was no need to assume any enrichment of Ca, Al, Sr, Ba, and U relative to Mg and Si in the source regions of basaltic magmas as suggested elsewhere by Gast (1972). These considerations indicate an approximately uniform distribution of incompatible elements throughout the primitive mantle, except in those regions which have suffered irreversible differentiation to form basaltic magmas and crustal rocks.

On the basis of the earth model developed in Sec. 5-3 and the above discussion, at least 30 to 50% of the mantle must have been irreversibly differentiated in order to account for the crustal abundances of U, K, and Rb and, probably, of Ba, La, Ta, and other incompatible elements.[4]

[1]Gast's primary argument is based upon a comparison of abundances of involatile elements between chondrites and the earth. The abundances of volatile elements (e.g., K and Rb) cannot be directly compared because of probable loss of these elements by the earth during or before its formation (Sec. 5-3). However, the depletions of Rb and K relative to involatile elements can be estimated quite closely by secondary isotopic and geochemical arguments (Gast, 1972) so that the total abundances of Rb and K in the earth can be derived, making possible an estimate of their degrees of concentration in the crust.

[2]E.g., models A and B of Gast (1972).

[3]Higher crustal abundances of incompatible elements than those given in Gast's model C would result if the composition of oceanic sediments and alkali basalts in the oceanic crust, together with amphibolitic-facies rocks in the lower continental crust, had been incorporated.

[4]Taylor (1964, 1967).

Conclusion

The combined weight of the evidence discussed previously strongly indicates that a large proportion of the earth's mantle has become irreversibly differentiated throughout geological time. The hypothesis[1] that the entire mesophere, amounting to 71% of the mass of the mantle, is composed of differentiated lithosphere is consistent with the evidence cited and provides a neat explanation of absence of deep-focus-earthquake activity below 700 km. Nevertheless, uncertainties in the data and assumptions relating to this interesting hypothesis are quite large, and a considerable body of further evidence will be necessary before it can be finally accepted.

We turn now to the less demanding hypothesis that a smaller proportion, some 30 to 50% of the mantle, has become irreversibly differentiated according to the mechanism shown in Fig. 8-6. This implies that the sinking lithosphere plates must descend considerably deeper than 650 km and that the deep mantle has participated to a considerable extent in the differentiation process.[2]

This hypothesis will be evaluated by considering the implications of its converse—that the sinking lithosphere plates descend only to 650 km,[3] where they are stopped on the major seismic discontinuity at this depth.[4] This would occur, for example, if the mantle beneath 650 km had a higher Fe-Mg ratio than the upper mantle,[5] or if the phase transformation at 650 km possessed a negative slope[4] dP/dT. In this case, accretion of differentiated lithosphere would have started at 650 km, gradually displacing the overlying layers upward. Thus the thickness of differentiated lithosphere lying on top of the 650-km discontinuity would grow with time, and the thickness of the overlying pyrolite layer would correspondingly decrease as it is displaced upward and further depleted by differentiation beneath oceanic ridges and above Benioff zones.

The reservoir of primitive pyrolite probably extends to at least 200 km. The highly alkalic magma series olivine nephelinite-olivine melilite nephelinite-olivine melilitite-kimberlite appear to have formed by extremely small degrees of partial melting of primary unfractionated pyrolite,[6] and the mineralogy of diamond-bearing kimberlite implies an origin at depths of at least 200 km.[7] Moreover, the concept of formation of normal basaltic magmas by partial melting during diapiric rise of magmas from the low-velocity zone implies the widespread existence of pyrolite at depths of 150 to 200 km.

A sketch of the structure of the mantle resulting from upward accretion of lithosphere upon the 650-km discontinuity is shown in Fig. 8-7. The total mass of differentiated lithosphere produced by sinking along Benioff zones and accumula-

[1]Dickinson and Luth (1971).
[2]Ringwood (1971, 1972).
[3]Rarely to 700 km.
[4]Isacks and Molnar (1971).
[5]Anderson and Jordan (1970).
[6]Chapter 4.
[7]Sobolev (1972).

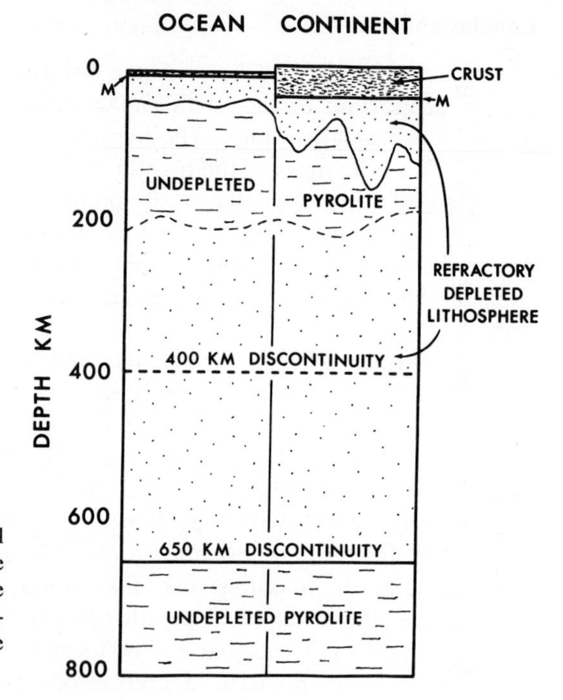

FIGURE 8-7
Structure of mantle based upon proposal that differentiated lithosphere is unable to sink beneath 650 km and therefore accretes outwards from this depth, displacing primitive unfractionated pyrolite upwards.

tion in the depth interval 200 to 650 km is 84×10^{25} g. If the rate of production of differentiated lithosphere (2 km^2/year, 40 km thick) had been constant for 3.7×10^9 years, the total mass produced would be 96×10^{25} g; in 4.5×10^9 years the corresponding mass would be 117×10^{25} g. These amounts exceed the mass of the zone between 200 and 650 km. If the rate of differentiation had been greater in the Precambrian, as seems likely, the discrepancy is increased.

We turn now to the arguments concerning the concentration of incompatible elements in the crust. It was concluded that these implied the occurrence of irreversible differentiation affecting at least 30 to 50% of the mantle. On the basis of Fig. 8-7, the depleted lithosphere occurring in the mantle above 650 km[1] amounts to 94×10^{25} g, or 23% of the mass of the mantle. If the degree of concentration of some incompatible elements in the crust is indeed a measure of the degree of differentiation of the mantle, then a discrepancy exists. We must conclude that either the amount of irreversibly differentiated mantle exceeds that which could be present in the outer 650 km or the abundances of involatile, incompatible elements in the upper mantle (relative to Si and Mg) were originally substantially higher than in Type 1 carbonaceous chondrites. This latter alternative was discussed and rejected in Sec. 5-3.

[1]Including the depleted zone immediately underlying the continental and oceanic crust ($\sim 10 \times 10^{25}$ g).

More detailed studies of the abundances of incompatible elements and the rates of crustal evolution in the past are needed before the above conclusions can be regarded as firmly established. Nevertheless, the evidence currently available is strongly suggestive. The implication that lithosphere plates descend much deeper than 700 km, displacing unfractionated pyrolite upward, is of considerable importance. The pyrolite now present in the low-velocity zone must ultimately have been derived from great depths in the mantle. Thus, the pyrolite-model composition derived in Chap. 5 is probably representative of the original composition of most or all of the mantle.[1,2] The fact that the composition range of basalts and alpine peridotites have been essentially unchanged from the early Precambrian to the present provides a similar indication of the major element homogeneity throughout the mantle source regions over this period.

The capacity for the pyrolite model to explain the observed physical properties of the mantle has been discussed extensively elsewhere[3] and is considered in Chap. 14. It is concluded that the model provides a self-consistent explanation of the elastic properties, seismic velocities, and density distribution throughout the entire mantle. In particular, an increase of FeO/MgO ratio at the 650-km discontinuity which would prevent subsidence of lithosphere below this level is not required. Moreover, the major phase transition occurring near 650 km probably has a positive slope which would facilitate transfer of lithosphere across the boundary.[4]

Finally, we consider possible reasons for the paucity or absence of continental crustal rocks more than 3.7 billion years old, suggesting that the processes of crustal evolution discussed in this chapter were not operative before then. Reasons which can be suggested for this nonuniformitarian behaviour are necessarily highly speculative.[5] It may well be significant that the formation of continental orogenic-type rocks appears to require a two-stage process. It is conceivable that, during the first billion years of earth history, extensive basaltic magmatism occurred over the earth, leading to the formation of a primitive basaltic crust. Because of high thermal gradients, the primitive basaltic crust was prevented from transforming to eclogite and thus was unable to sink into the mantle, thereby preventing operation of the second stage of partial melting required for the formation of orogenic-type magmas. A second speculation concerns the inference that the luminosity of the sun has increased by a factor of about 1.6 over the past 4.5 billion years.[6] Ringwood (1961) noted that, if this had occurred, the surface temperature over most of the earth before 3.5 billion years ago may have been below 0°C so that the normal geological cycle operating by erosional processes involving water would not have occurred. Moreover, the primitive oceans may

[1]Ringwood (1971, 1972).
[2]Dickinson and Luth (1971).
[3]Ringwood (1970).
[4]Chapter 14.
[5]Ringwood (1969).
[6]Schwarzschild (1958).

have been frozen. This would largely prevent the hydration of basaltic rocks at the earth's surface. We have previously discussed the key role that water, carried into the mantle by hydrated rocks in the sinking column beneath trenches, plays in the evolution of orogenic-type magmas. It is conceivable that during the first billion years there was no way of causing water to re-enter the mantle in sufficient concentration to cause the formation of these magmas.

REFERENCES

ANDERSON, D. L., and T. JORDAN (1970). The composition of the lower mantle. *Phys. Earth Planet. Interiors* **3,** 23–35.

ARMSTRONG, R. L. (1968). A model for the evolution of strontium and lead isotopes in a dynamic earth. *Rev. Geophys.* **6,** 175–199.

ARRIENS, P. A. (1971). The Archaean geochronology of Australia. *Geol. Soc. Australia Spec. Pub.* **3,** 11–23.

BARAGAR, W. R., and A. M. GOODWIN (1969). Andesites and Archaean volcanism on the Canadian Shield. In: A. R. McBirney (ed.), "*Proceedings of the Andesite Conference*," pp. 121–142. Bull. 65, Dept. Geology and Mineral Resources, State of Oregon.

BARAZANGI, M., and B. ISACKS (1971). Lateral variations of seismic-wave attenuation in the upper mantle above the inclined earthquake zone of the Tonga island arc: Deep anomaly in the upper mantle. *J. Geophys. Res.* **76,** 8493–8516.

———, ———, and J. OLIVER (1972). Propagation of seismic waves through and beneath the lithosphere that descends under the Tonga island arc. *J. Geophys. Res.* **77,** 952–958.

BENIOFF, H. (1955). Seismic evidence for crustal structure and tectonic activity. In: A. Poldervaart (ed.), "*The Crust of the Earth*," pp. 61–73. *Geol. Soc. Am. Spec. Paper* 62.

BIRCH, F. (1958). Differentiation of the mantle. *Bull. Geol. Soc. Am.* **69,** 483–486.

CLARK, S. P., and A. E. RINGWOOD (1964). Density distribution and constitution of the mantle. *Rev. Geophys.* **2,** 35–88.

CLEARY, J. (1967). Azimuthal variation of the Longshot source term. *Earth Planet. Sci. Letters* **3,** 29–37.

COATS, R. R. (1962). Magma type and crustal structure in the Aleutian arc. In: "*Crust of the Pacific Basin*," pp. 92–109. *Am. Geophys. Union, Geophys. Monograph* 6.

DEWEY, J. F. (1969). Continental margins: A model for the conversion of Atlantic type to Andean type. *Earth Planet. Sci. Letters* **6,** 189–197.

——— and J. M. BIRD (1970). Mountain belts and the New Global Tectonics. *J. Geophys. Res.* **75,** 2625–2647.

DICKINSON, W. R. (1969). Evolution of calcalkaline rocks in the geosynclinal system of California and Oregon. In: A.R. McBirney (ed.), "*Proceedings of the Andesite Conference*," pp. 151–156. Bull. 65, Dept. Geology and Mineral Resources, State of Oregon.

——— and T. HATHERTON (1967). Andesitic volcanism and seismicity around the Pacific. *Science* **157,** 801–803.

——— and W. C. LUTH (1971). A model for plate tectonic evolution of mantle layers. *Science* **174,** 400–404.

DRAKE, C. L., M. EWING, and G. H. SUTTON (1959). Continental margins and geosynclines: The east coast of North America north of Cape Hatteras. In: L.H. Ahrens, F. Press, K. Rankama, and S. K. Runcorn (eds.), "*Physics and Chemistry of the Earth,*" vol. 2, pp. 110-198. London.

ENGEL, A. E. J. (1963). Geologic evolution of North America. *Science* **140,** 143–152.

ESSENE, E. J., B. HENSEN, and D. H. GREEN (1970). Experimental study of amphibolite and eclogite stability. *Phys. Earth Planet. Interiors* **3,** 378–384.

FEDOTOV, S. A. (1963). The absorbtion of transverse seismic waves in the upper mantle and energy classification of near earthquakes of intermediate focal depth. *Izv. Acad. Sci. USSR, Geophys. Ser.* (Engl. Transl.) No. 6, 509–520.

——— (1968). On deep structure, properties of the upper mantle and volcanism of the Kuril-Kamchatka island arc according to seismic data. In: "*The Crust and Upper Mantle of the Pacific Area,*" pp. 94–111. *Am. Geophys. Union, Geophys. Monograph* 12.

FITTON, J. G. (1971). The generation of magmas in island arcs. *Earth Planet. Sci. Letters* **11,** 63–67.

FURUMOTO, A. S., G. P. WOOLLARD, J. F. CAMPBELL, and D. M. HUSSONG (1968). Variation in the thickness of the crust in the Hawaiian Archipelago. In: "*The Crust and Upper Mantle of the Pacific Area,*" pp. 94–111. *Am. Geophys. Union, Geophys. Monograph* 12.

GAST, P. W. (1960). Limitations on the composition of the upper mantle. *J. Geophys. Res.* **65,** 1287–1297.

——— (1968). Trace element fractionation and the origin of tholeiitic and alkaline magma types. *Geochim. Cosmochim. Acta* **32,** 1057–1086.

——— (1972). The chemical composition of the earth, the moon and chondritic meteorites. In: E. C. Robertson (ed.), "*The Nature of the Solid Earth,*" pp. 19–40. McGraw-Hill, New York.

GORSHKOV, G. S. (1962). Petrochemical features of volcanism in relation of the types of the earth's crust. In: "*Crust of the Pacific Basin,*" pp. 110–115. *Am. Geophys. Union, Geophys. Monograph* 6.

GREEN, D. H. (1971). Composition of basaltic magmas as indicators of conditions of origin: application to basaltic volcanism. *Phil. Trans. Roy. Soc. London* **A268,** 707–725.

——— (1972). Magmatic activity as the major process in chemical evolution of the earth's crust and upper mantle. *Tectonophysics* **13,** 47–71.

GREEN, T. H. (1972). Crystallization of calcalkaline andesite under controlled high pressure hydrous conditions. *Contr. Mineral. Petrol.* **34,** 150–166.

——— and A. E. RINGWOOD (1966). Origin of the calcalkaline igneous rock suite. *Earth Planet. Sci. Letters* **1,** 307–316.

——— and ——— (1967). Crystallization of basalt and andesite under high pressure hydrous conditions. *Earth Planet. Sci. Letters* **3,** 481–489.

——— and ——— (1968). Genesis of the calcalkaline igneous rock suite. *Contr. Mineral. Petrol.* **18,** 105–162.

——— and ——— (1969). High pressure experimental studies on the origin of andesite. In: A.R. McBirney (ed.), "*Proceedings of the Andesite Conference,*" pp. 21–32. Bull. 65, Dept. Geology and Mineral Resources, State of Oregon.

——— and ——— (1972). Crystallization of garnet-bearing rhyodacite under high pressure hydrous conditions. *J. Geol. Soc. Aust.* **19,** 203–212.

GRIGGS, D. T. (1972). The sinking lithosphere and the focal mechanism of deep earthquakes. In: E. C. Robertson (ed.), "*The Nature of the Solid Earth,*" pp. 361–384. McGraw-Hill, New York.

GUTENBERG, B., and C. F. RICHTER (1941). Seismicity of the earth. *Geol. Soc. Am. Spec. Paper* 34, 131.

——— and ——— (1954). "*Seismicity of the Earth and Associated Phenomena.*" Princeton Univ. Press, Princeton, N. J. 310 pp.

HAMILTON, W. (1969). The volcanic central Andes—a modern model for the Cretaceous batholiths and tectonics of western North America. In: A. R. McBirney (ed.), "*Proceedings of the Andesite Conference,*" pp. 175–184. Bull. 65, Dept. Geology and Mineral Resources, State of Oregon.

HILLS, E. S. (1953). Tectonic setting of Australian ore deposits. In: A. B. Edwards (ed.), "*Geology of Australian Ore Deposits,*" chap. 2, pp. 41–61. Aust. Inst. Mining Met. Inc., Melbourne.

HURLEY, P. M. (1968). Absolute abundance and distribution of Rb, K, and Sr in the Earth. *Geochim. Cosmochim. Acta* **32,** 273–283.

———, P. C. BATEMAN, G. W. FAIRBAIRN, and W. H. PINSON (1965). Investigation of initial Sr^{87}/Sr^{86} ratios in the Sierra Nevada plutonic province. *Bull. Geol. Soc. Am.* **76,** 165–173.

———, H. HUGHES, G. FAURE, H. FAIRBAIRN, and W. H. PINSON (1962). Radiogenic strontium-87 model of continent formation. *J. Geophys. Res.* **67,** 5315–5336.

ISACKS, B., and P. MOLNAR (1969). Mantle earthquake mechanisms and the sinking of the lithosphere. *Nature* **223,** 1121–1124.

——— and ——— (1971). Distribution of stresses in the descending lithosphere from a global survey of focal-mechanism solutions of mantle earthquakes. *Rev. Geophys. Space Phys.* **9,** 103–174.

———, J. OLIVER, and L. SYKES (1968). Seismology and the new global tectonics. *J. Geophys. Res.* **73,** 5855–5899.

JAKEŠ, P., and J. B. GILL (1970). Rare earth elements and the island arc tholeiite series. *Earth Planet. Sci. Letters* **9,** 17–28.

KANAMORI, H., and F. PRESS (1970). How thick is the lithosphere? *Nature* **226,** 330–331.

KARIG, D. (1971). Origin and development of marginal basins in the western Pacific. *J. Geophys. Res.* **76,** 2542–2561.

KITAHARA, S., S. TAKENOUCHI, and G. C. KENNEDY (1966). Phase relations in the system $MgO-SiO_2-H_2O$ at high temperatures and pressures. *Am. J. Sci.* **264,** 223–233.

KUNO, H. (1959). Origin of Cenozoic petrographic provinces of Japan and surrounding areas. *Bull. Volc.* Ser. II, **20,** 37–76.

MARKHININ, E. K. (1968). Volcanism as an agent of formation of the Earth's crust. In: "*The Crust and Mantle of the Pacific Area,*" pp. 413–422. *Am. Geophys. Union, Geophys. Monograph* 12.

MCBIRNEY, A. R. (1969). Compositional variations in Cenozoic calcalkaline suites of Central America. In: A. R. McBirney (ed.), "*Proceedings of the Andesite Conference,*" pp. 185–189. Bull. 65, Dept. Geology and Mineral Resources, State of Oregon.

MCELHINNY, W. (1973). "*Paleomagnetism and Plate Tectonics,*" Cambridge Univ. Press, London. 368 pp.

MCKENZIE, D. P. (1969). Speculations on the consequences and causes of plate motions. *Geophys. J.* **18,** 1–32.

MITRONOVAS, W., and B. L. ISACKS (1971). Seismic velocity anomalies in the upper mantle beneath the Tonga-Kermadec island arc. *J. Geophys. Res.* **76,** 7154–7180.

MOLNAR, P., and J. OLIVER (1969). Lateral variations of attenuation in the upper mantle and discontinuities in the lithosphere. *J. Geophys. Res.* **74,** 2648–2682.

MORGAN, W. J. (1971). Convection plumes in the lower mantle. *Nature* **230,** 42–43.

NICHOLLS, I. A., and A. E. RINGWOOD (1973). Effect of water on olivine stability in tholeiites and the production of SiO_2-saturated magmas in the island arc environment. *J. Geol.* **81,** 285–300.

OLIVER, J., and B. ISACKS (1967). Deep earthquake zones, anomalous structures in the upper mantle and the lithosphere. *J. Geophys. Res.* **72,** 4259–4275.

OXBURGH, E. R., and D. L. TURCOTTE (1968a). Problems of high heat flow and volcanism associated with zones of descending mantle convective flow. *Nature* **218,** 1041.

——— and ——— (1968b). Mid-ocean ridges and geotherm distribution during mantle convection. *J. Geophys. Res.* **73,** 2643–2661.

——— and ——— (1970). Thermal structure of island arcs. *Bull. Geol. Soc. Am.* **81,** 1665–1688.

RALEIGH, C. B., and W. H. LEE (1969). Sea-floor spreading and island arc tectonics. In: A. R. McBirney (ed.), "*Proceedings of the Andesite Conference,*" pp. 99–110. Bull. 65, Dept. Geology and Mineral Resources, State of Oregon.

RAMBERG, H. (1966). The Scandinavian Caledonides as studied by centrifuged dynamic models. *Bull. Geol. Inst. Univ. Uppsala* **43,** 1–72.

——— (1967). "*Gravity, Deformation and the Earth's Crust as Studied by Centrifuged Models.*" Academic, London. 214 pp.

——— (1970). Model studies in relation to intrusions of plutonic bodies. In: G. Newall and N. Rast (eds.), "*Mechanism of Igneous Intrusion,*" pp. 261–286. Galley Press, Liverpool.

——— (1972). Mantle diapirism and its tectonic and magmagenetic consequences. *Phys. Earth Planet. Interiors* **5,** 45–60.

RICHARDS, J. R., and J. EVERNDEN (1962). Potassium-argon ages in Eastern Australia. *J. Geol. Soc. Aust.* **9,** 1–49.

RINGWOOD, A. E. (1961). Changes in solar luminosity and some possible terrestrial consequences. *Geochim, Cosmochim, Acta* **21,** 295–296.

——— (1966). Discussion of paper by Sugimura. In: "*Continental Margins and Island Arcs,*" p. 346. *Geol. Surv. Canada, Paper* 66–15.

——— (1969). Composition and evolution of the upper mantle. In: "*The Earth's Crust and Upper Mantle,*" pp. 1–17. *Am. Geophys. Union Geophys. Monograph* 13.

——— (1970). Phase transformations and the constitution of the mantle. *Phys. Earth Planet. Interiors* **3,** 109–155.

——— (1971). Phase transformations and mantle dynamics. Publication 999, pp. 1–22, Dept. Geophysics and Geochemistry, Australian National University. Also presented at Pan-Pacific Science Congress, Canberra, August 1971.

——— (1972). Phase transformations and mantle dynamics. *Earth Planet. Sci. Letters* **14,** 233–241.

——— (1974). Petrological evolution of island arc systems. *J. Geol. Soc. Lond.* **130,** 183–204.

——— and D. H. GREEN (1966). An experimental investigation of the gabbro-eclogite transformation and some geophysical implications. *Tectonophysics* **3,** 383–427.

——— and A. MAJOR (1966). Synthesis of Mg_2SiO_4-Fe_2SiO_4 spinel solid solutions. *Earth Planet. Sci. Letters* **1,** 241–245.

——— and ——— (1967). High pressure reconnaissance investigations in the system Mg_2SiO_4-MgO-H_2O. *Earth Planet. Sci. Letters* **2,** 130–133.

RUBEY, W. W. (1951). Geologic history of sea water. *Bull. Geol. Soc. Am.* **62,** 1111–1147.

——— (1955). Development of the hydrosphere and atmosphere with special reference to the probable composition of the early atmosphere. In: A. Poldervaart (ed.), "*Crust of the Earth,*" pp. 631–650. *Geol. Soc. Am. Spec. Paper* 62.

SCHWARZSCHILD, M. (1958). "*Structure and Evolution of the Stars,*" p. 206. Princeton Univ. Press, Princeton, N. J.

SCLAR, C. B., L. C. CARRISON, and O. M. STEWART (1967). High pressure synthesis of a new hydroxylated pyroxene in the system MgO-SiO_2-H_2O. (Abstract.) *Trans. Am. Geophys. Union* **48,** 226.

SCLATER, J. G., and H. W. MENARD (1967). Topography and heat flow of the Fiji plateau. *Nature* **216,** 991.

SOBOLEV, N. V. (1972). Deep-seated inclusions in kimberlites and the problem of the composition of the earth's mantle. IGEM, Acad. Sci. USSR. Extended abstract, pp. 1–38. (Translated by D. A. Brown.)

SORGENFREI, T. (1971). On the granite problem and the similarity of salt and granite structures. *Geol. För. Stockholm. Forh.* **93**(2), 371–435.

STILLE, H. W. (1955). Recent deformations of the Earth's crust in the light of those of earlier epochs. In: A. Poldervaart (ed.), "*Crust of the Earth,*" pp. 171–192. *Geol. Soc. Am. Spec. Paper* 62.

SYKES, L. R. (1966). The seismicity and deep structure of island arcs. *J. Geophys. Res.* **71,** 2981–3006.

TAYLOR, S. R. (1964). Trace element abundances and the chondritic earth model. *Geochim. Cosmochim. Acta* **28,** 1989–1999.

——— (1967). The origin and growth of continents. *Tectonophysics* **4,** 17–34.

TOKSÖZ, M. N., J. W. MINEAR, and B. R. JULIAN (1971). Temperature field and geophysical effects of a downgoing slab. *J. Geophys. Res.* **76,** 1113–1138.

UYEDA, S., and V. VACQUIER (1968). Geothermal and geomagnetic data in and around the island arc of Japan. In: "*The Crust and Upper Mantle of the Pacific Area,*" pp. 349–366. *Am. Geophys. Union, Geophys. Monograph* 12.

WILSON, J. T. (1954). The development and structure of the crust. In: G. P. Kuiper (ed.), "*The Earth as a Planet,*" pp. 138–214. Univ. of Chicago Press, Chicago.

——— (1965). Submarine fracture zones, aseismic ridges and the ICSU line: Proposed western margin of the East Pacific ridge. *Nature* **207,** 907–911.

WYLLIE, P. J. (1971). Role of water in magma generation and initiation of diapiric uprise in the mantle. *J. Geophys. Res.* **76,** 1328–1338.

YAMAMOTO, K., and S. AKIMOTO (1974). High pressure and high temperature investigations in the system MgO-SiO_2-H_2O. *J. Solid State Chem.* **9,** 187–195.

——— and ——— (1975). High pressure and high temperature investigations of the phase diagram in the system MgO-SiO_2-H_2O. (Preprint)

PART TWO

The Deep Mantle

9
SOME INFERENCES FROM GEOPHYSICS

9-1 INTRODUCTION

The most important source of information on the physical properties of the mantle is provided by seismology and, in particular, by the variations of *P*- and *S*-wave velocities with depth. Until quite recently, nearly all discussions of the physics of the earth's interior were based ultimately upon the Jeffreys[1] and Gutenberg[2] depth-velocity curves (Fig. 9-1). As we saw in the introduction, these provided the primary justification for the subdivision of the mantle into three well-defined regions which have proved to be of a fundamental nature. It is reasonable to expect that a number of other physical properties will display overall patterns of variation in the mantle analogous to the seismic velocity variations. An ultimate explanation of the variation of gross physical properties with depth in the mantle requires an understanding of the nature of the individual mineral phases present, their properties, and the thermodynamic and crystal-chemical principles which govern their stabilities. Part Two will be mainly concerned with a review of

[1]Jeffreys (1937, 1939, 1959).
[2]Gutenberg (1958, 1959a, 1959b).

recent advances in this field, as applied to the constitution of the Transition Zone and Lower Mantle. It is sometimes convenient to refer to both these regions as the *Deep Mantle*.

The seismic velocity distributions of Jeffreys and Gutenberg were based upon first arrivals and upon smoothed travel-time data and, accordingly, are expected to vary more smoothly with depth than the real velocities in the mantle. Subsequent investigations[1,2,3,4,5,6,7,8,9] using seismic arrays and later arrivals have shown that this is indeed the case, particularly in the transition zone. These investigations show that there are major increases of velocity around 400 and 650 km (Fig. 9-1). Between these depths, the average velocity gradients appear to be much smaller than indicated by Jeffreys and Gutenberg. In the Lower Mantle, below 800 km, the revised velocity distributions are close, on the average, to those of Jeffreys and Gutenberg, but there is growing evidence of the presence of minor discontinuities and lateral inhomogeneities.[2,3,7,10]

Recent investigations suggest that the model possessing only two major discontinuities near 400 and 650 km, as shown in Fig. 9-1, may also prove to be an oversimplification. The 400-km discontinuity appears to have a complex structure, and there may be more than one discontinuity between 300 and 450 km.[11,12,13] There is now evidence for a small discontinuity around 550 km.[13,14,15] Reflection investigations[14,16,17] have shown that the major seismic discontinuity near 650 km is much sharper than those near 400 km, and this has been confirmed by a high-resolution array study.[13] Although one sharp and major discontinuity has been established near 650 km, it seems possible, even likely,[18] that future higher resolution seismic studies may also show this structure to be more complex than indicated in Fig. 9-1. These recent investigations suggest that the real seismic velocity distribution in the mantle may be somewhat intermediate between the Jeffreys-Gutenburg and Johnson models shown in Fig. 9-1—doubtless closer to the latter but with a more complex structure.

[1]Niazi and Anderson (1965).
[2]Chinnery and Toksöz (1967).
[3]Johnson (1967, 1969).
[4]Kanamori (1967).
[5]Hales et al. (1968).
[6]Julian and Anderson (1968).
[7]Archambeau et al. (1969).
[8]Nuttli (1969).
[9]Hales and Roberts (1970).
[10]Wright and Cleary (1972).
[11]Bolt et al. (1968).
[12]Bolt (1969).
[13]Simpson, Mereu, and King (1974).
[14]Whitcomb and Anderson (1970).
[15]Helmberger and Wiggins (1971).
[16]Engdahl and Flinn (1969).
[17]Adams (1971).
[18]Ringwood (1970).

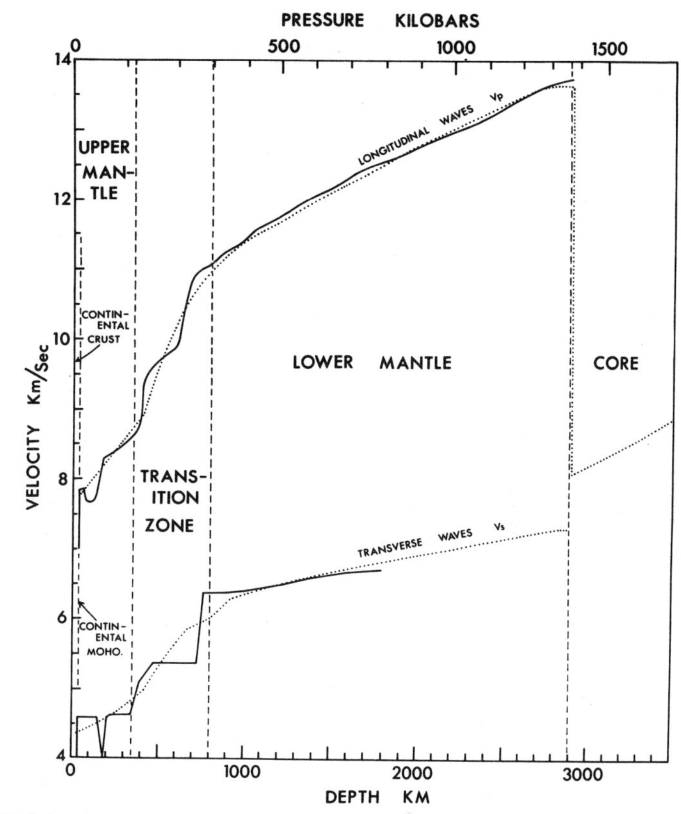

FIGURE 9-1
Seismic velocity distributions in the mantle. *P* waves—solid lines: Johnson (1967, 1969). *S* waves—solid line: Nuttli (1969). Broken line: Jeffreys (1937, 1939).

The outstanding features of the Jeffreys-Gutenberg mantle-velocity distributions are the high average gradients between 400 and 900 km and the region of uniform, relatively low gradients between 900 and 2900 km. Considered as *average* features, these may be considered to be firmly established, although as noted above, recent studies show that the 400 to 900 km region contains at least two major discontinuities separated by a region of low velocity gradient.

Most of the classical arguments concerned with the constitution of the mantle have been based upon the Jeffreys-Gutenberg curves. Some of these arguments—particularly, those of Bullen and Birch bearing upon the homogeneity of the mantle—were based particularly upon the interpretation of velocity gradients. Despite subsequent revisions to the velocity distributions, these arguments have not lost their essential validity. Although some of the conclusions reached by these earlier workers can now be shown to follow more directly from the newly established existence of seismic velocity discontinuities within the mantle and from the interpretation of free-oscillation data, it is worthwhile to follow some of the arguments of Bullen and Birch, both from the viewpoint of historical perspective and because of the lasting importance of the conclusions reached.

9-2 DENSITY DISTRIBUTION AND INHOMOGENEITY IN THE DEEP MANTLE

The density distribution in the upper mantle has been discussed in Chap. 3. In this section, we will consider the density distribution in the transition zone and lower mantle and the bearing of this information upon the homogeneity of the mantle. One of the most elegant arguments in geophysics deals with this subject and is due to Bullen (1936, 1940). Because of its fundamental importance to all considerations of the constitution of the mantle, it is given here at some length.

The velocities of longitudinal waves V_p and transverse waves V_s are given by

$$V_p^2 = \frac{K_s + \frac{4}{3}\mu}{\rho} \tag{1}$$

$$V_s^2 = \frac{\mu}{\rho} \tag{2}$$

where

$$K_s = \text{adiabatic bulk modulus} = \rho \left(\frac{\partial P}{\partial \rho}\right)_s \tag{3}$$

and μ = rigidity, ρ = density, P = pressure.

From (1), (2), and (3) we have

$$V_p^2 - \tfrac{4}{3}V_s^2 = \frac{K_s}{\rho} = \left(\frac{\partial P}{\partial \rho}\right)_s = \phi \tag{4}$$

where ϕ is defined as the elastic ratio (Birch, 1952).

Consider the variation of density (ρ) with depth (r) in a homogeneous, adiabatic region of the earth:

$$\frac{d\rho}{dr} = \left(\frac{\partial \rho}{\partial P}\right)_s \frac{dP}{dr} \tag{5}$$

From the hydrostatic relation,

$$\frac{dP}{dr} = -g\rho \qquad \text{where } g = \frac{Gm}{r^2} \tag{6}$$

Substituting in (5) from (4) and (6),

$$\frac{d\rho}{dr} = -\frac{g\rho}{\phi} = -\frac{Gm\rho}{\phi r^2} \tag{7}$$

Equation (7) is known as the *Williamson-Adams equation*, after those who first derived it.[1] This gives the relation between density and depth arising from self-compression in a homogeneous spherical shell in which the seismic velocities V_p and V_s are known. In the case under consideration, the initial value of m is the mass of the earth.

Bullen (1936, 1940) integrated equation (7) to obtain a density distribution throughout the mantle, taking the mean density ρ_0 of the mantle immediately beneath the Mohorovicic discontinuity as 3.32 g/cm^3 and making an appropriate small allowance for the mass of the overlying crust. From the resultant density distribution, Bullen calculated the moment of inertia of the mantle, I_M. Since the moment of inertia of the whole earth, I_E, was known, the moment of inertia of the core, I_C, could be obtained by subtraction. This was found to be 0.57 Mr^2. Now, the moment of inertia of a sphere of uniform density is 0.40 Mr^2 and that of a spherical shell is 0.67 Mr^2. The above procedure therefore implied that the density of the core decreased strongly from its outer margin to the centre. This was clearly an impossible situation for a liquid sphere (outer core) under self-compression. From this fundamental contradiction, it followed that one of the assumptions made in applying equation (7) to the mantle was wrong. These assumptions were those of adiabaticity, initial density, and chemical homogeneity.

It is certain that in the outer mantle the actual temperature gradient substantially exceeds the adiabatic. The appropriate correction[2] to equation (7) for a superadiabatic temperature gradient τ is

$$\frac{d\rho}{dr} = -\frac{g\rho}{\phi}\left(1 - \frac{\alpha\phi\tau}{g}\right) \tag{8}$$

where α is the coefficient of volume thermal expansion. The superadiabatic temperature gradient decreases the density gradient, resulting in a larger value of I_C. The net effect therefore is to compound the contradiction noted above.

Bullen chose 3.32 g/cm^3 as the initial density of the uppermost mantle because it corresponded to that of ultramafic rocks which were widely believed to predominate in the mantle. This choice can be strongly defended, as was argued in Chap. 3. However, earth models with higher upper mantle densities have been proposed.[3,4] These would tend to decrease the extent of the contradiction regarding I_C. Figure 9-2 shows the relationship between the ratio I/Mr^2 for the core and the initial density ρ_0 at the top of the mantle. It is seen that ρ_0 would need to be 3.7 in order to yield a core of uniform density. Allowance for self-compression in the real core, together with a superadiabatic temperature gradient in the mantle, would increase the required value of ρ_0 to about 3.8 g/cm^3. Many lines of evidence[4] can be called upon to show that this is an unacceptably high value.

[1]Williamson and Adams (1923).
[2]Birch (1952).
[3]Birch (1961b).
[4]See also Chap. 3.

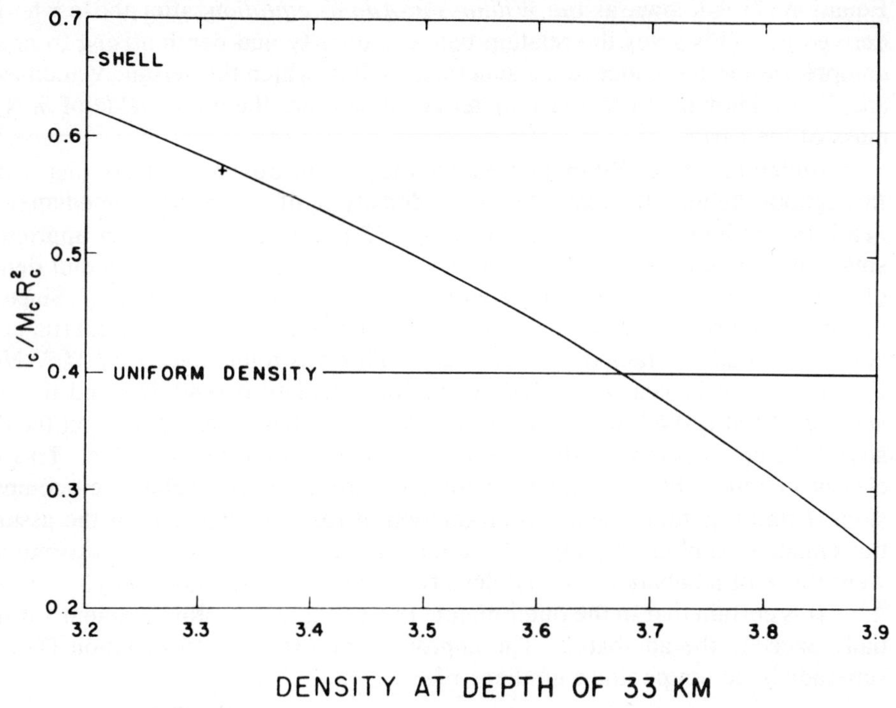

FIGURE 9-2
Moment of inertia of the central core as a function of density at the top of a uniform mantle; I_c, M_c, and R_c are the moment of inertia, mass, and radius of the core, respectively. (*From Birch, 1954, with permission.*)

Accordingly, Bullen concluded that the third assumption, i.e., homogeneity, implicit in the application of equation (7) is invalid. The mantle must therefore be inhomogeneous, and there must exist a region or regions in which density increases with depth at a rate which is much more rapid than would be caused by normal self-compression controlled by the bulk modulus K. The inhomogeneity might be caused by changes in chemical composition, by phase changes, or by a combination of these factors. In the further development of density models for the earth, Bullen made the reasonable assumption of a general correlation between seismic velocity and density and suggested that the inhomogeneity which he had demonstrated to be present was associated with Region C, i.e., the Transition Zone.

9-3 ELASTICITY

With the inhomogeneous nature of the mantle demonstrated by Bullen, Birch (1939, 1952) carried out a detailed investigation of the elastic properties of the mantle with the objective of defining the nature of the inhomogeneity. Birch's approach was based upon a comparison of the observed rate of change of seismic velocities with depth, with the rate of change which would be expected on theoretical grounds in a self-compressed homogeneous layer of the earth subject to an arbitrary temperature gradient. It should be mentioned that Birch's treatment was based upon the velocity distributions of Jeffreys and Gutenberg, which are much smoother than more recent distributions discussed in Sec. 9-1. Birch derived exact thermodynamic relationships for the parameter $1 - g^{-1}(d\phi/dr)$, which may also be obtained directly from the observed variation of seismic velocities in the mantle. After evaluating numerical coefficients and discarding insignificant terms, Birch showed that, to a sufficient approximation,

$$1 - g^{-1}\frac{d\phi}{dr} = \left(\frac{\partial K_T}{\partial P}\right)_T - 5T\alpha\gamma - \frac{2\tau\alpha\phi}{g} \tag{9}$$

(where γ is Grüneisen's ratio) and that, in equation (9), $(\partial K_T/\partial P)_T$ was by far the most important term. Evaluation of this term requires knowledge of the equation of state of the material. Birch used Murnaghan's (1937) theory of finite strain for this purpose. It is assumed that the Helmholtz free energy ψ or the strain energy can be expressed in terms of the negative of the strain (denoted by f) in the form

$$\psi = af^2 + bf^3 + cf^4 + \cdots \tag{10}$$

where a, b, and c are functions only of temperature.

It is assumed that higher order terms in (10) are small compared to the first two. This assumption might appear sweeping at first sight, but it should be remembered that the strain throughout the mantle falls within the rather restricted range 0 to 0.13.

From thermodynamics,

$$P = -\left(\frac{\partial\psi}{\partial v}\right)_T = -\frac{df}{dv}\left(\frac{\partial\psi}{\partial f}\right)_T \tag{11}$$

The relation between f, v (volume), and ρ is

$$\frac{v_0}{v} = \frac{\rho}{\rho_0} = (1 + 2f)^{3/2} \tag{12}$$

From (10), (11), and (12) Birch obtains the following relations, all for isothermal compressibility:

$$P = \tfrac{3}{2}K_0\left[\left(\frac{\rho}{\rho_0}\right)^{7/3} - \left(\frac{\rho}{\rho_0}\right)^{5/3}\right]\left\{1 - \xi\left[\left(\frac{\rho}{\rho_0}\right)^{2/3} - 1\right] + \cdots\right\} \tag{13}$$

$$P = 3K_0 f\,(1 + 2f)^{5/2}\,(1 - 2\xi f) \tag{14}$$

$$K = K_0\,(1 + 2f)^{5/2}\,[1 + 7f - 2\xi f(2 + 9f)] \tag{15}$$

$$\left(\frac{\partial K}{\partial P}\right)_T = \frac{12 + 49f - 2\xi(2 + 32f + 81f^2)}{3 + 21f - 6\xi f(2 + 9f)} \tag{16}$$

$$\phi = \phi_0\,(1 + 2f)\,[1 + 7f - 2\xi f(2 + 9f)] \tag{17}$$

In these expressions, ξ is a dimensionless parameter equal to $\frac{3}{4}(4 - K_0')$ and is a function of temperature alone.[1] It is clear that equations (13) to (17) would be greatly simplified if ξ were equal to zero. They would still be useful for $|\xi| < \frac{1}{2}$. Birch showed that this latter condition was satisfied for most materials for which there was adequate data and that, for some of these materials, ξ indeed appeared to be close to zero. The most decisive test of the above relationships was supplied by relatively compressible materials, such as the alkali metals, which, nevertheless, exhibit a fivefold range of initial compressibilities. Bridgman had measured the compressions of these up to $f \sim 0.3$ compared to $f \sim 0.13$ at the base of the mantle. Results compared to predictions from equation (13) are shown in Fig. 9-3. The agreement with the simplest form of (13) for $\xi = 0$ is most impressive. The implication of equation (13) is that *solids, regardless of initial compressibility, tend to follow a common law in which the most important variable is the degree of compression, rather than the pressure.*

[1] $K_0' = \left(\dfrac{dK}{dP}\right)_{P \to 0}$

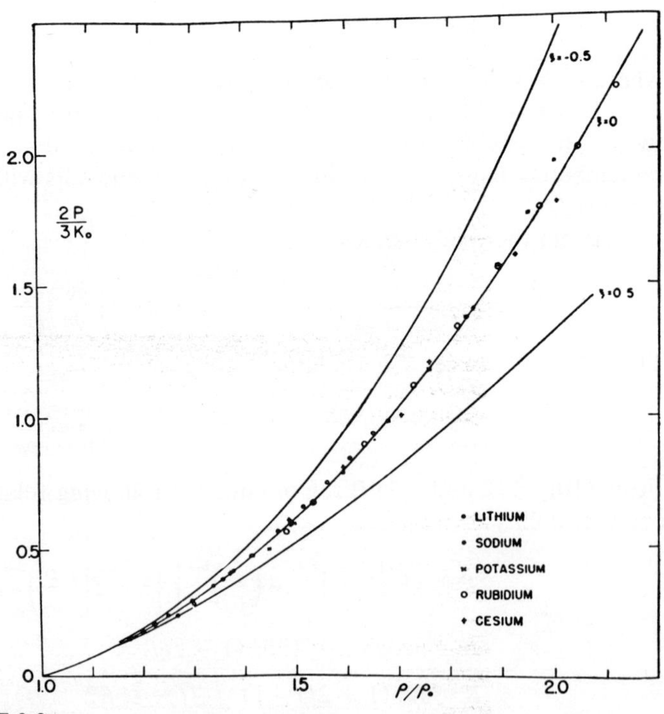

FIGURE 9-3
Experimental measurements of compressions of the alkali metals compared with compressions calculated from Birch-Murnaghan equation (13). (*From Birch, 1952, with permission.*)

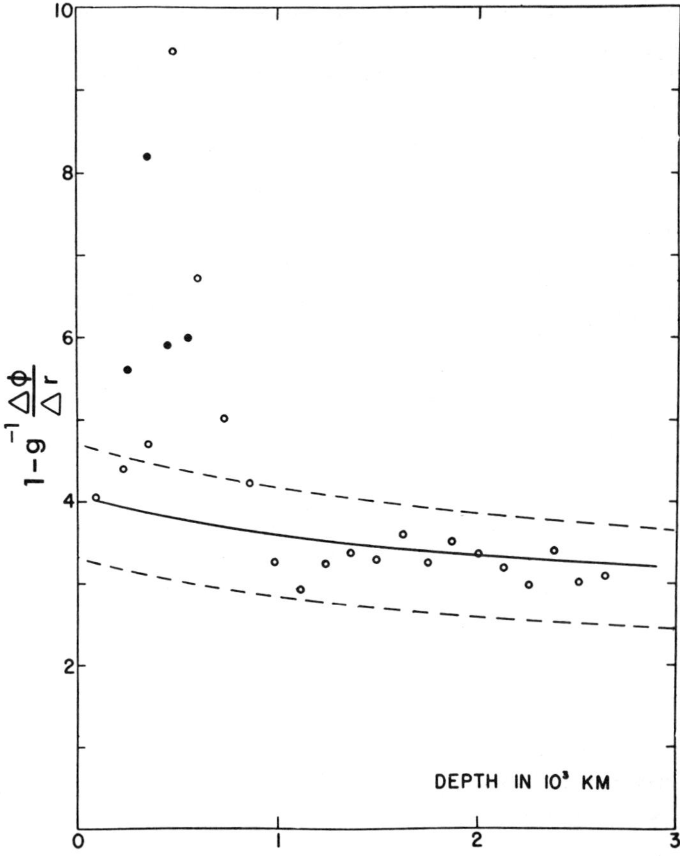

FIGURE 9-4
Comparison of observed values of the function $1 - g^{-1}\,(\Delta\phi/\Delta r)$ for the mantle, with calculated values of this function for compression of homogeneous material given by equations (9) and (18). Broken lines show calculated values of $1 - g^{-1}\,(\Delta\phi/\Delta r)$ obtained from equation (9) and the third-order Birch-Murnaghan equation (16) with $\xi = +\frac{1}{2}$ and $-\frac{1}{2}$ (*From Birch, 1952, with permission.*)

With $\xi = 0$, its most frequent value, equation (16), gives

$$\left(\frac{\partial K_T}{\partial P}\right)_T = \frac{12 + 49f}{3(1 + 7f)} \tag{18}$$

Thus $(\partial K_T/\partial P)_T$ is equal to 4 at zero pressure and diminishes to about 3 for the greatest compressions of the mantle. Birch (Fig. 9-4) compared values for the function $1 - g^{-1}(d\phi/dr)$ formed directly from the seismic velocities of Jeffreys and Gutenberg with $(\partial K_T/\partial P)_T$ from (18), which is by far the largest term in the theoretical expression for $1 - g^{-1}\,(d\phi/dr)$ in a homogeneous layer, given in (9). Finding that, between 1000 and 2900 km, the observed seismic values agreed well with those expected for a self-compressed homogeneous layer of the earth, Birch concluded that the lower mantle was indeed homogeneous; i.e., the variation of seismic velocities with depth was caused by self-compression and that important chemical changes and phase changes did not occur. Although this conclusion was reached initially upon the basis of $\xi = 0$, Birch showed that it was in fact valid for other values of ξ within the permissible range $|\xi| < \frac{1}{2}$.

However, between 300 and 900 km, it is seen from Fig. 9-4 that the observed values of $1 - g^{-1}(d\phi/dr)$ depart grossly from the values expected for a homogeneous layer. Birch observed that no reasonable adjustment in the parameters of his equation of state for a homogeneous layer could provide the observed high values of $1 - g^{-1}(d\phi/dr)$ and that, hence, this region must be inhomogeneous. Thus, the changes of composition and/or phase which were deduced by Bullen must be associated with this region of the earth between approximately 300 and 900 km.

By assuming the lower mantle to be adiabatic with $\xi = 0$, Birch used equations (13) to (17) to calculate ϕ_0 for the material of the lower mantle. He found that the elastic ratio ϕ_0 of the material of the lower mantle reduced to zero pressure averaged 51 $(km/sec)^2$. This value corresponded to an elevated temperature. Making an allowance for this factor Birch estimated that ϕ_0 for lower mantle material at atmospheric pressure and temperature was about 60 $(km/sec)^2$. The elastic ratios of common minerals such as olivines and pyroxenes which are believed to predominate in the upper mantle are between 30 and 40 $(km/sec)^2$, and clearly, these cannot be important constituents of the lower mantle. However, a number of simple oxide phases possess elastic ratios of the right order [e.g., Al_2O_3—69 $(km/sec)^2$, MgO—47 $(km/sec)^2$, and TiO_2—50 $(km/sec)^2$]. The high elastic ratios of these phases are primarily consequences of their crystal structures, which are much more closely packed than those of common silicates. Birch accordingly suggested that the lower mantle was composed of phases with elastic ratios and structures similar to those of close-packed oxide minerals. The transition zone was interpreted as a zone of major phase transformations accompanied perhaps by chemical changes resulting from instability of the principal minerals of the upper mantle at very high pressures. It was suggested that these minerals transformed to a new assemblage of close-packed polymorphs and phases, possessing structures similar to rutile, periclase, corundum, and spinel. The series of transformations was believed to be completed at 900 km, and the resulting new mineral assemblage was stable throughout the entire lower mantle, which was accordingly homogeneous.

Birch's (1952) paper is now recognized as a landmark in its field. During the 1950s, however, some of the principal conclusions were criticized, and alternative explanations of the properties of the transition zone were offered.[1,2,3,4,5,6,7,8] It

[1]Verhoogen (1953).
[2]Griggs (1954).
[3]Miki (1955).
[4]Shima (1956).
[5]Evernden (1958).
[6]Shimazu (1958).
[7]Magnitsky and Kalinin (1959).
[8]Wada (1960).

should be recalled that, in 1952, high-pressure phase transformations were all but unknown in oxides and silicates, and the conclusion that common dense silicate minerals such as pyroxenes and olivines might transform to entirely new and much denser phases at great depths in the mantle did not appear as intrinsically plausible as it does today.

It was argued that materials with complex structures such as silicates would be unlikely to follow the same compression law as compounds with very simple structures such as the alkali metals and that, accordingly, the conclusions regarding mantle inhomogeneity were highly uncertain. It was also suggested that a serious source of error might arise from ignoring the third- and higher order terms of equation (10) on which Birch's equation of state was based and that this would severely affect conclusions based on the extrapolation of lower mantle properties to atmospheric pressure conditions.

Birch's not unreasonable reply[1] was that the basic question of inhomogeneity had already been decided by Bullen's method and that his investigation was directed primarily towards finding the most probable location and cause of the inhomogeneity using an equation of state which, though simple in form, was nevertheless consistent with available data. Birch's identification of the 400 to 900 km region as the source of inhomogeneity was valid as long as $| \xi | < \frac{1}{2}$, which permitted the third-order term of (16) to be of substantial magnitude. However, his extrapolation of the elastic ratio of the lower mantle to yield ϕ_0 was based upon the assumption that $\xi = 0$. This was the most reasonable assumption which could be made at the time in the light of available data. Nevertheless, it was true that the presence of a significant third-order term in (13) within the range $| \xi | < \frac{1}{2}$ could, under some circumstances, result in values for ϕ_0 and ρ_0 which differed substantially from those inferred by Birch for $\xi = 0$. Thus Evernden (1958) argued that a lower mantle composed of common minerals (e.g., olivine and pyroxene) mixed with metallic iron was indicated by an equation of state in which higher order terms were retained.

Accurate measurements of $(\partial K_0/\partial P)_T$ for a number of oxides and silicates have since been made using the method of ultrasonic interferometry[2] (Table 9-1). It is seen that, for most of the relatively close-packed oxides and silicates which are relevant to the structure of the mantle (corundum, periclase, spinel, forsterite, and hematite), $(\partial K_0/\partial P)_T$ is between 3.9 and 4.9, corresponding to ξ between 0 and -0.7. However, garnet and rutile fall outside this range, and so do quartz, lime, and bromellite, which are, however, less relevant to mantle problems. The data in Table 9-1 thus imply that third-order terms in the Birch-Murnaghan equation of state are often substantial, with ξ generally negative. The third-order terms are not sufficiently large to shake Birch's basic conclusion concerning the inhomogeneity of the transition zone. However, the range of possible ξ values introduces a substantial uncertainty into estimates of ρ_0 and ϕ_0 for the lower mantle which were made on the assumption of $\xi = 0$. Thus, confirmation of Birch's interpretation of lower mantle properties by other methods would be necessary.

The availability of accurate measurements of dK/dP provide the justification

[1]Birch (1954).

[2]Anderson, Schreiber, Liebermann, and Soga (1968).

Table 9-1 $(\partial K_0/\partial P)_T$ FOR SOME OXIDES AND SILICATES*

Material		$(\partial K_0/\partial P)_T$
Corundum	Al_2O_3	3.9
Periclase	MgO	4.5
Spinel	$MgAl_2O_4$	4.2
Forsterite	Mg_2SiO_4	4.9
Garnet	$(Mg,Fe)_3Al_2Si_3O_{12}$	5.5
Hematite	Fe_2O_3	4.5
α-Quartz	SiO_2	6.4
Lime	CaO	5.3
Zincite	ZnO	4.8
Bromellite	BeO	5.5
Rutile	TiO_2	6.8

*Anderson, Schreiber, Liebermann, and Soga (1968).

for a more general equation of state[1] that that of Birch with zero ξ. The basic assumption (supported by measurement) is that the adiabatic and isothermal bulk moduli are linear with pressure, i.e.,

$$K_P = K_0 + K_0'P \qquad \text{where} \qquad K_0' = \left(\frac{dK}{dP}\right)_{P\to 0} \tag{19}$$

This embraces, but is a more general expression than, equation (18) of Birch since K_0' is unrestricted, whereas equation (18) implies $K_0' = 4$.

Murnaghan (1944) integrated (19) to obtain

$$P = \frac{K_0}{K_0'}\left[\left(\frac{v_0}{v}\right)^{K'_0} - 1\right] \tag{20}$$

Anderson[1] refers to (20) as the Murnaghan logarithmic equation, which leads to

$$\ln\left(\frac{v_0}{v}\right) = \frac{1}{K_0'}\ln\left[K_0'\left(\frac{P}{K_0}\right) + 1\right] \tag{21}$$

This equation is equivalent to the principal compression equation (13) of Birch in which third-order terms are retained and ξ is unrestricted. However, it has the advantage of simplicity over the full Birch equation and is superior to the restricted ($\xi = 0$) equations when $dK/dP \neq 4$.

An instructive example[2] of the application of equation (21) to calculate the compression of MgO over a range of pressures corresponding to those throughout the entire mantle and based only on experimental measurements of dK/dP up to 4 kbars is given in Fig. 9-5.

[1]O. Anderson (1965, 1966a, 1966b).
[2]O. Anderson (1966b).

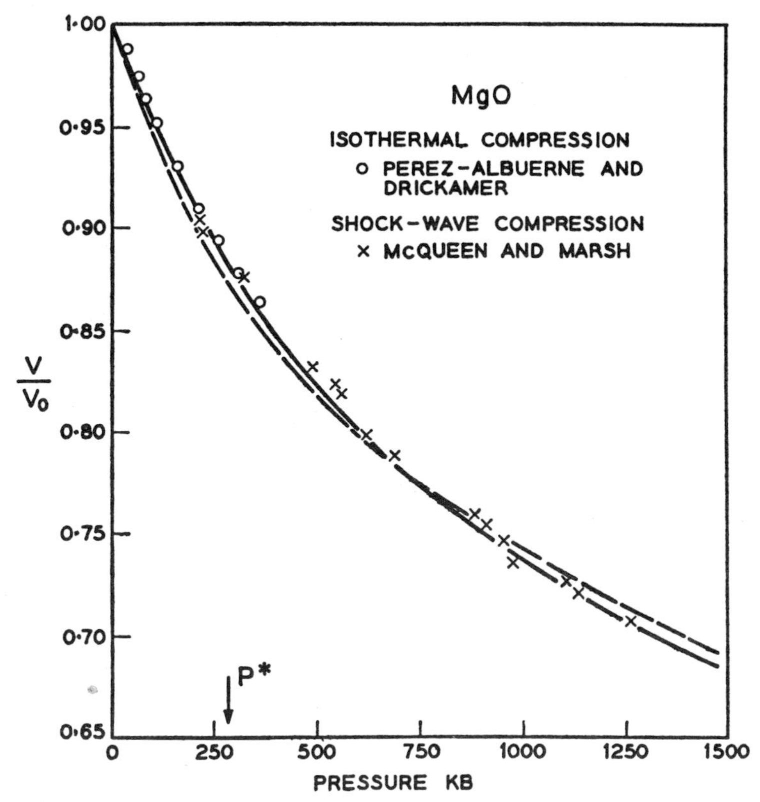

FIGURE 9-5
Experimental measurements of compression of MgO compared to theoretical curves given by equation (21) based upon ultrasonic measurements of K and dK/dP to 4 kbars. Broken line—single-crystal data. Full line—polycrystalline data. (*From Anderson, 1966b, with permission.*)

9-4 EMPIRICAL VELOCITY-DENSITY RELATIONSHIPS

Birch[1] has described an empirical relationship between P velocity and density which is displayed by a large number of rocks and minerals (Fig. 9-6). It is seen that, for samples with mean atomic weights $(\bar{M})^2$ of about 21, velocity is approximately proportional to density over a wide range. Samples with $\bar{M} > 21$ are systematically displaced with respect to the $\bar{M} = 21$ line in the direction of lower velocity. Indications of a linear velocity-density relationship exist for samples possessing similar but higher values of $\bar{M}$. These observations can be described in

[1]Birch (1960, 1961a).
[2]$\bar{M}$ = formula molecular weight divided by number of atoms in the molecular formula.

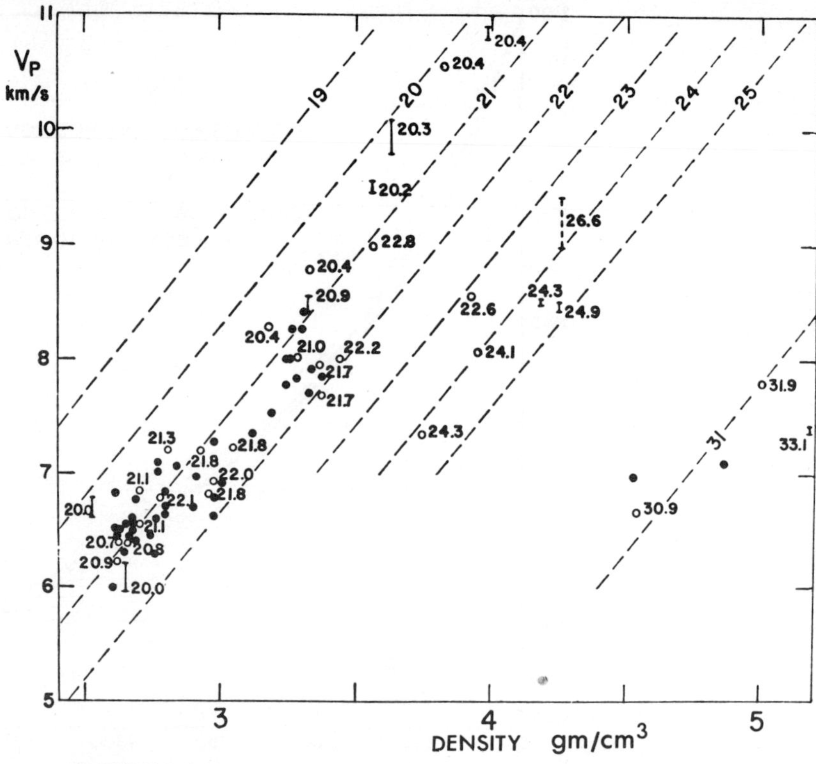

FIGURE 9-6
Velocity at 10 kbars versus density for silicates and oxides. The numbers attached to open circles are mean atomic weights. Dashed lines suggest variation for constant mean atomic weights. (*From Birch, 1961a, with permission. Copyright American Geophysical Union.*)

terms of a general rule:

$$V_p = A_{(\bar{M})} + B\rho \qquad (22)$$

where A is a constant determined approximately by the mean atomic weight, and B is the observed proportionality factor between velocity and density—commonly around 3(km/sec)/(g/cm³). Analogous rules have been proposed to describe empirical relationships between density and shear velocity[1] and bulk sound velocity,[2,3] respectively.

Birch pointed to three geophysical applications of the class of relationships exemplified by Fig. 9-6. Firstly, the lines of Fig. 9-6 may indicate the velocity-

[1]Simmons (1964a).
[2]McQueen, Fritz, and Marsh (1964).
[3]Wang (1969).

density path followed during isothermal compression of homogeneous materials. Secondly, and most important, the fact that equation (22) is applicable to a wide variety of oxides and silicates possessing different structures suggests that it might reasonably be applied to describe the velocity-density relationship of an oxide or silicate undergoing a phase transformation. Thirdly, the lines of Fig. 9-6 show the effect of substituting heavier atoms such as iron in a given structure. These relationships have proved of considerable value in estimating density distribution in the crust and upper mantle from known velocity distributions. They have also been widely applied to interpretations of the constitution of the deep mantle—particularly to the role of phase transformations and to possible changes in $\bar{M}$ with depth. Since we will be concerned with these applications in later chapters, it is desirable at this stage to examine further the basis of these velocity-density rules and their limitations.

With regard to the latter, it should be noted that, for different sets of rocks and minerals with $\bar{M}$ between 21 and 22 (which includes the best and most abundant primary data), the "constant" B in equation (22) is found to vary between 2.7 and 3.6 at low pressure and temperature.[1] Some rocks and minerals (granites and plagioclases) fall well outside this range. In the case of isothermal compression of some common minerals, B lies between 2.7 and 4; however, for homogeneous materials under isobaric conditions, the constant $B = (\partial V_p/\partial \rho)_P$ is much higher—between 4 and 6 (Table 9-2).

Birch's rule applied to the compression of homogeneous materials is implicit in Debye lattice theory, from which the following relationship has been derived:[2,3]

$$V = k(\bar{M})\,\rho^n \qquad \text{where } n = \gamma - \tfrac{1}{3} \tag{23}$$

γ is the Grüneisen parameter, which lies between 1 and 2 for most solids. Birch's rule represents the particular case where $n = 1$. Measurements of elastic properties on a wide range of oxide compounds actually suggest that n is closer to 1.5 than 1, at least for transverse and bulk sound-wave propagation. Birch's rule thus represents a linear approximation to the more fundamental power rule.[2,3] The velocity-density relationship for the power rule with $n = 1.5$ is shown in Fig. 9-7, which contains more recent observational data than were plotted on Fig. 9-6. In view of the scatter among the data, it is seen that the Birch rule represents an entirely adequate approximation within the density range over which it was originally derived.

It is sometimes more convenient to discuss velocity-density systematics in terms of relationships between bulk modulus K and density (or molecular volume v). From the Birch velocity-density relationship, assuming V_p/V_s constant, we have

$$V \propto \sqrt{\frac{K}{\rho}} \qquad \propto \rho \text{ for constant } \bar{M}$$

[1]Birch (1961a).
[2]Shankland (1972).
[3]Chung (1972).

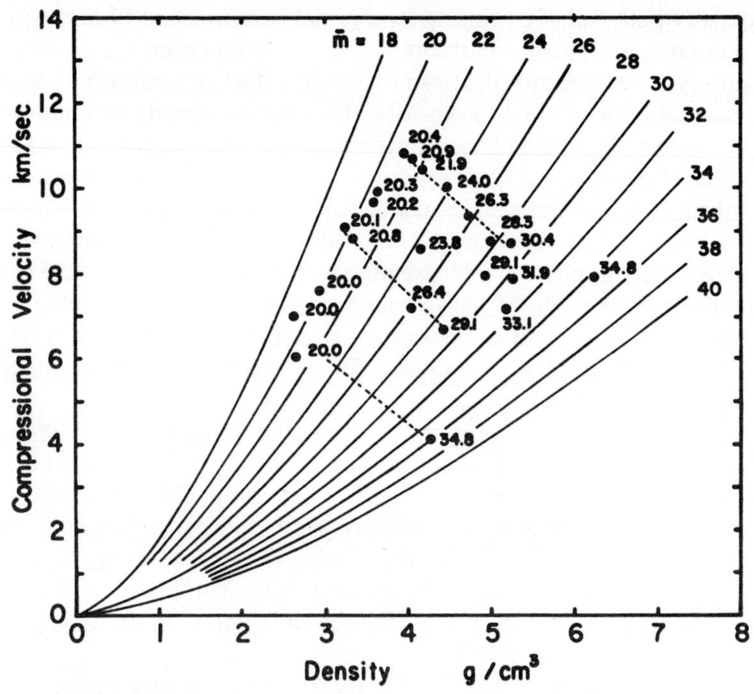

FIGURE 9-7
Compressional velocity versus density as a function of mean atomic weight $\overline{M}$ according to the power law $V = k(\bar{M})\rho^{1.5}$. *(From Soga, 1971, with permission. Copyright American Geophysical Union.)*

Therefore,

$$K \propto \rho^3 \tag{24}$$

On the other hand, O. Anderson (1966c) has reviewed data on K and ρ for oxide compounds[1] with $\bar{M} \sim 21$ to 22 and concluded that they are fitted better by $K \propto \rho^4$. This is equivalent to $V \propto \rho^{1.5}$, as discussed earlier and shown in Fig. 9-7.

The existence of a power-law relationship with exponent 4 follows from the fact that many relatively incompressible oxides have dK/dP values near 4[1] (Table 9-1). Consider the identity[1]

$$\frac{dK}{dv} = \frac{dK}{dP}\frac{dP}{dv}$$

Since $K = -v(dP/dv)$, the above is

$$\frac{dK}{dv} = -\frac{dK}{dP}\cdot\frac{K}{v}$$

[1]Anderson and Nafe (1965).

or

$$\frac{dK}{K} = -\frac{dK}{dP}\frac{dv}{v}$$

Taking dK/dP as constant and equal to 4, the above integrates to

$$\ln K = -4 \ln v + \text{constant}$$

which is equivalent to

$$K \propto v^{-4} \tag{25}$$

or, for constant $\bar{M}$,

$$K \propto \rho^4 \tag{26}$$

The effect of high pressure on the proportionality factors of equations (25) and (26) is probably not negligible. Equations (16) and (18) require that dK/dP decreases with pressure in the latter case, from 4 at zero pressure to 3 at very high pressures. This implies a change in n in the equation $K \propto \rho^n$ from 4 to 3 with increasing pressure.

D. Anderson (1967) derived a general form of the velocity-density relationship in which the quantity considered is ϕ (equal to K/ρ, rather than V_p). Anderson refers to this relationship as the seismic equation of state.

Consider an equation of state of the form

$$P = (N-M)^{-1} K_0 \left[\left(\frac{\rho}{\rho_0}\right)^N - \left(\frac{\rho}{\rho_0}\right)^M\right] \tag{27}$$

A variety of theoretical considerations lead to an equation of this form;[1] in particular, $N = \frac{7}{3}$, $M = \frac{5}{3}$ give the Birch-Murnaghan finite-strain equation of state (13). Differentiating (27) with respect to density, Anderson obtains for small compressions,

$$\frac{\rho_2}{\rho_1} = \left(\frac{\phi_2}{\phi_1}\right)^{1/(N+M-1)} \quad \text{or} \quad \rho = A\phi^n \tag{28a,b}$$

Equation (28) is useful since ϕ can be obtained directly from seismic- and shock-wave data. The exponent n is directly related to Grüneisen's ratio and to dK/dP and can be obtained empirically from shock-wave data, from observed relationships between ϕ and ρ for particular classes of oxides or silicates and from seismic data in restricted regions of the mantle. Equation (28) can be generalized to include the effect of composition changes by introducing the mean atomic weight $\bar{M}$, giving the seismic equation of state

$$\frac{\rho}{\bar{M}} = A\phi^n \tag{29}$$

Anderson shows that the above equation is applicable to a wide variety of selected

[1]Gilvarry (1957).

oxide compounds, with $\bar{M}$ varying between 18 and 90. For the data used, n is very close to $\frac{1}{3}$. Since $\phi = K/\rho$, this implies a relationship $K \propto \rho^4$ for constant $\bar{M}$.

Thus we have two "laws" proposed: $K \propto \rho^3$ and $K \propto \rho^4$. It is not quite clear which is to be preferred in the mantle. A degree of selection of preferred data is evident in different treatments.[1,2,3,4] The fourth-power law appears to be somewhat preferable in its applicability at relatively low pressures and temperatures to a wide range of oxide compounds possessing different structures. However, there is a substantial amount of scatter among the primary data. On the other hand, the third-power law appears to be superior in describing the compressions[5] of homogeneous materials to high densities, as for example, under shock-wave conditions.[6,7] There is also some evidence suggesting that the third-power law is more applicable in the cases of thermal expansion of homogeneous phases.[8,1,9]

These latter properties of the third-power law are likely to be particularly useful in describing the density variation throughout homogeneous regions of the mantle where both P- and S-wave velocities are reasonably well determined, thereby providing the bulk sound velocity V_c (where $V_c^2 = V_p^2 - \frac{4}{3}V_s^2$). Partial derivatives of P, S, and C velocities with respect to density for some mineral classes which are important constituents of the outermost 600 km of the mantle are shown in Table 9-2. In the cases of P and S waves, there are large differences between the two groups of corresponding partial derivatives, one at constant temperature and the other at constant pressure. On the other hand, the differences are relatively small in the case of C waves. Wang thus concludes that, for the empirical rule, $V_c \propto \rho$, applied to upper mantle minerals, "tempera-

[1]O. Anderson (1966c).
[2]D. Anderson (1967).
[3]Wang (1969).
[4]Simmons and England (1969) provide a revealing discussion of the uncertainties in the seismic equation of state when applied to a wider range of primary data.
[5]Corrected to adiabatic conditions.
[6]McQueen, Fritz, and Marsh (1964).
[7]Wang (1969, 1970).
[8]Clark and Ringwood (1964).
[9]Wang (1970).

Table 9-2 PARTIAL VELOCITY DERIVATIVES $\partial V_p/\partial\rho$, $\partial V_s/\partial\rho$, AND $\partial V_c/\partial\rho$ FOR CONSTANT PRESSURE AND CONSTANT TEMPERATURE FOR SOME MANTLE-TYPE MINERALS. *(After Wang, 1970)*

Mineral	$(\partial V_p/\partial\rho)_T$	$(\partial V_p/\partial\rho)_P$	$(\partial V_s/\partial\rho)_T$	$(\partial V_s/\partial\rho)_P$	$(\partial V_c/\partial\rho)_T$	$(\partial V_c/\partial\rho)_P$
Olivine (Fa_7)	4.0	5.9	1.4	4.3	3.9	3.5
Garnet (A1-Py)	3.3	4.3	0.9	2.4	3.5	3.4
Spinel	2.7	5.3	0.2	3.7	3.3	3.1

ture works in the opposite direction but probably in the same ratio as pressure." This is not the case for P and S waves.

The relationship of bulk sound velocity V_c to density for oxide systems of similar mean atomic weight also appears to be more regular than the corresponding V_p and V_s versus ρ rules. This may arise because the former relationship involves the bulk modulus through $V_c = \sqrt{K/\rho}$. K tends to be comparatively "well behaved," being rather simply related to repulsion potentials between atoms. However, V_p aand V_s also involve the shear modulus through $V_p = \sqrt{K + \frac{4}{3}\mu/\rho}$ and $V_s = \sqrt{\mu/\rho}$. The shear modulus and its derivatives vary in a complex manner and are structure sensitive.[1] The erratic behaviour of μ thus contributes to the rather wide variation of the proportionality factor in the Birch velocity-density relationship. On the other hand, these advantages of V_c tend to be offset by the circumstance that, in the mantle, this quantity is obtained by difference from V_p and V_s and so incorporates the combined errors of the latter. In certain parts of the mantle, particularly in the outer 1000 km, these are not negligible.

For families of isostructural compounds formed from atoms with similar valences, a very different relationship between K and v is observed[2,3] (Fig. 9-8). It is found that, for each family, K is proportional to v^{-1}. This relationship can be derived from considerations of the lattice energies of ionic solids.[3] It is not clear, however, why the expression applies also to groups of structures possessing covalent and metallic bonding. Possibly the repulsive terms in the lattice energies of ionic, covalent, and metallic compounds are dependent more upon the overall electronic structures of atoms than on the outermost electronic configurations.

There has been some confusion in the literature concerning the significance of the empirical power-law relationships discussed earlier: $K \propto \rho^n \propto v^{-n}$, where $3 < n < 4$ (Figs. 9-6 and 9-7), which apply to many oxycompounds, and the more general $K \propto v^{-1}$ relationship, which applies to groups of compounds with similar structures and valences (Fig. 9-8). It was noted by Birch (1961b) that the former relationship reflected crystallographic factors—particularly the closeness of the atomic packing. It appears that the rule is a consequence of the closeness of packing of the oxygen anions which occupy a far greater proportion than the cations of the volume of common oxides and silicates (Table 9-3). Thus, oxygen anions account for more than 90% of the volume of nearly all the minerals which obey Birch's rule ($\bar{M} \sim 21$). It appears probable that the reason that calcium silicates and other compounds rich in CaO do not follow the rule[4] closely is connected with the much larger relative proportion of the volume occupied by Ca^{++}. It is to be expected that other oxides possessing large cations, e.g., Sr, Ba, La, and Pb, may also behave anomalously. Accordingly, geophysical applications of these rules should be made only to materials in which oxygen anions occupy by far the

[1] Anderson and Liebermann (1970).
[2] Anderson and Nafe (1965).
[3] Anderson (1966c, 1972).
[4] Simmons (1964b).

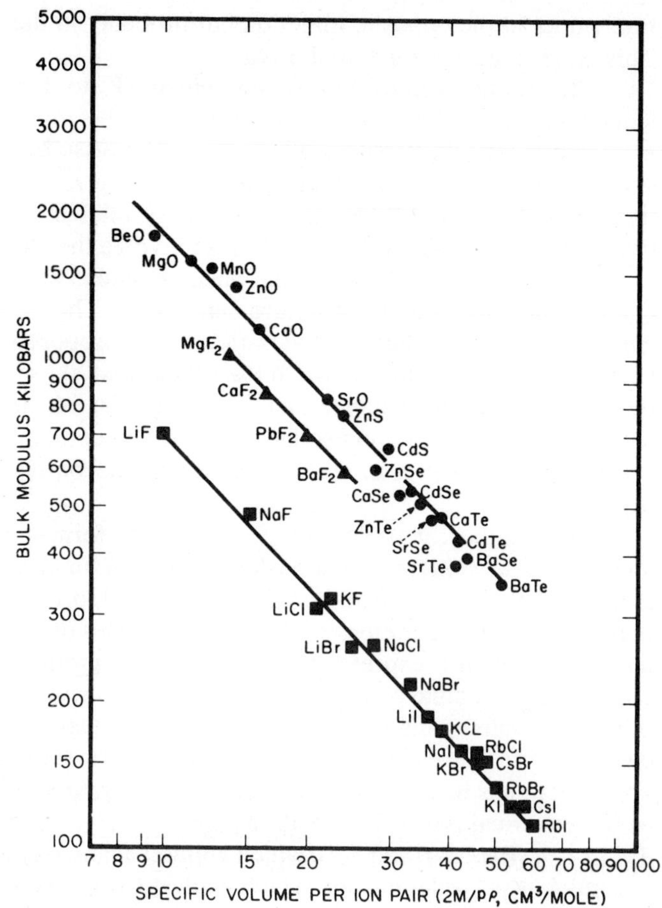

FIGURE 9-8
The bulk modulus versus specific volume for some diatomic solids. The slope of the solid lines is −1. (*From O. Anderson, 1972, with permission.*)

greatest proportion of the volume and in which the acoustic properties can be regarded to a first approximation as being determined by the closeness of packing of the oxygen anions. Fortunately, most of the probable major components of the mantle satisfy this condition.

Velocity-Density Rules and Phase Transformations

One of the principal applications of these rules has been to interpret observed velocity distributions in inhomogeneous regions of the mantle in terms of phase transformations and changes in mean atomic weight. Although the rules have

Table 9-3 PROPORTIONS OF THE VOLUMES OF SOME SIMPLE OXIDES OCCUPIED BY CATIONS*

Oxide	$V_{cations}/V_{oxide}$(%)
Na_2O	45
MgO	17
CaO	34
SrO	47
BaO	56
PbO	50
MnO	25
FeO	20
Fe_2O_3	8
Al_2O_3	5
La_2O_3	35
Cr_2O_3	7
SiO_2	1.3
GeO_2	2
TiO_2	5
ZrO_2	13

*The bonding in the oxides is assumed to be purely ionic, whereas in fact, varying degrees of covalent bonding are present. This will cause a general increase in the values tabulated. Nevertheless, proportional volumes on the ionic model are indicative in a relative sense.

been based upon best fits to somewhat scattered primary data for oxycompounds with different compositions, it has sometimes been implicitly assumed that velocity changes caused by phase changes may be rather more regular since no change in composition is involved. It is only relatively recently that sufficient data have become available to test this supposition[1,2,3] (Table 9-4). It is seen that the *P* velocity-density relationship existing between high-pressure and low-pressure polymorphs is considerably less regular than might have been anticipated.

Whilst there is no doubt of the value of empirical velocity-density relationships in providing a general guide to major problems of mantle constitution, these observations emphasize the need for caution in attempting to obtain detailed information regarding density changes and changes of mean atomic weight from observed seismic velocity distributions in regions of the mantle where phase changes are occurring.

The data in Table 9-4 show an interesting trend. For those phase transformations involving an increase of cation-oxygen coordination numbers, the values

[1]Mizutani et al. (1970, 1972).
[2]Liebermann (1970, 1972, 1973).
[3]Liebermann and Ringwood (1973).

Table 9-4 CHANGES IN *P* VELOCITY AND DENSITY ACROSS SOME PHASE TRANSFORMATIONS. (*After Liebermann and Ringwood, 1973, and Liebermann, 1974*)

Phase transformation	$\Delta V_p/\Delta\rho$ [(km/sec)/(g/cm^3)]
Quartz-coesite (SiO_2)	5.7
Quartz-rutile (SiO_2)	3.0
Coesite-rutile (SiO_2)	2.5
Quartz-rutile (GeO_2)	2.2
Olivine-spinel (Mg_2GeO_4)	3.5
Olivine-spinel (Fe_2SiO_4)	2.5
Olivine-spinel (Ni_2SiO_4)	2.2
Olivine-beta (Mn_2GeO_4)	3.2
Pyroxene-ilmenite ($MgGeO_3$)	1.6
Pyroxene-ilmenite ($MnGeO_3$)	1.0
"Pyroxene"-garnet ($CaGeO_3$)	1.5
"Pyroxene"-garnet ($CdGeO_3$)	1.2
Ilmenite-perovskite ($CdTiO_3$)	2.0
Felspar-hollandite ($NaAlGe_3O_8$)	1.3

of $\Delta V_p/\Delta\rho$ (average $B = 1.9$) are generally smaller than given by Birch's rule applied to compounds of similar mean atomic weight ($B = 3.1$). They are also smaller than the B values applying during compression or thermal expansion of homogeneous material. Similar behaviour has been noted by Davies (1974) in the case of transformations into isochemical mixed oxides; moreover, it extends to the relationship between bulk sound velocity and density.[1,2]

These results indicate that the velocity-density trajectories followed during the compression of homogeneous materials are likely to be offset if these materials transform to phase assemblages characterized by higher cation-oxygen coordinations. In general, these offsets will be similar to the offset caused by an increase of mean atomic weight amounting to about one unit in $\bar{M}$. It follows that inferred offsets of this magnitude in the bulk sound velocity-density relationship in the mantle at a depth of about 650 km,[3] where it is believed that phase changes involving an increase of silicon coordination from fourfold to sixfold occur,[4] do not necessarily constitute evidence favouring an increase in $\bar{M}$ at this depth. This topic is taken up again in Chap. 14.

[1]Liebermann (1974).
[2]Davies (1974).
[3]E.g., Press (1970).
[4]Chapter 14.

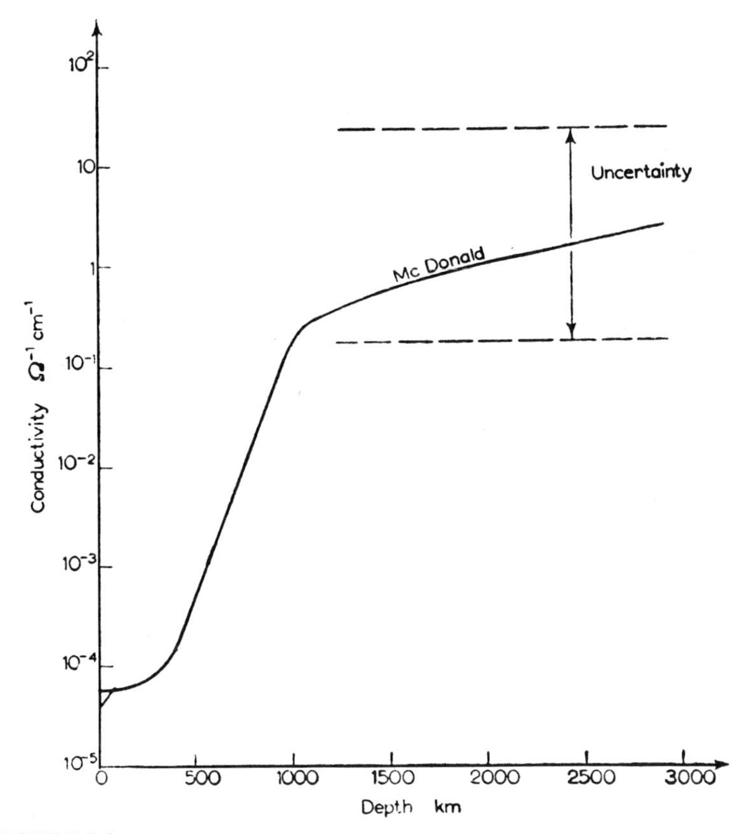

FIGURE 9-9
Electrical conductivity distribution for the mantle. (*From McDonald, 1957, with permission.*)

9-5 ELECTRICAL CONDUCTIVITY OF THE MANTLE

Lahiri and Price (1939) have studied the variation with depth of the electrical conductivity of the mantle. They found that conductivity in the upper mantle was about $10^{-4}\Omega^{-1}\text{cm}^{-1}$, but in the vicinity of 600 km it rose sharply to $10^{-2}\Omega^{-1}\text{cm}^{-1}$. The rate of increase of conductivity must flatten out below this depth in order to give an acceptable value near the core-mantle boundary. Studies of electrical conductivity distribution of the mantle have also been published by Rikitake (1951) and McDonald (1957).

McDonald's preferred conductivity distribution is given in Fig. 9-9. The strong qualitative resemblance to the seismic velocity distribution (Fig. 9-1) is to be remarked.

The conductivity of silicates is a result of several mechanisms—impurity electronic conduction, ionic conduction, and intrinsic electronic conduction.

The first mechanism is quantitatively unimportant in the mantle, whilst it is believed that the high pressures in the mantle will effectively inhibit the second mechanism, which relies on ionic diffusion.[1,2,3] Consequently, the conductivity of the mantle is attributed to intrinsic electronic conductivity σ, where

$$\sigma = \sigma_0 \exp\left(\frac{-E}{kT}\right)$$

where E is the excitation energy of the process, k is Boltzmann's constant, and σ_0 is a constant nominally equal to the conductivity at infinite temperature.

Runcorn[1] has examined the variation of conductivity in the mantle and has concluded that the rate of increase between 400 and 900 km (Fig. 9-9) cannot be explained if the sole mechanism responsible is intrinsic electronic conduction of homogeneous silicates subjected to the maximum temperature gradients which are permitted by thermal history considerations (~ 3°/km). Accordingly, Runcorn concluded that the conductivity distribution was more reasonably interpreted in terms of a phase transformation in which there is a sudden decrease in E, which is to be expected in a more close-packed lattice. Experimental measurement of the conductivity changes caused by the olivine-spinel transformation[4,5] support this interpretation. The conductivity of fayalite was found to increase by about two orders of magnitude when it transforms to the spinel structure. This change is similar to that observed in the mantle.[see also 2,3,6]

REFERENCES

ADAMS, R. D. (1971). Reflections from discontinuties beneath Antarctica. *Bull. Seism. Soc. Am.* **61,** 1441–1451.

AKIMOTO, S., AND H. FUJISAWA (1965). Demonstration of the electrical conductivity jump produced by the olivine-spinel transition. *J. Geophys. Res.* **70,** 443–449.

ANDERSON, D. L. (1967). A seismic equation of state. *Geophys. J. Roy. Astron. Soc.* **13,** 9–30.

ANDERSON, O. L. (1965). Two methods of estimating compression and sound velocity at very high pressures. *Proc. Nat. Acad. Sci.* **54,** 667–673.

——— (1966a). Seismic parameter ϕ: Computation at very high pressure from laboratory data. *Bull Seism. Soc. Am.* **56,** 725–731.

——— (1966b). The use of ultrasonic measurements under modest pressure to estimate compression at high pressure. *J. Phys. Chem. Solids* **27,** 547–565.

[1]Runcorn (1955).
[2]Tozer (1959).
[3]Bullard (1967).
[4]Bradley et al. (1962).
[5]Akimoto and Fujisawa (1965).
[6]Banks (1969).

——— (1966c). A proposed law of corresponding states for oxide compounds. *J. Geophys. Res.* **71,** 4963–4971.
——— (1972). Patterns in elastic constants of minerals important to geophysics. In: E. C. Robertson (ed.), "*The Nature of the Solid Earth,*" pp. 575–613. McGraw-Hill, New York.
——— and R. LIEBERMANN (1970). Equations for the elastic constants and their pressure derivatives for three cubic lattices and some geophysical applications. *Phys. Earth Planet. Interiors* **3,** 61–85.
——— and J. E. NAFE (1965). The bulk modulus-volume relationship for oxide compounds and related geophysical problems. *J. Geophys. Res.* **70,** 3951–3963.
———, E. SCHREIBER, R. LIEBERMANN, AND N. SOGA (1968). Some elastic constant data on minerals relevant to geophysics. *Rev. Geophys.* **6,** 491–524.
——— and N. SOGA (1967). A restriction to the law of corresponding states. *J. Geophys. Res.* **72,** 5754–5757.
ARCHAMBEAU, C. B., E. A. FLINN, AND D. G. LAMBERT (1969). Fine structure of the upper mantle. *J. Geophys. Res.* **74,** 5825–5866.
BANKS, R. J. (1969). Geomagnetic variations and the electrical conductivity of the upper mantle. *Geophys. J.* **17,** 457–487.
BIRCH, F. (1939). The variation of seismic velocities within a simplified earth model in accordance with the theory of finite strain. *Bull. Seism. Soc. Am.* **29,** 463–479.
——— (1952). Elasticity and constitution of the Earth's interior. *J. Geophys. Res.* **57,** 227–286.
——— (1954). The earth's mantle: Elasticity and constitution. *Trans. Am. Geophys. Union* **35,** 79–85, 97–98.
——— (1960). The velocity of compressional waves in rocks to 10 kilobars, 1. *J. Geophys. Res.* **65,** 1083–1102.
——— (1961a). The velocity of compressional waves in rocks to 10 kilobars, 2. *J. Geophys. Res.* **66,** 2199–2224.
——— (1961b). Composition of the earth's mantle. *Geophys. J. Roy Astron. Soc.* **4,** 295–311.
BOLT, B. A. (1969). PdP and PKiKP waves and diffracted PcP waves. (Preprint.)
———, M. O'NEILL, and A. QAMAR (1968). Seismic waves near 110°: Is structure in core or upper mantle responsible? Geophys. *J. Roy. Astron. Soc.* **16,** 475–487.
BRADLEY, R. S., A. K. JAMIL, and D. C. MUNRO (1962). Electrical conductivity of fayalite and spinel. *Nature* **193,** 965–966.
BULLARD, E. C. (1967). Electromagnetic induction in the earth. *Quart. J. Roy. Astron. Soc.* **8,** 143–160.
BULLEN, K. E. (1936). The variation of density and the ellipticities of strata of equal density within the earth. *Mon. Not. Roy. Astron. Soc., Geophys. Suppl.* **3,** 395–401.
——— (1940). The problem of the earth's density variation. *Bull. Seism. Soc. Am.* **30,** 235–250.
CHINNERY, M. A., and N. TOKSÖZ (1967). P-wave velocities in the mantle below 700 km. *Bull. Seism. Soc. Am.* **57,** 199–226.
CHUNG, D. H. (1972). Birch's law: Why is it so good? *Science* **177,** 261–263.
CLARK, S. P., and A. E. RINGWOOD (1964). Density distribution and constitution of the mantle. *Rev. Geophys.* **2,** 35–88.
CLEARY, J., and A. HALES (1966). An analysis of the travel times of P waves to North American stations in the distance range 32° to 100°. *Bull. Seism. Soc. Am.* **56,** 467–489.

DAVIES, G. F. (1974). The effects of polymorphic phase transitions on elasticity. *Earth Planet. Sci. Letters* (In press.)

ENGDAHL, E. R., and E. A. FLINN (1969). Seismic waves reflected from discontinuities within Earth's upper mantle. *Science* **163,** 177–179.

EVERNDEN, J. (1958). Finite strain theory and the earth's interior. *Geophys. J.* **1,** 1–8.

GILVARRY, J. J. (1957). Temperature-dependent equations of state of solids. *J. Appl. Phys.* **28,** 1252–1261.

GRIGGS, D. (1954). Discussion at symposium on the Earth's mantle. *Trans. Am. Geophys. Union* **35,** 93–94.

GUTENBERG, B. (1958). Velocity of seismic waves in the earth's mantle. *Trans. Am. Geophys. Union* **39,** 486–489.

——— (1959a). "*Physics of the Earth's Interior.*" Academic, New York. 240 pp.

——— (1959b). The asthenosphere low-velocity layer. *Ann. Geofisica* **12,** 439–460.

HALES, A. L., J. CLEARY, H. DOYLE, R. GREEN, and J. ROBERTS (1968). P-wave station anomalies and the structure of the upper mantle. *J. Geophys. Res.* **73,** 3885–3896.

——— and J. L. ROBERTS (1970). Shear velocities in the lower mantle and the radius of the core. *Bull. Seism. Soc. Am.* **60,** 1427–1436.

HELMBERGER, D., and R. A. WIGGINS (1971). Upper mantle structure of mid-western United States. *J. Geophys. Res.* **76,** 3229–3245.

JEFFREYS, H. (1937). On the materials and density of the earth's crust. *Mon. Not. Roy. Astron. Soc., Geophys. Suppl.* **4,** 50–61.

——— (1939). The times of P, S, and SKS and the velocities of P and S. *Mon. Not. Roy. Astron. Soc., Geophys. Suppl.* **4,** 498–533.

——— (1959). "*The Earth,*" 4th ed. Oxford Univ. Press, London. 420 pp.

JOHNSON, L. (1967). Array measurements of P velocities in the upper mantle. *J. Geophys. Res.* **72.** 6309–6325.

——— (1969). Array measurements of P velocities in the lower mantle. *Bull. Seism. Soc. Am.* **59,** 973–1008.

JULIAN, B., and D. L. ANDERSON (1968). Travel times, apparent velocities and amplitudes of body waves. *Bull. Seism. Soc. Am.* **58,** 339–366.

KANAMORI, H. (1967). Upper mantle structure from apparent velocities of P waves recorded at Wakayama microearthquake observatory. *Bull. Earthquake Res. Inst., Tokyo Univ.* **45,** 657–678.

LAHIRI, B. N., and A. T. PRICE (1939). Electromagnetic induction in non-uniform conductors, and the determination of the conductivity of the earth from terrestrial magnetic variations. *Phil. Trans. Roy. Soc. London* **A237,** 509–540.

LIEBERMANN, R. C. (1970). Velocity-density systematics for the olivine and spinel phases of Mg_2SiO_4-Fe_2SiO_4. *J. Geophys. Res.* **75,** 4029–4034.

——— (1972). Compressional velocities of polycrystalline olivine, spinel and rutile minerals. *Earth Planet. Sci. Letters* **17,** 263–268.

——— (1973). Elastic properties of polycrystalline SnO_2 and GeO_2: Comparison with stishovite and rutile data. *Phys. Earth Planet. Interiors* **7,** 461–465.

——— (1974). Elasticity of pyroxene-garnet and pyroxene-ilmenite phase transformations in germanates. *Phys. Earth Planet. Interiors* **8,** 361–374.

——— and A. E. RINGWOOD (1973). Birch's law and polymorphic phase transformations. *J. Geophys. Res.* **78,** 6926–6932.

MAGNITSKY, V. A., and V. A. KALININ (1959). The properties of the earth's mantle and the physical properties of the transition layer. *Bull. Acad. Sci. USSR, Geophys. Ser.* No. 1–6, 49–54.

MCDONALD, K. L. (1957). Penetration of the geomagnetic secular variation through a mantle with variable conductivity. *J. Geophys. Res.* **62,** 117–141.

MCQUEEN, R.G., J. FRITZ, and S. P. MARSH (1964). On the composition of the earth's interior, *J. Geophys. Res.* **69,** 2947–2965.

MIKI, H. (1955). Is the Layer-C (413-1000 km) inhomogeneous? *J. Phys. Earth* **3,** 1–6.

MIZUTANI, H., Y. HAMANO, and S. AKIMOTO (1972). Elastic-wave velocities of polycrystalline stishovite. *J. Geophys. Res.* **77,** 3744–3749.

———, ———, Y. IDA, and S. AKIMOTO (1970). Compressional wave velocities of fayalite, Fe_2SiO_4 spinel and coesite. *J. Geophys. Res.* **75,** 2741–2747.

MURNAGHAN, F. C. (1937). Finite deformations of an elastic solid. *Am. J. Math.* **59,** 235–260.

——— (1944). The compressibility of media under extreme pressures. *Proc. Nat. Acad. Sci.* **30,** 244–247.

NIAZI, M., and D. L. ANDERSON (1965). Upper mantle structure of western North America from apparent velocities of P waves. *J. Geophys. Res.* **70,** 4633–4640.

NUTTLI, O. W. (1969). Travel times and amplitudes of S waves from nuclear explosions in Nevada. *Bull. Seism. Soc. Am.* **59,** 385–398.

PRESS, F. (1970). Earth models consistent with geophysical data. Phys. Earth Planet. Interiors **3,** 3–22.

RIKITAKE, T. (1951). Electromagnetic induction within the earth and its relation to the electrical state of the earth's interior. *Bull. Earthquake Res. Inst., Tokyo* **28,** 45, 219.

RINGWOOD, A. E. (1970). Phase transformations and the constitution of the mantle. *Phys. Earth Planet. Interiors* **3,** 109–155.

RUNCORN, S. K. (1955). The electrical conductivity of the earth's mantle. *Trans. Am. Geophys. Union* **36,** 191–198.

SHANKLAND, T. J. (1972). Velocity-density systematics: Derivation from Debye theory and the effect of ionic size. *J. Geophys. Res.* **77,** 3750–3758.

SHIMA, M. (1956). On the variation in bulk modulus in the mantle. *J. Phys. Earth* **4,** 7–10.

SHIMAZU, Y. (1958). A chemical phase transition hypothesis on the origin of the C-layer within the mantle of the earth. *J. Earth Sci. Nagoya Univ.* **6,** 12–30.

SIMMONS, G. (1964a). Velocity of shear waves in rocks to 10 kilobars. *J. Geophys. Res.* **69,** 1123–1130.

——— (1964b). Velocity of compressional waves in various minerals at pressures to 10 kilobars. *J. Geophys. Res.* **69,** 1117–1121.

and A. W. ENGLAND (1969). Universal equations of state for oxides and silicates. *Phys. Earth Planet. Interiors* **2,** 69–76.

SIMPSON, D. W., R. MEREU, AND D. KING (1974). An array study of P wave velocities in the upper mantle and transition zone beneath northeastern Australia. *Bull. Seism. Soc. Am.* (In press.)

SOGA, N. (1971). Sound velocity of some germanate compounds and its relation to the law of corresponding states. *J. Geophys. Res.* **76,** 3983–3989.

TOZER, D. C. (1959). Electrical properties of the earth's interior. In: L. Ahrens, F. Press, K. Rankama, and S. Runcorn (eds.), "*Physics and Chemistry of the Earth,*" vol. 3, chap. 8, pp. 419–436. Pergamon, London.

VERHOOGEN, J. (1953). Elasticity of olivine and constitution of the earth's mantle. *J. Geophys. Res.* **58,** 337–346.

WADA, T. (1960). On the physical properties within the B-layer deduced from olivine model and on the possibility of polymorphic transition from olivine to spinel at the

20° discontinuity. Bull. 37, Disaster Prevention Institute, University of Kyoto, pp. 1–20.

WANG, C. Y. (1967). Phase transitions in rocks under shock compression. *Earth Planet. Sci. Letters* **3,** 107–113.

——— (1969). Equation of state of periclase and some of its geophysical implications. *J. Geophys. Res.* **74,** 1451–1456.

——— (1970). Density and constitution of the mantle. *J. Geophys. Res.* **75,** 3264–3284.

WHITCOMB, J. H., and D. L. ANDERSON (1970). Reflection of P′ P′ seismic waves from discontinuities in the mantle. *J. Geophys. Res.* **75,** 5713–5728.

WILLIAMSON, E. D., and L. H. ADAMS (1923). Density distribution in the earth. *J. Washington Acad. Sci.* **13,** 413–428.

WRIGHT, C., and J. CLEARY (1972). P wave travel-time measurements for the Warramunga seismic array and lower mantle structure. *Phys. Earth Planet. Interiors* **5,** 213–230.

10

EXPERIMENTAL METHODS OF INVESTIGATING MANTLE PHASE TRANSFORMATIONS

10-1 INTRODUCTION

Birch's hypothesis implied that, under P-T conditions equivalent to those existing at depths of 400 to 900 km, the major minerals of the upper mantle—olivines, pyroxenes, and garnets—should become unstable and transform to a new assemblage of close-packed "oxide" phases which possessed the elastic properties and densities required to explain the properties of the lower mantle. It will be recalled that, although Birch's hypothesis appeared to provide the most reasonable available explanation of the elastic properties of the mantle, its basic conclusions were nevertheless dependent upon the assumption of a rather simple equation of state. Although this equation described the compressions of the alkali metals remarkably well, it had not been demonstrated to be applicable to structurally more complex compounds such as oxides and silicates. Sceptics who questioned the equation of state accordingly preferred other interpretations of the constitution of the lower mantle. The most popular alternative hypothesis was that the lower mantle consisted of a mixture of metallic iron and olivine.[1] Other interpretations

[1]Evernden (1958).

held that the abnormal elastic properties of the lower mantle were caused by a change from ionic to covalent bonding,[1] by nonhydrostatic conditions,[2] or exclusively by changes in chemical composition.[3]

Clearly, it would be necessary to verify Birch's hypothesis by experimental methods before it could be regarded as finally established. Experimentally, this presented a difficult challenge at the time. The pressures in the 400 to 900 km depth interval vary from 130 to 340 kbars, and the temperatures are probably in the vicinity of 1500 to 3000°C. Experimental methods capable of attaining these conditions did not exist. Pressure transformations in mafic and ultramafic silicate minerals and in oxides were all but unknown, although Bridgman (1945) had expressed confidence that they would be found. The first encouragement came with the pioneering experiments by Coes (1953, 1955) and his successful synthesis of a new high-pressure polymorph of quartz, "coesite," possessing a density of 2.91 g/cm^3. Nevertheless, many sceptics remained, and diligent efforts were made by some to prove that certain phase transformations which had been previously suggested, e.g., the olivine-spinel transformation, were impossible on the grounds of lattice dynamics.[2,3,4]

It was clear that, during the 1950s, evidence bearing on Birch's hypothesis could be found only by indirect methods, using information obtained at relatively low pressures. Such methods included comparative crystal chemistry and studies of model systems by thermodynamic methods and by direct experiments within the limited pressure range then available. We will review in some detail the ways in which these methods were used. Although in more recent years, direct static experimental methods capable of reproducing the *P-T* conditions existing in the earth down to about 800 km have been developed, the use of these indirect techniques continues to be essential to an understanding of the role of phase transformations deeper than 800 km in the mantle.

10-2 INDIRECT METHODS

If phase transformations indeed play a major role in the mantle, it becomes important to understand the nature of the high-pressure mineral phases. Three lines of evidence bearing on this problem are discussed below.

As experimental observations on high-pressure transformations have expanded during recent years, it has been possible to reach some broad crystal chemical generalizations. An important result is the strong tendency of high-pressure phases to crystallize in structures which are already known. Of about 90 high-pressure transformations which have been discovered in the author's labora-

[1] Magnitsky and Kalinin (1959).
[2] Miki (1955).
[3] Shimazu (1958).
[4] Wada (1960a).

tory in silicates, germanates, and related oxycompounds, only about 10% have been to structures which were unknown, or not closely related to known structures. Bearing in mind the large number and complexity of existing silicate and germanate structures, this is a fortunate circumstance which could not have been confidently predicted before the advent of high-pressure experimentation. This situation greatly simplifies the discussion of possible structure types which might be adopted by a given compound when it undergoes a pressure transformation.

A second factor is the simple thermodynamic requirement that the high-pressure polymorph be denser than the lower pressure phase. This further limits the range of possible structures which might be adopted. Several methods exist for estimating the densities of possible high-pressure polymorphs of given compounds. These densities are found to be systematically related to such properties as mean cation-anion bond lengths and the densities of isochemical oxide mixtures. Examples of the application of these methods are discussed in Sec. 11-9 (Fig. 11-14) and Sec. 12-6 (Fig. 12-13).

A third factor also arises from crystal chemical considerations. Goldschmidt and his colleagues demonstrated the decisive role which the radius ratio, R_{cation}/R_{anion}, plays in determining the structures of inorganic compounds. Thus, in ternary oxide compounds $A_xB_yO_z$ of given stoichiometry, when the radius ratios R_A/R_O and R_B/R_O are plotted against one another, it is found that structures of a given type fall into well-defined fields (e.g., Fig. 12-4) which can be interpreted in terms of simple geometric packing of anions around cations, combined with charge neutralization or other bonding conditions. When a crystal is subjected to high pressure and compressed, the "effective" radii of the constituent ions may contract differentially, thus altering the radius ratios and packing requirements. Transformation into a new phase may occur when these radius ratios attain certain critical values.

In oxycompounds, it appears that the large oxygen anions (radius 1.40 Å)[1] tend to contract relatively more under pressure than the smaller cations such as Mg^{++}(0.72 Å) and Si^{4+} (0.26 Å).[1] Accordingly, the radius ratios R_{cation}/R_{oxygen} increase with pressure, and transformation into a new high-pressure phase occurs when these radius ratios reach the critical value.

The contraction is a more complex phenomenon than simple differential compression. It is probably connected with both deformation of the ions and pressure-induced changes in the nature of the chemical bonding. The polarizability[2] of the O^{--} ion is 3.1×10^{-24} cm³, compared to 0.043×10^{-24} cm³ for Si^{4+} and 0.12×10^{-24} cm³ for Mg^{++}. Thus the oxygen ion is much more readily deformed into nonspherical configurations, resulting in a reduction of its "effective radius" and an increase in the "effective radius ratio" of cation to oxygen anion. It has also been pointed out that compression may cause an increase in

[1]Ionic radii from Shannon and Prewitt (1969).
[2]Born and Heisenberg (1924).

the covalent component of the chemical bond in oxides.[1,2,3,4,5] This is manifested by an increased electron density between ions and a contraction of the bond length. For example, the ionic Mg^{++}-O^{--} bond length is 2.10 Å, compared to 1.95 Å for the covalent Mg-O bond.[1] More importantly, the position of the electron density minimum which marks the nominal point of contact of the ions or atoms moves closer to the oxygen atom as the covalent component of the bond increases. This causes an increase in the metal-to-oxygen radius ratio. Thus the Mg^{++}/O^{--} ideal ionic radius ratio is 0.5, compared to 2.7 for the corresponding ideal atomic radius ratio. Clearly, even a relatively small change in the nature of the metal-oxygen bond has a large influence on the effective radius ratio.

The consequence of the increase of effective metal-oxygen radius ratio with increasing pressure is that, when a high-pressure transformation occurs, the high-pressure polymorph nearly always belongs to a class of structures characterized by higher cation-to-anion radius ratios than the low-pressure structure. Since an increase in the cation-to-anion radius ratio also leads to higher metal-oxygen coordination numbers and thereby to structures of higher density, it is clear that the thermodynamic factor, through the $P\ \Delta v$ term in the free energy, also works in the same direction as the change in radius ratio with pressure. We have already noted that experience shows a strong probability that a new high-pressure phase will belong to a known crystal-structure type. The further inferences that the high-pressure structure is likely to belong to a class characterized by a higher cation-to-anion radius ratio and necessarily, from thermodynamic considerations, is likely to be denser than the low-pressure structure, provide most useful guides in predicting the types of structures to which a given phase may transform. Clearly, these relationships point to the desirability of studying the behaviour of a wide variety of oxide compounds under the available pressures so that the relationships of silicates with other sets of isostoichiometric oxide compounds can be adequately explored. In the following chapters we will be particularly concerned with studying the relationships between silicates and germanates. Sometimes, however, it will be found useful to proceed further afield and consider the crystal chemistry of manganates, titanates, stannates, and plumbates. We will find that high-pressure transformations such as those to be expected in the earth's interior do not belong to a mysterious and unpredictable class but can readily be interpreted and systematized in a framework based upon the vast amount of comparative crystal chemical data which has been obtained during the last 50 years, both at atmospheric pressure and at pressures up to 100 kbars.

Germanates as High-Pressure Models of Silicates

The crystal chemical relationships between germanates and silicates were first elucidated in a classical paper by Goldschmidt (1931). Both silicon and ger-

[1]Magnitsky and Kalinin (1959).
[2]Wada (1960b).
[3]Drickamer (1963).
[4]Neuhaus (1968).
[5]Amoros and San Miguel (1968).

Table 10-1 **EXAMPLES OF ISOSTRUCTURAL SILICATES AND GERMANATES.** (*After Ringwood and Seabrook, 1963*)

Si compound	Ge compound	Structure type
SiO_2	GeO_2	Quartz
SiO_2	GeO_2	Rutile
$MgSiO_3$	$MgGeO_3$	Orthopyroxene
$MgSiO_3$	$MgGeO_3$	Clinopyroxene
$CaMgSi_2O_6$	$CaMgGe_2O_6$	Diopside
Mg_2SiO_4	Mg_2GeO_4	Olivine
Ca_2SiO_4	Ca_2GeO_4	Olivine
Mn_2SiO_4	Mn_2GeO_4	Olivine
$CaMgSiO_4$	$CaMgGeO_4$	Olivine
Ni_2SiO_4	Ni_2GeO_4	Spinel
Co_2SiO_4	Co_2GeO_4	Spinel
Fe_2SiO_4	Fe_2GeO_4	Spinel
Zn_2SiO_4	Zn_2GeO_4	Phenacite
Be_2SiO_4	Be_2GeO_4	Phenacite
$ThSiO_4$	$ThGeO_4$	Zircon
$SrSiO_3$	$SrGeO_3$	Pseudowollastonite
Li_2SiO_3	Li_2GeO_3	
Li_4SiO_4	Li_4GeO_4	
Na_2SiO_3	Na_2GeO_3	
$Sc_2Si_2O_7$	$Sc_2Ge_2O_7$	Thortveitite
$Bi_4Si_3O_{12}$	$Bi_4Ge_3O_{12}$	Eulytite
$BaTiSi_3O_9$	$BaTiGe_3O_9$	Benitoite
$NaAlSi_3O_8$	$NaAlGe_3O_8$	Felspar
$KAlSi_3O_8$	$KAlGe_3O_8$	Felspar
$CaAl_2Si_2O_8$	$CaAl_2Ge_2O_8$	Felspar
$BaAl_2Si_2O_8$	$BaAl_2Ge_2O_8$	Felspar
$Ca_3Al_2Si_3O_{12}$	$Ca_3Al_2Ge_3O_{12}$	Garnet
$Ca_3Cr_2Si_3O_{12}$	$Ca_3Cr_2Ge_3O_{12}$	Garnet
$Ca_3Fe_2Si_3O_{12}$	$Ca_3Fe_2Ge_3O_{12}$	Garnet
$NaAlSiO_4$	$NaAlGeO_4$	Nepheline
$KAlSi_2O_6$	$KAlGe_2O_6$	Leucite
$3NaAlSiO_4 \cdot NaCl$	$3NaAlGeO_4 \cdot NaCl$	Sodalite
$Mg_3Si_2O_5(OH)_4$	$Mg_3Ge_2O_5(OH)_4$	Serpentine
$Ni_3Si_2O_5(OH)_4$	$Ni_3Ge_2O_5(OH)_4$	Serpentine
$Mg_3Si_4O_{10}(OH)_2$	$Mg_3Ge_4O_{10}(OH)_2$	Talc

manium readily form tetravalent ions which possess similar outer electronic structures and radii[1] (Si^{4+}—0.26 Å; Ge^{4+}—0.40Å). Accordingly, the crystal chemistry of silicates is very closely related to that of germanates. Although some exceptions occur, corresponding silicates and germanates are usually isostructural and, at elevated temperatures, display complete solid solutions with one another. To illustrate this point, a list of isostructural germanates and silicates, exhibiting a wide variety of different structures, is given in Table 10-1. From these observations it appears that if a germanate with a new structure should be synthesized, there would be a reasonable probability that, under some appropriate *P-T* conditions, a corresponding isostructural silicate would be stable. Thus the discovery

[1]The effective crystal radii are closer than given above owing to a large covalent component in the bonds. Metal-oxygen distances for fourfold coordination are Si-O—1.62 Å; Ge-O—1.74 Å (Shannon and Prewitt, 1969).

of a phase transformation in a given germanate would be a useful pointer towards the occurrence of a similar transformation in the corresponding silicate under appropriate *P-T* conditions.

Although this is a useful relationship to exploit, there is an even more important one which is of great value to the student of the earth's mantle. It appears that germanates often behave as high-pressure models of the corresponding silicates. If a germanate is found to display a given phase transformation at a particular pressure, the corresponding silicate often displays the same transformation, but at a much higher pressure. The reverse of this relationship has not been observed. Furthermore, in cases where high-pressure polymorphism is displayed by a silicate, the corresponding germanate, if it does not display the same polymorphism, is often found to display only the structure of the high-pressure silicate polymorph. Examples of this relationship are given in Tables 10-2 and 10-3.

This relationship can be interpreted in terms of the crystal chemical principles discussed in the previous section. During compression, we can expect that the "effective" radius ratios R_{Si}/R_O and R_{Ge}/R_O will increase. Transformation into a new phase occurs when these radius ratios attain some critical values. Since the zero-pressure radius of Ge^{4+} is already somewhat larger than that of Si^{4+}, germanates require smaller pressures to achieve the critical radius ratios required for given transitions than silicates do. Alternatively, because of their initially higher

Table 10-2 COMPARATIVE STABILITIES OF ISOSTRUCTURAL GERMANATE AND SILICATE PHASES. (IN EACH CASE, THE GERMANATE IS STABLE AT A LOWER PRESSURE THAN THE SILICATE AT THE SAME TEMPERATURE.) *(After Ringwood, 1972)*

Structure type	Germanate compound	Silicate compound
Rutile	GeO_2	SiO_2
Garnet	$Ca_3Al_2Ge_3O_{12}$	$Ca_3Al_2Si_3O_{12}$
	$Na_2CaTi_2Ge_3O_{12}$	$Na_2CaTi_2Si_3O_{12}$
	$Ca_3MgTiGe_3O_{12}$	$Ca_3MgTiSi_3O_{12}$
Spinel	Ni_2GeO_4	Ni_2SiO_4
	Co_2GeO_4	Co_2SiO_4
	Fe_2GeO_4	Fe_2SiO_4
	$(Mg_{0.8}Fe_{0.2})_2GeO_4$	$(Mg_{0.8}Fe_{0.2})_2SiO_4$
	$LiAlGeO_4$	$LiAlSiO_4$
Kyanite	Al_2GeO_5	Al_2SiO_5
Scheelite	$ZrGeO_4$	$ZrSiO_4$
	$HfGeO_4$	$HfSiO_4$
Jadeite	$NaAlGe_2O_6$	$NaAlSi_2O_6$
Perovskite	$CaGeO_3$	$CaSiO_3$
Hollandite	$KAlGe_3O_8$	$KAlSi_3O_8$
Pyrochlore	$Sc_2Ge_2O_7$	$Sc_2Si_2O_7$
	$In_2Ge_2O_7$	$In_2Si_2O_7$

Table 10-3 COMPARISON OF GERMANATE AND SILICATE DISPROPORTIONATION REACTIONS. (*After Ringwood, 1970*)

Reaction		*P* (kbars)	*T* (°C)
$2FeGeO_3$	→ $Fe_2GeO_4 + GeO_2$	10	700
$2FeSiO_3$	→ $Fe_2SiO_4 + SiO_2$	100	1000
(Pyroxene)	→ (spinel) + (rutile)		
$2CoGeO_3$	→ $Co_2GeO_4 + GeO_2$	10	700
$2CoSiO_3$	→ $Co_2SiO_4 + SiO_2$	100	1000
(Pyroxene)	→ (spinel) + (rutile)		
$NaAlGe_3O_8$	→ $NaAlGe_2O_6 + GeO_2$	15	1100
$NaAlSi_3O_8$	→ $NaAlSi_2O_6 + SiO_2$	28	1100
(Albite)	→ (jadeite) + (rutile, quartz)		
$3KAlGe_2O_6$	→ $2KAlGe_3O_8 + KAlO_2$	<90	1000
$3KAlSi_2O_6$	→ $2KAlSi_3O_8 + KAlO_2$	100	1000
(Leucite)	→ (hollandite)		

radius ratios, germanates may crystallize at zero pressure in a structure which is only attained by the silicate at high pressure.

For these reasons, the study of germanates as high-pressure models of silicates offers us the possibility of obtaining useful information about phase transformations which may occur in silicates at pressures beyond the range of currently available techniques. There are several ways in which these relationships may be taken advantage of. The first is purely qualitative. A systematic study at high pressure is made of the stabilities of germanate isotypes of the principal silicate minerals. In the vast majority of cases which have been investigated at pressures up to 100 kbars, the germanates have been found to transform to denser phases. From the considerations advanced above, there is a fairly high probability that silicates will transform to these phases at much higher pressures. This is a most valuable aid in exploration. Furthermore, from analogy with the germanates, various important properties, particularly density and probably also elastic properties, of the high-pressure silicates can be estimated and thus used to aid in the interpretation of the properties of the deep mantle.

Many transformations in germanates involve either a complete or a partial change in coordination of the germanium ions from fourfold to sixfold. In pure GeO_2, this change in coordination occurs at atmospheric pressure and at a temperature of 1007°C. However, in complex germanium oxycompounds, germanium almost invariably occurs in tetrahedral coordination at atmospheric pressure (excluding hydroxyl- and fluorine-bearing varieties). It appears that the combination of a basic oxide with GeO_2 tends to stabilize germanium in tetrahedral coordination, leading to the formation of "acid" GeO_4^{4-} groups.[1]

Because of the relative stabilization of GeO_4^{4-} groups in germanates as com-

[1]Weyl (1951, 1956).

pared with GeO_2, higher pressures are required to cause change in Ge coordination from 4 to 6 than in GeO_2. This generalization may reasonably be applied to silicates. It implies that high-pressure transformations in silicates which lead to octahedral coordination for silicon are likely to require pressures that are higher than those required for the 4 to 6 transformation in pure SiO_2. Thus, the equilibrium curve for the coesite-stishovite transformation (Fig. 13-4) might be expected to constitute a lower pressure limit for 4 to 6 transformations in silicates. Direct observations to date are consistent with this rule.

Prediction of Stabilities of New Phases from Thermodynamics of Germanate-Silicate Solid-Solution Equilibria

The relationships between silicates and germanates can also be used in a more quantitative manner than discussed above. Because of the similarity between germanium and silicon ions, and the complete solid solubility which is commonly displayed between isostructural germanates and silicates, it is probable that, at high temperatures, the thermodynamic behaviour of germanate-silicate solid solutions will not depart far from ideality.[1] This opens the possibility of using solid-solution equilibria between a germanate possessing a "high-pressure" structure (e.g., spinel) and a silicate possessing a "low-pressure" structure (e.g., olivine) to calculate the pressure at which the silicate transforms to the germanate structure.

This method was first applied by the author[1] to a consideration of olivine-spinel equilibria in the pseudobinary system Ni_2GeO_4-Mg_2SiO_4 at atmospheric pressure. It was found that, at 1500°C, Ni_2GeO_4 spinel dissolved 9 mol% Mg_2SiO_4. The spinel solid solution was found to be in equilibrium with an olivine solid solution of composition $(Mg_{0.40}Ni_{0.60})_2\,(Si_{0.55}Ge_{0.45})O_4$. The solid solution of Mg_2SiO_4 in Ni_2GeO_4 may be pictured in two stages. First, energy must be supplied to the stable form (olivine) of Mg_2SiO_4 in order to transform it to hypothetical spinel form. Secondly, both spinels, Ni_2GeO_4 and Mg_2SiO_4, undergo mixing and solid solution. There is a balance between the magnitude of the (free) energy required to transform the Mg_2SiO_4 from olivine to spinel and the (free) energy involved in the mixing of Mg_2SiO_4 and Ni_2GeO_4. If the energy of transformation is high, then only a small amount of Mg_2SiO_4 will be able to enter into solid solution. If the transformation energy is low, then an extensive range of solid solutions may be formed. It is clear from this qualitative argument that there is a relationship between the free energy of transformation of Mg_2SiO_4 olivine to the spinel structure and the amount of solid solution possible between Mg_2SiO_4 and Ni_2GeO_4.

Consider Mg_2SiO_4 as a component of both solid solutions in equilibrium at a given temperature—say, 1500°C. Let μ_1 represent the chemical potential of

[1]Ringwood (1956, 1958a).

Mg_2SiO_4 in the olivine phase and μ_2 represent the chemical potential of Mg_2SiO_4 in the spinel:

$$\mu_1 = \mu_1^\circ + RT \ln a_1$$

$$\mu_2 = \mu_2^\circ + RT \ln a_2$$

where μ_1° = chemical potential of pure Mg_2SiO_4 olivine at 1500°C

μ_2° = chemical potential of pure Mg_2SiO_4 spinel at 1500°C

a_1 = activity of Mg_2SiO_4 in the olivine solid solution

a_2 = activity of Mg_2SiO_4 in the spinel solid solution

At equilibrium, $\mu_1 = \mu_2$:

$$\therefore \quad \mu_2^\circ - \mu_1^\circ = RT \ln \frac{a_1}{a_2} = \Delta G^\circ$$

where ΔG° is the difference in chemical potential between the two pure modifications of Mg_2SiO_4 (i.e., ΔG° is the molar free energy of transformation of Mg_2SiO_4 between the olivine and spinel structures at 1500°C).

To obtain the activities required to calculate ΔG°, it would be possible to make a direct study of the thermodynamic properties of the solid solutions. However, this is not necessary if a host structure has been selected such that the replacing ions are very similar in crystal chemical properties. This is believed to be the case with the Si-Ge and Mg-Ni replacements under consideration. At high temperatures it is likely that these solid solutions will behave approximately ideally. In this case, according to Raoult's law, the activity of an ion becomes equal to its ionic fraction, which is the ratio of the number of specified ions to the total number of lattice sites available to them, i.e.,

$$a_{Mg} = \frac{N_{Mg}}{N_{Mg} + N_{Ni}} = [Mg]$$

whilst

$$a_{Si} = \frac{N_{Si}}{N_{Si} + N_{Ge}} = [Si]$$

In the case of a complex solid solution in which Ni and Mg ions mix at octahedral lattice sites, whilst Si and Ge mix independently at tetrahedral lattice sites, the activity[1] of Mg_2SiO_4 is given by $[Mg]^2\,[Si]$. Accordingly,

$$\Delta G^\circ = RT \ln \left\{ \frac{[Mg_1]^2\,[Si_1]}{[Mg_2]^2\,[Si_2]} \right\}$$

Substituting values for $[Mg_1]\,[Si_1]$ and $[Mg_2]\,[Si_2]$ from the measured compositions of olivine and spinel solid solutions in equilibrium, ΔG° for the

[1]Tempkin (1945).

olivine-spinel transition in Mg_2SiO_4 was found to be 70,000 ± 10,000 joules/mol at 1500°C. From a well-known thermodynamic relationship, ΔG° is also equal to $\int_0^P \Delta v \, dP$, where Δv is the molar volume difference between the two polymorphs and P is pressure. Δv was estimated by several crystal chemical methods, the most direct being an extrapolation of lattice parameters of spinel solid solutions. The values obtained were in the range 4.4 ± 1.0 cm^3. Because of the rather small compressibilities and thermal expansions of the phases concerned, the function $\int_0^P \Delta v \, dP$ is approximately equal to $P \, \Delta v$. It is possible to make a first-order correction using estimated compressibilities and thermal expansions. The appropriate mean value of Δv is found to be 4.0 + 1.0 cm^3/mol. Equating the $P \, \Delta v$ term to ΔG°, the pressure required to transform pure Mg_2SiO_4 from the olivine to the spinel structure at 1500°C was found to be 175,000 ± 55,000 bars, equivalent to depths of 500 ± 140 km. The calculated depths are thus within the transition zone and overlap the 20° discontinuity. This was the first quantitative evidence supporting the Jeffreys-Bernal specific hypothesis and Birch's more general hypothesis. Subsequent direct high-pressure experiments by Ringwood and Major (1966) have succeeded in synthesizing magnesia-rich silicate spinels within the pressure range predicted.

The above method has also been used successfully to predict the occurrence of an olivine-spinel transformation in Ni_2SiO_4.[1] Solid-solution equilibria between Ni_2GeO_4 (spinel) and Ni_2SiO_4 (olivine) were determined experimentally. Using the relationships derived above, the free-energy difference between the olivine and hypothetical spinel modifications of Ni_2SiO_4 was calculated as a function of temperature. The molar volume of Ni_2SiO_4 spinel was obtained by extrapolating the lattice parameters of $Ni_2(Ge, Si)O_4$ spinel solid solutions. Using this information, an equilibrium curve was calculated for the transformation. At the time when this work was done, the spinel modification of Ni_2SiO_4 was unknown, and some preliminary experiments[2] had failed to synthesize this phase at pressures of 95 kbars. The calculated equilibrium diagram indicated that Ni_2SiO_4 spinel should become stable at relatively low pressures, within the capabilities of simple available high-pressure apparatus. Accordingly, further attempts were made to synthesize this phase. These were successful, and the pressure required to stabilize Ni_2SiO_4 spinel was found to agree quite well with the predicted value.[1]

The success of this method in the cases of Ni_2SiO_4 spinel and in Mg-rich silicate spinels raises confidence in its applicability to other systems—particularly those which are beyond the range of presently available direct high P-T techniques. The observation that a given silicate displays substantial solid solubility in a structurally different and closer packed germanate compound is of considerable importance in determining the probable behaviour of the silicate at high pressure. Such observations will be exploited more than once in later chapters.

[1] Ringwood (1962a).
[2] Wentorf (1959).

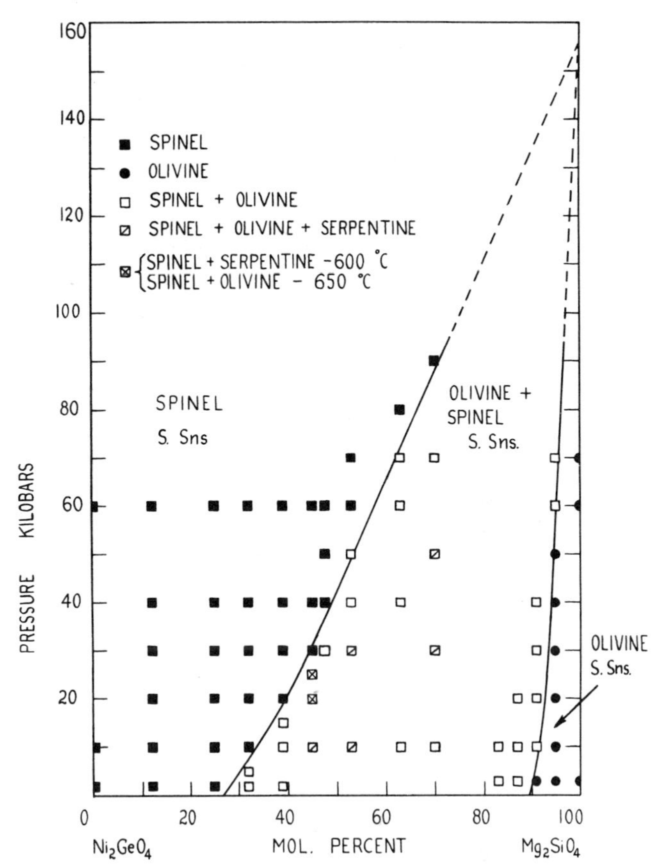

FIGURE 10-1
The system Ni_2GeO_4-Mg_2SiO_4 at 600°C and 0 to 90 kbars. (*From Ringwood and Seabrook 1962a, with permission.*)

Extrapolation Methods

Another method of using germanate-silicate phase equilibria to predict pressures at which transformations should occur in silicate end-members is by determining the solid-solubility boundaries in a given system over a wide range of pressures and then simply extrapolating the phase boundaries into higher pressure regions. Dachille and Roy (1960) used this approach to obtain a pressure for the olivine-spinel transformation in Mg_2SiO_4 by extrapolating phase boundaries determined in the system Mg_2GeO_4-Mg_2SiO_4 over a pressure range of 0 to 60 kbars. Their estimate for the olivine-spinel transformation in Mg_2SiO_4 was 100 ± 15 kbars at 530°C and 4.7% for the density change. These values are now known to be somewhat low. A more comprehensive but analogous investigation (Fig. 10-1) of the system Ni_2GeO_4-Mg_2SiO_4 over the pressure range 0 to 90 kbars by Ringwood (1958b) and Ringwood and Seabrook (1962a) led to the estimate 130 ± 20 kbars at 600°C and a density change ≥ 9%. These estimates are in good agreement with subsequent direct experiments. The general method used in the experiments is also of wide utility and has been applied in other investigations (Secs. 12-2 and 12-4).

10-3 SUMMARY OF RESULTS BY INDIRECT METHODS, 1952–1965

During the period 1952–1965, experimental static high *P-T* apparatus available for geophysical applications was unable to reproduce conditions in the mantle at depths much greater than 300 km. Accordingly, evidence relating to the occurrence of phase transformations in mantle silicates at depths greater than 400 km, as in Birch's hypothesis, could be obtained only by the indirect methods described previously (i.e., experiments conducted at less than 100 kbars). In the previous section, we reviewed some general aspects and achievements of these indirect methods. Specifically, they provided evidence that $(Mg,Fe)_2SiO_4$ olivine, probably the most abundant mineral in the upper mantle, was not stable at high pressure and was expected to transform to a spinel structure at a depth of 400 to 500 km in the mantle, with an increase in density of about 10%.[1,2,3] This conclusion was further supported by the discoveries that the minor and trace components of natural olivines Fe_2SiO_4,[4] Ni_2SiO_4,[5] and Co_2SiO_4[6] transformed directly to spinel structures at pressures between 20 and 70 kbars, at about 700°C, accompanied by density increases of 9 to 11%. These were the first direct transformations to be discovered in mantle-type silicates and removed any lingering doubts that silicates could crystallize in the spinel structure. Furthermore, the fact that Fe_2SiO_4 was a significant component of natural olivine in the mantle was particularly relevant in this context.

A key discovery was made by Stishov and Popova (1961), who found that quartz and coesite could be transformed at about 100 kbars, 1200°C into a new polymorph, "stishovite," possessing the rutile structure. The high density of this phase, 4.28 g/cm, was due to the fact that the silicon atoms were in octahedral coordination. This discovery greatly increased the range of transformations which might be possible in the deeper regions of the mantle. In particular, it suggested possible new modes of breakdown of the pyroxene and garnet families of minerals, probably second only to olivines in their abundance in the upper mantle.

[1]Ringwood (1956, 1958a,b,c).
[2]Dachille and Roy (1960).
[3]Ringwood and Seabrook (1962a).
[4]Ringwood (1958b,d).
[5]Ringwood (1962a).
[6]Ringwood (1963).

Ringwood and Seabrook (1962b, 1963) studied the stabilities of a number of germanate analogues of common silicates possessing pyroxene and pyroxene-like structures. All these phases were found to transform to denser phases or phase assemblages at modest pressures, suggesting similar behaviour of silicates at higher pressures. Thermodynamic calculations indicated that the transformation pressures for silicates corresponded to pressures in the transition zone.[1,2,3]

The bearing of these indirect experiments upon Birch's hypothesis was extensively discussed in a series of papers.[1,3,4,5,6] They showed that the common upper mantle minerals, olivine and pyroxenes, were indeed unstable at high pressures and should transform to new phases at depths between 400 and 1000 km. Furthermore, the properties of these new phases appeared to offer a satisfactory explanation of the gross physical properties of the deeper regions of the mantle.[5] Birch's hypothesis was thus essentially verified in a qualitative sense, and the role of phase transformations in the mantle was placed upon a firm basis.

10-4 DIRECT METHODS OF STUDYING MANTLE PHASE TRANSFORMATIONS

Despite the success of the indirect methods and results discussed in the previous sections, there remained a strong incentive to develop experimental methods which would be capable of reproducing the *P-T* conditions in the Transition Zone and Lower Mantle so that the predicted phases could be synthesized directly and their properties measured, thus leading to an understanding of the fine structure of the mantle. Furthermore, the prediction methods used are based upon the assumption that the high-pressure silicate phases will resemble germanates synthesized at relatively low pressure. Whilst this generalization is often valid, it is certainly not universally correct, and it is always possible that a high-pressure silicate phase may possess a structure which has no known germanate analogue (e.g., coesite) or that the sequence of transformations with increasing pressure in silicates may differ from that in germanates. There are also situations where germanate analogue studies suggest two or more possible types of high-pressure transformation for a given silicate. Clearly then, the model-system approach does not provide a final and sufficient answer, and ultimately, it is necessary to synthesize the predicted phases directly.

In recent years, important progress has been made towards this end. Thus,

[1]Ringwood (1962b).
[2]MacDonald (1962).
[3]Stishov (1962).
[4]Ringwood (1958c).
[5]Clark and Ringwood (1964).
[6]Ringwood (1966).

the development of shock-wave techniques[1] permits the generation of transient high pressures in the megabar region so that phase transformations in silicates can be observed directly and the equations of state of the high-pressure phases determined. A second field in which progress has been made has been in extending the pressure range of static high pressure-temperature apparatus. Equipment capable of developing up to 300 kbars at elevated temperatures is now in operation in a few laboratories. This is equivalent to a depth of about 800 km. Already a host of new transformations has been discovered, throwing much new light on the nature of the mantle. Activity in this field is likely to expand considerably during the next few years. Before discussing the experimental results, a brief description of some of the experimental methods currently in use appears appropriate.

Static High Pressure-High Temperature Apparatus

Reaction rates in most oxides and silicates are slow at temperatures less than 600°C. Accordingly, the achievement of transformations in these compounds necessitates that the samples be subjected to elevated temperatures whilst under high pressure. Conversely, the sluggish nature of reconstructive transformations in silicates and oxides often permits the high-pressure forms to be recovered by quenching or cooling to room temperature whilst under pressure, after which pressure is reduced, and the high-pressure modification in a metastable state may be removed from the apparatus and studied in the laboratory by conventional methods.

Several types of apparatus have been widely used in the pressure range below 100 kbars, e.g., the simple squeezer,[2] the piston cylinder apparatus,[3,4] the belt apparatus,[5] the tetrahedral anvil apparatus,[6] and the cube anvil apparatus.[7] Reviews of high-pressure apparatus and techniques have been published in books edited by Wentorf (1962), Giardini and Lloyd (1963), and Bradley (1969), and further descriptions of these types of apparatus, which have provided a wealth of experimental data relevant to the constitution of the upper mantle, are not required.

A pressure of 100 kbars corresponds to a depth of 300 km in the mantle. These classes of apparatus cannot therefore directly simulate the pressures in the transition zone. It is only during the last few years that apparatus capable of attaining 100 to 300 kbars simultaneously with temperatures of 1000°C and above have been developed. Brief descriptions of some of these follow. Before proceeding, however, we should consider the problem of pressure calibration.

Pressure calibration is ultimately dependent upon pressures assigned to certain key phase transformations—particularly those occurring in bismuth, thallium,

[1]E.g., McQueen, Marsh, and Fritz (1967).
[2]Griggs and Kennedy (1956).
[3]Coes (1962).
[4]Boyd and England (1960).
[5]Hall (1960).
[6]Hall (1958).
[7]Von Platen (1962).

barium, tin, and iron. These pressures have fluctuated alarmingly during recent years, but there are now heartening signs of convergence towards values for transformations below 100 kbars which are unlikely to be seriously in error. In the present review, we use values recommended at the Symposium on the Accurate Characterization of the High Pressure Environment—U.S. National Bureau of Standards, October 1968, as follows:

$Bi_{I\text{-}II}$, 25.5 kbars

$Tl_{II\text{-}III}$, 36.7 kbars

$Ba_{I\text{-}II}$, 55 kbars

$Bi_{III\text{-}V}$, 77 kbars

Sn, 96 kbars[1]

There is a dearth of reliable calibration points between 100 and 200 kbars. Most workers in this region have estimated their pressures by means of some kind of extrapolation procedure, using the fixed points below 100 kbars. However, since serious errors in some of these points were present until comparatively recently, the extrapolated pressures were sometimes grossly in error. The most common result has been for workers to overestimate the pressures obtained in their apparatus. In the following discussion, pressure estimates are based upon the NBS scale,[2] and pressures attributed to different kinds of apparatus may differ considerably from those claimed by their inventors. It is hoped that, in the future, the cross-checking of solid-solid transitions in the 100 to 300 kbar range against a widely accepted continuous pressure-density relationship for sodium chloride will alleviate calibration problems.

The measurement of pressures in internally heated apparatus of small working volume poses an additional problem. Use of the "standard" transitions determined at low temperature is of limited avail since the operation of the heater in the very small volume of the pressure cell has a substantial effect upon pressure distribution. Accordingly, it is necessary to use a series of secondary standards—transformations determined in large-volume apparatus at temperatures similar to those which are used in the small-volume apparatus. Useful transformations for this purpose are the olivine-spinel transitions in Fe_2SiO_4 and Co_2SiO_4 and the coesite-stishovite transition. At 1000°C, the pressures for these transitions, referred to the NBS pressure scale, are 52 kbars (Fe_2SiO_4),[3] 73 kbars (Co_2SiO_4),[4] and 98 kbars (SiO_2).[5]

The first ultrahigh-pressure apparatus to be described was the high-compression belt apparatus of Bundy (1963) (Fig. 10-2). Inward movement of the tapered pistons causes partial extrusion of the pyrophyllite gaskets, resulting in grada-

[1]See also Drickamer (1970).
[2]Lloyd (1971).
[3]Akimoto, Komada, and Kushiro (1967).
[4]Akimoto and Sato (1968).
[5]Akimoto and Syono (1969).

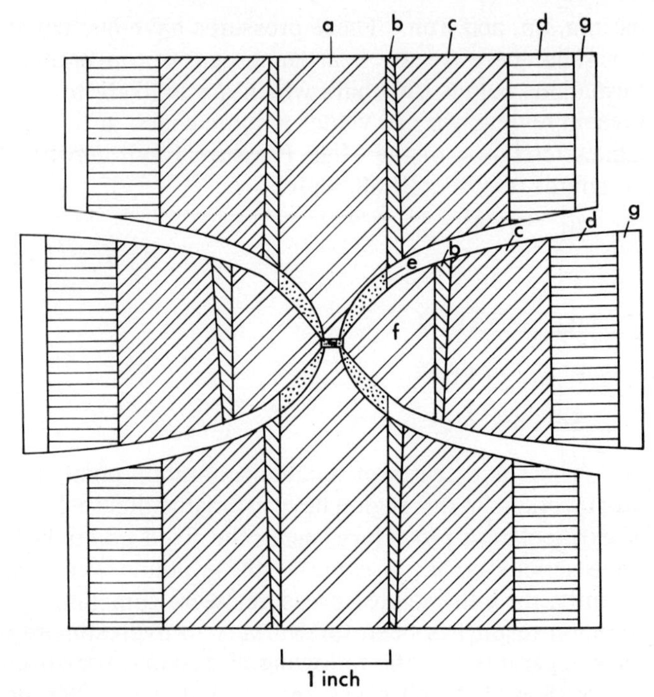

FIGURE 10-2
High-compression belt apparatus. (*a*) Carbide piston; (*b*) repair ring; (*c*) and (*d*) steel support rings; (*e*) extrudable pyrophyllite gasket; (*f*) carbide die; and (*g*) safety ring. (*From Bundy, 1963, with permission.*)

tional support to the tapered surfaces of pistons and pressure vessel, rising to a maximum where the ends of the pistons confront each other. In the high-pressure volume, a small resistance heater is placed. Samples may be heated to about 1000°C for extended periods of time or, transiently, to as much as 4000°C by means of a condenser discharge. This apparatus is capable of attaining up to about 150 kbars at elevated temperatures. The high-compression belt was used in the first successful synthesis of diamond directly from graphite (without use of a catalyst) and in many related investigations.[1] However, apart from some limited exploratory work on the olivine-spinel transformation,[2] it has not been used by experimental petrologists to the extent that its potential justifies.

Minomura et al.[3] have described a modification of Drickamer's apparatus[4] in which a graphite resistance heater is inserted between the faces of the pistons. Pressures in the vicinity of 200 kbars at 800°C have been claimed,[5] but in view of

[1]Bundy (1963).
[2]Sclar and Carrison (1966).
[3]Minomura et al. (1964).
[4]Drickamer and Balchan (1962).
[5]Akimoto and Ida (1966).

revisions in the pressure scale, it appears doubtful whether actual pressures have exceeded 150 kbars. A disadvantage of this apparatus is the minute amount of sample which is recovered after a run, which may not be sufficient to yield an x-ray-diffraction pattern of high quality.[1,2] Nevertheless, its possibilities were demonstrated by the achievement of a partial transformation in Mg_2SiO_4 olivine.[1]

Kawai[3] has described an interesting device with the potential for subjecting substantial volumes to high pressure. This is an extension of von Platen's cube, in which eight or more appropriately shaped pistons are arranged in a spherical configuration. The pistons are capable of taking electrical leads into the polyhedral central-pressure volume in which a sample may be heated and its temperature monitored by thermocouples. The sphere is placed in a standard piston cylinder apparatus, the pressure applied to the surface of the sphere being considerably magnified in the central region. More recently, Suito (1972) has described a double-staged modification of this apparatus in which a second set of eight cubic anvils are placed near the centre of the split-sphere anvils (Fig. 10-3). Pressures in excess of 200 kbars at temperatures up to 1200°C were achieved.

[1]Akimoto and Ida (1966).
[2]Ringwood and Major (1970).
[3]Kawai (1966).

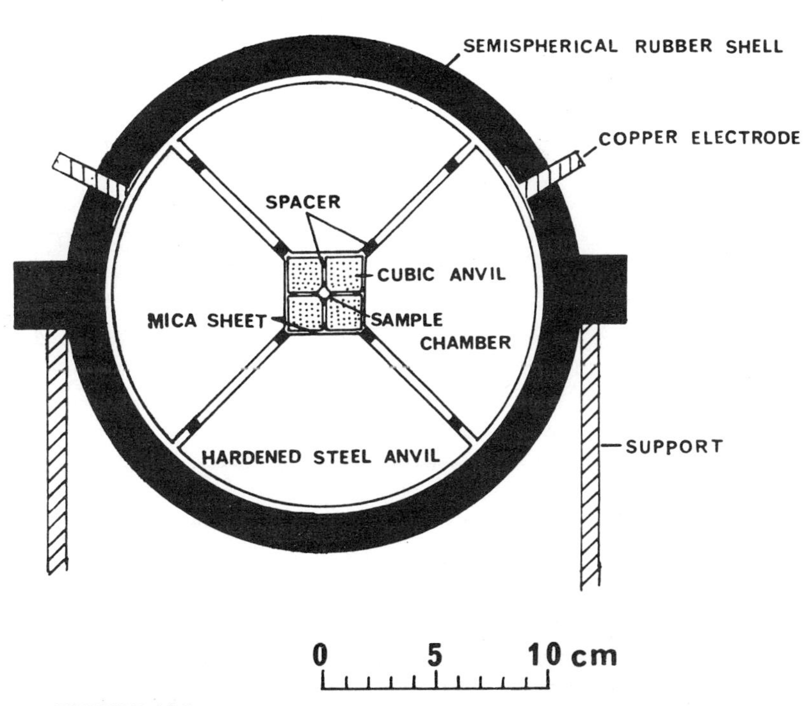

FIGURE 10-3
Split-sphere apparatus embodying double-stage anvil configuration. The assemblage is compressed in a piston-cylinder apparatus using oil as a pressure medium. (*From Suito, 1972, with permission.*)

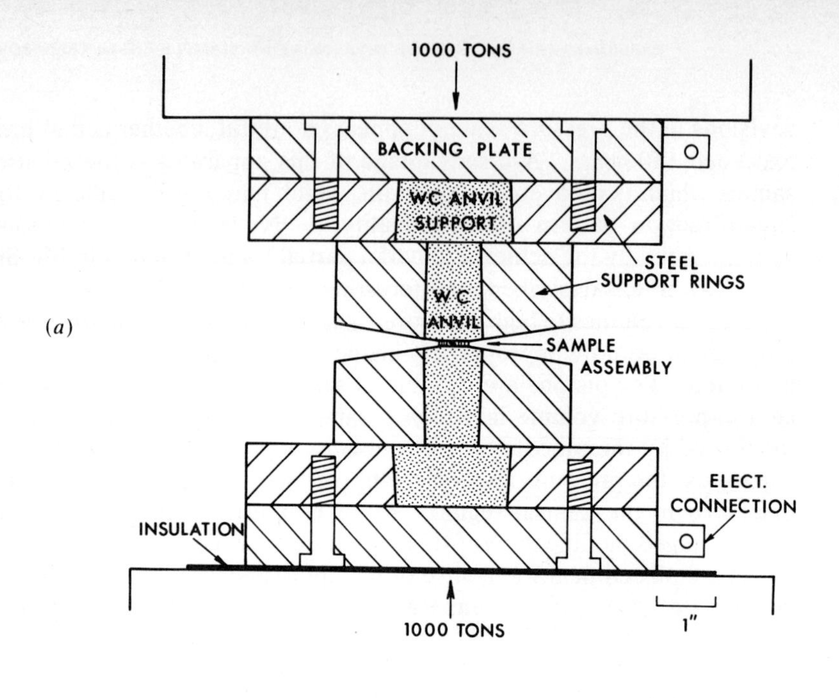

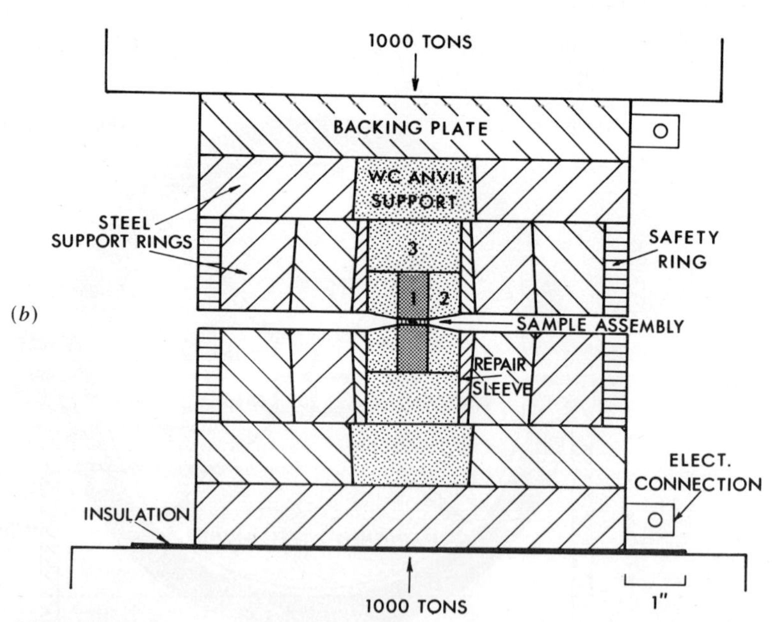

FIGURE 10-4
Bridgman anvils used in experiments by Ringwood and Major (1968, 1970). (*a*) Single support ring; (*b*) large compound anvil with compound support rings. The latter anvil is made up of a central cylinder (1), consisting of a material with extremely high compressive strength, e.g., binderless tungsten carbide or titanium carbide, surrounded by support rings and a backing plate. Components (2) and (3) are made from a softer grade of tungsten carbide. (*From Ringwood and Major, 1968, with permission.*)

A fourth type of high *P-T* apparatus has been described by Ringwood and Major (1968, 1970). Most of the high-pressure transformations to be described in later chapters have been discovered with this device. Accordingly, a more detailed description is supplied.

The apparatus consists of pairs of Bridgman anvils (Fig. 10-4) which have been provided with pressure cells containing internal heaters (Fig. 10-5). The design and construction of the pressure cell is the key to successful operation of this apparatus. The central alundum discs act effectively as a second set of pistons which confine the sample in an intermediate ring of alundum that serves as a die. The alundum components are supported by pyrophyllite rings which deform and extrude under pressure, supplying a high confining pressure to the alundum discs in the process. The central set of discs do not deform seriously under pres-

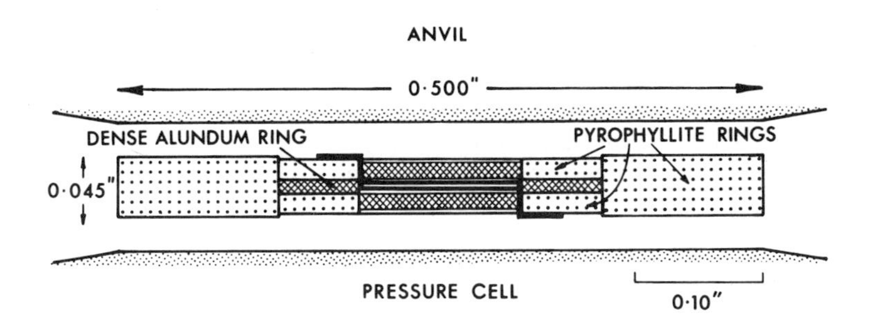

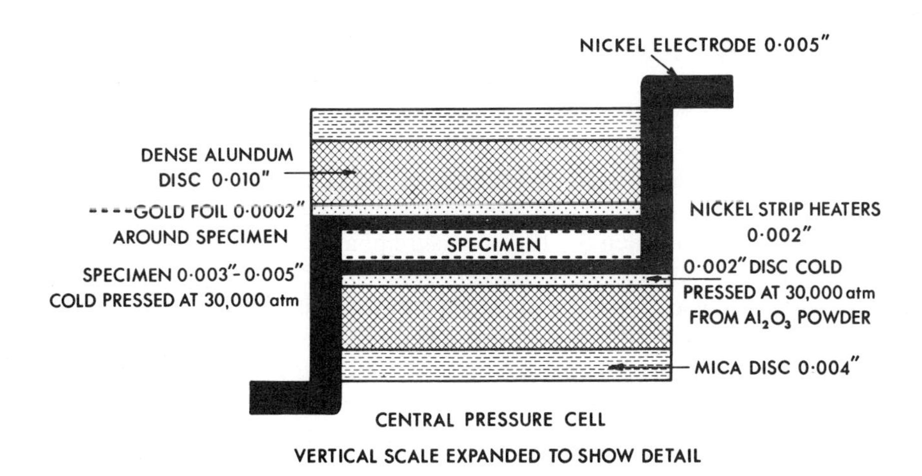

FIGURE 10-5
Internally heated pressure cell used in conjunction with Bridgman anvils. (*From Ringwood and Major, 1968, with permission.*)

sure, and it is possible to heat the nickel or platinum-rhodium strip furnace and the enclosed specimen reasonably uniformly and to recover the specimen after a run. Moreover, the arrangement acts as a pressure intensifier so that the pressure applied to the sample by the central alundum discs is about 60 kbars higher than the average pressure across the anvils. The combination of this pressure-intensification effect with the very high average pressures already obtainable from the Bridgman anvil apparatus results in the generation of pressures which exceed 200 kbars.

Between 80 and 150 kbars, the pressure on the sample in individual runs is believed to be correct to ± 10%, assuming a pressure of 98 kbars for the coesite-stishovite transition at 1000°C, which is the principal calibration point. With care and with repetition of runs, the uncertainty in determining a given phase boundary can be reduced to ± 5%. Outside the 80 to 150 kbar interval, errors in pressure become much larger. While under pressure, the sample is heated by passing a current through the nickel strip furnace. The temperature of the centre of the specimen is maintained at about 1000°C, which is sufficiently high to enable reactions to proceed within reasonable time intervals. Temperature is not employed as an independent variable in phase equilibria determined with this apparatus. After completion of an experimental run, temperature is swiftly lowered by terminating the power supply; pressure is then lowered and the sample extracted to be examined by optical and x-ray methods.

Because of the small sample space and consequent high temperature gradients, it is not possible to control temperature accurately with the previous apparatus. This problem has been solved recently with the development of new ultrahigh-pressure systems based on Bridgman anvils which are capable of containing much larger samples together with thermocouples.

Nishikawa and Akimoto (1971) have increased the diameter of the Bridgman anvils (Fig. 10-4) to 1 inch and have designed a new kind of pressure cell (Fig. 10-6) to be placed between the anvils. This permits a sample on the order of 10 to 20 mg to be heated over a range of temperatures up to 1200°C at pressures up to 150 kbars.

Finally, some ultimate limitations of the static techniques described above should be mentioned. These are applicable to transformations which can be quenched—i.e., in which the high-pressure phase is preserved intact after cooling under pressure followed by reduction to atmospheric pressure. The degree of metastability of these quenched phases increases with transformation pressure and volume change. At 200 kbars a quenched high-pressure phase 10% denser than the low-pressure phase is unstable relative to that phase by about 500 cal/cm^3. Clearly, there will be a relationship between the degree of metastability and a temperature T_c at which the reverse transformation runs spontaneously in a reasonable time (order of hours). In the case of stishovite, a substantial degree of reverse transformation to silica glass is observed at about 390°C in 6 hours.[1] Ultimately,

[1]Skinner and Fahey (1963).

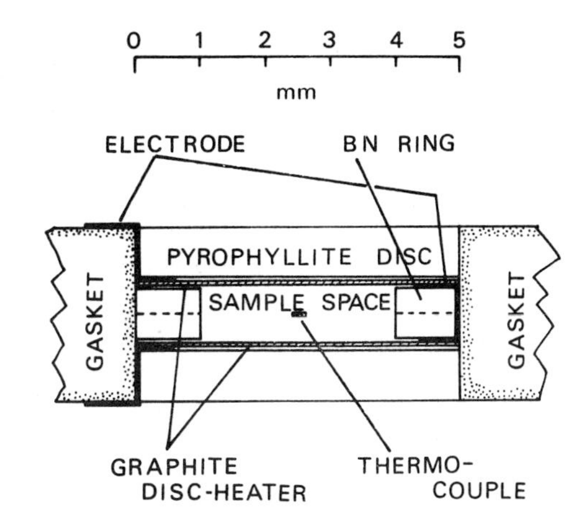

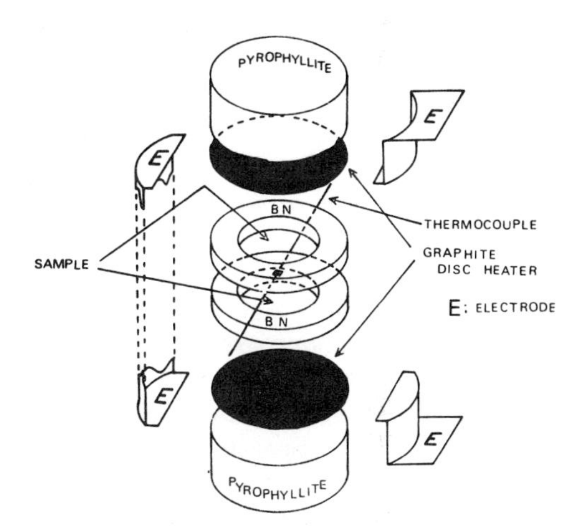

FIGURE 10-6
High pressure-temperature cell for use with 1-in.-diameter Bridgman anvils. Gasket is 1 in. in diameter and is constructed from semifired pyrophyllite. (*From Nishikawa and Akimoto, 1971, with permission.*)

with higher pressure transformations, T_c will become smaller than 15°C, and it will be impossible to preserve high-pressure phases by quenching to room temperature for periods sufficiently long to observe and study their properties. Already this stage has been reached in some systems where there is evidence that high-pressure phases formed between 150 and 200 kbars have reverted into low-pressure modifications when pressure was released.[1]

If this behaviour should become common above 200 kbars, as current investigations suggest, it will be necessary to abandon the quenching technique and to perform identification and measurements upon phases in situ whilst they are under pressure. The most promising technique here is x-ray diffraction by a sample

[1]Sections 12-4 and 12-5.

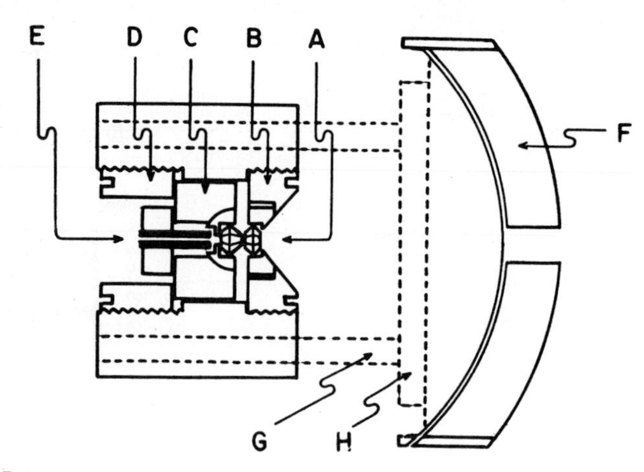

FIGURE 10-7
Cross section of diamond anvil press developed by Bassett et al., 1967. A—Diamond anvils; B—stationary piston; C—sliding piston; D—driver screw; E—collimator; F—film cassette; G—cassette mounting rods; H—cassette translating bars. (*From Bassett et al., 1967; with permission.*)

compressed between two diamond anvils. Bassett and coworkers (1967) have generated pressures above 300 kbars at room temperature by this method (Fig. 10-7). In order to activate the typically sluggish silicate transformations anticipated above 200 kbars, it would probably be necessary to heat the samples above 800°C. Important progress has already been made in this direction. In one device, an external furnace is placed around the diamond anvils as in the simple squeezer.[1,2] Temperatures of 800°C at pressures of about 200 kbars have been reached with this apparatus. In another system, a pulsed ruby laser beam has been used to heat the sample (but not the diamonds) to a temperature of about 3000°C whilst under pressure. More recently[3,4] a continuous yttrium aluminium garnet laser beam has been used to heat the sample to steady temperatures in the range 1000 to 2000°C at pressures exceeding 300 kbars. Some exciting new results have been obtained with this system (Fig. 10-8), as discussed in Secs. 11-5 and 12-4.

Shock-Wave Methods

Shocks produced directly or indirectly by high explosives enable the generation of pressures up to several million bars (megabars) in many metals, oxides, and sili-

[1]Bassett and Ming (1973).
[2]Mao and Bell (1971).
[3]Ming and Bassett (1974).
[4]Liu (1974).

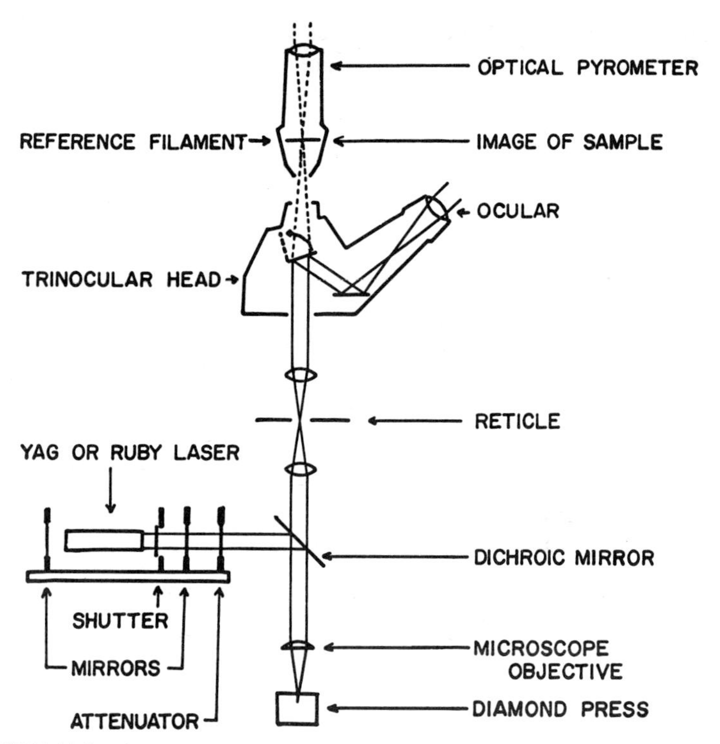

FIGURE 10-8
Schematic diagram showing optical system for heating a sample under pressure in a diamond anvil press by means of a focussed laser beam. Both pulsed ruby laser and continuous YAG laser can be used. Temperature is estimated by an optical pyrometer operating upon incandescent light emitted by sample. (*After Ming and Bassett, 1974, and Liu, 1974.*)

cates. Although these pressures are maintained only for very short intervals of time, usually less than a microsecond, they are sufficient to permit the measurement of some important properties. Since the pressure at the centre of the earth is about 3.6 Mbars, it is evident that shock-wave methods make the entire internal pressure regime within the earth accessible under laboratory conditions. Among the many recent reviews of shock-wave methods, the articles of Rice et al. (1958), Hamann (1966), Doran and Linde (1966), and McQueen, Marsh, and Fritz (1967) are especially useful.

The fundamental equations applying to the propagation of shock waves are known as the *Rankine-Hugoniot* relationships. These are obtained by applying the laws of conservation of mass, momentum, and energy across the shock-wave front, and they relate pressure P, internal energy E, specific volume V (or density ρ) behind the shock wave to these same quantities in front of the shock wave, in terms of the shock velocity U_s and the particle or mass velocity U_p. These equations lead to the following relationships[1] (the zero subscripts refer to initial state):

$$U_s = V_0 \left[\frac{P - P_0}{V_0 - V} \right]^{1/2}$$

$$U_p = [(P - P_0)(V_0 - V)]^{1/2}$$

$$E - E_0 = \frac{(P + P_0)(V_0 - V)}{2}$$

These relations allow the determination of pressure and energy in the shocked material as functions of the volume and define the Hugoniot equation of state of the material. This is the locus of all the P-V states which may be reached from a particular initial condition by means of single shock compressions. By carrying out a number of separate experiments using shocks of differing intensity, a series of P-V determinations is made, thus defining the Hugoniot for the material (e.g., Fig. 10-9). The key experimental measurements are of initial density (ρ_0), shock velocity (U_s), and particle velocity (U_p). Empirically, it is found that these latter quantities are linked by the relationship

$$U_s = C_0 + kU_p$$

where k is a constant determined by experiment, and C_0 is the hydrodynamical sound velocity of the material at atmospheric pressure:

$$C = \left(\frac{\partial P}{\partial \rho} \right)_s^{1/2}$$

The pressure P_H as a function of volume along the Hugoniot is then

$$P_H = \frac{C_0^2 (V_0 - V)}{[V_0 - k(V_0 - V)]^2}$$

This equation neglects the strength or rigidity of the material, assuming that this is small compared to the stresses developed in the shock. Whilst this is usually a reasonable assumption where P_H exceeds a few hundred kilobars, it is not always justified.[2]

Whilst the shock experiments provide a relatively direct and accurate determination of P and V along the Hugoniot, the determination of the corresponding temperature is less direct and more uncertain, requiring additional equation of state data for the material. The temperature gradient along the Hugoniot is higher than along the adiabat because the total energy involved is both internal and kinetic. The more compressible the starting material, or the higher the initial porosity, the higher are the shock temperatures.

McQueen, Marsh, and Fritz (1967)[3] determined the Hugoniots of a number of rocks and minerals of geophysical significance. These important experiments have provided invaluable raw material for many interpretative papers dealing with shock-wave-data reduction and its application to the constitution of the lower

[1] McQueen, Marsh, and Fritz (1967).
[2] E.g., the shock compression of corundum, Ahrens et al. (1969).
[3] See also McQueen and Marsh (1966).

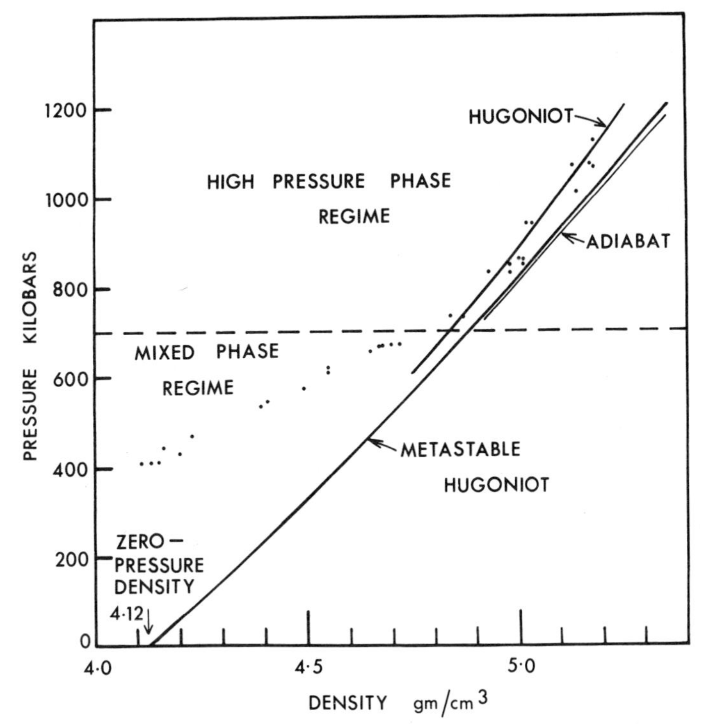

FIGURE 10-9
Pressure-density plot of Twin Sisters dunite under shock compression. Data points from McQueen, Marsh, and Fritz (1967). High-pressure regime represents Hugoniot of material which has been completely converted to high-pressure phase, with zero-pressure density obtained by extrapolation using finite strain theory and empirical ϕ-ρ relationship. Mixed-phase regime represents state of incomplete conversion to high-pressure phase. Metastable Hugoniot and adiabat for high-pressure phase are centered at state $P = 0$ and $T = 25°C$. Isotherm (not shown) is almost coincident with adiabat. (*After Ahrens et al., 1969.*)

mantle.[1,2,3,4,5,6,7,8,9,10] McQueen and coworkers demonstrated that many rocks and minerals displayed major phase transformations at pressures of a few hundred kilobars. Among these were the key upper mantle minerals—olivines and pyroxenes. A rock which was carefully studied, and for which good data exist, is the Twin Sisters dunite (Fig. 10-9). The scatter of experimental points gives an idea

[1]McQueen et al. (1964, 1967).
[2]Birch (1961, 1964).
[3]Takeuchi and Kanamori (1966).
[4]Wang (1967, 1968, 1969).
[5]Anderson and Kanamori (1968).
[6]Knopoff and Shapiro (1969).
[7]Shapiro and Knopoff (1969).
[8]Davies and Anderson (1971).
[9]Ahrens, Anderson, and Ringwood (1969).
[10]Davies and Gaffney (1973).

of experimental uncertainties. The area below 700 kbars is interpreted as a mixed-phase region in which phase transformations have not gone to completion. Above 800 kbars the data points are interpreted as referring to a single high-pressure phase. Clearly, it is important to determine the properties of this phase, particularly the density ρ_0 and hydrodynamical sound velocity C_0 which it would possess if it were possible to retain the high-pressure structure at atmospheric pressure.

Determination of these properties involves several steps, each of which introduces significant uncertainties. First, a curve representing the raw Hugoniot of the high-pressure phase is fitted to the experimental points in the single-phase region. Although the experimental scatter in Fig. 10-9 is not large, much greater scatter was observed in some other rocks and minerals. Further uncertainty in the location of the raw Hugoniot may also be caused by subjective judgments of the position of the boundary between mixed-phase and high-pressure-phase regimes and by the assumption of homogeneity in the high-pressure-phase regime. Phase transformations in this region accompanied by *small* density changes are very difficult to detect but have a strong influence on the slope of the Hugoniot. In the case of the Twin Sisters dunite, a reasonably large number of data points is available so that the uncertainties are minimized and the location of the raw Hugoniot is reasonably well established. This, however, is not the case for several of the other rocks and minerals studied.

As was noted earlier, temperature increases along the Hugoniot more rapidly than along the adiabat. For many geophysical purposes, it is desirable to convert the raw Hugoniot data to the metastable Hugoniot[1]—this is the locus of P-ρ-T points which the high-pressure phase would follow if it were possible to use it as a metastable specimen material for the shock experiments. Among other quantities, this conversion requires an estimate of the transformation energy between the high-pressure and low-pressure phases.

Having obtained a metastable P-ρ Hugoniot for the high-pressure phase, the next step is to transform this to adiabatic and isothermal conditions[2,3,4,5,6] (Fig. 10-9). These transformations require knowledge of the equation of state of the high-pressure phase, in particular, the volume dependence of the Grüneisen parameter. It has been pointed out[6] that this latter relationship requires knowledge of the pressure derivative of the shear modulus $d\mu/dP$. Most methods of shock-wave-data reduction have assumed a "regular" behaviour of $d\mu/dP$ rather analogous to the regular behaviour of dK/dP discussed in Sec. 9-4. Recent ultrasonic measurements have shown that, to the contrary, $d\mu/dP$ tends to vary rather widely and irregularly.[6,7] This introduces a further source of uncertainty in the calculation of adiabats and isotherms.

[1]McQueen, Marsh, and Fritz (1967).
[2]Ahrens, Anderson, and Ringwood (1969).
[3]Takeuchi and Kanamori (1966).
[4]Wang (1967, 1968, 1969).
[5]Anderson and Kanamori (1968).
[6]Knopoff and Shapiro (1969).
[7]Anderson and Liebermann (1970).

After transforming the experimental data in the high-pressure regime (Fig. 10-9) to isotherms or adiabats, the next step is to extrapolate the P-ρ relationship back to atmospheric pressure so as to obtain the zero-pressure density ρ_0. This requires the application of an equation of state. One method[1,4,5] is to fit the second-order Birch-Murnaghan equation to the data. It is found that the values of ρ_0 thus obtained are rather sensitive to small errors in the location of the isotherm. Accordingly, an additional constraint on ρ_0 is introduced. It is required that ρ_0 also obey the Birch velocity-density rule or Anderson's seismic equation of state[2] (Sec. 9-4).

Estimates of the zero-pressure densities of the high-pressure phases in shock experiments have been made by several authors[3,4,5,6,7,1] using different procedures. Agreement has been good in many cases. Thus, Wang estimates that the density of the high-pressure phase of the Twin Sisters dunite is 4.1 g/cm^3 at atmospheric pressure and 300°K, compared with 4.12 obtained by Ahrens et al. (1969) and 3.94 by Davies and Anderson (1971). These authors note that the derived density is very similar to the density which a mixture $MgO + FeO + SiO_2$ (as stishovite), equivalent in chemical composition to the dunite, would possess.

In the preceding discussion, some of the uncertainties involved in the treatment of shock-wave data have been mentioned. These should not be forgotten when geophysical applications of shock-wave data are made. At the same time, they should not be overemphasized. In some cases where it has been possible to test the methods used in reduction of shock data—e.g., estimating zero-pressure densities of MgO, Al_2O_3, and SiO_2 from high-pressure shock data—calculated densities have been in good agreement with observed densities.[5] Although the estimation of uncertainties is hazardous, it appears that, in favourable cases, the zero-pressure density is likely to be correct within $\pm$ 5%, whilst for most cases, the maximum error is unlikely to exceed 10%. These estimates assume the input of good-quality experimental shock data.

An alternative method of interpreting shock-wave data has been employed recently by Davies and Gaffney (1973). The procedure is essentially the reverse of that discussed above. The Hugoniots of a range of possible high-pressure polymorphs are calculated using crystal chemical estimates of ρ_0 (Secs. 11-9 and 12-6), together with semiempirical K-ρ relationships (Sec. 9-4). The calculated Hugoniots for each structure are then compared with the experimental data points. This method has advantages in interpreting the behaviour of incompletely transformed phases—e.g., the bronzitite studied by McQueen and coworkers.

Although shock-wave data are of great value in demonstrating the occurrence of phase transformations in silicates and oxides at pressures well beyond the range of static pressure apparatus and in estimating the densities, sound velocities,

[1] Davies and Anderson (1971).
[2] Anderson (1967).
[3] Wang (1967. 1968, 1969).
[4] Anderson and Kanamori (1968).
[5] Ahrens, Anderson, and Ringwood (1969).
[6] McQueen, Fritz, and Marsh (1963).
[7] McQueen, Marsh, and Fritz (1967).

and equations of state of the high-pressure phases, they do not serve to identify the nature of the high-pressure phases. The indirect methods based upon crystal chemistry and germanate analogue studies, described in Sec. 10-2, go far towards complementing the shock-wave data in this respect. In most cases where shock transformations occur, the probable nature of the high-pressure structure may be revealed by these indirect methods. Although they may sometimes suggest several alternative modes of transformation, the requirement that the proposed structure be consistent with the inferred zero-pressure density of the high-pressure phase, together with certain kinetic considerations, constitute strong constraints. A list of inferred structures and densities for high-pressure shock phases is given in Tables 10-4 and 10-5.

Under the usual experimental conditions employed, the properties of the specimen are measured as the shock propagates through it. This occurs in a period which is generally less than a microsecond. This period is extremely small compared to the times required for attainment of chemical equilibrium in most oxides and silicates under static high pressure-temperature conditions (usually periods of minutes to hours). Although the intense shear and high temperature associated with the shock undoubtedly speed up the reaction rate, it appears likely that chemical equilibrium is not reached under shock conditions in many cases.

Table 10-4 INTERPRETATION OF ZERO-PRESSURE DENSITIES AND STRUCTURES OF SOME SHOCKED MINERALS. (*After Ahrens, Anderson, and Ringwood, 1969*)*

Initial density (g/cm^3)	Mineral	Calculated zero-pressure density of shocked phases	Possible structure of shocked phases	Theoretical structure density
4.25	Rutile	5.71	Fluorite	5.5
2.65	Quartz	4.34	Stishovite	4.28
5.27	Hematite	5.96	Perovskite $Fe^{II} \cdot Fe^{IV}O_3$	5.8
3.58	Spinel	4.19	Calcium ferrite	4.13
5.20	Magnetite	6.30	Calcium ferrite	5.9
2.61	Albite	3.80	Hollandite	3.85
2.63	Oligoclase	3.69	Hollandite	3.86
2.56	Microcline	3.51	Hollandite	3.84
3.21	Forsterite	4.31	Calcium ferrite,	4.12†
3.32	Olivine (Fo_{90})	4.12	Sr_2PbO_4	4.04
3.72	Olivine (Fo_{45})	4.75	Sr_2PbO_4	4.64
4.39	Fayalite	5.31	Sr_2PbO_4	5.29
3.19	Enstatite	4.20	Perovskite	4.25
3.27	Bronzite (En_{90})	3.74	Ilmenite	3.87
3.25	Sillimanite	4.00	Pseudobrookite	3.81

*Davies and Anderson (1971) have revised these densities using a modified seismic equation of state to constrain the ρ versus ϕ relationships at zero pressure. In their revision, the initial densities shown above are decreased on the average by about 4%.

†Alternatively, K_2NiF_4.

Table 10-5 INTERPRETATION OF DENSITIES OF SHOCKED ROCKS. (*After Ahrens, Anderson, and Ringwood, 1969*)

Initial density (g/cm^3)	Rock	Calculated zero-pressure density of shocked phases	Inferred structure of shocked phases	Theoretical structure density
2.63	Granite	3.96	Stishovite, hollandites	3.94
2.75	Anorthosite	3.71	Hollandite (pyroxene)*	3.85
2.98	Diabase	3.69	Hollandite (pyroxene)*	3.70
3.02	Diabase	3.68	Hollandite (pyroxene)*	3.66

*In these rocks it is assumed that the relatively compressible felspars have transformed to hollandite structures, whereas the less compressible pyroxenes have not been transformed. Analogous behaviour is observed in naturally occurring shocked meteorites and terrestrial rocks.

For example, the known transformations of olivines to spinels have not been observed under experimental[1] shock conditions. Furthermore, the pressures at which many transformations, e.g., quartz-stishovite, are *complete* under shock conditions appear to be much higher than the equilibrium pressures as inferred from static high-pressure experiments, thermodynamic considerations, and geophysical data. (As an example of the latter, most iron-poor silicates, including olivine and pyroxene, require shock pressures of over 500 kbars to achieve densities of the order of those existing at a pressure of 250 kbars in the mantle.) Thus, in most shock experiments on oxides and silicates, it appears necessary to overstress the specimen far beyond the equilibrium transformation pressure before transformation becomes complete.

Because of this nonequilibrium situation, kinetic factors probably influence the nature of the high-pressure phase. For many compounds there may exist more than one structural state possible for a high-pressure phase. These states may be separated by relatively small energy differences. Which of these states is finally achieved under shock may be controlled dominantly by the kinetics. This applies particularly to cases where one possible structural state possesses an identical stoichiometry to the starting material, e.g., quartz → coesite → stishovite, whereas the other possible states represent a disproportionation into two or more phases, isochemical in bulk with the starting material, e.g., albite → jadeite + stishovite or pyroxene → spinel + rutile. In general, one might intuitively expect that the problems of nucleation and grain growth into two or more phases

[1]However, spinels transformed from olivines by shock have been recovered from a meteorite (Binns et al., 1969).

during the submicrosecond period of a shock would make this kind of transformation less probable than an isostoichiometric transformation in which the atoms need only move over distances of a few angstrom units in order to reach the lattice sites appropriate for the high-pressure phase. Nevertheless, in cases where a suitable isostoichiometric structure does not exist, or where it is very much less stable than an isochemical mixture of phases, disproportionation into two phases must still be considered a possibility.

An important property of the high-pressure phase which can be derived from shock-wave data is the hydrodynamical sound velocity C:

$$C = \left(\frac{\partial P}{\partial \rho}\right)_s^{1/2} = \phi^{1/2}$$

The essential step in deriving C is the differentiation of Hugoniot P-ρ data. However, since the Hugoniot does not lie along an adiabat, a further correction of the slope involving the Grüneisen ratio γ is required. McQueen at al. (1967) show that the sound velocity along the Hugoniot C_H is given by

$$C_H = C_0 \frac{1-x}{(1-kx)^{3/2}} \left[1 + kx \left(1 - \frac{\gamma x}{1-x}\right)\right]^{1/2}$$

where x is the compression $(V_0 - V)/V$.

As a result of the differentiation process, errors in C or ϕ values derived from shock-wave data are proportionally much higher than the corresponding errors in ρ. In addition, for some purposes, C_H values measured along the Hugoniot must be corrected to adiabatic and isothermal conditions. Because of these uncertainties and the different methods of treating the raw data used by different authors, the agreement in ϕ values for some geophysically important materials is not nearly as good as one might hope for.[1,2,3,4] More accurate determinations of this property are of key importance for interpretations of the constitution of the Lower Mantle (Sec. 14-6, Fig. 14-6). These uncertainties should not be ignored when shock-wave data are applied to problems of mantle constitution.

REFERENCES

AHRENS, T. J., D. L. ANDERSON, and A. E. RINGWOOD (1969). Equations of state and crystal structures of high pressure phases of shocked silicates and oxides. *Rev. Geophys.* **7,** 667–707.

——— and E. S. GAFFNEY (1971). Dynamic compression of enstatite. *J. Geophys. Res.* **76,** 5504–5513.

AKIMOTO, S., H. FUJISAWA, and T. KATSURA (1965). The olivine-spinel transition in Fe_2SiO_4 and Ni_2SiO_4. *J. Geophys. Res.* **70,** 1969–1977.

[1]McQueen et al. (1967).
[2]Wang (1967, 1968, 1969).
[3]Ahrens et al. (1969).
[4]Davies and Anderson (1971).

——— and Y. IDA (1966). High pressure synthesis of Mg_2SiO_4 spinel. *Earth Planet. Sci. Letters* **1,** 358–359.

———, E. KOMADA, and I. KUSHIRO (1967). Effect of pressure on the melting of olivine and spinel polymorphs of Fe_2SiO_4. *J. Geophys. Res.* **72,** 679–686.

——— and Y. SATO (1968) High pressure transformation in Co_2SiO_4 olivine and some geophysical implications. *Phys. Earth Planet. Interiors* **1,** 498–504.

——— and Y. SYONO (1969). Coesite-stishovite transition. *J. Geophys. Res.* **74,** 1653–1659.

AMOROS, J. L., and A. SAN MIGUEL (1968). The pressure field and the internal constitution of the earth. *Tectonophysics* **5,** 287–294.

ANDERSON, D. L. (1967). A seismic equation of state. *Geophys. J. Roy. Astron. Soc.* **13,** 9–30.

——— and H. KANAMORI (1968). Shock wave equations of state for rocks and minerals. *J. Geophys. Res.* **73,** 6477–6502.

ANDERSON, O. L., and R. LIEBERMANN (1970). Equations for the elastic constants and their pressure derivatives for three cubic lattices and some geophysical applications. *Phys. Earth Planet. Interiors* **3,** 61–85.

BASSETT, W. A., and L. MING (1973). Disproportionation of Fe_2SiO_4 to $2FeO + SiO_2$ at pressures up to 250 kilobars and temperatures up to 3000°C. *Phys. Earth Planet. Interiors* **6,** 154–160.

———, T. TAKAHASHI, and P. W. STOOK (1967). X-ray diffraction and optical observations on crystalline solids up to 300 kb. *Rev. Sci. Instr.* **38,** 37–42.

BINNS, R. A., R. J. DAVIS, and S. J. B. REED (1969). Ringwoodite, natural $(MgFe)_2SiO_4$ spinel in the Tenham meteorite. *Nature* **221,** 943–944.

BIRCH, F. (1961). Composition of the earth's mantle. *Geophys. J.* **4,** 205–311.

——— (1964). Density and composition of mantle and core. *J. Geophys. Res.* **69,** 4377–4388.

BORN, M., and W. HEISENBERG (1924). Über den Einfluss der Deformerarkeit der Ionen auf optische und chemische Konstanten I. *Z. Phys.* **23,** 388–410.

BOYD, F. R., and J. L. ENGLAND (1960). Apparatus for phase equilibrium measurements at pressures up to 50 kb and temperatures up to 1750°C. *J. Geophys. Res.* **65,** 741–748.

BRADLEY, C. C. (1969). "*High Pressure Methods in Solid State Research.*" Butterworth, London. 176 pp.

BRIDGMAN, P. W. (1945). Polymorphic transitions and geologic phenomena. *Am. J. Sci.* **243A** (Daly Volume), 90–97.

BUNDY, F. P. (1963). Direct conversion of graphite to diamond in static pressure apparatus. *J. Chem. Physics* **38,** 631–643.

CLARK, S. P., and A. E. RINGWOOD (1964). Density distribution and constitution of the mantle. *Rev. Geophys.* **2,** 35–88.

COES, L. (1953). A new dense crystalline silica. *Science* **118,** 131–132.

——— (1955). High pressure minerals. *J. Am. Ceram. Soc.* **38,** 298.

——— (1962). Synthesis of minerals at high pressures. In: R. Wentorf (ed.), "*Modern Very High Pressure Techniques*," pp. 137–150. Butterworth, London.

DACHILLE, F., and R. ROY (1960). High pressure studies of the system Mg_2GeO_4-Mg_2SiO_4 with special reference to the olivine-spinel transition. *Am. J. Sci.* **258,** 225–246.

DAVIES, G. F. , and D. L. ANDERSON (1971). Revised shock wave equations of state for high-pressure phases of rocks and minerals. *J. Geophys. Res.* **76,** 2617–2627.

——— and E. GAFFNEY (1973). Identification of high pressure phases of rocks and minerals from Hugoniot data. *Geophys. J.* **33,** 165–183.

DORAN, D. G., and R. K. LINDE (1966). Shock effects in solids. *Solid State Phys.* **19,** 229–290.

DRICKAMER, H. G. (1963). Electronic structure at high pressure. In: W. Paul and D. Warschauer (eds.), "*Solids Under Pressure*." McGraw-Hill, New York.

——— and A. S. BALCHAN (1962). High pressure optical and electrical measurements. In: R. H. Wentorf (ed.), "*Modern Very High Pressure Techniques*," pp. 25–50. Butterworth, London.

——— (1970). Revised calibration for high pressure electrical resistance cell. *Rev. Sci. Instrum.* **41,** 1667–1668.

EVERNDEN, J. (1958). Finite strain theory and the earth's interior. *Geophys. J.* **1,** 1–8.

GIARDINI, A. A., and E. C. LLOYD (eds.) (1963). "*High Pressure Measurement*." Butterworth, London. 409 pp.

GOLDSCHMIDT, V. M. (1931). Zur Kristallchemie des Germaniums. *Nachr. Ges. Wiss. Gottingen, Math. Physik* **XI,** 184–190.

GRIGGS, D. T., and G. KENNEDY (1956). A simple apparatus for high pressures and temperature. *Am. J. Sci.* **254,** 722–735.

HALL, H. T. (1958). Some high pressure-high temperature apparatus design considerations: Equipment for use at 100,000 atmospheres and 3000°C. *Rev. Sci. Inst.* **29,** 267–275.

——— (1960). Ultra high pressure, high temperature apparatus: the "Belt." *Rev. Sci. Inst.* **31,** 125–131.

HAMANN, S. D. (1966). Effects of intense shock waves. *Advances in High Pressure Res.* **1,** 85–141.

JAMIESON, J. C., and A. W. LAWSON (1962). X-ray diffraction studies in the 100 kilobar range. *J. Appl. Phys.* **33,** 776–780.

KAWAI, N. (1966). A static high pressure apparatus with tapering multi-pistons forming a sphere. I. *Proc. Japan Academy* **42,** 385–388.

KNOPOFF, L., and J. N. SHAPIRO (1969). Comments on the interrelationships between Grüneisen's parameter and shock and isothermal equations of state. *J. Geophys. Res.* **74,** 1439–1450.

LIU, L. (1974). Disproportionation of Ni_2SiO_4 to stishovite plus bunsenite at high pressures and temperatures. (In press.)

LLOYD, E. C. (ed.) (1971). "*The Accurate Characterization of the High Pressure Environment*." *National Bureau of Standards, Spec. Publ.* 326. 333 pp.

MACDONALD, G. J. F. (1962). On the internal constitution of the inner planets. *J. Geophys. Res.* **67,** 2945–2974.

MAGNITSKY, V. A., and V. A. KALININ (1959). The properties of the earth's mantle and the physical properties of the transition layer. *Bull. Acad. Sci. USSR, Geophys. Ser.* No. 1–6, pp. 49–54.

MAO, H. K., and P. M. BELL (1971). High pressure decomposition of spinel (Fe_2SiO_4). *Geophys. Lab. Washington, Annual Report* **70,** 176–178.

MCQUEEN, R. G., J. N. FRITZ, and S. P. MARSH (1963). On the equation of state of stishovite. *J. Geophys. Res.* **68,** 2319–2322.

———, ———, and ——— (1964). On the composition of the earth's interior. *J. Geophys. Res.* **69,** 2947–2965.

——— and S. P. MARSH (1966). In: S. P. Clark, Jr. (ed.), "*Handbook of Physical Constants*." *Geol. Soc. Am. Memoir* **97.**

———, ———, and J. N. FRITZ (1967). Hugoniot equation of state of twelve rocks, *J. Geophys. Res.* **72,** 4999–5036.

MIKI, H. (1955). Is the C-layer (413–984 km) inhomogeneous? *J. Phys. Earth* **3,** 1–6.

MING, L., and W. BASSETT (1974). Laser heating in the diamond anvil press up to 3000°C and 300 kilobars. (In press.)

MINOMURA, S., K. ITO, and B. OJAI (1964). Pressure and temperature measurements in Drickamer's resistance cell up to 161 kb and 4000°C. ASME Pub. 64–WA/PT6, High Pressure Technology Symposium of ASME, 1–4.

NEUHAUS, A. VON (1968). Über Phasen-und Materiezustände in den tieferen und tiefsten Erdzonen (Ergebrisse der Modernen Hochdruck-Hochtemperatur-Forschung zum geochemischen Erdbild). *Geol. Rundschau* **57**, 972–1001.

NISHIKAWA, M., and S. AKIMOTO (1971). Bridgman anvil with an internal heating system for phase transformation studies. *High temperatures-High pressures* **3**, 161–176.

PEREZ-ALBUERNE, E. A., K. F. FORSGREN, and H. G. DRICKAMER (1964). Apparatus for X-ray measurements at very high pressure. *Rev. Sci. Inst.* **35**, 29–33.

RICE, M. H., R. G. MCQUEEN, and J. M. WALSH (1958). Compression of solids by strong shock waves. *Solid State Phys.* **6**, 1–63.

RINGWOOD, A. E. (1956). The olivine-spinel transition in the earth's mantle. *Nature* **178**, 1303–1304.

——— (1958a). Constitution of the mantle, I. Thermodynamics of the olivine-spinel transition. *Geochim. Cosmochim. Acta* **13**, 303–321.

——— (1958b). Constitution of the mantle, II. Further data on the olivine-spinel transition. *Geochim. Cosmochim. Acta* **15**, 18–29.

——— (1958c). Constitution of the mantle, III. Consequences of the olivine-spinel transition. *Geochim. Cosmochim. Acta* **15**, 195–212.

——— (1958d). Olivine-spinel transition in fayalite. *Bull. Geol. Soc. Am.* **69**, 129–130.

——— (1962a). Prediction and confirmation of olivine-spinel transition in Ni_2SiO_4, *Geochim. Cosmochim. Acta* **26**, 457–469.

——— (1962b). Mineralogical constitution of the deep mantle. *J. Geophys. Res.* **67**, 4005–4010.

——— (1963). Olivine-spinel transformation in cobalt orthosilicate. *Nature* **198**, 79–80.

——— (1966). Mineralogy of the mantle. In: P. M. Hurley (ed.), "*Advances in Earth Science*," pp. 357–399, M.I.T. Press, Cambridge.

——— (1970). Phase transformations and the constitution of the mantle. *Phys. Earth Planet. Interiors* **3**, 89–108.

——— (1972). Mineralogy of the deep mantle: current status and future developments. In: E. C. Robertson (ed.), "*The Nature of the Solid Earth*." McGraw-Hill, New York. 677 pp.

——— and A. MAJOR (1966). Synthesis of Mg_2SiO_4-Fe_2SiO_4 solid solutions. *Earth Planet. Sci. Letters* **1**, 241–245.

——— and ——— (1968). Apparatus for phase transformation studies at high pressures and temperatures. *Phys. Earth Planet. Interiors* **1**, 164–168.

——— and ——— (1970). The system Mg_2SiO_4-Fe_2SiO_4 at high pressures and temperatures. *Phys. Earth Planet. Interiors* **3**, 109–155.

——— and M. SEABROOK (1962a). Olivine-spinel equilibria at high pressure in the system Ni_2GeO_4-Mg_2SiO_4. *J. Geophys. Res.* **67**, 1975–1985.

——— and ——— (1962b). High pressure transition of $MgGeO_3$ from pyroxene to corundum structure. *J. Geophys. Res.* **67**, 1690–1691.

——— and ——— (1963). High pressure phase transformations in germanate pyroxenes and related compounds. *J. Geophys. Res.* **68**, 4601–4609.

SCLAR, C. B., and L. C. CARRISON (1966). High pressure synthesis of a spinel on the join Mg_2SiO_4-Fe_2SiO_4. (Abstract.) *Trans. Am. Geophys. Union* **41**, 207.

SHANNON, R. D., and C. T. PREWITT (1969). Effective ionic radii in oxides and fluorides. *Acta Cryst.* **B25,** 925–946.

SHAPIRO, J. N., and L. KNOPOFF (1969). Reduction of shock-wave equations of state to isothermal equations of state. *J. Geophys. Res.* **74,** 1435–1438.

SHIMAZU, Y. (1958). A chemical phase transition hypothesis on the origin of the C-layer within the mantle of the earth. *J. Earth Sci., Nagoya University* **6,** 12–30.

SKINNER, B. J., and J. J. FAHEY (1963). Observations on the inversion of stishovite to silica glass. *J. Geophys. Res.* **68,** 5595–5604.

STISHOV, S. M. (1962). On the internal structure of the earth. *Geokhimiya* No. 8, 649–659.

——— and S. V. POPOVA (1961). New dense polymorphic modification of silica. *Geokhimiya* No. 10, 837–839.

SUITO, K. (1972). Phase transformations of pure Mg_2SiO_4 into a spinel structure under high pressures and temperatures. *J. Phys. Earth* **20,** 225–243.

TAKEUCHI, H., and H. KANAMORI (1966). Equations of state of matter from shock wave experiments. *J. Geophys. Res.* **71,** 3985–3994.

TEMPKIN, M. (1945). Mixtures of fused salts as ionic solutions. *Acta Physicochim. USSR* **20,** 411–420.

VON PLATEN, B. (1962). A multiple piston, high pressure, high temperature apparatus. In: R. H. Wentorf (ed.), "*Modern Very High Pressure Techniques*," pp. 118–136. Butterworth, London.

WADA, T. (1960a). On the physical properties within the B-layer deduced from olivine model and on the possibility of polymorphic transition from olivine to spinel at the 20° discontinuity. Bull. 37, Disaster Prevention Institute, University of Kyoto, pp. 1–20.

——— (1960b). On origins of region C and the core of the earth—ionic-intermetallic-metallic transition hypothesis. Bull. 38, Disaster Prevention Institute, University of Kyoto, pp. 1–64.

WANG, C. Y. (1967). Phase transitions in rocks under shock compression. *Earth Planet. Sci. Letters* **3,** 107–113.

——— (1968). Constitution of the lower mantle as evidenced from shock wave data for some rocks. *J. Geophys. Res.* **73,** 6459–6476.

——— (1969). Equation of state of periclase and some of its geophysical implications. *J. Geophys. Res.* **74,** 1451–1457.

WENTORF, R. H. (1959). Olivine-spinel transformation. *Nature* **183,** 1617.

——— (ed.) (1962). "*Modern Very High Pressure Techniques.*" Butterworth, London.

WEYL, W. A. (1951). "*Coloured Glasses.*" Society of Glass Technology, Sheffield, England.

——— (1956). Acid-base relationship in glass systems. *Glass Ind.* Nos. 5 and 6.

11

HIGH-PRESSURE TRANSFORMATIONS IN A_2BO_4 COMPOUNDS

We concluded in Chap. 3 that the predominant phase in the upper mantle is olivine, $(Mg,Fe)_2SiO_4$. Clearly, then a detailed knowledge of high-pressure transformations displayed by this phase is essential to an understanding of the constitution of the deep mantle. In the present chapter, results of direct experiments on olivines at pressures equivalent to those reached in the upper 600 km of the mantle are described, this limit representing (until recently) the capacity of static high pressure-high temperature apparatus. In order to infer the nature of transformations displayed by $(Mg,Fe)_2SiO_4$ at depths greater than about 600 km, it is necessary to consider results obtained from the application of shock-wave techniques and the indirect methods discussed in Chap. 10. Comparative studies of the crystal chemistry and high-pressure behaviour of a wide range of A_2BO_4 compounds, particularly germanates, titanates, and stannates, are particularly relevant in this latter connection.

Common olivine may be regarded as a binary compound formed between two components possessing sodium chloride (MgO, FeO) and rutile (SiO_2)[1] struc-

[1]The high-pressure polymorph of SiO_2, stishovite.

tures, respectively. Olivine is thus a particular example of a large class of A_2BO_4 compounds formed between AO (rocksalt structure) and BO_2 (rutile structure) components. Systematic studies of the stabilities of this class of A_2BO_4 compounds provide a useful basis for interpreting the crystal chemistry and stability relationships of high-pressure binary MgO-SiO_2 compounds. A list of compounds possessing NaCl (rocksalt) and rutile structures (Fig. 13-5) is given in Table 11-1.

11-1 THE OLIVINE-SPINEL-BETA PHASE TRANSFORMATIONS

Bernal (1936) first suggested that common olivine might transform in the mantle under a sufficiently high pressure to a new polymorph possessing the spinel structure, which would be about 9% denser than the olivine. This suggestion was based upon a previous observation[1] that the analogous compound Mg_2GeO_4 was dimorphous, displaying both olivine and spinel polymorphs at atmospheric pressure, the spinel being 9% denser. Bernal's suggestion was adopted by Jeffreys (1937) as the basis for an explanation of a rapid increase in seismic velocity which was believed to occur near 400 km—"the 20 degree discontinuity."

The general plausibility of Bernal's hypothesis was considerably enhanced

[1]Goldschmidt (1931).

Table 11-1 PARTIAL LIST OF OXIDE COMPOUNDS POSSESSING ROCKSALT AND RUTILE STRUCTURES. MOLAR VOLUMES (V) AND CATION RADII (r)* FOR SIXFOLD COORDINATION ARE ALSO TABULATED

Rocksalt			Rutile		
Compound	V (cm^3)	r (Å)	Compound	V (cm^3)	r (Å)
BaO	25.37	1.36	CrO_2	17.18	0.55
CaO	16.76	1.00	GeO_2	16.66	0.54
CdO	15.59	0.95	IrO_2	19.21	0.63
CoO	11.64	0.74	β-MnO_2	16.61	0.54
EuO	20.49	1.17	MoO_2‡	19.85	0.65
FeO	12.17	0.77			
MgO	11.25	0.72	β-PbO_2	25.02	0.78
MnO	13.22	0.82	RuO_2	19.04	0.62
NiO	10.97	0.70	SiO_2	14.01	0.40
SrO	20.69	1.16	SnO_2	21.55	0.69
(FeO)†	(9.90)	(0.61)*	TaO_2	20.46	0.66
			TeO_2	26.52	
			TiO_2	18.80	0.61
			VO_2	17.88	0.59

*Shannon and Prewitt (1969).
†Fe^{++} in low-spin configuration.
‡Distorted rutile structure.

Table 11-2 PARAMETERS OF OLIVINE-SPINEL TRANSFORMATIONS*

Compound	Transition pressure (kbars)	Temperature (°C)	$V_{olivine}$ ‡ (cm³)	V_{spinel} ‡ (cm³)	Density increase (%)	dP/dT (bars/°C)
Mg_2GeO_4	0	820	45.87	42.35	8.3	33¶
Fe_2SiO_4	49	1000	46.39	42.03	10.4	28
Ni_2SiO_4	31	1000	42.51	39.17	8.5	16
Co_2SiO_4	70	900	44.56	40.58	9.8	32
Mg_2SiO_4§	125	1000	43.79	39.58	10.6	(30–50)**
$MgMnGeO_4$	[35]†	1100	48.97	44.62	9.8	
$FeMnGeO_4$	[35]†	1100	49.85	45.72	9.0	
$CoMnGeO_4$	[35]†	1100	49.23	45.04	9.3	
$LiMgVO_4$			44.71	42.58	5.0	

*After Ringwood and Major (1970) and Ringwood and Reid (1971).
†Brackets denote synthesis pressure for spinel. Equilibrium pressures are probably smaller.
‡V denotes molar volume.
§Data obtained by extrapolation and refer to metastable Mg_2SiO_4.
¶Based on unpublished measurements by the author. (The transition was located at 19 kbars, 1400°C.)
**Estimated range.

by the discoveries[1,2,3] that olivines Fe_2SiO_4, Ni_2SiO_4, and Co_2SiO_4 transformed to spinel structures at pressures between 20 and 70 kbars (700°C) accompanied by an average density increase of 10%. Further examples of this class of transformation have since been uncovered. The olivines $MgMnGeO_4$,[4] $FeMnGeO_4$,[4] $CoMnGeO_4$,[4] and $LiMgVO_4$[5] all transform to spinels at high pressure. The related compounds $LiAlSiO_4$,[6] $LiAlGeO_4$,[7] and Zn_2GeO_4,[7,8] which possess phenacite structures at low pressures, also transform to spinels at high pressures. A list of olivine-spinel transformations is given in Table 11-2.

Because of apparatus limitations, studies of olivine-spinel transformations were mostly carried out on systems other than the key mantle system Fe_2SiO_4-Mg_2SiO_4 in the years prior to 1966. In that year, however, Ringwood and Major, using the apparatus described in Sec. 10-4, succeeded in synthesizing at 170 kbars and 1000°C a continuous series of spinel solid solutions between pure Fe_2SiO_4 and a magnesia-rich spinel containing 80 mol% Mg_2SiO_4 (Fig. 11-1). These were the first syntheses of spinels with Mg/Fe ratios close to those in the earth's mantle and thus confirmed to a considerable degree the predictions which had been made earlier.[9] The magnesia-rich spinels were found to be 10.6% denser than the corresponding olivines. The density change was also in accord with the earlier predic-

[1]Ringwood (1958b).
[2]Ringwood (1962a).
[3]Ringwood (1963).
[4]Ringwood and Reid (1971).
[5]Blasse (1963).
[6]Ringwood and Reid (1968b).
[7]Rooymans (1967).
[8]Ringwood and Major (1967).
[9]Ringwood (1956, 1958a).

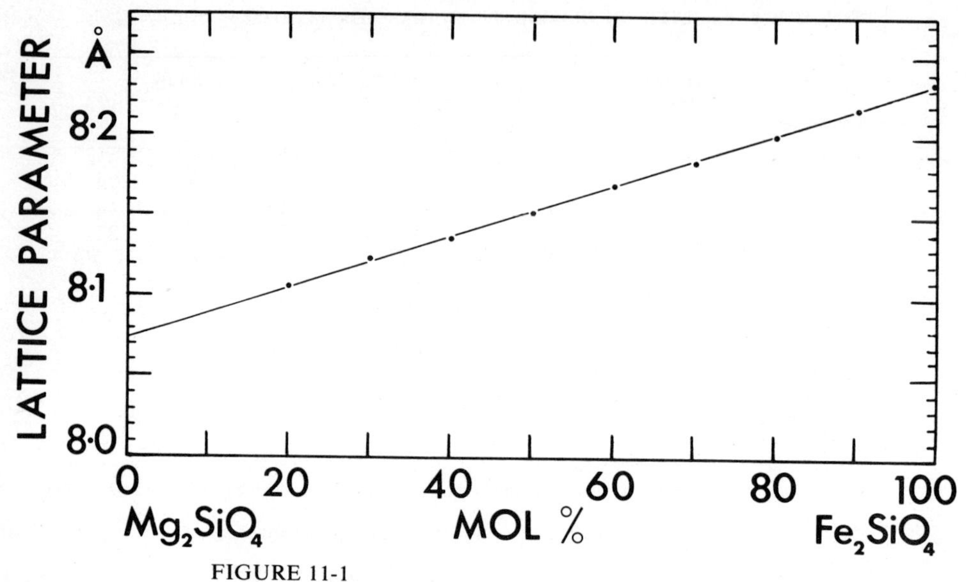

FIGURE 11-1
Lattice parameters of Mg_2SiO_4-Fe_2SiO_4 spinel solid solutions. Spinels were synthesized from olivine solid solutions at 170 kbars, 1000°C. (*From Ringwood and Major, 1966, with permission.*)

tions. By extrapolating the lattice parameters to pure Mg_2SiO_4 composition, both in the Mg_2SiO_4-Fe_2SiO_4 system and in the Mg_2SiO_4-Co_2SiO_4 and Mg_2SiO_4-Ni_2GeO_4 systems, the lattice parameter of pure Mg_2SiO_4 spinel was estimated to be 8.071 ± 0.005 Å, leading to a density of 3.56 g/cm^3 for the ideal Mg_2SiO_4 spinel.[1]

Only a short period elapsed between the synthesis of Mg-rich silicate spinels and the discovery of the first natural occurrence of the spinels by Binns, Davis, and Reed (1969). These workers observed numerous rounded purple isotropic grains up to 100 microns in diameter in thin sections of the Tenham chondritic meteorite. The grains were closely associated with veins thought to be produced by high-pressure shock waves during collisions between meteoritic parent bodies. Analyses of the grains showed that they were of the composition $(Mg_{0.74}Fe_{0.26})_2SiO_4$, which is identical to the composition of olivine occurring away from the veins. X-ray-diffraction studies showed that the grains possessed the spinel structure with a lattice parameter of 8.113 Å. The refractive index was 1.768. These properties agreed closely with those of the synthetic spinels for the relevant composition. A second occurrence of the natural $(Mg,Fe)_2SiO_4$ spinel also associated with veins of shock origin was found in the Coorara chondrite.[2]

[1]Ringwood and Major (1966, 1970).
[2]Smith and Mason (1970).

Although the transformations of olivines to spinels between Fe_2SiO_4 and $(Mg_{0.8}Fe_{0.2})_2SiO_4$ compositions closely followed predictions, the composition range between $(Mg_{0.8}Fe_{0.2})_2SiO_4$ and pure Mg_2SiO_4 provided a surprise. Ringwood and Major (1966, 1970) found that, at pressures above 150 kbars (1000°C), olivines in this composition range transformed not to spinels but into a birefringent phase with a complex x-ray-diffraction pattern which possessed some resemblance to a spinel but which had many extra lines and was clearly of a symmetry lower than cubic. The mean refractive index of this phase when produced by transformation of pure Mg_2SiO_4 was 1.702 ± 0.005. Based on the Gladstone-Dale rule, this indicated a density of 3.46 g/cm^3, which is 8% greater than that of forsterite.[1] The x-ray-diffraction pattern suggested that the new phase, which was called beta-Mg_2SiO_4, might have some kind of distorted spinel structure.[2]

Some preliminary observations[2] suggested that the new phase might be a retrogressive transformation product formed upon release of pressure. Evidence of such retrogressive transformations had been found at about the same time in other systems. Accordingly, it appeared possible that the new phase, actually stable under high pressure, may have been the true spinel. However, more extensive investigations[1,3] did not support this interpretation and provided strong evidence that beta-Mg_2SiO_4 was stable in its synthesis field. This conclusion was further supported by the results of Akimoto and coworkers,[4,5] who discovered new high-pressure modifications of Co_2SiO_4 and Mn_2GeO_4 which were isostructural with β-Mg_2SiO_4 and which were demonstrated to be thermodynamically stable phases. A high-pressure modification of Zn_2SiO_4 discovered by Ringwood and Major (1967) was also found to possess the β-Mg_2SiO_4 structure and to be thermodynamically stable.[6] The beta structure was found also for the compound Ni_2SiO_4-$NiAl_2O_4$ at high pressures.[7]

Relationships between the polymorphs of Co_2SiO_4 are shown in Fig. 11-2.[4] Above about 950°C, β-Co_2SiO_4 occupies a stable field between the olivine and spinel modifications, and moreover, the extent of this field broadens with increasing temperature. β-Co_2SiO_4 has a density of 5.05 g/cm^3, which is 7.1% higher than the olivine,[5] whereas Co_2SiO_4 spinel is 9.8% denser.[8] It has been agreed that the different polymorphs of A_2BO_4 compounds be designated as follows: α phase (olivine type), β phase (modified spinel type), γ phase (spinel type), and δ phase (strontium plumbate type). Relationships between different polymorphs of Mn_2GeO_4 are shown in Fig. 11-3.[5] The beta phase has a density of 5.13 g/cm^3 and

[1] Ringwood (1968).
[2] Ringwood and Major (1966).
[3] Ringwood and Major (1970).
[4] Akimoto and Sato (1968).
[5] Morimoto, Akimoto, Koto, and Tokonami (1969).
[6] Syono, Akimoto, and Matsui (1971).
[7] Ma and Hays (1972).
[8] Ringwood (1963).

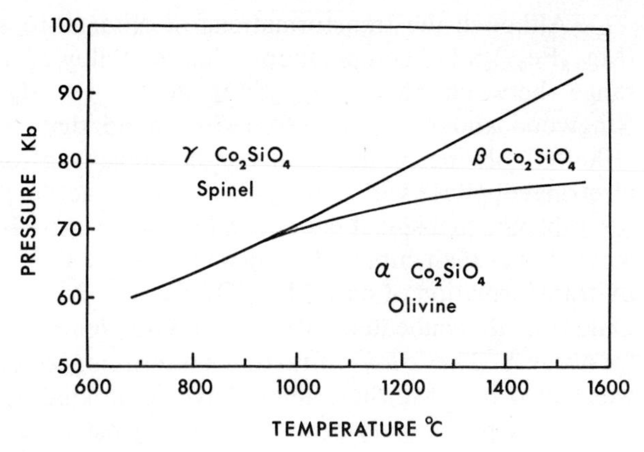

FIGURE 11-2
Relations between the three polymorphs of Co_2SiO_4. (*From Akimoto and Sato, 1968, with permission.*)

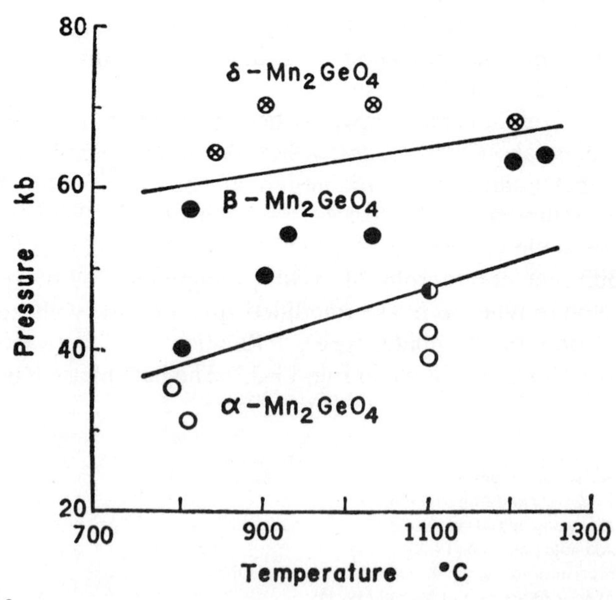

FIGURE 11-3
Relationships between the three polymorphs of Mn_2GeO_4. (*From Morimoto, Akimoto, Koto, and Tokonami, 1969, with permission. Copyright by the American Association for the Advancement of Science.*)

is 7.1% denser than the olivine modification (4.79 g/cm^3). The delta polymorph is discussed in a later section. It is noteworthy that all phase boundaries between α, β, γ, and δ polymorphs in Figs. 11-2 and 11-3 have positive slopes.

The transformations of Mg-rich olivines at high pressure point to both some of the strengths and the intrinsic weaknesses of the indirect methods of predicting stability fields of high-pressure phases.[1] Such methods make the assumption that the high-pressure phase will be one of a previously recognized class of structure. The systematics of crystal chemistry often make this a plausible assumption, but it is never certain. The prediction methods yield pressures which are required to transform the given phase to the new structure class. However, the results become irrelevant when a new or unsuspected phase becomes stable at a lower pressure than required for the anticipated phase. In the cases of the olivine-spinel transformations previously studied,[1] the existence of Ni_2SiO_4 and $(Mg_{0.8}Fe_{0.2})_2SiO_4$, and the pressures at which such spinels would become stable, were successfully predicted. On the other hand, the methods used were unable to predict that pure Mg_2SiO_4 would transform not to a spinel but to the beta phase. In this case, the prediction methods gave the pressure required for the olivine to *metastable* spinel transformation. This information is, nevertheless, of considerable value since it implies that olivine will not be a stable phase at high pressure and provides an estimate of the maximum pressure to which olivine may remain stable. It may break down to a different phase at a lower pressure, but if it does not, then the predicted transformation can be expected.

Crystal Structures

The olivine structure[2] is based upon a somewhat distorted hexagonal close packing of oxygen anions. The symmetry is orthorhombic with a tetramolecular unit cell. In Mg_2SiO_4 olivine, the Mg cations occupy octahedral holes in the oxygen sublattice, whereas the Si cations are situated in the tetrahedral holes. Thus the structure can be regarded as consisting of a packing together of MgO_6 octahedra and SiO_4 tetrahedra. The SiO_4 tetrahedra are isolated and linked only by O-Mg-O bonds.

The spinel structure (Fig. 11-4) consists of an approximate cubic close packing of oxygen anions, with the cations occupying some of the octahedral and tetrahedrally coordinated interstices. The oxygen sublattice is more closely packed than the somewhat distorted oxygen sublattice of the olivine structure, and hence, the spinel modification of a given compound is substantially denser than the olivine modification. As with the olivine structure, the spinel structure may be regarded as consisting of a packing of AO_6 octahedra and isolated BO_4 tetrahedra, with the tetrahedra and octahedra linked by A—O—B bonds. Crystallographic aspects of the olivine-spinel transformation have been discussed by

[1]Chapter 10.
[2]Bragg and Brown (1926).

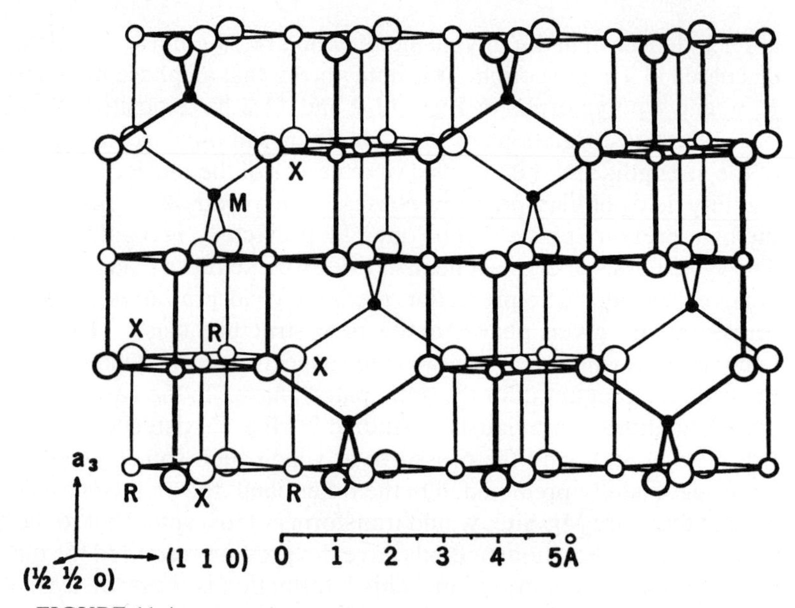

FIGURE 11-4
Cubic crystal structure of spinel R_2MX_4. (*From Morimoto, Akimoto, Koto, and Tokonami, 1969, with permission. Copyright by the American Association for the Advancement of Science.*)

Kamb (1968), who emphasized the importance of shortening of shared edges of coordination polyhedra in controlling the relative stabilities of olivine and spinel.[1]

Two principal classes of spinel structures are recognized. In the *normal* spinel structure, for compounds of the type $A_2^{++}B^{4+}O_4$, the A^{++} ions occupy octahedral sites, whereas the B^{4+} ion is tetrahedrally coordinated. The *inverse* spinel structure occurs when one of the A^{++} cations occupies the tetrahedral site, whilst the octahedral sites are shared by the remaining A^{++} and B^{4+} ions. Complete transitions between these configurations are possible, the degree of disorder tending to increase with temperature.[2,3] Detailed x-ray analyses of Ni_2SiO_4 and Fe_2SiO_4 spinels quenched from high pressures and temperatures showed that they were dominantly normal, with most of the silicon atoms in the tetrahedral sites.[4,5] However, a few percent of silicon atoms appeared to be octahedrally coordinated, indicating partial inverse character. It is uncertain to what extent high-temperature configuration disordering can be effectively

[1]See also Baur (1972).
[2]Stoll et al. (1964).
[3]Datta and Roy (1967).
[4]Ma, Hays, and Burnham (1971).
[5]Yagi, Marumo, and Akimoto (1974).

quenched, and thus the above analyses may underestimate the actual degree of inversion at high temperatures and pressures.[1]

The occurrence of an increasing degree of inversion in Fe_2SiO_4 at high temperatures[1] is suggested by the marked curvature[2] of the α-γ phase boundary, with the slope $dP/dT = \Delta S/\Delta V$ varying from ~28 bars/°C between 755 to 1170°C to <10 bars/°C above 1300°C. This decrease in slope cannot be explained by changes in ΔV but, rather, must be attributed to an anomalous increase of the entropy of the spinel phase with respect to fayalite. This may be a reflection of configurational entropy resulting from mixing of the Fe^{++} and Si^{4+} cations on equivalent structural sites and would require only 20% inversion in Fe_2SiO_4 spinel to explain the increase in entropy.

The x-ray crystal structure results on Ni_2SiO_4 and Fe_2SiO_4, combined with the observation that Fe_2SiO_4 - Mg_2SiO_4 spinel solid solutions obey Vegard's law (Fig. 11-1) suggests that Mg_2SiO_4-rich spinels are likely to be dominantly normal at low to moderate temperatures. However, Jackson et al. (1974) have pointed out that at high temperatures (e.g., >1500°C) some degree of Mg-Si disorder and partial inverse character may well occur. This would have important implications for mantle seismic velocity distributions and thermodynamic properties, topics which are considered in Chaps. 15 and 14.

A list of cations and anions which are capable of entering the spinel structure is given in Table 11-3. The spinel structure is tolerant towards a wide range of atomic substitutions, and spinel end-members often display extensive mutual solid solubility at high temperatures.

The first step towards solving the structure of β-Mg_2SiO_4 came when Akimoto and Sato (1968) synthesized a single crystal of Co_2SiO_4, the unit cell dimensions and space group of which were then determined by Morimoto and

[1]Jackson, Liebermann, and Ringwood (1974).
[2]Akimoto, Komada, and Kushiro (1967).

Table 11-3 LIST OF IONS WHICH CRYSTALLIZE IN SPINEL STRUCTURES

Charge →	I	II	III	IV	V	VI	Anions
	Ag	Cd	Al	Ge	Sb	Mo	O^{--}
	Na	Co	Co	Mn	V	W	S^{--}
	Li	Cu	Cr	Mo			Se^{--}
		Fe	Fe	Pb			Te^{--}
		Hg	Ga				
		Mg	In	Si			CN^{-}
		Mn	Mn	Sn			F^{-}
		Ni		Ti			
		Rh					
		Sn					
		Zn					

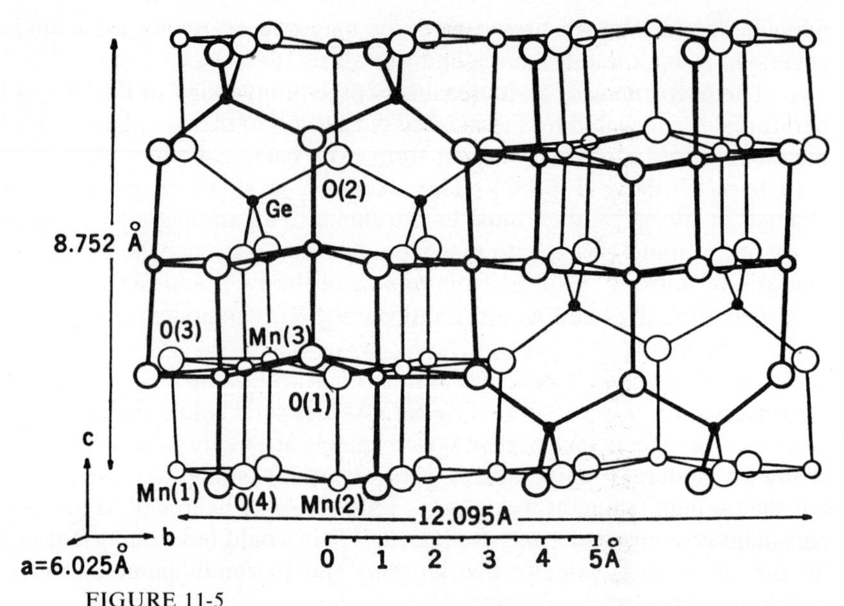

FIGURE 11-5
Orthorhombic crystal structure of β-Mn_2GeO_4. (*From Morimoto, Akimoto, Koto, and Tokonami, 1969, with permission. Copyright by the American Association for the Advancement of Science.*)

coworkers. The beta phase was found to be orthorhombic, with space group *Ibmm*. These workers then showed that the *d* spacings published for β-Mg_2SiO_4[1] could be successfully indexed on the basis of this space group with a similar unit cell to that of β-Co_2SiO_4.

The detailed crystal structure of beta phases was solved simultaneously by two groups working independently, and the results were first reported at the International Conference on Phase Transformations and the Earth's Interior in Canberra, January 1969. Working on a powder sample of β-$(Mg_{0.9}Ni_{0.1})_2SiO_4$ supplied by the author, Moore and Smith[2] indexed all 102 reflections on the basis of *Ibmm* and proposed a structural model which accounted well for all reflections and relative intensities. Meanwhile, Akimoto had discovered a high-pressure beta polymorph of Mn_2GeO_4 which yielded single crystals. The full single-crystal analysis of this phase[3] (Fig. 11-5) yielded a structure essentially identical to that proposed by Moore and Smith.

The oxygen anions in the beta phase (Fig. 11-5) are approximately in cubic close packing as in the spinel structure (Fig. 11-4). This accounts for the superficial resemblance between powder patterns of the two structures. The unit cell

[1]Ringwood and Major (1966).
[2]Moore and Smith (1969).
[3]Morimoto, Akimoto, Koto, and Tokonami (1969).

Table 11-4 PROPERTIES OF β-A_2BO_4 PHASES*

	Unit-cell dimensions (Å)			Density	Refractive indices		
	a	*b*	*c*	(g/cm³)	α	β	γ
Mg_2SiO_4	5.710	11.45	8.248	3.47	1.689	—	1.704
Co_2SiO_4	5.753	11.522	8.337	5.05	1.91	1.93	1.95
Mn_2GeO_4	6.025	12.095	8.752	5.13			
Zn_2SiO_4	5.740	11.504	8.395	5.34	1.75	—	1.77

*Morimoto et al. (1969); Ringwood and Major (1970); Syono, Akimoto, and Matsui (1971); Ringwood and Major (1967).

dimensions of the beta phase are simply related to those of the spinel (γ), with $a_\beta \sim a_\gamma$, $b_\beta \sim \sqrt{2}a_\gamma$, and $c_\beta \sim a_\gamma/\sqrt{2}$. The A and B cations possess similar coordinations to the spinel structure; however, their distributions are different. Whereas in spinel, the BO_4 tetrahedra are isolated, in the beta phase the BO_4 tetrahedra share one of their oxygen atoms, leading to the formation of B_2O_7 groups. As a result, some oxygen atoms are not bonded to any B atoms. The formula of this phase may thus be expressed as $A_4O \cdot B_2O_7$, where A = Mg, Mn, Co, Ni, Zn and B = Si, Ge. In Fig. 11-5, O(1) is not bonded to any Ge atom but is bonded to five Mn atoms; O(2) is bonded to two Ge atoms and one Mn atom because of the formation of the Ge_2O_7 groups; O(3) and O(4) are bonded to one Ge atom and three Mn atoms. Some properties of beta phases are given in Table 11-4.

11-2 THE SYSTEM Mg_2SiO_4-Fe_2SiO_4

Akimoto and Fujisawa (1966, 1968) carried out an extensive investigation of olivine-spinel equilibria in this key mantle system at pressures up to 95 kbars. Their results are shown in Fig. 11-6. Isothermal sections at 800, 1000, and 1200°C were constructed. These imply an average gradient dP/dT for the transformations over the range studied of approximately 30 bars/°C. In these experiments, homogeneous spinel solid solutions with composition ranging from pure fayalite to $(Fe_{0.6}Mg_{0.4})_2SiO_4$ were synthesized. The apparatus used was a tetrahedral anvil press.

A comprehensive investigation of this system over the pressure range 80 to 200 kbars at 1000°C was carried out by Ringwood and Major (1970).[1] Starting materials used were synthetic $(Mg,Fe)_2SiO_4$ olivines of known compositions. Experimental procedure was as described in Sec. 10-4. The nature and proportions of phases observed in the experiments are shown in Fig. 11-7. The experiments clearly define three fields of olivine (α) solid solutions, spinel (γ) solid solutions, and β-phase solid solutions, and two-phase regions of ($\alpha + \gamma$), ($\gamma + \beta$), and

[1]See also Ringwood (1969).

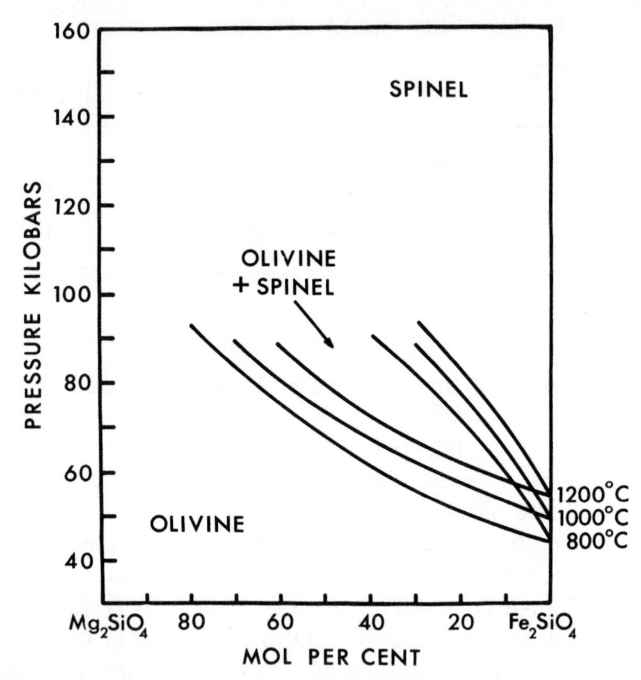

FIGURE 11-6
Isothermal sections at 800, 1000, and 1200°C through the system Mg_2SiO_4-Fe_2SiO_4 in the pressure range 0 to 95 kbars. (*After Akimoto and Fujisawa, 1968.*)

$(\alpha + \beta)$. Lattice parameters of spinel solid solutions in the two-phase $(\alpha + \gamma)$ field were determined and used, together with Fig. 11-1, to determine the compositions of spinels and accordingly the spinel solvus. These are shown as dots in Fig. 11-7.

The transformation of pure forsterite to the beta phase occurs at 120 kbars and 1000°C (Fig. 11-7). By extrapolating the olivine-spinel phase boundaries in Fig. 11-7, the corresponding pressure obtained for the metastable olivine-spinel transformation in pure Mg_2SiO_4 is 125 kbars. This small difference in pressures, combined with the small difference in molar volumes of the two polymorphs, in turn implies that the free-energy difference between the two high-pressure polymorphs of Mg_2SiO_4 is small. Accordingly, the positions of the $(\gamma\text{-}\beta)$ phase boundaries may be rather sensitively dependent upon temperature and possibly, in the more complex natural system in the mantle, upon the presence of other components which form solid solutions preferentially with one of these phases, thus increasing its relative stability. These circumstances complicate the application of the phase relations in the simple system to the mantle. The small difference in free energies between the gamma and beta phases is also reflected in the narrowness of the $(\gamma + \beta)$ field.

The $(\gamma + \beta)$ field boundaries in Fig. 11-7 are almost vertical. At atmospheric pressure the spinel is about 2% denser than the beta phase, and accordingly, it

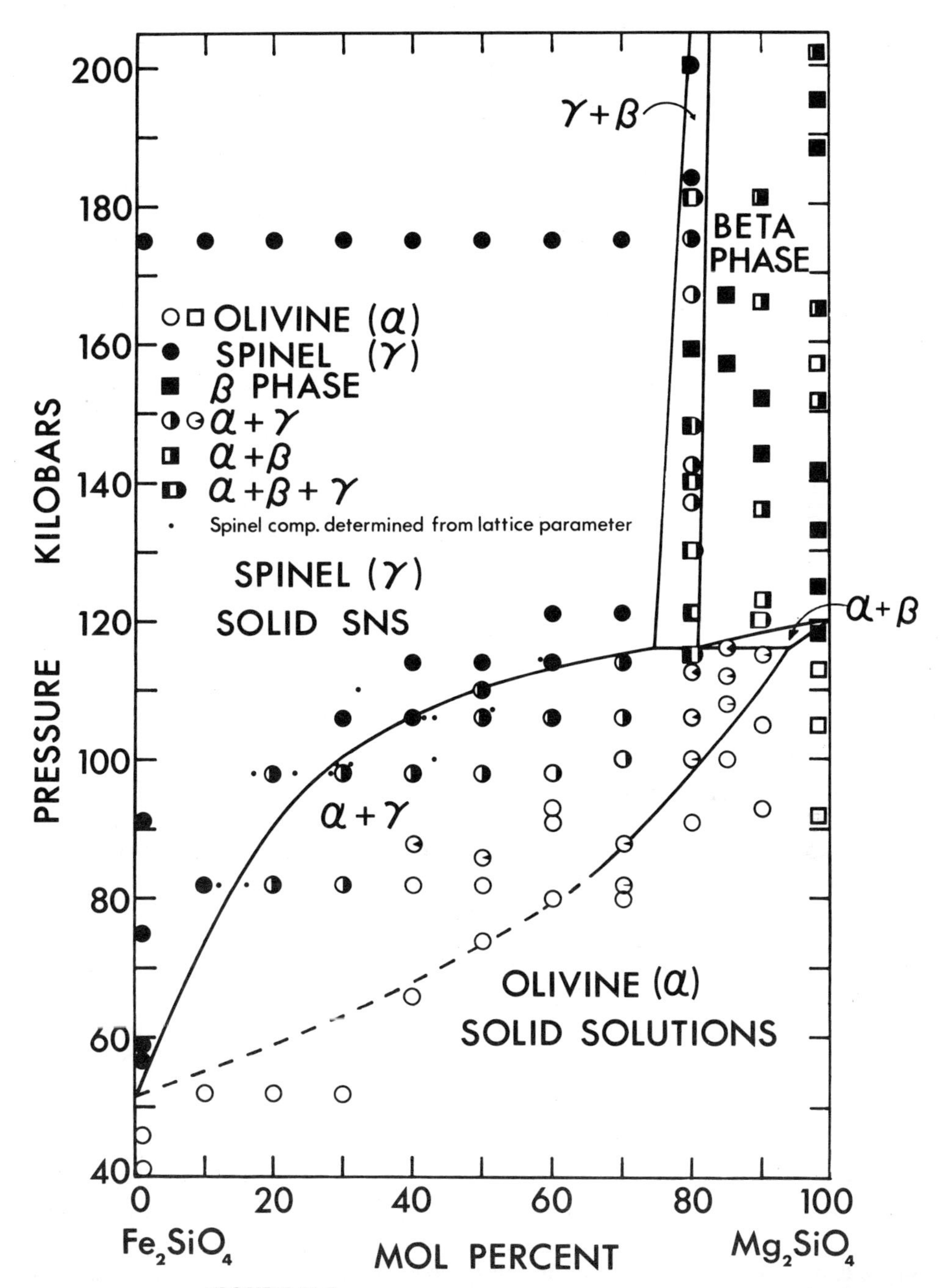

FIGURE 11-7
Phase relationships in the system Mg_2SiO_4-Fe_2SiO_4 at 50 to 200 kbars and at 1000°C. (*From Ringwood and Major, 1970, with permission.*)

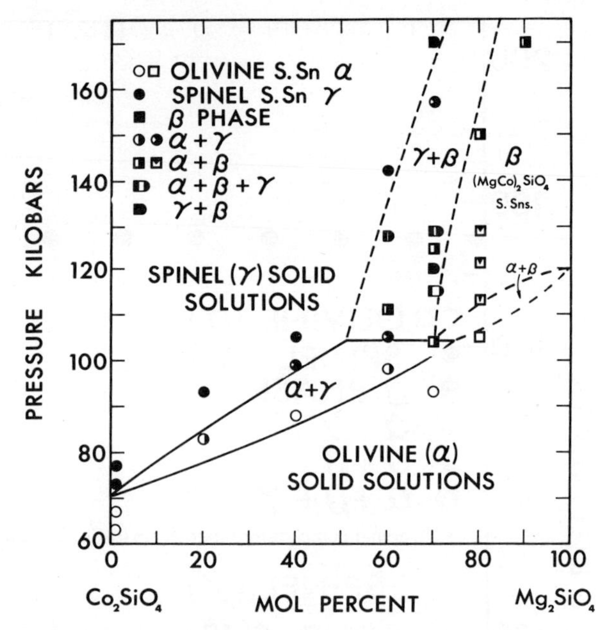

FIGURE 11-8
Phase relationships in the system Mg_2SiO_4-Co_2SiO_4 at 60 to 170 kbars, 1000°C. (*From Ringwood and Major, 1970, with permission.*)

might be expected that the $(\gamma + \beta)$ boundaries should have a positive slope and should intersect the Mg_2SiO_4 axis, corresponding to a transformation of β-Mg_2SiO_4 to Mg_2SiO_4 spinel at higher pressures. This is observed in the Co_2SiO_4 system (Fig. 11-2).

It appeared that an investigation of the system Co_2SiO_4-Mg_2SiO_4 might throw further light upon spinel-beta phase relationships. The results of experimental studies of this system[1] are shown in Fig. 11-8. The system is qualitatively similar to Fe_2SiO_4-Mg_2SiO_4, with fields of α, β, and γ phases. However, the extent of the beta field relative to γ and $(\beta + \gamma)$ fields is substantially increased, whilst the width of the olivine-spinel two-phase field in the Co_2SiO_4-Mg_2SiO_4 system is substantially smaller than for Fe_2SiO_4-Mg_2SiO_4. The olivine-spinel phase boundaries in the Co_2SiO_4-Mg_2SiO_4 system can be extrapolated to give a pressure of 125 kbars for the metastable olivine-spinel transformation in Mg_2SiO_4, agreeing with that obtained by extrapolations in the Fe_2SiO_4-Mg_2SiO_4 system. In contrast to results obtained for this latter system, the spinel-beta phase boundaries in the Co_2SiO_4-Mg_2SiO_4 system (Fig. 11-8) have a positive slope. Likewise, reconnaissance experiments[1] indicate that, in the analogous system Ni_2GeO_4-Mg_2SiO_4, the spinel-like phase boundaries also possess a positive slope.

[1]Ringwood and Major (1970).

The results of further investigations in the magnesia-rich region of the Fe_2SiO_4-Mg_2SiO_4 system have recently been published by Akimoto (1972) and Suito (1972). These confirm the essential nature of the phase diagram as established by Ringwood and Major (1966, 1970). However, the former investigators find a slight positive slope for the β-γ transition, implying that γ-Mg_2SiO_4 would become stable around 200 kbars at 1000°C (Fig. 11-9). This interpretation is based on only three runs which, however, demonstrate the occurrence of $(Mg_{0.9}Fe_{0.1})_2SiO_4$ spinel at pressures over 140 kbars (800 to 1000°), whereas the beta phase was found for this composition at lower pressures. When these results are considered in conjunction with the analogous results for the Mg_2SiO_4-Co_2SiO_4 and Mg_2SiO_4-Ni_2GeO_4 systems, the most likely conclusion is that the β-γ boundary indeed possesses a steep positive slope as in Fig. 11-9. Nevertheless, further confirmation remains desirable.[1]

Suito (1972) claimed to have transformed pure α-Mg_2SiO_4 into a mixture of β and γ phases in four runs above 200 kbars, 1000°C. The relative proportions of β and γ phases were not disclosed, nor was there any mention of the criteria used to establish the presence of γ phase in the presence of β phase. This is not necessarily a simple matter since nearly all x-ray reflections of γ phase overlap those of the β phase.

The gradients dP/dT for the α-γ, α-β, and β-γ transitions in Mg-rich compositions are of considerable geophysical importance. The gradients for olivine-spinel transitions in Mg_2GeO_4, Fe_2SiO_4, Co_2SiO_4, and Ni_2SiO_4 range between 16 and 33 bars/°C (average 27).[2] A corresponding gradient in the vicinity of 30 bars/°C appears to apply for compositions in the range Fe_2SiO_4-

[1] Ito et al. (1974) have recently provided convincing evidence of the synthesis of pure Mg_2SiO_4 spinel at 250 kbars, 1000°C.

[2] Table 11-2.

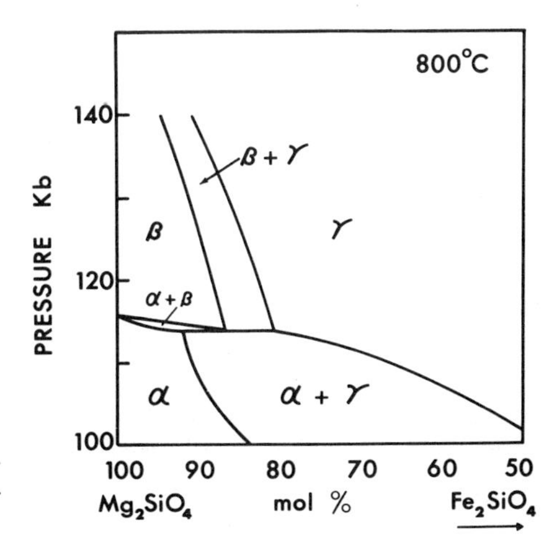

FIGURE 11-9
Mg-rich region of the system Mg_2SiO_4-Fe_2SiO_4 at 1000°C according to Akimoto (1972).

$(Mg_{0.5}Fe_{0.5})_2SiO_4$.[1] Thermodynamic considerations dictate that dP/dT for the α-γ transition should be greater than for the α-β transition.[2] Experimental measurements of the gradient for the α-β transition in pure Mg_2SiO_4 by Akimoto (1972) and Suito (1972) yielded values of 48 and 132 bars/°C respectively. The latter value seems much too high and implies an extremely anomalous entropy change. The former value also appears to be somewhat high when considered in relation to the gradients of the other olivine-spinel transitions discussed above. These gradients were measured in apparatus possessing very small pressurized working volumes, which were calibrated for pressure at room temperatures.

The operation of heaters within such small volumes usually causes a substantial pressure increase through the agency of thermal expansion. This is likely to lead to systematic overestimates of dP/dT slopes for phase transformations. Accordingly, dP/dT slopes determined in apparatus possessing relatively large working volumes are to be preferred. For this reason, the author believes that the slopes for the olivine-spinel transition depicted in Fig. 11-6 are likely to be more accurate and that dP/dT for magnesia-rich spinels is probably in the vicinity of 30 bars/°C, commensurate with other olivine-spinel transition slopes (Table 11-2; see also Ringwood and Major, 1970).

11-3 THE TRANSFORMATION OF OLIVINE IN THE MANTLE

It was concluded in Chap. 3 that olivine is a major constituent of the upper mantle and that the average Mg/Mg + Fe (molecular) ratio of this region, extending at least to 300 km, was close to 0.89. It will be assumed initially[3] that this ratio also holds at greater depths in the mantle. The extent to which transformations in olivine of this composition can account for the major seismic discontinuity[4] near 400 km is now examined.

As a first approximation to the problem, we will assume that the phase diagram for the system Mg_2SiO_4-Fe_2SiO_4 at 1000°C is applicable to the mantle as it stands, except for the appropriate temperature correction. The latter we will take as 30 bars/°C.

Referring to Fig. 11-7, we see that, for Mg/Mg + Fe = 0.89, olivine first begins to transform to a spinel of composition $(Mg_{0.46}Fe_{0.54})_2SiO_4$ at a pressure of 109 kbars. As pressure increases, the amount of spinel increases, whilst the spinel becomes richer in magnesia, reaching $(Mg_{0.75}Fe_{0.25})_2SiO_4$ at 116 kbars. At this point the spinel phase completely reacts to form a beta phase of composition $(Mg_{0.8}Fe_{0.2})_2SiO_4$. Above this pressure, spinel is absent, and we are in the field of olivine + beta phase. With a further slight increase of pressure, olivine finally

[1] Figure 11-6.

[2] Ringwood and Major (1970). See also Fig. 11-2.

[3] This assumption is discussed at length and justified in Chap. 14.

[4] Section 9-1, Fig. 9-1.

transforms to the beta phase by 118 kbars. The total transformation is thus spread over an interval of 9 kbars, with a median value of 114 kbars.

Now let us assume that the temperature near 400 km is 1600°C, which was estimated in an earlier study.[1] With a gradient of 30 bars/°C, the entire phase diagram would be raised in pressure by 18 kbars, and the median pressure for the transition would be 132 kbars. Thus the transformation of olivine through spinel into the beta phase would occur at a median depth of 397 km, with a width of about 27 km. This is in excellent agreement with the depth of the seismic discontinuity (Fig. 9-1), considering the uncertainties in the seismic data. Although the phase transformations extend over an interval of 27 km, the existence of the spinel-beta phase reaction point at 116 kbars (134 kbars or 403 km at 1600°C) causes a discontinuous change in mineralogy and density at this depth and a further rapid increase of density between 403 and 410 km. These effects will produce a first-order seismic discontinuity, rather than a second-order discontinuity as was suggested in earlier investigations.

It is most satisfactory that when we take a Mg/Mg + Fe ratio for the upper mantle, which is strongly supported by petrological and geochemical considerations, and a temperature at 400 km, which is consistent with geothermal considerations, the laboratory investigations are so closely consistent with seismic observations. This agreement suggests that the possible uncertainties in the experimental work and application are not very serious; nevertheless, their significance must be evaluated.

The principal experimental uncertainty relates to the effect of temperature upon the phase diagram. If the gradient dP/dT for the α-β transition were as high as 48 bars/°C, as maintained by Akimoto (1972), then the median depth for the transition would increase to 430 km (1600°C), which is within the depth range for the discontinuity permitted by uncertainties in seismic data. Alternatively, a reduction in temperature at 400 km from 1600 to 1400°C would bring the median depth close to 400 km. However, geothermal considerations suggest that this alternative is somewhat less probable.[1,2]

Chemical factors should also be considered. The possible effects of solid solution of other components present in the mantle environment upon the binary phase diagram (Fig. 11-8) have been discussed,[3] and it was concluded that they were probably unimportant. However, the partition of iron between orthosilicate and other phases in the mantle may well be a relevant factor.[3] As discussed in Chap. 14, the principal mineral coexisting with olivine immediately above the discontinuity is believed to be a complex garnet solid solution containing some octahedrally coordinated silicon. In garnet lherzolite xenoliths from diamond pipes, iron is strongly concentrated in garnet relative to olivine. Detailed cal-

[1]Clark and Ringwood (1964).
[2]Schatz and Simmons (1972).
[3]Ringwood and Major (1970).

culations by Ahrens[1] indicate that the distribution coefficient K, given by $(Fe/Mg)_{Ga}/(Fe/Mg)_{Ol}$ is about 3.7 at depths between 330 and 400 km. This would imply that, for an overall pyrolite composition, the olivine in equilibrium with garnet would have a Mg/Mg + Fe ratio of about 0.94. Referring to Fig. 11-7, we see that olivine of this composition would transform directly to the beta phase and that the total transformation interval would be much narrower than occurs for a Mg/Mg + Fe composition of 0.89. The phase diagram also shows that the depth of the discontinuity for the more magnesian olivine would be slightly increased.

The calculations presented by Ahrens possess substantial uncertainties in view of various thermodynamic assumptions made—particularly those concerning the ideality of garnet solid solutions. Moreover, the partition coefficient data refer to normal pyrope-almandine garnets, rather than to the complex high-pressure garnets containing some octahedral silicon, which are probably involved. Nevertheless, Ahrens' arguments possess considerable merit and emphasize the importance of direct experimental measurements of partition coefficients under the relevant P-T conditions.

Accepting the positive slope of the β-γ phase boundary, as shown in Fig. 11-9, we can expect that, at greater depths, possibly between 500 and 600 km, beta phase will transform to spinel, accompanied by a density increase of about 2%. An inferred minor seismic discontinuity at a depth of 520 to 550 km[2,3,4] might be connected with this transition.

11-4 TRANSFORMATION OF SPINEL AND BETA PHASES TO STRONTIUM PLUMBATE STRUCTURE

With the 400-km discontinuity largely explained by the olivine-spinel-beta phase transformation, it is natural to suggest that the major seismic discontinuity near 650 km (Fig. 9-1) might be explained by the further transformation of beta and gamma $(Mg,Fe)_2SiO_4$ to a still denser structure.[5,6] In both spinel and beta phases, the Si^{4+} ions are tetrahedrally coordinated. The observation that, at high pressures, the coordination of silicon in SiO_2 changes from 4 to 6, accompanied by a large increase in density,[7] and the observations of many such coordination changes in germanate analogues, suggest that silicate spinels and beta phases are likely to transform to denser structures characterized by octahedral coordination

[1]Ahrens (1972a).
[2]Whitcomb and Anderson (1970).
[3]Helmberger and Wiggins (1971).
[4]Simpson, Mereu, and King (1974).
[5]Ringwood (1966, p.390).
[6]Anderson (1967).
[7]Stishov and Popova (1961).

of silicon.[1,2] This is confirmed by the results of shock-wave experiments[3] on $(Mg,Fe)_2SiO_4$ olivine at pressures in the megabar range (Fig. 10-9). These experiments indicated that olivine had been shocked directly into modifications which considerably exceeded the densities anticipated for spinels. Under the nonequilibrium conditions which prevailed, the "spinel" field was apparently overshot, and the field of "post-spinel" transformation products was reached directly. The zero-pressure densities of the high-pressure phases were estimated to be very similar to those of isochemical mixed oxides:[4,5,6] $MgO + FeO + SiO_2$ (as stishovite).

Direct attempts to synthesize the post-spinel phase at pressures up to 200 kbars were unsuccessful.[7] Accordingly, studies of analogue compounds were indicated. Attempts to transform Mg_2GeO_4, Ni_2GeO_4, Co_2GeO_4, and Fe_2GeO_4 spinels at pressures up to 170 kbars were also unsuccessful.[8] However, Wadsley, Reid, and Ringwood (1968) found that Mn_2GeO_4, which possesses an olivine structure at atmospheric pressure, transformed to the strontium plumbate structure with an accompanying increase in density of 17.3% in runs at about 100 kbars. It was suggested that Mn_2GeO_4 might represent a high-pressure model for Mg_2SiO_4. The radius ratio Mn^{++}/Ge^{4+} is very similar to that of Mg^{++}/Si^{4+}. If, as discussed in Sec. 10-2, we regard the principal effect of pressure as shrinking the large oxygen anions relative to the small cations, then it appeared possible that $(Mg,Fe)_2SiO_4$ solid solutions might be capable of crystallizing in this structure at high pressure. From crystal chemical considerations, it was suggested that the likelihood of siliceous olivines forming the Sr_2PbO_4 structure under pressure was expected to increase in the order Fe_2SiO_4-Co_2SiO_4-Mg_2SiO_4-Ni_2SiO_4, corresponding to decreasing average divalent metal-oxygen distances. Subsequent investigations of the solid solubility of Fe, Co, Mg, Ni, and Si in the high-pressure Mn_2GeO_4 structure have generally confirmed these predictions.[9]

The subsequent discovery by Morimoto et al. (1969) that, at intermediate pressure, Mn_2GeO_4 displays a third polymorph isostructural with beta-Mg_2SiO_4 (Fig. 11-3) reinforces the applicability of the model relationship between Mn_2GeO_4 and Mg_2SiO_4. The sequence of transformations for Mn_2GeO_4 with increasing pressure is thus from the olivine to beta phase to strontium plumbate structures. An analogous sequence has been discovered in $FeMnGeO_4$, which transforms with increasing pressure from olivine to spinel to strontium plumbate structures.[8] The spinel Mn_2SnO_4 is also observed to transform to the Sr_2PbO_4

[1]Ringwood (1962b).
[2]Stishov (1962).
[3]McQueen, Marsh, and Fritz (1967).
[4]Anderson and Kanamori (1968).
[5]Wang (1967).
[6]Ahrens, Anderson, and Ringwood (1969).
[7]Ringwood (1970).
[8]Ringwood and Reid (1968b).
[9]Ringwood and Reid (1971, unpublished results).

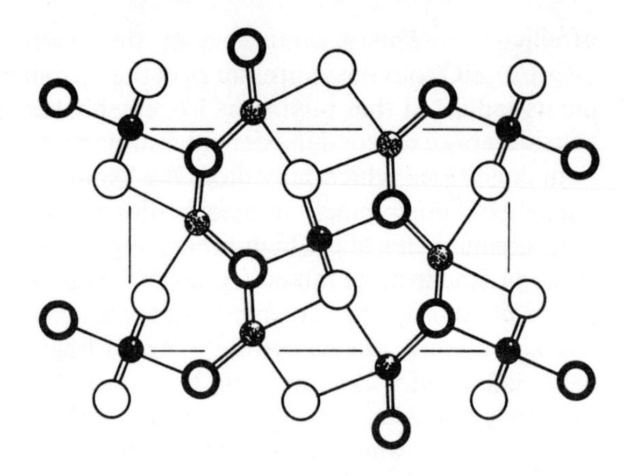

FIGURE 11-10
The structure of δ-Mn_2GeO_4 (Sr_2PbO_4 type) in projection onto (001). Small black circles, Ge; stippled circles, Mn; largest circles, oxygen appearing at two levels. (*From Wadsley, Reid, and Ringwood, 1968, with permission.*)

structure at pressures above 50 kbars.[1,2] (At still higher pressures, disproportionation into a mixture of MnO and $MnSnO_3$—corundum— is observed. The density of the latter assemblage is about 1% greater than that of the Sr_2PbO_4 modification.)

The orthorhombic strontium plumbate structure (space group *Pbam*) is also displayed by Sr_2PbO_4,[3] Ca_2PbO_4,[3] Ca_2SnO_4,[4] and Cd_2SnO_4, whilst Na_2CuF_4 is also closely related to it.[5] In the strontium plumbate modification of Mn_2GeO_4 (Fig. 11-10), the Mn^{++} ion is surrounded by six oxygen ions situated at the corners of a somewhat distorted trigonal prism, with an average Mn-O distance of 2.19 Å. A seventh oxygen at a distance of 2.73 Å through the centre of one prism is probably unbonded. The Ge^{4+} ions are situated within octahedra of oxygen anions at an average distance of 1.95 Å. Thus the transformation from olivine and beta structures to the strontium plumbate structure involves an increase of Ge^{4+} coordination from 4 to 6, which is responsible for the large increase in density.

Strontium plumbate isotypes are generally formed between end-members possessing rocksalt and rutile structures. It is instructive to compare the molar volumes of Sr_2PbO_4 structures with those of their isochemical rocksalt and rutile type end-members (Tables 11-6 and 11-7). It is seen that all members of this group are characterized by molar volumes which are very similar to the mixed oxides. This is an interesting and important property in view of the observation

[1]Ringwood and Reid (1968b).
[2]Syono, Sawamoto, and Akimoto (1969).
[3]Trömel (1965).
[4]Trömel (1967).
[5]Babel (1965).

previously noted that $(Mg_{0.9}Fe_{0.1})_2SiO_4$ olivine transforms under shock to a phase also possessing a molar volume similar to the isochemical mixed oxides.

Studies of the solubility of Mg_2SiO_4 in δ-Mn_2GeO_4 at about 150 kbars (1000°C) reveal a substantial degree of solid solubility—in the vicinity of 15%.[1] In view of the general solid-solubility relationships between high-pressure germanate phases and low-pressure silicates,[2] this is a most significant observation. It permits an approximate calculation of the pressure at which Mg_2SiO_4 may be expected to transform to the δ-Mn_2GeO_4 structure. This pressure is found to be in the vicinity of 200 to 250 kbars.[1] By comparison, the pressure at the second major seismic discontinuity in the mantle near 650 km is about 230 kbars. The results of this calculation, together with the general crystal chemical and density relationships discussed in this section, suggested that the 650-km discontinuity may be caused by transformation of $(Mg,Fe)_2SiO_4$ from the beta or gamma to the delta structure.

Baur (1972) investigated the crystallography of the hypothetical strontium plumbate-type (δ) polymorph of Mg_2SiO_4. He demonstrated that δ-Mg_2SiO_4 could be constructed from appropriately linked MgO_6 and SiO_6 polyhedra in which Mg-O, Si-O, and O-O bond lengths agreed well with those observed in stishovite and in magnesium oxycompounds. It was concluded that the δ structure satisfied the crystallographic requirements for a high-pressure polymorph of Mg_2SiO_4.

11-5 DISPROPORTIONATION OF $A_2^{++}B^{4+}O_4$ SPINELS AT HIGH PRESSURE

As an alternative to the β-δ transformation discussed in the previous section, it has been suggested that Mg_2SiO_4 might disproportionate in the deep mantle into two phases. One possibility is the reaction[3,4,5]

$$A_2BO_4 \text{ (spinel)} \rightarrow ABO_3 \text{ (ilmenite str.)} + AO \text{ (rocksalt str.)}$$

Studies of the $MgGeO_3$-$MgSiO_3$ system (Sec. 12-2) have shown that an ilmenite form of $MgSiO_3$ may become stable between 200 and 300 kbars so that the above reaction must be seriously considered for Mg_2SiO_4. Analogue reactions have not been discovered in germanates. However, the titanate spinels Mg_2TiO_4,[6] Co_2TiO_4,[6] Fe_2TiO_4,[6] Zn_2TiO_4,[7] $ZnMgTiO_4$,[7] and Mn_2TiO_4[8] are observed to disproportionate into $MTiO_3$ ilmenite phases + MO oxides at high pressure. Zn_2SnO_4 and Mn_2SnO_4 behave analogously.[8] The densities of the ilmenite-oxide

[1] Ringwood and Reid, unpublished results.
[2] Section 10-2.
[3] Birch (1952).
[4] Ringwood (1962b).
[5] Ringwood (1966).
[6] Akimoto and Syono (1967).
[7] Ringwood and Reid (1968b).
[8] Reid and Ringwood, unpublished results.

mixtures are usually from 1 to 4% smaller than the mean densities of the isochemical component oxides.[1] The density of the appropriate mixture of $MgSiO_3$ ilmenite and MgO would be 3.72,[1] which is about 3.4% smaller than that of isochemical mixed oxides $2MgO + SiO_2$ stishovite (3.85 g/cm^3).

It has also been suggested by several workers[2,3,4,5,6] that, under the high pressures existing in the lower mantle, silicate minerals may actually disproportionate into physical mixtures of their constituent oxides, principally SiO_2 (as stishovite), MgO, FeO, CaO, and Al_2O_3. Such reactions are theoretically possible if the density of the oxide mixture is greater than that of the mineral or compound. The free-energy change ΔG in these reactions is equal to $\Delta G_0 + \int_0^P \Delta v \, dP$, where ΔG_0 is the free energy of formation of the compound from its constituent oxides at atmospheric pressure, and Δv is the difference in molar volume at pressure P between the compound and its isochemical oxide mixture. The condition for disproportionation is for ΔG to be zero; hence, $\int_0^P \Delta v \, dP = -\Delta G_0$. If we consider Δv to be constant (a fair approximation in many cases), then $P \, \Delta v = -\Delta G_0$, and hence, $P = -\Delta G_0/\Delta v$. Many calculations of the pressures at which Mg_2SiO_4 "spinel" might be expected to disproportionate into oxide components have been carried out.[6,7,8,9,10] The calculated pressures are in the vicinity of 300 kbars at 1000 to 2000°C.

Examples of these transformations have been discovered[11] in the spinels Mg_2SnO_4 and Co_2SnO_4, which disproportionate into mixtures of MgO, CoO (rocksalt) + SnO_2 (rutile) at high pressures, and in Li_2NiF_4 spinel, which disproportionates into LiF (rocksalt str.) + NiF_2 (rutile str.) under pressure. Several $A^{++}B_2^{3+}O_4$ spinels were similarly found to disproportionate into constituent oxides at high pressures.[11] Another compound which disproportionated into oxides was Al_2GeO_5 (kyanite).

More recently, two groups[12,13,14] have described experiments using diamond anvils in the range 150 to 250 kbars, 800 to 3000°C in which disproportionation of fayalite into mixed oxides (FeO + SiO_2) was achieved. This important result

[1] Section 12-2.
[2] Birch (1952, 1964).
[3] MacDonald (1956).
[4] Shimazu (1958).
[5] McQueen, Fritz, and Marsh (1964).
[6] Ahrens and Syono (1967).
[7] Ringwood (1962b).
[8] MacDonald (1962).
[9] Stishov (1963).
[10] Anderson (1967).
[11] Ringwood and Reid (1968b). See also Sec. 11-6.
[12] Bassett and Takahashi (1970).
[13] Bassett and Ming (1973).
[14] Mao and Bell (1971).

directs attention to the possible disproportionation of Mg_2SiO_4 at higher pressures.

Discussion

In this, and in the preceding section, three possible modes of transformation for spinels and related structures have been discussed. All transformations lead to densities similar to the mixed oxides and involve change of coordination of one cation from fourfold in spinel to sixfold in the post-spinel phases. The bearing of these observations on the probable transformation mode of Mg_2SiO_4 may be considered. The nature of the post-spinel transformation is strongly influenced by (1) the free energy ΔG_0 of formation of the spinel from constituent oxides and (2) stereochemical factors controlling the stability of dense binary phases such as ABO_3 ilmenite and A_2BO_4 (strontium plumbate). It is readily seen that, where ΔG_0 is relatively small, a corresponding small pressure may be sufficient to cause dissociation according to the relationship $P = -\Delta G_0/\Delta v$. On the other hand, where ΔG_0 is large (i.e., a strong compound-forming tendency exists between the oxides), a correspondingly higher pressure will be required to cause dissociation into oxides, and transformation, if it occurs, is more likely to result in the formation of a new single phase (providing that this is permitted by stereochemical considerations).

The experimental results are generally in accord with these expectations. The spinels which have been observed to disproportionate into oxides, e.g., some of the stannates and $A^{++}B_2^{3+}O_4$ spinels, were characterized by relatively low ΔG_0 values. On the other hand, none of the germanate spinels possessing higher ΔG_0 values have been observed to disproportionate into simple oxides. Where transformations occurred, they were to isochemical denser phases possessing A_2BO_4 stoichiometry. The titanate spinels, which would be expected to possess ΔG_0 values intermediate between the stannates, on the one hand, and the silicates and germanates, on the other,[1] tended to display an intermediate transformation behaviour, mostly disproportionating into mixtures of ABO_3 (ilmenite) + AO (rocksalt). The energy required for the latter transformation is much smaller than that for complete disproportion into oxides. Fayalite also has a relatively low free energy of formation, and disproportionation would not be unexpected.

In view of the strong compound-forming tendency of MgO and SiO_2 as shown by the comparatively high free energy of formation of Mg_2SiO_4, and also because of the crystal chemical and thermodynamic considerations discussed in Sec. 11-4, Ringwood (1970) concluded that Mg_2SiO_4 was more likely to transform to the strontium plumbate structure than to dissociate into MgO + SiO_2.

Whilst this book was in press, two groups[2,3] reported that they had achieved

[1]Ringwood and Reid (1968b).
[2]Ming and Bassett (1974).
[3]Kumazawa et al. (1974).

the disproportionation of Mg_2SiO_4 into $2MgO + SiO_2$ (stishovite) at pressures exceeding 200 kbars and at high temperatures.[1] This important achievement represents a milestone in mantle phase transformation studies. It is, however, not necessarily inconsistent with the above conclusion regarding the probable stability of a strontium plumbate polymorph of Mg_2SiO_4. The disproportionation into mixed oxides was observed in runs which had not been quenched under pressure. It is quite possible that this technique would not discover a strontium plumbate modification of Mg_2SiO_4 since this may be unquenchable, spontaneously reverting to low-pressure phases on release of pressure.[2]

Accordingly, it remains possible that beta- or gamma-Mg_2SiO_4 transforms first to the Sr_2PbO_4 structure, as suggested by the earlier arguments, before finally disproportionating to $MgO + SiO_2$ (stishovite). This interpretation remains to be tested in the future by means of in situ x-ray-diffraction studies carried whilst the sample is under pressure (Sec. 10-4). It requires that the Sr_2PbO_4 polymorph is slightly less dense (e.g., $\leq 2\%$) than the isochemical oxide mixture. This relationship is indeed displayed by several Sr_2PbO_4-type isomorphs (Table 11-7).

11-6 TRANSFORMATIONS IN $A^{++}B_2{}^{3+}O_4$ SPINELS

Another important class of spinels includes those formed between oxides of divalent elements (usually rocksalt structure) and trivalent elements (usually corundum structure). Transformations are known in these spinels and may be of importance in determining the behaviour of trivalent oxides in the lower mantle. This family of spinels also displays normal structures $A^{IV}B_2{}^{VI}O_4$ and inverse structures $(AB)^{VI}B^{IV}O_4$. (The superscripts denote the oxygen coordinations.)

The behaviour of a large number of these spinels at high pressure (~ 120 kbars, 1000°C) was investigated.[3,4] A list of transformations observed in spinels is given in Table 11-5. The most common transformation observed was disproportion into constituent oxides:

$$AB_2O_4\text{spinel} \rightarrow AO\text{ (rocksalt)} + B_2O_3\text{ (corundum)}$$

This transformation was displayed by $MnAl_2O_4$, $FeAl_2O_4$, $NiAl_2O_4$, $CoAl_2O_4$, MgV_2O_4, MnV_2O_4, and $CdIn_2O_4$. This type of transformation was not unexpected in view of the medium to low free energies of formation of these spinels. On the other hand, a number of other AB_2O_4 spinels which probably possessed comparable free energies of formation failed to transform at all. This was proba-

[1]Liu (personal communication) has recently succeeded in disproportionating Ni_2SiO_4 and Co_2SiO_4 spinels into their constituent oxides at pressures in the vicinity of 200 kbars and temperatures of 1400 to 1800°C.

[2]This conclusion is based upon the author's experimental results on solid solubility relations in the systems Mg_2SiO_4-Mn_2GeO_4 and $Mg_3Al_2Si_3O_{12}$-$Mg_3Al_2Ge_3O_{12}$. See also the discussions relating to this point in Secs. 10-4, 12-4, and 12-5.

[3]Ringwood and Reid (1968b).

[4]Reid and Ringwood, unpublished results.

Table 11-5 HIGH-PRESSURE TRANSFORMATIONS IN OXIDE SPINELS AND CLOSELY RELATED STRUCTURES

A. *Transformation into single dense phases* $\Delta\rho/\rho \sim 10\%$:
Mn_2GeO_4 ("beta phase") transforms to Sr_2PbO_4 structure
Mn_2SnO_4 (spinel) transforms to Sr_2PbO_4 structure
$FeMnGeO_4$ (spinel) transforms to Sr_2PbO_4 structure
Mn_3O_4 (tetragonal spinel) transforms to calcium manganite structure
Transformations also observed in $CdCr_2O_4$, $CdFe_2O_4$, $LiFeTiO_4$, and Fe_3O_4

B. *Disproportionation into oxide plus binary ilmenite (or corundum) structure* $\Delta\rho/\rho \sim 8\%$:
A_2BO_4 (spinel) $\rightarrow$ ABO_3 (ilmenite) + AO (rocksalt)
Mg_2TiO_4, Co_2TiO_4, Fe_2TiO_4 $MgZnTiO_4$, Zn_2TiO_4, Mn_2TiO_4,
Mn_2SnO_4, and Zn_2SnO_4

C. *Complete disproportionation into oxides* $\Delta\rho/\rho \sim 10\%$:
$A_2^{++}B^{4+}O_4$ (spinel) $\rightarrow$ AO (rocksalt) + BO_2 (rutile)
Mg_2SnO_4, Co_2SnO_4, (Li_2NiF_4)
$A^{++}B_2^{3+}O_4$ (spinel) $\rightarrow$ AO rocksalt + B_2O_3 (corundum)
$MnAl_2O_4$, $FeAl_2O_4$, $NiAl_2O_4$, $CoAl_2O_4$, $MgAl_2O_4$, MnV_2O_4,
MgV_2O_4, and $CdIn_2O_4$

bly due to kinetic difficulties. Thus, pure $MgAl_2O_4$ spinel failed to transform in a number of runs at pressures up to 170 kbars. However, a spinel with the composition $MgAlVO_4$ disproportionated completely at 120 kbars to a mixture of $MgO + Al_2O_3 + V_2O_3$. This strongly suggests that the assemblage $MgO + Al_2O_3$ is stable relative to $MgAl_2O_4$ under these conditions.

At much higher pressures, under shock-wave conditions, $MgAl_2O_4$ is observed to transform[1] to a phase which is substantially denser than the isochemical mixed oxides (Fig. 11-11). Apparently the stability field of the assemblage $MgO + Al_2O_3$ has been "overshot," perhaps for kinetic reasons, with the formation of a phase with an estimated zero-pressure density[2] of 4.19 g/cm^3 compared to 3.86 g/cm^3 for an isochemical mixture of $MgO + Al_2O_3$.

Possible structures for the high-pressure modification of $MgAl_2O_4$ are those of calcium ferrite ($CaFe_2O_4$) and the closely related calcium titanite ($CaTi_2O_4$) and calcium manganite ($CaMn_2O_4$). These structures are commonly formed between end-members possessing rocksalt and corundum structures and are characteristically about 4 to 6% denser than the mixed oxides (Tables 11-6 and 11-7). The likelihood of $MgAl_2O_4$ transforming to the calcium ferrite structure is enhanced by the observations that $CaAl_2O_4$ also transforms to the calcium ferrite structure at high pressures, whilst $MgSc_2O_4$ possesses the calcium ferrite structure at atmospheric pressure.[3]

Reid and Ringwood (1969a) have shown that Mn_3O_4 which possesses a hausmanite structure (i.e., a tetragonal spinel) transforms to the $CaMn_2O_4$ structure at high pressure, with a 10% increase in density. A transformation has also

[1] McQueen and Marsh (1966).
[2] Ahrens, Anderson, and Ringwood (1969).
[3] Reid and Ringwood (1970).

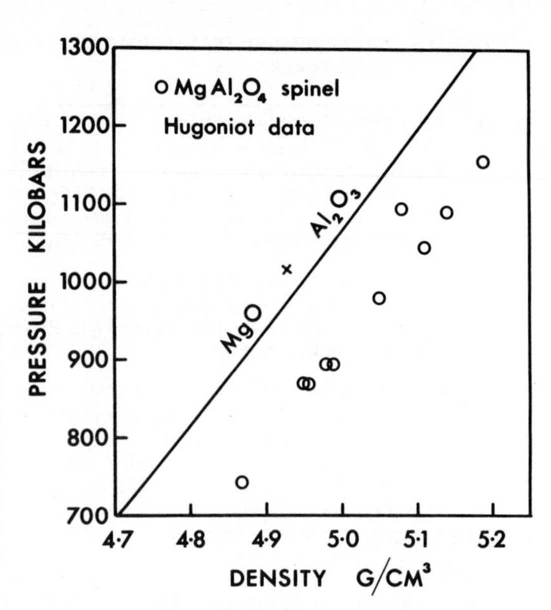

FIGURE 11-11
Shock-wave compression of $MgAl_2O_4$ spinel compared to compression of isochemical mixture of $MgO + Al_2O_3$ based upon weighted average of data for pure MgO and pure Al_2O_3. Note that $MgAl_2O_4$ spinel attains substantially higher densities than the isochemical oxide mixture at corresponding pressures. Hugoniot data from McQueen and Marsh (1966).

been found in magnetite at about 250 kbars.[1] The structure has not been solved but it may possibly be related to calcium ferrite or manganite. The transformation of magnetite is also observed under shock-wave conditions. The zero-pressure density of the shocked phase is estimated to be about 6.3 g/cm³, compared to a density of 5.54 g/cm³ for a mixture of $FeO + Fe_2O_3$.[2] The density of a calcium ferrite modification of Fe_3O_4 would be about 5.8 g/cm³ (Tables 11-6 and 11-7), which is somewhat low. However, if the Fe^{3+} ions in the inferred calcium ferrite structure were in the low spin state,[3] the density would be about 6.3 g/cm³, which is in satisfactory agreement, considering the various uncertainties.

The calcium ferrite structure type[4,5,6,7] (Fig. 11-12) and the closely related $CaTi_2O_4$[8] and $CaMn_2O_4$[9] types are among the densest AB_2O_4 structures so far known. They are built up of pairs of edge-shared octahedra which, at any one level, corner-share to produce eight-coordinate sites that may be occupied by Ca^{++}, Eu^{++}, Mn^{++}, Mg^{++}, or Na^{+}. The pairs of octahedra repeat at successive levels, again by edge-sharing, forming a dense array in which each metal-oxygen octahedron edge-shares to four others. The octahedrally coordinated sites can be

[1]Mao, Bassett, and Takahashi (1970).
[2]Ahrens, Anderson, and Ringwood (1969).
[3]See also Davies and Gaffney (1973).
[4]Hill, Peiser, and Rait (1956).
[5]Bertaut, Blum, and Magnano (1956).
[6]Decker and Kasper (1957).
[7]Reid, Wadsley, and Sienko (1968).
[8]Bertaut and Blum (1956).
[9]Lepicard and Protas (1966).

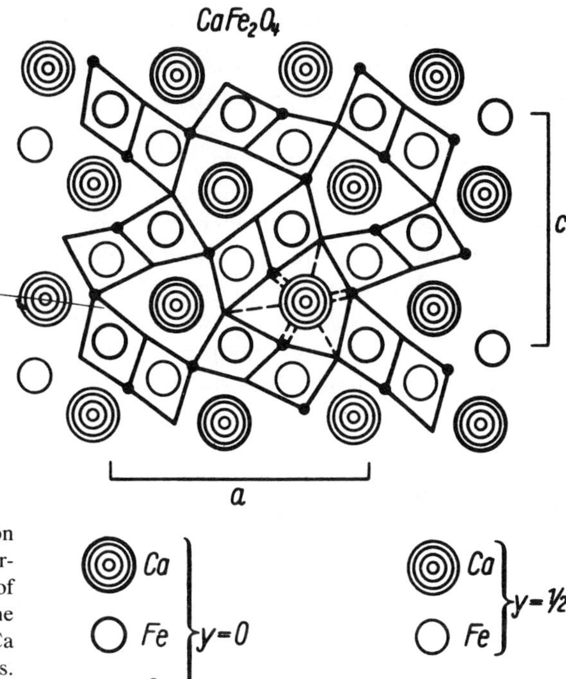

FIGURE 11-12
The structure of $CaFe_2O_4$ projected on an *a-c* plane showing pairwise edge-sharing by FeO_6 octahedra with formation of "tunnels" occupied by Ca ions. The eight- to ninefold coordination of the Ca ions is indicated by the broken lines. (*From Gorter, 1960, with permission.*)

occupied by the following trivalent ions: Fe^{3+}, Cr^{3+}, V^{3+}, Eu^{3+}, Mn^{3+}, Al^{3+}, and Sc^{3+}.

Although, above, we have been considering mainly calcium ferrite structures of the type $A^{++}B_2^{3+}O_4$, it is important to note that the same structure is displayed by a number of $A^{+}B^{3+}C^{4+}O_4$ compounds, e.g., $NaScTiO_4$,[1] $NaFeSnO_4$,[1] and high-pressure $NaAlGeO_4$.[2] In these compounds, the sodium ions occupy the eightfold site, whereas the octahedral sites are shared between the trivalent and quadrivalent ions. More complex cation distributions are also observed. In $MgSc_2O_4$,[3] there is a substantial degree of disorder, with Mg and Sc cations populating both the eightfold and sixfold sites. In $NaCo_{0.5}Ti_{1.5}O_4$, the sixfold sites are shared between divalent and quadrivalent cations.[4] Furthermore, extensive degrees of solid solutions are possible between different end-members. These properties of the calcium ferrite structure may be of considerable geochemical significance in connection with a possible transformation of Mg_2SiO_4 at extreme pressures and are discussed in Sec. 11-8.

[1]Reid, Wadsley, and Sienko (1968).
[2]Reid, Wadsley, and Ringwood (1967).
[3]Müller-Buschbaum (1966).
[4]A.F. Reid (personal communication).

11-7 FURTHER TRANSFORMATIONS IN OLIVINES

The olivine Ca_2GeO_4 transforms to the K_2NiF_4 structure at high pressure.[1,2] This transformation is accompanied by a density increase of 25%. The K_2NiF_4 structure is commonly formed between end-members possessing rocksalt and rutile structures and is between 2 and 7% (average 3.5%) *denser* than isochemical mixtures of these end-members.[3] Such compounds will therefore be stabilized relative to the isochemical oxide mixture by pressure. Examples of compounds displaying this structure are Sr_2TiO_4, Sr_2SnO_4, Ca_2MnO_4, and Ba_2PbO_4. The K_2NiF_4 structure may be regarded as being made up of alternate layers possessing rocksalt and perovskite structures. The large cation is surrounded by nine oxygen atoms, whereas the small cation is octahedrally coordinated. Most of the oxides capable of crystallizing in the K_2NiF_4 structure are also capable of crystallizing in the perovskite structure when reacted in 1:1 molecular ratios. Perovskites are also substantially denser than their isochemical oxide mixtures.[4] From the germanate-silicate relationships discussed previously, it appears likely that Ca_2SiO_4 would be capable of adopting the K_2NiF_4 structure under very high pressure. The possibility that Mg_2SiO_4 might ultimately be capable of adopting this structure is considered in Sec. 11-8.

In contrast to Ca_2GeO_4, the closely related olivine Cd_2GeO_4 disproportionates under high pressure:[1]

$$Cd_2GeO_4 \text{ (olivine)} \rightarrow CdGeO_3 \text{ (perovskite)} + CdO \text{ (rocksalt)}$$

Analogous reactions are displayed by several other olivines:[1,5]

$$CaMgGeO_4 \text{ (olivine)} \rightarrow CaGeO_3 \text{ (perovskite)} + MgO$$

$$CaMgSiO_4 \text{ (olivine)} \rightarrow CaSiO_3 \text{ (perovskite)} + MgO$$

$$CaMnGeO_4 \text{ (olivine)} \rightarrow CaGeO_3 \text{ (perovskite)} + MnO$$

$$Mn_2SiO_4 \text{ (olivine)} \rightarrow MnSiO_3 \text{ (garnet)} + MnO$$

The density increases accompanying the above disproportionations average about 25%. The compounds Ca_2SnO_4, Cd_2SnO_4, and Cd_2TiO_4 also disproportionate into mixtures of perovskite and rocksalt-type phases under pressure.[5]

11-8 POSSIBLE TRANSFORMATIONS OF Mg_2SiO_4 TO STRUCTURES DENSER THAN ISOCHEMICAL MIXTURES OF $MgO + SiO_2$

We have already discussed in some detail the transformations by which A_2BO_4 spinels may attain densities similar to those of the isochemical mixtures of rocksalt and rutile end-members. There is evidence that the silicate phases occurring

[1] Ringwood and Reid (1968a).
[2] Reid and Ringwood (1970).
[3] Section 11-10.
[4] Section 12-5.
[5] Reid and Ringwood, unpublished results.

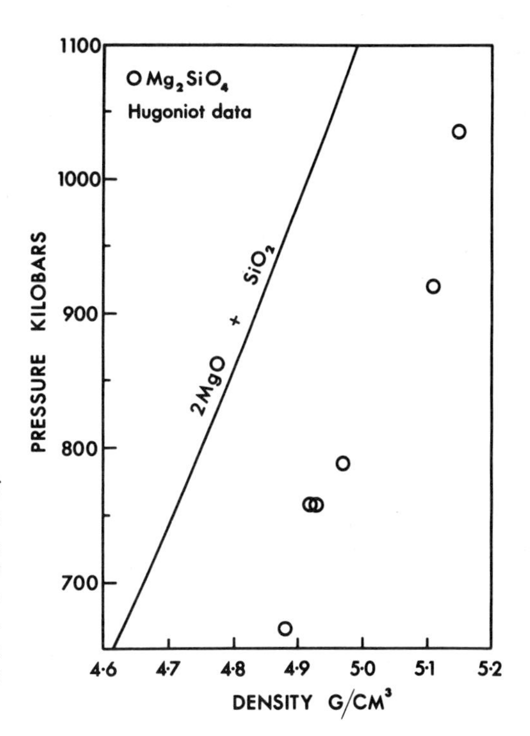

FIGURE 11-13
Compression of a mixture of $2MgO + SiO_2$ based on shock-wave Hugoniot data for pure MgO and SiO_2 compared with corresponding data for pure forsterite, Mg_2SiO_4. Note that forsterite attains substantially higher densities than the isochemical oxide mixture at corresponding pressures. *(Based on data of McQueen and Marsh, 1966.)*

in the lower mantle may attain densities which are about 5% higher than those of the equivalent oxide mixtures.[1]

Shock-wave data[2] strongly suggest that such phases indeed exist. The density of the high-pressure polymorph of forsterite produced by shock compression is several percent higher than that of the mixed oxides (Fig. 11-13). Davies and Gaffney (1973) have reanalyzed the Hugoniot data for hortonolite dunite and fayalite. In the case of fayalite, the compressions observed at pressures between 600 and 850 kbars corresponded to a phase possessing the density of a Sr_2PbO_4-type polymorph or of an oxide mixture. However, the densities obtained in two experiments between 1000 and 1300 kbars were substantially higher, indicating the existence of a structure which was more closely packed. Experiments carried out on the hortonolite dunite, approximately $MgFeSiO_4$, were also interpreted in terms of transformation into a phase which was about 5% denser than the mixed oxides. On the other hand, the Twin Sisters dunite $(Mg_{0.9}Fe_{0.1})_2SiO_4$ appeared to transform to a phase with a density similar to isochemical mixed oxides. The reason for the differing behaviour is uncertain and may be connected with kinetic factors.

In order to achieve densities substantially greater than those of isochemical mixed oxides, phases with iron and magnesium coordinations greater than 6 are

[1]Discussed in Chap. 14.
[2]McQueen and Marsh (1966).

required. A possible structure for these densest modifications of Mg_2SiO_4 and Fe_2SiO_4 is that of calcium ferrite.[1] Other possibilities are the closely related $CaTi_2O_4$ and $CaMn_2O_4$ structures.[1] These structures contain two nonequivalent octahedral sites and an eightfold coordinated site. It would be possible for one Mg^{++} cation to occupy the eightfold site as in pyrope garnet, whereas the octahedral sites might be shared by Mg^{++} and Si^{4+} as in the $Mg_3(Mg,Si)Si_3O_{12}$ high-pressure garnet end-member.[2] The calcium ferrite structure is stable for a wide variety of ions possessing different charges and radii, and extensive solid solutions are possible. A calcium ferrite-type polymorph of forsterite with the formula $Mg^{VIII}(Mg,Si)^{VI}O_4$ would be about 5% denser than the isochemical mixture of $2MgO + SiO_2$.[3] Such a phase might well be stable in the lower mantle, where it would probably form solid solutions with other components possessing calcium ferrite structures, principally $MgAl_2O_4$ and $NaAlSiO_4$.[4] The ability to form complex solid solutions may well be an important factor stabilizing this phase in the lower mantle.

Another A_2BO_4 structure which might be considered for Mg_2SiO_4 is that of Ca_2IrO_4—again, a binary compound formed between components possessing rocksalt and rutile structures. Ca_2IrO_4 possesses a complex hexagonal structure[5] which is 7% denser than the isochemical mixture of $CaO + IrO_2$.

The K_2NiF_4 structure is also very dense, consisting of alternative subunits possessing perovskite and rocksalt structures. However, a crystallographic analysis by Baur (1972) suggests that this structure is unlikely to be adopted by a high-pressure form of Mg_2SiO_4. Alternatively, Mg_2SiO_4 might disproportionate under extreme pressures into a mixture of $MgSiO_3$ perovskite + MgO. The conditions under which $MgSiO_3$ might exist in the perovskite structure are discussed in Sec. 12-5. Such an assemblage would be 3 to 5% denser than isochemical mixed oxides. Possible models for these transformations are the olivines, $MgCaSiO_4$, $MgCaGeO_4$, $MnCaGeO_4$, and Cd_2GeO_4, which transform to perovskite plus rocksalt mixtures as described in the previous section.

From the above discussion, it is evident that several structures or assemblages exist which are 3 to 7% denser than isochemical mixed oxides and to which Mg_2SiO_4 might ultimately transform in the deeper regions in the mantle. On the basis of some recent results,[6] it is now believed that an assemblage of $MgSiO_3$ perovskite plus MgO is the most promising candidate. However, a calcium ferrite-type polymorph of Mg_2SiO_4 remains a distinct possibility.

It should be noted that the recently observed disproportionation of Mg_2SiO_4 into mixed oxides at high pressure in no way conflicts with the possibility that these oxides may recombine to form denser compounds at still higher pressures. As an example we may cite the case of $MgAl_2O_4$ spinel, which disproportionates into $MgO + Al_2O_3$ around 200 kbars but which, at shock pressures exceeding

[1] Reid and Ringwood (1970).
[2] Section 12.3
[3] Section 11.9
[4] Ringwood (1970).
[5] Babel, Rudorf, and Tschopp (1966).
[6] Reid and Ringwood (1975).

700 kbars, transforms to a phase which is substantially denser than the mixed oxides (Sec. 11-6, Fig. 11-11).

11-9 DENSITY RELATIONSHIPS AMONG A_2BO_4 STRUCTURE TYPES

An interpretation of the constitution of the deep mantle in terms of the probable nature of the minerals present requires accurate methods of estimating the densities of possible high-pressure polymorphs. This information may be obtained from a systematic study of the densities or molar volumes of different classes of A_2BO_4 compounds relative to their isochemical mixed oxides. The two principal classes of A_2BO_4 (or AB_2O_4) compounds are related to their constituent oxides by the equation

$$A_2BO_4 = 2AO \text{ (rocksalt str.)} + BO_2 \text{ (rutile str.)}$$

$$AB_2O_4 = AO \text{ (rocksalt str.)} + B_2O_3 \text{ (corundum str.)}$$

When the volumes of different types of A_2BO_4 compounds are compared to their oxide volumes, some important systematic relationships emerge.[1] These are

[1]Reid and Ringwood (1970).

Table 11-6 DENSE A_4BO_4 COMPOUNDS, FORMULA VOLUMES, Å^3.
(From Reid and Ringwood, 1970)

Compound	V(Å^3)	Vol. compound (V)† / Vol. oxides (V_0)
1. Olivine (α) type		
Ni_2SiO_4	70.58	1.186
Mg_2SiO_4	72.68	1.199
Co_2SiO_4	73.96	1.191
Mg_2GeO_4	76.34	1.175
Fc_2SiO_4	76.81	1.206
Mn_2SiO_4	80.70	1.201
$MgMnGeO_4$	81.29	1.191
$CoMnGeO_4$	81.72	1.185
$FeMnGeO_4$	82.76	1.186
Mn_2GeO_4	84.84	1.186
$CaMgSiO_4$	85.27	1.222
γ-Ca_2SiO_4	96.75	1.226
Ca_2GeO_4	102.0	1.226
2. β-Mg_2SiO_4 type		
Mg_2SiO_4*	67.41	1.112
Co_2SiO_4*	69.08	1.112
Mn_2GeO_4*	79.72	1.115
Zn_2SiO_4*	69.29	

*High-pressure phases.

†V is the volume per formula unit. V_0 is the sum of the formula volumes of the constituent oxides of rocksalt, rutile, or corundum type.

Table 11-6 (*continued*)	$V(\text{Å}^3)$	V/V_0†
3. A_2BO_4 spinel (γ) type		
Ni_2SiO_4*	65.04	1.093
Co_2SiO_4*	67.37	1.085
Ni_2GeO_4	69.45	1.088
Fe_2SiO_4*	69.78	1.096
Mg_2GeO_4	70.09	1.079
Li_2NiF_4	71.81	1.092
Co_2GeO_4	71.94	1.083
$NiMnGeO_4$*	73.03	1.079
Mg_2VO_4	73.71	1.103
Fe_2GeO_4	74.38	1.094
$CoMnGeO_4$*	74.78	1.084
Co_2TiO_4	75.13	1.073
Mg_2TiO_4	75.15	1.096
$FeMnGeO_4$*	75.90	1.088
Fe_2TiO_4	77.74	1.085
Mg_2SnO_4	80.59	1.103
Co_2SnO_4	80.73	1.082
4. Sr_2PbO_4 (δ) type		
$FeMnGeO_4$*	70.21	1.006
Mn_2GeO_4*	71.90	1.005
Mn_2SnO_4	80.63	1.012
Na_2CuF_4‡	85.28	1.022
Cd_2SnO_4	87.85	1.004
Ca_2SnO_4	90.95	0.995
Ca_2PbO_4	96.00	0.990
Sr_2PbO_4	108.30	0.984
5. $CaMn_2O_4$ type		
Mn_3O_4*	70.93	0.979
$CaMn_2O_4$	77.00	0.984
6. K_2NiF_4 type		
Ca_2GeO_4*	81.30	0.976
Ca_2MnO_4	81.35	0.975
Sr_2TiO_4	94.85	0.949
Sr_2RuO_4	95.38	0.951
Sr_2IrO_4	97.74	0.972
Sr_2MoO_4	98.64	0.970
Sr_2SnO_4	102.10	0.978
K_2NiF_4	104.92	0.956
K_2MgF_4	107.20	0.982
Ba_2SnO_4	113.95	0.950
Ba_2PbO_4	123.0	0.980
7. $CaFe_2O_4$ type		
$CaAl_2O_4$*	66.03	0.939
β-$CaCr_2O_4$	71.90	0.944
CaV_2O_4	73.80	0.953
$CaFe_2O_4$	74.55	0.954
$CaIn_2O_4$	87.45	0.962
8. $CaTi_2O_4$	76.08	0.951
9. Ca_2IrO_4	81.90	0.935

*High-pressure phases.

†V is the volume per formula unit, V_0 the sum of the formula volumes of the constituent oxides of rocksalt, rutile, or corundum type.

‡Monoclinic distortion of Sr_2PbO_4 type.

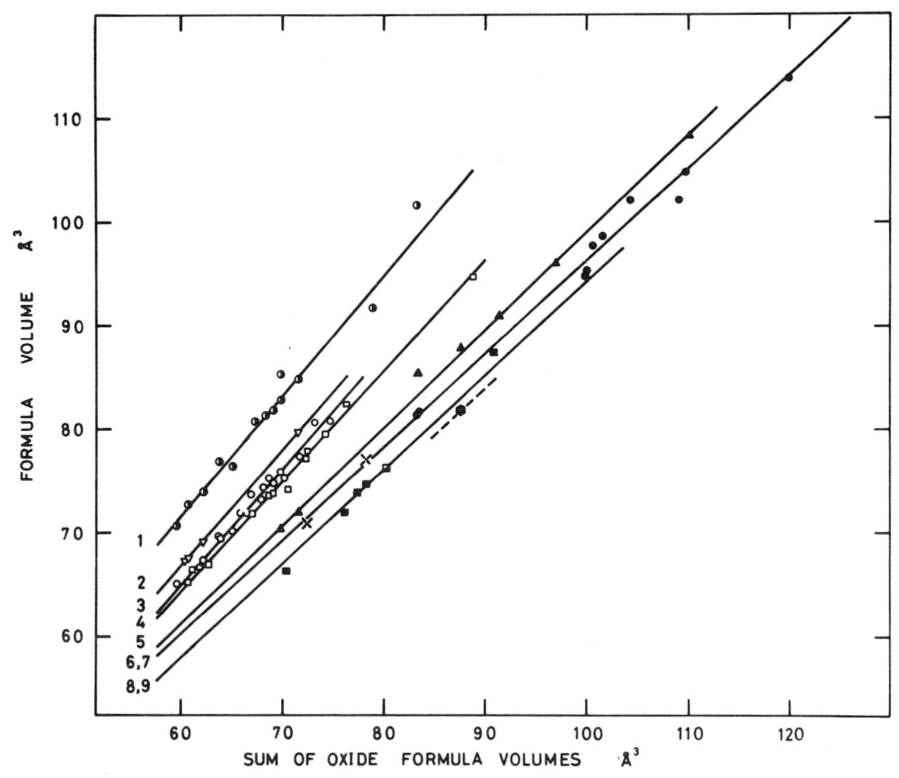

FIGURE 11-14
The formula volumes of A_2BO_4 and AB_2O_4 polymorphs versus the volumes of their isostructural mixtures of 2AX (rocksalt) + BX_2 (rutile) or AO (rocksalt) plus B_2O_3 (corundum). The structure types are shown:

1.	Olivine (α)	vertically shaded circles
2.	β-Mg_2SiO_4 type	inverted triangles
3.	A_2BO_4 spinel (γ)	open circles
4.	AB_2O_4 spinel	open squares
5.	Sr_2PbO_4 (δ)	shaded triangles
6.	$CaMn_2O_4$	crosses
7.	K_2NiF_4	black circles
8.	$CaFe_2O_4$	shaded squares
9.	Ca_2IrO_4	shaded hexagon

(*From Reid and Ringwood, 1970, with permission.*)

shown in Tables 11-6 and 11-7 and in Fig. 11-14. It is seen that the volumes of isotypes of a given structure class, relative to the isochemical mixed oxides, are very similar and are indeed characteristic of the structure class. The average volume for each class is compared with that of the isochemical mixed oxides (2AO rocksalt + BO_2 rutile or AO rocksalt + B_2O_3 corundum) in Table 11-7. These systematic relationships make it possible to estimate quite accurately the density which any given A_2BO_4 (or AB_2O_4) compound would possess if it transformed to

Table 11-7 RELATIVE VOLUMES OF A_2BO_4 POLYMORPHS. (*From Reid and Ringwood, 1970*)

Structure type	Examples, with coordination numbers,* of metals and anions	V/V_0†	$\Delta V/V_0$(%)‡
Olivine [α]	$^{[6]}Mg_2{}^{[4]}Si^{[4]}O_4$	1.20	20
β-Mg_2SiO_4 type	$^{[6]}Co_4{}^{[4]}Si_2{}^{[4]}O_6{}^{[5]}O_1{}^{[3]}O_1$	1.11_5	11.5
Spinel A_2BO_4[γ]	$^{[6]}Fe_2{}^{[4]}Si^{[4]}O_4$	1.09	9
Spinel AB_2O_4	$^{[4]}Mg^{[6]}Al_2{}^{[4]}O_4$	1.07_5	7.5
$2AO + BO_2$	$2^{[6]}Mg^{[6]}O + {}^{[6]}Ti^{[3]}O_2$	1.00	0
$AO + B_2O_3$	$^{[6]}Mg^{[6]}O + {}^{[6]}Al_2{}^{[4]}O_3$	1.00	0
Sr_2PbO_4 [δ]	$^{[6]}Mn_2{}^{[6]}Ge^{[5]}O_2{}^{[4]}O_2$	1.00	0
$CaMn_2O_4$	$^{[8]}Ca^{[6]}Mn_2{}^{[5]}O_2{}^{[6]}O_1{}^{[4]}O_1$	0.98	− 2
Defect NiAs§	$^{[6]}Fe^{[6]}Cr_2{}^{[5]}S_2{}^{[4]}S_2$	0.98	− 2
K_2NiF_4	$^{[9]}Ca_2{}^{[6]}Ge^{[6]}O_4$	0.96_5	− 3.5
$CaFe_2O_4$	$^{[8]}Ca^{[6]}Al_2{}^{[5]}O_4$	0.94_4	− 5.5
$CaTi_2O_4$	$^{[8]}Ca^{[6]}Ti_2{}^{[5]}O_2{}^{[6]}O_1{}^{[4]}O_1$	0.94_5	− 5.5
Ca_2IrO_4	$^{[9]}Ca_2{}^{[7]}Ca_3{}^{[6]}Ca_1{}^{[6]}Ir_3{}^{[6]}O_3{}^{[5]}O_9$	0.93_5	− 6.5

*Coordination number given by superscripts in brackets.

†Structure volumes relative to sum of constituent oxide volumes, averaged from Fig. 11-14.

‡$\Delta V = V - V_0$.

§A_2BO_4 type not as yet obtained for oxides.

another structure class. This information has been repeatedly utilized in previous sections. Also given in Table 11-7 are the coordination numbers of both cations and oxygen anions for each structure type. In several instances, one kind of atom is seen to occupy more than one type of lattice site possessing differing coordinations. As might be expected, there is a general increase in average coordination numbers as the densities of structures increase. Since, in any structure, the total number of bonds from metal to oxygen atoms must be equal to the number of bonds from oxygen to metal atoms, the sum of the products (number of metal atoms of given coordination number) × (coordination number) must be equal to the sum of the products (number of oxygen atoms of given coordination number) × (coordination number), as is seen to be the case in Table 11-7. This requirement leads, in phases as dense or denser than the corresponding mixtures of simple oxides, to oxygen coordinations as high as 5 or 6.

The defect nickel arsenide structure class has been included in Table 11-7, although this is not formed by oxides. Transformations of several chalcogenide spinels ($FeCr_2S_4$, $CoCr_2S_4$, $CuCr_2Se_4$, $CuCr_2Te_4$, $CuCr_2Te_4$, and Fe_3S_4) to the defect nickel arsenide structure under high pressure have been reported.[1] The density increase associated with the transformation is about 8%, and it has been suggested[2] that this structure may be important in the lower mantle. However, this does not appear likely unless the nature of the bonding in oxides changes rather drastically to covalent-metallic under pressures. This type of bonding occurs in nickel-arsenide structures, which permits the metallic atoms to have two

[1]Rooymans (1967).

[2]Ahrens and Syono (1967).

other metallic atoms as near neighbours. Such a structure would be improbable in an oxide as long as the metallic atoms retained any polar (i.e., ionic) character. Considerations of the electron orbitals of the oxygen atoms suggests that they cannot meet the rather elaborate orbital configuration required for the bonding in the nickel arsenide structure.[3]

REFERENCES

AHRENS, T. J. (1972a). The mineralogic distribution of iron in the upper mantle. *Phys. Earth Planet. Interiors* **5**, 267–281.

——— (1972b). The state of mantle minerals. *Tectonophysics* **13**, 189–219.

———, D. L. ANDERSON, and A. E. RINGWOOD (1969). Equations of state and crystal structures of high pressure phases of shocked silicates and oxides. *Rev. Geophys.* **7**, 667–707.

——— and Y. SYONO (1967). Calculated mineral reactions in the Earth's mantle. *J. Geophys. Res.* **72**, 4181–4188.

AKIMOTO, S. I. (1972). The system MgO-FeO-SiO_2 at high pressures and temperatures—phase equilibria and elastic properties. *Tectonophysics* **13**, 161–187.

——— and H. FUJISAWA (1966). Olivine-spinel transition in system Mg_2SiO_4-Fe_2SiO_4 at 800°C. *Earth Planet. Sci. Letters* **1**, 237–240.

——— and ——— (1968). Olivine-spinel solid solution equilibria in the system Mg_2SiO_4-Fe_2SiO_4. *J. Geophys. Res.* **73**, 1467–1479.

——— and Y. IDA (1966). High-pressure synthesis of Mg_2SiO_4 spinel. *Earth Planet. Sci. Letters* **1**, 358–359.

———, E. KOMADA, and I. KUSHIRO (1967). Effect of pressure on the melting of olivine and spinel polymorphs of Fe_2SiO_4. *J. Geophys. Res.* **72**, 679–686.

——— and Y. SATO (1968). High pressure transformation in Co_2SiO_4 olivine and some geophysical implications. *Phys. Earth Planet. Interiors* **1**, 498–505.

——— and Y. SYONO (1967). High pressure decomposition of some titanate spinels. *J. Chem. Physics* **47**, 1813–1817.

ANDERSON, D. L. (1967). Phase changes in the upper mantle. *Science* **157**, 1165–1173.

——— and H. KANAMORI (1968). Shock wave equations of state for rocks and minerals. *J. Geophys. Res.* **73**, 6477–6502.

BABEL, D. (1965). Die Struktur des Na_2CuF_4. *Z. anorg. allgem. Chem.* **336**, 200–206.

———, W. RUDORF, and R. TSCHOPP (1966). Erdalkaliridium (IV)-oxide. Struktur von Dicalciumiridium (IV)-oxide, Ca_2IrO_4. *Z. anorg. allgem. Chem.* **347**, 282–288.

BASSETT, W. A., and L. MING (1973). Disproportionation of Fe_2SiO_4 to $2FeO + SiO_2$ at pressures up to 250 kilobars and temperatures up to 3000°C. *Phys. Earth Planet. Interiors* **6**, 154–160.

——— and T. TAKAHASHI (1970). Disproportionation of Fe_2SiO_4 to $2FeO + SiO_2$ at high pressure and temperature. (Abstract.) *Trans. Am. Geophys. Union* **51**, 828.

BAUR, W. H. (1972). Computer-simulated crystal structures of observed and hypothetical Mg_2SiO_4 polymorphs of high and low density. *Am. Min.* **57**, 709–731.

BERNAL, J. D. (1936). Discussion. *Observatory* **59**, 268.

[3]Wang (1970).

BERTAUT, E. F., and P. BLUM (1956). Determination of the structure of Ti_2CaO_4. *Acta Cryst.* **9,** 121–126.

———, ———, and G. MAGNANO (1956). Structure des vanadite, chromite et ferrite monocalcique. *Bull. Soc. Franc. Min. Crist.* **79,** 536–561.

BINNS, R. A., R. J. DAVIS, and S. B. J. REED (1969). Ringwoodite, natural $(MgFe)_2SiO_4$ spinel in the Tenham meteorite. *Nature* **221,** 943–944.

BIRCH, F. (1952). Elasticity and constitution of the Earth's interior. *J. Geophys. Res.* **57,** 227–286.

——— (1964). Density and composition of mantle and core. *J. Geophys. Res.* **69,** 4377–4388.

BLASSE, G. (1963). Crystal structures of some compounds of the type $LiMe^{3+}$ $Me^{4+}O_4$ and $LiMe^{2+}Me^{5+}O_4$. *J. Inorg. Nuclear Chem.* **25,** 230–231.

BRAGG, W. L., and G. B. BROWN (1926). The structure of olivine. *Z. Krist.* **64,** 538–556.

——— and G. F. CLARINGBULL (1965). "*Crystal Structures of Minerals.*" Bell and Sons, London. 409 pp.

CLARK, S. P., and A. E. RINGWOOD (1964). Density distribution and constitution of the mantle. *Rev. Geophys.* **2,** 35–88.

DACHILLE, F., and R. ROY (1960). High pressure studies of the system Mg_2GeO_4-Mg_2SiO_4 with special reference to the olivine-spinel transition. *Am J. Sci.* **258,** 225–246.

DATTA, R. K., and R. ROY (1967). Equilibrium order-disorder in spinels. *J. Am. Ceram. Soc.* **50,** 578–583.

DAVIES, G. F., and E. GAFFNEY (1973). Identification of high pressure phases of minerals and rocks from Hugoniot data. *Geophys. J.* **33,** 165–183.

DECKER, B. F., and J. S. KASPER (1957). The structure of calcium ferrite. *Acta Cryst.* **10,** 332–337.

GAFFNEY, E. S., and D. L. ANDERSON (1973). The effect of low-spin Fe^{2+} on the composition of the lower mantle. *J. Geophys. Res.* **78,** 7005–7014.

GOLDSCHMIDT, V. M. (1931). Zur Kristallchemie des Germaniums. *Nachr. Ges. Wiss. Göttingen, Math.-Physik. Kl.* 184–190.

GORTER, E. W. (1960). Some structural relationships of ternary transition metal oxides. In: "*Proc. 17th Internat. Congress for Pure and Applied Chemistry, Munich,* 1959": vol. 1, Inorganic Chem. pp. 303–328.

HELMBERGER, D., and R. A. WIGGINS (1971). Upper mantle structure of mid-western United States. *J. Geophys. Res.* **76,** 3229–3245.

HILL, P. M., H. S. PEISER, and J. R. RAIT (1956). The crystal structure of calcium ferrite and β calcium chromite. *Acta Cryst.* **9,** 981–986.

ITO, E., Y. MATSUI, K. SUITO, and N. KAWAI (1974). Synthesis of γ-Mg_2SiO_4. *Phys. Earth Planet. Interiors* **8,** 342–344.

JACKSON, I. N., R. C. LIEBERMANN, AND A. E. RINGWOOD (1974). Disproportionation of spinels to mixed oxides: Effect of inverse character and implications for the mantle. *Earth Planet. Sci. Letters* **24,** 203–208.

JEFFREYS, H. (1937). On the materials and density of the earth's crust. *Mon. Not. Roy. Astron. Soc., Geophys. Suppl.* **4,** 50–61.

KAMB, B. (1968). Structural basis of the olivine-spinel stability relation. *Am. Min.* **53,** 1439–1455.

KAWAI, N., S. ENDOH, and S. SAKATA (1966). Synthesis of Mg_2SiO_4 with a spinel structure. *Proc. Japan. Acad.* **42,** 626–628.

KUMAZAWA, M., H. SAWAMOTO, E. OHTANI, and K. MASAKI (1974). Postspinel phase of forsterite and evolution of the earth's mantle. *Nature* **247,** 356–358.

LEPICARD, G., and J. PROTAS (1966). Structural study of the double oxide of manganese and calcium, orthorhombic $CaMn_2O_4$ (marokite). *Bull. Soc., Franc. Mineral. Crist.* **89,** 318–324.

MA, C. B., and J. F. HAYS (1972). New phases on the join $NiAl_2O_4$-Ni_2SiO_4. Geol. Soc. Am. Annual Meeting. (Abstract.)

———, ———, and C. W. BURNHAM (1971). Octahedral co-ordination of silicon in silicate spinels (revised abstract). *EOS* **52,** 536.

MACDONALD, G. J. F. (1956). Quartz-coesite stability relations at high temperatures and pressures. *Am. J. Sci.* **254,** 713–721.

——— (1962). On the internal constitution of the inner planets. *J. Geophys. Res.* **67,** 2945–2974.

MAO, H., W. BASSETT, and T. TAKAHASHI (1970). High pressure transformation in magnetite. *Carnegie Inst. Washington Yearbook* **68,** 249–251.

——— and P. BELL (1971). High pressure decomposition of spinel (Fe_2SiO_4). *Carnegie Inst. Washington Yearbook* **70,** 176–178.

MCQUEEN, R. G., J. N. FRITZ, and S. P. MARSH (1964). On the composition of the earth's interior. *J. Geophys. Res.* **69,** 2947–2978.

——— and S. P. MARSH (1966). In: S. P. Clark, Jr. (ed.), "*Handbook of Physical Constants.*" *Geol. Soc. Am. Memoir* 97.

———, ———, and J. N. FRITZ (1967). Hugoniot equation of state of twelve rocks. *J. Geophys. Res.* **72,** 4999–5036.

MING, L., and W. BASSETT (1974). Postspinel phases in Mg_2SiO_4-Fe_2SiO_4 system up to 80% Mg_2SiO_4. (Abstract.) *EOS* **55,** 416–417.

MOORE, P. B., and J. V. SMITH (1969). High pressure modification of Mg_2SiO_4: Crystal structure and crystallochemical and geophysical implications. *Nature* **221,** 653–655.

MORIMOTO, N., S. AKIMOTO, K. KOTO, and M. TOKONAMI (1969). Modified spinel, beta-manganous orthogermanate: Stability and crystal structure. *Science* **165,** 586–588.

MÜLLER-BUSCHBAUM, H. (1966). Uber oxoscandate II zue Kenntnis des $MgSc_2O_4$. *Z. anorg. allgem. Chem.* **343,** 113–224.

REID, A. F., and A. E. RINGWOOD (1969a). Newly observed high pressure transformations in Mn_3O_4, $CaAl_2O_4$ and $ZrSiO_4$. *Earth Planet. Sci. Letters* **6,** 205–208.

——— and ——— (1969b). High pressure scandium oxide and its place in the molar volume relationships of dense structures of M_2X_3 and ABX_3 type. *J. Geophys. Res.* **74,** 3238–3252.

——— and ——— (1970). The crystal chemistry of dense M_3O_4 polymorphs: High pressure Ca_2GeO_4 of K_2NiF_4 structure type. *J. Solid State Chem.* **1,** 557–565.

——— and ——— (1975). High pressure modification of $ScAlO_3$ and some geophysical implications. *J. Geophys. Res.* (In press.)

———, A. D. WADSLEY, and A. E. RINGWOOD (1967). High pressure $NaAlGeO_4$, a calcium ferrite isomorph and model structure for silicates at depth in the earth's mantle. *Acta Cryst.* **23,** 736–739.

———, ———, and M. SIENKO (1968). Crystal chemistry of sodium scandium titanate $NaScTiO_4$, and its isomorphs. *Inorg. Chem.* **7,** 112–118.

RINGWOOD, A. E. (1956). The olivine-spinel transition in the earth's mantle. *Nature* **178,** 1303–1304.

——— (1958a). The constitution of the mantle I. Thermodynamics of the olivine-spinel transition. *Geochim. Cosmochim. Acta* **13,** 303–321.

——— (1958b). The constitution of the mantle II. Further data on the olivine-spinel transition. *Geochim. Cosmochim. Acta* **15,** 18–29.

——— (1958c). Constitution of the mantle III. Consequences of the olivine-spinel transition. *Geochim. Cosmochim. Acta* **15,** 195–212.
——— (1958d). Olivine-spinel inversion in fayalite. *Bull. Geol. Soc. Am.* **69,** 129–130.
——— (1962a). Prediction and confirmation of olivine spinel transition in Ni_2SiO_4. *Geochim. Cosmochim. Acta* **26,** 457–469.
——— (1962b). Mineralogical constitution of the deep mantle. *J. Geophys. Res.* **67,** 4005–4010.
——— (1963). Olivine-spinel transformation in cobalt orthosilicate. *Nature* **198,** 79–80.
——— (1966). Mineralogy of the mantle. In: P. Hurley (ed.), "*Advances in Earth Sciences*," pp. 357–399. M.I.T. Press, Cambridge.
——— (1968). High pressure transformations in A_2BO_4 compounds. (Abstract.) *Trans. Am. Geophys. Union* **49,** 355.
——— (1969). Phase transformations in the mantle. *Earth Planet. Sci. Letters* **5,** 401–412.
——— (1970). Phase transformations and the constitution of the mantle. *Phys. Earth Planet. Interiors* **3,** 109–155.
——— and A. MAJOR (1966). Synthesis of Mg_2SiO_4-Fe_2SiO_4 solid solutions. *Earth Planet. Sci. Letters* **1,** 241–245.
——— and ——— (1967). High pressure transformations in zinc germanates and silicates. *Nature* **215,** 1367–1368.
——— and ——— (1970). The system Mg_2SiO_4-Fe_2SiO_4 at high pressures and temperatures. *Phys. Earth Planet. Interiors* **3,** 89–108.
——— and A. F. REID (1968a). High pressure polymorphs of olivines: the K_2NiF_4 type. *Earth Planet. Sci. Letters* **5,** 67–70.
——— and ——— (1968b). High pressure transformations of spinels, I. *Earth Planet. Sci. Letters* **5,** 245–250.
——— and ——— (1971). Olivine-spinel transformations in $MgMnGeO_4$, $FeMnGeO_4$ and $CoMnGeO_4$. *J Phys. Chem. Solids* **31,** 2791–2793.
ROOYMANS, C. J. M. (1967). Structural investigations on some oxides and other chalcogenides at normal and very high pressures. Doctoral thesis, University of Amsterdam.
SCHATZ, J. F., and G. SIMMONS (1972). Thermal conductivity of earth materials at high temperatures. *J. Geophys. Res.* **77,** 6966–6983.
SHANNON, R. D., and C. T. PREWITT (1969). Effective ionic radii in oxides and fluorides. *Acta Cryst.* **B25,** 925–946.
SHIMAZU, Y. (1958). A chemical phase transition hypothesis on the origin of the C-layer within the mantle of the earth. *J. Earth Sci. Nagoya University* **6,** 12–30.
SIMPSON, D. W., R. F. MEREU, and D. KING (1974). An array study of P wave velocities in the upper mantle and transition zone beneath northeastern Australia. *Bull. Seism. Soc. Am.* (In press.)
SMITH, J. V., and B. MASON (1970). Pyroxene-garnet transformation in Coorara meteorite. *Science* **168,** 832–834.
STISHOV, S. M. (1962). On the internal structure of the earth. *Geokhimiya* No. 8, 649–659.
——— (1963). Equilibrium line between coesite and the rutile-like modification of silica. (In Russian.) *Dokl. Akad. Nauk. SSSR* **148,** no. 5, 1186–1188.
——— and S. V. POPOVA (1961). New dense polymorphic modification of silica. *Geokhimiya* No. 10, 837–839.

STOLL, E., P. FISCHER, W. HAELG, and G. MAIER (1964). Redetermination of cation distribution of spinel ($MgAl_2O_4$) by means of neutron diffraction. *J. Phys. (Paris)* **25,** 447–448.

SUITO, K. (1972). Phase transformations of pure Mg_2SiO_4 into a spinel structure under high pressures and temperatures. *J. Phys. Earth* **20,** 225–243.

SYONO, Y., S. A. AKIMOTO, and Y. MATSUI (1971). High pressure transformations in zinc silicates. *J. Solid State Chem.* **3,** 369–380.

———, H. SAWAMOTO, and S. A. AKIMOTO (1969). Disordered ilmenite $MnSnO_3$ and its magnetic property. *Solid State Communications* **7,** 713–716.

TARTE, P., and A. E. RINGWOOD (1962). Infra-red spectra of the spinels Ni_2SiO_4, Ni_2GeO_4 and their solid solutions. *Nature* **193,** 971–972.

TRÖMEL, M. (1965). Zur Structur der Verbindungen vom Sr_2PbO_4-Typ. *Naturwiss.* **17,** 492–493.

——— (1967). Kristallstrukturdaten für Ca_2SnO_4 und Cd_2SnO_4. *Naturwiss.* **54,** 17–18.

WADSLEY, A. D., A. F. REID, and A. E. RINGWOOD (1968). The high pressure form of Mn_2GeO_4, a member of the olivine group. *Acta Cryst.* **B24,** 740–742.

WANG, C. Y. (1967). Phase transitions in rocks under shock compression. *Earth Planet. Sci. Letters* **3,** 107–113.

——— (1970). Can mantle minerals have the NiAs structure? *Phys. Earth Planet. Interiors* **3,** 213–217.

WHITCOMB, J. H., and D. L. ANDERSON (1970). Reflection of $P'P'$ seismic waves from discontinuities in the mantle. *J. Geophys. Res.* **75,** 5713–5728.

WYCKOFF, R. W. G. (1965). "*Crystal Structures*," vol. 3, 2d ed. Interscience, a division of Wiley, New York. 981pp.

YAGI, T., F. MARUMO, and S. I. AKIMOTO (1974). Crystal structure of spinel polymorphs of Fe_2SiO_4. *Am. Min.* **59,** 486–490.

12

HIGH-PRESSURE TRANSFORMATIONS IN ABO_3-TYPE COMPOUNDS

Next to olivine, the major mineral constituents of the upper mantle are pyroxenes and garnets.[1] The fundamental chemical formulae of these two mineral families are similar, having two cations to three oxygen anions. Although characterized by complex compositions and the formation of extensive solid solutions, the end-members of these groups usually have simple formulae of the type $A^{++}B^{4+}O_3$ and $C_2^{3+}O_3$. Because of the abundance of these minerals in the upper mantle, their behaviour at high pressures is of vital importance to an understanding of the constitution of the deep mantle. As we shall see, they display a wide variety of transformations, most of which have been discovered only recently. We shall commence with a description of transformations in pyroxenes and pyroxenoids.

[1]Ref. Chaps. 3, 5, and 6.

Table 12-1 DISPROPORTIONATION TRANSFORMATIONS DISPLAYED BY PYROXENES AT 700 TO 1000°C

$2ABO_3 \rightarrow A_2BO_4 + BO_2$ (Pyroxene) → (spinel) + (rutile)	$\Delta\rho/\rho \sim 15\%$	Ref.
Pyroxene	Disproportionation pressure (approx.) (kbars)	
$FeGeO_3$	10	1
$CoGeO_3$	10	1
$(MgNi)GeO_3$	10	1
$FeSiO_3$	100	2
$CoSiO_3$	100	3
$FeSiO_3$-$MgSiO_3$ solid solutions as far as $(Mg_{0.43}Fe_{0.57})SiO_3$	100–200	4

References:
1. Ringwood and Seabrook (1963).
2. Ringwood and Major (1966a).
3. Ringwood and Major (1966b).
4. Ringwood and Major (1968).

12-1 DISPROPORTIONATION OF PYROXENES INTO SPINEL + RUTILE STRUCTURES

Some pyroxenes are observed to disproportionate at high pressures into mixtures of spinel + rutile-type phases according to the following reaction:

$$2ABO_3 \text{ (pyroxene)} \rightarrow A_2BO_4 \text{ (spinel)} + BO_2 \text{ (rutile)}$$

Transformations of this type were discovered[1] in the germanate pyroxenes $FeGeO_3$, $CoGeO_3$, and $(Mg_{0.75}Ni_{0.25})GeO_3$. They occurred at pressures between 10 and 25 kbars at 700°C and were accompanied by density increases of about 15% (Table 12-1).

Later, it was found that the pyroxenes $FeSiO_3$ and $CoSiO_3$ likewise transformed to Fe_2SiO_4 and Co_2SiO_4 spinels plus SiO_2 (stishovite), respectively, at pressures close to 100 kbars and around 1000°C.[2,3] It is interesting to observe the similarity in the modes of transformation of $FeSiO_3$ and $FeGeO_3$, on one hand, and $CoSiO_3$ and $CoGeO_3$, on the other, and also that the germanates transform at only 10 kbars compared to 100 kbars for the silicates.

Having obtained a transformation in $FeSiO_3$, a component of natural pyroxenes, the next step was to attempt to extend this towards more magnesian compositions by means of a study of the system $FeSiO_3$-$MgSiO_3$. Reconnaissance results[4] in this system are shown in Fig. 12-1. The position of the spinel boundary

[1] Ringwood and Seabrook (1963).
[2] Ringwood and Major (1966a,b,c).
[3] The pressures cited are somewhat smaller than those given in the original reference because of revisions in the pressure scale.
[4] Ringwood and Major (1968).

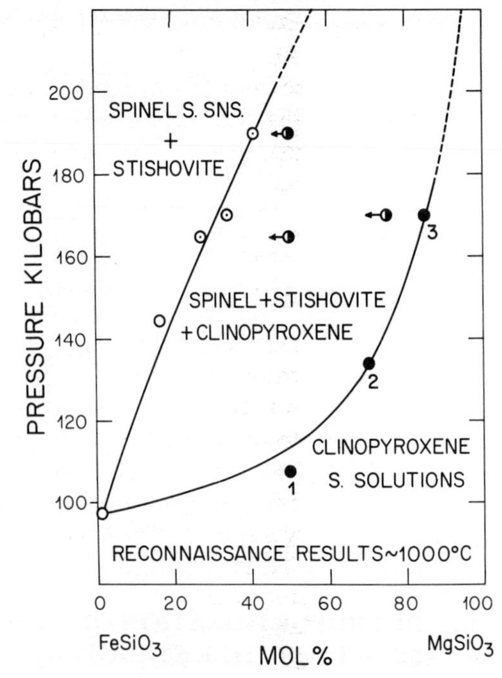

FIGURE 12-1
Phases observed in the system $FeSiO_3$-$MgSiO_3$ at pressures up to 200 kbars and at approximately 1000°C. Symbols as follows:

○ Complete transformation to spinel + stishovite.
⊙ Boundary determined by spinel composition obtained from lattice parameter.
←○ Experimental run in three-phase field in which composition of spinel was determined.
● (1,2,3) Inferred points on boundary of pyroxene stability field.

(*From Ringwood and Major, 1968, with permission.*)

is reasonably well established from measurements of the lattice parameters of the spinels; however, the boundary for initial breakdown of pyroxene is not fixed so well because of problems in obtaining equilibrium during crystallization of $MgSiO_3$-$FeSiO_3$ glasses. Nevertheless, the general nature of the phase diagram is clearly established. As pressure increases from 100 to 190 kbars, the composition of the pyroxenes which can be completely transformed to $(Mg, Fe)_2SiO_4$ spinel + SiO_2 (stishovite) rises from $FeSiO_3$ to $(Mg_{0.43}Fe_{0.57})SiO_3$. The field of spinel, stishovite, and untransformed pyroxene is wide so that a broad spectrum of pyroxene compositions ranging to about $(Mg_{0.8}Fe_{0.2})SiO_3$ can be transformed partially according to this equilibrium. An interesting feature is that, at the same Fe/Mg ratio, pyroxenes are stable to much higher pressures than the corresponding olivines.[1] It is possible to extrapolate the phase boundaries in Fig. 12-1 to obtain the transformation pressure for pure $MgSiO_3$. The uncertainties in the location of the phase boundaries make extrapolation hazardous. Nevertheless, it appears reasonable to infer that $MgSiO_3$ would transform[2] to β-Mg_2SiO_4 plus SiO_2 (stishovite) at some pressure between about 200 and 300 kbars, *providing that no other major phase transformation intervenes.*

12-2 THE PYROXENE-ILMENITE TRANSFORMATION

The possibility that $MgSiO_3$ might ultimately transform to the corundum structure (essentially a disordered ilmenite strucure) was first suggested by J. B. Thompson.[3] Experimental evidence supporting this suggestion was obtained by Ringwood and Seabrook,[4] who showed that the orthopyroxenes $MgGeO_3$ and $MnGeO_3$ were transformed to ilmenite structures at 25 to 30 kbars, 700°C, accompanied by density increases of 15 and 18%, respectively.[5] These large density increases were due to the fact that the germanium ion changed from fourfold to sixfold coordination through the transition. In view of the model relationships between germanates and silicates previously discussed, it was concluded that ultimate transformation of $MgSiO_3$ to the ilmenite structure was probable. Here then, was an alternative transformation to the disproportionation inferred in the previous section.

Further information upon the probable mode of transformation of $MgSiO_3$ was sought through a study of the system $MgGeO_3$-$MgSiO_3$ at high pressures.[6] Results of reconnaissance work in this system are shown in Fig. 12-2. $MgSiO_3$ is observed to display substantial solid solubility in $MgGeO_3$ ilmenite, and the degree of solid solubility increases with pressure, reaching about 25% at 170 kbars.

The observation of a substantial solid solubility of $MgSiO_3$ in the ilmenite structure at 170 kbars is important in showing that the free-energy difference between the ilmenite and pyroxene forms of $MgSiO_3$ is of small magnitude ($\sim$ 2.6 kcal/mol) at this pressure. A formal calculation along the lines discussed in Sec. 10-2 leads to an estimated pressure of 213 kbars for the transformation pressure in pure $MgSiO_3$. Because of the possibility that departures from ideality of the solid solutions may increase at high pressure, this estimate has a substantial margin of uncertainty. Nevertheless, it is indicative.

[1]Compare Figs. 12-1 and 11-7.

[2]*Note added in press*: Ito et al. (1972) have confirmed this transformation occurring at pressures above 200 kbars and 1000°C.

[3]Ref. Birch (1952, p. 234).

[4]Ringwood and Seabrook (1962, 1963).

[5]An ilmenite-type modification of $ZnGeO_3$ was also synthesized by Ringwood and Major (1967c) and Syono, Akimoto, and Matsui (1971).

[6]Ringwood and Major (1966c, 1968).

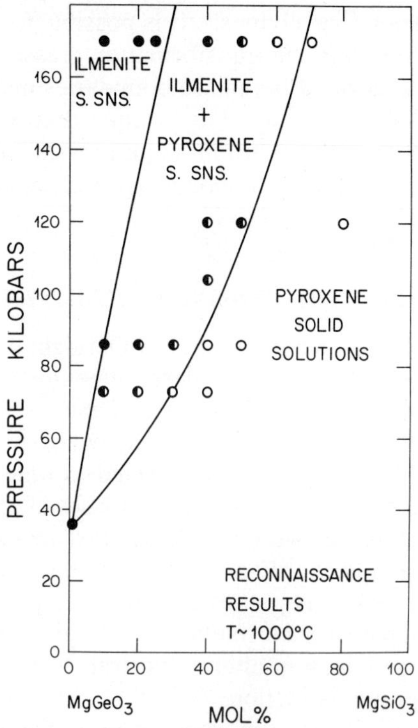

FIGURE 12-2
Phases present in the system $MgGeO_3$-$MgSiO_3$ at pressures up to 170 kbars and at approximately 1000°C. Symbols as follows:

- ● Homogeneous ilmenite type solid solutions.
- ◑ Two-phase field of ilmenite + pyroxene solid solutions.
- ○ Pyroxene solid solutions.

(From Ringwood and Major, 1968, with permission.

Alternatively, one may simply extrapolate the phase diagram (Fig. 12-2) across to the $MgSiO_3$ side, in which case it would be reasonable to conclude that $MgSiO_3$ is likely to transform to an ilmenite structure at some pressure between 200 and 300 kbars, *providing that transformations to other phases do not intervene.*

Incorporating the results of the previous section, we see that two alternative modes of transformation between pressures of 200 and 300 kbars are indicated by the experimental data. Enstatite may either transform directly to an ilmenite structure, or alternatively, the stability field of β-Mg_2SiO_4 + SiO_2 (str.) may be reached at a lower pressure, in which case disproportionation to this assemblage would intervene. The uncertainties in the experimental data (Figs. 12-1 and 12-2) do not permit a choice to be made as to the most probable mode of transformation.[1]

The density of the ilmenite form of $MgSiO_3$ is a matter of some importance.

[1]*Note added in press*: This problem has been clarified in recent experiments by Ito et al. (1972) and Kawai et al. (1974). These workers found that $MgSiO_3$ disproportionated to Mg_2SiO_4 (spinel) plus stishovite at about 200 kbars, 1000°C. However, at still higher pressures (~500 kbars), $MgSiO_3$ transformed to a corundum- or ilmenite-type structure. The density of this new polymorph (Kawai et al., 1974) was 3.81 g/cm^3, which is very similar to the value predicted above.

Table 12-2 COMPARATIVE MOLAR VOLUMES OF COMPOUNDS POSSESSING ILMENITE STRUCTURES AND FORMED FROM END-MEMBERS POSSESSING ROCKSALT AND RUTILE STRUCTURES

Compound	Molar volume V (cm^3)	Isochemical oxide volume v (cm^3)	$\Delta v/v$ (%)*
$CdTiO_3$	35.69	30.39	3.8
$CoTiO_3$	31.05	30.44	2.0
$MgGeO_3$	29.14	27.91	4.4
$MgTiO_3$	30.86	30.05	2.7
$MnGeO_3$	31.28	29.88	4.7
$MnTiO_3$	32.76	32.02	2.3
$MnVO_3$	32.01	31.10	2.9
$FeTiO_3$	31.69	30.84	2.8
$NiTiO_3$	30.56	29.77	2.7
$NiMnO_3$	28.42	27.58	3.0
$CoMnO_3$	29.43	28.25	4.2
$MgSiO_3$	26.40	25.26	4.5 (assumed)
$FeSiO_3$†	24.87	23.91	4.0 (assumed)

*$\Delta v = V - v$.

†Fe^{++} in low-spin configuration with radius of 0.61 Å (Shannon and Prewitt, 1969).

The molar volumes of ilmenite compounds compared to the volumes of their component rocksalt- and rutile-structure end-members are given in Table 12-2.

The densities of the ilmenite compounds are observed to be between 2.0 and 4.7% smaller than the isochemical oxide mixtures. This is apparently caused by the slight distortion involved in introducing two cations of differing size into the octahedral interstices of a hexagonal close packing of oxygen anions. The density differential of $MgSiO_3$ ilmenite compared with its oxide components is probably close to that of $MgGeO_3$ (4.4%) and $MnGeO_3$ (4.7%). Assuming a figure of 4.5%, the density of $MgSiO_3$ (ilmenite) would be 3.80 g/cm^3. An alternative method based upon extrapolation of bond lengths for A_2O_3 and ABO_3 compounds gives a value of 3.76 g/cm^3 for $MgSiO_3$ ilmenite.[1,2] An average value of 3.78 g/cm^3 is unlikely to be far wrong.

The corundum structure may be regarded as consisting of a hexagonal close-packed array of oxygen atoms, with metal atoms occupying two-thirds of the octahedrally coordinated interstices. Each metal atom is thus surrounded by an octahedron of oxygen atoms, whilst the oxygen atoms, in turn, are surrounded by four metal atoms. The ilmenite structure is essentially an ordered corundum structure, in which the layers of aluminium atoms are replaced successively by layers of iron and titanium atoms, as shown in Fig. 12-3. Ilmenite structures are commonly formed between end-members possessing rutile and rocksalt structures (Table 12-2).

[1] Reid and Ringwood (1969).

[2] Ref. Section 12-6, Fig. 12-13.

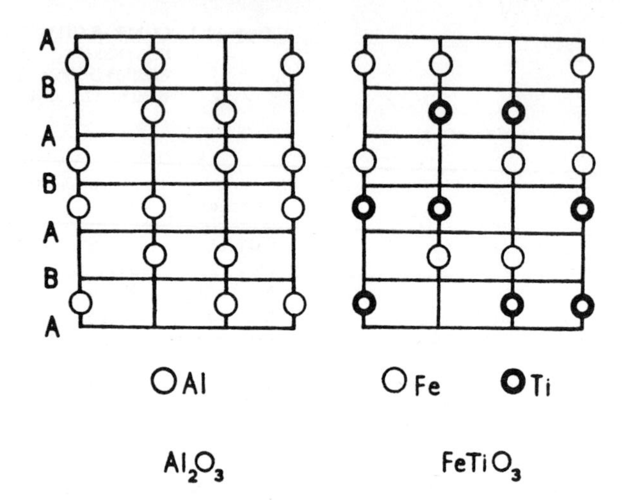

FIGURE 12-3
Diagrammatic comparison of the distribution of metal atoms in corundum and ilmenite. The structures are viewed on $(2\bar{1}\bar{1}0)$. Horizontal lines A, B indicate the positions of the close-packed layers of oxygen atoms. (*From Wells, 1962, with permission. Copyright Oxford University Press. By permission of the Clarendon Press, Oxford.*)

12-3 THE PYROXENE-GARNET TRANSFORMATION

The basic chemical formulae of the pyroxene (e.g., $MgSiO_3$) and garnet (e.g., $Mg_3Al_2Si_3O_{12}$) families are closely related, the ratio of cations to anions being 2:3. This relationship is easily seen by expressing the formula of pyrope garnet as $3MgSiO_3 \cdot Al_2O_3$. It has long been known that pyroxenes are able to take Al_2O_3 into solid solution under appropriate P-T conditions. Such aluminous pyroxenes can be regarded as consisting of solid solutions of garnet in nonaluminous pyroxene, e.g.,

$$\underset{\text{aluminous enstatite}}{3MgSiO_3 \cdot xAl_2O_3} = \underset{\text{pyrope garnet}}{xMg_3Al_2Si_3O_{12}} + \underset{\text{enstatite}}{3(1-x)MgSiO_3}$$

This type of equilibrium is of considerable importance in the lower crust and upper mantle.[1] Because the right-hand side is denser than the left-hand side, it follows that high pressures tend to break down aluminous pyroxenes into low-alumina pyroxenes plus garnets. A considerable amount of experimental data upon equilibria of this type now exist and have been applied to petrological and geophysical problems. The above equilibria may be considered to represent an important class of pyroxene-garnet transformations. However, a second and opposite class of these equilibria is also conceivable. These involve solid solutions of pyroxenes in garnet structures. Because of the large density differences in-

[1]Section 6-2.

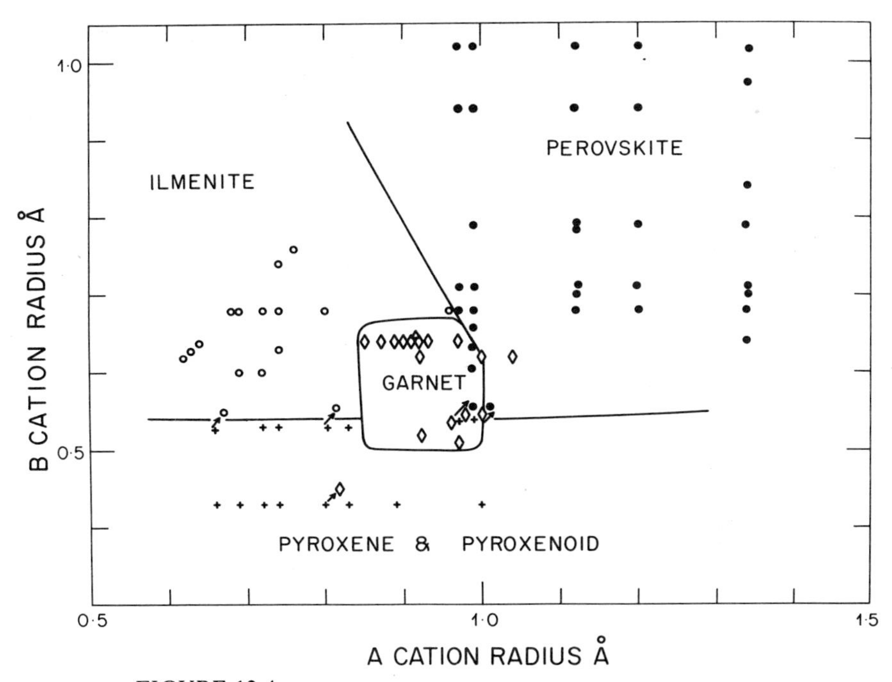

FIGURE 12-4
Plot of A and B ionic radii for $A^{++}B^{4+}O_3$ compounds possessing pyroxene, ilmenite, perovskite, and garnet structures and for $A_3^{3+}B_5^{3+}O_{12}$ garnets. Arrows indicate high-pressure transformations. (*From Ringwood, 1970, with permission.*)

volved, such equilibria would also be expected to be strongly pressure dependent. This section will be concerned with this latter class.

A Goldschmidt diagram showing the structure fields of some ABO_3 and $A_3B_5O_{12}$ compounds in relation to the ionic radii of A and B cations is shown in Fig. 12-4. The central and transitional position of the garnet field relative to the pyroxene, ilmenite, and perovskite fields is to be noted. This suggests the possibility that garnet structures may represent a third mode of transformation from pyroxenes and pyroxenoids.

The general oxygarnet formula is $A_3^{VIII}B_2^{VI}C_3^{IV}O_{12}$, where the superscripts denote the coordination of the specified cation by oxygen. The structure of pyrope garnet $Mg_3Al_3Si_3O_{12}$ was refined by Gibbs and Smith (1965), to whom the following description is due. The structure consists of independent SiO_4 and AlO_6 polyhedra which share corners to form an aluminosilicate framework within which each magnesium atom is surrounded by an irregular polyhedron of eight oxygen atoms, which may be described as a distorted cube (Figs. 12-5 and 12-6). Two edges of the silicon tetrahedron and six edges of the aluminium octahedron are shared with the magnesium cube, leaving unshared four edges in the tetrahedron, six in the octahedron, and six in the cube. The high density of garnets can be attributed to the large percentage of shared edges, which leads to a

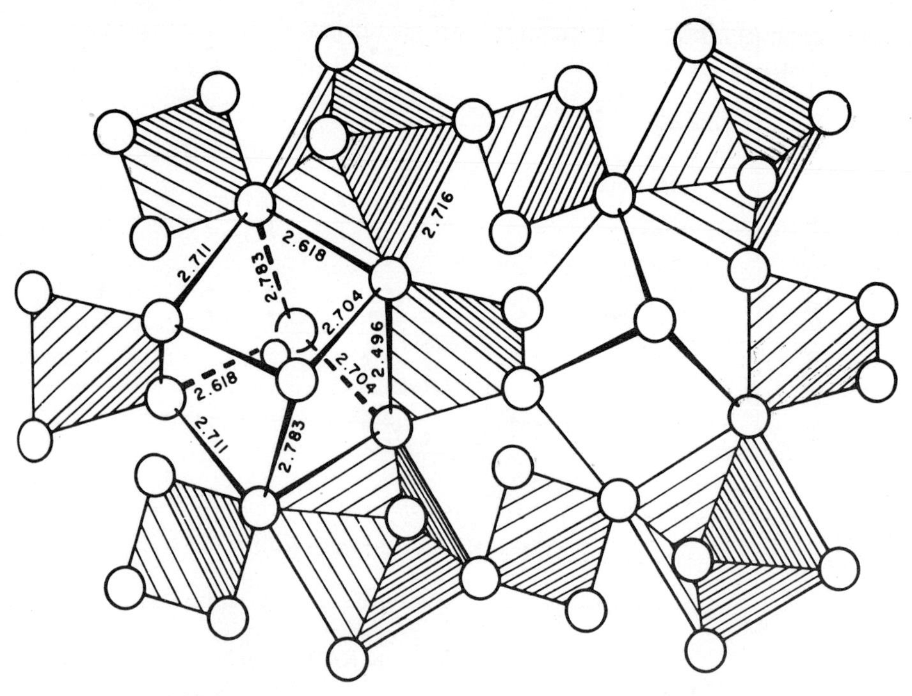

FIGURE 12-5
Part of the pyrope structure projected down the *z* axis, showing corner sharing of SiO_4 tetrahedra and AlO_6 octahedra. The large open circles represent oxygen atoms, and the smaller one within the open cube is magnesium. Aluminium and silicon atoms at the centres of the octahedra and tetrahedra are not shown, nor is the magnesium atom at the centre of the closed cube. The values along the edges of the polyhedra refer to the O-O distances. (*From Gibbs and Smith, 1965, with permission.*)

tightly packed arrangement. The shared edges are shorter than the unshared edges, leading to considerable distortion of the polyhedra. In particular, four of the Mg-O distances in the MgO_8 cube are significantly shorter than the remaining four Mg-O bonds. In the pyrope structure, each oxygen atom is coordinated by one silicon atom, one aluminium atom, and two magnesium atoms (Fig. 12-6).

Until about 20 years ago, the garnet group was thought to be essentially restricted to a rather small number of rockforming silicates $A_3B_2Si_3O_{12}$, where $A = Mg$, Ca, Fe, Mn and $B = Al$, Fe, Cr. Since then, systematic studies of the crystal chemistry of garnets have revealed that a remarkable range of elements are capable of entering this structure, and an extremely complex series of ionic substitutions and charge balances are possible. In Table 12-3, the cations which are capable of entering the A, B, and C sites are shown, together with examples of various general and specific types of ionic substitutions. Notice particularly that it is possible to replace the characteristic trivalent cations in the B position (e.g.,

Al^{3+}) by a combination of a divalent and tetravalent atom, e.g., $Mg^{++} + Ti^{4+}$. The most general case of this replacement would be where the B atoms are replaced by an (A + C) couple. In such a case, the garnet has the formula $A_3(AC)C_3O_{12}$ or, more simply, $(ACO_3)_4$, where A is divalent and C is tetravalent.

An example of this type of structure was discovered by Ringwood and Seabrook (1963), who found that $CaGeO_3$, which has a wollastonite-like structure at atmospheric pressure, was transformed completely into a garnet structure at 40 kbars and 700°C, with an inferred formula $Ca_3^{VIII}(CaGe)^{VI}Ge_3^{IV}O_{12}$. In some runs, presence of line splitting indicated that the garnet was slightly distorted with a symmetry lower than cubic. The same workers found that $CdGeO_3$, which has a complex structure of low density at atmospheric pressure, transformed to a distorted garnet structure at pressures greater than about 10 kbars, 700°C. Detailed studies of the structures of these germanate garnets[1] confirmed that the octahedral

[1]Prewitt and Sleight (1969).

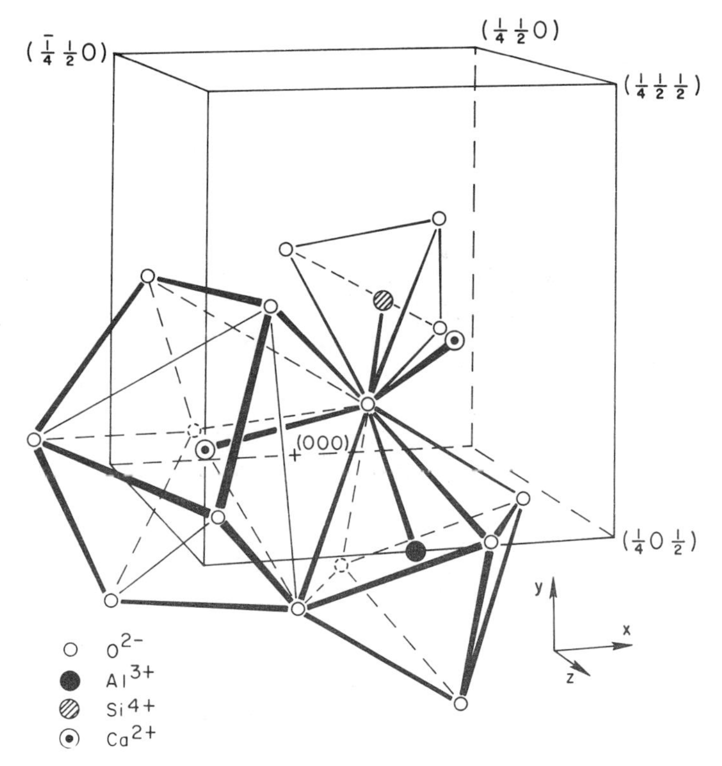

FIGURE 12-6
The coordination polyhedra of oxygen atoms about Al, Ca, and Si in grossularite. The pyrope structure is similar. (*From Abrahams and Geller, 1958, with permission.*)

Table 12-3 STRUCTURAL FORMULAE, IONIC SUBSTITUTIONS, AND EXAMPLES OF GARNETS*

$A_3^{VIII}B_2^{VI}C_3^{IV}O_{12}$
$A_3^{VIII}C_2^{VI}C_3^{IV}O_{12}$
$A_3^{VIII}(AC)^{VI}C_3^{IV}O_{12}$

			Examples
A cations	+	Na, K	$Mg_3Al_2Si_3O_{12}$
	++	Ca, Mg, Fe, Mn, Cd, Co, Cu, Zn, Sr, Ba, Mn	$Ca_3(TiMg)Si_3O_{12}$
	3+	Rare earths, Bi, Y	$(CdGd_2)Mn_2Ge_3O_{12}$
	4+	Zr, Hf	$(NaCa_2)Ni_2V_3O_{12}$
B cations	+	Li	$Na_3Al_2Li_3F_{12}$
	++	Mn, Co, Mg, Ni, Zn, Fe, Ca, Cd	$(CaNa_2)Ti_2Si_3O_{12}$
	3+	Fe, Ga, Al, Cr, V, Sc, In, Rh, Mn, Co	$Y_3Al_2Al_3O_{12}$
	4+	Sn, Ge, Si, Ti, Zr, Hf, Ru	$Ca_3(CaGe)Ge_3O_{12}$
	5+	Nb, Ta, Sb	$Mn_3(MnSi)Si_3O_{12}$
C cations	+	Li	$(NaCa_2)Mg_2P_3O_{12}$
	++	Co	$(Na_2Ca)Si_2Si_3O_{12}$
	3+	Fe, Ga, Al, Co	
	4+	Si, Ge, Sn, Ti	
	5+	As, V, P	
Anions		O^{--}, F^-, OH^-	

*General references: Wyckoff (1965), Geller (1967).

sites were shared by Cd(Ca) and Ge. It was found that the Cd and Ge were ordered into distinct crystallographic sites, resulting in a reduction in symmetry from cubic to tetragonal.

An analogous transformation was observed in rhodonite, $MnSiO_3$.[1] This mineral has a pyroxenoid structure at atmospheric pressure. At pressures of about 60 kbars, 700°C, transformation to a true pyroxene structure was observed. Finally, in runs between 120 and 170 kbars (1000°C), transformation into a garnet structure occurred. This had a density of 4.27 g/cm^3 compared to 3.71 g/cm^3 for rhodonite. The structural formula of the garnet can be expressed as $Mn_3^{VIII}(MnSi)^{VI}Si_3^{IV}O_{12}$, where the superscripts denote coordination numbers. The fact that two cations varying in radius by as much as Mn^{++} and Si^{4+} share the octahedral sites almost certainly implies ordering at these sites. It is notable that this inferred transformation implies a change in coordination of one-quarter of the silicon atoms from fourfold to sixfold. A further study of the high-pressure behaviour of $MnSiO_3$ was reported by Akimoto and Syono (1972). The transition from clinopyroxene structure (density 3.82 g/cm^3) to garnet occurred close to 120 kbars and was relatively insensitive to temperature. The x-ray-diffraction

[1]Ringwood and Major (1967a).

pattern of the garnet displayed line splitting and was indexed with a tetragonal unit cell, analogous to $CdGeO_3$.

The occurrence of these transformations suggested the possibility that $MgSiO_3$ and $CaSiO_3$ might perhaps transform to garnet structures at high pressure. This would represent a third mode of transformation for $MgSiO_3$ in addition to those previously considered. High-pressure experiments upon compositions along the join $MgSiO_3$-$Mg_3Al_2Si_3O_{12}$ supported this possibility.[1,2] A glass of composition $MgSiO_3 \cdot 10\%\ Al_2O_3$ (wt%) was crystallized over a wide range of pressures at about 1000°C. This composition is equivalent to a mixture of $MgSiO_3$ (enstatite) 60% and $Mg_3Al_2Si_3O_{12}$ (pyrope) 40%. Results are shown in Figs. 12-7 and 12-8. At pressures up to 90 kbars the glass crystallized to a mixture of clinoenstatite 40 and pyrope 60, as expected. However, between 90 and 100 kbars, a remarkable increase in the proportion of garnet occurred, and by 150 kbars the glass had transformed completely to a garnet. The composition of this garnet would be, according to previous discussion and interpretation,

$$[Mg_3(MgSi)Si_3O_{12}]_{60}\ [Mg_3Al_2Si_3O_{12}]_{40}$$

Similar results were obtained on glasses with compositions $MgSiO_3 \cdot 13.5\%\ Al_2O_3$ and $MgSiO_3 \cdot 5\%\ Al_2O_3$. In the latter case, complete transformation to garnet was not obtained, the lowest Al_2O_3 content of a homogeneous garnet at the limit of

[1]Ringwood and Major (1966c).
[2]Ringwood (1967).

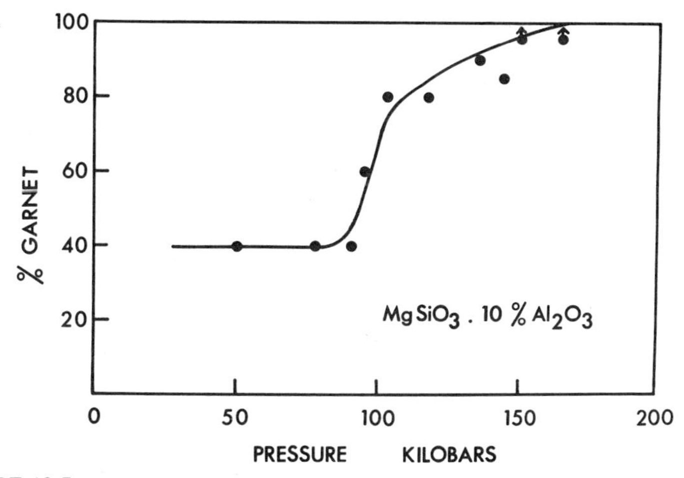

FIGURE 12-7
Proportions of garnet crystallizing from an $MgSiO_3 \cdot 10\%\ Al_2O_3$ glass as a function of pressure at approximately 1000°C. (*From Ringwood, 1967, with permission.*)

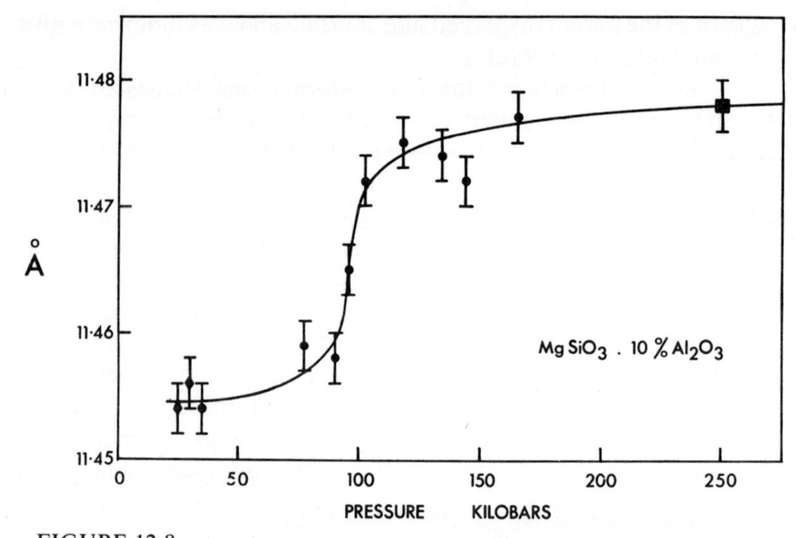

FIGURE 12-8
Lattice parameters of garnets crystallized from a glass of composition $MgSiO_3 \cdot 10\% \ Al_2O_3$ as a function of pressure at (approx.) 1000°C. Run at 250 kbars (square symbol) was on an $MgSiO_3 \cdot 5\% \ Al_2O_3$ glass. Runs at 25 and 30 kbars (diamond symbols) were on a glass of pyrope composition. (*From Ringwood, 1967, with permission.*)

experimental pressures being about 7%, corresponding to a composition $[Mg_3(MgSi)Si_3O_{12}]_{75}[Mg_3Al_2Si_3O_{12}]_{25}$.

These results show that, at high pressures, enstatite is capable of dissolving extensively in the garnet structure. The fact that solid solubility increases so rapidly at about 100 kbars suggests that this is not a normal solid solution but may be more closely related to compound formation. The composition of the garnet formed at about 100 kbars is very nearly $Mg_3(AlMg_{0.5}Si_{0.5})Si_3O_{12}$. It appears that ordering of Al, Mg, and Si in the 2:1:1 proportion in the B sites may, in fact, define a compound which displays solid solubility with pure $MgSiO_3$ at higher pressures (Figs. 12-7 and 12-8). The transformation involves an effective increase in density of about 10% in the $MgSiO_3$ component as it transforms from pyroxene to garnet. The experimental results also suggest that, at even higher pressures than were employed, garnets with still smaller amounts of Al_2O_3 would be stable, and ultimately, at very high pressures, a pure $MgSiO_3$ garnet might occur.[1]

This possibility has been strongly supported by the discoveries of garnets in chondritic meteorites[2,3,4] formed by shock waves. In the Coorara and Tenham

[1]Ringwood (1967).
[2]Mason, Nelen, and White (1968).
[3]Binns, Davis, and Reid (1969).
[4]Smith and Mason (1970).

Table 12-4 COMPOSITIONS OF NATURAL AND SYNTHETIC MAJORITE

	Natural occurrence*	Synthetic†
SiO_2	52.0	52.43
Al_2O_3	2.6	2.46
Cr_2O_3	0.7	0.62
Fe_2O_3	—	0.65
FeO	16.9	17.14
MgO	27.5	26.07
Na_2O	0.7	0.63
Total	100.4	100.00

*Smith and Mason (1970). Average of "best" analyses. All iron calculated as Fe^{++}.
†Ringwood and Major (1971).

chondrites, microcrystalline garnet occurs associated with the $(Mg, Fe)_2SiO_4$ spinel as a major component of thin veins believed to have been produced by shock and is interpreted as a high-pressure modification of the orthopyroxene present in these meteorites. Chemical analysis of the Coorara garnet showed that it was indeed of pyroxene composition and could be recast into a garnet formula, providing that a substantial proportion of the silicon was placed in octahedral coordination (Tables 12-4 and 12-5). The chemical analysis showed small but significant differences in minor element compositions between the garnet and the associated orthopyroxene. This suggested that the garnet may not have been transformed directly from solid orthopyroxene but may, rather, have crystallized from melted vein material during adiabatic cooling and solidification under pressure as the shock pressure relaxed.

The meteoritic garnet (majorite) was synthesized by Ringwood and Major (1971) from a glass of appropriate composition at pressures between 200 and 300 kbars using the anvil configuration shown in Fig. 10-4*b*. The compositions and structural formulae of natural and synthetic majorite are shown in Tables 12-4 and 12-5. Complete conversion of the glass to garnet was achieved.

A wide range of alumina-deficient silicate garnets has been synthesized at pressures exceeding 100 kbars. These include $(Fe_{0.5}Mn_{0.5})SiO_3$,[1] $(Ca_{0.7}Fe_{0.3})SiO_3$,[2] and $(Mg_{0.75}Fe_{0.25})SiO_3 \cdot 5\% \ Al_2O_3$.[1] Glasses with the compositions $CaMgSi_2O_6 \cdot 10\% \ Al_2O_3$, $CaSiO_3 \cdot 10\% \ Al_2O_3$, and $FeSiO_3 \cdot 10\% \ Al_2O_3$ also crystallized to form homogeneous garnet phases at high pressures,[2,3,4] and these are inferred to possess analogous structures to that of $MgSiO_3 \cdot 10\% \ Al_2O_3$. The aluminous diopside glass behaved very similarly to $MgSiO_3 \cdot 10\% \ Al_2O_3$. Solid solubility of diopside in garnet was low up to about 90 kbars, after which it increased very sharply (see Figs. 12-7 and 12-8).

[1]Ringwood and Major (1971).
[2]Ringwood (unpublished observations).
[3]Ringwood and Major (1966c).
[4]Ringwood (1967).

Table 12-5 ANALYSES OF NATURAL AND SYNTHETIC MAJORITE RECAST TO GARNET FORMULAE, $A_3B_2C_3O_{12}$ (CATIONS PER 12 OXYGEN ANIONS)

	Natural majorite*			Synthetic majorite†		
C group	Si	3.00		Si	3.00	
B group	Si	0.78	2.06	Si	0.788	2.000
	Al	0.23		Al	0.210	
	Fe^{++}	1.02		Cr	0.036	
	Cr	0.03		Fe^{3+}	0.035	
				Mg	0.931	
A group	Mg	2.98	3.08	Mg	1.877	3.001
	Na	0.10		Fe^{++}	1.036	
				Na	0.088	
$\frac{Mg}{Mg + Fe^{++}} = 0.75$				$\frac{Mg}{Mg + Fe^{++}} = 0.73$		
$a_0 = 11.524 \pm 0.002$ Å				$a_0 = 11.529 \pm 0.005$ Å		

*Smith and Mason (1970). Average of "best" analyses. All iron calculated as Fe^{++}.
†Ringwood and Major (1971).

From these studies on Mg-, Fe-, and Ca-bearing pyroxenes, it appears that the pyroxene-garnet transformation is of rather general occurrence, particularly in the presence of some Al_2O_3. This was confirmed by some experiments[1] upon more complex, eclogitic systems. The results of high-pressure experiments on the "alkali-poor olivine tholeiite" (Table 1-1) are given in Fig. 12-9. This composition crystallizes to an eclogite containing about 50% garnet and 50% pyroxene at pressures over 20 kbars[2] and possesses a density of 3.55 g/cm³. The behaviour of this complex eclogitic system at higher pressures is seen to be very similar to the simple systems. There is a drastic increase in the solubility of pyroxene in the garnet structure at about 100 kbars. At a pressure of 120 kbars, crystallization of the glass to a "garnetite" possessing a density of 3.72 g/cm has occurred. The glass starting material was formed by the fusion of pyrope garnet and omphacite extracted from a kimberlite xenolith, and accordingly, it may be held to represent the composition of a typical mantle eclogite xenolith. Analogous behaviour was displayed in the case of a quartz tholeiite.[1]

The pyroxene-garnet transformation is probably of considerable importance in the earth's mantle. The above experimental results imply that, in the presence of more than 3% Al_2O_3 (relative to pyroxene), the entire pyroxene component of the mantle would ultimately be transformed to the garnet structure. An alternative statement is that the pyroxene component will form extensive solid solutions

[1]Ringwood (1967).
[2]Ringwood and Green (1966).

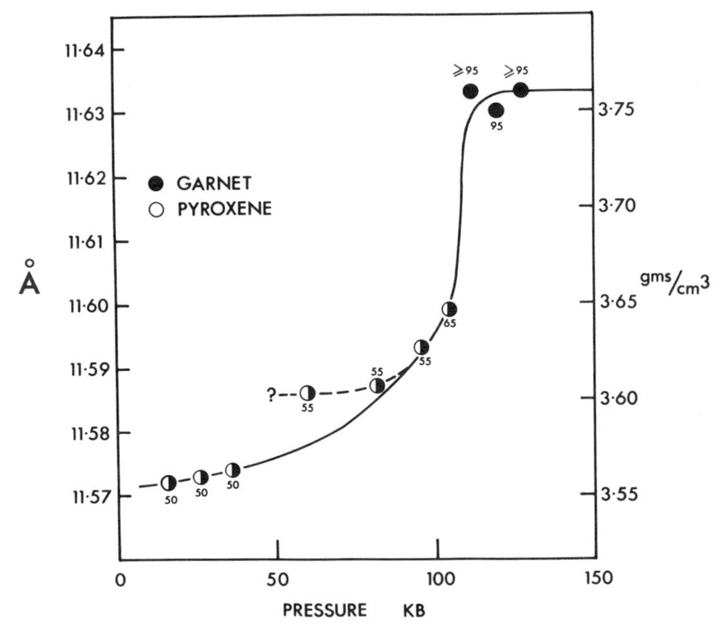

FIGURE 12-9
Lattice parameters of garnets, proportions of garnets, and mean densities for an eclogite of alkali-poor olivine-tholeiite composition (Table 1-1) in the pressure range 16 to 130 kbars, 1000°C. (*From Ringwood, 1967, with permission.*)

with existing normal garnet and that this reaction will set in suddenly at a pressure of about 100 kbars, 1000°C and will be extended to somewhat higher pressures by solid-solution effects. The effective increase in density of the pyroxene as it enters the garnet structure is about 10%. Since pyroxenes are abundant components of the upper mantle, and almost any geochemically reasonable model of the mantle[1] contains sufficient $R_2O_3(Al_2O_3 + Cr_2O_3 + Fe_2O_3)$ in relation to pyroxene for this transformation to proceed to completion, it is probably of considerable geophysical importance and will strongly influence the seismic velocity and density distribution between about 300- and 400-km depth, depending upon the temperature gradient which is not yet known. It is of interest that, at 1000°C, the transformation proceeds nearly to completion at a somewhat smaller pressure than is required for the transformation of mantle olivine to spinel and beta phase (Figs. 11-7 and 12-7).

The gradient dp/dt of the pyroxene-garnet transformation in the mantle is a quantity of considerable geophysical importance but is not yet known. If it is small (e.g., similar to the corresponding gradient exhibited by $MnSiO_3$), then there would be a substantial difference between the depths of the pyroxene-garnet tran-

[1]Chapters 5 and 14.

sition (perhaps 300 km) and the olivine-spinel-β-Mg_2SiO_4 transition ($\sim$ 400 km). This may give rise to a double seismic discontinuity. On the other hand, if the gradient is very large ($\geq$ 50 bars/°C), then the garnet-pyroxene transition could overlap the olivine transformation near 400 km, forming a single seismic discontinuity.

One further transformation of a pyroxene may be noted. Germanium diopside $CaMgGe_2O_6$ was found to disproportionate at 70 kbars, 700°C into a mixture of $CaGeO_3$ (garnet) and $MgGeO_3$ (ilmenite).[1]

12-4 THE GARNET-ILMENITE TRANSFORMATION

In Sec. 12-2 we concluded that an ilmenite polymorph of $MgSiO_3$ would be stable relative to $MgSiO_3$ pyroxene at a pressure between 200 and 300 kbars. However, in Sec. 12-3 we found that $MgSiO_3 \cdot xAl_2O_3$ compositions transformed initially to a garnet structure and that there was also evidence that pure $MgSiO_3$ would transform first to this structure (Sec. 12-3). The garnet structure has most of its silicon atoms in fourfold coordination and is less dense than the ilmenite structure in which all cations are octahedrally coordinated. Consequently, it appears likely that $MgSiO_3$ and $MgSiO_3 \cdot xAl_2O_3$ garnets may transform to the ilmenite structure at even higher pressures.[2,3,4] This would essentially be a solid solution between $MgSiO_3$ (ilmenite) and Al_2O_3 (which already has the corundum structure).

As attempts to transform silicate garnets at high pressure were unsuccessful, recourse was made to germanate analogue methods described earlier. An investigation[5] of the system $Mg_3Al_2Ge_3O_{12}$-$Mg_3Al_2Si_3O_{12}$ at 0 to 170 kbars and 1000°C revealed results which are shown in Fig. 12-10. Above 70 kbars, the system is binary, and it is clear that Ge-Si pyrope garnets indeed transform to ilmenite structures at high pressure. Some of the points in Fig. 12-10 (squares) illustrate a problem which is likely to arise increasingly in experiments at the highest pressures using quenching methods. Exsolution and retrogressive transformation of ilmenite solid solutions have apparently occurred when pressure was released.

At pressures above 70 kbars at 1000°C, the garnet composition $Mg_3Al_2Ge_3O_{12}$ crystallizes as an ilmenite-type solid solution. In view of the model relationships between germanate and silicate systems, this result supports the suggestion that silicate garnets may transform to the ilmenite structure deep within the mantle. The existence of a homogeneous series of ilmenite solid solutions containing up to at least 20 mol% $Mg_3Al_2Si_3O_{12}$ were synthesized during the experiment but compositions richer in pyrope were retrogressively transformed on

[1]Ringwood and Seabrook (1963).
[2]Ringwood (1962, 1966).
[3]Clark, Schairer, and de Neufville (1962).
[4]Boyd (1964).
[5]Ringwood and Major (1967b).

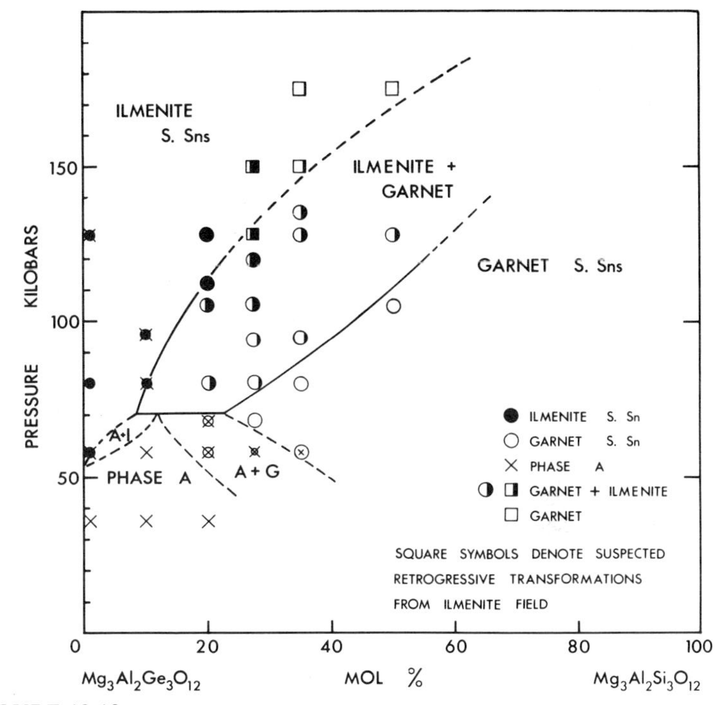

FIGURE 12-10
Phases present in the system $Mg_3Al_2Ge_3O_{12}$-$Mg_3Al_2Si_3O_{12}$ as a function of pressure at about 1000°C. (*From Ringwood and Major, 1967b, with permission.*)

release of pressure. The extensive degree of solid solution of pyrope in the germanium ilmenite structure at high pressure further supports the view that pure pyrope will ultimately transform to an ilmenite polymorph at sufficiently high pressure (Sec. 10-2). Extrapolation of phase boundaries in Fig. 12-10 suggests that pressures of 200 to 300 kbars would be required to transform pure Si pyrope to an ilmenite structure. Calculations based upon the solid-solution relationships indicate a transition pressure of only 180 kbars. This is probably too low because of nonideality of the solid solutions at high pressure. Nevertheless, the result is considered indicative. The ilmenite structure is about 8% denser than the corresponding garnet. The density of $Mg_3Al_2Si_3O_{12}$ ilmenite is expected to be 3.85 g/cm^3.

Shock-wave data[1,2] on the compression of bronzitite suggest the occurrence of a series of successive phase transformations, each extending over a broad pressure interval. For this reason, the method of analysis used by Davies and Gaffney (1973) is preferable to that used by Ahrens et al. (1969). At pressures between

[1]McQueen, Marsh, and Fritz (1967).
[2]Ahrens and Gaffney (1971).

350 and 750 kbars, the observed compressions are consistent with transformation into an ilmenite structure. At higher pressures, the densities progressively increase beyond those expected for ilmenite and, in the vicinity of 1 Mbar, are close to that expected for a perovskite structure. This is discussed further in Sec. 12-5.

Graham and Ahrens (1973)[1] report experiments on the compression of an almandine garnet ($Al_{0.74}Py_{0.14}Ca_{0.04}Sp_{0.03}$) to a pressure of 650 kbars. Evidence of transformation into another phase was observed at about 170 kbars. The zero-pressure density of this phase was estimated to be about 6% greater than garnet, suggesting the possibility of transformation into an ilmenite structure. Some very weak new x-ray reflections were observed in the shocked material and were attributed to a residual "distorted ilmenite" phase which had escaped retrogressive transformation. This phase could not be identified microscopically, however, and the x-ray evidence presented is considered to be rather equivocal.

Although transformation of almandite into a distorted ilmenite-type structure appears conceivable because of the stabilizing role of Al_2O_3, it is doubtful whether $FeSiO_3$ with iron in its normal high-spin state would transform to this structure. The difference in octahedral radii between Fe^{++} (0.77Å) and Si^{4+} (0.40Å) is greater than is observed in ilmenite-type structures. This would lead to distortion of the hexagonal close-packed array of oxygen anions in the ilmenite lattice which would necessarily be destabilizing. For this reason, it appears more likely that $FeSiO_3$ (high-spin iron) would transform ultimately to a perovskite-related structure than to an ilmenite structure. This topic is discussed further in Sec. 12-5.

It has been suggested[2,3] that, at very high pressures, the ferrous ion might adopt the low-spin configuration which would involve contraction to a radius of 0.61Å.[4] In the low-spin state, an ilmenite modification of $FeSiO_3$ appears entirely reasonable from crystal chemical considerations. The density of such a modification[5] is estimated to be 5.31 g/cm^3. On the other hand, the density of $FeSiO_3$ (high-spin Fe) in the perovskite structure is estimated to be higher at 5.45 g/cm^3 (Sec. 12-5). Even if $FeSiO_3$ (ilmenite-type, low-spin iron) became stable at a given pressure, it appears that higher pressures would favour transformation to the perovskite-type (high-spin iron) structure. The higher density of the perovskite structure is caused by the higher Fe^{++} coordination, which outweighs the effect of the radius contraction accompanying the low-spin state.

Transformation of Mg-Pyroxene in the Mantle: Summary

The experiments described previously indicate three possible modes of transfor-

[1]See also Ahrens and Graham (1972).
[2]Fyfe (1960).
[3]Strens (1969).
[4]Shannon and Prewitt (1969).
[5]Table 12-2.

mation of Mg-rich pyroxenes:

1 Disproportionation to β-Mg_2SiO_4 + stishovite
2 Transformation to garnet structure
3 Transformation to ilmenite structure

The results of Ito et al. (1972) showed that pure $MgSiO_3$ disproportionated above 200 kbars into β-Mg_2SiO_4 + stishovite. This assemblage would have a density of 3.68 g/cm^3, intermediate between those of $MgSiO_3$ garnet (3.52 g/cm^3) and $MgSiO_3$ ilmenite (3.81 g/cm^3). The above results were obtained by a quenching method. They do not preclude the possibility that a garnet modification of $MgSiO_3$ is stable under high pressure but is not retained during quenching because of retrogressive transformation. This behaviour has been observed in other systems.[1,2] Irrespective of whether a stability field for pure $MgSiO_3$ (garnet) intervenes between the pyroxene and β-Mg_2SiO_4 + stishovite fields, there is no doubt that in the mantle, in the presence of some Al_2O_3, pyroxene would transform initially to a garnet phase. This is shown by the synthesis of majorite containing only 2.6% Al_2O_3 and also of other Al-deficient garnets (Sec. 12-3).

At still higher pressures, it appears most likely, in the light of earlier discussion, that majorite would transform to an ilmenite or closely related structure. This is strongly supported by Kawai et al.'s (1974) successful synthesis of an ilmenite-type polymorph of $MgSiO_3$. The presence of Al_2O_3 (also Cr_2O_3 and Fe_2O_3) would probably have an important stabilizing effect upon this transformation. The unit cell dimensions of Al_2O_3 and $MgSiO_3$ (ilmenite) are similar and complete solid solubility would be expected (cf. $3MgGeO_3 \cdot Al_2O_3$ ilmenite). We have already noted that $MgSiO_3$ ilmenite would be about 4% less dense than the isochemical mixed oxides. Thus, the possibility of further transformations of $MgSiO_3$-Al_2O_3 ilmenites to denser phases or phase assemblages at higher pressures might be considered.

Liu (personal communication) has recently carried out some experiments upon a range of pyrope-almandine garnets using a diamond-anvil squeezer with laser-heating of the samples. Pressures employed were estimated to be in the range 150 to 350 kbars, whilst the sample was heated to 1500 to 2000°C. Samples were removed from the anvils after runs and examined by x-ray diffraction and optical methods at room pressure and temperature.

These experiments are currently being analyzed. At pressures between 150 and 300 kbars, the garnets displayed a complex series of transformations which, as yet, are incompletely understood. The possible role of retrogressive transformation has yet to be clarified. At still higher pressures, above about 300 kbars, Liu observed that pyrope garnet transformed into a new phase plus corundum. The new phase could be indexed as an orthorhombic perovskite form of $MgSiO_3$. The role of perovskite-type phases in the lower mantle is discussed in the next section.

[1]Sections 12-3 and 12-4.
[2]Ringwood and Major (1971).

12-5 TRANSFORMATION OF GARNETS AND ILMENITES TO PEROVSKITE-TYPE STRUCTURES

It was concluded in Sec. 12-3 that calcium-bearing pyroxenes in the upper mantle would transform to garnet structures at depths of 300 to 400 km. It is of interest therefore to study the behaviour of the calcium-rich component of garnets at high pressure. We turn first to a study of germanium analogues.

Ringwood and Major (1967a) showed that $CaGeO_3$ garnet, which is stable at moderate pressures, is further transformed to a perovskite structure above 100 kbars at 1000°C. Similarly, $CdGeO_3$ garnet was transformed into a distorted perovskite structure. These results suggest that $CaSiO_3$ in the mantle is likely to transform ultimately to a perovskite structure. Studies of the system $CaGeO_3$-$CaSiO_3$ at high pressure strongly support this. In a series of runs at 170 kbars, $CaGeO_3$ perovskite was found to take at least 35 mol% $CaSiO_3$ into solid solution[1] (Fig. 12-11). Almost certainly, the amount of $CaSiO_3$ dissolved substantially exceeded 35%. However, on releasing pressure, exsolution of the solid solution apparently occurred and proceeded nearly to completion in some cases.

A further study of high-pressure solid-solubility relations was carried out in the system $CaSiO_3$-$CaTiO_3$.[2] Perovskite solid solutions containing up to 83 mol% $CaSiO_3$ were synthesized at pressures above 100 kbars (Fig. 12-11). Formal

[1]Ringwood and Major (1967a).
[2]Ringwood and Major (1971).

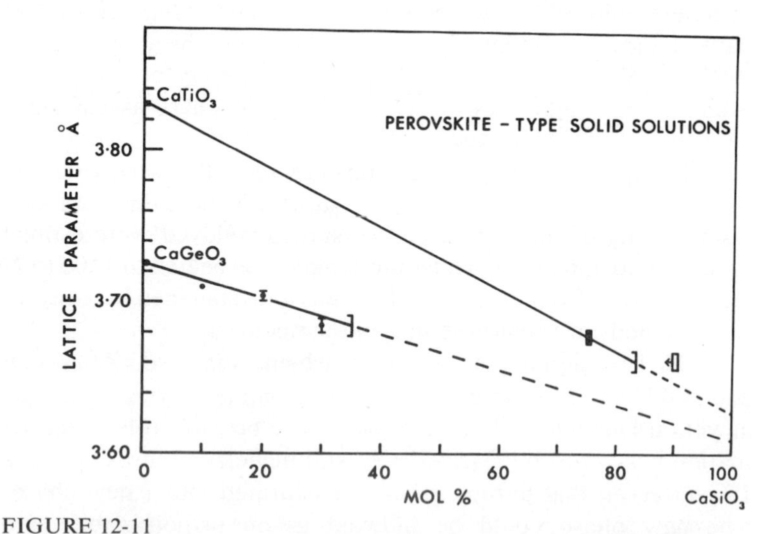

FIGURE 12-11
Lattice parameters of $CaSiO_3$-$CaTiO_3$ (rectangles) and $CaSiO_3$-$CaGeO_3$ (circles) perovskite solid solutions synthesized at high pressures. Solid lines represent solid solution ranges observed after release of pressure. Actual solid solution ranges under pressure are believed to have extended across to pure $CaSiO_3$ perovskite (see text). (*After Ringwood and Major, 1971, with permission.*)

calculations along the lines presented in Sec. 10-2 indicated that the pressure needed to form pure $CaSiO_3$ perovskite could hardly exceed 120 kbars, which is easily within the reach of static high-pressure apparatus. However, in runs on $Ca(Si_{0.9}Ti_{0.1})O_3$ and pure $CaSiO_3$ compositions at pressures exceeding 100 kbars, unexpected behaviour was displayed. Even when crystalline starting material was used, the products recovered usually consisted of *glass*, sometimes associated with a weakly diffracting phase possessing diffuse lines and with a density of about 3.45 g/cm³. The observations[1] clearly indicated that these were retrogressive transformation products of a still denser phase which had actually been synthesized in the apparatus but which could not be quenched in the high-pressure state when pressure was released. What was the nature of this high-pressure phase?

The presence of glass as a retrogressive transformation product implied that the phase contained octahedrally coordinated silicon.[1] Clearly, its density had to be greater than the 3.45 g/cm³ exhibited by one of the metastable retrogressive transformation products. Thermodynamic calculations showed that the additional pressure needed to synthesize $CaSiO_3$ perovskite over that needed for $Ca(Si_{0.83}Ti_{0.17})O_3$ perovskite was only 2.5 kbars. These considerations, combined with crystal chemical limitations, led to the conclusion that the $CaSiO_3$ phase, stable above about 100 kbars (1000°C), almost certainly possessed the perovskite or a closely related structure.[1,2]

The perovskite structure is based upon a cubic close packing of the large A cations (e.g., Ca^{++}) and oxygen anions. The smaller B cations, e.g., Ti^{4+}, Ge^{4+}, Si^{4+}, occur in interstices in octahedral coordination with respect to oxygen. The structure is shown in Fig. 12-12. The B atoms (e.g., Ti) are seen to be surrounded

[1]Discussed in detail by Ringwood and Major (1971).

[2]Since confirmed using diamond anvils with x-ray diffraction studies under pressure (Liu and Ringwood, in press).

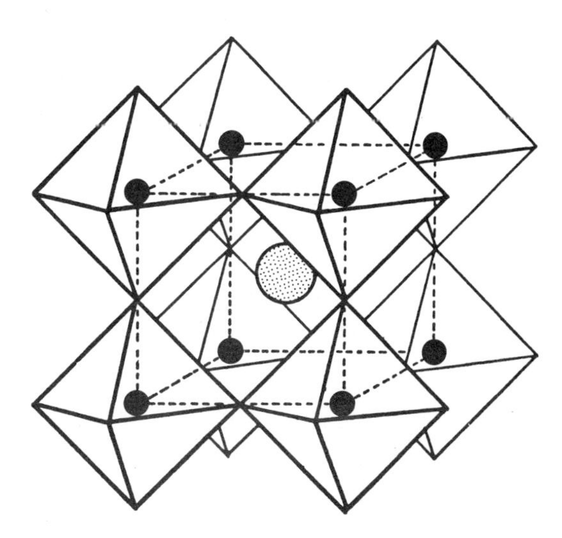

FIGURE 12-12
The idealized structure of perovskite $CaTiO_3$. A Ca atom is at the cube centre, and the Ti-O octahedra are shown diagrammatically. (*After Naray-Szabo, 1943.*)

by regular octahedra of oxygen atoms which are linked by sharing corners to form a three-dimensional framework. The large A atoms occur in holes between the octahedra and in the ideal structure are surrounded by 12 oxygen atoms. Geometrical packing considerations dictate that the ideal oxide perovskite structure can be formed only for a rather restricted range of sizes of A and B atoms. The A atoms have to be sufficiently large to form a close-packed array with oxygen atoms ($r = 1.40$Å), whilst the B atoms must fit within the octahedral holes of the A-O close-packed array. When these requirements are not met, the simple perovskite lattice may become distorted, giving rise to several closely related structures and superstructures based on the perovskite lattice but with lower symmetry. The coordination numbers of the A atoms in these distorted perovskites may be smaller than 12. One such class possesses the orthorhombic $GdFeO_3$ structure, in which each unit cell is built up of four distorted perovskite units oriented diagonally to the a_0 and b_0 axes.[1] Relatively small cations are capable of entering this structure, which is displayed by $YAlO_3$ at atmospheric pressure and by $ScAlO_3$, $InGaO_3$, and $InCrO_3$ at high pressures.[2,3,4] The (octahedral) ionic radii of these cations are Sc^{3+}—0.73Å, In^{3+}—0.79Å, Y^{3+}—0.89Å, Al^{3+}—0.53Å, Ga^{3+}—0.62Å, and Cr^{3+}—0.62Å. In this class of "distorted perovskites," the larger cation (e.g., Sc^{3+}, In^{3+}, Y^{3+}) is actually in eightfold coordination with respect to oxygen.[3,4]

The perovskite structure is formed by a large number of binary ABO_3 compounds, including many of those where AO has a rocksalt structure and BO_2 has a rutile structure. Where perovskites are formed between these end-members, their densities are characteristically between 5 and 10 % greater than the isochemical mixed oxides (Table 12-6). Thus the structure is relatively close packed. The density of $CaGeO_3$ perovskite is 5.17 g/cm^3 compared to 4.81 g/cm^3 for an isochemical mixture of $CaO + GeO_2$. For comparison, the densities of $CaGeO_3$ wollastonite and garnet are 3.98 and 4.44 g/cm^3, respectively. The density of $CaSiO_3$ perovskite may be obtained from the extrapolated lattice parameter (3.62 ± 0.02Å, Fig. 12-11) and is found to be 4.07 ± 0.04 g/cm^3. This is 7.2% denser than an isochemical mixture of CaO + stishovite.

The calcium-rich component of natural garnet will generally have the formula $CaSiO_3 \cdot xAl_2O_3$, where for grossularite, $x = 25$ wt%. In a perovskite dominantly composed of Ca^{++} and Si^{4+}, it would be difficult to incorporate much aluminium, which is obliged to enter as an $RAlO_3$ component, where R is a large trivalent cation—e.g., Y^{3+}. The low abundance of such large trivalent cations will accordingly severely restrict the amount of aluminium which can enter a $CaSiO_3$ perovskite structure in the mantle. Thus, the transformation of Ca garnet in the mantle into a perovskite will also involve disproportionation according to the equation

[1]Wyckoff (1964).
[2]Reid and Ringwood (1975).
[3]Geller (1969).
[4]Marezio (1969).

$$CaSiO_3 \cdot xAl_2O_3 \text{ (garnet)} \rightarrow CaSiO_3 \text{ (perovskite)} + xAl_2O_3$$

This transformation has been demonstrated in a germanium analogue system.[1] Analogous transformations[2] have been reported to be displayed by the garnets $Y_3Al_5O_{12}$ and $Y_3Fe_5O_{12}$, which disproportionated into mixtures of perovskite- and corundum-type phases at high pressures, e.g.,

$$Y_3Al_5O_{12} \text{ (garnet)} \rightarrow 3YAlO_3 \text{ (perovskite)} + Al_2O_3 \text{ (corundum)}$$

The possibilities that $MgSiO_3$ and $FeSiO_3$ may ultimately transform to perovskite-type structures[3,4] may now be considered. The most likely transformations would be to the $GdFeO_3$-type modification. This would require eightfold coordination of Mg and Fe and sixfold coordination of Si. The former coordinations are observed in garnets, whilst the latter is seen in stishovite. This structure would therefore appear reasonable on crystal chemical grounds. In the deep regions of the mantle, the Fe-O and Mg-O bonds will be shortened by about 10%, a large proportion of these contractions being borne by the oxygen anions, as discussed in Sec. 10-2, so that the effective radius ratios Mg/O and Fe/O are substantially increased. The tendency of Fe and Mg to enter eightfold coordin-

[1]Ringwood (1970).
[2]Marezio, Remeika, and Jayaraman (1966).
[3]Ringwood (1962, 1966, 1970).
[4]Reid and Ringwood (1969).

Table 12-6 MOLAR VOLUMES OF PEROVSKITES FORMED FROM ROCKSALT- AND RUTILE-TYPE END-MEMBERS

Compound	Molar volume V (cm^3)	Isochemical oxide volume (v) (cm^3)	$\Delta v/v$* (%)
$BaSnO_3$	42.03	46.92	−10.4
$SrSnO_3$	39.52	42.24	−6.4
$CaSnO_3$	36.28	38.31	−5.3
$CdSnO_3$	35.73	37.14	−3.8
$BaMoO_3$	39.73	45.22	−12.1
$SrMoO_3$	37.83	40.54	−6.7
$BaTiO_3$	38.89	44.17	−12.0
$SrTiO_3$	35.87	39.49	−9.2
$EuTiO_3$	35.65	39.29	−9.3
$CaTiO_3$	34.11	35.56	−4.1
$CdTiO_3$	31.76	34.39	−7.6
$CaMnO_3$	31.32	33.37	−6.1
$CaVO_3$	32.02	34.64	−7.6
$MnVO_3$	29.91	31.10	−3.8
$BaPbO_3$	46.99	50.39	−6.7
$CdGeO_3$	30.44	32.26	−5.6
$CaGeO_3$	31.08	33.43	−7.0
$CaSiO_3$	28.56	30.77	−7.2

*$\Delta v = V - v$.

ated sites will therefore become stronger with increasing pressures. Looking at the matter from a slightly different perspective, we note that the cation radius ratios Mg^{++}/Si^{4+} and Fe^{++}/Si^{++} are within the range of A^{++}/B^{4+} cation radius ratios displayed by known perovskite structures. If we regard the principal effect of pressure as the shrinking of large oxygen anions relative to the small cations,[1] it appears likely that $MgSiO_3$ and $FeSiO_3$ will attain, at high pressures, the range of relative ionic sizes required for perovskite structures.

A perovskite modification of $MnVO_3$ has been produced at a pressure of 50 kbars.[2] The Mn^{++} cation (0.82 Å) is the smallest divalent cation so far observed to form a perovskite structure and encourages the view that slightly smaller divalent cations (Fe^{++}—0.77 Å, Mg^{++}—0.72 Å) may adopt this structure at much higher pressures. Recently, Reid and Ringwood (1975) synthesized at high pressures a perovskite modification of $ScAlO_3$. This is an important model structure since the (octahedral) radius of Sc^{3+} (0.73 Å) lies between those of Mg^{++} and Fe^{++}, whilst the radius of octahedral Al^{3+} (0.53 Å) does not greatly exceed that of Si^{4+} (0.40 Å), as exemplified by the observed substitution of Al by Si in octahedral crystal sites. $ScAlO_3$ has the smallest perovskite volume so far observed. An important conclusion to be drawn from the details of its structural determination[3] is that the orthorhombic ($GdFeO_3$-type) perovskite lattice allows a movement of oxygen atoms such that relatively small A ions can be accommodated in the network of corner-joined BO_6 octahedra while still maintaining normal eight-coordinate bond lengths. We have already noted that Mg^{++} and Fe^{++} exhibit eightfold coordination in the garnet structure. Moreover, Si^{4+} is known to occur as the B ion in $CaSiO_3$ perovskite solid solutions.

Limited shock-wave results[4] on pure enstatite indicated that it had transformed to a structure which was substantially denser than the isochemical oxides (analogously to Mg_2SiO_4, Fig. 11-13). The estimated zero-pressure density[5] of the high-pressure form of $MgSiO_3$ was about 4.2 g/cm^3. From Table 12-6 we see that perovskites are from 4 to 12% denser than isochemical mixtures of rutile- and rocksalt-type end-members. A perovskite modification of $MgSiO_3$ would be expected to be 4 to 5% denser than the mixed oxides,[3] giving it a density close to 4.2 g/cm^3, in agreement with the shock-wave value. Shock compression results on bronzitite[4] also indicated that, at pressures in the vicinity of 1 Mbar, the density attained was compatible with a perovskite structure.[6] In the light of the shock-wave data and general crystal chemical considerations, it was concluded that (Mg, Fe)SiO_3 would probably transform ultimately to the perovskite structure in the deep mantle.[7] Because of the larger ionic radius of Fe^{++}, and its marked prefer-

[1] Section 10-2.
[2] Syono, Akimoto, and Endoh (1971).
[3] Reid and Ringwood (1975).
[4] McQueen and Marsh (1966).
[5] Ahrens, Anderson, and Ringwood (1969).
[6] Davies and Gaffney (1973).
[7] *Noted added in press:* A large measure of confirmation of this prediction has been provided by Liu (1974), who successfully synthesized an (orthorhombic) perovskite modification of $MgSiO_3$ at pressures exceeding 300 kbars. The calculated density of $MgSiO_3$ perovskite was 4.12 g/cm^3, some 4% denser than the isochemical mixed oxides.

ence for eightfold coordination as displayed in upper mantle garnets, it is likely that the pressure required to form $FeSiO_3$ perovskite will be smaller than for $MgSiO_3$ perovskite.

The compounds $CdTiO_3$ and $CdSnO_3$ are dimorphous, displaying both perovskite- and ilmenite-type polymorphs at atmospheric pressure, the perovskites being about 6% denser than the ilmenite modifications. The *P-T* relations for the ilmenite-perovskite transition in $CdTiO_3$ have been studied.[1] A notable property is the negative slope dP/dT of this transition. An ilmenite to perovskite-type transition is also displayed by the compound $MnVO_3$.[2,3] This transition involves a 6.6% increase in density and has a small positive gradient (<10 bars/°C).

An ilmenite-perovskite transformation in the mantle would involve the disproportionation of a complex solid solution between $MgSiO_3$ and $(Al, Cr, Fe)_2O_3$:

$$(Mg, Fe)SiO_3 \cdot x(Al, Cr, Fe)_2O_3 \text{ (ilmenite)} \rightarrow$$

$$(Mg, Fe)SiO_3 \text{ (perovskite)} + x(Al, Cr, Fe)_2O_3 \text{ (corundum)}$$

It is likely that the corundum phase would react with MgO to form $MgAl_2O_4$ in the dense structure (calcium ferrite?) inferred to occur under shock-wave conditions (Sec. 11-6).

12-6 RELATIVE VOLUMES OF ABO_3 STRUCTURES

The importance of being able to predict the volume which an $A_xB_yO_z$ compound would possess if it were transformed to a given crystal structure has been emphasized in previous sections. We have extensively discussed systematic relationships which exist between the volumes of binary A_2BO_4 and ABO_3 compounds and the combined volumes of their isochemical AO (rocksalt str.) and BO_2 (rutile str.) components. In this section, an alternative method[4] of obtaining these relative volumes is outlined. This is based upon an empirical plot of average octahedral bond lengths (R, Å) versus unit cell or molar volumes (V) for different series of isostructural compounds. In a given ABO_3 or A_2BO_4 compound, the average octahedral bond length is obtained simply by averaging the distances of the A-O and B-O bonds in the rutile- and rocksalt-type end-member components or in the ABO_3 and A_2BO_4 structures in which octahedral coordination prevails. When this procedure is carried through, it is found that members of each crystal structure class fall upon characteristic *V-R* curves. These curves clearly define the relative volumes or packing densities of different structure types and permit extrapolation to predict the volumes of new polymorphs of given compounds.

[1]Rooymans (1967).

[2]Syono, Akimoto, and Endoh (1971).

[3]King (1974) has recently studied the compression of $FeTiO_3$ (ilmenite) under shock-wave conditions. At pressures above 300 kbars, ilmenite transformed to a phase estimated to possess a zero-pressure density zone 10 to 15% higher than ilmenite. In view of the corresponding transitions found in $MnVO_3$, $CdTiO_3$, and $CdSnO_3$, it seems possible that this phase might possess a perovskite-related structure.

[4]Reid and Ringwood (1969).

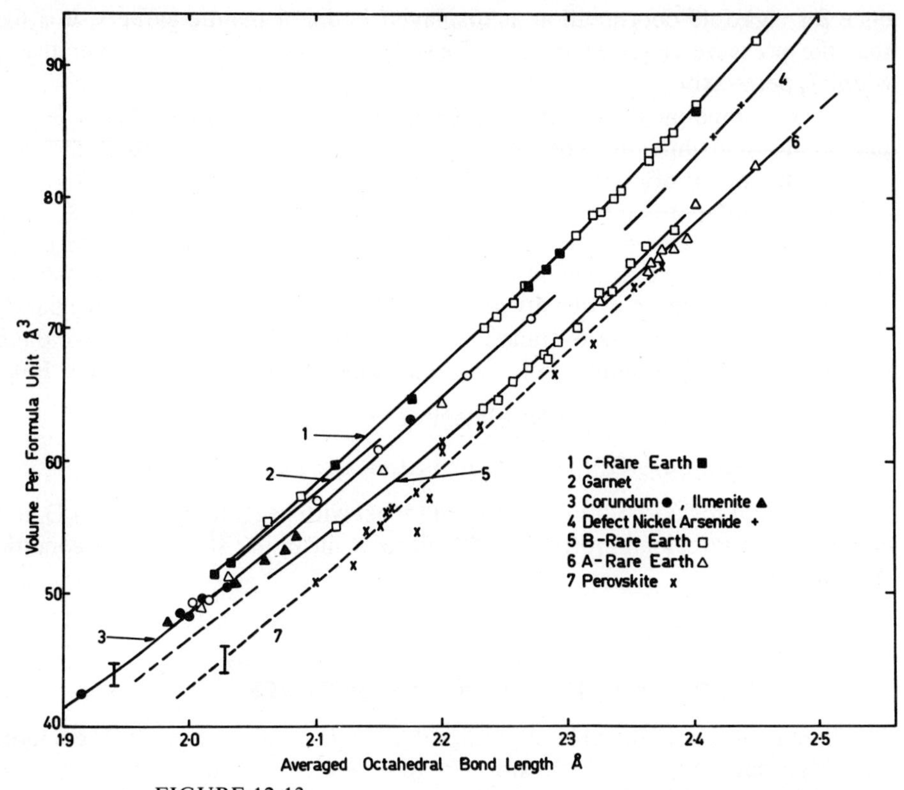

FIGURE 12-13
Averaged octahedral bond lengths versus volume per formula unit for M_2X_3 and ABX_3 structure types. Solid symbols refer to bond lengths determined from atomic coordinates, open symbols to bond lengths estimated from simple oxides. (*From Reid and Ringwood, 1969, with permission. Copyright American Geophysical Union.*)

The application of this method[1] to M_2X_3 and ABX_3 compounds is shown in Fig. 12-13. It is seen that the relative volumes of these structures decrease in the sequence defect spinel, C rare earth, garnet, rare earth manganite, corundum (ilmenite), metal-deficient nickel arsenide, B rare earth, A rare earth, and perovskite. These curves led to the conclusion[1] that $MgSiO_3$ (ilmenite) would have a density close to 3.76 g/cm³.

The application of comparative crystal chemistry to geophysical problems is not limited to the estimation of volumes of high-pressure silicate polymorphs in the deep mantle. Many physical properties which are of vital importance to an understanding of the earth's interior are dominantly or substantially controlled by

[1]Reid and Ringwood (1969).

crystal structure. Examples are elastic properties, thermodynamic properties (particularly entropy and specific heats), rheological properties, and electrical conductivity. Extrapolations based upon comparative crystal chemistry offer perhaps the most direct and promising means of estimating and understanding these properties in the deep mantle. This is a field of research which is relatively untouched and yet has tremendous potential.

12-7 SIGNIFICANCE OF PYROXENE-ILMENITE INTERGROWTHS AMONG KIMBERLITE XENOLITHS

Williams (1932) drew attention to the occurrence of xenoliths in South African kimberlites consisting of oriented ilmenite plates in a matrix of pyroxene (Fig. 12-14). The structure strongly resembles an exsolution texture. This interpretation directs attention to the nature of the homogeneous parental phase. In view of the deep origin of kimberlites, there is a possibility that the intergrowths represent a high-pressure phase which was unstable when transported to high levels in the crust and, hence, exsolved. Chemical analyses of two such intergrowths are given in Table 12-7. The resemblance is very close.

The fact that the cation-to-anion ratio in both pyroxene and ilmenite is 2:3, which is the same as garnets, immediately suggests the possibility that the high-pressure phase might have been a garnet. This possibility appeared all the more plausible in view of the wide flexibility of cation distributions permitted in the garnet structure (Sec. 12-3, Table 12-3).

To test this hypothesis, a portion of the xenolith (Table 12-7, column 1) was fused to a glass and recrystallized over a range of *P-T* conditions.[1] In runs below 90 kbars (900 to 1500°C), the glass crystallized to an ilmenite-pyroxene assemblage essentially identical to the starting material. However, runs at pressures higher than 100 kbars (1000°C) produced up to 90% garnet, with small quantities of other phases.

These experiments provide strong evidence that the predecessor to the intergrowth shown in Fig. 12-14 was indeed a garnet. This hypothesis was supported[2] by experiments upon a simplified model composition $Ca_3(MgTi)Si_3O_{12}$ or $3CaSiO_3 \cdot MgTiO_3$. A glass of this composition crystallized to a mixture of clinopyroxene and $CaTiO_3$ (perovskite) at pressures less than 60 kbars (1000 to 1200°C). Between 70 to 120 kbars, mixtures of pyroxene and garnet were observed. Finally, at 150 kbars (1000°C), complete conversion to garnet was obtained. The garnet displayed sharply resolved back reflections, yielding a lattice parameter of 12.085Å.

The composition of the xenolith of Table 12-7 has been recast into a garnet

[1]Ringwood and Lovering (1970).
[2]Ringwood and Major (1971).

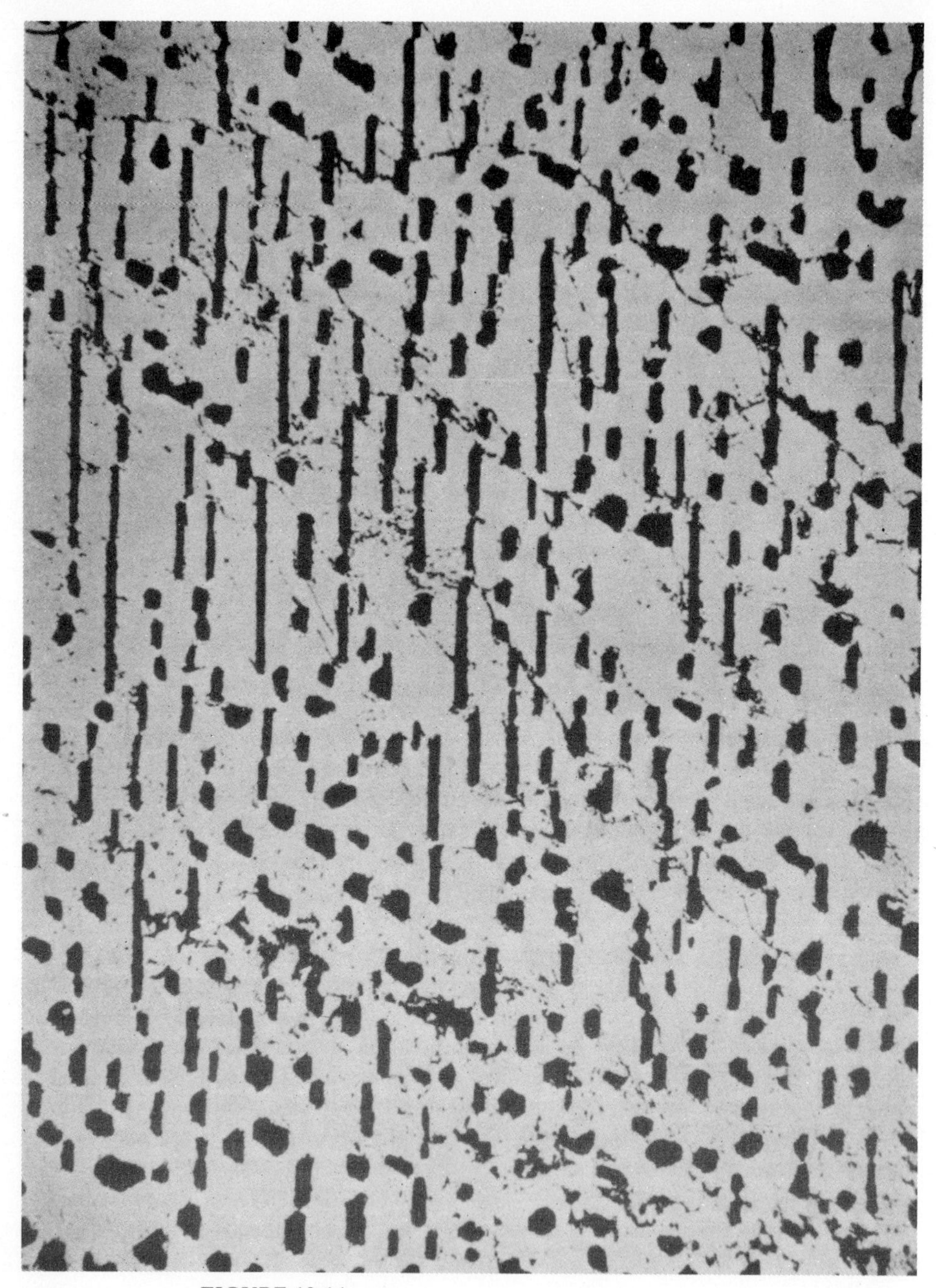

FIGURE 12-14
Xenolith from kimberlite at Monastery Mine, South Africa, consisting of an oriented intergrowth of ilmenite plates in a matrix of diopsidic pyroxene. (*After Williams, 1932.*)

Table 12-7 COMPOSITIONS OF PYROXENE-ILMENITE NODULES FROM THE DIAMOND PIPE AT THE MONASTERY MINE, SOUTH AFRICA

	(1)	(2)
SiO_2	36.92	33.56
TiO_2	15.40	15.50
Al_2O_3	1.89	4.01
Fe_2O_3	4.47	6.72
Cr_2O_3	0.10	0.25
V_2O_3	0.10	
FeO	11.91	9.07
MnO	0.14	0.39
MgO	16.11	15.62
NiO	0.10	
CaO	10.82	11.20
Na_2O	1.21	0.09
P_2O_5	0.013	0.05
$H_2O(+)$	0.65	1.91
$H_2O(-)$	0.18	0.20
CO_2	0.09	1.98

Explanation:
(1) Analyst E. Kiss, Australian National University.
(2) Williams (1932).

formula in Table 12-8. It is seen that ilmenite enters the garnet structure via a coupled replacement of two Al^{3+} ions in the octahedral site by $(FeMg)^{++} + Ti^{4+}$, analogously to the germanate garnet $Ca_3MgTiGe_3O_{12}$ which is stable at atmospheric pressure.

It has not been possible to synthesize the new garnet at pressures less than 100 kbars, which is equivalent to a depth of 100 kbars. Because of kinetic difficulties and apparatus limitations, it was not possible to reverse the reaction and thus to conclusively demonstrate that 100 kbars is the equilibrium pressure at 1000°C. Nevertheless, the synthesis experiments suggest[1] that the equilibrium pressure is not far from 100 kbars.

If we accept this evidence, it implies that kimberlites containing these xenoliths are derived from depths greater than 300 km. This is about double the depth of origin previously inferred for these rocks by virtue of the presence of diamonds in them. The xenolith suite occurring in kimberlites takes on new significance. Previously, the peridotite-eclogite series of xenoliths had been regarded as samples of the mantle extending to a depth of 150 km. It now appears that peridotite-eclogite may be representative of the mantle down to depths exceeding 300 km.

[1]Ringwood and Lovering (1970).

Table 12-8 ANALYSIS OF PYROXENE-ILMENITE XENOLITH (TABLE 12-7, COLUMN 1) RECAST TO A GARNET FORMULA, $A_3B_2C_3O_{12}$

Cations per 12 oxygen anions:						
C group	Si^{4+}	2.904	3.000			
	Al^{3+}	0.096				
B group	M^{3+}	0.356	2.000			
	Ti^{4+}	0.911				
	M^{++}	0.733				
A group	Na^{+}	0.184	3.051			
	Ca^{++}	0.912				
	M^{++}	1.955				
M^{++}	Fe^{++}	0.784		M^{3+}	Al^{3+}	0.079
	Mn^{++}	0.010			Fe^{3+}	0.265
	Mg^{++}	1.889			Cr^{3+}	0.006
	Ni^{++}	0.006			V^{3+}	0.006
Total (atoms)		2.689		Total (atoms)		0.356

Unit cell size of garnet synthesized at 130 kbars: 11.842 Å

Sodium Garnets

The analysis (Table 12-7) shows 1.2% Na_2O which is apparently accommodated in the garnet. Likewise, the natural occurrence of majorite contained about 0.7% Na_2O (Table 12-4). In the former case, the possibility of sodium entry into garnet via a $(CaNa_2)Ti_2Si_3O_{12}$ component was suggested.[1] In the latter case, insufficient titanium was available, and a sodium garnet component $(NaCa_2)(AlSi)Si_3O_{12}$ was hypothesized.[1] This suggestion was made independently by Sobolev and Lavrentev (1971), who found that garnets from diamond-bearing eclogites and garnets occurring as inclusions in natural diamonds commonly contain abnormal amounts of sodium—up to 0.26%.

A systematic examination of these possibilities was made by Ringwood and Major (1971). At pressures exceeding 100 kbars (1000°C), a garnet series extending from grossularite $Ca_3Al_2Si_3O_{12}$ through $(Ca_2Na)(AlTi)Si_3O_{12}$ to $(CaNa_2)Ti_2Si_3O_{12}$ was synthesized. An analogous series extending from grossularite through $(Ca_2Na)(AlSi)Si_3O_{12}$ to $(CaNa_2)Si_2Si_3O_{12}$ was also synthesized.[2] Composition-lattice parameter relationships of these garnet series are shown in Fig. 12-15. The preceding interpretations were thus fully substantiated.[2]

These results have a bearing on the occurrence of sodium in the earth's

[1]Ringwood and Lovering (1970).
[2]Ringwood and Major (1971).

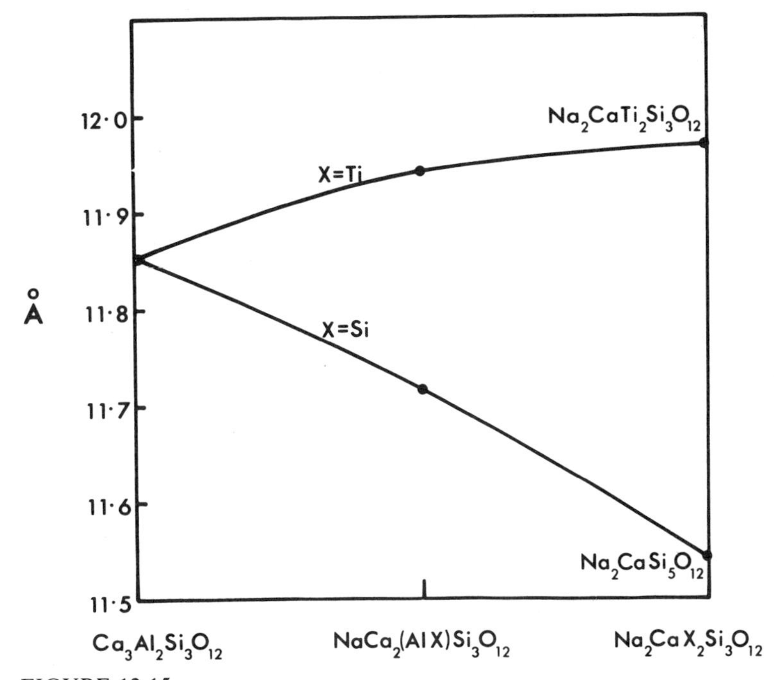

FIGURE 12-15
Relations between lattice parameters and compositions of some high-pressure sodium garnets. (*From Ringwood and Major, 1971, with permission.*)

mantle. In an earlier study[1] it was suggested that, in the depth interval between 350 and 650 km, sodium might be accommodated in jadeite. This mineral appeared to be stable up to very high pressures. The above syntheses of $(NaCa_2)(AlSi)Si_3O_{12}$ and $(Na_2Ca)Si_2Si_3O_{12}$ garnets strongly indicate, however, that sodium would enter the complex garnet solid solutions which are present in this depth interval, and jadeite is probably not present.

REFERENCES

ABRAHAMS, S. C., and S. GELLER (1958). Refinement of the structure of a grossularite garnet. *Acta Cryst.* **11,** 437–441.

AHRENS, T. J., D. L. ANDERSON, and A. E. RINGWOOD (1969). Equations of state and crystal structures of high pressure phases of shocked silicates and oxides. *Rev. Geophys.* **7,** 667–707.

——— and E. GAFFNEY (1971). Dynamic compression of enstatite. *J. Geophys. Res.* **76,** 5504–5513.

——— and E. GRAHAM (1972). A shock-induced phase change in iron-silicate garnet. *Earth planet. Sci. Letters* **14,** 87–90.

[1] Ringwood (1970).

AKIMOTO, S., and Y. SYONO (1972). High pressure transformations in $MnSiO_3$. *Am. Min.* **57,** 76–84.

BINNS, R. A., R. J. DAVIS, and S. B. J. REED (1969). Ringwoodite, natural $(MgFe)_2SiO_4$ spinel in the Tenham meteorite. *Nature* **221,** 943–944.

BIRCH, F. (1952). Elasticity and constitution of the Earth's interior. *J. Geophys. Res.* **57,** 227–286.

BOYD, F. (1964). Geological aspects of high-pressure research. *Science* **145,** 13–20.

——— (1969). Electron probe study of diopside inclusions in kimberlite. *Am. J. Sci.* **267A,** 50–59.

BRAGG, L., and G. F. CLARINGBULL (1965). "*Crystal Structures of Minerals.*" G. Bell and Sons, London. 409 pp.

CLARK, S. P., F. SCHAIRER, and J. DE NEUFVILLE (1962). Phase relations in the system $CaMgSi_2O_6$-$CaAl_2SiO_6$-SiO_2 at low and high pressure. *Carnegie Inst. Washington Yearbook* **61,** 61.

DAVIES, G. F., and E. GAFFNEY (1973). Identification of phases of rocks and minerals from Hugoniot data. *Geophys. J.* **33,** 165–183.

FYFE, W. S. (1960). The possibility of d-electron coupling in olivine at high pressures. *Geochim. Cosmochim. Acta* **19,** 141–143.

GELLER, S. (1967). Crystal chemistry of the garnets. *Z. Krist.* **125,** 1–47.

——— (1969). Discussion. *Trans. Am. Cryst. Assoc.* **5,** 36–37.

GIBBS, G. V., and J. V. SMITH (1965). Refinement of the crystal structure of synthetic pyrope. *Am. Min.* **50,** 2023–2039.

GRAHAM, E. K., and T. J. AHRENS (1973). Shock-wave compression of iron-silicate garnet. *J. Geophys. Res.* **78,** 375–392.

ITO, E., T. MATSUMOTO, K. SUITO, and N. KAWAI (1972). High pressure breakdown of enstatite, *Proc. Japan Acad.* **48,** 412–415.

KAWAI, N., M. TACHIMORI, and E. ITO (1974). A high pressure hexagonal form of $MgSiO_3$. *Proc. Japan Acad.* **50,** 378–380.

KING, D. A. (1974). A Hugoniot for ilmenite and its implications on possible pyroxene high pressure phase transitions. *EOS* **55,** 557.

LIEBERTZ, J., and C. J. M. ROOYMANS (1965). Die ilmenit-perowskit Phasenumwandlung von $CdTiO_3$ unter hohem Druck. *Z. Phys. Chem. Neue Folge* **44,** 242–249.

LIU, L. (1974). Silicate perovskite from phase transformations of pyrope-garnet at high pressure and temperatures. *Geophys. Res. Letters* **1,** 277–280.

MAREZIO, M. (1969). Oxides at high pressure. *Trans. Am. Cryst. Assoc.* **5,** 29–37.

———, J. P. REMEIKA, and A. JAYARAMAN (1966). High pressure decomposition of synthetic garnets. *J. Chem. Phys.* **45,** 1821–1824.

MASON, B., J. NELEN, and J. S. WHITE (1968). Olivine-garnet transformation in a meteorite. *Science* **160,** 66–67.

MCQUEEN, R. G., and S. P. MARSH (1966). In: S. P. Clark, Jr. (ed.), "*Handbook of Physical Constants,*" *Geol. Soc. Am. Memoir* 97.

———, ———, and J. N. FRITZ (1967). Hugoniot equation of state of twelve rocks. *J. Geophys. Res.* **72,** 4999–5036.

NARAY-SZABO, ST. V. (1943). Der Structurtyp des Perovskits ($CaTiO_3$). *Naturwiss* **31,** 202–203.

PREWITT, C. T., and A. W. SLEIGHT (1969). Garnet-like structures of high pressure cadmium germanate and calcium germanate. *Science* **163,** 386–387.

REID, A. F., and A. E. RINGWOOD (1969). High pressure scandium oxide and its place in the molar volume relationships of dense structures of M_2X_3 and ABX_3 type. *J. Geophys. Res.* **74,** 3238–3252.

——— and ——— (1975). High pressure modification of $ScAlO_3$ and some geophysical implications. *J. Geophys. Res.* (In press.)

RINGWOOD, A. E. (1962). Mineralogical constitution of the deep mantle. *J. Geophys. Res.* **67,** 4005–4010.

——— (1966). Mineralogy of the mantle. In: P. M. Hurley (ed.), "*Advances in Earth Science*," pp. 357–399. M.I.T. Press, Cambridge.

——— (1967). The pyroxene-garnet transformation in the earth's mantle. *Earth Planet. Sci. Letters* **2,** 255–263.

——— (1970). Phase transformations and the constitution of the mantle. *Phys. Earth Planet. Interiors* **3,** 109–155.

——— and D. H. GREEN (1966). An experimental investigation of the gabbro-eclogite transformation and some geophysical implications. *Tectonophysics* **3,** 383–427.

——— and J. F. LOVERING (1970). Significance of pyroxene-ilmenite intergrowths among kimberlite xenoliths. *Earth Planet. Sci. Letters* **7,** 371–375.

——— and A. MAJOR (1966a). High pressure transformation of $FeSiO_3$ pyroxene to spinel plus stishovite. *Earth Planet. Sci. Letters* **1,** 135–136.

——— and ——— (1966b). High pressure transformation in $CoSiO_3$ pyroxene and some geochemical implications. *Earth Planet. Sci. Letters* **1,** 209–211.

——— and ——— (1966c). High pressure transformations in pyroxenes. *Earth Planet. Sci. Letters* **1,** 351–357.

——— and ——— (1967a). Some high pressure transformations of geophysical interest. *Earth Planet. Sci. Letters* **2,** 106–110.

——— and ——— (1967b). The garnet-ilmenite transformation in Ge-Si pyrope solid solutions. *Earth Planet. Sci. Letters* **2,** 331–334.

——— and ——— (1967c). High pressure transformation in zinc germanates and silicates. *Nature* **215,** 1367–1368.

——— and ——— (1968). High pressure transformations in pyroxenes II. *Earth Planet. Sci. Letters* **5,** 76–78.

——— and ——— (1971). Synthesis of majorite and other high pressure garnets and perovskites. *Earth Planet Sci. Letters* **12,** 411–418.

——— and M. SEABROOK (1962). High pressure transition of $MgGeO_3$ from pyroxene to corundum structure. *J. Geophys. Res.* **67,** 1690–1691.

——— and ——— (1963). High pressure phase transformations in germanate pyroxenes and related compounds. *J. Geophys. Res.* **68,** 4601–4609.

ROOYMANS, C. J. M. (1967). Structural investigations on some oxides and other chalcogenides at normal and very high pressures. Doctoral thesis, University of Amsterdam.

SHANNON, R. D., and C. T. PREWITT (1969). Effective ionic radii in oxides and fluorides. *Acta Cryst.* **B25,** 925–946.

SMITH, J. V., and B. MASON (1970). Pyroxene-garnet transformation in Coorara meteorite. *Science* **168,** 832–833.

SOBOLEV, V., and J. LAVRENTEV (1971). Isomorphic sodium admixture in garnets formed at high pressures. *Contr. Mineral. Petrol.* **31,** 1–9.

STRENS, R. G. (1969). The nature and geophysical importance of spin pairing in minerals of

iron II. In: S.K. Runcorn (ed.), "*The Application of Modern Physics to the Earth and Planetary Interiors,*" pp. 213–220. Interscience, a division of Wiley, New York.
SYONO, Y., S. AKIMOTO, and Y. ENDOH (1971). High pressure synthesis of ilmenite and perovskite type $MnVO_3$ and their magnetic properties. *J. Phys. Chem. Solids* **32,** 243–249.
———, ———, and Y. MATSUI (1971). High pressure transformations in zinc silicates. *J. Solid State Chem.* **3,** 369–380.
WANG, C. Y. (1967). Phase transitions in rocks under shock compression. *Earth Planet. Sci. Letters* **3,** 107–113.
WELLS, A. F. (1962). "*Structural Inorganic Chemistry,*" 2d ed. Clarendon, Oxford.
WILLIAMS, A. F. (1932). "*The Genesis of the Diamond*" (2 vols.). Ernest Benn, London. 636 pp.
WYCKOFF, R. W. G. (1964). "*Crystal Structures,*" vol. 2, 2d ed. Interscience, a division of Wiley, New York.
——— (1965). "*Crystal Structures,*" vol. 3, 2d ed. Interscience, a division of Wiley, New York.

13
MISCELLANEOUS TRANSFORMATIONS

13-1 TRANSFORMATIONS IN ALKALI ALUMINOSILICATES AND ALUMINOGERMANATES

The geochemistry of the alkali metals in the crust has been the subject of a vast amount of research. In contrast, much less is known about the behaviour and distribution of these elements in the mantle. In the crust, the alkali metals occur dominantly as felspar and other aluminosilicates. Since these open structures are unstable under high pressures, it is clear that the geochemistry of alkali metals in the mantle will be strongly influenced by the nature of the phases into which alkali aluminosilicates transform at depth. A well-known example is the transformation of albite into jadeite plus quartz.[1] Another more general example is of the transformation of felspars to the hollandite structure.[2]

[1]Birch and Le Comte (1960).
[2]Ringwood, Reid, and Wadsley (1967a,b).

The Felspar-Hollandite Transformation

Kume, Matsumoto, and Koizumi (1966) showed that, at 35 kbars and 1100°C, the germanium analogue of orthoclase, $KAlGe_3O_8$, transformed to a hollandite structure. Ringwood, Reid, and Wadsley (1967a) later succeeded in transforming natural orthoclase to the hollandite structure. The transformation pressure is about 100 kbars at 1000°C. The new phase has a density of 3.84 g/cm^3 and is thus about 50% denser than orthoclase. This large density change is caused by the circumstance that all the silicon and aluminium atoms are in octahedral coordination.

This transformation was also observed in the felspars $NaAlGe_3O_8$ and $RbAlGe_3O_8$.[1] Germanium albite first transformed at 15 kbars, 1100°C according to the reaction

$$NaAlGe_3O_8 \text{ (albite)} \rightarrow NaAlGe_2O_6 \text{ (jadeite)} + GeO_2 \text{ (rutile)}$$

At pressures above 25 kbars (1100°C), the jadeite and rutile phases recombined to form the hollandite. Analogous behaviour might be expected for common albite, which is known to transform to jadeite + quartz at relatively low pressures. In runs up to 150 kbars, albite crystallizes to jadeite + stishovite. However, at still higher pressures, these phases might be expected to recombine to form $NaAlSi_3O_8$ hollandite. Evidence for this is provided by the shock-wave studies on an albitite,[2] which yielded evidence of major phase transformation. The zero-pressure density[3] of the shocked phase was estimated as 3.80 g/cm^3. This is in good agreement with the density expected for a hollandite phase, and it appears probable that albite was indeed transformed to hollandite in the shock.

The hollandite structure[4,5,6] (Fig. 13-1) is tetragonal and closely related to the α-MnO_2 structure. The small cations are octahedrally coordinated, and the octahedra are linked together by edge-sharing to form double strings running parallel to the c axis. The double strings of octahedra, in turn, share corners to form a three-dimensional framework which contains tunnels parallel to the c axis, in which the large cations are situated. The latter are surrounded by eight oxygen atoms at the corners of a slightly distorted cube, with another four oxygen atoms substantially further away. The network of MO_6 octahedra is also closely related to the arrangement in the rutile structure, in which, however, the absence of the large cations permits a closer packing.

The general formula[4,5,6] of the hollandite group is $A_x(B_yC_{8-y})O_{16}$, where A represents the large cations occupying the tunnels, and B and C represent the small M cations forming the centres of the edge- and corner-linked MO_6 octahedra. The following ions are observed to occur in the A, B, and C positions:

[1]Ringwood, Reid, and Wadsley (1967a,b).
[2]McQueen, Marsh, and Fritz (1967).
[3]Ahrens, Anderson, and Ringwood (1969).
[4]Byström and Byström (1950).
[5]Bayer and Hoffman (1966).
[6]Wadsley (1964).

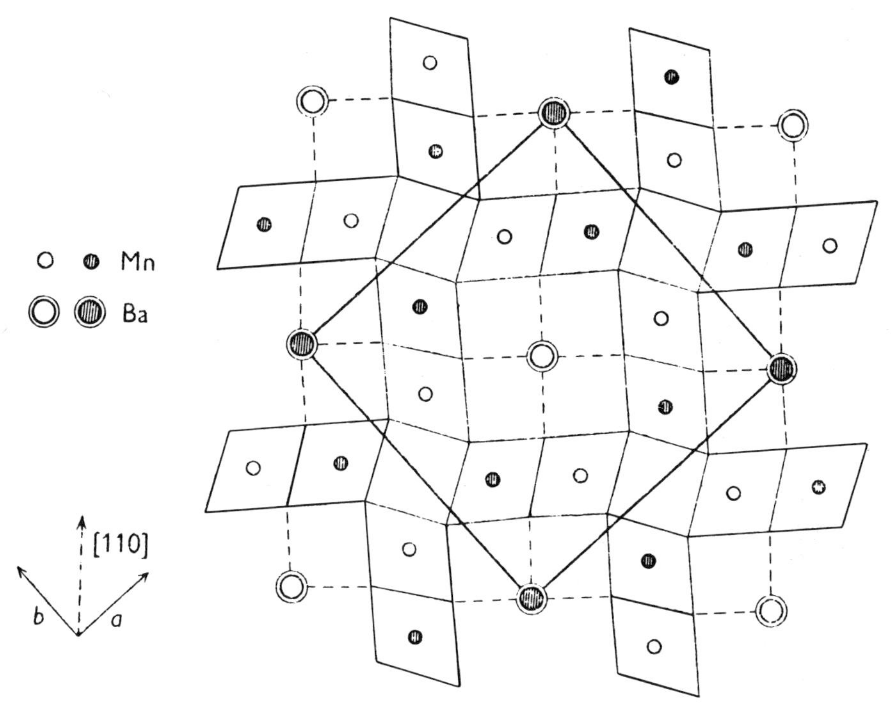

FIGURE 13-1
The network of oxygen octahedra in an ideal $BaMn_8O_{16}$ hollandite projected onto the basal plane. Diagram shows double strings of edge-linked MnO_6 octahedra defining tunnels occupied by Ba atoms. (*From Byström and Byström, 1950, with permission.*)

A	Na, K, Rb, Cs, Ba, Sr
B^{++}	Mg, Co, Ni, Cu, Zn
B^{3+}	Al, Ti, Cr, Fe, Ga, In
C^{4+}	Mn, Ti, Ge, Si, Sn

There is a considerable degree of flexibility in the ionic substitutions and charge balances permitted, and in addition, a substantial degree of non-stoichiometry often exists, corresponding to deficiencies of A and B cations compared to an ideal formula represented, for example, by $K_2Al_2Ti_6O_{16}$. (This particular compound represents the case of a titanium analogue serving as a high-pressure model for the corresponding germanate and silicate.) A list of felspar-hollandite transformations is given in Table 13-1.

The flexibility in ionic substitutions and the observation that Ba^{++} is capable of entering the hollandite structure suggests the possibility that Ba, Sr, and Ca felspars may be capable of forming hollandites at high pressure. Under static conditions, anorthite is known to break down to a mixture of grossularite, kyanite,

Table 13-1 HIGH-PRESSURE HOLLANDITE PHASES

$\frac{\Delta\rho}{\rho} \sim 50\%$ (References—see text)
A. Obtained under static high pressures:
$K_2Al_2Si_6O_{16}$
$K_2Al_2Ge_6O_{16}$
$Rb_2Al_2Ge_6O_{16}$
$Na_2Al_2Ge_6O_{16}$
$BaAl_2Si_6O_{16}$
$SrAl_2Si_6O_{16}$
B. Inferred to occur under shock conditions:
$Na_2Al_2Si_6O_{16}$
$K_2Al_2Si_6O_{16}$
$NaCaAl_3Si_5O_{16}$

and quartz.[1] This type of breakdown continues to 150 kbars, except that stishovite occurs instead of quartz.[2] However, under shock conditions[3] an anorthosite of composition $Ab_{51}An_{49}$ transformed to a phase with an estimated zero-pressure density[4] of 3.72 g/cm^3. It is possible that transformation to a hollandite structure has occurred. Such a transformation would be favoured kinetically during the few microseconds duration of the shock, rather than breakdown into a mixture of $NaAlSi_3O_8$ hollandite, grossularite, kyanite, and stishovite, involving complex nucleation processes and diffusion.

Support for this interpretation comes from studies of barium and strontium felspars.[5] When subjected to pressures of 100 to 150 kbars, at 1000°C, $BaAl_2Si_2O_8$ and $SrAl_2Si_2O_8$ were observed to disproportionate into hollandite phases plus unknown phases. Under similar conditions, glasses with compositions $Ba_{0.5}AlSi_3O_8$ and $Sr_{0.5}AlSi_3O_8$ (compare $KAlSi_3O_8$) crystallized to single-phase hollandite. However, an attempt to synthesize $Ca_{0.5}AlSi_3O_8$ hollandite was unsuccessful. This behaviour may be interpreted in terms of solid-solution formation in the series

$$Ba_2Al_4Si_4O_{16} \rightarrow BaAl_2Si_6O_{16} \rightarrow Ba_0Si_8O_{16}$$

Entry of barium (or strontium) ion is accompanied by the replacement of two silicon ions by aluminium in order to preserve electroneutrality. It is found that the homogeneous hollandite composition extends as far as about $Ba_{1.5}Al_3Si_5O_{16}$ for barium and $SrAl_2Si_6O_{16}$ for strontium, whilst for calcium, the formation of a hollandite does not occur. The contrasting behaviour apparently arises from the repulsion energies of the large cations in the hollandite structure. These cations

[1] Boyd and England (1961).
[2] Ringwood (unpublished observations).
[3] McQueen and Marsh (1966).
[4] Ahrens, Anderson, and Ringwood (1969).
[5] Reid and Ringwood (1969b).

occur as neighbours occupying channels in the structure, and the number of cations which can be accommodated is evidently governed both by the space available and by the repulsion potential between these ions. For this reason, the hollandite field does not extend as far as $Ba_2Al_4Si_4O_{16}$. The conclusion that anorthosite $Ab_{51}An_{49}$ transformed to a hollandite structure under shock conditions is consistent with this interpretation. This composition could be written $CaNaAl_3Si_5O_{16}$ and is analogous to the barium-rich hollandite $Ba_{1.5}Al_3Si_5O_{16}$ which was synthesized. Although 1.5 Ba (radius 1.36 Å) atoms are replaced by one atom each of Na^+ and Ca^{++} (radii 1.0 Å), the repulsion potential in the latter case may be comparable.

Several other disproportionation reactions involving hollandite phases have also been observed.[1] Leucite disproportionated at high pressure into hollandite plus potassium aluminate at 100 kbars, 1000°C. Germanium leucite and germanium kalsilite displayed analogous transformations. The fact that, at high pressure, a silica-rich phase such as $KAlSi_3O_8$ hollandite is stable in association with the extremely basic $KAlO_2$ shows that K hollandite would be stable in the conditions of silica saturation which occur in the mantle:

$$3KAlSi_2O_6 \rightarrow 2KAlSi_3O_8 + KAlO_2$$
$$\underset{\text{leucite}}{3KAlGe_2O_6} \rightarrow \underset{\text{hollandite}}{2KAlGe_3O_8} + KAlO_2$$
$$\underset{\text{kalsilite}}{3KAlGeO_4} \rightarrow \underset{\text{hollandite}}{KAlGe_3O_8} + 2KAlO_2$$

It is possible that $KAlSi_3O_8$ hollandite will be discovered as a naturally occurring mineral in impactites associated with meteorite craters. The conditions under which stishovite was formed should also have resulted in the transformation of potassium felspar grains present in the parent rock. However, since $KAlSi_3O_8$ hollandite is probably soluble in hydrofluoric acid, the solution method used to recover stishovite may not be successful in the case of $KAlSi_3O_8$.

Transformation of Nepheline and Jadeite

$NaAlGeO_4$, which has a nepheline-related structure, transformed at high pressure into a new dense phase possessing the calcium ferrite structure.[2,3] The sodium atoms were in eightfold coordination, whilst the aluminium atoms were in sixfold coordination. This structure is displayed by a large number of $NaA^{3+}B^{4+}O_4$ compounds, and it appears likely that common nepheline $NaAlSiO_4$ will ultimately transform to a calcium ferrite structure which would have a density of about 3.9 g/cm³ (Fig. 13-2). Efforts to synthesize this phase directly were unsuccessful since nepheline was found to disproportionate initially into jadeite plus

[1]Ringwood, Reid, and Wadsley (1967a,b).
[2]Ringwood and Major (1967a).
[3]Reid, Wadsley, and Ringwood (1967).

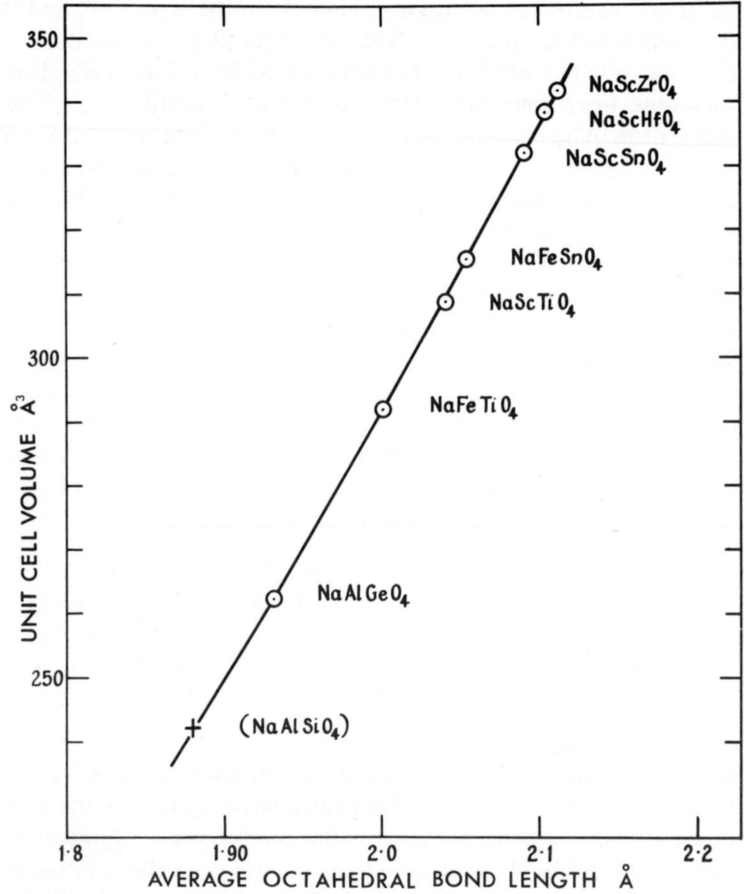

FIGURE 13-2
Plot of unit-cell volumes versus average octahedral bond lengths for $NaA^{3+}B^{4+}O_4$ calcium ferrite-type isomorphs. The density of the hypothetical calcium ferrite isomorph of $NaAlSiO_4$ is determined by extrapolation to be 3.9 g/cm³. (*From Reid, Wadsley, and Ringwood, 1967, with permission.*)

a new high-pressure phase of $NaAlO_2$ possessing the sodium ferrite structure:[1]

$$2NaAlSiO_4 \text{ (nepheline)} \rightarrow NaAlSi_2O_6 \text{ (jadeite)} + NaAlO_2$$

Jadeite is remarkably stable at high pressures. Nevertheless, it appears probable that, at pressures above 200 kbars, the jadeite and sodium aluminate will recombine to form $NaAlSiO_4$ (calcium ferrite structure).

Experiments were carried out[2,3] on the germanium analogue of jadeite. This

[1]Reid and Ringwood (1968).
[2]Ringwood and Major (1967a).
[3]Reid, Wadsley, and Ringwood (1967).

was found to disproportionate at high pressures as follows:

$$NaAlGe_2O_6 \text{ (jadeite)} \rightarrow NaAlGeO_4 \text{ (calcium ferrite)} + GeO_2 \text{ (rutile)}$$

It appears plausible that natural jadeite may ultimately disproportionate analogously into $NaAlSiO_4$ (calcium ferrite) + SiO_2 (stishovite).

The fact that sodium is able to form high-pressure phases of the type $NaAlSi_2O_6$, $NaAlGe_2O_6$, $NaAlGeO_4$, and probably $NaAlSiO_4$, whereas potassium appears unable to form these phases, leads to a significant difference in the geochemical behaviour of these elements in the deep mantle. Potassium probably occurs as the hollandite phase $KAlSi_3O_8$, which, as was noted, is stable in extremely Si-undersaturated environments. However, the sodium hollandite $NaAlSi_3O_8$ would probably be unstable under the Si-saturation conditions of the mantle because of the preferential formation of $NaAlSi_2O_6$, $Na_2CaSi_5O_{12}$ garnet, and at the highest pressures, $NaAlSiO_4$ (calcium ferrite). Thus sodium and potassium probably reside in different phases and, accordingly, may become fractionated from each other.

13-2 HIGH-PRESSURE HYDRATED MAGNESIUM SILICATES

Ringwood and Major (1967b) carried out some reconnaissance work on the system MgO-SiO_2-H_2O in compositions where the MgO/SiO_2 ratio varied from 1 to 5 at pressures between 100 and 170 kbars and at 1000°C. Starting materials consisted of hydrous gels containing 2 to 10% H_2O and mixtures of periclase with silicic acid (5% H_2O). In the high-pressure regime, familiar phases such as talc and serpentine were not encountered and were replaced by a new series of hydrated phases. The principal x-ray *d* spacings of these spacings were established, but it was not possible to characterize these phases completely. Accordingly, they were named phases A, B, and C. Phase A had a mean refractive index of about 1.65 and a low to medium birefringence. It was believed to be hydrated, with a MgO/SiO_2 ratio of 1.5 to 2. This phase was not formed from runs at pressures below 100 kbars, its place being taken by forsterite.

Phase B was almost ubiquitous in runs above 110 kbars but always occurred in association with other phases including phases A, C, MgO, and stishovite. This rendered optical identification and characterization difficult. It appeared to correspond to one of two optically identifiable phases possessing mean refractive indices of about 1.71 and 1.77, respectively. From the preferred synthesis field, it appeared that phase B was hydrated and that the MgO/SiO_2 ratio was between 2 and 3. The refractive indices indicate that this phase possesses a high density, possibly between 3.5 and 3.8 g/cm^3.

Phase C occurred sporadically in runs on water-rich gels, with MgO/SiO_2 ratios between 3 and 5, and was believed to be a layer-lattice mineral, perhaps related to the chondrodite-humite series.

The importance of these phases lies in the possibility that they may serve as hosts for OH^- ions in the mantle.[1] Phases A and B are both stable at relatively high temperatures, and phase B in particular possesses a high density.

Sclar, Carrison, and Stewart (1967) have carried out investigations of the system MgO-SiO_2-H_2O, with MgO/SiO_2 in the 2:1 ratio. Their results thus far have been reported in abstract form. Three phases at this composition were recognized and were believed to represent different polymorphs of hydroxylated pyroxenes, isostructural with orthopyroxene and clinopyroxene. The deduced formula was $MgSi_{0.5}(H_4)_{0.5}O_3$, in which only half the tetrahedral sites are occupied by silicon, charge compensation occurring by substitution of $(H_4O_4)^{4-}$ for SiO_4^{4-}. The principal x-ray *d* spacings for this phase published by Sclar et al. show that it is identical with the phase A previously discussed. These workers also drew attention to the geochemical significance of the postulated hydroxylated pyroxenes for the storage and release of large amounts of water in the upper mantle, according to the reaction

$$Mg_2SiO_4 + 2H_2O \rightleftharpoons 2MgSi_{0.5}(H_4)_{0.5}O_3$$

Sclar et al. did not present any crystallographic evidence to back up their claim that phase A possessed a hydroxylated pyroxene structure, and their interpretation was criticized by Martin and Donnay (1972).

A further study of the MgO-SiO_2-H_2O system was reported by Yamamoto and Akimoto (1974). They succeeded in establishing its formula and working out its stability relationships with clinoenstatite and forsterite (Fig. 13-3). The composition of phase A was found to be $H_6Mg_7Si_2O_{14}$ or $2Mg_2SiO_4 \cdot 3Mg(OH)_2$, and its density was 2.95 g/cm^3. These workers also discovered a fourth hydrated mag-

[1] Ringwood and Major (1967b).

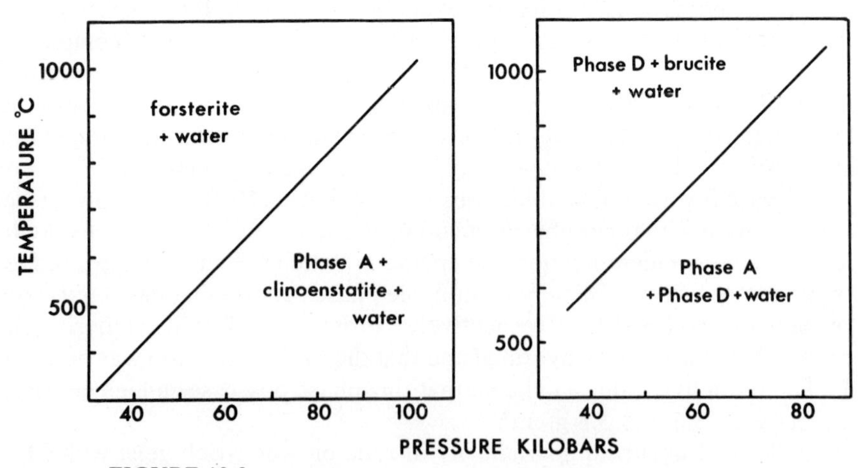

FIGURE 13-3
Stability relationships of phases A and D under conditions of $P_{H_2O} = P_{total}$. (*From Yamamoto and Akimoto, 1974.*)

nesium silicate "D phase," with an Mg/Si ratio around 5:2. The lattice constants and space group of phase D were close to those of chondrodite, but its diffraction patterns and refractive indices were clearly different.[1] Unlike phase A, phase D coexisted with brucite but was not observed to coexist with clinopyroxene (Fig. 13-3).

The observation that phase A coexists with clinoenstatite at high pressures is important in connection with the dehydration of serpentinite in the sinking lithosphere of subduction zones. It appears probable that phase A plays an important role in the dehydration sequence and serves as a vehicle for transporting water deep into the mantle. This topic was discussed further in Sec. 8-3. For convenience, in that discussion, phase A was given the nonspecific acronym DHMS (dense hydrated magnesium silicate).

13-3 TRANSFORMATIONS IN Al, Sc, In, Zr, AND Hf SILICATES

Attempts to transform sillimanite and kyanite to denser forms at pressures up to about 200 kbars have so far been unsuccessful.[2] This may well be a result of kinetic difficulties. The germanium analogue of kyanite was observed to disproportionate into GeO_2 (rutile) + Al_2O_3 (corundum) at relatively modest pressures.[3]

An aluminium silicate phase, $Al_2Si_2O_7$, has recently been found to be stable above about 100 kbars.[4] At lower pressures, this composition crystallizes as a mixture of kyanite and coesite. The mean refractive index of this phase is 1.757, and its molar refractivity indicates that most or all silicon atoms may be octahedrally coordinated.[4] Its density is approximately 4 g/cm^3, slightly smaller than that of the isochemical mixed oxides. X-ray structural studies are in progress. The major low-angle reflections suggest that the structure may be based upon a subcell with oxygen atoms arranged as in the corundum lattice, i.e., in hexagonal close packing. The aluminium and silicon atoms would then be likely to occupy octahedral interstices in this lattice. The possibility that this phase may contain essential $(OH)^-$ ions in its structure cannot yet be eliminated.

A series of phases of the type $M_2O_3 \cdot nTiO_2$ ($1 \leq n < 4$; $M = Cr^{3+}$, Fe^{3+}) has been synthesized and their crystallography investigated.[5,6] These are also based upon hexagonal close-packed oxygen lattices with the M^{3+} and Ti^{4+} ions ordered in octahedral interstices. The structures can also be related to that of rutile via crystallographic shear operations.[6] Although the structure of the high-pressure $Al_2O_3 \cdot 2SiO_2$ phase differs from that of $Cr_2O_3 \cdot 2TiO_2$, it is possible that there may be a general relationship to the extent that both are based upon a

[1] Yamamoto and Akimoto (1975) have since identified phase D as hydroxyl-chondrodite.
[2] Ringwood (unpublished observations).
[3] Ringwood and Reid (1969).
[4] Ringwood (1975).
[5] Grey and Reid (1972).
[6] Grey, Reid, and Allpress (1973).

corundum-type oxygen sublattice. The structures probably differ primarily in the way in which Cr^{3+}, Al^{3+} and Ti^{4+}, Si^{4+} are ordered in the octahedral sites. It is possible that additional related $Al_2O_3 \cdot nSiO_2$-type phases remain to be discovered and that they play a role in the transition zone. Their densities are expected to be high; for example, the density of Cr_2TiO_5 is 1% greater[1] than that of the isochemical mixture of Cr_2O_3 (corundum) and TiO_2 (rutile).

At atmospheric pressure the silicates $Sc_2Si_2O_7$ and $In_2Si_2O_7$ crystallize in the thortveitite structure. This is built upon Si_2O_7 groups which share one oxygen ion and Sc^{3+} cations in octahedral coordination. At high pressures these compounds transform to the pyrochlore structure.[2] The high-pressure polymorphs are of interest since the silicon ions are octahedrally coordinated as in stishovite, whilst scandium is in eightfold coördination. As a result, the structure is remarkably dense.

In_2O_3 possesses a corundum structure, whilst Sc_2O_3 possesses a C rare earth structure at low pressures, transforming into B rare earth type at high pressure.[3] Relationships between these forms and the hypothetical corundum polymorph of Sc_2O_3 have been studied, and the density of Sc_2O_3 (corundum) determined.[3] It is intermediate between the C and B rare earth forms. The densities of $In_2Si_2O_7$ and $Sc_2Si_2O_7$ may thus be compared with isochemical oxide mixtures of In_2O_3, Sc_2O_3 (corundum) + SiO_2 (stishovite). The pyrochlore-type polymorphs are respectively 4 and 5% denser than the mixed oxides.[4]

The pyrochlore structure is of potential importance in the deep mantle because of the possibility that it might be adopted by trivalent elements—e.g., in $Al_2Si_2O_7$ and $Cr_2Si_2O_7$. Such polymorphs would be denser than the isochemical mixtures of corundum plus rutile-type end-members.

Although zircon, $ZrSiO_4$, is only a minor to trace constituent of crustal and upper mantle rocks, its geochemical importance is enhanced since it serves as a host for an appreciable part of the uranium and thorium in these regions. The compounds $USiO_4$ and $ThSiO_4$ are both isostructural with zircon. Accordingly, the stability of zircon at high pressure is a matter of interest.

The compounds $ZrGeO_4$ and $HfGeO_4$ possess scheelite structures at atmospheric pressures,[5] and following the discussion of germanate-silicate model relationships (Sec. 10.2), this suggested that zircon might transform to the scheelite structure at high pressures. This was indeed found to be the case—when $ZrSiO_4$ and $HfSiO_4$ (also zircon structure) were subjected to a pressure of 120 kbars at 1000°C, complete transformations to the scheelite structure were observed,[6] thus

[1] Reid (personal communication).

[2] Reid, Ringwood, and Li (1974).

[3] Reid and Ringwood (1969a).

[4] $Sc_2Si_2O_7$ is also 1.4% denser than a mixture of Sc_2O_3 (B rare earth) and SiO_2 (stishovite).

[5] Wyckoff (1965).

[6] Reid and Ringwood (1969c).

providing further examples of the germanate-silicate model relationship. The scheelite polymorph of zircon may yet be found to occur as a natural mineral in diamond pipes of deep-seated origin or in impact breccias. Efforts to transform $USiO_4$ and $ThSiO_4$ zircons to the scheelite structure under similar conditions were unsuccessful.

The zircon structure consists of discrete SiO_4 tetrahedra interspersed by ZrO_8 groups. The latter consist of two interpenetrating tetrahedra of different sizes with four oxygens at 2.15 Å and another four oxygens further away at 2.29 Å.[1] The metal-oxygen coordinations in the $ZrSiO_4$ scheelite polymorph are essentially similar to those in zircon. However, the SiO_4 and ZrO_8 groups are more compactly arranged, and the Zr-O distances are more nearly equal than in the zircon polymorph.

At still higher pressures, it appears possible that $ZrSiO_4$ may transform to the $KAlF_4$ structure. This is formed by a number of fluorides, including $NaAlF_4$ in which the cation-anion sizes and radius-ratios resemble those in $ZrSiO_4$. This structure is very closely packed, with the Al ion in octahedral coordination and the Na (and K) ions in eight-coordinated square prismatic sites.

13-4 TRANSFORMATIONS IN SOME SIMPLE OXIDES

Quartz-Coesite-Stishovite Transformations

The first high-pressure polymorph of quartz "coesite" was discovered by Coes (1957) in the course of his pioneering investigations in high-pressure mineralogy. The equilibrium between quartz and coesite has been studied by many workers.[e.g.,2,3,4,5,6,7,8] Differences in the positions of equilibrium curves reported by various workers are mainly due to the different types of apparatus used and to accompanying problems of pressure calibration. The quartz-coesite equilibrium has proved to be useful for interlaboratory comparison of high pressure-temperature apparatus. There is now a pleasing convergence of results from several laboratories[3,4,8,5] where careful attention has been paid to the problems of estimating the role of friction (hysteresis) and nonuniform distribution of pressure in the pressure cell due to strength of the pressure-transmitting medium.

[1]Wyckoff (1965).
[2]MacDonald (1956).
[3]Khitarov (1964).
[4]Kitahara and Kennedy (1964).
[5]Boettcher and Wyllie (1968).
[6]Boyd and England (1960).
[7]Boyd (1964).
[8]Green, Ringwood, and Major (1966).

The density of coesite is 2.91 g/cm^3, and its structure shows that the silicon atoms remain in fourfold coordination.[1] Coesite is a rare terrestrial mineral, sometimes occurring in impactites associated with meteoritic craters, where it has been formed under transient high pressures generated by shock waves derived from the impact of the meteorite.[2,3] Coesite has also been found as a very rare inclusion in diamonds from kimberlite pipes.[4]

Stishov and Popova (1961) made the important discovery that coesite could be transformed to a new phase possessing the rutile structure at pressures above about 100 kbars[5] and at 1200 to 1600°C. The density of the new phase, "stishovite," was 4.28 g/cm^3. The large increase in density was caused by the change in silicon coordination from fourfold to sixfold. Successful syntheses of stishovite were reported shortly afterwards by other laboratories,[6,7,8] and the first natural occurrence of stishovite, associated with coesite in an impact breccia from Meteor Crater, Arizona, was discovered at about the same time.[9]

The standard entropy of stishovite was estimated by extrapolation of an empirical density-entropy relationship displayed by the rutile-type compounds PbO_2-SnO_2-GeO_2.[10] Using this estimate, together with available equilibrium data on the coesite-stishovite and quartz-coesite transformations, Stishov[10] calculated an equilibrium *P-T* curve for the coesite-stishovite transformation. Subsequently, the equilibrium has been determined directly.[11,12] When all results are reduced to the same pressure scale, they are found to be in reasonable agreement. In particular, the entropy change estimated for the coesite-stishovite transition, largely upon the basis of crystal chemical considerations,[10] agrees well with the entropy change derived directly from the experimental equilibrium curves and from solution calorimetry.[13] This points to yet another possible application of systematic crystal chemical relationships to obtain thermodynamic properties of high-pressure silicate phases. Phase relationships between quartz, coesite, and stishovite are shown in Fig. 13-4.

High-Pressure Transformations of Some Rutile-Type Phases

The rutile structure has played a prominent role in the crystal chemical discussions of previous chapters. In this structure (Fig. 13-5) each titanium atom is sur-

[1] Zoltai and Buerger (1959).
[2] Chao, Shoemaker, and Madsen (1960).
[3] Pecora (1960).
[4] H. O. A. Meyer (1968), personal communication.
[5] The pressure of 160 kbars originally cited has been corrected downward in accordance with revision in the high-pressure scale.
[6] Ringwood and Seabrook (1962).
[7] Wentorf (1962).
[8] Sclar, Young, Carrison, and Schwartz (1962).
[9] Chao, Fahey, Littler, and Milton (1962).
[10] Stishov (1963).
[11] Ostrovsky (1965, 1967).
[12] Akimoto and Syono (1969).
[13] Holm, Kleppa, and Westrum (1967).

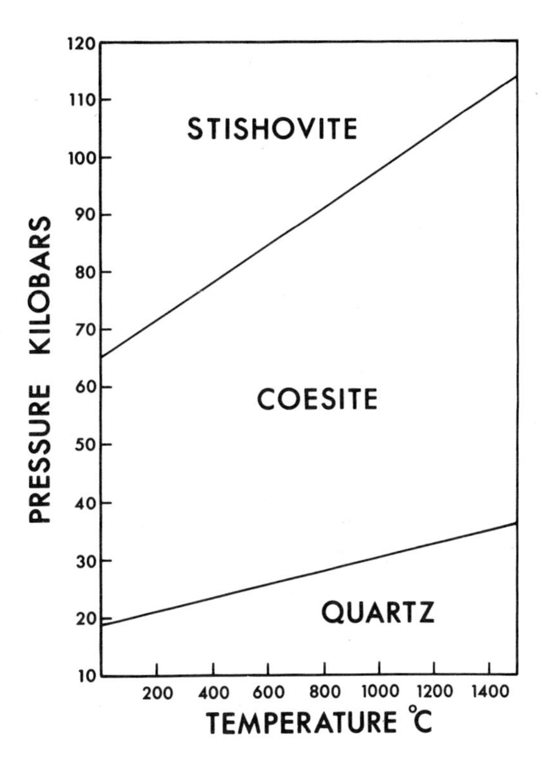

FIGURE 13-4
Quartz-coesite and coesite-stishovite equilibrium curves. The former is based upon the results of Khitarov (1964), Green, Ringwood, and Major (1966), and Boettcher and Wyllie (1968). The latter is based upon the results of Akimoto and Syono (1969) referred to the NBS pressure scale. (See Sec. 10-4.)

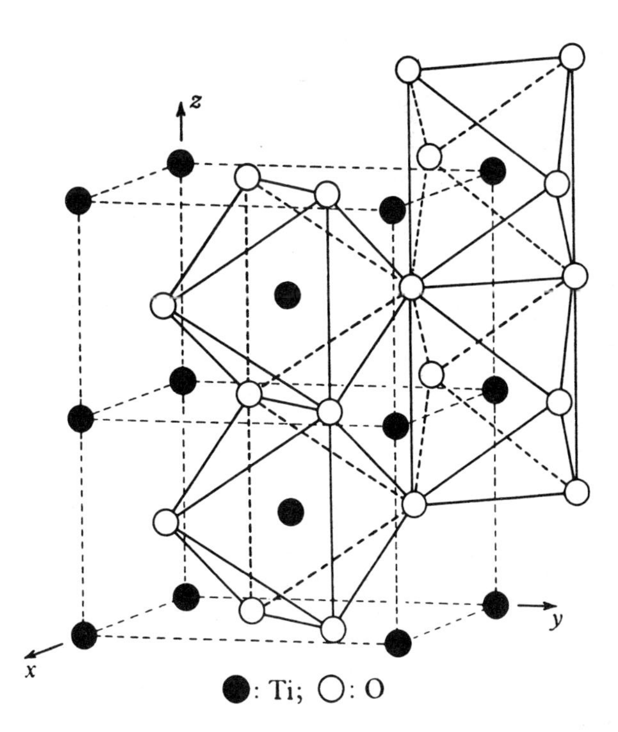

FIGURE 13-5
The tetragonal structure of rutile, showing the coordinating octahedra of oxygen anions around the titanium cations and the way in which these octahedra are linked in bands by sharing horizontal edges. (*From Evans, 1966, with permission.*)

rounded by six oxygen atoms at the corners of a slightly distorted octahedron, whilst each oxygen atom is surrounded by three titanium atoms lying in a plane at the corners of an approximately equilateral triangle. If several unit cells of this structure are considered, it is seen that each of the coordinating octahedra of oxygen atoms shares its two horizontal edges with adjacent octahedra. The octahedra are thus linked in bands, which run vertically through the structure and pass through the centre and corners of the unit cell.[1]

The edge-shared SiO_6 octahedra forming chains of SiO_4 composition are a common feature of both stishovite and strontium plumbate-type Mg_2SiO_4.[2] In stishovite, each SiO_4 chain is connected via common corners to four neighbouring chains, whilst in strontium plumbate-type Mg_2SiO_4, the SiO_4 chains are isolated from each other by intervening edge-shared chains of Mg coordination polyhedra. In both structures, the chains are parallel to the *c* axis.

Several rutile-type phases, including TiO_2 and PbO_2, are known to transform to a slightly denser (~2%) orthorhombic modification with a complex structure under high pressure.[3,4,5] At pressures above 70 kbars, PbO_2 was observed to transform to yet another polymorph possessing the fluorite structure and about 8% denser than the rutile-type modification.[6] Under shock compression, TiO_2 was observed to transform to a phase estimated to possess a density between 20 and 30% greater than rutile.[5,7] It seems very likely that this phase also possessed the fluorite structure.[5,7] The large difference in density changes associated with the rutile-fluorite transformations in TiO_2 and PbO_2 may be connected with the smaller ionic radius of Ti^{4+} (0.61 Å) compared to Pb^{4+} (0.78 Å). The transformation involves an increase in metal-oxygen coordination from 6 to 8. The strong evidence that this change of coordination is possible for a small cation such as Ti^{4+} makes it even more plausible that larger cations such as Fe^{++} and Mg^{++} will also adopt eightfold coordination in the deep mantle as was proposed in Secs. 11-8 and 12-5.

Other rutile-type structures have also been shocked at pressures exceeding 1 Mbar.[8,9] MnO_2 and SnO_2 did not give any evidence of phase transformation. The estimated zero-pressure densities of these phases existing above 0.5 Mbar were similar to the densities of the starting material.[7] However, the possibility of transformations involving small increases in density, for example, to the orthorhombic modification, could not be eliminated. Silica transforms under shock to a phase which possesses a zero-pressure density similar to that of

[1]Evans (1966).
[2]Baur (1972).
[3]White, Dachille, and Roy (1961).
[4]Azzaria and Dachille (1961).
[5]McQueen, Jamieson, and Marsh (1967).
[6]Syono and Akimoto (1968).
[7]Ahrens, Anderson, and Ringwood (1969).
[8]McQueen and Marsh (1966).
[9]Al'tschuler, Trunin, and Simakov (1965).

stishovite.[1,2,3] This observation, together with the successful recovery of small amounts of stishovite in shock experiments,[4] strongly indicates that the phase formed under shock was indeed stishovite.

Fujisawa (1968) has suggested that silica might be transformed from the rutile to the fluorite structure under the pressures which exist in the deep mantle and that this transformation may be of considerable geophysical significance. The suggestion appears implausible on several grounds. Firstly, the shock data on SiO_2 which extend to pressures much greater than those at the core-mantle boundary imply that a phase with the density of stishovite is formed over the entire pressure range. The pressures required to induce a small atom like Si^{4+} (radius 0.26 Å) to transform from sixfold to eightfold coordination with respect to oxygen are likely to be very much higher than those required for the transformation in TiO_2 and PbO_2 (Ti^{4+}—0.61 Å, Pb^{4+}—0.78 Å). Furthermore, it should be noted that neither SnO_2 or MnO_2 (rutile str.) transformed to the fluorite structure at pressures up to about 1 Mbar. Both of these, because of their larger cation radii (Sn^{4+}—0.69 Å, Mn^{4+}—0.54 Å) are likely to require smaller pressures to transform to the fluorite structure than stishovite (Sec. 10-2). Finally, the density increase associated with the rutile-fluorite transformation, together with the corresponding increases[5] in seismic velocity and elastic ratio, are so large as to be inconsistent with the observed velocity-density relationship in the lower mantle,[6] unless the postulated fluorite polymorph of SiO_2 were present in only very small amounts.

Stability of Al_2O_3 and MgO under High Pressure

The Hugoniots of periclase and corundum have both been determined[7] up to pressures of 1.3 Mbars, which is the pressure at the core-mantle boundary. No evidence of phase transformation was observed. The zero-pressure density of the phase stable above 500 kbars was estimated and found to be similar to the zero-pressure densities of periclase and corundum. However, the experimental uncertainties do not preclude the possibility of phase transformations accompanied by small volume changes ($<5\%$). The transformation of MgO to the cesium chloride structure thus remains a possibility, although not favoured by the data. The shock experiments are of value in providing evidence as to the stability under extreme pressures of these two geochemically important minerals.

A high-pressure transition of BaO (rocksalt) to a distorted cesium chloride structure accompanied by a density increase of about 12% has also been reported.[8] However, because of the much greater size and polarizability of

[1] Ahrens, Anderson, and Ringwood (1969).
[2] Wackerle (1962).
[3] McQueen, Fritz, and Marsh (1963).
[4] DeCarli and Milton (1965).
[5] Estimated from systematic velocity-density relationships (Sec. 9-4).
[6] Chapter 14.
[7] McQueen and Marsh (1966).
[8] Liu (1971).

Ba^{++}(1.36Å) as compared to Mg^{++}(0.72Å), this transition has little relevance to the high-pressure behaviour of MgO.

Transformations in Some R_2O_3 Compounds

Relationships between corundum (ilmenite) structures and B and C rare earth structures[1] are shown in Fig. 12-13. Most of the C rare earths, which possess a type of defect fluorite structure, have been found to transform to the B rare earth structure at pressures of a few tens of kilobars.[2] The transformation has also been observed in Sc_2O_3,[1] accompanied by a density increase of 8.1%. The B rare earth structure is monoclinic, and the cations are distributed among three crystallographically distinct sites, two with sevenfold coordination and one with sixfold coordination. By comparison, the cation coordination in the C rare earth structure is sixfold.

The B rare earth structure is about 5% denser than the corresponding ilmenite or corundum types and must be regarded as a possible structure into which corundums may transform at very high pressure. No direct examples have yet been found, although some general relationships suggest the probability of this behaviour. Thus, Mn_2O_3 possesses the C rare earth structure, and a modification of Fe_2O_3 with the C rare earth structure is also known.[3] (Fe_2O_3 displays extensive solid solubility in C rare earth Mn_2O_3.) In view of the fact that most other C rare earths are known to transform to the B rare earth structure at modest pressure, it can be expected that, with still higher pressures, Fe_2O_3 (which normally possesses the corundum structure) might also transform to the B rare earth structure. A few C rare earths (In_2O_3, Tl_2O_3, and $GaFeO_3$), however, transform not to B rare earth but to the corundum or ilmenite structure, which is of intermediate density.[4]

Hematite is observed to display a major phase transformation under shock-wave compression at pressures greater than 900 kbars.[5] The zero-pressure density of this phase was estimated[6] to be 5.96 g/cm^3 compared with 5.27 g/cm^3 for hematite. The increase in density of 13% is substantially higher than is estimated for a B rare earth modification ($\Delta\rho/\rho \sim 5\%$). However, the shock data do suggest that an intermediate phase with density appropriate to the B rare earth structure may be formed around 700 kbars.

Reid and Ringwood (1969a) suggested that the highest pressure phase of hematite might possess the perovskite structure with the formula $Fe^{++}Fe^{4+}O_3$. This structure would require an electron transfer and would yield a density of 5.8 g/cm^3, which is within the uncertainty of the shock-wave estimate. This suggestion is admittedly somewhat ad hoc, and moreover, the density of this structure

[1]Reid and Ringwood (1969a).
[2]Hoekstra and Gingerich (1964).
[3]Wyckoff (1964).
[4]Shannon (1966).
[5]McQueen and Marsh (1966)
[6]Ahrens, Anderson, and Ringwood (1969).

does not appear to explain well the highest pressure (1.2 to 1.4 Mbars) data points.[1]

Perhaps a more plausible interpretation is that of Syono et al. (1971), who suggest that the dense form of Fe_2O_3 might be a consequence of a high spin-low spin transition in Fe^{3+}. The radius of Fe^{3+} would be decreased from 0.65 to 0.55 Å by this transition,[2] and the zero-pressure density of Fe_2O_3 (corundum, low spin) would be 6.05 g/cm^3, which agrees well with the shock-wave data. A similar interpretation has been proposed by Davies and Gaffney (1973).

REFERENCES

AHRENS, T. J., D. L. ANDERSON, and A. E. RINGWOOD (1969). Equations of state and crystal structures of high pressure phases of shocked silicates and oxides. *Rev. Geophys.* **7,** 667–707.

AKIMOTO, S., and Y. SYONO (1969). Coesite-stishovite transition. *J. Geophys. Res.* **74,** 1653–1659.

AL'TSCHULER, L. G., R. F. TRUNIN, and G. SIMAKOV (1965). Shock-wave compression of periclase and quartz, and the composition of the Earth's lower mantle. *Bull. Acad. Sci. USSR, Phys. Solid Earth* (English translation) **10,** 657–661.

AZZARIA, L., and F. DACHILLE (1961). The new high pressure polymorph of MnF_2. *J. Phys. Chem.* **65,** 889.

BAUR, W. H. (1972). Computer-simulated crystal structures of observed and hypothetical Mg_2SiO_4 polymorphs of high and low density. *Am. Min.* **57,** 709–731.

BAYER, G., and W. HOFFMAN (1966). Complex alkali titanium oxides $A_x(B_yTi_{8-y})O_{16}$ of the α-MnO_2 structure-type. *Am. Min.* **51,** 511–516.

BIRCH, F., and P. LE COMTE (1960). Temperature-pressure plane for albite composition. *Am. J. Sci.* **258,** 209–217.

BOETTCHER, A. L., and P. J. WYLLIE (1968). The quartz-coesite transition measured in the presence of a silicate liquid and calibration of piston-cylinder apparatus. *Contr. Mineral. Petrol.* **17,** 224–232.

BOYD, F. (1964). Geological aspects of high-pressure research. *Science* **145,** 13–20.

——— and J. L. ENGLAND (1960). The quartz-coesite transition. *J. Geophys. Res.* **65,** 749–756.

——— and ——— (1961). Melting of silicates at high pressures. *Carnegie Inst. Washington Yearbook* **60,** 120.

BYSTRÖM, A., and A. M. BYSTRÖM (1950). The crystal structure of hollandite, the related manganese oxide minerals and α-MnO_2. *Acta Cryst.* **3,** 146–154.

CHAO, E. C. T., J. FAHEY, J. LITTLER, and D. MILTON (1962). Stishovite, SiO_2, a very high pressure new mineral from Meteor Crater, Arizona. *J. Geophys. Res.* **67,** 419–421.

———, E. M. SHOEMAKER, and B. M. MADSEN (1960). The first natural occurrence of coesite from Meteor Crater, Arizona. *Science* **132,** 220–222.

COES, L. (1953). A new dense crystalline silica. *Science* **118,** 131–133.

[1] Davies and Gaffney (1973).
[2] Shannon and Prewitt (1969).

DAVIES, G. F., and E. GAFFNEY (1973). Identification of high pressure phases of rocks and minerals from Hugoniot data. *Geophys. J.* **33,** 165–183.

DE CARLI, P., and D. MILTON (1965). Stishovite: synthesis by shock waves. *Science* **147,** 144.

EVANS, R. C. (1966). "*An Introduction to Crystal Chemistry*," 2d ed. Cambridge Univ. Press, London. 410 pp.

FUJISAWA, H. (1968). Temperature and discontinuities in the transition layer in the earth's mantle: Geophysical application of the olivine-spinel transition in the $Mg_2SiO_4-Fe_2SiO_4$ system. *J. Geophys. Res.* **73,** 3281–3294.

GREEN, T. H., A. E. RINGWOOD, and A. MAJOR (1966). Friction effects and pressure calibration in a piston-cylinder apparatus at high pressures and temperature. *J. Geophys. Res.* **71,** 3589–3594.

GREY, I. E., and A. F. REID (1972). Shear structure compounds $(Cr,Fe)_2\,(Ti_{n-2})_{2n-1}$ derived from the α-PbO_2 structural type. *J. Solid. State Chem.* **4,** 186–194.

———, ———, and J. ALLPRESS (1973). Compounds in the system Cr_2O_3-Fe_2O_3-TiO_2-ZrO_2 based on intergrowth of the α-PbO_2 and V_3O_5 structure types. *J. Solid State Chem.* **8,** 86–99.

HOEKSTRA, H., and K. GINGERICH (1964). High pressure B-type polymorphs of some rare earth sesquioxides. *Science* **152,** 1163–1164.

HOLM, J. L., O. J. KLEPPA, and E. F. WESTRUM (1967). Thermodynamics of polymorphic transformations in silica. Thermal properties from 5 to 1070°K and pressure-temperature stability fields for coesite and stishovite. *Geochim. Cosmochim. Acta* **31,** 2289–2307.

KHITAROV, N. I. (1964). New experimental work in the field of deepseated processes. *Geochem. Inter.* **3,** 532–535.

KITAHARA, S., and G. C. KENNEDY (1964). The quartz-coesite transition. *J. Geophys. Res.* **69,** 5395–5400.

KUME, S., T. MATSUMOTO, and M. KOIZUMI (1966). Dense form of germanium orthoclase ($KAlGe_3O_8$). *J. Geophys. Res.* **71,** 4999–5001.

LIU, L. (1971). A dense modification of BaO and its crystal structure. *J. Appl. Phys.* **42,** 3702–3704.

MACDONALD, G. J. F. (1956). Quartz-coesite stability relations at high temperatures and pressures. *Am. J. Sci.* **254,** 713–721.

MARTIN, R. F., and G. DONNAY (1972). Hydroxyl in the mantle. *Am. Min.* **57,** 554–570.

MCQUEEN, R. G., J. N. FRITZ, and S. P. MARSH (1963). On the equation of state of stishovite. *J. Geophys. Res.* **68,** 2319–2322.

———, J. C. JAMIESON, and S. P. MARSH (1967). Shock-wave compression and X-ray studies of titanium dioxide. *Science* **155,** 1401–1404.

——— and S. P. MARSH (1966). In: S. P. Clark, Jr. (ed.), "*Handbook of Physical Constants.*" Geol. Soc. Am. Memoir **97.**

———, ———, and J. N. FRITZ (1967). Hugoniot equation of state of twelve rocks. *J. Geophys. Res.* **72,** 4999–5036.

OSTROVSKY, I. A. (1965). Experimental fixation of the position of the coesite-stishovite equilibrium curve. (In Russian.) *Izv. Akad. Nauk SSSR, Geol. Ser.* no. 10, 132–135.

——— (1967). On some sources of errors in phase equilibria investigations at ultra-high pressures, 2, Phase diagram of silica. *Geol. J.* **5,** 321–328.

PECORA, W. T. (1960). Coesite, craters and space geology. *Geotimes* **5,** 16.

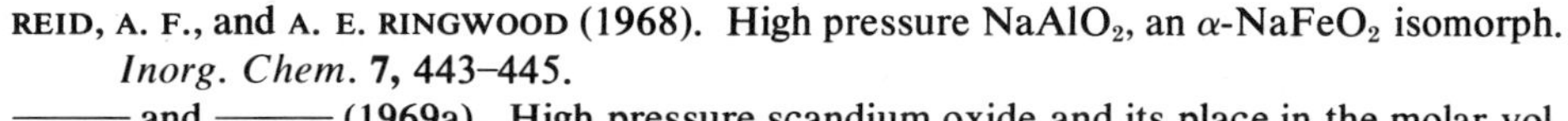

REID, A. F., and A. E. RINGWOOD (1968). High pressure $NaAlO_2$, an α-$NaFeO_2$ isomorph. *Inorg. Chem.* **7,** 443–445.

——— and ——— (1969a). High pressure scandium oxide and its place in the molar volume relationships of dense structures of M_2X_3 and ABX_3 type. *J. Geophys. Res.* **74,** 3238–3252.

——— and ——— (1969b). Six-coordinate silicon: High pressure strontium and barium aluminosilicates with the hollandite structure. *Solid State Chem.* **1,** 6–9.

——— and ——— (1969c). Newly observed high pressure transformations in Mn_3O_4, $CaAl_2O_4$ and $ZrSiO_4$. *Earth Planet. Sci. Letters* **6,** 205–208.

———, ———, and C. LI (1974). (In press.)

———, A. D. WADSLEY, and A. E. RINGWOOD (1967). High pressure $NaAlGeO_4$, a calcium ferrite isomorph and model structure for silicates at depth in the earth's mantle. *Acta Cryst.* **23,** 736–739.

RINGWOOD, A. E. (1975). (In press.)

——— and A. MAJOR (1967a). Some high pressure transformations of geophysical interest. *Earth Planet. Sci. Letters* **2,** 106–110.

——— and ——— (1967b). High pressure reconnaissance investigation in the system Mg_2SiO_4-MgO-H_2O. *Earth Planet. Sci. Letters* **2,** 130–133.

——— and A. F. REID (1969). High pressure transformations of spinels (1). *Earth Planet. Sci. Letters* **5,** 245–250.

———, ———, and A. D. WADSLEY (1967a). High pressure $KAlSi_3O_8$, an aluminosilicate with 6-fold coordination. *Acta Cryst.* **23,** 1093–1095.

———, ———, and——— (1967b). High pressure transformations of alkali aluminosilicates and aluminogermanates. *Earth Planet. Sci. Letters* **3,** 38–40.

——— and M. SEABROOK (1962). Some high pressure transformations in pyroxenes. *Nature* **196,** 883–884.

SCLAR, C. B., L. C. CARRISON, and O. M. STEWART (1967). High pressure synthesis of a new hydroxylated pyroxene in the system MgO-SiO_2-H_2O. (Abstract.) *Trans Am. Geophys. Union* **48,** 226.

———, A. P. YOUNG, L. C. CARRISON, and C. M. SCHWARTZ (1962). Synthesis and optical crystallography of stishovite, a very high pressure polymorph of SiO_2. *J. Geophys. Res.* **67,** 4049–4054.

SHANNON, R. D. (1966). New high pressure phases having the corundum structure. *Solid State Comm.* **4,** 629–630.

——— and C. T. PREWITT (1969). Effective ionic radii in oxides and fluorides. *Acta Cryst.* **B25,** 925–946.

STISHOV, S. M. (1963). Equilibrium line between coesite and the rutile-like modification of silica. (In Russian.) *Dokl. Akad. Nauk SSSR* **148,** no. 5, 1186–1188.

——— and S. V. POPOVA (1961). New dense polymorphic modification of silica. *Geokhimiya* No. 10, 837–839.

SYONO, Y., and S. AKIMOTO (1968). High pressure synthesis of fluorite-type PbO_2. *Mat. Res. Bull.* **3,** 153–158.

———, ———, and Y. ENDOH (1971). High pressure synthesis of ilmenite and perovskite type $MnVO_3$ and their magnetic properties. *J. Phys. Chem. Solids* **32,** 243–249.

WACKERLE, L. (1962). Shock-wave compression of quartz. *J. Appl. Phys.* **33,** 922–937.

WADSLEY, A. D. (1964). In: L. Mandelcorn (ed.), "*Non-Stoichiometric Compounds,*" chap. 3, p. 108 et seq. Academic, New York.

WANG, C. Y. (1968). Equation of state of periclase and Birch's relationship between velocity and density. *Nature* **218,** 74–76.

WENTORF, R. H. (1962). Stishovite synthesis. *J. Geophys. Res.* **67,** 3648.

WHITE, W. B., F. DACHILLE, and R. ROY (1961). High-pressure–high temperature polymorphism of the oxides of lead. *J. Am. Ceram. Soc.* **44,** 170–174.

WYCKOFF, R. W. G. (1964). "*Crystal Structures,*" vol. 2, 2d ed. Interscience, a division of Wiley, New York.

——— (1965). "*Crystal Structures,*" vol. 3, 2d ed. Interscience, a division of Wiley, New York.

YAMAMOTO, K., and S. AKIMOTO (1974). High pressure and high temperature investigations in the system MgO-SiO_2-H_2O. *J. Solid State Chem.* **9,** 187–195.

——— and ——— (1975). High pressure and high temperature investigations of the phase diagram in the system MgO-SiO_2-H_2O. (Preprint.)

ZOLTAI, T., and M. J. BUERGER (1959). The structure of coesite, the dense high-pressure form of silica. *Z. Krist.* **111,** 129–141.

14

CONSTITUTION OF THE DEEP MANTLE

14-1 CHEMICAL COMPOSITION

The chemical composition of the upper mantle was discussed in Chap. 5. It was concluded that this region is chemically zoned, being composed in most places of a layer of peridotite, perhaps 30 to 200 km thick, passing downward into a more primitive rock type, *pyrolite*, which is intermediate between peridotite and basalt in composition but closer to peridotite. A key property of pyrolite is that, on partial melting, it yields a basalt magma and leaves behind a residual refractory peridotite. The major element composition of pyrolite as estimated by several different methods is given in Table 5-2.

An important question is whether the pyrolite composition is representative of the entire mantle or whether significant changes in chemical composition with depth occur in the transition zone and lower mantle. Relevant evidence must be sought from cosmochemical considerations.

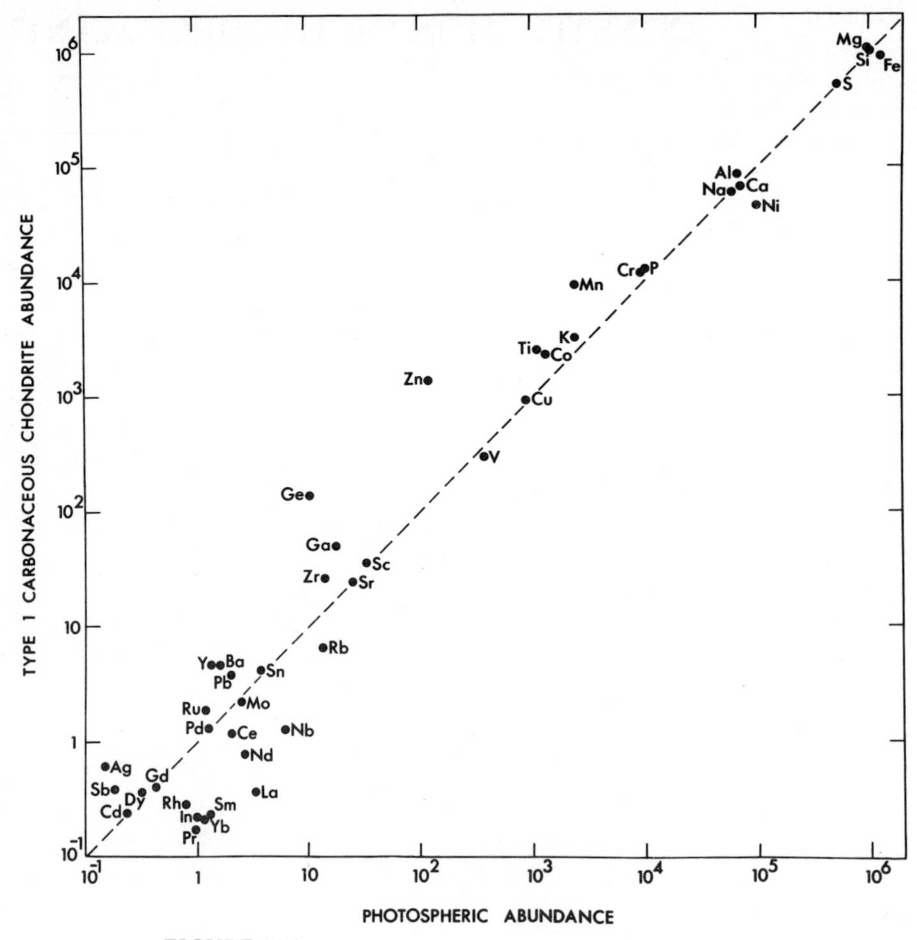

FIGURE 14-1
Comparison of elemental abundances in Type 1 carbonaceous chondrites with those in the solar photosphere. All abundances are normalized on the basis of Si = 10^6. Chondrite data from Urey (1972). Photosphere data from Aller (1968) and Müller (1968) for Ge, Ga, and Sn and from Garz et al. (1969) for Fe.

It is now generally agreed[1,2,3,4,5] that, except for highly volatile elements such as H, He, C, N, O, and the inert gases, abundances in Type 1 carbonaceous chondrites closely approach the primordial abundances of the parental solar nebula from which the earth and other planets were formed. This conclusion is based upon several lines of evidence—nucleosynthesis theory and systematics, studies

[1]Mason (1960).
[2]Ringwood (1961, 1962a, 1966a,b,c).
[3]Anders (1964, 1971).
[4]Cameron (1968).
[5]Urey (1972).

of the chemical and genetic relationships between different classes of chondrites, abundances of elements in cosmic rays, and finally, comparison with solar abundances. The abundances of elements in Type 1 carbonaceous chondrites are plotted against the solar photosphere abundances in Fig. 14-1. When due allowances are made for experimental uncertainties, which may be very large for the less abundant elements, the agreement is seen to be remarkably good. This applies particularly to the more abundant elements, for which the data are most reliable.

The relationship between these primordial abundances, the overall composition of the earth, and the pyrolite composition was discussed in Secs. 5-3 and 5-4. It was found that the relative abundances of involatile, oxyphile elements (e.g., Mg, Si, Ca, Al, Ti, U, Th, Ba, Sr, and rare earths) in pyrolite were similar to the primordial relative abundances of these elements. This similarity is of considerable genetic significance. The lack of marked fractionation between upper mantle pyrolite and the primordial abundances of this group of major rock-forming and trace elements, possessing diverse crystal chemical properties, suggests that the pyrolite composition may be applicable to the entire mantle.

We will explore this hypothesis by investigating the sequence of phase changes, which, in the light of evidence reviewed earlier, would be expected to occur in material of pyrolite composition with increasing depth in the mantle. We will then enquire to what extent phase changes in pyrolite are capable of explaining the observed and inferred distribution of physical properties with depth. Because iron is a siderophile element and not a member of the involatile, oxyphile group, we will treat the Mg/Mg + Fe ratio as an independent variable in order to discuss the possibility of changes in this ratio with depth, as have been suggested by several authors. The Mg/Mg +Fe ratio of pyrolite is 0.89 (atomic).[1] Initially, we will assume this ratio to be constant throughout the mantle. The effects of varying the ratio will then be explored.

Before commencing, however, a brief discussion of seismic velocity distributions in the mantle is required.

14-2 SEISMIC VELOCITY DISTRIBUTIONS

The seismic velocity distributions of Jeffreys and Gutenberg (Fig. 9-1) were based upon first arrivals and upon smoothed travel-time data, and hence, the derived velocity distributions between 400 and 1000 km were also smooth. This was held by some to be an objection against the proposal that phase changes occurred in this region.[2] Early attempts to explain this velocity distribution in terms of phase transformations attempted to use the smearing-out effects associated with solid-solution formation.[3] In the light of evidence then available, this explanation did

[1]If the other transition elements present in pyrolite (Ni^{++}, Mn^{++}, Fe^{3+}, Cr^{3+}, and Ti^{4+}) are included with Fe^{++} to make a "simplified pyrolite" composition, the effective Mg/Mg + Fe ratio would be 0.87 to 0.88.

[2]Verhoogen (1953).

[3]Ringwood (1958, 1962b). See also Meijering and Rooymans (1958).

not appear unreasonable. However, subsequent experimental investigations in relevant systems[1] have shown that the two phase loops were generally narrower than anticipated and that some important phase transitions in the mantle would occur within limited depth intervals, producing relatively sharp velocity increases.

Fortunately, parallel developments in seismology have demonstrated that the velocity distribution between 400 and 1000 km is not smooth and that two or more seismic discontinuities are present, as shown, for example, in Figs. 6-3 and 9-1. The first hint of this structure arose from surface-wave investigations which indicated the presence of abnormally large increases in velocity concentrated in two relatively narrow zones 50 to 100 km thick around 400 km and 700 to 800 km.[2,3] Ringwood (1966d, p. 390) accordingly suggested that the high velocity gradient between 400 and 500 km might be caused by the olivine-spinel transformation and by the transformation of pyroxene, whilst the high gradients around 700 to 800 km might be caused by the further transformation of spinel into ilmenite and oxide structures. A similar suggestion was later made by Anderson (1967a).

The velocity distributions proposed by Anderson and coworkers have been confirmed in general by a number of subsequent investigations of mantle body waves.[4] There is general agreement concerning the presence of a major seismic discontinuity near 400 km and of a further discontinuity or, alternatively, a zone of high velocity gradient, near 650 km. This structure, in which nearly all the velocity increase between 300 and 800 km is concentrated in two major discontinuities, is depicted in Figs. 6-3 and 9-1.

More recent high-resolution seismic investigations suggest that the above double discontinuity structure, in turn, represents an oversimplified interpretation of the actual velocity distribution. The *P*-velocity distribution in the mantle beneath northern Australia as determined in a detailed array study by Simpson (1973) is shown in Fig. 14-2. Note that the "400-km discontinuity" is resolved into two regions of high gradient at depths centred around 330 and 390 km. This feature is also displayed in the solution of Masse et al. (1972) for the mantle beneath western United States, which showed sharp increases in velocity at 315 km and at 422 km. Both of these studies suggest that the velocity contrast across the deeper discontinuity is about twice as large as the contrast across the shallower discontinuity. Evidence of a complex structure in the vicinity of 300 to 400 km was also obtained by Bolt et al. (1968).

Referring to Fig. 14-2, we see that a substantial increase in seismic velocities occurs between depths of 490 and 550 km. This feature is also displayed on a profile obtained by Helmberger and Wiggins (1971) but is absent on earlier

[1] Akimoto and Fujisawa (1968), Ringwood and Major (1970), Ringwood (1970).
[2] Anderson and Toksöz (1963).
[3] Kovach and Anderson (1964).
[4] Reviewed in Sec. 9-1.

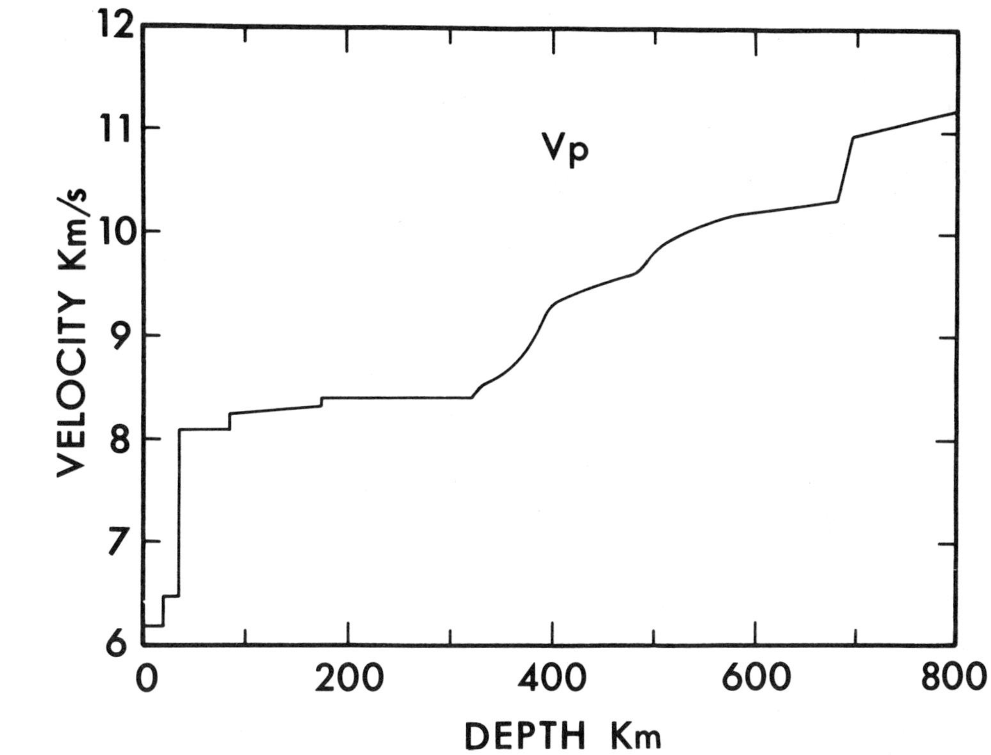

FIGURE 14-2
P-wave distribution for the upper mantle and transition zone beneath northern Australia based on data obtained by the Warramunga array, Tennant Creek. (*From Simpson, 1973.*)

profiles exemplified by Figs. 6-3 and 9-1. Seismic reflections from a horizon in this depth interval have also been reported.[1]

Reflection investigations[1,2,3] provided evidence that the discontinuity near 650 to 700 km is much sharper than those near 300 to 400 km and occurs over a depth interval of only a few kilometres.[3] Well-defined reflections were also obtained from 420 km and indicated that the velocity change across this boundary was comparable with that near 650 to 700 km but was spread out over a broader depth interval.[3] Confirmation that the deeper discontinuity was much sharper than those near 300 to 400 km was provided by the array results[4] depicted in Fig.

[1]Whitcomb and Anderson (1970).
[2]Engdahl and Flinn (1969).
[3]Adams (1971).
[4]Simpson (1973).

14-2. However, in the latter investigation, the discontinuity was placed at 680 km, rather than 650 km. Likewise, a recent study of *S* velocity in the mantle placed this discontinuity close to 690 km.[1] In all seismic velocity-depth models, there are substantial uncertainties in the actual depths of the discontinuities implied by the data. Accordingly, in subsequent discussions, it is both realistic and convenient to round off the depths of the discontinuities discussed above. We will take the depths of these major discontinuities to be centred at 330, 400, and 700 km, with a zone of high velocity gradient between 500 to 550 km.

The presence of seismic discontinuities or regions of anomalously high velocity gradient in the mantle below 800 km is also indicated by the investigations of Archambeau et al. (1969) and Johnson (1969). The former authors suggest that a substantial increase in velocity (0.4 km/sec) may occur near 1050 km (Fig. 6-3). This feature is not regarded as uniquely established: "A single change or perhaps several small but rapid changes in velocity gradient occur in the range from 1000 km to 1200 km." They believe that the region of uniform gradient between 700 and 1000 km is reliably determined from the combined data of all their profiles.

Johnson has carried out a detailed investigation of *P* velocity in the lower mantle using array data. He inferred the presence of anomalously large velocity gradients near 830, 1000, 1230, 1540, and 1910 km (Fig. 14-3). The high-gradient regions are spread over depth intervals of about 50 km. These regions account for an integrated velocity increase of 2.7%, which is about 20% of the total velocity increase between 800 and 1800 km. Johnson suggests the possibility that phase

[1]Worthington, Cleary, and Anderssen (1974).

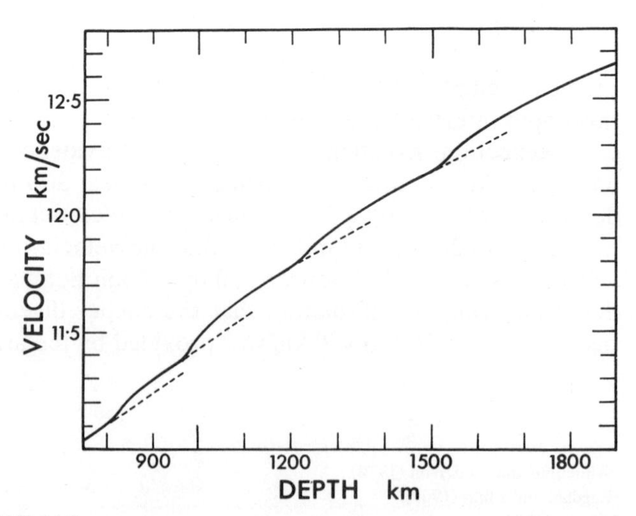

FIGURE 14-3
P-wave velocity distribution in the lower mantle. (*After Johnson, 1969.*)

changes might be responsible for these features. There appears to be a significant difference between the Johnson and Archambeau et al. solutions in the region between 800 and 1200 km. The latter authors obtain a much larger increase of velocity from discontinuities or anomalously high gradients (~0.4 km/sec) in this region than does Johnson (~ 0.2 km/sec). This is seen to result from the higher average gradient for "normal" regions between 700 and 1000 km obtained by Johnson, as compared with the gradient for this interval obtained by Archambeau et al. and considered to be well established. This region warrants further study. Kanamori (1967), using an array, also finds evidence of an abnormally high gradient around 950 km.

Taken at face value, the general pattern of the Johnson and Archambeau et al. velocity distributions is strongly suggestive of the presence of further phase transformations to denser and seismically faster structures occurring below 800 km, although the combined magnitude of these phase changes is 5 to 10 times smaller than the major phase changes occurring between 330 and 700 km. This interpretation must be treated with caution, however, until the seismic velocity distributions in the lower mantle obtained by different seismologists show a greater degree of concordance. Chinnery and Toksöz (1967) have derived a velocity distribution in the lower mantle which is somewhat different from Johnson's. Wright and Cleary's (1972) results agree substantially with those of Johnson's; nevertheless, there are some significant differences. Apparently, the problems of local heterogeneity beneath different arrays have not yet been fully resolved. There is also the possibility of lateral heterogeneity in the lower mantle. Nevertheless, although detailed interpretations vary, it seems likely that the lower mantle is not strictly homogeneous and that several regions of abnormal velocity gradient due to phase changes and/or chemical changes are present.

14-3 MINERALOGY IN A PYROLITE MANTLE AS A FUNCTION OF DEPTH

We now proceed to set up a model which depicts the most probable variation of mineralogy with depth for the pyrolite composition (Table 5-2) based upon the experimental data described in previous chapters. *It is emphasized that we are dealing with a model subject to future changes as new and improved experimental data become available.* The mineralogy down to 600 km is based directly upon the results of high-pressure experiments which reproduce the *P-T* conditions down to this depth.[1] Below 600 km, the model is based largely upon indirect evidence such as germanate analogue systems combined with shock-wave data.

In the pyrolite composition, the atomic ratio of the major M^{++} cations ($Mg + Fe + Ca + \overline{NaAl}$) to Si^{4+} falls between the 1:1 and 2:1 ratios. Thus the mineralogy of the mantle will be dominated by compounds of $MSiO_3$ and M_2SiO_4

[1]Recent reconnaissance investigations using diamond anvils have extended this limit to 800 km.

stoichiometry, as discussed in Chaps. 11 and 12. The possibility of intermediate compounds within these limits, e.g., $M_3Si_2O_7$, should perhaps be considered. It is significant, however, that the only binary compounds formed between the oxides AO (MgO, FeO, CoO, NiO, and MnO) and BO_2 (GeO_2, TiO_2, and SnO_2) within the 2:1 and 1:1 range of stoichiometries are of the types A_2BO_4 and ABO_3. Intermediate compounds between these compositions are not formed.[1] In view of the general crystal chemical relationships existing at high pressure between silicates, on the one hand, and germanates, titanates, and stannates, on the other,[2] it does not appear likely that such intermediate compounds will play a significant role in the deep mantle, at least in the cases of the major components MgO, FeO, and SiO_2. Accordingly, it appears reasonable to discuss the mineralogy of the deep mantle mainly in terms of the crystal chemistry of A_2BO_4 and ABO_3 compounds as reviewed in Chaps. 11 and 12.

The Region between Depths of Approximately 150 to 330 km

The *P*-velocity distributions (Figs. 6-3, 9-1, and 14-2) show a uniform and zero to small rate of increase of velocity with depth in this region. The velocity gradients, in the light of Birch's (1952) analysis, suggest that this region is essentially homogeneous. This inference is supported by experimental data. High-pressure investigations show that, throughout the *P-T* conditions in this region, pyrolite would crystallize to a mineral assemblage comprising olivine, orthopyroxene, clinopyroxene (both low in alumina), and garnet. No further phase transformations which might cause velocity anomalies have yet been discovered, despite intensive search.

The detailed mineralogy for the pyrolite composition is as follows:

		wt%
Olivine (Fo_{89})	$(Mg, Fe)_2SiO_4$	57
Orthopyroxene	$(Mg, Fe)SiO_3$	17
Omphacitic clinopyroxene	$(Ca, Mg, Fe)_2Si_2O_6$-$NaAlSi_2O_6$	12
Pyrope-rich garnet	$(Mg, Fe, Ca)_3(Al, Cr)_2Si_3O_{12}$	14

The density (ρ_0) of this mineral assemblage, reduced to the value which it would display at atmospheric pressure and temperature, is 3.38 g/cm^3. The minerals are characterized by fourfold-coordinated Si and by six- and eightfold-coordinated Mg, Fe, and Ca (with respect to oxygen).

Detailed studies of the elastic properties and densities of olivines and of mineral assemblages close to the pyrolite composition at high pressures and temperatures have been made by Graham (1970) and Ahrens (1972a). They demonstrate

[1]Nor are they formed at *high pressure* between CaO and GeO_2. However, intermediate compounds are known in the system CaO-TiO_2. Their structure is closely related to those of the Ca_2TiO_4 and $CaTiO_3$ compositions on either side (Roth, 1958).

[2]Chapters 10, 11, 12, and 13.

that the pyrolite composition with Mg/Mg + Fe = 0.88 ± 0.03 provided a quantitative explanation of the *P*- and *S*-velocity distributions and of the density distribution in the upper mantle.

Seismic Discontinuities and High Velocity Gradients between Approximately 320 and 400 km

This depth interval contains one of the two major seismic discontinuities within the mantle (Figs. 6-3 and 9-1). As was noted in the previous section, recent investigations have shown that this feature possesses a complex structure and can be resolved into a double discontinuity, consisting of a smaller and possibly smeared-out velocity increase commencing around 330 km and a larger and more distinct discontinuity centred around 400 km (Fig. 14-2).

The experimental investigations reviewed previously demonstrated that two major transformations would occur in pyrolite within this pressure range: (1) the pyroxene-garnet transformation and (2) the olivine-spinel-β-Mg_2SiO_4 transformation. The former involves the solid solution of pyroxene in the garnet structure, which would set in rather suddenly at about 100 kbars (1000°C) (Figs. 12-7, 12-8, and 12-9). The resultant complex garnet solid solution is characterized by partial octahedral coordination of silicon and possesses the general formula $M_3Al_2Si_3O_{12}$-$M_3(MSi)Si_3O_{12}$, where M = Mg, Fe, Ca.

The gradient dP/dT of the pyroxene-garnet transformation in the mantle is not yet known. If, however, it falls within the "normal" range exhibited by most solid-solid phase transitions (say, 20 bars/°C), then the pyroxene-garnet transition might be expected to occur around 330 km for a temperature of about 1500°C. The transformation involves an effective density increase of 10% for the pyroxene component, which comprises some 29% of pyrolite, so that the total density increase would be about $2\frac{1}{2}$%,[1] corresponding to a *P*-velocity increase of about 0.3 km/sec, which is identical to the velocity increase around 330 km as shown in Fig. 14-2. Figures 12-7, 12-8, and 12-9 show that a large proportion of this velocity increase would occur within a narrow depth interval, with the remainder being smeared out over greater depths, also in agreement with the profile shown on Fig. 14-2. It may be concluded, therefore, that the pyroxene-to-garnet transition in pyrolite provides a good explanation of this feature.

The next transition to occur at greater depths is that of olivine through spinel to the beta phase or, perhaps, directly to the beta phase, depending upon the Fe/Fe + Mg ratio of the olivine, which in turn is influenced by iron-partition relations with coexisting garnet.[2] The details of this transformation in the mantle were discussed in Sec. 11-3. Taking a gradient of 30 bars/°C and a temperature of 1600°C at 400 km, this transition for the pyrolite Fe/Fe + Mg ratio would occur over a depth interval of about 30 km, centred upon 400 km. Within this interval a

[1]Under the *P*-*T* conditions existing at 330 km.
[2]Ahrens (1972b).

first-order discontinuity could arise from the presence in the phase diagram (Fig. 11-7) of a reaction point between the spinel and beta phases.[1] Alternatively, and perhaps more probably, the preferential partition of iron in coexisting garnet would cause the olivine to have a composition of Fo_{94}. This would transform directly to beta phase at slightly greater (~ 10 to 20 km) depth, but the total transition interval would be quite small—perhaps about 10 km. Likewise, a higher gradient than 30 bars/°C would place the discontinuity below 400 km. For reasonable ranges of uncertainty concerning temperature, composition, gradient, and nature of the phase diagram, the depth of this major transformation is likely to lie between about 370 and 430 km. The transformation of olivine into the beta phase involves a density increase of about 6.5% in the olivine, which constitutes 57% pyrolite, corresponding to a velocity increase of 0.43 km/sec.[2]

The experimental data discussed above thus imply the occurrence of two seismic discontinuities in the 320 to 430 km region, together with boundary regions of high velocity gradient. The overall agreement between seismological observations and experimental results for a pyrolite model is most satisfactory.

High Velocity Gradients in the 500 to 550 km Region

In the depth interval between 400 and 500 km (approx.) the experimental observations discussed in Chaps. 11 and 12 imply that a mantle of pyrolite composition would crystallize as follows:

	wt%
β-$(Mg, Fe)_2SiO_4$	57
Complex garnet solid solution: $M_3(Al, Cr, Fe)_2Si_3O_{12}$-$M_3(MSi)Si_3O_{12}$ plus $Na_2CaSi_5O_{12}$ (M = Mg, Fe, Ca)	43

The density ρ_0 of this mineral assemblage, reduced to atmospheric pressure and temperature, would be 3.63 g/cm³. Coordination of silicon atoms by oxygen is mainly fourfold, with some octahedrally coordinated silicon in the garnet.

Between 500 and 550 km, a zone of abnormally high velocity gradient may occur (Fig. 14-2). Helmberger and Wiggins (1971) recognize a similar feature centred upon 550 km. This region is clearly inhomogeneous, and the P-velocity component attributable to inhomogeneity[3] is about 0.3 km/sec. Two distinct phase transitions may be responsible for this feature. There is some evidence that the beta phase may transform to spinel around this depth[4] (Sec. 11-2, Fig. 11-9). This would cause a velocity increase[5] of about 0.14 km/sec in pyrolite

[1] Section 11-3.

[2] This refers to the P-T conditions pertaining in the mantle. The density difference at normal P and T is 8.2%. Velocity is related to density by $dV_p = 3.2\ d\rho$ (Birch, 1961a).

[3] As distinct from the velocity increase caused by homogeneous self-compression.

[4] Whitcomb and Anderson (1970), since confirmed; ref. Chap. 11.

[5] Taking the bulk moduli of β- and γ-Mg_2SiO_4 to be equal, as indicated in preliminary observations by Takahashi (personal communication).

composition. The transition would be practically discontinuous and might explain the seismic reflections observed from a horizon near 520 km.[1]

A second transformation which can be expected to occur between 400 and 600 km is that of the calcium component of garnet into the perovskite structure, as discussed in Sec. 12-5. Pure $CaSiO_3$ is believed to transform into the perovskite structure at about 100 kbars, 1000°C. However, in the region above 500 km, $CaSiO_3$ occurs not as a free phase but as a component of the complex garnet solid solution, and the pressure required to partially transform the garnet to form $CaSiO_3$ perovskite would be substantially increased. The reaction essentially is

$$(Mg, Fe, Ca) \text{ garnet} \rightarrow (Mg, Fe) \text{ garnet} + CaSiO_3 \text{ (perovskite)}$$

The pyrolite composition contains 6.5% potential $CaSiO_3$ perovskite, and this reaction would account for a velocity increase of about 0.1 km/sec.[2] If combined with the β-γ Mg_2SiO_4 transition, the total velocity increase would be 0.24 km/sec, slightly smaller than indicated on Fig. 14-2. The discrepancy could readily be accounted for by the entry of some Fe^{++} into the perovskite-type phase—i.e., $(CaFe)SiO_3$ perovskite.

The Region between 600 and 800 km, Including the 700-km Discontinuity

Figures 6-3, 9-1, and 14-2 show a major seismic discontinuity occurring around depths of 650 to 700 km at a pressure of about 250 kbars. The germanate model studies suggested that garnets, spinels, and beta phases would transform at higher pressures into new phases possessing ilmenite, strontium plumbate, and perovskite structures. The density of this assemblage of phases for the pyrolite composition would be similar to that of the isochemical mixed oxides. Thermodynamic investigations based upon the observed solid solubilities of silicates in the high-pressure germanate structures yielded calculated pressures in the range 200 to 300 kbars for the above transformations in silicates.[3] These estimates, although admittedly imprecise, are in good agreement with the pressures around 600 to 800 km, suggesting that these transformations may play an important role in this region of the mantle.

Experimental results described in Secs. 12-2 and 12-4 provided strong evidence that the $(Mg,Fe)SiO_3$-Al_2O_3 components of the garnet solid solution would transform to the ilmenite structure.

On the other hand, recent investigations reviewed in Sec. 11-5 showed that pure Mg_2SiO_4 disproportionated into a mixture of $MgO + SiO_2$ (stishovite) at pressures between 200 and 300 kbars. There can be little doubt that this transformation plays a key role in the corresponding region of the mantle, although, as

[1]Whitcomb and Anderson (1970).
[2]Estimate based upon use of the velocity-density relationship (Sec. 9-4) and density data of Chap. 12.
[3]Chapters 11 and 12.

noted earlier, the very limited experimental data so far obtained by the quenching technique do not preclude the occurrence of a strontium plumbate modification of Mg_2SiO_4 occurring either before or after the disproportionation reaction, depending upon relative densities.

It is not possible to specify the detailed nature of the phase transformations occurring between 650 and 800 km in the light of presently available data. This will require the application of high-temperature diamond-anvil techniques employing in situ x-ray-diffraction analysis whilst the samples are under pressure, a technique now known to be feasible, but not yet applied to these problems. Important advances in this area may be anticipated during the next few years. Nevertheless, although the detailed nature and sequence of the transformations is not known, their general nature seems reasonably clear.

In the light of the discussion of phase relationships in Chaps. 11, 12, and 13, the most probable series of transformations occurring around 650 to 800 km is believed to be as follows:

1 $(Mg, Fe)_2SiO_4$ disproportionates to an isochemical mixture of $(Mg, Fe)O + SiO_2$ (stishovite).

2 Pyrope-rich component of garnet transforms to the ilmenite structure.

3 Calcium-rich component of garnet transforms to the perovskite structure (see previous discussion).

4 Sodium aluminosilicate component of garnet transforms to $NaAlSiO_4$ (calcium ferrite structure).

An alternative to (1) would be for $(Mg,Fe)_2SiO_4$ to transform first to the strontium plumbate structure, which at still higher pressures, would disproportionate into the mixed oxides.[1] This sequence might also be reversed, depending upon the relative densities of the different forms. These possibilities simply serve to emphasize the fact that the $P\ \Delta v$ term in the free energy is predominant in this region of the mantle and that the nature of the mineral assemblage present is largely controlled by small differences in intrinsic density.

It is unlikely that these transformations would occur at exactly the same depth. By far the most important transformation would be that of $(Mg,Fe)_2SiO_4$. This transformation may well occur within a relatively narrow depth range and would be primarily responsible for the occurrence of the 700-km seismic discontinuity. The transformations of garnets to ilmenite and related hexagonal close-packed structures involve the equilibria of very complex solid solutions and may occur at somewhat greater depths. They would probably be smeared out much more than the $(Mg, Fe)_2SiO_4$ transformation, and it would be coincidental if they were to occur at exactly the same pressure. It appears that they may contribute to a rather complex fine structure in this region which has yet to be resolved by seismology.

The observations of Adams (1971) imply that a large proportion of the ve-

[1]Another possibility would be for $(Mg,Fe)_2SiO_4$ to disproportionate into $MgSiO_3$ (ilmenite) + $(Mg,Fe)O$.

locity increase near 700 km occurs within a few kilometres (~ 1 to 2 kbars). The sharpness of the transformation suggests that the composition of the pure phase is close to pure Mg_2SiO_4. This may be explicable in terms of Mg, Fe partition relationships between octahedral lattice sites in β, γ (and δ) $(Mg, Fe)_2SiO_4$ and also in (Mg, Fe)O, with eightfold coordinated sites in coexisting garnet- and perovskite-type[1] phases. The preferential entry of Fe^{++} into eightfold coordinated sites in garnet in equilibrium with olivine has already been noted.[2] If this tendency is increased by pressure, then the transitions may occur in nearly pure Mg_2SiO_4 and would therefore be first order.

Although it is plausible to link the Mg_2SiO_4 transition with the major discontinuity near 700 km, seismic investigations have not yet disclosed another discontinuity which can be linked with the Mg garnet-to-ilmenite transformation which is believed to occur in this region.[3] This is probably a consequence of two factors. Firstly, the garnet-ilmenite solid solution is probably smeared out over a broad depth interval so that a sharp discontinuity is not caused. Secondly, the detailed seismic structure in this region has not yet been finally resolved, and regions with abnormally high velocity gradients may well be present, particularly between 700 and 800 km.[4]

Presently available data are inadequate to settle these questions, and further experimental work aimed at establishing the relative sequences of β-γ-δ Mg_2SiO_4 transitions, the disproportionation of Mg_2SiO_4 into mixed oxides, and the disproportionation of the complex garnet solid solution into perovskite and hexagonal close-packed phases is required. Evidence relating to the partition of iron between these phases will be particularly important. For the present, we will accept the experimental evidence reviewed in Chaps. 11, 12, and 13 that transformation of pyrolite into a mineral assemblage consisting most probably of periclase-, stishovite-, ilmenite-, and perovskite-type structures is attained by a depth of 800 km (~ 300 kbars) and will also assume that below 800 km, Fe^{++} is preferentially partitioned into the eightfold coordinated sites presented by a distorted perovskite structure.[5]

According to this view, the mineral assemblage attained at this depth would consist of

	wt%
MgO (periclase)	29
SiO_2 (stishovite)	22
$MgSiO_3$-(Al, Cr, Fe)AlO_3 (ilmenite-type s. soln.)	24
(Ca, Fe)SiO_3 (perovskite s. soln.)	23
$NaAlSiO_4$ (calcium ferrite structure)	2

[1] I.e., $GdFeO_3$ modification.
[2] Section 11-3. See also Ahrens (1972b).
[3] Section 12-4.
[4] Preliminary observations by D. Simpson (personal communication).
[5] Section 12-5.

The zero-pressure density of this assemblage is 4.03 g/cm^3, with a small uncertainty arising from the unknown distribution of Fe^{++}, mainly partitioned between the perovskite and periclase phases. This density is almost identical to the mean density (4.02 g/cm^3) of an isochemical mixture of component oxides. The mineral assemblage is characterized by octahedrally coordinated Mg^{++}, Si^{4+}, and Al^{3+} and by eightfold coordinated Ca^{++} and (perhaps) Fe^{++}.

Phase Transitions in the Lower Mantle

The general nature of the phase transformations responsible for increasing the zero-pressure density of pyrolite up to that of the equivalent isochemical mixed oxides appears to be fairly well understood. If we accept the above interpretation that these transformations are complete by about 800 km, then the possibility of further transformations to denser states at greater depths must be considered. The seismic velocity distribution in Fig. 6-3 shows a sharp increase of 0.4 km/sec at 1050 km. If verified, this may well represent a further phase transformation. The regions of anomalously high velocity gradient located by Johnson (Fig. 14-3) are likewise readily interpreted in terms of phase transformations, as previously noted. It should be pointed out on completely general grounds that further phase transformations in the deep mantle over the pressure range 300 to 1300 kbars are not unlikely, when considered in relation to the number of transformations which occur at pressures less than 300 kbars. Perhaps the most surprising aspect is the small total magnitude of the density changes which is permitted for possible further transformations by constraints arising from the density distribution and elasticity of the deep mantle.

Possible transformations leading to densities higher than the isochemical mixed oxides have already been discussed in Secs. 11-8 and 12-5. These result in mineralogies characterized largely by eightfold coordinated Mg and Fe. A possible mineral assemblage for pyrolite composition which is suggested by analogue studies discussed in earlier chapters is as follows:

$(Ca, Mg, Fe)SiO_3$	Perovskite structure
$(Mg, Fe)O$	Rocksalt structure
$NaAlSiO_4$	Calcium ferrite structure
$(Mg, Fe)(Al, Cr, Fe)_2O_4$	Calcium ferrite structure

Alternatively, $(Mg, Fe)_2SiO_4$ may crystallize directly in the calcium ferrite structure. The zero-pressure density of this assemblage would be about 4% higher than that of the mixed oxides.

Fyfe (1960) has suggested that the Fe^{++} ion in silicates may undergo a contraction in radius (by about 0.15 Å) under very high pressures due to a transition from the paramagnetic to the diamagnetic state caused by coupling of *d* electrons.[1] This high spin-low spin transition would cause a density increase of about 1% in

[1]See also Strens (1969) and Gaffney and Anderson (1973).

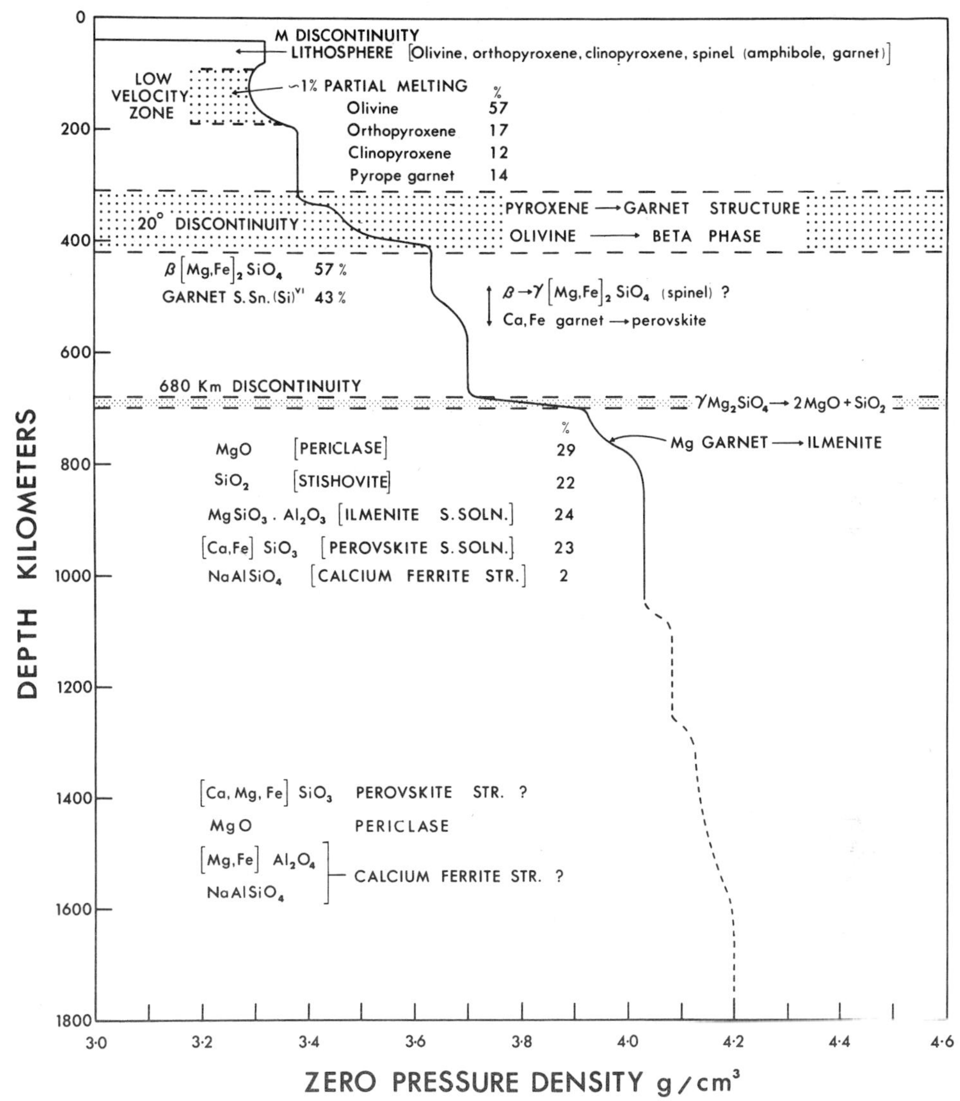

FIGURE 14-4
Possible mineral assemblages and corresponding zero-pressure densities for a model mantle of pyrolite composition. Mineralogy to a depth of 700 km is based mainly upon direct static experiments. Mineralogy at greater depths is inferred on the basis of indirect methods and shock-wave methods.

$(Mg, Fe)_2SiO_4$ of mantle composition. Shock-wave investigations suggest that the high spin-low spin transitions in Fe^{++} and Fe^{3+} may occur in magnetite and hematite at pressures around 1 Mbar.[1,2]

[1]Gaffney and Anderson (1973).
[2]Section 13-4.

An illustration of the mineralogical zoning in the deep mantle as outlined in the preceding discussion is given in Fig. 14-4.

14-4 MAGNITUDE OF VELOCITY CHANGES CAUSED BY MANTLE PHASE TRANSFORMATIONS

We have seen that phase transformations which are known or expected to occur in a mantle of pyrolite composition provide an excellent qualitative explanation of the principal features of the seismic velocity distribution in the mantle. Moreover, as shown in Sec. 14-3, they account quantitatively[1] for the magnitudes of the velocity changes at the 320, 400, and 500 to 550 km discontinuities (Fig. 14-2). We now pursue this latter topic further and examine the extent to which phase transformations can account for the overall velocity increases in the transition zone and in inhomogeneous regions of the lower mantle (Fig. 14-3).

The basis for this comparison rests upon the empirical Birch velocity-density relationship $V_p = a + b\rho$, as discussed in Sec. 9-3. The constant b[2] is taken as 3.16 (km/sec)/(g/cm³). In the previous section, the zero-pressure densities of the mineral assemblages believed to occur in different regions of the mantle were calculated. Using Birch's relationship, the density changes caused by transformations from one mineral assemblage to another can be converted into equivalent changes in seismic velocities. Before these can be compared with the changes in seismic velocities inferred at the discontinuities, a correction for homogeneous self-compression within the mantle must be applied so that the density changes occurring under the P-T conditions existing in the mantle at the discontinuities may be estimated.

This correction may be applied using the Birch-Murnaghan equation

$$P = \tfrac{3}{2}K_0 \left[\left(\frac{\rho}{\rho_0}\right)^{7/3} - \left(\frac{\rho}{\rho_0}\right)^{5/3} \right]$$

where P is pressure, and K_0 is the bulk modulus of the particular mineral assemblage at zero pressure. Values of K_0 for the various mineral assemblages can be estimated if the zero-pressure densities are known, using the Birch compressional velocity-density relationship, which has been extended by Wang (1967) to a corresponding relationship between density and hydrodynamical sound velocity V_c. Since V_c is equal to $\sqrt{K/\rho}$ and is proportional to ρ, we have K proportional to ρ^3. Known bulk moduli are therefore plotted against the cube of density for a variety of compounds and minerals of constant mean atomic weight and the constant of proportionality determined. This relationship then serves to determine the bulk

[1]See also Graham (1970) and Ahrens (1972a).

[2]Birch (1961a, table 15, no. 3). This is the most relevant solution and is based upon accurately measured single-crystal values.

Table 14-1 ESTIMATED ZERO-PRESSURE DENSITIES, BULK MODULI, AND ELASTIC RATIOS (ϕ) FOR PYROLITE MINERAL ASSEMBLAGES

Mineral assemblages	ρ_0 (g/cm^3)	K_0 (kbars)	$\phi_0 = K_0/\rho_0$ (km/sec)^2
1. Olivine + pyroxenes + garnet	3.38	1280	38
2. β-$(Mg,Fe)_2SiO_4$ + garnet	3.63	1700	47
3. Periclase + stishovite + $MgSiO_3 \cdot (Al, Cr, Fe)_2O_3$ ilmenite str. + $(Ca, Fe)SiO_3$ perovskite str. + $NaAlSiO_4$ calcium ferrite str.	4.03	2350	60
4. $(Mg, Fe, Ca)SiO_3$ perovskite str. + $(Mg, Fe)O$ rocksalt str. + $Mg(Al, Cr, Fe)_2O_4$ calcium ferrite str. + $NaAlSiO_4$ calcium ferrite str.	4.20	2800	67

modulus of an unknown structure if the density is known. Alternative empirical methods may be employed with essentially similar results.[1,2] The estimated zero-pressure bulk moduli and densities obtained in this manner for different pyrolite mineral assemblages are given in Table 14-1. Using these values, the variation of density with pressure for each assemblage is calculated using the Birch-Murnaghan equation, thus yielding the density differences between assemblages as a function of pressure. This is converted into an equivalent velocity change by means of the Birch V-ρ relationship. In addition, a small correction to the density changes to allow for the high temperature of the mantle is applied.[3]

These methods were applied to estimate the velocity changes arising from the combined pyroxene-garnet and α-β Mg_2SiO_4 phase transitions near 400 km, on the one hand, and the combined "post-spinel" transitions near 680 km, on the other.[4] The total velocity increase caused by phase changes around 400 km was 0.7 km/sec, which may be compared with 0.65 km/sec found by Johnson (Fig. 9-1) and 0.9 km/sec obtained by Archambeau et al. (Fig. 6-3). Considering the uncertainties, the agreement is satisfactory. For the combined "post-spinel" transition, a velocity increase of 1.0 km/sec was calculated, which may be compared with 0.9 km/sec (Johnson) and 1.1 km/sec (Archambeau et al.).

Because the seismic velocity structure between 300 and 800 km has not been finally resolved, it is perhaps preferable to attempt to combine all the velocity increases due to phase changes between 320 and 800 km. From Figs. 6-3, 9-1, and 14-2, these may be estimated to lie between 1.5 and 2.0 km/sec. The calculated velocity increase for phase transitions from assemblages 1 to 3 (Table

[1]O. Anderson and Nafe (1965).
[2]D. Anderson (1967b).
[3]Using the procedure of Clark and Ringwood (1964).
[4]Ringwood (1970).

14-1) is 1.7 km/sec. Considering the several sources of uncertainty,[1] the agreement may be considered satisfactory.[2]

The conclusion following from the previous discussion is that increases of velocity arising out of phase transformations between 320 and 800 km in a mantle of pyrolite composition are quantitatively consistent with the density changes caused by the transformations. In particular, they do not demand any increase in the iron-to-magnesium ratio with depth in order to maintain consistency. This differs from a conclusion reached by others[3,4] that a large increase in Fe/Fe + Mg ratio is required through the transition zone if the changes in seismic velocities are to be reconciled with density changes arising out of phase transformations.

It is emphasized that the present arguments, to this stage, do not preclude the possibility of a small change in iron content with depth. The situation is that a change in iron content is not *required* in order to achieve consistency between phase changes and accompanying density changes in a pyrolite mantle when adequate allowance is made for the various sources of uncertainty arising from the data and the simplifying assumptions made in treating this data as discussed in Sec. 9-4. In this respect, it is revealing that conclusions requiring an increase of iron content with depth were based upon application of an earlier form of the seismic equation of state.[5] More recently, a revised form of this equation has been proposed,[6] the application of which does not require an increase of iron content with depth.

A second argument supporting an increase in Fe/Fe + Mg ratio near 400 km arose out of attempts to match an extrapolated phase diagram for the system Mg_2SiO_4-Fe_2SiO_4[7] with the velocity distribution and probable temperature near 400 km.[8,9] However, a direct determination of equilibria in the Mg-rich portion of this system[10] showed that the postulated diagram was in error. When the new phase diagram for this system is used, a satisfactory explanation of the 400-km discontinuity emerges without any need to alter the Fe/Fe + Mg ratio in order to achieve consistency.[11]

Finally, we consider the magnitude of density increases which are associated with discontinuities or abnormally high velocity gradients in the deep mantle. Between 800 and 1700 km, these abnormal regions account for a velocity change

[1] Particularly those involving the velocity-density coefficient applicable to phase changes (Sec. 9-4, Table 9-4).
[2] A minor change in Table 14-1 and in the calculated velocity increase from the results given by Ringwood (1970) results from the present assumption that most Fe^{++} occurs in the perovskite phase in mineral assemblage 3.
[3] Anderson (1967a, 1968).
[4] Press (1968).
[5] Anderson (1967b).
[6] Anderson and Jordan (1970).
[7] Akimoto and Fujisawa (1968).
[8] Anderson (1967a).
[9] Fujisawa (1968).
[10] Ringwood and Major (1970).
[11] See also Graham (1970) and Ahrens (1972a).

of 0.28 km/sec (Fig. 14-3). From the Birch velocity-density relation, the corresponding density increase is 0.09 g/cm^3. Making a correction for differential self-compression of the two states, taking the 0.09 g/cm^3 differential to apply at an average depth of 1200 km, and taking the bulk modulus of the dense assemblage as 2800 kbars (Table 14-1), this leads to an initial density ρ_0 which is 0.16 g/cm^3 higher than that of isochemical mixed oxides. If these velocity increases are attributed to phase transformations, as seems reasonable, this implies that the stable mineral assemblage in the deep regions of the mantle is about 4% denser than the isochemical mixed oxides.

14-5 DENSITY OF THE LOWER MANTLE

The assumption of a chemical and mineralogical model for the upper mantle determines the mean density and moment of inertia of this region, and these in turn constrain the density distribution of the lower mantle. The density distribution within the earth on the assumption that the upper mantle is of pyrolite composition was thereby obtained by Clark and Ringwood (1964).

If the lower mantle is assumed to be approximately homogeneous,[1] following Birch's (1952) conclusion, then the density distribution over the pressure range 400 to 1200 kbars can be extrapolated back to zero pressure in order to obtain the equivalent zero-pressure density and elastic ratio of the lower mantle mineral assemblage. This procedure was followed by Clark and Ringwood, who fitted the Birch-Murnaghan equation (below) based on the third-order theory of finite strain to their density distribution for the lower mantle:

$$P = 3K_0 f(1 + 2f)^{5/2}(1 - 2\xi f)$$

where f is strain, and ξ is a dimensionless parameter proportional to the coefficient of the third-order terms. The density ρ_0 which the mineral assemblage in the lower mantle would possess at zero pressure and low temperature was found to be 4.25 g/cm^3. The corresponding values for K_0 and ϕ_0 were 2700 kbars and 63 km/sec^2, respectively. The values of ρ_0 were found to be insensitive to the presence of superadiabatic gradients in the lower mantle on the order of 1 to 2°C/km.[2]

The calculated zero pressure-room temperature density of the lower mantle was found to be about 5% greater than that of a mixture of oxides (SiO_2 as stishovite) with the pyrolite composition. Suggested explanations for this discrep-

[1] I.e., as a first approximation, the small velocity abnormalities in the lower mantle inferred by Johnson (1969) and others are ignored.

[2] Analogous calculations were also carried out for a range of earth models by Birch (1964), Wang (1970), and Anderson and Jordan (1970). The zero pressure-low temperature densities for the lower mantle obtained by these authors varied between 4.1 and 4.3 g/cm^3.

ancy were as follows:

1 Bearing in mind the uncertainties of the various procedures employed, it might not be significant.
2 The 5% discrepancy might be caused by the occurrence of phase transformations in the deep mantle, leading to the formation of phases which were intrinsically denser than the isochemical mixed oxides.
3 The Fe/Fe + Mg ratio may increase from about 0.1 in the upper mantle to 0.2 in the lower mantle.

Studies of seismic velocity distribution in the lower mantle (Fig. 14-3) suggest the occurrence of further phase transformations and thus contradict the assumption of homogeneity in this region, which is frequently adopted. Because of the relatively small magnitude of these velocity changes, the contradiction is not severe, and the homogeneity assumption remains useful as a first approximation to the properties of this region. Nevertheless, the presence of these small inhomogeneous regions will have a significant effect upon estimates of zero-pressure parameters for the lower mantle to the extent that these are obtained by an extrapolation procedure. The inhomogeneities will cause the values of ρ_0 and ϕ_0 to be systematically underestimated. This strengthens the conclusion that the zero-pressure density of the lower mantle may be significantly higher than mixed oxides of pyrolite composition.

The invaluable shock-wave data on the Hugoniot equations of state of a number of rocks and minerals of geophysical interest obtained by McQueen and coworkers[1,2] have provided the raw material for many interpretative papers dealing with the density and elastic properties of the lower mantle.[2,3,4,5,6,7,8,9,10,11,12,13,14,15] The Hugoniots of forsterite, Twin Sisters dunite, hortonolite dunite, and fayalite indicated complete transformation above 500 to 700 kbars into high-pressure phases.

The densities which these high-pressure phases would have if it were possible to retain them metastably at atmospheric pressure and temperature have been

[1]McQueen and Marsh (1966).
[2]McQueen, Marsh, and Fritz (1967).
[3]Birch (1961b, 1964).
[4]McQueen, Fritz, and Marsh (1964).
[5]Takeuchi and Kanamori (1966).
[6]Wang (1967, 1968, 1970).
[7]Anderson and Kanamori (1968).
[8]Knopoff and Shapiro (1969).
[9]Shapiro and Knopoff (1969).
[10]Ringwood (1969, 1970).
[11]Ahrens, Anderson, and Ringwood (1969).
[12]Davies and Anderson (1971).
[13]Al'tschuler (1972).
[14]Davies and Gaffney (1973).
[15]Gaffney and Anderson (1973).

estimated[1,2,3,4,5,6,7] by several workers, as discussed in Secs. 10-4 (Fig. 10-9) and 11-8. The zero-pressure density of the Twin Sisters dunite was found to be very similar to those of the isochemical mixed oxides MgO, FeO, and SiO_2 (as stishovite). Likewise, at pressures between 600 and 800 kbars, the hortonolite dunite and fayalite attained densities similar to the isochemical mixed oxides. However, at higher pressures, both these minerals apparently transformed to phases which were significantly denser than the isochemical mixed oxides.[7] Pure forsterite also transformed to a denser phase than the mixed oxides (Fig. 11-13).

The mean atomic weight ($\bar{M}$) for the Twin Sisters dunite is 20.9, compared to 21.3 for pyrolite.[8] The shock-wave P-ρ data for the hortonolite dunite ($\bar{M} = 24.3$) and fayalite ($\bar{M} = 29.1$) can be used to interpolate the P-ρ relations for dunites of intermediate composition.[9] In Fig. 14-5, the Hugoniot P-ρ relationship for a dunite of mean atomic weight equal to pyrolite is shown. The correction to the Twin Sisters dunite data is small and does not introduce a significant error. Also shown in Fig. 14-5 is an envelope enclosing recently determined density distributions[10] for the lower mantle, most of which, in addition to satisfying the standard constraints, also incorporate refinements imposed by free-oscillation studies.

It is seen that the P-ρ curve for the dunite ($M = 21.3$) falls within the lower part of this envelope. In view of the earlier interpretation of the state reached by the Twin Sisters dunite under shock compression, this would suggest that the P-ρ distribution in the lower mantle is consistent with the P-ρ curve for pyrolite in a mineral assemblage equivalent to mixed oxides. However, there is a further complication. The temperatures attained by the dunite under shock compression (Fig. 14-5) are probably 500 to 1000°C lower than the actual temperatures in the lower mantle. A correction for this factor would displace the dunite curve downward by about 0.1 g/cm³, significantly outside the mantle envelope. This might be considered to constitute evidence, albeit somewhat marginal, that the density of the lower mantle is indeed slightly higher than mixed oxides, isochemical with pyrolite, under the same P-T conditions.

A rather clearer picture emerges if other model systems are used to represent the lower mantle. In Fig. 14-5 the density of an MgO, FeO, SiO_2 (stishovite) mixture with the components in the same ratios as in pyrolite and with a mean atomic weight of 21.3 is compared with the density envelope for the lower mantle. The density of the model mixture, which was obtained by interpolation[9] from

[1]Takeuchi and Kanamori (1966).
[2]Wang (1967, 1968, 1970).
[3]Anderson and Kanamori (1968).
[4]Ahrens, Anderson, and Ringwood (1969).
[5]Davies and Anderson (1971).
[6]Al'tschuler (1972).
[7]Davies and Gaffney (1973).
[8]Uncertainties in the pyrolite composition permit a range of $\bar{M}$ values between 21.2 and 21.4.
[9]Ringwood (1969, 1970).
[10]The density distributions falling within this envelope are those of Wang (1970, 1972), Haddon and Bullen (1969), Birch (1964), and Jordan (1973).

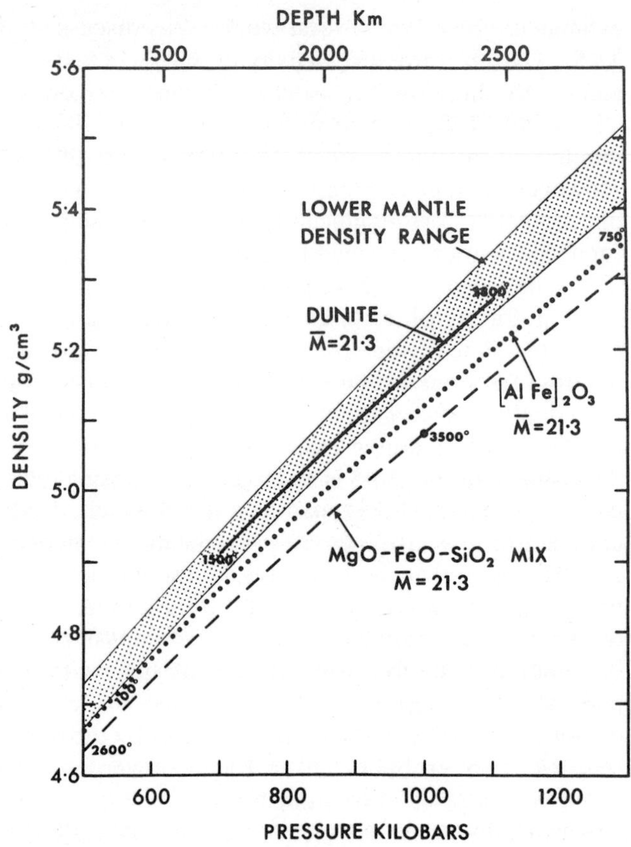

FIGURE 14-5
Comparison of density distribution in lower mantle (see text) with densities of some model mantle materials derived from shock-wave and ultrasonic data. Density-pressure relationships for dunite and $(Al, Fe)_2O_3$ are along the respective Hugoniots (McQueen et al., 1967), and the estimated temperatures (Ahrens et al., 1969) are shown. In the case of the MgO-FeO-SiO_2 mix, the Hugoniot densities have been corrected to allow for the probable higher temperatures in the lower mantle. The indicated temperatures are notional.

shock-wave data on MgO, SiO_2, and Fe_2SiO_4, lies substantially below that of the lower mantle. This conclusion has been confirmed by further detailed studies of the equations of state of SiO_2, FeO, and MgO using more recent data on elastic and thermodynamic properties.[1,2] A third model[3] shown in Fig. 14-5 is based upon Al_2O_3-Fe_2O_3 corundum-type solid solutions with mean atomic weight 21.3.

[1]Graham (1973).
[2]Davies (1973).
[3]Ringwood (1970).

In this case, also, the model densities fall below those in the lower mantle. Allowance for the large temperature difference between the model and the lower mantle would increase the discrepancy substantially.

Considering all the relevant evidence, it appears probable that the density of the lower mantle is indeed slightly higher than that of mixed oxides isochemical with pyrolite under corresponding *P-T* conditions. It should be emphasized that the difference is quite small, probably less than 5% and closer to 3%.

14-6 ELASTICITY OF THE LOWER MANTLE

If we accept the previous evidence relating to the density of the lower mantle, we must next enquire whether the increased density over the "mixed-oxide" state is caused by further phase transformations leading to denser structures, by an increase in iron content, or perhaps, by a combination of these factors. Although the density differential under discussion is small, it is nevertheless of considerable genetic and dynamical significance. The occurrence of chemical zoning in the deep mantle would have an important bearing upon the nature of convective circulations in this region and would provide evidence relating to chemical and physical processes operative during the formation of the earth.

The seismic evidence for small velocity discontinuities occurring below 800 km (Fig. 14-3) suggests the occurrence of further transformations. However, this evidence cannot be considered decisive until the differences between existing velocity distributions in the lower mantle are resolved.

In principle, the property which is likely to be of most use in distinguishing between the two alternatives is the elastic ratio ϕ. A proportional change in density is related to the corresponding proportional change in elastic ratio by

$$\frac{\Delta\phi}{\phi} = n\,\frac{\Delta\rho}{\rho}$$

For an increase in density caused by a phase change or by normal thermal contraction, n is between $+2$ and $+3$.[1,2,3] However, for an increase in density resulting in an increase of iron content, n is negative, with a value of approximately -1.

The fact that both phase transformations to denser states and thermal contraction change ϕ in the opposite direction to the change caused by an increase in iron content provides an important constraint upon proposed solutions. It should be possible to compare ϕ values for the mantle obtained directly from the observed seismic velocities ($\phi = V_p^2 - \frac{4}{3}V_s^2$), with ϕ values derived from shock-wave experiments on rocks under pressures similar to those in the lower mantle.

The problem is that ϕ equal to $(\partial P/\partial\rho)_s$ must be obtained by differentiat-

[1]Clark and Ringwood (1964).
[2]Anderson (1967b).
[3]Wang (1970).

ing the primary shock P-ρ data after correction to adiabatic conditions (Fig. 10-9). This involves a number of assumptions about the nature of the equation of state and the transformation energy associated with phase changes. These assumptions lead to corrections which can be carried out in different ways. As discussed in Sec. 10-4, there are also some additional uncertainties arising from the interpretation of the primary Hugoniot data. Accordingly, the uncertainties in obtaining ϕ from shock data are much larger than in obtaining ρ, and results of different investigators[1,2,3] in deriving ϕ from the same set of shock data differ substantially. Nevertheless, the results are indicative (Fig. 14-6).

The ϕ values obtained by McQueen et al.[1] for the high-pressure phase of the Twin Sisters dunite (adjusted to $\bar{M} = 21.3$) fall substantially below the mantle ϕ curve, as do those calculated by Ahrens et al.[3] at pressures greater than 800 kbars. If the 3% discrepancy in density between a pyrolite lower mantle with a mixed oxide density and the actual density in the lower mantle (Fig. 14-5) is to be resolved by an increase in iron content, the mean atomic weight $\bar{M}$ of the lower mantle must be increased from 21.3 to 21.8. This would cause a further reduction of the ϕ values in Fig. 14-6 by about 3 $(\text{km/sec})^2$, and the discrepancies between shock and mantle ϕ values become much more marked. Thus the treatments of shock-wave data by McQueen et al. and Ahrens et al. do not favour an increase of iron content with depth. They would be consistent with the occurrence of further minor phase transformations.

On the other hand, Wang's ϕ-P curve for dunite $\bar{M} = 21.3$ falls slightly above the mantle ϕ curve. If we increase the iron content of the dunite sufficiently to bring its density-pressure relationship (Fig. 14-5) into agreement with the lower mantle, the calculated ϕ values are decreased and agree with the mantle ϕ values. The hypothesis of a slightly higher iron content ($\bar{M} = 21.8$) in the lower mantle is thus apparently supported. However, differences in temperature distribution between the mantle and shock-wave ϕ-P curves have not been taken into account. The temperatures along Wang's adiabat are right at the bottom of the wide range of temperature distributions which have been proposed by many authors for the lower mantle. The actual temperature distribution in the lower mantle may well be 1000°C or more above Wang's adiabat. The appropriate correction would then bring Wang's calculated ϕ-P curve substantially below the mantle ϕ-P curve.[4] In this case, the density discrepancy of Fig. 14-6 would be resolved by further phase transformations to denser states, rather than by an increase in iron content. On the other hand, if the lower mantle is indeed relatively cool ($\sim$ 2700°C at the core-mantle boundary), an increase in iron content would be favoured by Wang's ϕ-P calculation but not by those of McQueen and Ahrens et al. Reviewing the shock-wave evidence discussed above, Ringwood (1970) concluded that, on balance, the case for a constant $\bar{M}$ value throughout the

[1]McQueen, Marsh, and Fritz (1967).
[2]Wang (1968).
[3]Ahrens, Anderson, and Ringwood (1969).
[4]Ringwood (1970).

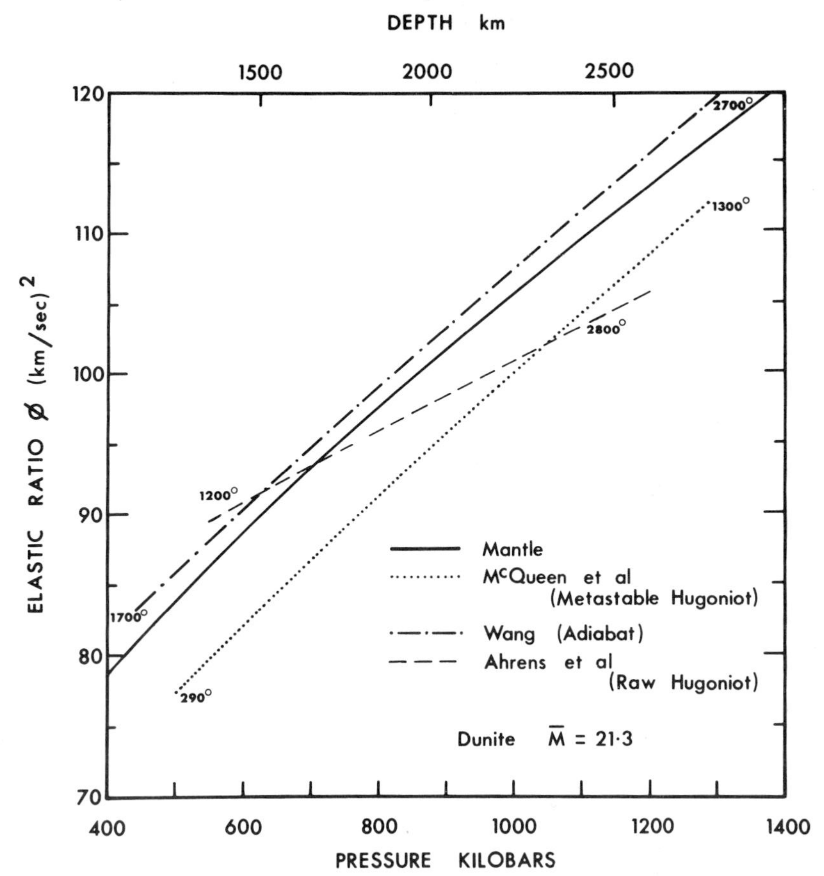

FIGURE 14-6
Comparison of mantle ϕ-P relationship (obtained from Wang 1970, 1972 models) with ϕ-P relationship for Twin Sisters dunite (corrected to $\bar{M} = 21.3$) derived from shock-wave data by McQueen et al. (1967), Wang (1968), and Ahrens et al. (1969). The temperatures calculated by these authors are indicated.

mantle, accompanied by transformations to states slightly denser than mixed oxides, was favoured:

> However, in view of possible uncertainties in the seismic and shock wave data, this conclusion cannot be regarded as firmly established and the possibility of a small increase in the iron content of the lower mantle may warrant reconsideration as further shock and seismic data become available.

Velocity-Density Relationships

The application of empirical velocity-density relationships in oxides and silicates (Sec. 9-4) may provide further evidence on the iron content of the lower mantle.

Birch (1961b) first applied this method using the relation $V_p = a(\bar{M}) + 3.31\rho$ (where a is a constant related to mean atomic weight) in conjunction with the P-wave velocities of Jeffreys and Gutenberg and constrained by the observed mass and moment of inertia of the earth. Birch concluded that, if the upper mantle is of peridotite composition with an initial density of 3.32 g/cm^3, then an increase of iron content corresponding to 1.5 units of $\bar{M}$ would be required through the transition zone. However, he pointed out that the estimated change of iron content would be reduced or even eliminated if the effects of temperature gradients had been considered. In a subsequent investigation, Birch (1964) applied the velocity-density relationship to the upper mantle and transition zone and used the Adams-Williamson equation to obtain the density distribution in the lower mantle. He considered two cases, with initial densities below the Moho of 3.22 and 3.43 g/cm^3. His estimates of the zero-pressure density of the lower mantle based upon extrapolation of inferred lower mantle densities with the second-order Birch-Murnaghan equation were 3.90 to 3.96 g/cm^3. Birch concluded that the lower mantle contained about 10% FeO, implying a mean atomic weight similar to that of the pyrolite.

An alternative procedure, employed by Wang (1970), is to apply the empirical relationship[1,2] between hydrodynamical sound velocity $V_c = \phi^{1/2}$ and density $V_c = a + b\rho$ to obtain the density distribution in various regions of the mantle. The V_c-ρ relationship has an advantage over the V_p-ρ relationship since it applies equally to density changes caused both by temperature and pressure.[2,3] However, this advantage is partially offset by the situation that the V_c distribution in the mantle is substantially more uncertain than either V_p or V_s since it is obtained by subtraction of these quantities: $V_c^2 = V_p^2 - \frac{4}{3}V_s^2$. Thus, uncertainties in both V_p and V_s tend to be compounded in V_c. This problem is most acute in the transition zone and upper mantle.

Wang (1970) applied the V_c-ρ relationship to the outer 1000 km of the mantle and the Adams-Williamson equation to the lower mantle to obtain the density distribution throughout the mantle constrained by the earth's mass, moment of inertia, and free-oscillation periods. His results indicate that the entire mantle may have a uniform iron content, with mean atomic weight of 21.3 to 21.5, and that the density at the top of the mantle is about 3.30 g/cm^3. These results are thus consistent with a mantle of pyrolite composition throughout.

In a second paper, Wang (1972) calculated the density and P- and S-velocity distributions for a pyrolite upper mantle using the petrological model of Green and Ringwood (1967) and the measured elastic properties and densities of upper mantle minerals. The density distribution throughout the transition zone was then estimated using Birch's empirical V_p–ρ relationship and, in the lower mantle, using Wang's empirical V_c–ρ relationship, both for $\bar{M} = 21$. The Adams-William-

[1]McQueen, Fritz, and Marsh (1964).
[2]Wang (1968, 1970).
[3]Section 9-4.

son equation was used to obtain the density distribution in the core. This preliminary density model for the earth was then successively perturbed to obtain agreement with free-oscillation data, subject to the additional standard constraints imposed by mass, moment of inertia, and seismic body-wave travel times.[1] An interesting feature was that only minor perturbations to the original simple earth model were necessary in order to obtain consistency with the other boundary conditions. This investigation again showed that a wide range of geophysical data relating to density distribution within the earth were consistent with a mantle characterized by a constant value of $\bar{M}$ close to 21.3.

In contrast to the above conclusion, Anderson[2] and others have argued in favour of a large increase in iron content in the lower mantle, based upon application of an assumed phase diagram for the system Mg_2SiO_4-Fe_2SiO_4 in conjunction with shock-wave data and use of the "seismic equation of state." Anderson concluded that the lower mantle possessed a mean atomic weight $\bar{M}$ between 22.3 and 23.4, corresponding to an FeO/FeO + MgO (mol) ratio of about 0.3 to 0.4 (compare pyrolite: $\bar{M} = 21.3$, FeO/FeO + MgO = 0.11). It has already been pointed out,[3] however, that the phase diagram used by Anderson was incorrect, whilst the applicability of the seismic equation of state to the deep mantle is of dubious validity.[4]

More recently, Anderson and Jordan (1970) produced a revised seismic equation of state which was based upon elastic-property and shock-wave data for close-packed minerals and hence considered more applicable to the lower mantle. Using the revised seismic equation of state, they concluded that the lower mantle was characterized by an $\bar{M}$ value in the range 21.0 to 22.0, i.e., similar to pyrolite. Curiously, they were reluctant to accept the most simple and obvious interpretation of this result—i.e., that the upper mantle and lower mantle were chemically similar. Instead, a new hypothesis was proposed.[5,6,7] The phases in the lower mantle were assumed equivalent (in elastic properties) to a mixture of MgO + FeO + SiO_2 (stishovite). However, the ratios of all three components were arbitrarily varied from the pyrolite ratios. By increasing the percentages of FeO and SiO_2 simultaneously, whilst decreasing MgO, a satisfactory match to lower mantle properties was claimed.[8] The effect of a modest increase in FeO (smaller than was proposed in earlier models) is to provide a higher density than that of pyrolite in the "mixed oxide" state, as indeed is required (Fig. 14-5). However, an increase in ϕ is also needed (Fig. 14-6). This is provided by a large

[1] *P*-wave data of Johnson (1967, 1969) and *S*-wave data of Hales and Roberts (1970) were employed.
[2] Anderson (1967a, 1968).
[3] Section 14-4.
[4] The seismic equation of state implies $K \propto \rho^4$, which is reasonable for low pressures and temperatures. However, under the very high pressures and temperatures of the deep mantle, a more appropriate relation is $K \propto \rho^3$—ref. Sec. 9-4. See also Simmons and England (1969).
[5] Anderson and Jordan (1970).
[6] Anderson (1970).
[7] Anderson, Sammis, and Jordan (1971, 1972).
[8] The procedures followed were criticized by Wang and Simmons (1972).

increase in SiO_2 possessing the very high ϕ value of 78 $(km/sec)^2$, which more than offsets the opposing effect on ϕ of increased FeO. The resultant composition favoured for the lower mantle[1,2] possessed a pyroxene stoichiometry and was close to $(Mg_{0.64}Fe_{0.36})\ SiO_3$.

The model advocated by these authors has been investigated by others[3,4] using more accurate data on the elastic properties of MgO, FeO, and SiO_2. With some important qualifications, they concluded that such a model, in which the proportions of mixed oxides MgO, FeO, and SiO_2 can be varied arbitrarily, may be capable of satisfying the observed data on the elasticity and density of the lower mantle. However, a serious uncertainty is introduced by the assumption that the elastic ratio ϕ of the oxide mixture can be obtained from the molar average of the ϕ values for the individual oxides. This is denied by Thomsen (1972).

Thus we have two alternative models to consider: (1) the lower mantle is similar in composition to the upper mantle (pyrolite) and displays a mineral assemblage slightly denser than isochemical mixed oxides; (2) the lower mantle possesses the physical properties of an oxide mixture in which both FeO and SiO_2 are substantially more abundant relative to MgO than in the upper mantle. When realistic account is taken of the several sources of uncertainty, it must be concluded that evidence based upon shock-wave data, empirical velocity-density relationships, and equations of state, combined with present knowledge of the ϕ, ρ, and T distributions in the lower mantle, is not sufficiently decisive to permit a choice between these alternative hypotheses. More accurate and more abundant elastic-property data and shock-wave data on relevant minerals, oxides, and germanate analogue compounds will be necessary before this question can be settled solely from considerations of elasticity.

There are, however, other kinds of arguments which bear upon these hypotheses. The first hypothesis rests upon the proposition that phase transformations to states slightly denser than mixed oxides are possible, and indeed likely, under the P-T conditions of the lower mantle. This proposition is supported by the evidence of such transformations in Mg_2SiO_4, $MgFeSiO_4$, Fe_2SiO_4, $(Mg_{0.86}Fe_{0.14})SiO_3$, $MgAl_2O_4$, and Fe_3O_4 under shock-wave conditions, as discussed in Chaps. 11 and 12, by the known occurrence of many compounds crystallizing in structures which are intrinsically denser than the mixed oxides (Secs. 11.9, 12.6, and 13.3) and by the indications of several small discontinuities in the lower mantle believed to be caused by phase transformations.[5,6] It is supported also by the discussions of Chaps. 11 and 12, which show that crystal structures denser than the mixed oxides, which might reasonably be adopted by $(Mg, Fe)SiO_3$ and $(Mg, Fe)_2SiO_4$ under very high pressures, indeed exist and are plausible on crystal chemical grounds.

[1]Anderson (1970).
[2]Anderson, Sammis, and Jordan (1971, 1972).
[3]Graham (1973).
[4]Davies (1973).
[5]Figure 14-3.
[6]Johnson (1969).

The second hypothesis makes the implicit assumption that phase transformations to states significantly denser than isostructural mixed oxides do *not* occur in the lower mantle. This assumption is challenged by the considerable body of evidence, mentioned above, which supports or suggests the occurrence of these transformations. Moreover, shock-wave data[1] on the compression of bronzitite ($En_{86}Fs_{14}$) show that it attains a density of 5.15 to 5.3 g/cm³ at pressures of 950 to 1100 kbars, which are centrally placed within the mantle density-pressure band shown in Fig. 14-5. However, the pyroxene-like composition inferred for the lower mantle[2,3] is close to $(Mg_{0.64}Fe_{0.36})SiO_3$. Assuming that the more iron-rich pyroxene transformed to the same phase as the bronzitite, its density at these pressures would be about 0.3 g/cm³ higher and well above the mantle density range.

In the author's opinion, the combined weight of the latter evidence favours the hypothesis that the mineral assemblage in the lower mantle is a few percent denser than the "mixed oxide" state.[4] This hypothesis, in turn, permits the upper mantle and lower mantle to possess similar mean chemical compositions. Certainly, presently available evidence does not justify the contrary conclusion to be drawn.

Some recent theories of the origin of the earth[5] imply that the lower mantle is devoid of FeO. Although there might be some question as to whether the lower mantle contains *more* FeO than pyrolite, the proposition that the lower mantle contains *no* FeO can be firmly rejected. Referring to Fig. 14-5, an oxide mixture of MgO, SiO_2, Al_2O_3, CaO, and Na_2O in primordial proportions would yield a density curve lying about 5% below the lower mantle density curve; yet the ϕ distribution for this mixture would be similar to, or even somewhat higher than, the ϕ distribution in the lower mantle. If the oxide mixture were permitted to transform to a more closely packed mineral assemblage with a sufficiently high density to match the lower mantle density distribution, the corresponding ϕ values for this assemblage would lie at least 10% higher than the observed ϕ distribution in the lower mantle.

14-7 CONCLUSION

As a result of advances in our understanding of the role of phase transformations in the deep mantle and of parallel advances in seismology, a satisfying and widely self-consistent interpretation of the physical constitution of the mantle in terms of its mineralogical constitution has emerged.

[1]McQueen, Marsh, and Fritz (1967).
[2]Anderson (1970).
[3]Anderson, Sammis, and Jordan (1971).
[4]*Note added in press.* This interpretation has been confirmed in large measure with the synthesis by Liu (1974) of a perovskite modification of $MgSiO_3$, some 4% denser than the isochemical mixed oxides.
[5]E.g., the "heterogeneous accumulation theory" of Clark, Turekian, and Grossman (1972).

Directly discovered and indirectly inferred phase transformations in a mantle of pyrolite composition quantitatively explain the occurrences, depths, and velocity changes of the major seismic discontinuities in the transition zone. The density of the mineral assemblage present in the lower mantle is probably 3 to 5% higher than the density of isochemical mixed oxides of pyrolite composition. In principle, this density increment could be caused by the occurrence of additional small transformations below 800 km or by an increase of iron content with depth or by some combination of these effects.

Available evidence from several directions, whilst not finally conclusive, nevertheless favours the first alternative. There are no valid grounds at present for maintaining that the mean atomic weight of the lower mantle is higher than that of the upper mantle or that the FeO/MgO ratio of the lower mantle is higher than that of the upper mantle. In the absence of significant evidence to the contrary, and in view of the close similarity of the pyrolite abundances of involatile oxyphile elements to the primordial relative abundances of these elements, we are entitled to conclude that the *mean* composition of the deep mantle is probably similar to that of the upper mantle.[1] This conclusion is supported by evidence discussed in Sec. 8-7 and constitutes a fundamental boundary condition for theories of the earth's origin and evolution.

REFERENCES

ADAMS, R. D. (1971). Reflections from discontinuities beneath Antarctica. *Bull. Seism. Soc. Am.* **61,** 1441–1451.

AHRENS, T. J. (1972a). The state of mantle minerals. *Tectonophysics* **13,** 189–219.

——— (1972b). The mineralogic distribution of iron in the upper mantle. *Phys. Earth Planet. Interiors* **5,** 267–281.

———, D. L. ANDERSON, and A. E. RINGWOOD (1969). Equations of state and crystal structures of high pressure phases of shocked silicates and oxides. *Rev. Geophys.* **7,** 667–707.

AKIMOTO, S., and H. FUJISAWA (1968). Olivine-spinel solid solution equilibria in the system Mg_2SiO_4-Fe_2SiO_4. *J. Geophys. Res.* **73,** 1467–1479.

ALLER, L. H. (1968). The chemical composition of the sun and solar system. *Proc. Astron. Soc. Australia* **1,** 133–135.

AL'TSCHULER, L. V. (1972). Composition and state of matter in the deep interior of the Earth. *Phys. Earth Planet. Interiors* **5,** 295–300.

ANDERS, E. (1964). Origin, age and composition of meteorites. *Space Sci. Rev.* **3,** 583–714.

——— (1971). How well do we know "Cosmic" abundances? *Geochim. Cosmochim. Acta* **35,** 516–522.

ANDERSON, D. L. (1967a). Phase changes in the upper mantle. *Science* **157,** 1165–1173.

——— (1967b). A seismic equation of state. *Geophys. J.* **13,** 9–30.

[1] I.e., second-order effects due to irreversible differentiation of pyrolite into residual eclogite plus peridotite are averaged. This topic is further discussed in Secs. 8-5, 8-7, and 15-6.

——— (1968). Chemical inhomogeneity of the mantle. *Earth Planet. Sci. Letters* **5,** 89–94.
——— (1970). Petrology of the mantle. *Mineral Soc. Am. Spec. Paper* **3,** 85–93.
——— and T. JORDAN (1970). The composition of the lower mantle. *Phys. Earth Planet. Interiors* **3,** 23–35.
——— and H. KANAMORI (1968). Shock wave equations of state for rocks and minerals. *J. Geophys. Res.* **73,** 6477–6502.
———, C. SAMMIS, and T. JORDAN (1971). Composition and evolution of the mantle and core. *Science* **171,** 1103–1113.
———, ———, and ——— (1972). Composition of the mantle and core. In: E. C. Robertson (ed.), "*The Nature of the Solid Earth,*" pp. 41–66. McGraw-Hill, New York.
——— and N. TOKSÖZ (1963). Surface waves on a spherical earth I, Upper mantle structure from Love waves. *J. Geophys. Res.* **68,** 3483–3500.
ANDERSON, O. L., and J. E. NAFE (1965). The bulk modulus-volume relationship for oxide compounds and related geophysical problems. *J. Geophys. Res.* **70,** 3951–3962.
ARCHAMBEAU, C. B., E. A. FLINN, and D. G. LAMBERT (1969). Fine structure of the upper mantle. *J. Geophys. Res.* **74,** 5825–5865.
BIRCH, F. (1952). Elasticity and constitution of the Earth's interior. *J. Geophys. Res.* **57,** 227–286.
——— (1961a). The velocity of compressional waves in rocks to 10 kilobars, 2. *J. Geophys. Res.* **66,** 2199–2224.
——— (1961b). Composition of the earth's mantle. *Geophys. J.* **4,** 295–311.
——— (1964). Density and composition of mantle and core. *J. Geophys. Res.* **69,** 4377–4388.
BOLT, B. A., M. O'NEILL, and A. QAMAR (1968). Seismic waves near 110°: is structure in core or upper mantle responsible? *Geophys. J.* **16,** 475–487.
CAMERON, A. G. W. (1968). A new table of abundances of the elements in the solar system. In: L. H. Ahrens (ed.), "*Origin and Distribution of the Elements,*" pp. 125–143. Pergamon, London.
CHINNERY, M. A., and M. N. TOKSÖZ (1967). P-wave velocities in the mantle below 700 km. *Bull. Seism. Soc. Am.* **57,** 199–226.
CLARK, S. P., and A. E. RINGWOOD (1964). Density distribution and constitution of the mantle. *Rev. Geophys.* **2,** 35–88.
———, K. TUREKIAN, and L. GROSSMAN (1972). Model for the early history of the Earth. In: E. C. Robertson (ed.), *"Nature of the Solid Earth,"* pp. 3–18. McGraw-Hill, New York.
DAVIES, G. F. (1973). Elasticity of solids at high pressures and temperatures: theory, measurement and geophysical interpretation. Ph.D. Thesis, California Inst. Technology.
——— and D. L. ANDERSON (1971). Revised shock wave equations of state for high pressure phases of rocks and minerals. *J. Geophys. Res.* **76,** 2617–2627.
——— and E. S. GAFFNEY (1973). Identification of phases of rocks and minerals from Hugoniot data. *Geophys. J.* **33,** 165–183.
ENGDAHL, E. R., and E. A. FLINN (1969). Seismic waves reflected from discontinuities within Earth's upper mantle. *Science* **163,** 177–179.
FUJISAWA, H. (1968). Temperature and discontinuities in the transition layer within the earth's mantle: Geophysical application of the olivine-spinel transition in the Mg_2SiO_4-Fe_2SiO_4 system. *J. Geophys. Res.* **73,** 3281–3294.

FYFE, W. S. (1960). The possibility of d-electron coupling in olivine at high pressures. *Geochim. Cosmochim. Acta* **19,** 141–143.

GAFFNEY, E. S., and D. L. ANDERSON (1973). The effect of low-spin Fe^{2+} on the composition of the lower mantle. *J. Geophys. Res.* **78,** 7005–7014.

GARZ, T., M. KOCK, J. RICHTER, B. BASCHEK, H. HOLWEGER, and A. UNSOLD (1969). Abundances of iron and some other elements in the sun and in meteorites. *Nature* **223,** 1254–1255.

GRAHAM, E. K. (1970). Elasticity and composition of the upper mantle. *Geophys. J.* **20,** 285–302.

——— (1973). On the compression of stishovite. *Geophys. J.* **32,** 15–34.

GREEN, D. H., and A. E. RINGWOOD (1967). The stability fields of aluminous pyroxene peridotite and garnet peridotite and their relevance in upper mantle structure. *Earth Planet. Sci. Letters* **3,** 151–160.

HADDON, R. A. W., and K. E. BULLEN (1969). An earth model incorporating free earth oscillation data. *Phys. Earth Planet. Interiors* **2,** 35–49.

HALES, A. L., and J. L. ROBERTS (1970). Shear velocities in the lower mantle and the radius of the core. *Bull. Seism. Soc. Am.* **60,** 1427–1436.

HELMBERGER, D., and R. A. WIGGINS (1971). Upper mantle structure of mid-western states. *J. Geophys. Res.* **76,** 3229–3245.

JOHNSON, L. (1967). Array measurements of P-velocities in the upper mantle. *J. Geophys. Res.* **72,** 6309–6325.

——— (1969). Array measurements of P-velocities in the lower mantle. *Bull Seism. Soc. Am.* **59,** 973–1011.

JORDAN, T. H. (1973). Estimation of the radial variation of seismic velocities and density in the earth. Ph.D. Thesis, California Institute of Technology.

KANAMORI, H. (1967). Upper mantle structure from apparent velocities of P-waves recorded at Wakayama microearthquake observatory. *Bull. Earthquake Res. Inst. Tokyo Univ.* **45,** 657–678.

KNOPOFF, L., and J. N. SHAPIRO (1969). Comments on the interrelationships between Grüneisen's parameter and shock and isothermal equations of state. *J. Geophys. Res.* **74,** 1439–1450.

KOVACH, R. L., and D. L. ANDERSON (1964). Higher mode surface waves and their bearing on the structure of the earth's mantle. *Bull. Seism. Soc. Am.* **54,** 161–182.

LIU, L. (1974). Silicate perovskite from phase transformation of pyrope garnet at high pressure and temperature. *Geophys. Res. Letters* **1,** 277–280.

MASON, B. (1960). The origin of meteorites. *J. Geophys. Res.* **65,** 2965–2970.

MASSE, R. P., M. LANDISMAN, and J. B. JENKINS (1972). An investigation of upper mantle compressional velocity distribution beneath the Basin and Range province. *Geophys. J. Roy. Aston. Soc.* **30,** 19–36.

MCQUEEN, R. G., J. N. FRITZ, and S. P. MARSH (1964). On the composition of the earth's interior. *J. Geophys. Res.* **69,** 2947–2978.

——— and S. P. MARSH (1966). In: S. P. Clark (ed.), "*Handbook of Physical Constants,*" pp. 153–159. *Geol. Soc. Am. Memoir* 97.

———, ———, and J. N. FRITZ (1967). Hugoniot equation of state of twelve rocks. *J. Geophys. Res.* **72,** 4999–5036.

MEIJERING, J., and C. J. M. ROOYMANS (1958). On the olivine-spinel transition in the earth's mantle. *Koninkl. Ned. Akad. Wetenschap. Proc. Ser.* **B61,** 333–344.

MÜLLER, E. A. (1968). The solar abundances. In: L. H. Ahrens (ed.), "*Origin and Distribution of the Elements,*" pp. 155–176. Pergamon, London.

PRESS, F. (1968). Earth models obtained by Monte Carlo inversion. *J. Geophys. Res.* **73,** 5223–5234.

——— (1970). Earth models consistent with geophysical data. *Phys. Earth Planet. Interiors* **3,** 3–22.

RINGWOOD, A. E. (1958). Constitution of the mantle, III: Consequences of the olivine-spinel transition. *Geochim. Cosmochim. Acta* **15,** 195–212.

——— (1961). Chemical and genetic relationships among meteorites. *Geochim. Cosmochim. Acta* **24,** 159–197.

——— (1962a). Present status of the chondritic earth model. In: C. B. Moore (ed.), "*Researches on Meteorites,*" pp. 198–216. Wiley, New York.

——— (1962b). Mineralogical constitution of the deep mantle. *J. Geophys. Res.* **67,** 4005–4010.

——— (1966a). Genesis of chondritic meteorites. *Rev. Geophys.* **4,** 113–175.

——— (1966b). The chemical composition and origin of the earth. In: P. M. Hurley (ed.), "*Advances in Earth Sciences,*" pp. 287–356. M.I.T. Press, Cambridge.

——— (1966c). Chemical evolution of the terrestrial planets. *Geochim. Cosmochim. Acta* **30,** 41–104.

——— (1966d). Mineralogy of the Mantle. In: P. M. Hurley (ed.), "*Advances in Earth Sciences,*" pp. 357–398. M.I.T. Press, Cambridge.

——— (1969). Phase transformations in the mantle. *Earth Planet. Sci. Letters* **5,** 401–412.

——— (1970). Phase transformations and the constitution of the mantle. *Phys. Earth Planet. Interiors* **3,** 109–155.

——— and A. MAJOR (1970). The system Mg_2SiO_4-Fe_2SiO_4 at high pressures and temperatures. *Phys. Earth Planet. Interiors* **3,** 89–108.

ROTH, R. S. (1958). Revision of the phase equilibria diagram of the binary system calcia-titania showing the compound $Ca_4Ti_3O_{10}$. *J. Res. Nat. Bur. Standards* **61,** 437–440.

SHAPIRO, J. N., and L. KNOPOFF (1969). Reduction of shock wave equations of state to isothermal equations of state. *J. Geophys. Res.* **74,** 1435–1438.

SIMMONS, G., and A. W. ENGLAND (1969). Universal equations of state for oxides and silicates. *Phys. Earth Planet. Interiors* **2,** 69–76.

SIMPSON, D. W. (1973). P wave velocity structure of the upper mantle in the Australian region. Ph.D. Thesis, Australian National University.

STRENS, R. G. (1969). The nature and geophysical importance of spin pairing in minerals of iron (II). In: S. K. Runcorn (ed.), "*The Application of Modern Physics to the Earth and Planetary Interiors,*" pp. 213–220. Interscience, a division of Wiley, New York.

SYONO, Y., S. AKIMOTO, Y. ISHIKAWA, and Y. ENDOH (1969). A new high pressure phase of $MnTiO_3$ and its magnetic property. *J. Phys. Chem. Solids* **30,** 1665–1672.

TAKEUCHI, H., and H. KANAMORI (1966). Equations of state of matter from shock wave experiments. *J. Geophys. Res.* **71,** 3985–3994.

THOMSEN, L. (1972). Elasticity of polycrystals and rocks. *J. Geophys. Res.* **77,** 315–327.

UREY, H. C. (1972). Abundances of the elements. *Annals New York Acad. Sci.* **194,** 35–44.

VERHOOGEN, J. (1953). Elasticity of olivine and constitution of the earth's mantle. *J. Geophys Res.* **58,** 337–346.

WANG, C. Y. (1967). Phase transitions in rocks under shock compression. *Earth Planet. Sci. Letters* **3,** 107–113.

——— (1968). Constitution of the lower mantle as evidenced from shock wave data for some rocks. *J. Geophys. Res.* **73,** 6459–6476.

——— (1970). Density and constitution of the mantle. *J. Geophys. Res.* **75,** 3264–3284.

——— (1972). A simple earth model. *J. Geophys. Res.* **77,** 4318–4329.

WANG, H., and G. SIMMONS (1972). FeO and SiO_2 in the lower mantle. *Earth Planet. Sci. Letters* **14**, 83–86.

WHITCOMB, J. H., and ANDERSON D. L. (1970). Reflection of P′P′ seismic waves from discontinuities in the mantle. *J. Geophys. Res.* **75**, 5713–5728.

WORTHINGTON, M. H., J. R. CLEARY, and R. S. ANDERSSEN (1974). Upper and lower mantle shear velocity modelling by Monte Carlo inversion. *Geophys. J.* **36**, 91–103.

WRIGHT, C., and J. R. CLEARY (1972). P wave travel-time measurements from the Warramunga seismic array and lower mantle structures. *Phys. Earth Planet. Interiors* **5**, 213–230.

15

PHASE TRANSFORMATIONS AND MANTLE DYNAMICS

In my opinion, the transitional layer is the key to the problem of what is going on in the mantle. When we understand its nature, we shall be well on the way to a grasp of the dynamics of the earth's interior.

(Birch, 1954)

15-1 INTRODUCTION

The role of phase transformations in the earth's mantle has been a principal theme throughout this volume. Several different kinds of transformations have been discussed. In addition to the major polymorphic transitions covered in Chaps. 11 to 14, they include also changes in mineral assemblages in response to variations of P, T, P_{H_2O}, and f_{O_2} in the earth's interior and changes of state such as occur during the formation of magmas by partial melting. All these categories are of importance both to considerations of the mantle's present constitution and to the dynamical behaviour of the mantle throughout geological time.

Referring to the major geologic cycle depicted in Fig. 8-6, we recall that formation of the observed range of basaltic magmas beneath mid-oceanic ridges by partial melting of pyrolite constitutes a particular class of phase transformations which was discussed in Chaps. 4 and 5. The outward motion of the resultant differentiated lithosphere plates away from the ridge appears to be made possible by another phase transformation (pressure-induced dehydration of amphibole) which causes incipient melting to occur at depths of 70 to 80 km and provides the plates with low-viscosity bearings on which to slide (Chaps. 4 and 6). As the plates subside into the mantle beneath oceanic trenches, a further set of transformations occurs. In the mafic crust, greenschists transform to amphibolite, basalt and amphibolite to eclogite, whilst serpentinites undergo a range of dehydration reactions. These are accompanied or followed by a second stage of partial melting, both in the former oceanic crust and in the overlying mantle wedge, leading to the generation of orogenic-type magmas (Chaps. 7 and 8).

These examples testify to the pervasive role played by phase transformations during the generation and consumption of lithosphere plates. But what of the engine which transports the plates? The possible role of phase transformations in regulating and driving this engine deserves examination. In Chaps. 11 to 14 we reviewed evidence relating to the occurrence of major solid-solid transformations in the transition zone referred to by Birch in the opening quotation. As lithosphere plates sink below 300 km or so, these major transformations will proceed, accompanied by corresponding changes in physical properties. In the present chapter, we will consider the possibility that these changes influence the driving mechanism for plate motions. Other aspects of the role of phase transformations in a dynamic mantle will also be discussed.

Although the theories of sea-floor spreading and plate tectonics have enjoyed spectacular success in explaining the behaviour and evolution of the earth's crust, the nature of the engine which drives the sea-floor-spreading process is far from being understood. It is generally agreed that the engine is a form of "thermal convection" operating within the mantle-crust system, although it must be understood that this term is used in its broadest sense.

The literature contains numerous statements to the effect that phase transformations will enhance or inhibit mantle convection. Either of these statements may be correct, depending upon the particular convective mechanism which is envisaged.[1,2] Two major types of convection theories have been widely discussed.[3] The oldest maintains that large-scale convection currents occur throughout the mantle. These currents transport the passive lithosphere plates by viscous coupling and ultimately drag them down into the mantle beneath oceanic trenches. A more recent hypothesis is that the cold lithosphere plates sink into the mantle because of their higher density and thus provide the primary driving force for the sea-floor-spreading cycle. We will consider these in turn.

[1]Knopoff (1964).
[2]Verhoogen (1965).
[3]E.g., McKenzie (1969, 1972).

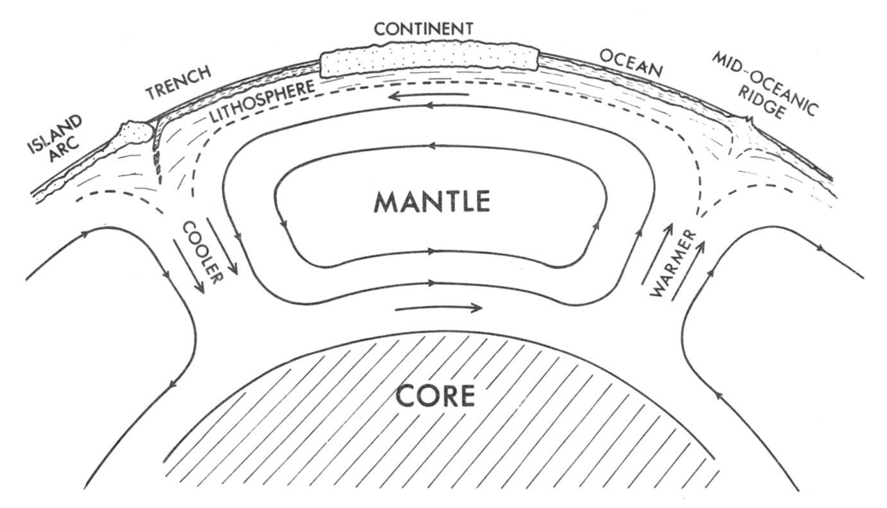

FIGURE 15-1
Diagrammatic representation of passive plate theory according to which plates of lithosphere are transported on the backs of mantle-wide convection currents occurring under near-adiabatic conditions.

15-2 CONVECTION CURRENTS AND THE PASSIVE PLATE THEORY

Theories in this category have been proposed by Holmes.[1] Hales,[2] Pekeris,[3] Vening Meinesz,[4] Hess,[5] Runcorn,[6] Griggs,[7] and many others. A sketch of the mechanism is given in Fig. 15-1. In most of these models[8] it is assumed that the mantle beneath the lithosphere displays a constant Newtonian viscosity and that convection occurs under the marginal stability criterion whenever the adiabatic gradient is exceeded by more than a very small amount. The convection cells are usually assumed to extend between the base of the lithosphere and the core-mantle boundary. Mid-oceanic ridges mark the loci of rising and diverging currents and the creation of new oceanic crust. Passive plates of lithosphere are transported on the backs of currents by viscous coupling and are finally dragged down into the mantle beneath trenches which mark the sites of sinking limbs. Runcorn[6] argued that, for convection systems of this kind, negative gravity anomalies would be produced above the rising limbs and positive anomalies above sinking limbs. He

[1]Holmes (1931).
[2]Hales (1936).
[3]Pekeris (1935).
[4]Vening Meinesz (1947, 1962).
[5]Hess (1955, 1962).
[6]Runcorn (1963, 1965).
[7]Griggs (1939).
[8]However, Griggs (1939) proposed a theory of cyclic convection currents based upon the assumption of a small finite strength in the mantle.

concluded from the low harmonics of the earth's gravity field that the difference in temperature between ascending and descending limbs was very small, on the order of 1°C. However, as McKenzie (1968) pointed out, Runcorn did not allow for the gravitational effect of the elevation of the crust above a rising current, which produces a positive anomaly, thereby tending to cancel out the negative anomaly caused by the density deficiency in the rising plane. McKenzie estimated that, when an appropriate allowance is made for surface elevation (e.g., over mid-oceanic ridges), the temperature difference between rising and sinking currents is larger and in the vicinity of 10 to 20°C. Nevertheless, in terms of the discussion which follows, this is still to be regarded as a very small difference. Moreover, McKenzie did not consider the subsurface structure of mid-oceanic ridges, which shows that the elevated topography is isostatically compensated at depths smaller than 100 km by a shallow, low-density root (Figs. 2-8 and 2-9).[1] An adequate allowance for this effect would substantially reduce McKenzie's estimate of 20 degrees superheat for a convection current rising from great depths beneath oceanic ridges.

Convection currents of this type, driven by superadiabatic temperature gradients smaller than 0.01°C/km, face considerable difficulties in penetrating the transition zone,[2,3] particularly in the neighbourhood of the 400-km discontinuities. The gradient dP/dT for the olivine-spinel-beta phase transformation is probably about 30 bars/°C.[4] Quantitative data for the pyroxene-garnet transformation are lacking, but the combined results of two groups indicate that the gradient is at least not significantly negative.[5,6,7] For simplicity in subsequent discussion, we will assume that these transformations occur at the same depth and are equivalent to a single first-order transformation with a positive gradient of 30 bars/°C.

The situation in the neighbourhood of the 650 to 800 km discontinuities is more complicated. The gradient for disproportionation of Mg_2SiO_4 into oxides has been estimated to be -13 bars/°C.[8] However, this estimate is probably incorrect since it is based partly upon an overestimate for the gradient of the α-γ transition in Mg_2SiO_4. Taking a more realistic value for the latter (Sec. 11-1) and using more recent thermodynamic data, the gradient dP/dT for disproportionation of Mg_2SiO_4 (normal spinel) into $MgO + SiO_2$ (stishovite) is found to be very close to zero.[9] This value may be too small, however. An experimental study[9] of the disproportionation of Mg_2SnO_4 (spinel) into $MgO + SnO_2$ (rutile str.) which provides a useful model for Mg_2SiO_4 gave a gradient of +40 bars/°C. The large size of this gradient was attributed to additional configurational entropy in

[1] See also Sclater and Francheteau (1970).
[2] Knopoff (1964).
[3] Verhoogen (1965).
[4] Sections 11-2 and 11-3.
[5] Ringwood (1967).
[6] Kushiro, Syono, and Akimoto (1967).
[7] Akimoto and Syono (1972).
[8] Ahrens and Syono (1967).
[9] Jackson, Liebermann, and Ringwood (1974).

Mg_2SnO_4 spinel arising from its inverse character.[1] This characteristic may well be present in Mg_2SiO_4 spinel. Mg_2TiO_4 is inverse, whilst Mg_2GeO_4 appears to be partly inverse.[2] Evidence has been cited that Fe_2SiO_4 spinel becomes increasingly disordered and inverse above 1200°C.[1,3] Many other spinels are known which have normal cation distributions at low temperatures but which become increasingly disordered at high temperatures, and it seems quite likely that Mg_2SiO_4 spinel would behave likewise at high temperatures.[1] A modest degree of disorder (e.g., 20%) would have a large effect upon the slope of the disproportionation reaction, making it more positive. All things considered, a slope of about +20 bars/°C for this reaction does not appear unreasonable. It is also possible that a transformation involving δ-Mg_2SiO_4 may either precede or follow the oxide disproportionation reaction.[4] The β-δ transformation in Mn_2GeO_4 has a gradient of 22 bars/°C,[5] and the corresponding transformation in Mg_2SiO_4 would probably have a generally similar slope. The gradient of the other major transformation around 700 to 800 km involving garnet is unknown. Recognizing the considerable uncertainties, we will assume for the sake of simplicity in subsequent discussion that all the transformations in the 650 to 800 km region are combined into a single first-order transformation at 650 km with a gradient of +20 bars/°C.

The latent heats associated with these transformations at 400 and 650 km would cause difficulties for convection currents.[6,7,8] Applying the Clapeyron equation, $dP/dT = \Delta H/T\Delta v$, the latent heats ($\Delta H$) associated with the transformations at the 400- and 650-km discontinuities would both be in the vicinity of 25 cal/g, giving a total of about 50 cal/g over the entire transition zone.

The adiabatic gradient may be readily calculated in a phase-transition region.[9,10,7] In the limiting case of a single phase displaying a univariant transformation, the adiabatic gradient is equal to the gradient of the transformation. For the principal transformations in the mantle, this would be 10 to 15°C/km. In a multicomponent system and in the presence of other phases, the adiabatic gradient, given by $(\partial T/\partial r)_s = \alpha g T/C_p$, is reduced ($\alpha$ is the coefficient of thermal expansion, g the acceleration of gravity, and C_p the specific heat at constant pressure). If ΔT is the temperature interval over which the transformation occurs, we may substitute in the above equation an effective thermal expansion $\alpha' = \Delta v/v\Delta T$ and an effective specific heat $C_p' = \Delta H/\Delta T$. Estimates of these quantities yield an adiabatic gradient of about 4°/km in the vicinity of the major phase transformations, which is about 10 times higher than the "normal" adiabatic gradient. The

[1] Jackson, Liebermann, and Ringwood (1974).
[2] Dachille and Roy (1960).
[3] Section 11-1.
[4] Sections 11-4 and 14-3.
[5] Morimoto et al. (1969).
[6] Knopoff (1964).
[7] Verhoogen (1965).
[8] A contrary view is argued by Schubert et al. (1975).
[9] Ringwood (1958).
[10] Tozer (1959).

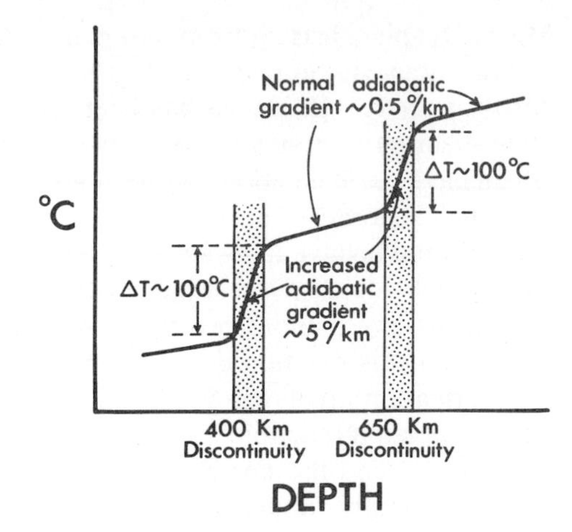

FIGURE 15-2
Sketch showing variation of adiabatic gradient throughout transition zone. (Not to scale.) (*From Ringwood, 1972, with permission.*)

adiabatic temperature distribution in the transition zone thus has the form shown in Fig. 15-2. For convection to occur under the marginal stability criterion, a superadiabatic gradient is required, and this can be established and maintained only by thermal conduction. Continuity of heat flux demands a thermal conductivity distribution qualitatively similar to Fig. 15-2, with the conductivity varying by a factor of 10 across the phase transformations. This is exceedingly improbable.

To avoid this problem, the marginal stability criterion must be abandoned and the assumption made that a substantial superadiabatic gradient exists throughout a large depth either above or below the transition layers. Under these conditions, the latent heat barrier can be overcome.[1,2] With the latent heat estimated for the mantle transformations (total 50 cal/g), the product of the superadiabatic gradient T°C/km and the depth r km of the convecting layer must be at least 200°C. For a convection cell extending from the upper mantle to the base of the lower mantle, a superadiabatic gradient of 0.1°C/km would be required throughout the lower mantle. For smaller dimensions, the superadiabatic gradient would be correspondingly higher.

Convection currents of the type depicted in Fig. 15-1 driven by substantial superadiabatic gradients as in this case face two difficulties. Firstly, convection under these conditions would involve a mean temperature difference of 200°C between ascending and descending currents. This would cause regional gravity anomalies at the earth's surface which are far greater than observed.[3,4,5]

[1]Verhoogen (1965.
[2]Schubert and Turcotte (1971).
[3]Runcorn (1963, 1965).
[4]McKenzie (1968).
[5]See discussion on page 518.

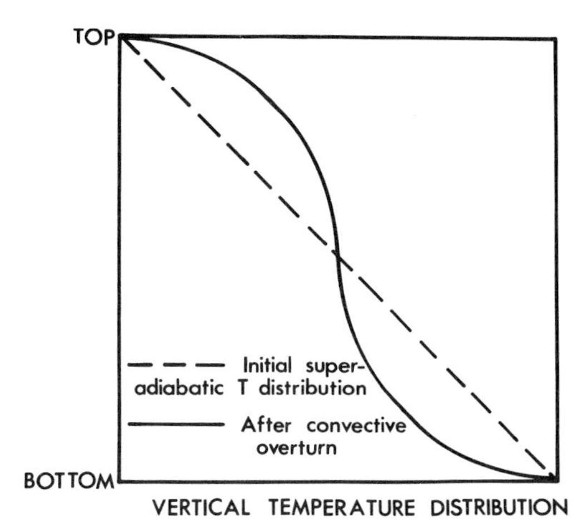

FIGURE 15-3
Sketch showing vertical temperature distribution in mantle after convective overturn from the initial superadiabatic temperature distribution required to initiate convection. (*After Brooks, 1941.*)

Secondly, there is the problem of re-establishing a superadiabatic temperature gradient by thermal conduction after completion of half a turn of the cycle (Fig. 15-3). The time t required for this process[1] is given by $t = r^2/4\pi k$. Taking k = thermal diffusivity $= 0.01$ and $r = 3000$ km, $t = 2.5 \times 10^{10}$ years, or over 5 times the age of the earth. Thus, only one convective overturn could have occurred throughout the history of the earth.[1] Verhoogen (1954) suggested that it might be possible to avoid the difficulty by appealing to heat supplied by conduction from the core or by radioactive decay within the deep mantle. This does not appear feasible. If convection currents are to drive lithosphere plates at the observed rates (1 to 10 cm/year), the time scale for convective overturn must be on the order of 10^8 years. Increasing the temperature of the lower mantle by 200°C over this period by radioactive heating would require the abundances of U, Th, and K to be over 10 times the chondritic levels. This seems most implausible on several geochemical grounds. Furthermore, if heat were generated at this rate in the lower mantle and transported by convection to the upper mantle, the near-surface heat-flux should be about 10 times the observed value. Likewise, if the lowermost mantle is to be heated by 200°C over 10^8 years by heat sources in the core, the temperature of the entire core must be raised by the same amount. This would also require mean abundances of U, Th, and K throughout the core which are more than tenfold higher than chondritic abundances. Besides raising the problem of why U, Th, and K had been so efficiently concentrated in the core despite their lithophile nature,[2] heat generation of this magnitude throughout the core,

[1]Brooks (1954).
[2]Hall and Murthy (1971) and Lewis (1971) have indeed speculated that most of the earth's potassium has been partitioned into the core because of the alleged chalcophile nature of potassium. However, Oversby and Ringwood (1972) have measured experimentally the partition coefficient of potassium between iron sulphide and silicate melts and demonstrated that the amount of potassium entering the sulphide is negligible.

if transferred to the upper mantle by convection, would produce a near-surface heat-flux of approximately 3 times the observed value.

It appears therefore that these rather simple models of mantle convection must be abandoned and more complex, but probably more realistic models, considered. Turcotte and Oxburgh (1967)[1] have proposed a model for laminar, cellular convection in which significant temperature gradients are restricted to thin thermal boundary layers adjacent to the horizontal cell boundaries and to thin thermal plumes on the vertical boundaries between cells. The core of each convection cell is assumed to be highly viscous and isothermal.[2] Models of this kind nevertheless represent oversimplifications because of their neglect of temperature-dependent viscosity and/or the possible plastic behaviour of the mantle. Qualitatively, both of these effects are likely to have the effect of concentrating zones of movement into even narrower vertical plumes than those considered by Oxburgh and Turcotte. Models of this kind can be devised which escape the time-scale difficulties mentioned earlier since the volumes in motion are small compared to the volume of the mantle. Thus, adequate flow velocities (of a few centimetres per year) can be maintained even if only a limited volume of the entire mantle has passed through the convective cycle during the past 4.5 billion years.

Nevertheless, if convective motions of this kind are to break through the phase transitions in the transition zone, the temperature differences between rising plumes and surrounding stagnant mantle must still be on the order of 200°C, following the previous arguments. This may be compared with the corresponding difference of 20°C obtained by McKenzie (1968) from an analysis of relations between mantle convection, surface elevation, and the earth's gravity field.[3]

In order to account for the thickness and structure of the lithosphere plates formed on either side of the ridges, the width of the rising plumes can hardly be smaller than 300 km. If a plume of this width and 200°C warmer than surrounding mantle extended deep into the lower mantle, a large regional negative gravity anomaly would be created. One of the striking features of the earth's gravity field is the small magnitude of free-air anomalies over oceanic ridges as shown both by the short-wavelength gravity anomaly field obtained from ship-based surveys (Fig. 2-8) and by the long-wavelength gravity anomaly field obtained from satellite data (Fig. 15-10). In contrast, substantial positive free-air anomalies occur over Benioff zones both at short and long wavelengths (Figs. 2-10 and 15-10).

In the light of the above discussion, the absence of large negative anomalies over oceanic ridges may be taken as evidence that the rising plumes do not extend deep into the lower mantle. Most probably they do not descend further than a few hundred kilometres, or less. This interpretation is in harmony with the hypothesis that the plumes beneath ridges are fed mainly by horizontal flow within the low-velocity zone (Fig. 8-6).

[1]See also Oxburgh and Turcotte (1968).
[2]See also Wilson (1963).
[3]As noted on page 518, the 20°C of superheat obtained by McKenzie probably represents an overestimate because of neglect of near-surface isostatic compensation beneath mid-oceanic ridges.

Elsasser (1963) has proposed that convection is restricted to the upper mantle and that the cells are characterized by very large ratios of horizontal to vertical dimensions. Cells of this kind, however, are very difficult to reconcile with the evidence of chemical and petrological zoning in the upper mantle which has been discussed in earlier chapters. Moreover, the interpretation of Benioff zones as representing the sinking of lithosphere to depths of at least 700 km is in conflict with this model.

15-3 THERMAL STRUCTURE OF PLATES

An alternative hypothesis is that plate motions are driven not by viscous coupling with a convecting mantle but by gravitational body forces acting directly on the lithosphere plates themselves.[1] This hypothesis appeals to common sense, although it may represent an oversimplification.[2] Before considering it further, a diversion is necessary in order to consider the thermal structure of plates.[1,2]

Following McKenzie,[2] we note that plates are formed initially at high temperatures along the axes of mid-oceanic ridges. As they move outward, cooling occurs, resulting in regions of high heat-flow through the ocean floors on either side of the ridge axes and a corresponding increase in the thickness of the plates. The width of the high-heat-flow region is controlled by the time t taken for a slab of hot material to cool and this is governed by the thickness L of the plate that is formed.

The relationship is given by[3]

$$t = \frac{\rho C_P L^2}{\pi^2 K}$$

where ρ = density, C_P = specific heat, and K = thermal conductivity. If the plate is being formed at velocity v on either side of the ridge, the high-heat-flow region will extend outward approximately vt from the ridge axis. An analysis of heat-flow observations in relation to topography and spreading rates yields a value of 75 to 100 km for the thickness of the plate[4]—which agrees well with the lithosphere thickness obtained from seismology.[5]

When the cool lithosphere plates plunge into the mantle beneath trenches, they are heated primarily by conduction from the mantle. The time t required for temperature changes in boundary layers to be felt appreciably at the center of the slab is given by the above equation, except that L is now the half-thickness of the plate.

For $K = 0.005$ cgs and $L = 4 \times 10^6$ cm, $t = 10^7$ years, which is identical to

[1]Elsasser (1967, 1969).
[2]McKenzie (1969, 1972).
[3]Carslaw and Jaeger (1959).
[4]Sclater and Francheteau (1970).
[5]Kanamori and Press (1970).

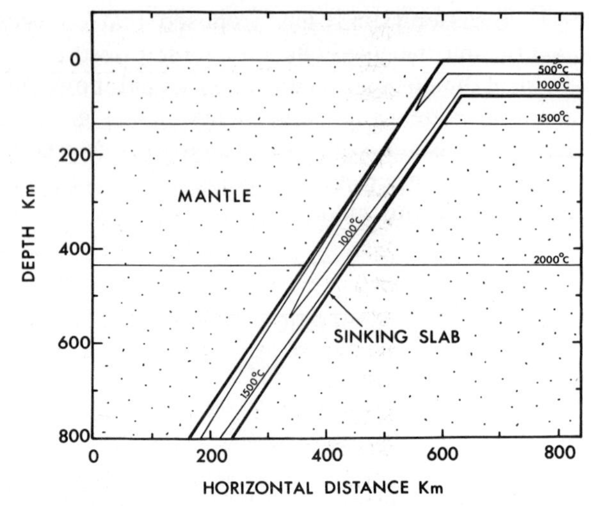

FIGURE 15-4
Temperature distribution in the sinking lithosphere compared to surrounding mantle. Thickness of lithosphere is 80 km, velocity of sinking is 8 cm/year, and thermal conductivity of slab = 0.005 cgs. Adiabatic heating included. (*After Griggs, 1972.*)

the time required for the slab to reach 700 km[1] (Fig. 15-5). The key result as emphasized by Griggs[2] is that, at this depth, the interior of the slab remains much cooler than the surrounding mantle.

Detailed studies of the thermal structure of sinking slabs, taking into account other sources of heat such as adiabatic compression (including latent heat associated with phase changes), viscous dissipation in the boundary layer, and radioactivity, have been undertaken by several authors.[3,2,4,5,6] Griggs' results,[2] which agree essentially[7] with those of McKenzie,[3] are shown in Fig. 15-4. It is seen that the temperature at 300 km in the slab's interior is about 1000°C cooler than the surrounding mantle, whilst at 700 km, the difference exceeds 500°C. Toksöz et al.[5] agree that the temperature in the slab at 400 km is much cooler than in the surrounding mantle but maintain that, by 700 km, thermal equilibrium has been reached. Griggs[2] points out that this is primarily a result of excessive shear-strain heating assumed by Toksöz et al. The latter authors also assume rather high thermal conductivities resulting from radiant heat transfer. Whilst the role of

[1] Isacks, Oliver, and Sykes (1968).
[2] Griggs (1972).
[3] McKenzie (1969, 1972).
[4] Minear and Toksöz (1970).
[5] Toksöz, Minear, and Julian (1971).
[6] Oxburgh and Turcotte (1970).
[7] I.e., for models with similar thicknesses and conductivities.

the latter factor is debatable, it seems likely that Minear et al. have overemphasized the role of shear-strain heating at slab boundaries and that the internal slab temperatures reached in their models are excessive. Thus, Griggs' and McKenzie's conclusion that the interior of the slab is substantially and possibly much cooler than surrounding mantle at a depth of 700 km appears more soundly based, although the magnitude of this temperature difference remains in doubt.

Their conclusion is strongly supported by the results of Pascal et al. (1973) and Fitch (1975), who find that seismic velocities in the slabs at depths of 600 to 700 km beneath the New Hebrides and beneath the Tonga trench are 3 to 6% faster than in "normal" mantle at the same depth. This implies the occurrence of considerably lower temperatures in the slabs in this depth interval, perhaps combined with phase changes as discussed in the next section.

Isacks, Oliver, and Sykes (1968) demonstrated that an approximately linear relationship existed between the length of seismic zones in island arcs and the calculated slip rate perpendicular to the arc (Fig. 15-5). The slope of the line corresponded to an isochron of 10 million years. One explanation which they considered was that 10 m.y. might be the time constant for "assimilation" of lithosphere by mantle. However, the thermal calculations referred to above suggest that it is improbable. A more plausible explanation was suggested by Griggs (1972), who argued that the limiting depth of earthquakes is that point at which the lithosphere becomes too hot for seismic failure (caused perhaps by shear instability or brittle fracture?) to occur. Below this limiting depth, which is given by a critical ratio of minimum slab temperature to melting temperature, all deformation occurs by flow. This model would be consistent with the slab sinking below 700 km as suggested by its thermal properties.

The question as to whether the slabs sink below the 650 to 800 km discontinuities is of considerable importance. In the discussion of irreversible petrological differentiation of the mantle given in Secs. 8-5 and 8-7, evidence favouring the interpretation that sinking lithosphere plates probably penetrated the discontinuities and entered the deep mantle was cited. This would not be possible if the 650-km discontinuity were characterized by a marked increase in Fe/Mg ratio as had earlier been argued by some. However, the evidence discussed in Chap. 14 did not support this interpretation. More recently, Kumazawa et al. (1974) concluded that the disproportionation of Mg_2SiO_4 to $MgO + SiO_2$ stishovite at about 650 km would function as a brake and will prevent the plates sinking through the discontinuity. This conclusion, however, was based upon earlier thermodynamic data suggesting that the disproportionation reaction had a negative slope dP/dT. In the light of later data, we concluded in Sec. 15-2 that the slope of this reaction was most probably positive. It will be seen in the next section that this would facilitate sinking of the plate through the discontinuity. Indeed, the occurrence of a positive gradient for phase transitions in this interval when combined with the evidence earlier cited for marked velocity contrasts between slab and mantle at depths of 600 to 650 km would strongly suggest that the plates indeed sink through the discontinuity.

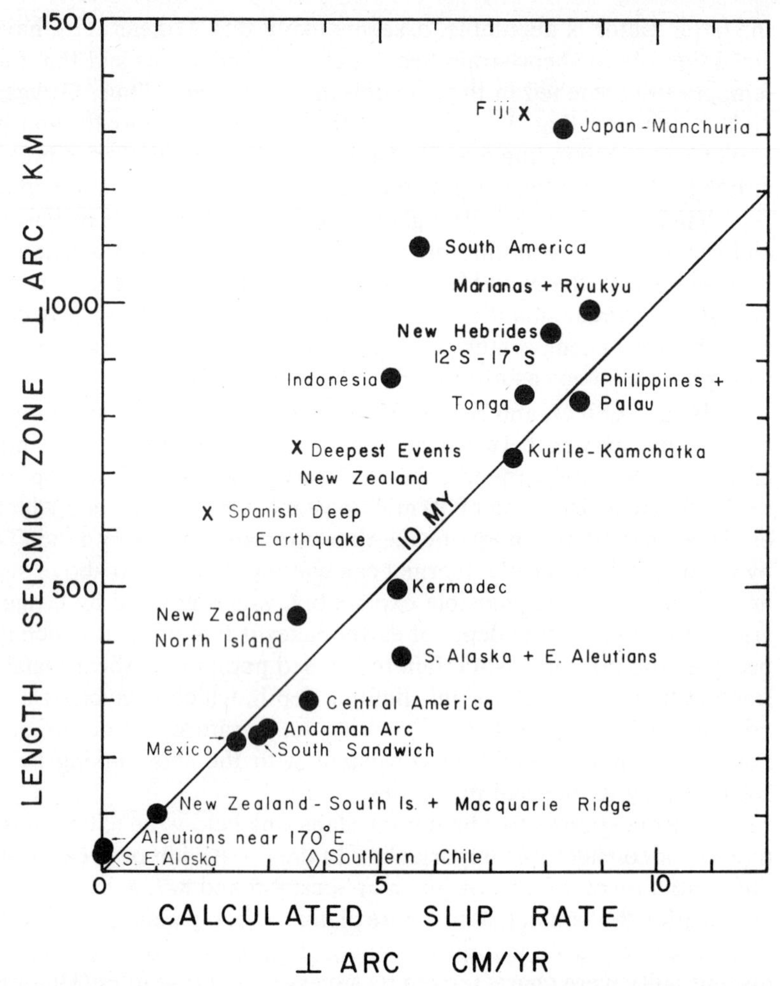

FIGURE 15-5
Calculated rates of underthrusting versus length of Benioff zones for various island arcs and arc-like features (solid circles), for several unusual deep events (crosses), and for Southern Chile (diamond). The solid line indicates the theoretical locus of points for uniform spreading over a 10 m.y. interval. (*From Isacks, Oliver, and Sykes, 1968, with permission. Copyright American Geophysical Union.*)

15-4 ACTIVE PLATE THEORY

We return now to the hypothesis[1] that plate motions are driven by gravitational body forces acting directly upon the lithosphere plates which may be regarded as

[1]Elsasser (1967, 1969).

the "active" elements, the role of the asthenosphere being essentially passive. It has been suggested that plate motions of this kind are caused by a combination of two effects: (1) a thrust directed outward from the mid-oceanic ridges and (2) a tensional force exerted by the sinking of dense lithosphere into the mantle beneath trenches. The lithosphere between these two regions behaves as a stress guide. Variations in plate velocities from place to place are interpreted as being caused by the combined effects of forces of differing magnitudes exerted at the extremities.

A sketch of this model is shown in Fig. 15-6. The mid-oceanic ridges are elevated due to high temperatures and their partially molten state whilst the thickness of the lithosphere increases outward from the ridge for a distance comparable with the zone of high heat-flow so that the base of the plate is significantly inclined to the horizontal.[1,2] Clearly, the plate will tend to slide down the inclined plane and exert a force on the remaining lithosphere, tending to push it outward over the weak, incipiently molten asthenosphere. The configuration has been analyzed by Hales[1] and Jacoby,[2] who have shown that, for reasonable assumptions regarding asthenosphere viscosity and plate structures, the forces generated may be sufficient to push plates against the viscous resistance of the asthenosphere with velocities on the order of a few centimetres per year. Lliboutry

[1]Hales (1969).
[2]Jacoby (1970).

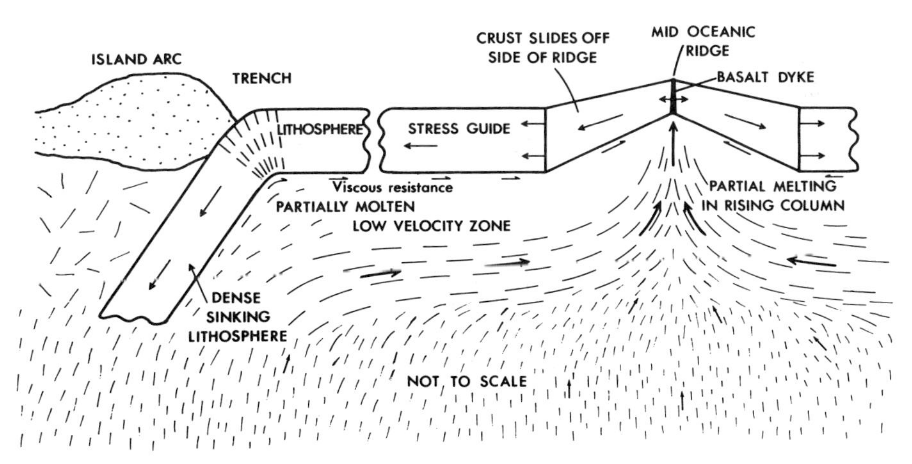

FIGURE 15-6
Sketch of plate motions driven by gravitational body forces—a combination of sinking of dense lithosphere beneath trenches and sliding of lithosphere plates down the slopes of mid-oceanic ridges. Injection of magma along the axes of the ridges contributes to the outward thrust. Between ridges and trenches, the lithosphere behaves as a stress guide. The sketch illustrates ideas proposed by Elsasser, Hales, Jacoby, and Lliboutry.

(1969) pointed to another factor which increases the thrust from mid-oceanic ridges. Diapiric uprise of pyrolite beneath ridges accompanied by partial melting leads to the continuous injection of basaltic magma along deep cracks which are formed at the axis of the ridge. These enormous elevated dykes full of basaltic magma under hydrostatic conditions exert an outward thrust on the plates at either side. Hales, Jacoby, and Lliboutry estimate that the global energy generated by these mechanisms exceeds that released by earthquakes.

Whilst the forces directed outward from the ridges may suffice to push plates over the asthenosphere, they do not explain the subsidence of lithosphere into the mantle beneath trenches. This process is crucial if plate motions over the earth's surface are to be sustained, and it seems possible that the lithosphere subsidence mechanism may provide the main rate-determining control for plate motions. Without lithosphere subsidence, there would be no sea-floor spreading or continental drift.

We will discuss the proposition[1,2] that lithosphere subsidence is also governed by the operation of gravitational body forces. According to this view, the lithosphere sinks into the mantle beneath trenches simply because it is appreciably denser than surrounding mantle. In this section we will be primarily concerned with the role of phase transformations in contributing to the density contrast between subsiding slab and surrounding mantle and thereby providing a principal cause of lithosphere subsidence.

We have already noted that, because of high thermal inertia in relation to sinking velocity, the interior of the slab is probably at least 1000°C cooler than surrounding mantle at a depth of 400 km and may well be 500°C or more cooler at a depth of 700 km (Fig. 15-7). Because of this large temperature contrast, the slab is easily able to overcome the latent heat barriers caused by phase transformations in the transition zone, as discussed previously. Paradoxically, if the descending material is cool enough to break through the phase transformations, the latter will then increase the driving force on the sinking plates by enlarging the density contrast, providing the slopes of the transformations are positive, as appears to be mostly the case. This effect has been noted by a number of authers, although appreciation of its full significance did not generally arise until the magnitude of temperature difference between slab and surrounding mantle was realized.[3,4,5,6,7,8,9]

The effect is shown in Fig. 15-7, where it is seen that the principal phase transformations in the mantle will occur at much shallower depths in the sinking slab because of lower temperatures. With the temperature differences shown, and

[1]Elsasser (1967, 1969).
[2]See also Ringwood and Green (1966, fig. 10).
[3]Vening Meinesz (1962).
[4]Verhoogen (1965).
[5]Isacks and Molnar (1969, 1971).
[6]Schubert, Turcotte, and Oxburgh (1970).
[7]Schubert and Turcotte (1971).
[8]Ringwood (1971, 1972).
[9]Griggs (1972).

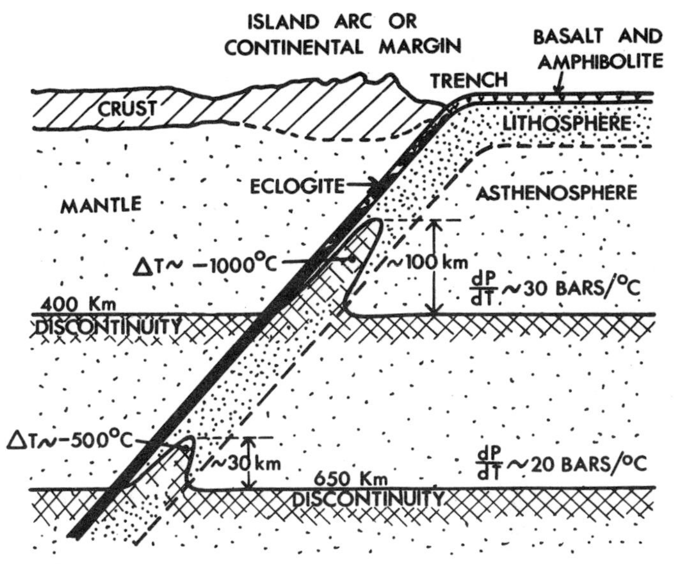

FIGURE 15-7
Simplified sketch showing possible relationships between sinking lithosphere plate and phase transformations. ΔT is the maximum temperature difference between the interior of the sinking slab and the normal mantle at given depth. (Not to scale.) (*From Ringwood, 1971, 1972, with permission.*)

taking the gradient of the phase transformations occurring near 400 km as 30 bars/°C and of those near 650 km as 20 bars/°C, as discussed earlier, the 400-km discontinuity would be displaced upward in the sinking slab by about 100 km, whilst the 650-km discontinuity would be displaced upward by about 30 km. In the displaced zones, the density contrasts between sinking slab and surrounding mantle caused by phase transformations will be about 6 to 8%. This may be compared with the density contrasts of 1 to 2% between 300 and 650 km caused by simple thermal contraction as a result of the lower mean temperature of the slab. Thus, phase transformations play the major role in driving the lithosphere into the mantle at depths greater than 300 km.

Transformation of the mafic oceanic crust to eclogite in the sinking slab would also play a significant role since eclogite ($\rho = 3.45$ to $3.60\ g/cm^3$) is intrinsically denser than pyrolite or mantle peridotite ($\rho = 3.28$ to $3.40\ g/cm^3$). The role of this transformation has been extensively discussed by Ringwood and Green (1966), who suggested that it might function as a tectonic engine driving the plates into the mantle. It was shown that basalt was thermodynamically unstable compared to eclogite, even near the surface. However, because of kinetic reasons, transformation to eclogite might not occur until a temperature of a few hundred degrees was reached. At a depth in the vicinity of 300 km, eclogite would transform to a denser garnetite,[1] which would maintain the density contrast with the

[1]Ringwood (1967). See also Sec. 12-3.

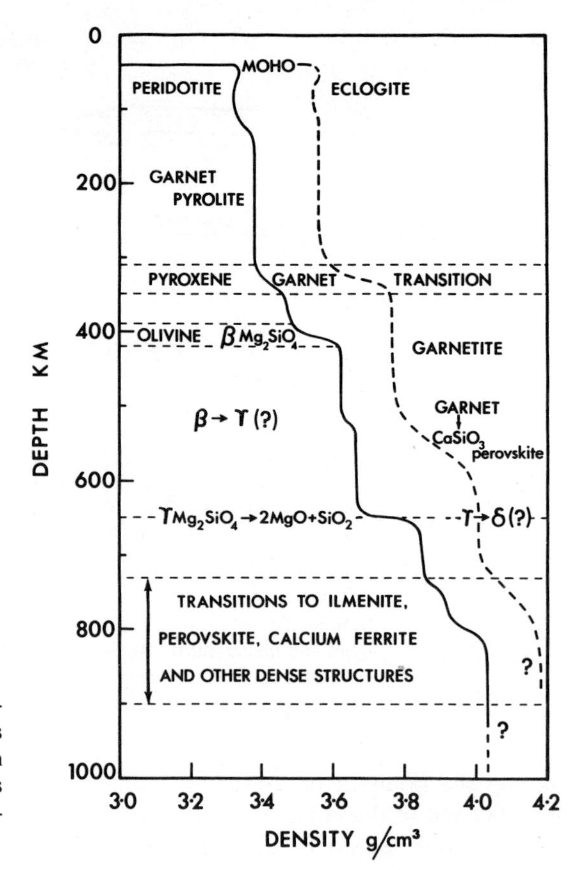

FIGURE 15-8
Comparison of "zero-pressure densities" of possible mineral assemblages displayed by pyrolite and eclogite with increasing depth in the mantle, which is assumed to be isothermal at 0°C. (*Modified after Ringwood, 1967.*)

surrounding mantle (Fig. 15-8). As discussed in Chap. 8, the basalt-eclogite transformation may play an important role in initiating the sinking of a lithosphere slab into the mantle.

The structure of a sinking lithosphere slab is shown in Fig. 15-7. Because of its high mean density caused jointly by phase transformations and relatively low temperature, this structure,[2,3,4] gives rise to a surface positive gravity anomaly of about 100 mgal (Fig. 2-10), which is a characteristic of island arc-trench systems.[1] Another important effect is that the velocity of seismic waves along the slab will be substantially higher than through neighboring normal mantle. Seismic velocities in the tongues of dense phases extending upward in the lithospheric plates from the 400- and 650-km discontinuities should be on the average about 1 km/sec faster than in neighbouring mantle. Seismic observations reviewed in

[1]Hatherton (1969).
[2]Oxburgh and Turcotte (1970).
[3]Morgan (1965).
[4]Griggs (1972).

Sec. 8-2 indeed demonstrated that *P* and *S* velocities along the axis of the sinking plate are much higher than in surrounding "normal" mantle. It is an exciting prospect that accurate measurements of travel times within the lithosphere plates from earthquakes at different depths may, in conjunction with measured slopes of relevant phase transformations, make it possible to deduce the temperature-depth distributions in the slabs.[1]

15-5 PHASE TRANSFORMATIONS AND DEEP EARTHQUAKES

Studies of earthquake source mechanisms in sinking lithosphere plates imply that the principal stresses are developed within the plates and parallel to the direction of dip.[2,3,4] Many intermediate-focus earthquakes indicate conditions of tension, whereas earthquakes occurring at depths greater than 300 km show that the plate is in compression (Fig. 15-9). At depths of 300 to 500 km, between the intermediate- and deep-focus earthquakes, zones of low earthquake frequency are

[1]Griggs (1972).
[2]Sykes (1966).
[3]Isacks and Molnar (1969, 1971).
[4]Isacks, Sykes, and Oliver (1969).

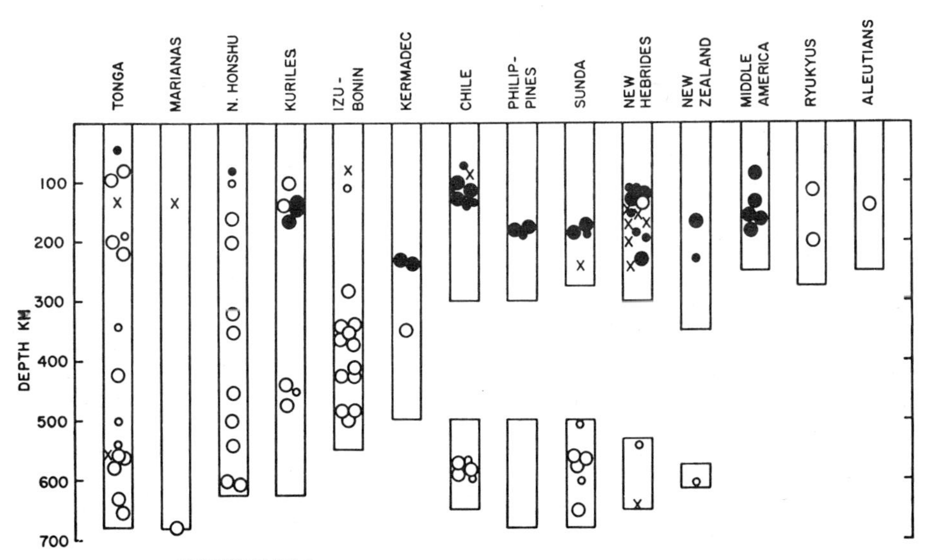

FIGURE 15-9
Downdip stresses from focal mechanism studies as a function of the depth for various arcs. Most arcs in which the seismic activity is discontinuous with depth show extension in the intermediate depth range. O—compressional; ●—extensional; X—other. (*From Isacks and Molnar, 1969, with permission.*)

sometimes found. Isacks and Molnar[1] have interpreted these observations as providing support for the model according to which plates descend because of gravitational body forces acting on their excess density. Referring to Fig. 15-9, it is seen that the tensional earthquakes occur immediately above the tongue of high-density phases, extending upward from the 400-km discontinuity. The large excess density associated with this feature may well be responsible for tension above 300 km.[1]

Compressional stresses within the slab at depths below 300 km might also be caused by the large excess densities within the slabs, caused by phase transitions below 300 km.[1] The establishment of compressive stresses also requires that appropriate resistance to downward movement be supplied. Isacks and Molnar (1971) consider two alternative sources of this resistance:

1 If the slab were unable to penetrate the 650-km discontinuity, then an upward force would be applied to the slab from this level, throwing it into compression. Inability to penetrate the transition could be due to either a change in chemical composition (e.g., an increase in Fe/Mg ratio) at this depth or the occurrence of a negative dP/dT gradient at the transition which would cause a buoyancy effect (analogous but opposite to that illustrated in Fig. 15-7). We have earlier considered the evidence relating to these possibilities and rejected them (Chaps. 11, 12, and 14). Moreover, Griggs (1972) has pointed out that a thin lithosphere tongue retarded at its extremity would hardly be expected to transmit compressive stresses upward for several hundred kilometres but, rather, would be subject to short-wavelength folding, for which there is little or no evidence. The fact that compression is frequently observed in slabs which do not appear to have reached the 650-km discontinuity is also hard to reconcile with this explanation. Finally, we have already cited geochemical and petrological evidence which suggests that slabs descend below 650 km (Sec. 8-7). These difficulties appear to warrant rejection of this explanation, and so we turn to the second explanation suggested by Isacks and Molnar (1971).

2 This maintains that the resistance is simply the result of strength in the mantle increasing with depth (down to 650 km). The resistance is thus applied over the surfaces of the slab by viscous coupling with surrounding normal mantle. This interpretation appears preferable at the present time.

Phase changes may play an indirect role in influencing the strength distribution since mechanical properties are partly controlled by the crystal structures of the mineral phases present at different depths. We have seen that the phase changes around 650 to 800 km are caused by transformation of spinels and garnets into denser mineral assemblages, most probably containing 15 to 30% periclase (Fig. 14-4). The presence of this amount of periclase in the lower mantle

[1]Isacks and Molnar (1969, 1971).

would probably have a major effect upon the mechanical properties of this region. Periclase is a weaker and more ductile material than upper mantle silicates at low pressure,[1] and this may also be true at the high pressures of the mantle below 700 km. The lower mantle may thus possess a smaller viscosity and ultimate strength than the overlying region between about 200 and 700 km. This factor would facilitate the descent of slabs into this region and may be connected with the absence of deep focus earthquakes below 700 km. It may also be relevant in connection with the possible occurrence of convection in the deep mantle.

The possibility that intermediate- and deep-focus earthquakes may be caused by phase transformations has frequently been suggested.[2,3,4,5] Two difficulties, however, have hitherto prevented general acceptance of this hypothesis. Firstly, it has been hard to understand how a phase transition can occur sufficiently quickly in the mantle to cause an earthquake. Secondly, analyses of first motions from earthquake seismograms have indicated a prevalence of double-couple mechanisms, implying some kind of shear failure, rather than monopolar implosion of explosion mechanisms as were believed to be required by the phase change hypothesis. Upon further examination, neither of these objections appears compelling.

In a static mantle, where temperatures below 300 km would probably exceed 1400°C,[6] it has not been possible to suggest a plausible model for rapidly running phase transitions. These require an initial state of metastability so that, once nucleated, the transformations are able to proceed rapidly towards completion. However, at average mantle temperatures, reaction rates are so fast that phase changes occur essentially under equilibrium conditions and without appreciable hysteresis.

The situation is radically different in a dynamic mantle. The sinking lithosphere slab may be 1000°C or more cooler than surrounding mantle (Fig. 15-4). Reaction rates vary exponentially with temperature. Experimental investigations show that most reconstructive transformations in silicates of the kind which occur in the mantle proceed extremely slowly under anhydrous conditions at 600°C but proceed at measurable rates at 800 to 900°C.[7] Moreover, under laboratory conditions, reaction rates may be increased by orders of magnitude owing to small crystal size and shear stress within the pressure apparatus. In the mantle, large, relatively unstrained crystals may be present and reaction rates may be very much lower than observed in the laboratory. Although relevant quantitative kinetic data are sparse,[8] it appears likely that, at inferred rates of lithosphere de-

[1]E.g., compare periclase (Paterson and Weaver, 1970) and olivine (Carter and d'Allemant, 1970).
[2]Bridgman (1945).
[3]Ringwood (1956, 1967).
[4]Evison (1963, 1967).
[5]Dennis and Walker (1965).
[6]Clark and Ringwood (1964).
[7]Author's unpublished observations.
[8]See, however, Kasahara and Tsukahara (1971).

scent, low-pressure mantle minerals (e.g., olivine) may be carried some distance into the equilibrium field of the corresponding high-pressure phases, providing that the temperature in the sinking slab is less than about 700 to 800°C. Once a condition of metastability has been reached, phase transitions may proceed very rapidly (< 1 sec) after they are nucleated, thus causing a corresponding rapid volume change capable of producing an earthquake. In general, the greater the degree of metastability, the more rapidly a transformation will run, once nucleated. Thus, phase transformations may proceed during microsecond intervals when overstressed by the propagation of shock waves.[1] A more familiar example is the sudden "explosive" devitrification of a metastable silicate glass near room temperature.

Thermal studies have shown that temperatures of parts of sinking slabs may be smaller than 700°C at a depth of 400 km (Fig. 15-4). Reaction rates at these temperatures are probably sufficiently slow for the required metastabilities to develop for the basalt-eclogite and eclogite-garnetite transformations between 100 and 300 km and for the olivine-spinel-β-Mg_2SiO_4 and pyroxene-garnet transformations between 300 and 400 km. Whether the temperature within the slab at a depth of 600 km can be sufficiently low to permit metastability of β-Mg_2SiO_4 and garnet is less clear. Nevertheless, in view of uncertainties concerning the kinetics of transformations and temperature distribution within the slab, this possibility cannot be dismissed.[2]

There is a further effect which has received inadequate consideration. It is not necessary that the phase change should proceed sufficiently rapidly to be the prime source of the earthquake. The contraction of a volume of mantle in the sinking slab due to a phase change may occur relatively slowly but, nevertheless, may yet cause the development of large stresses in the region surrounding the contracting transforming volume. Earthquakes may thereby be caused by failure in response to these stresses in the surrounding envelope.[3,4]

The argument that phase transitions appear unable to produce the seismic radiation field exhibited by deep earthquakes is weakened by the observations of large shear-wave components in seismic waves generated by nuclear explosions.[5,6] These have been explained as resulting from the relaxation of pre-existing stress in the region where the bombs were detonated. This explanation, however, is directly applicable to the sinking slab, which as discussed earlier, is in a state of uniaxial stress as a result of gravitational body forces acting on excess densities primarily caused by the phase transformations themselves (Fig. 15-7). Another explanation of the *S*-wave radiation from nuclear explosions has been in terms of failure in the walls surrounding the bomb-produced cavity owing to the stress field produced.[5,6] This is directly analogous to the situation suggested

[1]McQueen, Marsh, and Fritz (1967).
[2]Griggs (1972).
[3]Bridgman (1945).
[4]Ringwood (1967).
[5]Brune and Pomeroy (1963).
[6]Toksöz, Harkrider, and Ben-Menaheim (1965).

above for the generation of earthquakes owing to failure under stresses developed in the region surrounding a volume which is contracting owing to phase transformations. These considerations suggest that earlier objections to phase transformations as the ultimate causes of deep and intermediate earthquakes should be re-examined. A recent detailed study of earthquake source mechanisms indeed indicates the frequent occurrence of a component in the seismic radiation field which is not of a double couple type and which is consistent with a phase-transformation origin.[1,2] If deep-focus earthquakes should prove to be direct or indirect consequences of phase transformations, the temperature in the sinking slabs at a depth of 700 km must be much smaller than the surrounding mantle temperatures, as suggested earlier in this chapter.

15-6 OTHER EFFECTS ASSOCIATED WITH PHASE TRANSFORMATIONS

Vertical Movements of the Crust

Investigations of relationships between phase transformations and dynamical processes in the mantle are still in the exploratory stage, and many fields of enquiry remain to be pursued. One of these concerns possible relations between phase transformations and vertical movements of the crust. This problem excited considerable interest when it was believed that the Mohorovicic discontinuity was caused by a phase change from gabbro to eclogite. When a thick pile of sediments is dumped on a model crust with a gabbro-eclogite transformation at its lower boundary, pressure is transmitted comparatively rapidly to the boundary, whereas changes in temperature at this boundary resulting from thermal blanketing by sediments develop on a time scale which is many orders of magnitude longer. Thus the crust first subsides as pressure increases and gabbro turns to eclogite, permitting deposition of a greater thickness of sediments which enhances the subsidence. This is ultimately halted as temperature rises due to sediment blanketing; eclogite then transforms to gabbro and the crust rises.

Numerous investigations have been made of variations of the kind of instability mentioned above.[3,4,5,6,7] Whilst the hypothesis that the Moho is generally caused by an isochemical phase change is no longer viable (Sec. 2-4), the above analyses have not lost their value. Rather than attempt to explain vertical movements in terms of a single phase-transition boundary characterized by a large density change, they could be applied to the case of a series of phase changes charac-

[1]Knopoff and Randall (1970).
[2]Randall and Knopoff (1970).
[3]MacDonald and Ness (1960).
[4]Wetherill (1961).
[5]Broecker (1962).
[6]Joyner (1967).
[7]O'Connell and Wasserburg (1967).

terized by relatively small density changes, but spread over a thickness of many tens of kilometres. This is more relevant to the evolution of the crust as changes occur in metamorphic facies in response to variations of P, T, and P_{H_2O} and are, in turn, accompanied by changes in density (generally small but affecting large volumes), resulting in variations of crustal thickness.

The elevation of mid-oceanic ridges is another effect which may be attributed in part to phase changes. Increase of temperature in the convection plume causes pyroxene pyrolite to transform to less dense plagioclase pyrolite in the upper 40 km (Sec. 6-2, Table 6-1), resulting in elevation of the surface.[1] The effect is augmented by partial melting and normal thermal expansion.

With a static earth, phase transformations at depths greater than a few hundred kilometres do not appear to provide plausible sources of vertical crustal movements because of the long time constant ($> 10^9$ years) for temperature changes at these depths caused by thermal conduction. However, the dynamical behaviour of the mantle as envisaged by the plate theory is constantly causing large perturbations of temperature at depths of several hundred kilometres, on a time scale of 10^7 to 10^8 years, particularly where plates sink into the mantle. These temperature perturbations will propagate outward by conduction, causing variations of the levels at which major phase transformations occur. The accompanying volume changes are likely to be reflected at the surface in terms of vertical uplifts or depressions of large areas of the crust. The motions of continental blocks relative to these deep perturbations will increase the complexity of epeirogenic movements observed at the surface of the earth.

Smaller temperature perturbations at depth may also be caused by variations in the circulation pattern within the low-velocity zone and by deep radioactive heating, particularly beneath continents which may have been stationary relative to underlying mantle for long periods. Such perturbations may diffuse downward to the 400-km discontinuity on a longer time scale (10^8 to 10^9 years) and may cause small variations in crustal elevation over regions of continental extent.

Phase Transformations and Gravity Anomalies

The role of phase transformations in contributing to the positive gravity anomalies over Benioff zones was mentioned in Sec. 15-4. Bott (1971) discussed the origin of the earth's long-wavelength global gravity anomalies in analogous terms. As seen in Fig. 15-10, these are not obviously related to the present distribution of continents and oceans. Although some of the positive anomalies tend to correlate with the circum-Pacific belt, where cool, dense lithosphere is sinking into the mantle, there are other large anomalies which are unrelated to surface features or known deep structure[2] or to the global patterns of heat-flow and seismic delay times. These observations led Bott to conclude that the sources of the anomalies

[1]See also Sclater and Francheteau (1970) and Miyashiro et al. (1970).

[2]E.g., the negative anomalies just south of India and west of southwestern Australia.

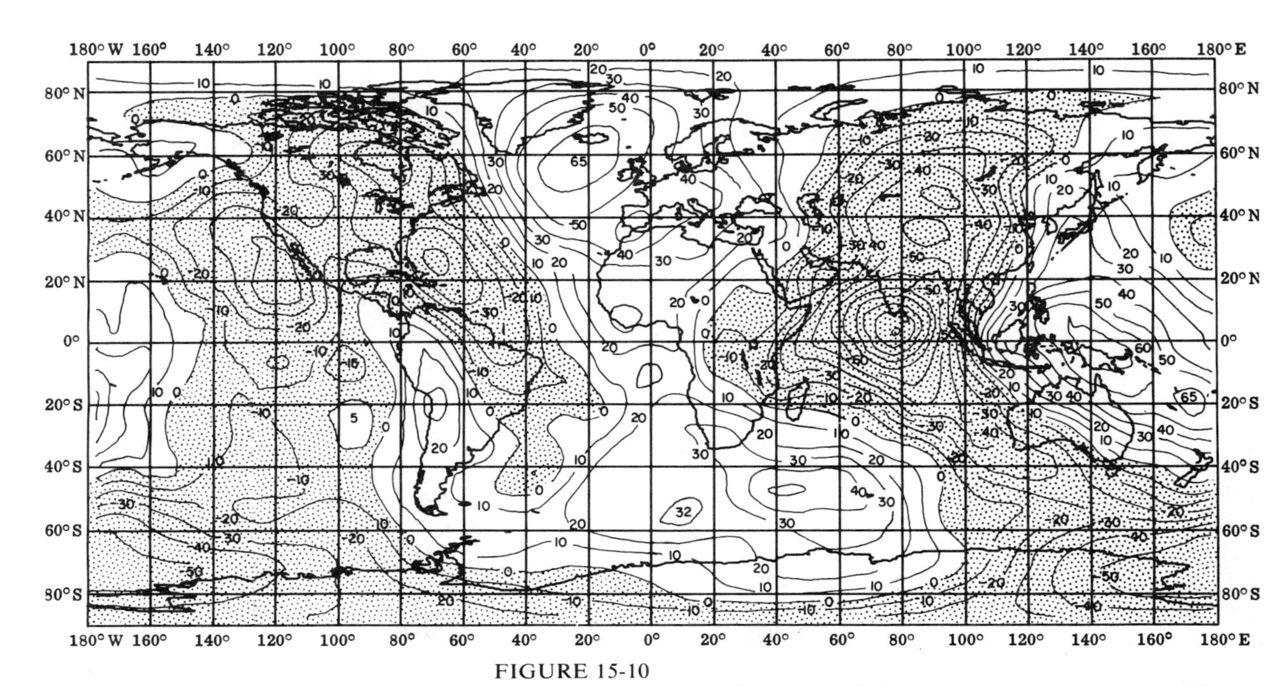

FIGURE 15-10
Global geoid undulations from a spherical harmonic expansion to degree 20, contour interval 10 meters, reference flattening 1/298.256. (*From Rapp, 1973.*) Note positive anomalies in circum-Pacific regions of lithosphere subsidence and absence of significant anomalies correlated with mid-oceanic-ridge systems.

lay deeper than the asthenosphere and were possibly connected with the transition zone. He demonstrated that small temperature perturbations, up to 20°C at a depth of 400 km, would, by virtue of their effect upon the depth of the 400-km discontinuity, provide an explanation of the amplitude and scale of these anomalies. The generation of such perturbations in a dynamic mantle appears entirely plausible. Bott also reverses the argument, showing that the observed gravity anomalies place quite strict limits on the allowable fluctuation in depth of the olivine-β Mg_2SiO_4 transition, which in turn imply that lateral temperature fluctuations of more than a few hundred kilometres wavelength at this depth cannot exceed a few tens of degrees.

Development of Chemical Inhomogeneity in the Mantle

In Sec. 14-7 we concluded that there was no compelling evidence that the average mean atomic weight of the lower mantle differed significantly from that of the upper mantle. Thus, considered on this largest of possible scales, there is no evidence that the mean major element chemical composition of the lower mantle differs substantially from that of the upper mantle. On the other hand, we emphasized in Secs. 8-5 and 8-7 that the fundamental mechanism of upper mantle evolution involved irreversible differentiation of pyrolite ($\bar{M} = 21.3$) into residual peridotite ($\bar{M} = 21.0$ to 21.2) and eclogite ($\bar{M} \sim 22.0$). Thus, considered on a smaller scale, this kind of petrologic evolution results in the development of a large degree of heterogeneity, even though the mean major element composition of the differentiated peridotite plus eclogite is similar to that of pyrolite.[1]

Although the mantle heterogeneity caused by differentiation of pyrolite into peridotite + eclogite + continental crust is primarily caused by solid-liquid phase transformations, a significant role may possibly be played by solid-solid transformations in the sinking lithosphere plates. These plates consist of layers of basaltic crust, residual harzburgite and lherzolite (Fig. 5-1). As they sink into the mantle, phase transformations in the respective mafic, harzburgite, and lherzolite layers will occur at different depths because of the differences in chemical compositions. The depths to which the lithospheric layers sink may be controlled by the sequence and properties of the phase transformations which occur.

For example, consider the sinking mafic crust, which transforms successively to eclogite ($\rho \sim 3.5$ g/cm^3) at depths less than 100 km and then to garnetite ($\rho \sim 3.8$ g/cm^3) at about 300 km (Fig. 15-8). At greater depths, further transformations occur. In the light of the discussion in Chaps. 12 and 14, it appears likely that the former mafic crust will attain states which remain denser than surrounding pyrolite. This is largely a consequence of the relatively important role of the

[1]Qualification is necessary because of segregation of the crust. Although this has a major effect upon the distribution of some incompatible trace elements, its effect on major element differentiation between (peridotite + eclogite) versus pyrolite is negligible because of its small mass.

garnet-perovskite transition and the higher Fe/Mg ratio. Thus, the former mafic crust is likely to sink very deeply into the lower mantle.

The situation is different in the case of the residual peridotite layer underlying the oceanic crust (Fig. 5-1). This may consist partly of harzburgite having a higher ratio of normative olivine to pyroxene (80:20) than underlying lherzolite and pyrolite (55:45). If Mg_2SiO_4 indeed transforms to the "post-spinel" state at a depth which is less than that at which Mg garnet transforms to denser phases (Fig. 14-4), then the differentiated part of the sinking lithosphere will become substantially denser than surrounding material of pyrolite composition below 650 km and will therefore penetrate this discontinuity. But, when the garnet component of the undifferentiated mantle transforms at greater depths, the differentiated former harzburgitic component of the sinking slab may or may not be able to break through, depending upon the sequence of phase transformations in the slab and in the surrounding mantle and upon the density changes associated with these transformations. If unable to penetrate a particular transformation, the differentiated material would spread out to form a distinct layer. Thus, phase transformations may act as a kind of filter, leading to significant degrees of lateral and vertical chemical heterogeneity in the mantle. Perhaps some of the differences between seismic velocity distributions observed by workers in different parts of the world can be attributed to this cause. These heterogeneities may also be responsible for some of the seismic reflections which have been observed from deep mantle horizons.[1]

An important factor controlling the nature of ultimate differentiation of the mantle will be the relative densities of orthosilicate $(Mg,Fe)_2SiO_4$ versus metasilicate $(Mg,Fe)SiO_3$ phases under the pressures which exist in the deep mantle. If the orthosilicate phases transform to intrinsically denser states than metasilicates, then the differentiated former harzburgitic component of the lithosphere, which is enriched in orthosilicate relatively to metasilicate, may sink initially to the bottom of the mantle and accumulate by successive outward growth as advocated by Dickinson and Luth (1971). On the other hand, if the densities are reversed, the harzburgitic component will accumulate at an intermediate depth.

REFERENCES

AHRENS, T. J., and Y. SYONO (1967). Calculated mineral reactions in the earth's mantle. *J. Geophys. Res.* **72,** 4181–4188.

AKIMOTO, S., and Y. SYONO (1972). High pressure transformations in $MnSiO_3$. *Am. Min.* **57,** 76–84.

BIRCH, F. (1954). The earth's mantle: elasticity and constitution. *Trans. Am. Geophys. Union* **35,** 79–85.

[1]Whitcomb and Anderson (1970).

BOTT, M. H. (1971). The mantle transition zone as possible source of global gravity anomalies. *Earth Planet. Sci. Letters* **11,** 28–34.

BRIDGMAN, P. W. (1945). Polymorphic transitions and geological phenomena. *Am. J. Sci.* **243A,** 90–97.

BROECKER, W. S. (1962). The contribution of pressure-induced phase changes to glacial rebound. *J. Geophys. Res.* **67,** 4837–4842.

BROOKS, H. (1941). Cyclic convection currents. *Trans. Am. Geophys. Union* **22,** 548–551.

——— (1954). Discussion. *Trans. Am. Geophys. Union* **35,** 92–93.

BRUNE, J., and P. POMEROY (1963). Surface wave radiation patterns for underground nuclear explosions and small magnitude earthquakes. *J. Geophys. Res.* **68,** 5005–5028.

CARSLAW, H. S., and J. C. JAEGER (1959). "*Conduction of Heat in Solids.*" Oxford, London. 496 pp.

CARTER, N., and H. D'ALLEMANT (1970). High temperature flow of dunite and peridotite. *Bull. Geol. Soc. Am.* **81,** 2181–2202.

CLARK, S. P., and A. E. RINGWOOD (1964). Density distribution and constitution of the mantle. *J. Geophys. Res.* **2,** 35–88.

DACHILLE, F., and R. ROY (1960). High pressure studies of the system Mg_2GeO_4 - Mg_2SiO_4 with special reference to the olivine spinel transition. *Am. J. Sci.* **258,** 225–246.

DENNIS, J. C., and C. WALKER (1965). Earthquakes resulting from metastable phase transitions. *Tectonophysics* **2,** 401–407.

DICKINSON, W. R., and W. C. LUTH (1971). A model for plate tectonic evolution of mantle layers. *Science* **174,** 400–404.

ELSASSER, W. M. (1963). Early history of the earth. In: J. Geiss and E. Goldberg (ed.), "*Earth Science and Meteoritics,*" chap. 1, pp. 1–30. North-Holland, Amsterdam.

——— (1967). Convection and stress propagation in the upper mantle. Technical Report No. 5, June 15, 1967. Princeton, New Jersey.

——— (1969). Convection and stress propagation in the upper mantle. In: S.K. Runcorn (ed.), "*The Application of Modern Physics to the Earth and Planetary Interiors,*" pp. 223–246. Interscience, a division of Wiley, New York.

EVISON, F. (1963). Earthquakes and faults. *Bull Seism. Soc. Am.* **53,** 873–891.

——— (1967). On the occurrence of volume change at the earthquake source. *Bull. Seism. Soc. Am.* **57,** 9–26.

FITCH, T. (1975). Compressional velocity and velocity contrast in the source region of deep earthquakes. *Earth Planet. Sci. Letters.* (In press.)

GRIGGS, D. T. (1939). A theory of mountain building. *Am. J. Sci.* **237,** 611–650.

——— (1972). The sinking lithosphere and the focal mechanism of deep earthquakes. In: E. C. Robertson (ed.), "*The Nature of the Solid Earth,*" chap. 14, pp. 361–384. McGraw-Hill, New York.

HALES, A. L. (1936). Convection currents in the earth. *Mon. Not. Roy. Astron. Soc., Geophys. Supp.* **3,** 372–379.

——— (1969). Gravitational sliding and continental drift. *Earth Planet. Sci. Letters* **6,** 31–34.

HALL, H. T., and V. R. MURTHY (1971). The early chemical history of the earth: some critical elemental fractionations. *Earth Planet. Sci. Letters* **11,** 239–244.

HATHERTON, T. (1969). Gravity and seismicity of asymmetric active regions. *Nature* **221,** 353–355.

HESS, H. H. (1955). Serpentines, orogeny and epeirogeny. In: A. Poldervaart (ed.), "*Crust of the Earth*," pp. 391–407. *Geol. Soc. Am. Spec. Paper* 62.

——— (1962). History of ocean basins. In: A. E. J. Engel, H. L. James, and B. F. Leonard (ed.), "*Petrologic Studies*," Buddington Volume, pp. 599–620. Geol. Soc. Am., New York.

HOLMES, A. (1931). Radioactivity and earth movements. *Trans. Roy. Soc. Glasgow* **18,** 559–606.

ISACKS, B., and P. MOLNAR (1969). Mantle earthquake mechanisms and the sinking of the lithosphere. *Nature* **223,** 1121–1124.

——— and ——— (1971). Distribution of stresses in the descending lithosphere from a global survey of focal mechanism solution of mantle earthquakes. *Rev. Geophys. Space Phys.* **9,** 103–174.

———, J. OLIVER, and L. R. SYKES (1968). Seismology and the new global tectonics. *J. Geophys. Res.* **73,** 5855–5899.

———, L. R. SYKES, and J. OLIVER (1969). Focal mechanisms of deep and shallow earthquakes in the Tonga-Kermadec region and the tectonics of island arcs. *Bull. Geol. Soc. Am.* **80,** 1443–1470.

JACKSON, I. N., R. C. LIEBERMANN, and A. E. RINGWOOD (1974). Disproportionation of spinels to mixed oxides: Effect of inverse character and implications for the mantle. *Earth Planet. Sci. Letters* **24,** 203–208.

JACOBY, W. R. (1970). Instability in the upper mantle and global plate movements. *J. Geophys. Res.* **75,** 5671–5680.

JOYNER, W. B. (1967). Basalt-eclogite transition as a cause for subsidence and uplift. *J. Geophys. Res.* **72,** 4997–4998.

KANAMORI, H., and F. PRESS (1970). How thick is the lithosphere? *Nature* **226,** 330–331.

KASAHARA, J., and J. TSUKAHARA (1971). Experimental measurements of reaction rate at the phase change of nickel olivine to spinel. *J. Phys. Earth* **19,** 79–88.

KAULA, W. M. (1972). Global gravity and tectonics. In: E. C. Robertson (ed.), "*The Nature of the Solid Earth*," chap. 15, pp. 385–405. McGraw-Hill, New York.

KNOPOFF, L. (1964). The convection current hypothesis. *Rev. Geophys.* **2,** 89–122.

——— and M. RANDALL (1970). The compensated linear-vector dipole: A possible mechanism for deep focus earthquakes. *J. Geophys. Res.* **75,** 4957–4963.

KUMAZAWA, M., H. SAWAMOTO, E. OHTANI, and K. MASAKI (1974). Postspinel phase of forsterite and evolution of the earth's mantle. *Nature* **247,** 356–358.

KUSHIRO, I., Y. SYONO, and S. AKIMOTO (1967). Effect of pressure on garnet-pyroxene equilibrium in the system $MgSiO_3$-$CaSiO_3$-Al_2O_3. *Earth Planet. Sci. Letters* **2,** 460–464.

LEWIS, J. S. (1971). Consequences of the presence of sulphur in the core of the earth. *Earth Planet. Sci. Letters* **11,** 130–134.

LLIBOUTRY, L. (1969). Sea-floor spreading, continental drift and lithosphere sinking with an asthenosphere at melting point. *J. Geophys. Res.* **74,** 6525–6540.

MACDONALD, G. J. F., and N. F. NESS (1960). Stability of phase transitions in the earth. *J. Geophys. Res.* **65,** 2173–2190.

MCKENZIE, D. P. (1968). The influence of the boundary conditions and rotation on convection currents in the earth's mantle. *Geophys. J. Roy. Astron. Soc.* **15,** 457–500.

——— (1969). Speculations on the consequences and causes of plate motions. *Geophys. J.* **18,** 1–32.

——— (1972). Plate tectonics. In: E.C. Robertson (ed.), "*The Nature of the Solid Earth*," chap. 13, pp. 323–360. McGraw-Hill, New York.

MCQUEEN, R. G., S. MARSH, and J. FRITZ (1967). Hugoniot equation of state of twelve rocks. *J. Geophys. Res.* **72,** 4999–5036.

MINEAR, J. W., and M. N. TOKSÖZ (1970). Thermal regime of a downgoing slab and new global tectonics. *J. Geophys. Res.* **75,** 1397–1419.

MIYASHIRO, A., F. SHIDA, and M. EWING (1970). Petrologic models for the Mid-Atlantic Ridge. *Deep Sea Res.* **17,** 109–123.

MORGAN, W. J. (1965). Gravity anomalies and convection currents, 2. *J. Geophys. Res.* **70,** 6189–6204.

MORIMOTO, N., S. AKIMOTO, K. KOTO, and M. TOKONAMI (1969). Modified spinel, beta-manganous orthogermanate: Stability and crystal structure. *Science* **165,** 586–588.

O'CONNELL, R. J., and G. J. WASSERBURG (1967). Dynamics of the motion of a phase change boundary to changes in pressure. *Rev. Geophys.* **5,** 329–410.

OVERSBY, V. M., and A. E. RINGWOOD (1972). Potassium distibution between metal and silicate and its bearing on the occurrence of potassium in the earth's core. *Earth Planet. Sci. Letters* **14,** 345–347.

OXBURGH, E. R., and D. L. TURCOTTE (1968). Mid-ocean ridges and geotherm distribution during mantle convection. *J. Geophys. Res.* **73,** 2643–2661.

——— and ——— (1970). Thermal structure of island arcs. *Bull. Geol. Soc. Am.* **81,** 1665–1688.

PASCAL, G., J. DUBOIS, M. BARAZANGI, B. ISACKS, and J. OLIVER (1973). Seismic velocity anomalies beneath the New Hebrides island arc: evidence for a detached slab in the upper mantle. *J. Geophys. Res.* **78,** 6998–7004.

PATERSON, M. S., and C. W. WEAVER (1970). Deformation of polycrystalline MgO under pressure. *J. Am. Ceram. Soc.* **53,** 463–471.

PEKERIS, C. L. (1935). Thermal convection in the earth's interior. *Mon. Not. Roy. Astron. Soc., Geophys. Supp.* **3,** 343–367.

RANDALL, M., and L. KNOPOFF (1970). The mechanism at the focus of deep earthquakes. *J. Geophys. Res.* **75,** 4965–4976.

RAPP, R. H. (1973). Numerical results from the combination of gravimetric and satellite data using the principles of least squares collocation. Dept of Geodetic Science Report No. 200, The Ohio State University, Columbus, Ohio, 53 pp.

RINGWOOD, A. E. (1956). Ph.D. Thesis, University of Melbourne.

——— (1958). The constitution of the mantle III. *Geochim. Cosmochim. Acta* **15,** 195–212.

——— (1967). The pyroxene-garnet transformation in the earth's mantle. *Earth Planet. Sci. Letters* **2,** 255–263.

——— (1971). Phase transformations and mantle dynamics. Presented at Pan Pacific Science Congress, Canberra, August 1971. Publication 999, Department of Geophysics and Geochemistry, Australian National University.

——— (1972). Phase transformations and mantle dynamics. *Earth Planet. Sci. Letters* **14,** 233–241.

——— and D. H. GREEN (1966). An experimental investigation of the gabbro-eclogite transformation and some geophysical implications. *Tectonophysics* **3,** 383–427.

——— and A. MAJOR (1970). The system Mg_2SiO_4-Fe_2SiO_4 at high pressures and temperatures. *Phys. Earth Planet. Interiors* **3,** 89–108.

RUNCORN, S. K. (1963). Satellite gravity measurements and convection in the mantle. *Nature* **200,** 628–630.

——— (1965). Changes in the convection pattern in the earth's mantle and continental drift: evidence for a cold origin of the earth. *Phil. Trans. Roy. Soc. London* **A258,** 228–251.

SCHUBERT, G., and D. L. TURCOTTE (1971). Phase changes and mantle convection. *J. Geophys. Res.* **76,** 1424–1432.

———, ———, and E. R. OXBURGH (1970). Phase change instability in the mantle. *Science* **169,** 1075–1077.

———, D. YUEN, and D. L. TURCOTTE (1975). Role of phase transitions in a dynamic mantle. (In press.)

SCLATER, J. G., and J. FRANCHETEAU (1970). The implications of terrestrial heat flow observations on current tectonic and geochemical models of the crust and upper mantle of the earth. *Geophys. J.* **20,** 509–542.

SYKES, L. R. (1966). The seismicity and deep structure of island arcs. *J. Geophys. Res.* **71,** 2981–3006.

TOKSÖZ, M. N., D. HARKRIDER, and A. BEN-MENAHEIM (1965). Determination of source parameters by amplitude equalization of seismic waves, 2. *J. Geophys. Res.* **70,** 907–922.

———, J. MINEAR, and B. JULIAN (1971). Temperature field and geophysical effects of a downgoing slab. *J. Geophys. Res.* **76,** 1113–1138.

TOZER, D. C. (1959). The electrical properties of the earth's interior. In: L. H. Ahrens, F. Press, K. Rankana and S. K. Runcorn (eds.), "*Physics and Chemistry of the Earth,*" vol. 3, pp. 414–436. Pergamon, London.

TURCOTTE, D. L., and E. R. OXBURGH (1967). Finite amplitude convective cells and continental drift. *J. Fluid Mech.* **28,** 29–42.

VENING MEINESZ, F. A. (1947). Major tectonic phenomena and the hypothesis of convection currents in the earth. *Quart. J. Geol. Soc. London* **103,** 191–207.

——— (1962). Thermal convection in the earth's interior. In: S.K. Runcorn (ed.), "*Continental Drift,*" pp. 144–176. Academic, New York.

VERHOOGEN, J. (1954). Discussion. *Trans. Am. Geophys. Union* **35,** 96–97.

——— (1965). Phase changes and convection in the earth's mantle. *Phil. Trans. Roy. Soc. London* **A258,** 276–283.

WETHERILL, G. W. (1961). Steady state calculations bearing on geological implications of a phase-transition Mohorovicic Discontinuity. *J. Geophys. Res.* **66,** 2983–2993.

WHITCOMB, J. H., and D. L. ANDERSON (1970). Reflection of P′P′ seismic waves from discontinuities in the mantle. *J. Geophys. Res.* **75,** 7513–5728.

WILSON, J. T. (1963). Hypothesis of earth's behaviour. *Nature* **198,** 925–929.

16
MANTLE COMPOSITION AND THE EARTH'S ORIGIN

16-1 INTRODUCTION

This book has been concerned chiefly with reviewing evidence relating to the composition and constitution of the mantle. Evidence of this nature necessarily has an important bearing upon theories of the earth's origin and provides several boundary conditions which must be satisfied by any acceptable theory. Currently, there is a widespread trend to regard meteorites as providing the key evidence relating to the chemical and physical conditions in the solar system which obtained during planetary formation.[1] It appears, however, that meteorites have undergone a complex evolution which is still poorly understood and controversial. It has always seemed to the author that greater use should be made of information derived directly from the study of the earth in the formulation of general hypotheses of planet formation. Curiously enough, as we shall see, certain widely discussed theories of planetary origin based largely upon the interpretation of meteoritic evidence provide inadequate explanations of the earth's composition and constitution.

[1]E.g., as in the models of Anders, Urey, Cameron, Wood, Grossman, and many others.

In Chap. 5 we developed the rationale for the pyrolite model of the primitive *upper mantle* composition and reviewed several sources of evidence which rather tightly constrained this primitive composition (Table 5-2). In Chap. 8 we saw how pyrolite was becoming irreversibly differentiated throughout geological time, leading to the formation of lithosphere plates of peridotite and eclogite which subsided to great depths in the mantle, displacing unfractionated primitive pyrolite upward to serve as the source of fresh basaltic magmas (Fig. 8-6). We reviewed evidence indicating that a large proportion of the mantle, perhaps 30 to 60%, had become irreversibly differentiated in this manner, implying that the pyrolite now in the low-velocity zone was *ultimately* derived from deep within the mantle (Sec. 8-7). These considerations, in conjunction with the relative constancy of the basaltic composition spectrum over the last 3 billion years suggested rather strongly that the mantle was originally rather well mixed and that the pyrolite composition as obtained for the present *upper* mantle may originally have applied to a large proportion of the deep mantle also. Perhaps, before the irreversible differentiation process began, the pyrolite composition was representative of the entire mantle.

In Chap. 14 we reviewed the extent to which the pyrolite composition could satisfy the known physical properties of the entire mantle when due account was taken of the effects of phase transformations occurring under the high pressures of the earth's interior. We concluded that, to a first approximation, the pyrolite composition undergoing phase changes required by experimental data provided a satisfactory explanation of the radial distribution of physical properties—principally density and elasticity—throughout the mantle. This did not preclude the possibility of a degree of local heterogeneity in the deep mantle caused by the irreversible differentiation processes mentioned above. The first-order conclusion was that the lower mantle appeared to be *approximately* uniform and homogeneous in major element composition, except perhaps for the lowermost 200 km or so.

A suggestion that the deep mantle may contain a few percent more iron oxide than the pyrolite upper mantle was examined and found to be unnecessary and without adequate justification, although it could not be absolutely precluded. It could be firmly concluded, however, that the lower mantle does not contain substantially *less* iron than pyrolite.

Perhaps the most important single boundary condition for the earth's origin arising from study of the mantle is the approximate chemical homogeneity of the mantle.[1,2] The absence of any substantive evidence for large-scale radial chemical zoning below 200 km suggests either that the silicate component of the material from which the earth accreted was uniform in composition or that, if this were

[1]I.e., over volumes which are sufficiently large to average the compositions of residual peridotite and eclogite (and their high-pressure derivatives) in the subducted lithosphere. According to the discussion of Sec. 8-7, a closer approach to ideal homogeneity would have been realized very early in the earth's history before the mantle's structure had been modified by plate subduction processes.

[2]Excluding the lowermost 200 km or so of the mantle (Sec. 16-9).

not so, an efficient stirring process occurred subsequently to accretion, resulting in a close approach to homogeneity.

The above conclusion regarding the original approximate homogeneity of the mantle is further supported by a comparison of the pyrolite composition with primordial abundances as discussed in the next section.

16-2 COSMOCHEMICAL ASPECTS OF THE PYROLITE COMPOSITION

Some relationships between the pyrolite composition and the primordial chondritic or solar elemental abundances were discussed in Sec. 5-3. It was noted that virtually all current theories of formation of the planets hold or imply that processes involving differential volatility or condensation played an important role in establishing the present composition of planets. In the case of the earth, it is known that a wide range of metals and nonmetals which are volatile under high-temperature reducing conditions were lost during, or prior to, accretion. An important question is whether or not there was any substantial fractionation among *involatile* elements during the earth's formation from primordial material. In other words, can we construct a geochemically and geophysically self-consistent earth model from the primordial relative abundances of a group of elements which are characterized by the property of involatility under the probable conditions of earth formation?

The classification of elements as "volatile" or "involatile" in the above context necessarily requires the employment of rather arbitrary criteria, relating to the assumed conditions of earth formation. We will take elements in the volatile group to be those which are readily volatilized from silicate melts and oxycompounds at temperatures of 1000 to 1500°C in the presence of significant partial pressures of hydrogen and/or in the presence of carbon. Elements of the involatile group do not volatilize under these conditions. It is useful to divide the involatile group further into oxyphile and siderophile subgroups, according to the way in which they are partitioned when silicate-oxide phases are equilibrated with metallic iron. The siderophile elements are those which preferentially enter the metal phase, and vice versa for the oxyphile elements. Examples of elements in the above groups and subgroups are given in Table 16-1.

It was demonstrated in Sec. 5-3 that an earth model possessing the observed core-to-mantle mass ratio could be constructed from the primordial abundances of involatile elements (group I, Table 16-1), providing several percent of SiO_2 were reduced to elemental silicon and assigned to the core.[1] Of greater importance to

[1]Alternatively, the core may contain 10 to 15% sulphur, whilst some silicon (about 20% of the total) must be assumed to have been lost from the earth as a volatile element during its formation. This possibility is discussed further in Sec. 16-9. It is consistent with the above model and classification since the volatility of Si is intermediate between those of Mg and Na (Grossman, 1972a). In Table 16-1, sodium is the least volatile member of group II, whereas silicon is the most volatile member of group IA.

Table 16-1 CLASSIFICATION OF SOME ELEMENTS* ACCORDING TO THEIR RELATIVE VOLATILITIES FROM MAFIC SILICATE MELTS UNDER HIGH-TEMPERATURE REDUCING CONDITIONS

I. Involatile group	II. Volatile group
A. Oxyphile subgroup	*A. Nonmetals*
Li, Mg, Al, Si, Ca, Sc, Ti, Sr, Y, Zr, Nb, Ba, rare earths, Hf, Ta, Th, U	He, Ne, Ar, Kr, Xe H, C, N, S, F, Cl, Br, I
B. Siderophile subgroup	*B. Metals*
Fe, Co, Ni, Cu, Ag, Au, Mo, W, Ru, Rh, Pd, Re, Os, Ir, Pt	Na, K, Rb, Cs, Zn, Cd, Hg, Tl, In, Pb, Bi

*This list is not intended to be complete.

the present discussion was the conclusion from Sec. 5-3 that the relative abundances of involatile oxyphile elements (group IA of Table 16-1) present in pyrolite were similar to the primordial relative abundances of these elements as observed in Type I carbonaceous chondrites and in the solar photosphere. The absence of marked fractionation of involatile elements between these different environments appears to represent a pattern of major cosmochemical significance, and it seems reasonable to assume that it extends also to the lower mantle. Certainly, any contrary assumption would be contrived and entirely arbitrary in the light of the geophysical evidence discussed in Chap. 14. These considerations strengthen the case for believing that the pyrolite composition may be representative of the entire mantle.

Whilst a self-consistent earth model can be constructed from the primordial abundances of involatile elements, there is no doubt that the earth has been strongly depleted in volatile elements (group II, Table 16-1) relative to the primordial abundances.[1,2,3,4] The depletions seem to fall into two subgroups. The metals (IIB of Table 16-1) appear depleted mostly by factors of 4 to 20 compared to subgroup IA, whereas the depletion factors of the nonmetals IIA are more variable and mostly much greater than those of subgroup IIB.[3] The pattern of depletions of volatile elements relative to involatile elements (Table 16-1) provides an important boundary condition for theories of the earth's origin.

Another key boundary condition is provided by the observed abundances of siderophile elements (subgroup IB of Table 16-1) in pyrolite. Where data on partition coefficients between metal and silicate phases exist (e.g., in the cases of Ni, Co, Cu, Au, Re),[3,5] the abundances of siderophile elements in pyrolite appear to

[1]Gast (1960, 1972).
[2]Ringwood (1962).
[3]Ringwood (1966a, b).
[4]Oversby and Ringwood (1972).
[5]Kimura, Lewis, and Anders (1973).

be much greater (10- to 100-fold) than would be expected if pyrolite had once equilibrated with metallic iron, which subsequently had segregated into the earth's core.[1] This behaviour probably extends to the remainder of the siderophile subgroup IB of Table 16-1.[1] The comparatively high abundances of siderophile elements in pyrolite seem to require that equilibration with metallic iron has not occurred.

This inference is supported by evidence from another direction. Primary pyrolite contains substantial amounts of ferric iron. In order to explain the Fe^{3+}/Fe^{++} ratios of fresh oceanic basaltic glasses and of unaltered primary mantle minerals (from xenoliths and peridotites), an Fe^{3+}/Fe^{++} ratio of 0.05 to 0.1 is required for primary pyrolite. On the other hand, the Fe^{3+}/Fe^{++} ratios of basalts, pyroxenes, and spinels which have equilibrated with metallic iron at high temperatures are much lower, probably by at least an order of magnitude[1] (Table 16-2).

This difference in oxidation states is reflected in the nature of volatiles developed by degassing (Table 16-2). It is well known that the volatiles degassed from the earth are dominantly composed of H_2O and CO_2, rather than H_2 and CO. Thus, the gas phase in equilibrium with an average Hawaiian tholeiite[2] at 1200°C would have $H_2O/H_2 \sim 120$ and $CO_2/CO \sim 35$. These ratios agree closely with Rubey's (1951) estimates for the composition of volatiles degassed from the earth. The observed composition of the gas phase is thus compatible with the oxidation state of the pyrolite mantle, which indeed constitutes the ultimate redox-state buffer. On the other hand, if the mantle had equilibrated with metallic iron, as in the lunar interior, the gas phase evolved would be dominated[3,1] by CO and H_2, rather than CO_2 and H_2O (Table 16-2).

[1]Ringwood (1966b, 1971).
[2]Holland (1963).
[3]Mueller (1964).

Table 16-2 COMPOSITIONS OF GAS PHASES IN EQUILIBRIUM AT 1200°C WITH MINERAL ASSEMBLAGES IN THE EARTH'S MANTLE AND IN THE LUNAR INTERIOR†

	Simplified mineral assemblage	f_{O_2} (atm)	CO_2/CO	H_2O/H_2
Earth	$(Mg, Fe)_2SiO_4 + Fe_3O_4$* $Fe^{++}/Fe^{3+} \sim 10$	10^{-8}	35	120
Moon	$(Mg, Fe)_2SiO_4$ + Fe metal $Fe^{++}/Fe^{3+} > 200$	10^{-14}	0.05	0.1
	Simplified equilibria			
Earth	$Fe_2O_3 + CO = 2FeO + CO_2$			
Moon	$FeO + CO = Fe\ (metal) + CO_2$ The same for H_2O, H_2			

*Fe_3O_4 as component of spinels and pyroxenes.
†Holland (1963).

We proceed now to discuss the extent to which various hypotheses of the earth's origin are capable of accounting for the characteristics of the pyrolite composition discussed earlier in this section.

16-3 HOMOGENEOUS ACCRETION HYPOTHESES

The most widely supported class of hypotheses regarding the earth's origin during the last 25 years has maintained that the earth accreted from an intimate mixture of silicate particles and metal particles, generally resembling ordinary chondritic meteorites.[e.g.,1,2,3,4,5,6,7,8] The chondritic material is assumed to have been formed in the solar nebula by a complex series of chemical and physical processes which occurred prior to the accretion of planets. There are some difficult problems in explaining the nature of these processes and, for example, the way in which they lead to explanations of density differences between planets. These, however, are beyond the scope of this chapter. We are primarily concerned with evaluating the hypothesis in relation to evidence obtained directly from the earth.

According to this class of hypotheses, accretion of the earth occurred over a sufficiently long period (10^7 to 10^8 years) so that its gravitational potential energy was efficiently radiated away and it formed in an initially "cool" and unmelted condition with an average temperature less than 1000°C. Subsequently, heating by long-lived radioactive elements occurred, leading to melting of the metal phase and its segregation into bodies which were large enough to sink through the plastic mantle to form the core. Thus, according to these models, not only did the earth accrete initially from relatively homogeneous, well-mixed material, but also, for a considerable period thereafter, the earth itself was approximately homogeneous, consisting of an intimate mixture of metal and silicate phases.[9] The major internal differentiation resulting in formation of the core occurred much later (e.g., 10^9 years) than the primary accretion of the earth.

Homogeneous accretion hypotheses account simply in principle for one key property of the mantle as discussed in Sec. 16-1—namely, its approximately uniform chemical composition, which reflects the composition of the well-mixed silicate component of the accreting material. This is an important attribute. Nevertheless, a formidable array of difficulties has arisen.

[1]Urey (1952, 1956, 1957, 1958, 1962, 1963).
[2]Kuiper (1952).
[3]Wood (1962).
[4]Birch (1964, 1965a).
[5]Vinogradov (1961).
[6]Runcorn (1962, 1965).
[7]Elsasser (1963).
[8]Lubimova (1958).
[9]Because of these characteristics, this class of theories may be described as "homogeneous accretion" theories. This represents a restriction of the usage of Turekian and Clark (1969), who used this term to represent hypotheses in which the earth accreted from uniform material without reference in its state immediately after accretion.

The arguments[1] based largely upon the distribution and abundances of volatile elements, which convinced most geochemists and cosmogonists during the fifties and sixties that the earth accreted in a cool, unmelted condition, are now known to be invalid, being based upon incorrect estimates of the primordial abundances of volatile metals and their distribution within the earth.[2,3] It is now widely accepted that much of the accretion of the earth could have occurred under sustained high-temperature conditions, providing that a small proportion of a distinct low-temperature "carrier," rich in volatile components were suitably incorporated.[4] Many variants of this idea have since been suggested.[e.g.,3,5,6,7,8,9,10,11]

Parallel to this development has been the conclusion based upon lead-uranium fractionation between metal and silicates and its effect upon the isotopic composition of mantle lead that the time interval between accretion of the earth and segregation of the core was short, most probably less than 10^8 years and almost certainly less than 5×10^8 years.[4,12,13] Core formation may well have proceeded simultaneously with accretion. Since core formation is a highly energetic event, evolving about 600 cal/g for the entire earth,[14,15] enough to heat it by 2000°C, and moreover, it could hardly commence and proceed unless the mean temperature of the earth after its formation was at least about 1000°C.[16] it is difficult to avoid the conclusion that either during or very soon after its formation most of the earth's interior attained high temperatures, with an average in the vicinity of 2500°C.

This is a very different picture of the earth's primitive state than was originally presented by advocates of homogeneous accumulation models. As we saw in Sec. 5-5, however, it agrees with considerations of the earth's thermal history based upon the K, U, and Th abundances in pyrolite.

We turn now to some specific chemical difficulties of the homogeneous accretion hypothesis arising from the pyrolite mantle composition. Firstly, we note the deficiency of alkali metals in the earth as compared to chondrites, as demonstrated by Gast (1960). Likewise, the FeO/FeO + MgO ratio and SiO_2 content

[1] Urey (1954).
[2] Ringwood (1966a, b).
[3] Anders (1968, 1971a).
[4] Ringwood (1960).
[5] Turekian and Clark (1969).
[6] Clark, Turekian, and Grossman (1972).
[7] Larimer and Anders (1967, 1970).
[8] Ganapathy et al. (1970).
[9] Cameron (1972, 1973a).
[10] Gast (1972).
[11] Anderson and Hanks (1972).
[12] Armstrong (1968).
[13] Oversby and Ringwood (1971).
[14] Urey (1952).
[15] Flasar and Birch (1973).
[16] Even allowing for the possible presence of a substantial amount of sulphur.

of the silicate phases of ordinary chondrites are dissimilar to those of pyrolite. Accordingly, it is necessary to make some ad hoc assumptions about the nature of the chemical fractionation processes which occurred in the nebula prior to accretion if the model is to be sustained.

A major difficulty concerns the abundances of siderophile elements in pyrolite, as detailed in Sec. 16-2. If the earth had once consisted of an intimate mixture of iron and silicate phases which had been slowly heated until the metal melted and segregated, it is inevitable that local chemical equilibrium would have been established between metal and silicate phases. Yet the pyrolite abundances of Ni, Co, Cu, Au, and probably, also many other siderophile elements are one to two orders of magnitude higher than can be explained by equilibrium partition between metal and silicate. It appears certain that the core has formed under conditions in which chemical equilibrium with the mantle was not maintained.[1]

A corresponding problem arises in connection with the redox state of the mantle and the nature of the volatiles which are observed to be degassed from pyrolite to form the hydrosphere and atmosphere. As noted in Sec. 16-2, the Fe^{3+}/Fe^{++} ratio of the mantle is at least 10 times higher than would occur if the mantle had equilibrated with metallic iron which subsequently separated into the core. The difference in redox states between the earth's mantle and the lunar interior reflects this situation. In consequence, the volatiles degassed from the mantle are dominantly composed of CO_2 and H_2O, whereas if the mantle had once equilibrated with metallic iron, the gases evolved should be composed dominantly of CO and H_2 (Table 16-2).[2] This evidence supports the previous conclusion that the earth's core and mantle were formed under conditions during which chemical equilibrium was not maintained.

A related difficulty faced by homogeneous accretion hypotheses is the origin of the earth's hydrosphere and atmosphere. According to these hypotheses, small amounts of volatile components such as H_2O and CO_2 were chemically bound, and/or adsorbed in the silicate component of the material from which the earth accreted.[3,4] According to a more recent version,[5] the volatiles were introduced via a small proportion of a well-mixed "carrier," similar to Type 1 carbonaceous chondrites. It is assumed that, with sufficiently slow accretion under cool conditions, the volatiles were trapped within the earth. Subsequent heating of the earth's interior by radioactivity led to magmatism accompanied by partial degassing—the principal species degassed being H_2O, CO_2, and N_2. These species accumulated to form the hydrosphere and atmosphere.

However, even if it were possible to trap some H_2O and CO_2 in this manner, these molecules would be decomposed in the presence of excess metallic iron. As

[1]Ringwood (1966a, b).

[2]See Ringwood (1966b, p. 69; 1971) for further discussion of this point.

[3]Brown (1952).

[4]Rubey (1951).

[5]Larimer and Anders (1967).

the intimate mixture of iron and silicates is heated above 1000°C, leading to melting and core formation, any trapped water and carbon dioxide would be reduced according to the following simplified reactions:[1]

$$Fe + H_2O \rightarrow FeO \text{ (in silicate)} + H_{2(\text{Fe soln.})}$$

$$2Fe + CO_2 \rightarrow 2FeO \text{ (in silicate)} + C_{(\text{Fe soln.})}$$

The hydrogen and carbon produced are both soluble in excess iron, and the free energies of formation of these solutions will drive the equilibria to the right. This effect is accentuated by reaction of FeO to form ferromagnesian silicates. Moreover, the volume changes of these reactions when the H_2 and C enter the metal phase to form interstitial solutions are markedly negative so that the high pressures in the earth's interior also drive the equilibria strongly to the right. The net effect is the total decomposition of any trapped CO_2 and H_2O, the oxygen entering mantle silicates as FeO, whilst the C and H_2 are removed from the mantle with the iron as it sinks into the core. The homogeneous accretion model thus appears to offer an inadequate explanation of the earth's hydrosphere and atmosphere.

16-4 MODIFIED HOMOGENEOUS ACCRETION HYPOTHESIS

In an attempt to escape some of the difficulties mentioned previously, Murthy and Hall (1970)[2] have proposed that the earth accreted homogeneously from a mixture of meteoritic material consisting of 40% Type I carbonaceous chondrites, 45% ordinary chondrites, and 15% iron meteorites. This mixture contains a total of about 5% sulphur, mainly in the carbonaceous chondrite component. Because of the low eutectic temperature in the Fe-FeS system (990°C), accretion and core segregation is able to proceed at a lower temperature than in the case of the previous model, permitting early formation of the core, perhaps even during accretion. According to this model, the additional light element in the core is mainly sulphur.

A serious constraint on this model is imposed by the high volatility of sulphur in the environment provided by the solar nebula. Sulphur readily forms the volatile species H_2S, COS, and CS_2, and the conditions required for the condensation of sulphur are highly restrictive. For example, troilite (FeS) can be condensed only at quite low temperatures (~ 400°C—Table 16-3). Yet the model requires that half of the primordial sulphur was retained by the earth during accretion, whereas most of the rubidium, potassium, and sodium were lost. These latter elements are much less volatile than sulphur (Table 16-3) under most conditions, and it is difficult to define a chemical environment during the earth's formation

[1] Mueller (1964).
[2] See also Hall and Murthy (1971), Murthy and Hall (1972), and Lewis (1971).

which would permit the accretion of sulphur but not of Na, K, and Rb. In the case of the model advocated by Murthy and Hall, in which accretion of much of the earth is required to proceed at relatively *low* temperatures ($< 1000°C$), the difficulty appears insuperable.[1] The problem is highlighted by the compositions of meteorites. In the chemical processing by which ordinary chondrites were formed from primordial material, about 80% of the sulphur was lost, but Na, K, and Rb remained essentially intact. Likewise, iron meteorites contain on the average only about 1% sulphur, yet, they evidently formed in an environment (prior to metal segregation) where Na, K, and Rb were undepleted, as indicated by the compositions of silicate inclusions found in iron meteorites.

Lewis (1971) and Hall and Murthy (1971) have recognized this problem[2] and have proposed an interesting solution. They claim that K and Rb are actually chalcophile under core-formation conditions and, accordingly, become preferentially partitioned into the sulphide phase which enters the core. Thus the earth as a whole possesses a chondritic composition and is not depleted in volatile metals.

This proposal was based upon rather inadequate[3] thermodynamic data. A detailed experimental investigation of the partition of potassium between Fe-FeS melts and a wide range of silicates demonstrated that potassium did not possess the claimed chalcophile properties and would not preferentially enter the core.[3] Thus, there can be no reasonable doubt that earlier conclusions regarding the earth's depletion in alkali metals and other volatiles compared to primordial abundances are correct. The problem of retaining most of the sulphur whilst losing most of the alkali metals remains a formidable objection to the Hall-Murthy-Lewis hypothesis.

This is by no means the only serious problem. The model necessarily results in local chemical equilibrium between a predominantly (85%) metallic iron phase and surrounding silicates. Siderophile elements would thus become preferentially partitioned into the metal phase and transported into the core. The high abundances of siderophile elements in pyrolite are thus unexplained. Indeed, compared to primordial abundances, pyrolite contains at least 10 times more nickel, cobalt, and copper than sulphur. It is scarcely credible that the core-formation process extracted sulphur 10 times more efficiently than these other metals. It seems much more likely that sulphur has suffered a net depletion during formation of the earth comparable to that displayed by other volatile elements. (This leaves open the possibility that a limited amount, say, 20%, of the primordial sulphur was retained and may now be present in the core—Sec. 16-9.)

Another problem concerns the iron content of the mantle. The Hall-Murthy model leads to a Fe/Fe + Mg (mol) ratio of 0.3, which is much higher than that of

[1]In Sec. 16-9, the possibility of condensing a small proportion of the cosmic complement of sulphur in a liquid Fe-S alloy at *high* temperature (1300 to 1600°C) is considered. The conditions required are incompatible with Murthy and Hall's model.

[2]However, they appear to have ignored the depletion of Na in the mantle.

[3]Oversby and Ringwood (1972, 1973).

pyrolite and, indeed, is higher than permitted by the geophysical constraints discussed in Chap. 14. A further aspect concerns the redox state of the mantle and the nature of the equilibrium gas phases. The model leads to a redox state close to that of earlier versions of homogeneous accretion hypotheses (Sec. 16-3), thereby encountering similar difficulties.

Alternative Approaches

The difficulties encountered by homogeneous accretion models have caused several authors to abandon them and to explore other possibilities. Ringwood (1960, 1966a, b) developed a "single-stage" model according to which the earth formed by accretion in a cold solar nebula from primitive planetesimals resembling Type I carbonaceous chondrites. According to this model, chemical reduction to produce metallic iron, evaporation of volatiles, and core formation occurred during accretion, and the primary energy source was the earth's gravitational potential energy. A contrasting "inhomogeneous accretion" model has been suggested by Turekian and Clark (1969) and further developed by several others.[1,2,3,4,5,6,7] According to this model, the earth formed by accretion from a hot, condensing solar nebula and developed a primary chemical zoning corresponding to the sequence of condensation of solids from the nebula.

In most aspects, these hypotheses are diametrically opposed. However, they have one important element in common. Both maintain that the siderophile and volatile elements now present in the mantle were introduced in the form of a primitive low-temperature component generally resembling Type I carbonaceous chondrites and that this component was introduced under such conditions that equilibration with metallic iron did not occur. The "single-stage" model introduces this component at the earliest stage of accretion, whereas the "heterogeneous accumulation" model plasters them on as a veneer at the final stage of accretion.

16-5 THE HETEROGENEOUS ACCUMULATION HYPOTHESIS

One of the earliest hypotheses in this category was proposed by Eucken (1944), who investigated the condensation behaviour of hot solar gases which, for example, might have been torn out of the sun by tidal interaction with a passing star, as proposed in Jeans' hypothesis of solar-system formation. Eucken found that

[1]Clark, Turekian, and Grossman (1972).
[2]Grossman (1972a, b).
[3]Anders (1968, 1971a).
[4]Ganapathy et al. (1970).
[5]Cameron (1973a).
[6]Anderson (1973).
[7]Anderson and Hanks (1972).

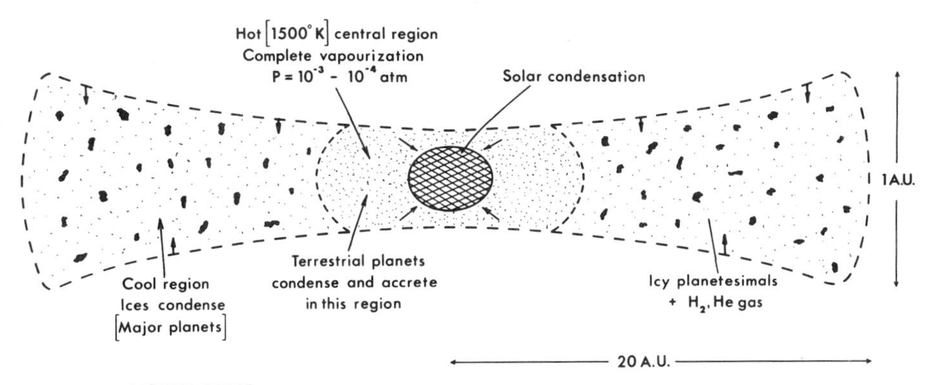

FIGURE 16-1
Parental solar nebula according to Cameron's "stellisk" model. Nebula amounts to 2 solar masses.

molten metallic iron would condense first, followed by silicates. He suggested that the gross structure of the earth, with its metallic core surrounded by silicate mantle, was a consequence of direct sequential condensation in this manner. However, with the demise of Jeans' hypothesis and with the evidence produced by Latimer (1950), Urey (1952), and others that at least *part* of the earth had accreted at low temperatures, most probably in a *cool* solar nebula, Eucken's proposals were generally discarded in favour of the homogeneous accumulation models discussed in Sec. 16-3.

In recent years, the heterogeneous accumulation model for the formation of the earth has been revived[1,2,3,4,5,6,7] in the context of a specific theory of solar system origin developed by Cameron and coworkers.[8,9,10] According to this theory, the solar system developed from a rotating discoidal nebula (stellisk) amounting to two solar masses. Cameron shows that the contraction of a nebula of this mass will produce high temperatures (1000 to 1700°C) in the vicinity of the terrestrial planets (Fig. 16-1). Moreover, the dissipation time of the nebula is very short (10^3 to 10^4 years), and Cameron argues that most of the formation of planets must have occurred over this brief time interval, before dissipation was complete.

[1]Turekian and Clark (1969).
[2]Clark, Turekian, and Grossman (1972).
[3]Grossman (1972a, b).
[4]Anders (1968, 1971a).
[5]Ganapathy et al. (1970).
[6]Anderson (1973).
[7]Cameron (1973a).
[8]Cameron (1962, 1963, 1969, 1970, 1972, 1973a, b, c).
[9]Ezer and Cameron (1963).
[10]Cameron and Pine (1973).

Table 16-3 CONDENSATION SEQUENCE FROM NEBULA OF SOLAR COMPOSITION AT 10^{-3} ATM. (*After Grossman*, 1972*a*)

Phase	Temperature (°C)
CaO, Al_2O_3, TiO_2, ZrO_2	1570–1230
Metallic iron	1200–1030
Forsterite (Mg_2SiO_4), enstatite ($MgSiO_3$)	1170–1030
Alkali felspar [$(Na, K)AlSi_3O_8$]	~730
Troilite (FeS)	430
FeO (as component of ferromagnesian silicates, e.g., olivine [$(Fe_{0.1}Mg_{0.9})_2SiO_4$]	300
Magnetite	130
Carbonaceous compounds*	100–200
Hydrated magnesium silicates*	0–100
Ice	<0

*Larimer and Anders (1967).

Under the *P-T* conditions reached in the inner regions of the solar nebula, the dust grains originally present in the parental cloud would have been completely evaporated. On cooling of the gas, components would condense over a wide range of temperatures, allowing for the possibility of complex chemical and physical fractionations. Condensation sequences for hot solar gases have been calculated by several authors[1,2,3,4] using modern thermodynamic data. In general, metallic iron and silicates condense at high temperatures and oxidized iron, iron sulphide and hydrated silicates at low temperatures (Table 16-3). It has been demonstrated that many chemical differences between the several classes of chondritic meteorites can be explained in principle by selective condensation from the gas phase followed by remixing of fractions condensed in specific temperature ranges in appropriate proportions.[3,5,6] These processes are held by some to have occurred in the environment provided by Cameron's model and, thus, to provide direct support for the reality of the model.[5,6,3] It should be mentioned in passing, however, that there are a sizable number of students of meteorites who, whilst recognizing the influence upon their compositions of selective condensation and evaporation accompanied by remixing of various fractions, nevertheless do not believe that *T-P* conditions required for these processes were *representative* of the inner solar nebula (Fig. 16-1) in the sense of defining the *average T-P* condi-

[1]Lord (1965).
[2]Larimer (1967).
[3]Grossman (1972a, b, 1973).
[4]Wood (1963).
[5]Larimer and Anders (1967, 1970).
[6]Anders (1968, 1971a).

tions followed by this region in the course of its evolution. A more likely environment for evaporation-condensation fractionations and chondrule formation is believed by some to be provided by highly localized impact phenomena at the surfaces of parent bodies of asteroidal (and perhaps larger) dimensions.[1,2,3,4,5,6,7,8,9,10] Collisions between comets, resulting in the formation of rapidly cooling gas balls of primordial composition (but depleted in H and He), might also have provided appropriate environments for chondrule formation and for the formation of the Ca-Al rich inclusions found in some carbonaceous chondrites.

According to Larimer's (1967) calculations, iron would condense at higher temperatures than magnesium silicates in a cooling solar gas ($P \sim 10^{-3}$ atm), as in the earlier results by Eucken. Turekian and Clark (1969) applied these results in the context of Cameron's model to propose that accretion of the earth occurred simultaneously with condensation of the solar nebula over a period of about 10^4 years, resulting in gross stratification of the earth, with an iron core surrounded by a lower mantle dominantly of magnesium silicates. The later stages of accretion occurred after the nebula had cooled to relatively low temperatures so that the equilibrium condensate contained oxidized iron, iron sulphide, hydrated magnesium silicates, plus other volatiles—i.e., similar to Type I carbonaceous chondrites. Volatile-rich material of this nature, mixed with earlier high-temperature condensates which had failed to accrete previously are believed to have accreted upon the earth over a longer time scale (10^5 to 10^7 years) to produce the upper mantle-crust system.[11]

An important feature[12,13] of this heterogeneous accumulation model is that the upper mantle has never been in contact with the core: "It is a direct consequence of the short time scale of the main phase of accretion, which enables the core to settle to the center and be surrounded by the lower mantle before the outer layers are added."[13] In this way, an acceptable explanation of the boundary conditions discussed in Sec. 16-2 is provided—i.e., the high abundance of siderophile elements in the upper mantle, the high oxidation state of this region, and its retention of volatiles, dominated by H_2O and CO_2.

[1] Urey and Craig (1953).
[2] Urey (1956).
[3] Fredriksson (1963, 1969).
[4] Ringwood (1966c).
[5] Kurat, Keil, Prinz, and Nehru (1972).
[6] Wlotzka (1969).
[7] Kurat (1970).
[8] King, Carman, and Butler (1972).
[9] Dodd (1971).
[10] Fredriksson, Noonan, and Nelen (1973).
[11] See also Anders (1968).
[12] Turekian and Clark (1969).
[13] Clark, Turekian, and Grossman (1972).

Some General Considerations

Before considering the capacity of this model to account specifically for the composition and constitution of the mantle, as discussed in the preceding chapters of this book, some general aspects should be considered.

In order to provide the required high temperatures in the nebula and the rapid accretion of planets, the model seems rather firmly committed to Cameron's specific hypothesis of a massive parental nebula amounting to two solar masses. It should be recalled that there are several alternative and competing hypotheses dealing with the development of the solar nebula[1] and that these often differ in major respects from Cameron's. It is not appropriate in the present chapter to discuss their astrophysical merits and demerits, simply to note that this is a highly controversial and speculative area. There is, however, one obvious astrophysical difficulty facing Cameron's hypothesis which is of such a magnitude that it cannot be ignored.[2,3]

The hypothesis[4] requires that nearly one solar mass was lost during the formation of the sun and planets. The loss of perhaps one-fifth of a solar mass may have occurred in the form of an intense solar wind emitted during the T Tauri phase. However, the remaining material must have been lost from the region where planets accumulated. The model explicitly leads to rapid cooling of the nebula on a time scale of 10^3 to 10^4 years, and this is accompanied by precipitation of solid metal, oxide, and silicate planetesimals in the region of the terrestrial planets. Likewise, the accretion of the planets from condensing material is maintained to be highly efficient since it is synchronized with cooling of the nebula on this rapid time scale. According to the hypothesis, there should be about 100 times more solid material in the region of the terrestrial planets than is now present.

Bearing in mind the highly efficient condensation and accretion processes which were proposed, why did the terrestrial planets capture such a small proportion of the available solid condensates? How was the immense amount of unaccreted solid condensate removed from the solar system? The answers so far provided are unconvincing.[4] Much of the hot, solid condensate must surely have collected into planetesimals which in turn should have formed additional planets. The problems attached to ejection of large aggregates of solids from the solar system are immeasurably greater than for gases.

Returning to the heterogeneous accumulation hypothesis, subsequent thermochemical studies have disclosed complications in the simple scheme proposed by Turekian and Clark. For example, using the latest solar abundance and thermodynamic data, Grossman (1972a) found a large degree of overlap in the con-

[1]Section 16-6.
[2]Kaula (1968).
[3]Öpik (1973).
[4]Cameron (1973a).

densation intervals of iron and magnesium silicates. At 10^{-3} atm, 46% of the iron condensed before forsterite appeared, after which magnesium silicates and iron crystallized simultaneously. However, at 3×10^{-4} atm, which is taken to be the pressure in the region where the earth accreted,[1] only 30% of total iron precipitated before being joined by magnesium silicates. The uncertainties in these estimates are substantial. Thus the appealing early picture[2] of iron-core accumulation being cleanly differentiated from mantle accumulation is considerably blurred.[3] Another problem concerns the condensation of highly involatile oxides—principally CaO and Al_2O_3. These are almost totally precipitated well before the condensation of iron, and the model therefore implies that they should be clearly separated from the magnesium silicates of the lower mantle (Table 16-3). Theoretically, the earth should possess an inner core of involatile oxides dominated by CaO and Al_2O_3.[4]

Application to the Earth—Some Specific Difficulties

(i) Zoning in the mantle The heterogeneous accretion model leads to an earth zoned radially according to volatility (Fig. 16-2). If we ignore the difficulty that Ca-Al oxides should form an inner core and permit them to rise through the surrounding metal,[4] they should have segregated at the base of the mantle to form a layer several hundred kilometres thick. Such a layer should have highly distinctive seismic properties which have not been observed.

Likewise, the upper mantle, representing a low-temperature lately accreted component, might be expected to be strongly depleted in Ca, Al, and other highly involatile elements. Yet it is certain that, in the upper mantle source regions of basalt, Ca, Al, Ti, Zr, etc., are *not* depleted compared to Mg + Si in relation to the primordial abundances. Indeed, the most striking inference (Secs. 5-3 and 5-4) is that the relative abundances of these elements in pyrolite resemble their primordial relative abundances.

(ii) Iron content of the lower mantle Equilibrium calculations[5] show that significant amounts of FeO do not begin to enter the magnesium silicates condensing from solar gases until the temperature has fallen below 400°C (Fig. 16-3). This is far below the temperature at which the lower mantle is believed to condense according to the heterogeneous accretion model (1000 to 1200°C). Accordingly, the model requires that the magnesium silicates in the lower mantle are essentially

[1]Cameron (1973c).
[2]Turekian and Clark (1969).
[3]Clark, Turekian, and Grossman (1972).
[4]Anderson and Hanks (1972) proposed that the earth indeed formed in this manner. However, the U + Th content of the oxide inner core produced sufficient heat to melt the surrounding iron core and permitted the Ca, Al-rich material to rise into the lower mantle. The model grows in complexity.
[5]Grossman (1972a).

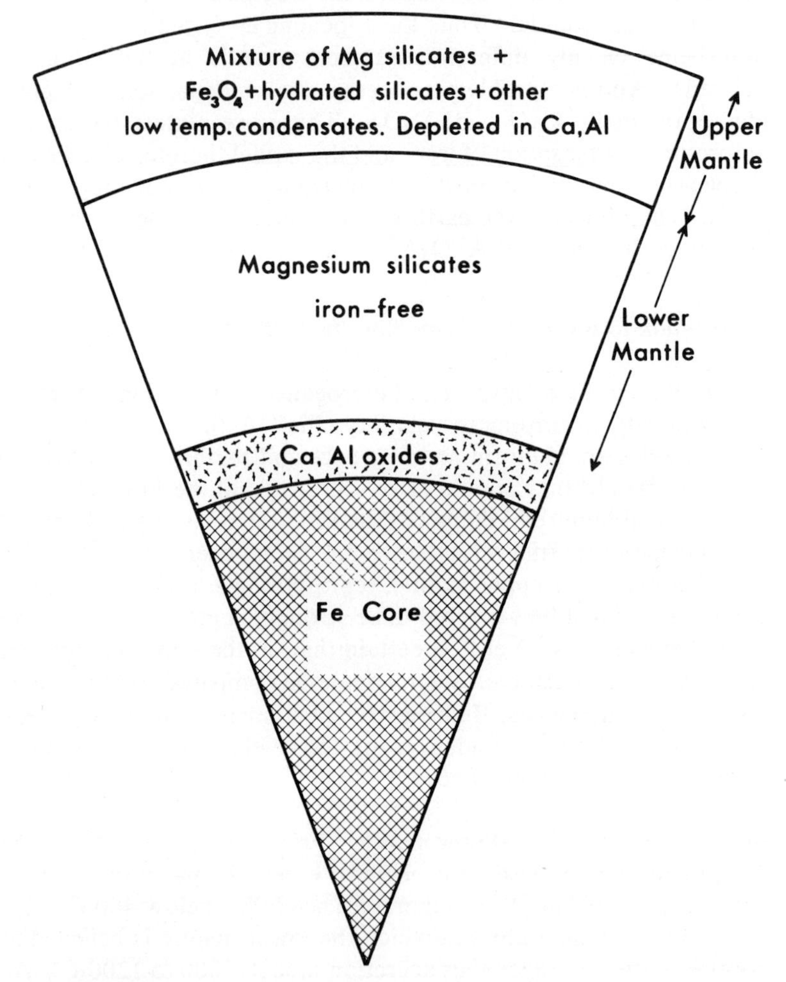

FIGURE 16-2
Zonal structure of earth suggested by heterogeneous accretion hypothesis of Turekian and Clark (1969). The Ca, Al oxide layer is assumed to have segregated from its original location in the inner core. (*Anderson and Hanks. 1972.*)

free of FeO (Fig. 16-3). The iron content of the lower mantle was extensively discussed in Chap. 14. It was concluded that the FeO/FeO + MgO ratio of this region almost certainly lies between 0.1 and 0.2 and most probably is similar to that of upper mantle pyrolite (0.12). If the lower mantle contained no iron, its

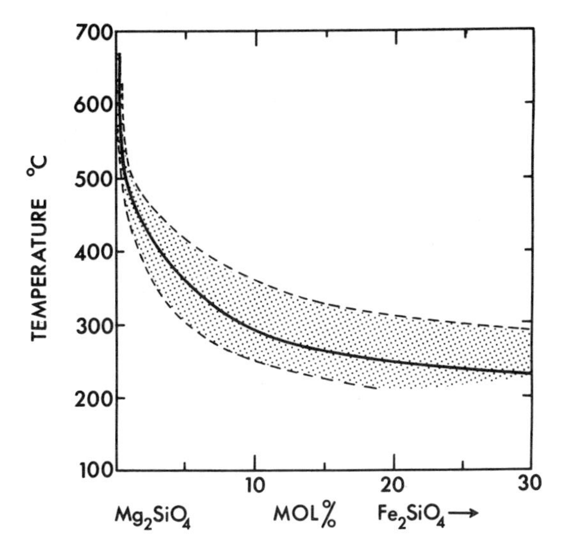

FIGURE 16-3
Equilibrium fayalite content of olivine as a function of temperature in a partially condensed system of solar composition. Shaded area represents uncertainty limits. (*Based on data of Grossman, 1972a.*)

elastic ratio ($\phi = K/\rho$) distribution would be far too high in relation to its density distribution.

We conclude that the heterogeneous accumulation hypothesis offers a totally unsatisfactory explanation of the inferred properties of the earth's mantle. It requires strong vertical zoning of Ca + Al relative to Mg + Si, iron relative to magnesium, and other volatile components relative to involatiles (see below). Evidence for these consequences is lacking. Indeed, the first-order conclusion arrived at in this and previous chapters has been that the mantle is *approximately* chemically homogeneous, on the large scale.

(iii) Volatiles in the upper mantle According to the model under consideration,[1,2] the *upper* mantle accreted relatively slowly under cool conditions and is a mixture of two components: high-temperature condensates *A*, consisting mainly of silicates and metallic iron which had failed to accrete during the main phase of accumulation, and low-temperature condensates *B*, rich in volatiles and oxidized iron and broadly similar to Type I carbonaceous chondrites. This latter component is held to be responsible ultimately for the atmosphere, the hydrosphere, and the content of volatile metals in the upper mantle. From mass balance and Rb-Sr isotopic considerations[3] it may be estimated that the low-temperature component *B* would amount to about 10% of the late "veneer."

A problem with this model is that the volatile-rich material falls upon the

[1]Turekian and Clark (1969).
[2]Clark, Turekian, and Grossman (1972).
[3]Gast (1972).

earth with such high velocities that most of it is evaporated or intensely heated and, therefore, rather thoroughly degassed (Fig. 16-5). Only a very small proportion of infalling material survives atmospheric entry in a cool state. Thus, late accretion of volatiles provides an inefficient manner of trapping them within the solid earth, later to be released by degassing. Yet, large amounts of volatiles have nevertheless been trapped in pyrolite which contains about 0.1% H_2O and 16 ppm N_2.[1,2] The amounts of these trapped volatiles in the upper mantle exceed those now in the atmosphere and hydrosphere.[3] This is a difficult situation to achieve if the volatiles are added as a late-stage veneer. The problem is acute in the cases of gases such as nitrogen and, probably, the inert gases.

(iv) Sulphur in the upper mantle Condensation equilibria show that FeS should be a major component of a low-temperature (<400°C) condensate from the solar nebula. According to the heterogeneous accumulation model, the upper mantle should contain about 0.1 of the primordial abundance of sulphur in the form of FeS. Actually, however, pyrolite is extremely depleted in sulphur, the fraction present amounting to less than 1% of the primordial abundance.[3]

(v) Composition of the core Density and elasticity data imply that the earth's core does not consist of pure iron-nickel but contains, in addition, substantial amounts (10 to 20%) of lighter elements.[4,5] Opinions differ as to the most likely nature of the light element(s)—sulphur, silicon, and carbon are considered to be the most promising candidates. The presence of these components cannot be explained in terms of the heterogeneous accumulation model since the metal which condenses early consists of nearly pure nickel-iron, devoid of S, Si, and C. The model does not offer any plausible means of subsequently introducing these elements into the core.[6]

Discussion

In its original form,[7] the heterogeneous accumulation model necessarily produced an earth characterized by strong radial chemical zoning. As we have seen, this structure cannot be reconciled with observations and inferences concerning the actual composition of the mantle, unless some very efficient subsequent stirring process leading to homogeneity were postulated. Once a large-scale stratification is established, it is extremely difficult to remove. Even if initially unstable—i.e., a

[1]Ringwood (1960, 1966a,b).
[2]Section 4-6.
[3]Ringwood (1966a, b).
[4]E.g., Birch (1963, 1964).
[5]E.g., Stewart (1973).
[6]Levin (1972).
[7]Turekian and Clark (1969).

dense layer overlying a less dense layer—the most likely result would be a simple convective overturn, leading to reversal of the layers and a stable stratification.

The only process which seems adequate to produce a thorough stirring of an inhomogeneous mantle would be the segregation of an initially dispersed metal phase into the earth's core, which is a highly energetic event. This difficulty seems to have been recognized since, in later developments of the model,[1,2] in which much of the metallic iron accretes simultaneously with magnesium silicates, the lower mantle is rather thoroughly stirred by the core-separation process and thereby presumably homogenized. The simplicity of the model is thereby lost, but with negligible compensations in other directions—apart from mixing the Ca-Al rich lowermost layer more uniformly throughout the lower mantle. The revised model still requires that the lower mantle is free of FeO and maintains that the upper mantle did not participate in the stirring and homogenization process since it accreted after this had been completed. Thus the model remains open to all the other objections cited in the previous paragraphs.

16-6 PLANET ACCRETION IN A COOL SOLAR NEBULA

It is generally believed that the planets formed by accretion in a parental discoidal nebula surrounding the sun. The minimum mass of the nebula is obtained from the present masses of the planets plus the complementary volatile components, mainly hydrogen and helium, which were lost from the solar system. The proportions of lost volatiles are readily obtained from solar abundances (Table 16-4). The minimum mass of the nebula is found to be about 1% of the mass of the sun. Allowing for uncertainties in the amounts of gases lost from the outer planets, in the solar abundances, and also for the amount of material present in Oort's comet cloud (Öpik, 1973), this primitive nebula may have amounted to as much as several percent of a solar mass. This is the simplest and most economical hypothesis which can be proposed and has been adopted by most cosmogonists, e.g., Schmidt

[1]Clark, Turekian, and Grossman (1972).
[2]Grossman (1972a).

Table 16-4 MASSES OF MAJOR COMPONENTS OF PRIMORDIAL SOLAR NEBULA

		wt %
"Gases"	H, He	98
"Ices"	C, N, O, Ne, S, A, Cl (as hydrides, except Ne, A)	1.5
"Rock"	Na, Mg, Al, Si, Ca, Fe, Ni (as silicates and oxides)	0.5

(1958), Urey (1952), Kuiper (1952), Safronov (1972), Schatzman (1967), Hoyle (1960), Öpik (1973), and Prentice (1974).

The mechanisms by which such a small mass became distributed into a disc with the dimensions of the solar system are not fully understood. Hoyle (1960) and others suggested rotational instability (which would have occurred when the collapsing solar condensation had contracted to about the orbit of Mercury) accompanied by hydromagnetic coupling of the sun to the spun-off matter. This resulted in transfer of angular momentum from the sun to the nebula, thereby braking the sun's rotation and causing the nebula to be pushed far outward to form the parental disc. Another, and perhaps more plausible, model due to Schatzman (1967) and Prentice (1974) envisaged a major role for turbulent convection during collapse of the solar cloud. The turbulent pressure in the outer regions greatly exceeded the normal gas pressure, resulting in a very low moment of inertia coefficient ($I/MR^2 \sim 0.01$) for the contracting system. This permitted rotational instability to commence at the orbits of the outer planets so that matter was shed continuously or episodically as the solar cloud contracted, whilst angular momentum was transferred from the solar condensation to the surrounding nebula via turbulent viscosity. The residual matter rapidly collapsed to form a disc (Schatzman) or a series of rings (Prentice) analogous to the classical configuration proposed by Laplace (Fig. 16-4).

Regardless of the mechanism of formation, a fundamental result is that a nebula with this small mass would be relatively cold (<0°C) by the stage that it had contracted into a disc or ring system,[1] except for the region closest to the sun, within the orbit of Mercury. Accordingly, for this class of models, the separation of solids from gases in the nebula leading to the formation of planetesimals from which ultimately the planets accreted must have occurred under relatively cool or cold conditions. This is the viewpoint which in the past has been adopted by most cosmogonists.[2] It is also the viewpoint adopted by the author in his attempts to interpret the chemistry of planet formation.

In general terms, we might expect the solid particles in our cold solar nebula to consist of a mixture of "rock" and "ices," as specified in Table 16-4. At the low temperatures under consideration, total condensation of metallic oxides would occur, accompanied by variable degrees of condensation of ices. Of particular importance is the oxidation state of iron. This is controlled by the equilibrium:

$$\tfrac{1}{4}Fe_3O_4 + H_2 = \tfrac{3}{4}Fe + H_2O \qquad K = \frac{H_2O}{H_2}$$

[1]This result does not conflict with the possibility that the material of the nebula may have passed through a transitory high-temperature stage as part of the contracting solar condensation *before* being spun out to a disc or ring system.

[2]Cameron's model of a massive nebula (two solar masses), leading to high temperatures in the inner solar system, represents a notable exception—Sec. 16-5.

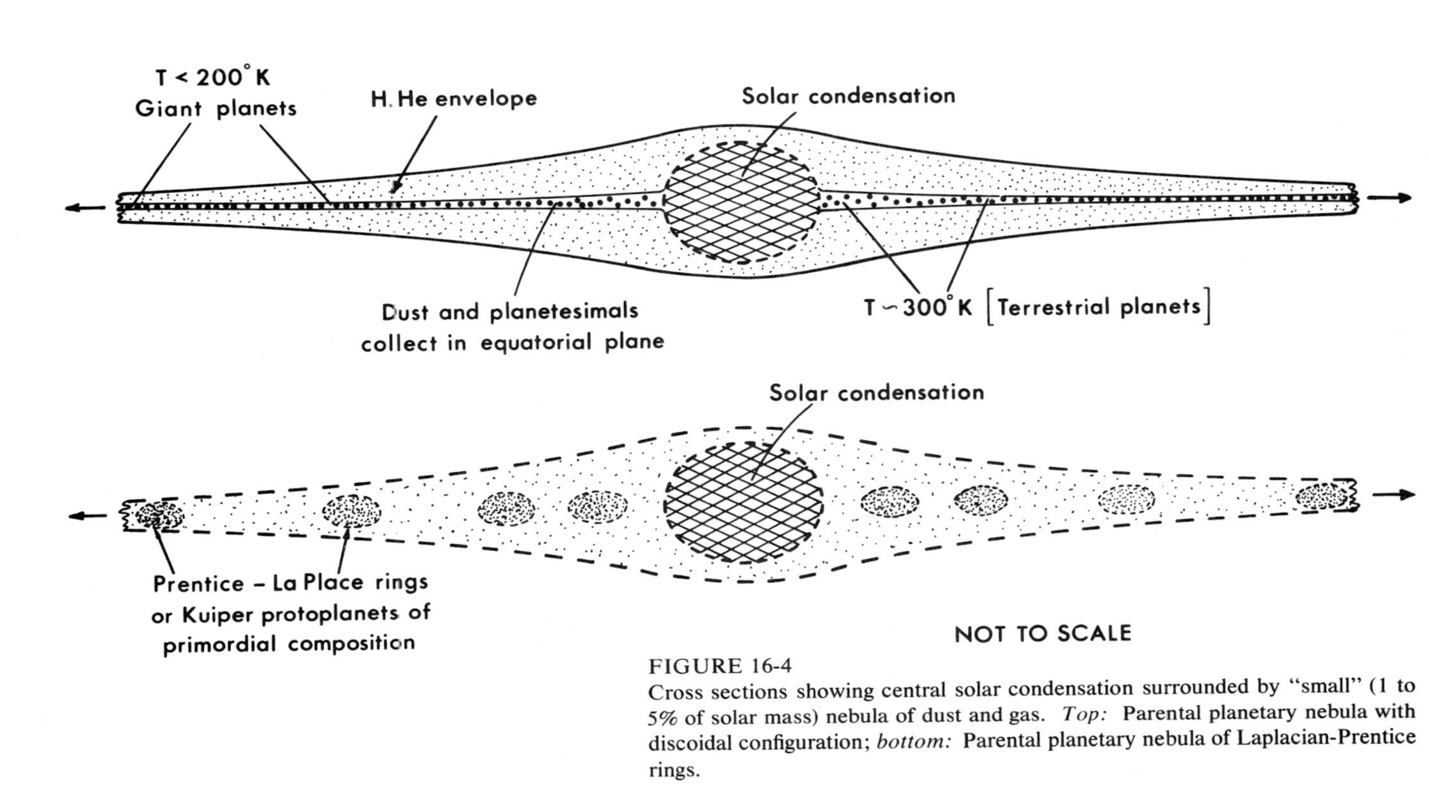

FIGURE 16-4
Cross sections showing central solar condensation surrounded by "small" (1 to 5% of solar mass) nebula of dust and gas. *Top:* Parental planetary nebula with discoidal configuration; *bottom:* Parental planetary nebula of Laplacian-Prentice rings.

In the solar nebula, K is fixed by the relative abundances of hydrogen and oxygen, which yield a H_2O/H_2 ratio of 10^{-3}. The equilibrium constant K is also related to the free-energy change ΔG for the above reaction by the well-known expression

$$\Delta G_T = -RT \ln K$$

With ΔG_T determined from thermochemical data and K fixed by the H_2O/H_2 ratio of the nebula, the equilibrium temperature for the reaction is obtained and found to be 130°C. At temperatures below this, iron would occur oxidized, as magnetite,[1] whilst above 130°C, magnetite would be reduced to metallic iron.

This important result, due to Latimer (1950), shows that, in the cool primordial solar nebula, all iron would be present originally in the *oxidized state*. Since we know that the Earth, Moon, meteorites, and almost certainly Venus and Mercury contain *metallic iron*, it follows that accretion of dust to form planets was either preceded by, or accompanied by, partial reduction of oxidized iron to form a metallic phase.

Some years ago, Urey (1953) conjectured that a particular class of meteorites, the *Type I carbonaceous chondrites*, might be closely related to the primitive dust particles in the solar nebula. An analysis of such a chondrite is given in Table 16-5. Notice that all the iron is *oxidized*, as would be expected for such primitive material. Notice also the very large amounts of volatile components present—components related to the "ices" of Table 16-4. Detailed chemical analyses have shown that the Type I carbonaceous chondrites have retained, to a considerable

[1]A large proportion of iron would also occur as FeO in solid solution in olivines and pyroxenes. The temperatures needed to reduce the FeO component of these minerals to metallic iron (for the solar H_2O/H_2 ratio) are higher, ranging up to about 500°C (Fig. 16-3).

Table 16-5 COMPOSITION OF TYPE I CARBONACEOUS CHONDRITES. (*Orgueil-Wiik, 1956*)

	wt %
SiO_2	22
FeO*	23
MgO	15
Al_2O_3	1.6
CaO	1.2
NiO	1.2
Na_2O	0.7
Cr_2O_3	0.4
H_2O	19
S	5.7
Carbonaceous material	9.7

*Fe_2O_3 and FeO all calculated as FeO.

degree, the primordial abundances of most elements, except for those which are highly volatile.[1,2] They have experienced a very simple chemical and thermal history and have not been heated to more than about 100°C after accreting into parent bodies. A considerable body of evidence now supports the view that these are extremely primitive objects.

One of the remarkable features of these chondrites is the large amount of carbonaceous material which they contain, including a wide range of complex organic compounds. These organic compounds probably formed by Fischer-Tropsch reactions[3,4] when gas mixtures consisting principally of carbon monoxide and hydrogen cooled in the presence of silicate-oxide grains which functioned as catalysts. Instead of reacting to form the thermodynamically stable methane, a crop of metastable complex organic compounds is readily produced under these conditions[3,4] and may be retained indefinitely, providing cooling is sufficiently rapid.

The discovery of a wide range of interstellar organic molecules,[5,6] evidently forming on the surfaces of grains in cool dust-gas clouds, testifies to the cosmochemical importance of these processes, particularly in connection with the condensation and trapping of carbon in association with oxidized iron. It can be expected that such molecules would also be formed in the solar system during the rapid cooling and condensation of initially hot solar gases as they were spun out to form the cool, parental planetary nebula.[7]

Although carbonaceous chondrites amount to only 4% of the chondrites falling on the earth, it is likely that they are much more abundant than these figures would indicate. Because of their fragility, they have a much lower chance of surviving atmospheric entry than other stronger meteorites. Most meteorites are believed to contain a dispersed component related to carbonaceous chondrites,[4] and discrete carbonaceous chondrite xenoliths are found in most other classes of meteorites. The major component of interplanetary material which has fallen on the moon during the last 3 billion years chemically resembles carbonaceous chondrite material.[8] Moreover, the most abundant component of the meteoroid population presently entering the earth's atmosphere is a fragile, low-density material, most probably related to carbonaceous chondrites and comets.[9,10] Finally, the comets themselves are believed to be composed of a primitive mixture of the "rock" and "ice" components of Table 16-4, possessing a mean molecular weight

[1]Ringwood (1966c).
[2]Anders (1971a, b).
[3]Studier, Hayatsu, and Anders (1968).
[4]Anders (1971a).
[5]Buhl and Ponnamperuma (1971).
[6]Brown (1974).
[7]Herbig (1970).
[8]Ganapathy et al. (1970).
[9]McCrosky (1971).
[10]Millman (1972).

of about 21.[1] Öpik believes that comets represent material formed by accretion originally in the vicinity of Jupiter, after which they were ejected by gravitational interactions to their present reservoir in the outermost solar system (Oort's sphere). Their total mass may be quite large (~ 1.5 earth masses), and they may represent the original material from which the outer planets commenced to accrete.[1]

The preceding discussion, although something of a diversion, has aimed at emphasizing the following points which are believed to be of far-reaching cosmochemical significance.

1 Chemical and thermodynamic considerations indicate that the solid condensate in a cold solar nebula would consist of the primordial abundances of metals mostly, *including iron*, in the form of *oxides*, accompanied by carbonaceous compounds and substantial amounts of other volatiles—particularly water and sulphur.
2 Material with compositions resembling that of the expected low-temperature condensate appears to have been abundant and widely distributed throughout the inner solar system[2] during the past 4.5 billion years.
3 In the light of these circumstances, it is reasonable, as a working hypothesis, to assume that the planetesimals from which the terrestrial planets[3] accumulated were dominantly composed of this primitive oxidized, volatile-rich material.

These considerations have formed the basis of the author's approach to the problem of planet formation over a number of years.[4] To prevent confusion, one point should be made clear. It is not proposed that the primitive material from which planets formed was *identical* to Type 1 carbonaceous chondrites. The essence of the model has been that these provide us with our closest approach to the primitive material.[5] The latter are believed to have formed when primitive material accreted into a small parent body which was heated to about 100°C, mildly metamorphosed and reconstituted, possibly accompanied by some fractionation of volatiles. On the other hand, the primitive planetesimals are not believed to have passed through this phase. Moreover, in the case of the earth, the dust from which the primitive planetesimals were formed may have separated from the nebula at significantly higher temperatures than in the case of the Type I carbonaceous chondrites, resulting in rather different proportions of trapped volatiles—particularly with regard to the net H_2O/C ratio. The distribution of trapped inert gases may also have differed.[6]

[1] Öpik (1973).
[2] In the region between Jupiter and Venus.
[3] Possibly excepting Mercury.
[4] Ringwood (1959, 1960, 1961a, 1962, 1966a, b, c, 1970a, 1974).
[5] This qualification is discussed in some detail by Ringwood (1966a, p. 332).
[6] Recent discoveries of "solar" inert gases in oceanic tholeiites (Dymond, 1973, Fisher, 1973) suggest that the primitive material from which the earth formed may have contained solar-type gases rather than the "planetary"-type gases found in carbonaceous chondrites.

16-7 SINGLE-STAGE HYPOTHESIS FOR THE ORIGIN OF THE EARTH

Some of the difficulties confronting homogeneous accumulation hypotheses were mentioned in Sec. 16-3. Not the least of these was their complexity and contrived nature, particularly regarding the multistage chemical and physical fractionation processes which were assumed to have occurred in the solar nebula prior to accretion of planets. In an attempt to avoid these difficulties, Ringwood (1959, 1960, 1966a,b) proposed a simpler "single-stage" hypothesis of the earth's origin, which was also extended to cover the origin of terrestrial planets,[1,2] chondritic meteorites,[3] and the moon.[4]

This hypothesis proposed that the earth formed by accretion in an initially cool solar nebula from planetesimals appropriate in composition to the condensate to be expected on chemical grounds in such a nebula—i.e., highly oxidized primitive material containing carbonaceous compounds, water, and other volatiles and resembling Type I carbonaceous chondrites, as discussed in Sec. 16-6. It was maintained, furthermore, that reduction of oxidized iron to metal, loss of volatiles, and differentiation occurred essentially simultaneously and as a direct result of the primary accretion process. The assumptions of intermediate stages of reduction and fractionation in the solar nebula prior to accretion were avoided so that that hypothesis amounted to a single-stage process.

It was also assumed that accretion of most of the earth occurred over an interval of 1 million years or less. According to current theories, this period is estimated to range between 10^3 and 10^8 years.[5,6,7,8,9,10] The present imperfect understanding of the accretion process permits considerable freedom in the assumption of a time scale for accretion of the earth. A further assumption was that accretion of most of the earth was completed *before* the sun passed through the T Tauri phase. This assumption is shared with other theories of planetary origin.[11,5]

Formation of the earth under the above boundary conditions is strongly influenced by the gravitational energy dissipated during accretion, which in turn controls the chemical equilibria in the accreting material. The gravitational potential energy E released when a sphere of mass M and radius r accretes from a highly dispersed dust cloud is

$$E = -\frac{3}{5}\frac{GM^2}{r}$$

[1]Ringwood (1959, 1966b, 1974).
[2]Ringwood and Clark (1971).
[3]Ringwood (1961a, 1966c).
[4]Ringwood (1966b, 1970a, b, 1972).
[5]Cameron (1973a).
[6]Clark, Turekian, and Grossman (1972).
[7]Hanks and Anderson (1969).
[8]Safronov (1959).
[9]Levin (1972).
[10]MacDonald (1959).
[11]Kuiper (1957).

where G is the gravitational constant. For the earth, this amounts to about 9000 cal/g, a large quantity, sufficient to vapourize the entire earth. The above relationship implies that the energy per unit mass increases approximately with the square of the radius of the accreting earth, reaching 15,000 cal/g in the terminal stages (Fig. 16-5).

During accretion, most of the gravitational energy is converted to thermal energy and largely radiated away. The equilibrium condition at the surface of the earth (without an atmosphere) growing by accretion in a *cold* nebula is approximately

$$\rho \frac{GM(r)}{r} \frac{dr}{dt} = \sigma T^4$$

where ρ = density, t = time during accretion process, $M(r)$ = mass enclosed within a sphere of radius r at time t, r = radius of body at time t, dr/dt = accretion rate, σ = Stefan-Boltzmann constant, and T is the surface temperature at radius r. The equation can be solved for T if we know dr/dt. However, the details of the accretion process are so poorly understood that a wide range of assumptions concerning the form of dr/dt are permissible. It is generally assumed[1,2] that, throughout most of the accretion process, dr/dt increases sharply with r. However, at the beginning and end of accretion, dr/dt should approach zero.[3] Clearly, the time interval t_{acc} over which the major part of the accretion of the earth occurs is of fundamental importance. For a wide range of assumptions concerning the form of dr/dt, a value of 10^5 years for t_{acc} leads to high temperatures (>1500°C) being developed during the later stages of accretion,[1,2,3,4,5] whereas a value for t_{acc} of 10^8 years leads in most cases to a "cool" earth with surface temperatures smaller than a few hundred degrees C.[5,6,7,8]

The author's models have assumed that the accretion time was so short that the surface temperature throughout the greater part of the earth's accretion exceeded 1000°C, rising to 1500°C or more during the later stages of accretion (Fig. 16-5). These models led to the development of a massive primitive atmosphere consisting mainly of CO and H_2, together with several percent of volatilized silicates. The latter are precipitated in the outer, cooler regions of the atmosphere, leading to a high opacity. This has the effect of insulating the earth and drastically decreasing the rate at which gravitational energy released deep within the atmosphere is radiated away from the earth. Thus the time scale for accretion may be extended substantially and still lead to temperatures exceeding 1500°C at the sur-

[1] Hoyle (1946).
[2] Ter Haar and Wergeland (1948).
[3] Hanks and Anderson (1969).
[4] Benfield (1950).
[5] Urey (1952).
[6] Safronov (1959, 1972).
[7] Levin (1972).
[8] MacDonald (1959).

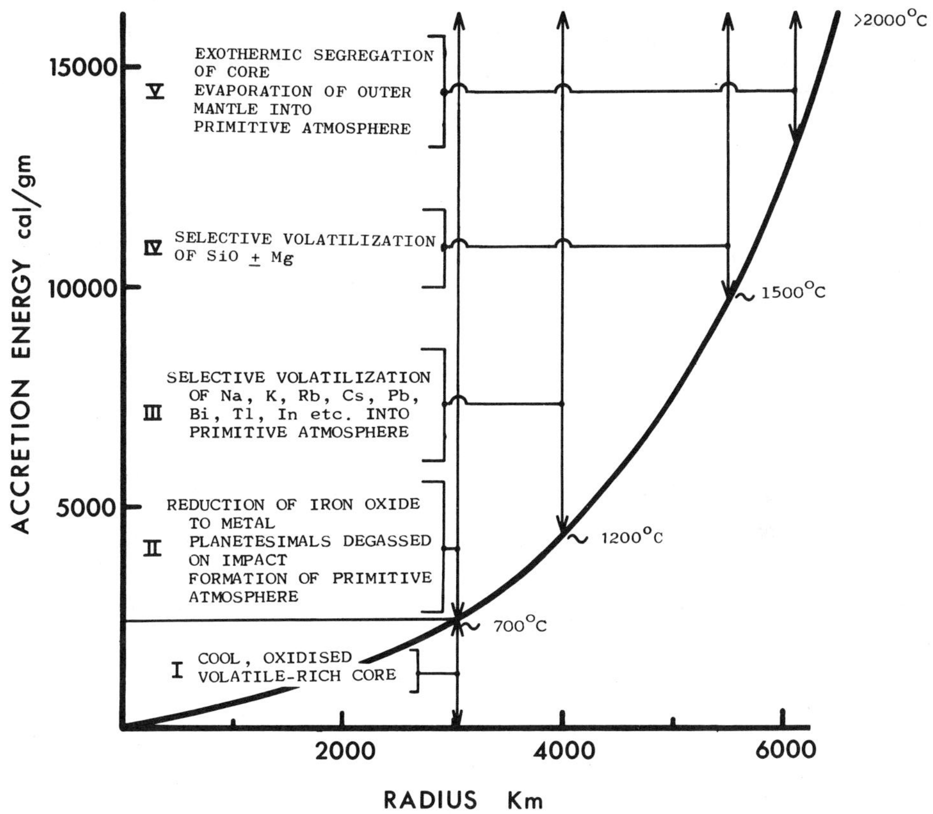

FIGURE 16-5
Relationship between energy of accretion and radius of a growing earth-sized terrestrial planet. The principal stages of accretion are also shown in relation to the energy of accretion and approximate surface temperatures. (*After Ringwood, 1970a, with permission.*)

face. Öpik[1] has shown that, with reasonable assumptions about the opacity of the atmosphere, the required high surface temperatures may be attained even though the accretion time is on the order of 1 million years. The development of the earth according to the boundary conditions discussed above is shown in Fig. 16-5. The stages referred to below are those depicted in this diagram.

Stage I During the early stages of accretion, the energy evolved is small and accretion is relatively slow. The temperature is accordingly low and is buffered by the latent heat of evaporation of volatile components (especially water) in the accreting material. Thus, during this stage, a cool oxidized, volatile-rich nucleus

[1]E. J. Öpik (personal communication).

of primordial material is formed. The size of this nucleus might be about 5 to 10% of the mass of the earth, and it contains the primordial abundances of most metals, with iron present in the oxidized state. However, partial degassing at low temperatures might be expected. A key equilibrium is

$$CH_4 + H_2O = CO + 3H_2$$

The equilibrium constant $K_p = P_{CO}P_{H_2}{}^3/P_{CH_4}P_{H_2O}$ is 10^{-25} at 25°C and 2×10^7 at 1700°C. The equilibrium is thus driven strongly to the left at low temperatures and to the right at high temperatures. Accordingly, in stage I, partial loss of trapped carbon as methane might be expected. Most of the primordial complement of sulphur is also trapped during this stage so that an (Fe, Ni)S phase would coexist with a silicate assemblage consisting of magnetite, olivine, and hydrated Fe-Mg silicates also containing NiO in solid solution. Stage I is of key importance to the earth's subsequent geochemical development since it represents the stage at which volatile components are securely trapped within the deep interior of the growing planet.

The size of this nucleus may be similar to Mars. Indeed, Mars has been interpreted as representing a planet in which accretion did not proceed beyond stage I. The observed chemical and physical properties of Mars are in excellent agreement with this interpretation.[1]

Stage II As the mass of the nucleus increases, the energy of infall of planetesimals becomes sufficient to cause strong transient heating on impact, leading to reduction of oxidized iron to metal by accompanying carbonaceous material. The heating also causes complete degassing, resulting in the generation of a primitive reducing atmosphere. The surface temperatures during stage II are probably in the range 700 to 1200°C.

Stage III When the mass of the nucleus has increased to about one-fifth of the mass of the earth, the accretion temperature has risen to about 1200°C. Reduction of iron oxide to metal now proceeds within the primitive reducing atmosphere (mainly CO and H_2), rather than at the solid surface of the nucleus. With further increase of temperature in the interval 1200 to 1500°C, reduction and volatilization of a number of relatively volatile elements (Na, K, Rb, Cs, Zn, Cd, Hg, T1, In, Bi, Ga, Ge, Pb, S, C1, and others) into the primitive atmosphere occur. The material accreting thus consists of a mixture of metallic iron plus iron-free silicates, strongly depleted in volatile components.

Stage IV As the mass of the nucleus and the rate of accretion increase still further, the temperature becomes sufficiently high (>1500°C) to reduce and volatilize major components of the silicate phase of the infalling planetesimals.

[1]Ringwood and Clark (1971).

Thermodynamic calculations[1,2,3] show that silicon (as SiO) is the most volatile major component and will be preferentially evaporated into the primitive atmosphere at this stage. Consequently, the enstatite component of the accreting material loses silicon and is replaced by forsterite, which becomes by far the dominant accreting phase. At still higher temperatures, some of the MgO will be reduced to Mg and evaporated into the primitive atmosphere. At this stage, also, some silica will be reduced to elemental silicon which enters the metal phase, forming a ferrosilicon alloy.[4] The less volatile components—calcium and aluminium silicates, together with ferrosilicon—continue to accrete upon the growing nucleus. Thus the primitive atmosphere becomes selectively enriched, primarily in silicon and to a lesser degree in magnesium.

Stage V According to the model, the earth develops "inside out," with a cool, oxidized nucleus and becoming successively more reduced and metal-rich towards the surface. This state is gravitationally unstable. As melting occurs near the surface, the metal segregates into bodies which are large enough to sink through the solid interior into the core. This process is highly exothermic, liberating 600 cal/g for the entire earth. As a result, the outer part of the earth's mantle is evaporated in toto into the primitive atmosphere. The atmosphere, coupled by turbulent viscosity, is spun out into a disc and cools to form a sediment ring of planetesimals from which the moon ultimately accretes. These processes are examined in greater detail below.

Formation of the Core

Except for the oxidized, volatile-rich nucleus, the earth accretes as an intimate mixture of metal and silicate phases. Segregation of dispersed metal to form the core represents a change of gravitational potential energy to a more stable state, most of the energy being liberated as heat via viscous dissipation as metal bodies sink through the mantle. The importance of this heat source to the earth's thermal regime was first recognized by Urey[1] and was further examined by Tozer[5] and Birch.[6] The heat evolved amounts to 600 cal/g for the entire earth,[7] enough to raise its temperature by 2000°C.[8] This heat considerably exceeds that which is trapped within the earth's interior during accretion, according to the temperature distribution shown in Fig. 16-5.

[1]Urey (1952).
[2]Ringwood (1966b).
[3]Grossman (1972a).
[4]Ringwood (1959, 1961b).
[5]Tozer (1965).
[6]Birch (1965a, b).
[7]Flaser and Birch (1973).
[8]For the present period of revolution (24 hours), the energy evolved would be 640 cal/g. For a more probable initial period of about 5 hours, this is reduced to 600 cal/g.

Urey[1] argued that the earth formed in a cool state and that segregation of much of the core occurred gradually,[2] throughout geological time, thus driving mantle convection and providing a tectonic engine. This model was further developed by others.[3,4] However, Ringwood (1960) drew attention to the intrinsic instability of the core-forming process and pointed out that, once started, it would proceed to completion in a very short time. The evolution of heat resulting from metal segregation would cause the effective viscosity of the mantle to fall exponentially with temperature, thereby increasing the rate of core formation, leading to a runaway process. A similar conclusion was reached by Elsasser (1963), Tozer (1965), and Birch (1965a). Ringwood (1960) also argued from lead-uranium distributions and lead isotopic considerations that core formation occurred either during accretion or very soon afterward.[5] This argument was supported by experimental studies of the partition of lead between silicate and metallic iron phases.[6]

The temperature distribution within the earth immediately after accretion represents the combined effects of adiabatic heating under self-compression (including phase changes) plus the temperature distribution resulting from accretion as obtained from Fig. 16-5. These contributions are shown separately in Fig.

[1]Urey (1952, 1954, 1957, 1962).
[2]However, Urey (1952) also recognized the difficulties for this model caused by the heat of core formation and expressed some reservations.
[3]Runcorn (1962, 1965).
[4]Munk and Davies (1964).
[5]See also Hanks and Anderson (1969).
[6]Oversby and Ringwood (1971).

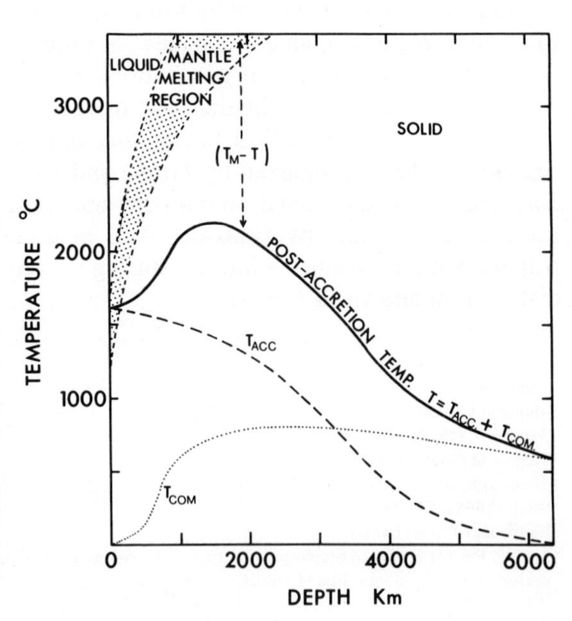

FIGURE 16-6
Temperature distribution for present model at post-accretion stage immediately prior to core formation. The temperature distribution T is the sum of surface accretion temperatures (T_{acc}) given by Fig. 16-5 and a heating term T_{com} resulting from adiabatic self-compression including phase changes. Also shown is an envelope containing several estimates of the variation of mantle melting temperatures (T_M) with depth. The difference ($T_M - T$) between the melting temperatures and post-accretional temperatures is seen to increase strongly with depth below 1000 km.

16-6 and combined to provide the net post-accretional temperature distribution. The adiabatic heating is obtained by integrating the relationship[1,2,3]

$$\frac{dT}{dr} = \frac{\alpha g T}{C_p}$$

where α is the thermal expansion coefficient, and C_p is specific heat. Note that the adiabatic gradient is proportional to absolute temperature, which in turn is given by the accretional contribution T_{acc} of Fig. 16-6. Because T_{acc} is highest in the outer mantle, the adiabatic gradient is high there also so that adiabatic temperatures increase rapidly with depth in the first 1000 km. This characteristic is accentuated by latent heat associated with the phase changes in the Transition Zone.

The net post-accretional temperature distribution (Fig. 16-6) is seen to rise sharply to 2200°C at a depth of 1400 km, before turning over to approach the earth's centre at 600°C. Also shown on Fig. 16-6 is an envelope of mantle melting temperatures as estimated by several authors.[4,5,6,7,8] Note that the upper 100 to 200 km is above the solidus and that, between 200 to 1000 km, the mantle temperatures are relatively close to the solidus. Below 1000 km, this trend changes drastically, and the temperature difference between post-accretional temperature T and the melting temperature T_M increases strongly with increasing depth.

The dynamics of core formation have been studied by Elsasser (1963), and his discussion is applicable to the present model. Near the earth's surface, the temperature exceeds the melting point of the metal which is depressed because of the presence of silicon and/or sulphur,[9] whilst the silicate phase is also molten or partly molten. Below 200 km, the silicates are solidified by the increase of pressure, but because of the proximity to the melting temperature, this outer layer will possess very low mechanical strength to a depth of some hundreds of kilometres (Fig. 16-6). The metal phase is thereby able to segregate into sizable bodies near the surface and to sink downward through this weak layer.

Below this layer, however, the strength or effective viscosity of the silicates begins to increase drastically with increasing depth. This is a consequence of the elevation of the melting point by high pressure (Fig. 16-6). Elsasser points out that the effective viscosity of the silicate layer increases exponentially with the difference between melting temperature T_M and real temperature T. Since, as shown in Fig. 16-6, $T_M - T$ increases strongly with depth below 500 to 1000 km, a corresponding exponential increase of viscosity would occur. On the other hand, because of the lower melting point caused by the presence of silicon

[1]Benfield (1950).
[2]Birch (1952).
[3]Verhoogen (1956).
[4]Uffen (1952).
[5]MacDonald (1959).
[6]Clark (1963).
[7]Green and Ringwood (1967).
[8]Kennedy and Higgins (1973).
[9]Section 16-9.

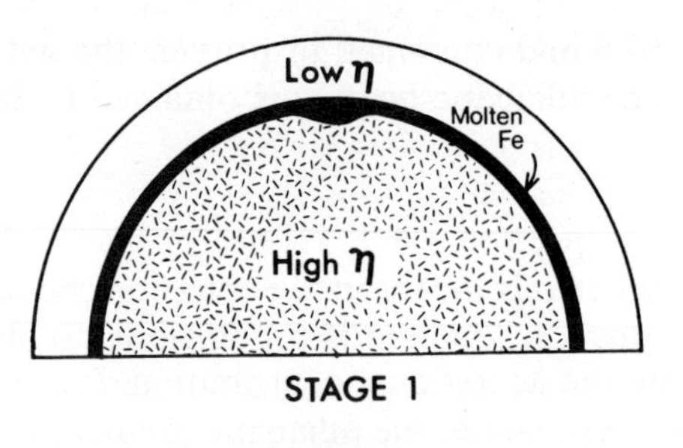

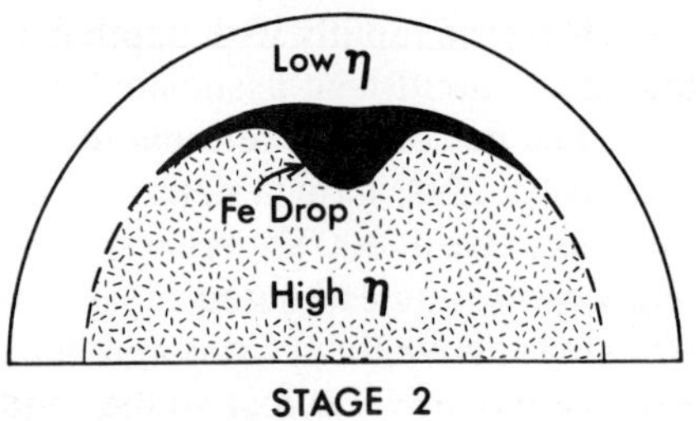

FIGURE 16-7
Early stages of core formation. Drop formation from a homogeneous layer. (Not to scale.) (*After Elsasser, 1963.*)

and/or sulphur,[1] the sinking metal phase remains partly molten and mobile in the outer mantle. Large metal segregations sink more rapidly than small ones, and hence, the latter are "cleaned out" of the weak layer. However, sinking of metal bodies becomes greatly impeded as they encounter layers of vastly increased viscosity at greater depths. Ultimately, the sinking is almost brought to a standstill. At this stage, the metal collects to form a continuous metallic stratum which is highly unstable. Elsasser shows that the metallic stratum will collapse rapidly to form a huge "drop," perhaps some hundreds of kilometres in diameter (Fig. 16-7).

Applying Stokes' law, $V = 2gr^2(\rho_1 - \rho_2)/\eta$, where V = velocity, r = radius of drop, ρ_1 = density of metal, ρ_2 = density of silicate, and η = viscosity, it can readily be shown that, for a viscosity of 10^{21} poises, the drop of this size would sink through the mantle to the core on a time scale of 10^4 to 10^5 years. Actually, the time scale is much shorter because the Newtonian viscosity assumption implicit in Stokes' law probably represents a gross oversimplification of the rheological properties of the mantle. Intense stress is concentrated in the silicate mantle near the boundaries of the drops, and the velocity of sinking is probably related to the marginal stress by a power law. Moreover, the gravitational energy associated with sinking is liberated by viscous dissipation in the boundary layer. Accompanying heating causes local "viscosity" to fall exponentially with temperature, leading to a corresponding increase in sinking velocity. Altogether, it does not appear too unreasonable to expect that the first large drops

[1]Section 16-9.

formed when the earth's deep interior was relatively cool might sink to the centre on a time scale of 1000 years or so.

Sinking drops of this size cause considerable stirring of the mantle and displace unsegregated material upward to participate in the process. Further evolution of gravitational energy causes overall mantle temperatures to rise above the distribution shown in Fig. 16-6. This would be accompanied by a corresponding large increase in the rate of sinking, leading to further heating and even more rapid drop segregation, perhaps attaining catastrophic proportions. In the light of these considerations, it is conceivable that the entire core-forming episode was completed within an interval of about 10^4 years, accompanied by an input of 600 cal/g for the entire earth, as noted earlier. Birch's (1965a) recognition that this was the most decisive event in the earth's thermal history is amply warranted.

If the 600 cal/g evolved by core formation were retained within the earth, the resultant mean temperature increase of 2000°C, superimposed upon the initial temperature distribution (Fig. 16-6), would have led to extensive melting throughout much or most of the earth's interior. This was the view adopted earlier by the author.[1] Upon further consideration, this appears less likely. Core formation will cause strong convection in the mantle and mixing of the oxidized volatile-rich nucleus (Fig. 16-5) into the mantle, leading to approximate homogeneity. Accordingly, the temperature distribution throughout the earth is not likely to depart greatly from adiabatic. The temperature at the surface of the earth cannot exceed that at which silicates of the outer mantle evaporate into the primitive atmosphere under the relevant redox conditions. This is probably in the vicinity of 2000°C.[2] Thus the temperature throughout the earth immediately after core formation would follow an adiabat inward from a surface temperature of 2000°C (Fig. 16-8). The energy needed to heat the earth from the initial post-accretional temperature distribution (Fig. 16-6) to the post-core formation distribution (Fig. 16-8) amounts to 270 cal/g, which is 45% of the core-forming energy. The remaining 330 cal/g will be available to support evaporation of silicates from the outer mantle into the primitive atmosphere and also radiation losses. If all this energy was used for evaporation, about 17% of the earth's mass would enter the primitive atmosphere.[3] However, because of radiation losses during this brief interval of core formation, the actual amount evaporated would be smaller, possibly a few percent of the earth's mass.

According to the above model, the uppermost mantle would be completely molten immediately after core formation. Because of the rapid increase of melting point with pressure, the pyrolite solidus[3] would intersect the adiabat at a depth of 400 km (Fig. 16-8). Below this depth, and extending to the core, the mantle would be solid since the adiabatic gradient remains much smaller than the melting-point gradient (Fig. 16-8).

[1]Ringwood (1960, 1966a,b).
[2]Ringwood (1966b).
[3]Taking the latent heat of evaporation of silicates at 2000°C as about 2000 cal/g (Ahrens and O'Keefe, 1972).
[4]Green and Ringwood (1967).

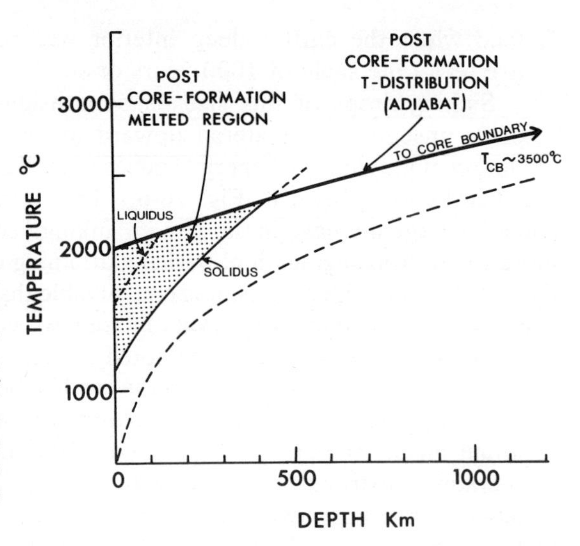

FIGURE 16-8
Temperature distribution in mantle following core formation, in relation to pyrolite solidus and liquidus. The mantle temperature distribution follows an adiabat from 2000°C, based upon data of Birch (1952) and Verhoogen (1956). Note extensive molten and partly molten zone extending to depth of 400 km. Also shown is an estimate of present mantle temperature distribution (broken line).

Because of the relationship between adiabatic and melting-point gradients, crystallization of the melted region commences at the base and proceeds upward, as heat is lost by radiation from the surface of the liquid. The liquid would crystallize completely within about 10^4 years. Because of the phase relationships described in Chap. 4, fractional crystallization in the depth interval 60 to 400 km is confined within the volume olivine-clinopyroxene-orthopyroxene-garnet of Fig. 4-5. Cumulates would consist of an ultramafic assemblage of these minerals, grading up to olivine eclogite near 60 km, whilst residual liquids would be ultramafic, ranging from komatiitic to picritic. At shallower depths (15 to 60 km) the residual liquid attains the composition of olivine tholeiite, and crystallization occurs within the olivine-orthopyroxene-clinopyroxene-plagioclase volume of Fig. 4-1. Cumulates consisting of these minerals are dominantly of olivine gabbro composition. Only within the uppermost 15 km would residual liquids become quartz-normative, thereby ultimately permitting the iron-enriched quartz tholeiite magma to differentiate to form a small volume, perhaps 10% of silica-rich granophyre. Thus, the phase relations dictate that only a minor amount of silicic crustal material would be produced by fractional crystallization of such a large thickness of the upper mantle. However, the residual silicic material would be strongly enriched in incompatible elements.

After solidification, a vertical section would consist of a layer of gabbro (~ 40 km) grading through garnet granulite (40 to 60 km) into eclogite.[1] At greater depths (80 to 100 km), eclogite would grade into ultramafic cumulates. With further cooling, the gabbro-eclogite phase boundaries would migrate

[1]Chapter 1.

upward, and rigid lithosphere plates would develop. As discussed in Chap. 8 (Fig. 8-5), a large thickness of eclogite overlying ultramafics leads to gravitational instability. This would be likely to initiate the process of plate subsidence into the mantle.

After many cycles of plate creation and consumption near the surface, accompanied by an enormous loss of heat, the entire upper 400 km would have been "processed," and following discussions in Sec. 8-7, we suspect that the differentiated products subsided into the deep mantle, perhaps causing a substantial degree of local heterogeneity in this region. The disposal of such a volume of cold material into the deep mantle would cause general cooling, probably by a few hundred degrees. Ultimately, through the agency of convective cooling operating through plate creation and consumption, the temperature distribution throughout the mantle approached more closely to the present distribution (Fig. 16-8), thereby leading to crustal evolution processes of the kind we see today.

Core-Mantle Disequilibrium and the Incorporation of Volatiles in the Mantle

According to the above model, the earth accretes in a state which is grossly out of chemical equilibrium. The deep interior (nucleus) is highly oxidized and rich in volatiles, whereas the outer regions are progressively more reduced and poor in volatile components. After melting near the surface, the metal phase collects into bodies which are large enough to sink into the core. Equilibrium between metal and silicates can be attained only by diffusion across the interfaces of sinking metal bodies. If the rate of sinking of metal is high compared to the rate of attainment of equilibrium by diffusion, the core which separates will not be in equilibrium with the mantle.

These are precisely the conditions which arise according to the previous discussion of core formation. The dimensions of sinking metal "drops" are on the order of hundreds of kilometres. Moreover, the drops will freeze[1] as they sink below 500 to 700 km. The time scale for equilibration of such drops with the surrounding solid mantle is many, many orders of magnitude greater than the descent time.

The sinking drops displace the oxidized, volatile-rich material of the central nucleus upward into the mantle and also drive a forced convection in the latter. As a result of this strong stirring caused by core formation, the oxidized volatile-rich nucleus is thoroughly mixed with the overlying degassed zone of iron-poor silicates. In this manner, oxidized iron ($FeO + Fe_2O_3$) becomes uniformly distributed throughout the mantle, accompanied by the siderophile elements—e.g., Ni, Co, Cu, Au, Re, Pt—present in the nucleus. The segregation of metal drops into

[1]This situation would be realized even if the additional light component in the core were sulphur rather than silicon—see discussion in Sec. 16-9.

the core, as outlined above, occurs too rapidly to permit bulk equilibrium to be achieved, and accordingly, a large proportion of the siderophile elements (and also Fe^{3+}) remains in the mantle, thereby accounting for the relationships discussed in Sec. 16-2. The occurrence of silicon as a component of the earth's core, whilst the mantle contains oxidized iron, can also be understood on these grounds.

A significant property of the single-stage model is the explanation provided of the occurrence of volatiles (group II elements of Table 16-1) in the mantle. We have already noted[1] the problems encountered by models in which these elements are "plastered" on to the earth as a veneer during the late stages of accretion. In the present model, the volatiles are incorporated into the earth at an early stage of accretion as components of the cool, oxidized nucleus. Once incorporated in this manner, they are securely trapped. During core segregation, the volatiles become mixed into the mantle and homogenized (together with oxidized iron and siderophile elements). Their distribution then becomes controlled by crystal-chemical factors, e.g., OH^- ions would enter silicate phases replacing O^{--} or form separate hydrous phases, whilst inert gas atoms would become dispersed within crystals, occupying lattice defects. Indeed, the term *volatile* has little descriptive significance under such conditions. The partial pressures of these components are insignificant, compared to the load pressures. Leakage to the atmosphere and hydrosphere can occur only via exsolution from magmas within the uppermost few kilometres of the crust. This is a very slow process—it was estimated in Sec. 8-7 that perhaps 30 to 60% of the mantle's trapped volatiles might have been degassed in this manner since the earth's formation.

The role of water merits special attention. Type I carbonaceous chondrites contain about 20% water, and a comparable proportion may have been present in the earth's cool nucleus. When introduced into the mantle, the water may have caused a substantially decreased viscosity which would have expedited the core-forming process, as well as the homogenization of the cool nucleus with the overlying mantle, leading to a relatively uniform pyrolite composition. A significant proportion of water from the nucleus may have been consumed by the oxidation of small amounts of metal phase in the mantle which had escaped the core-segregation process.[2]

Dissipation of Primitive Atmosphere

The single-stage model for the formation of the earth involves the production of an enormous atmosphere (0.1 to 0.2 M_E) composed chiefly of H_2, H_2O, and CO, together with a few percent of volatilized metals. A critical requirement of the model is that the primitive atmosphere was completely lost at a very early stage of

[1]Section 16-5.
[2]See also Gast (1972).

the earth's history. Most of the criticisms directed at the model have maintained that atmospheric loss on this scale was not possible.[1,2,3,4]

These criticisms apparently rest upon application of the Jeans-Spitzer theory of selective escape of gases from a gravitational field. This theory applies to the selective escape of gas molecules from an exosphere of extremely low density where the mean free path is on the same order as the scale height. It has no relevance to the present model in which a relatively dense, high molecular weight terrestrial atmosphere is surrounded by, and continuous with, a tenuous, hydrogen-rich solar nebula. In the latter case, an exosphere with the required low density is not present, and Öpik (1963a,b) has pointed out that under such conditions, escape will not be selective with respect to molecular weight. If the temperature is sufficiently high, the atmosphere may "blow off." The critical parameter for this process is the *mean* molecular weight of the atmosphere. If this can be lowered sufficiently by introduction of hydrogen from the solar nebula, escape by blowing off becomes correspondingly easier (see below).

Geochemical considerations indicate rather strongly that escape of a primitive atmosphere from the earth actually occurred at an early stage. It is implicit in earlier discussions (Chap. 8) that most of the earth's present atmosphere and hydrosphere is of secondary origin, having developed gradually over geological time via degassing processes.[5] The abundances of H_2O and N_2 in primitive basalt magmas and ultramafics imply that volatiles presently locked in the mantle exceeded the amounts released to the atmosphere and hydrosphere.[6] Immediately after its formation, before the present atmosphere and hydrosphere had developed substantially, the proportion of volatiles, chiefly H_2O, CO_2, and N_2, locked inside the earth would have been far greater.

It is most difficult to understand how the earth was formed in this condition. Referring to Fig. 16-5, we see that, when the mass of the earth has grown beyond one-tenth of its present value, accreting solids arrive with such high velocities that they are subjected to strong transient heating, accompanied by melting and degassing. By the time the earth has reached one-quarter of its present mass, accreting matter is largely volatilized during impact. Beyond this stage, nearly all the accreting material is strongly degassed.[7] At the conclusion of accretion, the amount of degassed volatiles present in the atmosphere and hydrosphere exceeds, on any reasonable assumptions, the amount of volatiles trapped in the

[1]Urey (1960, 1962).
[2]Levin (1972).
[3]Lewis (1972).
[4]Harris and Rowell (1960).
[5]Rubey (1951).
[6]Ringwood (1966a), Sec. 4-5.
[7]Even if accretion is very slow, so that the equilibrium surface temperature of the earth is low, nitrogen and the inert gases will be irreversibly released into the atmosphere.

interior. These considerations strongly indicate that a primitive atmosphere much larger than the present secondary atmosphere and hydrosphere was present after the earth was formed and that some mechanism for the escape of this primitive atmosphere must have existed. In recent years, many authors have explicitly or implicitly assumed that this was possible.[1,2,3,4,5,6]

The recent discovery[7,8,9] of solar-type inert gases in oceanic tholeiites promises to add a new dimension to the question of atmosphere escape. If these gases prove to be characteristic of the primordial gases locked in the earth, and therefore of the gases subsequently released to the atmosphere, the observation that present atmospheric gases are strongly mass-fractionated relative to the solar pattern is highly significant. It may reasonably be interpreted in terms of Suess' suggestion[1] that the strongly mass-fractionated inert gases of the atmosphere represent a residuum from a more massive primitive atmosphere which was lost from the earth early in its history, owing to a decreased surface gravitational field resulting from a rapid rotation.

In earlier versions of the single-stage model,[10] the author envisaged that the massive primitive atmosphere accumulated until a late stage of accretion, when it was removed by some combination of high earth-rotation rate, solar T-Tauri particle radiation, and turbulent interaction with the solar nebula. It must be admitted, however, that it is difficult to dissipate a massive atmosphere once it is allowed to accumulate.

A more satisfactory approach is to establish the conditions required for the gaseous reduction products to escape continuously back into the solar nebula whilst the earth is accreting, so that a steady state is reached. Under such conditions, the accumulation of a massive atmosphere would not occur. Öpik (1963b) has pointed out that blowing-off of a planetary atmosphere is governed by an escape parameter $\bar{B}$ which represents the ratio of gravitational potential energy to thermal energy of the molecules,

$$\bar{B} = \frac{GM\bar{m}}{RkT}$$

where G is the gravitational constant, M is the mass of the planet inside the spherical surface with radius R, $\bar{m}$ is the mean mass of the gas molecules, k is Boltzmann's constant, and T is absolute temperature. For a diatomic gas, the thermal energy equals 2.5 kT per molecule; if this exceeds the gravitational energy, the top

[1]Suess (1949).
[2]Kuiper (1957).
[3]Hoyle and Fowler (1964).
[4]Cameron (1963, 1973a).
[5]Sagan (1967).
[6]Öpik (1963a, b).
[7]Dymond (1973).
[8]Dymond and Hogan (1973).
[9]Fisher (1973).
[10]Ringwood (1960, 1966a,b).

of the atmosphere blows off into interplanetary space. Thus, for escape of this kind, which is indiscriminate as to molecular species and depends only upon the *mean* properties of the gas

$$\frac{GM\bar{m}}{RkT} = \bar{B} < 2.5$$

In the case of the primitive earth rotating with angular velocity ω, the effective gravitational energy is reduced, and the condition for blow-off becomes

$$\bar{B} = \frac{\bar{m}}{kT}\left(\frac{GM}{R} - \omega^2 R^2\right) < 2.5$$

Clearly, blow-off is favoured by increasing R, T, and ω and decreasing $\bar{m}$.

Consider the situation in the later stages of accretion of the earth. The rotation period was probably between 4 and 5 hours, assuming that the moon was derived from the earth, as argued in Sec. 16-8. The primitive atmosphere would convect strongly. Since turbulent convective mixing creates a large turbulent viscosity, it is reasonable to expect that such an atmosphere would co-rotate with the earth out to a considerable distance, possibly amounting to several earth radii (Prentice, personal communication).

Following the discussion in Sec. 16-9, we assume that the mean molecular weight $\bar{\mu}$ of the gaseous reduction products was lowered to about 3.5 by turbulent mixing of hydrogen from the solar nebula. Then, for $T \sim 1900°K$ as appropriate for late stages of accretion (Fig. 16-5), blow-off of atmospheric gases into the nebula would occur at about 3 earth radii. At earlier stages of accretion, blow-off according to the above equation would occur at correspondingly lower temperatures and smaller distances from the earth.[1] It follows that, providing the temperatures in the earth's environment can be kept sufficiently high during accretion, blow-off of atmospheric gases will occur continuously as the earth accretes. The required high temperatures in turn imply a rapid rate of accretion and a high atmospheric opacity (supplied by infalling material). The model under discussion seems adequate to provide these (Fig. 16-5).

For $T = 1900°K$ and a mean molecular weight of 3.5, the scale height of the atmosphere is about 450 km. Taking the pressure in the nebula as about 10^{-5} bar (for a "small" nebula), the pressure at the earth's surface would probably be on the order of 0.1 bar. The mass of the "steady-state" primitive atmosphere (mainly CO, H_2O, and H_2) is seen to be far smaller than that of the present terrestrial atmosphere. Thus, under the physical conditions assumed in the model, reduction

[1]If blow-off were to occur preferentially, tangentially, in the direction of the earth's rotation, the *excess* escape velocity required over the free circular motion would be reduced by a factor $\sqrt{2} - 1$. The corresponding temperatures required for this kind of blow-off would be reduced by a factor of about 4 (Kuiper, 1952). Escape of gases in this direction would remove excess angular momentum from the system, thereby tending to reduce ω. On the other hand, accretion of solid planetesimals by the growing earth would provide angular momentum, tending to offset this factor. A steady state might thereby have been reached, resulting in a rotation period of about 5 hours.

products would be continually "washed" back into the nebula during accretion, and at no stage would a massive primitive atmosphere be formed (Fig. 16-12).

During contraction on to the main sequence, it is believed that solar-type stars pass through the T-Tauri phase, during which they may lose matter at a rate up to a solar mass per million years in the form of high energy particle radiation.[1,2] Most current theories of evolution of the solar nebula invoke the T-Tauri phase of the sun to dissipate the unaccreted gases and finely particulate solids. It appears probable that following the turbulent mixing phase described above, the T-Tauri "solar hurricane" may have been responsible for the final dissipation of the attenuated primitive terrestrial atmosphere.

Post Core-Formation Accretion and Origin of the D″ Layer

The development of the earth as previously discussed corresponds to the stage of rapid accretion during which most of the available planetesimals were captured by the earth. However, the final stages of accretion must have been much slower owing to depletion of planetesimals in the earth's neighbourhood. This low-intensity accretional tail must have extended for a considerable period after the time of core formation when the primitive atmosphere had been lost and the earth's surface had cooled. Indeed, the lunar highlands provide striking evidence of an intense bombardment between 4.6 and 3.9 billion years ago which must have been shared by the earth.

The plastering-on to the earth of a late stage veneer of primitive volatile-rich planetesimals under cool conditions constitutes an important tenet of heterogeneous accumulation hypotheses which were critically discussed in Sec. 16.5. Nevertheless, accepting the view that a large amount of such material fell upon the earth during the first half billion years, what became of it?

In the early stages, immediately after the outer region of the earth had crystallized and cooled, the rate of infall of planetesimals would have been relatively high. The material would have been degassed upon impact but because of the relatively cool ambient conditions, iron would have remained oxidized or have become reoxidized. Thus a cool, rigid crust of oxidized chondritic material would have collected. Because of its high iron content, it would be much denser than the mantle. After a reasonably thick (10 to 20 km) cold plate of this material had collected, the consequent gravitational instability would cause the plate to subside into the mantle which was hotter than now and much weaker (Fig. 16-8). Because of the large density contrast between mantle and plate, arising from the high iron content of the latter, the plate would sink directly to the base of the mantle and would accumulate above the core-mantle boundary. In such a way, a dense layer

[1]Herbig (1962).
[2]Kuhi (1966).

of iron-enriched material would be built up at the base of the mantle. Such a layer would possess a lower seismic velocity than pyrolite.

It has long been known that the lower 200 km or so of the mantle possesses anomalous seismic properties and that seismic velocities appear to decrease with depth. Bullen (1963) pointed out that this region which he called the D″ layer must be chemically inhomogeneous and characterized by abnormally high density. The hypothesis advanced above may explain these characteristics.

16-8 ORIGIN OF THE MOON

Why is this topic discussed in a book dealing primarily with the earth's mantle? The reason is the author's belief that the moon was ultimately derived from the mantle, being formed by recondensation of material evaporated into the earth's primitive atmosphere during core formation.[1]

We have seen that the single-stage model for the earth's formation leads to generation of a massive (0.1 to 0.2 M_E) primitive atmosphere composed mainly of carbon monoxide and hydrogen together with smaller quantities (10 to 20%) of volatilized silicates. After accretion, it was argued that this atmosphere was dissipated by a combination of the processes discussed in the previous section. As the atmosphere expanded and cooled, a proportion of the volatilized silicates was precipitated to form a "sediment ring" of planetesimals surrounding the earth. This sediment ring subsequently became unstable and coagulated to form the moon. Öpik (1955, 1961) had earlier argued on dynamical grounds that the moon formed from a terrestrial sediment ring[2] but did not explain its origin or composition. These follow rather directly from the present model which complements Öpik's hypothesis.

As the primitive atmosphere expanded and cooled, precipitation of solids occurred, with the less volatile components being precipitated first, at relatively high temperatures close to the earth, whereas the more volatile components were precipitated at lower temperatures and further from the earth. It appears likely that the components precipitating at high temperatures may have grown into relatively large (10^2 to 10^7 cm) planetesimals. Their size would be further enhanced if, as implied by the model, a significant proportion of the precipitate (or condensate) were partially molten and very "sticky." Planetesimals of large size would tend to be left behind by the escaping terrestrial atmosphere. However, the more volatile components precipitating at relatively low temperatures are likely to have formed fine, micron-sized particles or smoke. This material would be sufficiently fine to be carried away with the escaping atmospheric gases by viscous drag, and accord-

[1]Ringwood (1960, 1966b, 1970a, b, 1972).
[2]See also Schmidt (1950, 1958).

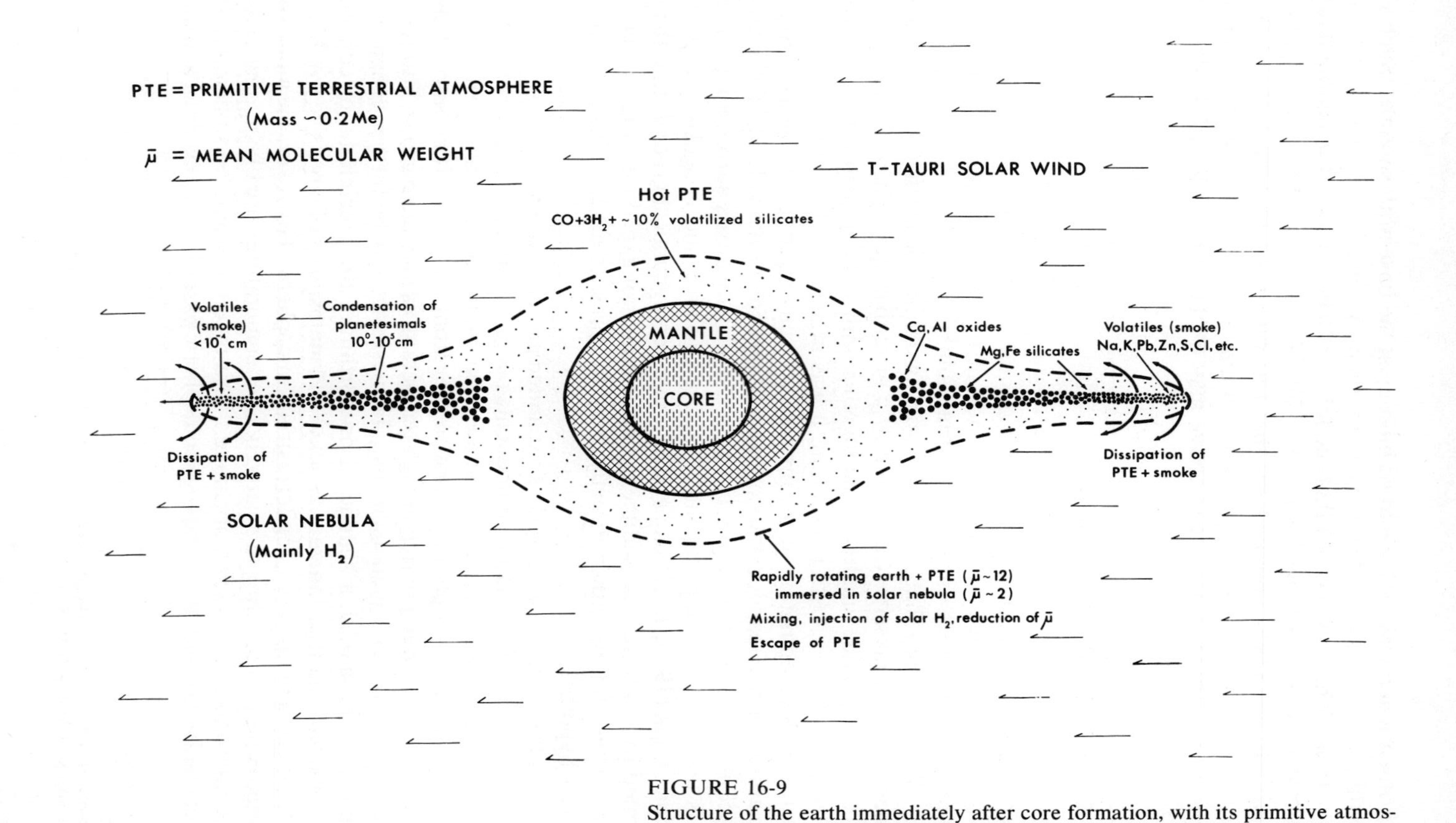

FIGURE 16-9
Structure of the earth immediately after core formation, with its primitive atmosphere and sediment ring. (*From Ringwood, 1972, with permission.*)

ingly would be lost from the earth-moon system. Thus the processes envisaged would lead to a selective fractionation, causing the planetesimals in the sediment ring to be comprised mainly of relatively involatile components. An impression of the proposed structure of the earth with its primitive atmosphere and sediment ring is given in Fig. 16-9.

The prime requirement of a hypothesis of lunar origin is an explanation of the density difference between moon and earth and of the circumstance that the density of the moon is similar to the density of the upper mantle. The present model, according to which the moon formed from material evaporated from the upper mantle after the accompanying metal had segregated into the core, provides a straightforward explanation of the moon's density.

Chemical Boundary Conditions for Theories of Lunar Origin

Studies of returned lunar materials have revealed several important differences between the chemistry of the moon and of the earth's mantle.

1 From investigations of high-pressure melting relationships displayed by maria basalts, combined with major and trace element compositional data, it has become possible to define the nature of the source region within the lunar interior from which the maria basalts were generated by partial melting.[1,2,3,4,5] The methods used are closely analogous to those discussed in Chaps. 4 and 5 in connection with terrestrial basalts and their source regions in the earth's mantle. These methods indicate that the most abundant mineral in the lunar mantle is pyroxene, as compared to a preponderance of olivine in the terrestrial mantle.

2 The same investigations also imply that the pyroxenes in the lunar basalt source regions are poorer in calcium and aluminium than the corresponding pyroxenes in the earth's mantle. This appears to be a consequence of the larger amount of total pyroxene in the moon (item *1* above) since the absolute abundances of calcium and aluminium are probably not smaller in the lunar interior than in the mantle.

3 Experimental investigations also show that the $FeO/FeO + MgO$ ratio in the source region of maria basalts is between 0.20 and 0.25[1,2,3,4,5] compared to a corresponding value of 0.12 for pyrolite.

4 We have already noted that elements which are volatile under high-temperature reducing conditions (e.g., K, Rb, Pb, Tl, Bi, In) are depleted in the earth (by factors of 5 to 20) compared to the primordial abundances (Table

[1]Ringwood and Essene (1970).
[2]Ringwood (1970b).
[3]Green et al. (1971).
[4]Green and Ringwood (1973).
[5]Ringwood and Green (1975).

16-1). A conclusion of key importance is that the moon is also depleted in these elements, but by larger factors. The volatile elements appear to be depleted in the moon relative to the earth by factors of 3 to 100.[e.g.1,2,3]

5 Lunar basalts and their source regions are characterized by a redox state which is very different from the earth's mantle. Lunar rocks contain significant quantities of metallic iron whilst ferric iron is virtually absent, whereas in pyrolite, metallic iron is absent and significant quantities of ferric iron are present. Thus the lunar interior is much more reduced than the earth's mantle, indeed, its oxygen fugacity at 1200°C is 6 orders lower than that of the earth's mantle (Table 16-2).

6 Lunar basalts are strongly depleted in siderophile elements compared to terrestrial basalts.[3]

A satisfactory theory of lunar origin must be capable of explaining these chemical and mineralogical differences between earth and moon. It is believed that the present hypothesis may meet this requirement; indeed, some of the characteristics referred to in the above paragraphs were predicted on the basis of the hypothesis long before material was returned from the moon.[4]

The high proportion of pyroxene relative to olivine is interpreted as the result of the addition of SiO_2 selectively evaporated from the accreting earth during stage IV (Fig. 16-5), to the pyrolite which was evaporated in toto, without fractionation, during stage V (Fig. 16-5). The same factor also accounts for the low ratio of Ca and Al to total pyroxene. The differences in redox states and FeO/FeO + MgO ratios are attributed to the fundamentally different processes and chemical environment attending the accretion of the earth on the one hand and the moon on the other. As discussed earlier, the redox state and FeO/FeO + MgO ratio of pyrolite are believed to have resulted from a specific nonequilibrium terrestrial accretion model, whereas the moon accreted under different P, T, f_{O_2} conditions from a mixture of silicates containing some dispersed metallic iron. Equilibration between silicates and iron thus occurred within the moon, but not in the earth. Besides accounting for the moon's lower oxygen fugacity, separation of the metal during basalt genesis accounts for the relative depletion of siderophile elements in lunar basalts (item *6* above).

The depletion of volatile elements in the moon compared to the earth can be understood on the basis of the model depicted in Fig. 16-9. The volatiles are condensed at low temperature as fine smoke particles which remain coupled to the gases of the primitive atmosphere and hence are ultimately lost from the system. The moon accretes from volatile-depleted, large planetesimals of the sediment ring.

[1]Ringwood and Essene (1970).
[2]Ringwood (1970b).
[3]Ganapathy et al. (1970).
[4]Ringwood (1966b, pp. 84-87).

Sediment Ring and Tidal Evolution

Several authors[1] have argued on dynamical grounds that the moon could not have formed by coagulation of a terrestrial sediment ring. The objection is based upon Goldreich's (1966) conclusion that the stable orbits of planetesimals within about 10 earth radii should lie in the equatorial plane because of the perturbing influence of the earth's rotational bulge. If the moon formed from a sediment ring, it should have occupied the same plane, whereas tidal evolution studies show that the lunar orbit possessed a substantial inclination to earth's equatorial plane when it was much closer to the earth.[2]

This problem no longer appears formidable. O'Keefe (1972a,b) has drawn attention to some results by Darwin (1908) which show that if the earth possessed a low viscosity soon after it was formed, "then the effect of the tides would have been to wrench it [i.e., the moon] out of the earth's equatorial plane." In reply, Goldreich (1972) agreed that "the inclination of the moon's orbit to the earth equator might have increased with time if the [viscoelastic] properties of the earth were very different then from what they are now." According to the present model (Sec. 16-7), this was indeed the case. As shown in Fig. 16-8, the outer 400 km of the earth is believed to have been molten immediately after accretion, whilst below 400 km, the temperatures were 200 to 500°C higher than now exist. The mean viscosity and also the viscosity distribution within the earth were accordingly very different then from today.

The moon's nonzero inclination could also have been caused by external angular momentum contributed by impacts on the moon by a swarm of heliocentric planetesimals, the tail end of which was responsible for the cratering in the highlands.[3,4] Alternatively, the moon may have captured a small proportion of heliocentric planetesimals during its accretion.[3,4]

The view that the moon accreted from earth-orbiting matter is supported by recent interpretations of lunar thermal history. In order to account for evidence of extensive near-surface differentiation in connection with formation of the lunar highlands prior to 4 billion years ago, several authors[5,6,7,8,9] have concluded that acceptable thermal history models require an early high-temperature pulse in the outer regions, which might be supplied most plausibly by partial conservation of gravitational potential energy during accretion. Thermal studies[10] show that in order to be strongly heated in the outer regions by gravitational energy, the moon

[1] E.g., Singer (1971).
[2] Goldreich (1966).
[3] Kaula (1971).
[4] Shoemaker (1972).
[5] Ringwood (1966b).
[6] Ringwood (1970a, b).
[7] Wood (1970, 1972).
[8] Hubbard and Gast (1971).
[9] Toksöz et al. (1972).
[10] Mizutani et al. (1972).

FIGURE 16-10
An impression of the primitive earth surrounded by its "sediment ring" of dust and planetesimals. See text for further details.

must accrete within an interval shorter than about 1000 years. Öpik[1] has pointed out that accretion of the moon from earth-orbiting material may have occurred over a period of about 100 years, which would permit the inferred gravitational heating. On the other hand, if the moon formed as an independent planet from sun-orbiting material, the accretional period would be orders of magnitude higher.[1] With the longer time scale, the rather small amount of gravitational potential energy liberated during formation of the moon ($\sim$ 400 cal/g) would be effectively radiated away.

An impression of the appearance of the primitive earth is given in Fig. 16-10. This shows an incandescent central condensation surrounded by a cooler "sediment ring" which was parental to the moon. This impression might equally be of the primordial major planets surrounded by their protosatellite rings or of the contracting sun surrounded by its discoidal nebula which was parental to the planets. Actually, the picture is of the Sombrero Hat Galaxy (NGC 4594), from the Mt. Wilson and Palomar Galaxy Catalogue, and is reproduced by courtesy of the Director of the Hale Observatories. It appears that the processes responsible for the

[1]Öpik (1961, 1969).

evolution of many galaxies, stars, and planet-satellite systems may conform to a unified pattern.

16-9 MODIFIED SINGLE-STAGE HYPOTHESIS

One of the most striking characteristics of alpine ultramafics and of lherzolite xenoliths in alkali basalts is their uniformity of FeO/FeO + MgO ratios, which mostly lie within the range 0.08 to 0.11. This is particularly true of residual high-temperature peridotites which have been subjected to only small degrees of partial melting. Likewise, the FeO/FeO + MgO ratios of oceanic tholeiites which have not been subjected to high-level crystallization differentiation are also relatively uniform and close to 0.34. The derived model composition of pyrolite is accordingly characterized by a uniform FeO/FeO + MgO ratio of approximately 0.11.[1]

According to the single-stage model (Sec. 16-7), the FeO content of pyrolite is the result of physically mixing FeO + Fe_2O_3 from the cool, oxidized nucleus (stage I, Fig. 16-5) with iron-free silicates formed during the later stages (III, IV, Fig. 16-5). Extremely effective mixing is required in order to account for the uniformity of FeO/FeO + MgO ratios in the upper mantle. It was suggested that this might have been achieved by intense convection during core formation, facilitated by the introduction of water into the mantle from the cool nucleus. The water would have two functions. Firstly, it would depress the mantle solidus causing a small degree of widespread partial melting and greatly reducing the viscosity, thereby leading to turbulent convection and homogenization. Secondly, the interstitial H_2O-rich fluid phase might provide an internal diffusional-exchange medium of high mobility which would facilitate equilibration and homogenization throughout large volumes.

Whilst the operation of these processes appears plausible in the context of the model, it is difficult to gauge just how effective they would be in achieving complete homogeneity. It appears quite possible that this objective would be realized. At the same time, there is an incentive to explore alternative mechanisms which might account more directly for the homogeneity of iron distribution. This incentive is increased by the author's impression, based as yet upon incomplete evidence,[2] that the variances of siderophile elements (e.g., Ni, Co, Au) in alpine ultramafics and in primary basalts may be substantially larger than the corresponding variance of FeO. If verified by future detailed investigations, this would suggest that FeO (on the one hand) and siderophile elements (on the other) were introduced into the mantle via two distinct mechanisms. The view is retained herein that the siderophile and volatile elements were introduced from the

[1] If other transition elements (e.g., Ni, Co, Mn, Fe^{3+}, Cr^{3+}, Ti) are included with Fe^{++}, as in the construction of "simplified" pyrolite compositions (Chap. 14), the effective FeO/FeO + MgO ratio would be 0.12.

[2] See, for example, Goles (1967).

cool, oxidized nucleus, as described in Sec. 16-7. We proceed to explore the possibility that oxidized iron was introduced in another way.

Two possibilities might be considered. Firstly, in the primitive, oxidized material from which the earth accreted, the amount of carbon may have been insufficient[1] to reduce all the $FeO + Fe_2O_3 + H_2O$ (i.e., in atomic abundances, $C < O_{Fe2} + O_{Fe3} + O_{H_2O}$). Thus a significant amount of FeO would have remained unreduced in the silicates which fell upon the earth throughout the entire accretion process. This variant would preclude the possibility of reducing silicates to elemental silicon (as ferrosilicon) during the final stages of accretion. Apart from this aspect (which is discussed later), the hypothesis of development of the earth remains essentially the same as described in Sec. 16-7.

The second variant embodies some more significant differences. It is still assumed that the earth accreted directly from primitive planetesimals in a cool solar nebula and that, accordingly, all the iron in the planetesimals was present in the fully oxidized state (Sec. 16-6). It is suggested, however, that the planetesimals were substantially poorer[2] in water and carbonaceous compounds than Type I carbonaceous chondrites. This might result if they had separated from the nebula at slightly higher temperatures than Type I carbonaceous chondrites, but still below 130°C, at which temperature magnetite becomes stable (Sec. 16-6).

As the earth accreted from primitive planetesimals of this composition, it followed essentially the same path as depicted in Fig. 16-5, with one important exception. After the accreting earth exceeded a radius of, say, 3000 km (stage II), it would capture a hydrogen-rich atmosphere from the solar nebula in which it was immersed.[3] Oxidized planetesimals falling into this hydrogenous primitive atmosphere would undergo partial reduction as before, e.g.,

$$\tfrac{1}{4}Fe_3O_4 + H_2 \rightleftharpoons \tfrac{3}{4}Fe + H_2O$$

According to the model (Fig. 16-5) about four-fifths of the earth accretes at surface temperatures between 1200°C (at a radius a little short of 4000 km) and 1600°C (at completion). As the mass grows sharply in this size range, the size of the primitive hydrogen-rich atmosphere captured from the nebula would increase correspondingly. Moreover, the atmosphere would have been highly turbulent.

Consider the redox equilibria as primitive oxidized planetesimals fall into this atmosphere in the stated temperature range. The reduction of iron and the composition of coexisting silicates are governed by the simplified equilibrium[4]

$$\tfrac{1}{2}Fe_2SiO_4 + \tfrac{1}{2}Mg_2SiO_4 + H_2 = MgSiO_3 + Fe + H_2O \qquad \text{``}K\text{''} = \frac{H_2O}{H_2} \qquad \text{(I)}$$

Using Mueller's thermodynamic data[4] the "K" values for the above equilibrium,

[1]In the model discussed in Sec. 16-7, it was assumed that there was sufficient carbon present to reduce *all* the oxidized iron and water (after stage I) and also some of the silica (in stage IV).

[2]E.g., by a factor of about 5.

[3]Cameron (1973a).

[4]Mueller (1964)

corresponding to an FeO/(FeO + MgO) ratio of 0.11 in the silicates of the mantle. vary from 0.12 at 1200°C to 0.07 at 1600°C. Thus it is seen that, providing the H_2/H_2O ratio of the atmosphere is in the vicinity of 10, the composition of accreting silicates is not strongly temperature-dependent and is close to that which now prevails in the earth's mantle.

If the effective H_2/H_2O ratio near the earth's surface is to be maintained at 10 during stages III and IV, it is evident that a continual supply of hydrogen must be introduced from the solar nebula, whilst the H_2O resulting from reduction of iron oxide must be carried away. This would be a consequence of turbulent convection in the primitive atmosphere, combined with the introduction and mixing of hydrogen-rich ($H_2/H_2O \sim 2000$)[1] gases from the solar nebula. As a result, the distribution of H_2O would be governed by a kinetic equilibrium. The mean molecular weight of the primitive terrestrial atmosphere would not exceed about 3.5 (for $H_2/H_2O \sim 10$) so that, at the elevated temperatures under consideration, the terrestrial atmosphere would continually "blow off" into the nebula.[2,3,4] The principal reduction product, H_2O, together with smaller quantities of CO resulting from the presence of carbon in the primitive planetesimals, would thus be continually "washed" back into the solar nebula during accretion. The departing atmospheric gases would carry with them the metals (e.g., K, Rb, Pb, etc.) which are volatilized during these stages. The abundances of these metals in the atmosphere are low so that they make an insignificant contribution to its mean molecular weight. During stage IV, a marked selective volatilization of SiO into the atmosphere is required, thereby causing an increase in the $Mg_2SiO_4/MgSiO_3$ ratio of accreting solids. The loss of SiO at this stage is permitted by equilibria relevant to the model.[1]

During the final, core-formation, stage V (Fig. 16-5), the combination of substantially increased surface temperatures plus a decreased rate of accretion would result in more efficient mixing of the atmosphere with the solar nebula so that the H_2/H_2O ratio of the terrestrial atmosphere may have been considerably higher, thereby lowering the temperatures[5] required for evaporation of the outer part of the earth directly into the primitive atmosphere. Formation of the moon from this evaporated outer mantle proceeds as in the previous model (Sec. 16-8).

Composition of the Core

The oxygen fugacities attained in the present model are insufficiently low to permit the reduction of silicates to silicon. We are thus confronted by the necessity to

[1]Grossman (1972a).
[2]Öpik (1963a,b).
[3]Cameron (1973a).
[4]I am indebted to Dr. A. G. W. Cameron (personal communication) for pointing out to me some of the advantages of this revised version of the single-stage hypothesis.
[5]Ringwood (1966a, fig. 5).

explain the occurrence of the light element believed to occur in the earth's outer core.[1] A currently popular candidate is sulphur.[1] The conditions under which sulphur may enter the earth are examined below.

For the model under consideration, accretion temperatures are fixed according to Fig. 16-5, which requires that most of the earth accreted at sufficiently high temperatures (1200 to 1600°C) to evaporate alkali metals. Moreover, the redox conditions (i.e., H_2/H_2O ratios) required to produce mantle silicates of observed composition[2] in this temperature range are fixed by equation (I) above and are in the vicinity of 10.

The condensation of sulphur is governed by the equilibrium within the terrestrial atmosphere:

$$FeS + H_2 \rightleftharpoons H_2S + Fe \qquad K = \frac{H_2S}{H_2} \tag{II}$$

If we reduce the H_2 abundance by a factor of 200 from the solar abundance[3] in order to produce mantle silicates of observed compositions, as required by equation (I), it can readily be shown[4] that pure FeS cannot be condensed above 1000°C.

If, however, the activity of FeS could be lowered by solution in molten iron, some condensation would be possible. This is controlled by the Fe-FeS phase diagram (Fig. 16-11). In order to reduce the sulphur activity adequately, only a limited amount of FeS can be permitted to enter the metallic liquid, and this is only possible at relatively high temperatures. Calculations assuming ideal solution of FeS in molten Fe show that 6 to 8 wt% sulphur would condense in a metallic Fe-S liquid at temperatures above 1400°C.[4] Condensation of sulphur would not be possible below 1400°C under ideal solution conditions. However, making reasonable allowances for nonideality and for the presence of some carbon and nickel in solution, which would lower the liquidus temperatures of Fig. 16-11, it appears likely that the metallic liquid condensing under the accretion conditions discussed earlier would contain 6 to 10% sulphur at temperatures in the range 1300 to 1600°C. This would be marginally sufficient to satisfy current estimates of the proportion of light elements in the core.[5] Incorporation into the core of FeS accreted in the cold oxidized nucleus during stage I (Fig. 16-5) could increase the sulphur content of the core by about 5% so that the total amount of sulphur (11 to 15%) would be quite compatible with current estimates within their uncertainty limits.[5,6] Thus, the present model, in which sulphur is regarded as the principal light element in the core, marks a return to the classical views of Washington and Goldschmidt as well as to the more recent models of Murthy, Hall, and Lewis. It

[1] Section 16-4.
[2] I.e., $(Mg_{0.89}Fe_{0.11})_2SiO_2$ coexisting with metallic iron.
[3] Corresponding to $H_2/H_2O \sim 10$ and $H_2S/H_2 \sim 8 \times 10^{-3}$.
[4] Author's unpublished calculations.
[5] Stewart (1973).
[6] King and Ahrens (1973).

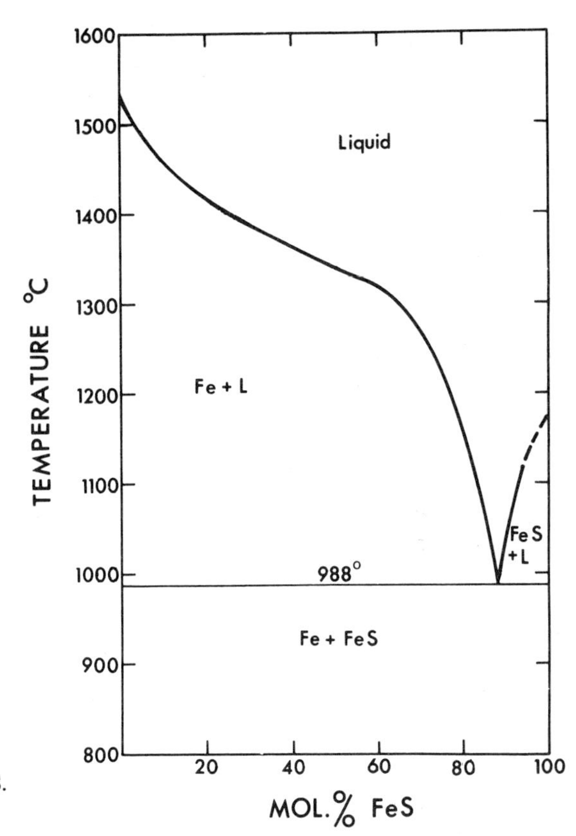

FIGURE 16-11
Phase diagram for the system Fe-FeS. (*After Hansen and Anderko, 1958.*)

must be emphasized, however, that the physicochemical conditions according to which sulphur enters the earth in the present model differ fundamentally from those envisaged by the latter authors.

Because of the deep eutectic trough (Fig. 16-11), the presence of sulphur has an important effect in facilitating the segregation of metal phase into the core, as described in Sec. 16-7. Much has been made[1] of the small pressure effect[2] on the eutectic temperature of this system in this connection. It should be noted that this effect is anomalous and directly connected with the rather low density of FeS. As a result, the density of the liquid at the eutectic point is almost identical to that of an isochemical mixture of FeS and Fe, and following the Clapeyron equation, dP/dT is nearly zero.[3] This situation will persist only as long as iron and FeS do not transform to denser phases. Such transformations occur at 100 to 130 kbars

[1]Murthy and Hall (1970, 1972).
[2]Brett and Bell (1969).
[3]Verhoogen (1973).

both for iron and FeS.[1] The density increase for FeS is about 12%,[1] and it is possible that even denser polymorphs may be stable at higher pressures. Thus, above about 100 kbars, the volume change during melting in the Fe-FeS system will be substantial, and the eutectic temperature will increase normally with pressure.

The above relationships imply that the presently observed low eutectic temperatures of the Fe-FeS system are expected to persist only in the upper few hundred kilometres of the mantle. In this region, this property will greatly facilitate core segregation, as discussed in Sec. 16-7. In the deeper regions of the mantle, however, a large increase in minimum melting temperature can be expected because of the influence of high pressures so that sinking "drops" of metal (cf Fig. 16-7) would freeze as they sink, following the discussion in Sec. 16-7.

Conclusion

A considerable volume of new data upon the abundances of siderophile elements in alpine ultramafics, lherzolite xenoliths, and unfractionated primary basalts in relation to their FeO/FeO + MgO ratios is required before a firm preference for either of the two single-stage models discussed above[2] can be justified. Nevertheless, the author is influenced by present evidence for constancy of FeO/FeO + MgO ratios within the upper mantle and by the evidence that this applies also in the deep mantle.[3] On the other hand, there is suggestive evidence that siderophile elements may possess greater variances, at least in the upper mantle. If finally confirmed, these inferences would favour the "modified single-stage hypothesis" discussed in this section.

It is worth noting that the modified hypothesis encounters lesser difficulties in disposing of the primitive atmosphere and is also closer to some variants of homogeneous accumulation models than the model discussed in Sec. 16-7. The principal differences are the substitutions of hydrogen for carbon as the reducing agent and of sulphur for silicon in the core.

Nevertheless, the modified hypothesis still possesses all the essential elements of the single-stage model. It maintains that the earth accreted in a single stage from primitive, oxidized, volatile-containing planetesimals in a cold solar nebula, that volatile elements (and siderophiles) were trapped in a cool nucleus during the earliest stage of accretion, and that chemical reduction of iron and evaporation of volatile elements occurred during the primary accretion process, the energy source being provided by the gravitational energy liberated during accretion. Finally, according to the model, an intense burst of thermal energy produced by segregation of the core caused strong convection and mixing of the volatile and siderophile elements into the mantle under nonequilibrium conditions,

[1] King and Ahrens (1973).
[2] Sections 16-7 and 16-9.
[3] Chapter 14.

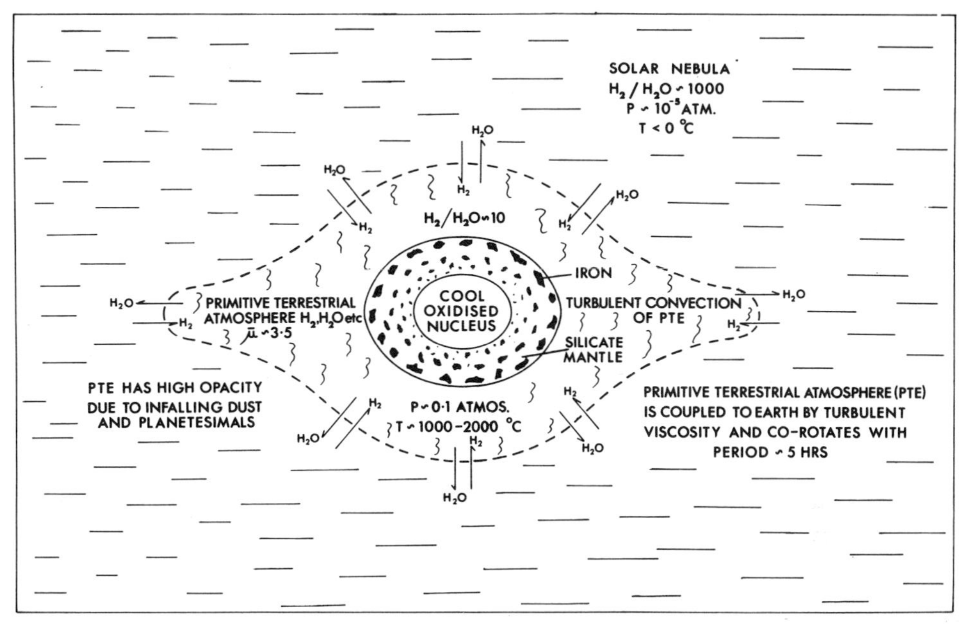

FIGURE 16-12
Sketch of accreting earth prior to core-formation stage as envisaged in the modified single-stage hypothesis.

accompanied by evaporation of the outermost mantle into a primitive atmosphere which was spun off and recondensed to form, ultimately, a sediment ring, the parent of the moon.

A sketch of the accreting earth immediately prior to core formation according to the modified model discussed in this section is given in Fig. 16-12.

REFERENCES

AHRENS, T. J., and J. D. O'KEEFE (1972). Shock melting and vapourization of lunar rocks and minerals. *The Moon* **4,** 214–249.

ANDERS, E. (1968). Chemical processes in the early solar system as inferred from meteorites. *Acc. Chem. Res.* **1,** 289–298.

——— (1971a). Meteorites and the early solar system. *Ann. Rev. Astron. Astrophys.* **9,** 1–34.

——— (1971b). How well do we know "Cosmic Abundances"? *Geochim. Cosmochim. Acta* **35,** 516–522.

ANDERSON, D. L. (1973). The composition and origin of the moon. *Earth Planet. Sci. Letters* **18,** 301–316.

——— and T. C. HANKS (1972). Formation of the Earth's core. *Nature* **237,** 387-388.

ARMSTRONG, R. L. (1968). A model for Sr and Pb isotope evolution in a dynamic earth. *Rev. Geophys.* **6,** 175–199.

BENFIELD, A. E. (1950). The temperature of an accreting earth. *Trans. Am. Geophys. Union* **31,** 53–57.

BIRCH, F. (1952). Elasticity and constitution of the earth's interior. *J. Geophys. Res.* **57,** 227–286.

——— (1963). Some geophysical applications of high pressure research, In: W. Paul and D. Warschauer (eds.), "*Solids Under Pressure*," pp. 137-162. McGraw-Hill, New York.

——— (1964). Density and composition of mantle and core. *J. Geophys. Res.* **69,** 4377–4388.

——— (1965a). Speculations on the earth's thermal history. *Bull. Geol. Soc. Am.* **76,** 133–154.

——— (1965b). Energetics of core formation. *J. Geophys. Res.* **70,** 6217-6221.

BRETT, R. (1971). The earth's core: speculations on its chemical equilibrium with the mantle. *Geochim. Cosmochim. Acta* **35,** 203–221.

——— and P. BELL (1969). Melting relations in the Fe-rich portion of the system Fe-FeS at 30 Kbar pressure. *Carnegie Inst. Washington Yearbook* **67,** 198-199.

BROWN, H. (1952). Rare gases and the formation of the earth's atmosphere. In: G. P. Kuiper (ed.), "*The Atmospheres of the Earth and Planets,*" 2d ed., pp. 258–266. Univ. Chicago Press, Chicago.

BROWN, R. D. (1974). Organic matter in interstellar space. In: J. P. Wild (ed.), "*In The Beginning,*" chap. 1, pp. 1-14. Copernicus 500th Birthday Symposium. Austral. Acad. Sci. 133 pp.

BUHL, D., and C. PONNAMPERUMA (1971). Interstellar molecules and the origin of life. *Space Life Sciences* **3,** 157–164.

BULLEN, K. E. (1963). "*Introduction to the Theory of Seismology,*" 3d ed. Cambridge Univ. Press, London. 381 pp.

CAMERON, A. G. W. (1962). The formation of the sun and planets. *Icarus* **1,** 13-69.

——— (1963). Formation of the solar nebula. *Icarus* **1,** 339-342.

——— (1969). Physical conditions in the primitive solar nebula. In: P. Millman (ed.), "*Meteorite Research,*" pp. 7-15. Reidel, Dordrecht, Holland.

——— (1970). Formation of the earth-moon system. *Trans. Am. Geophys. Union* **51,** 628–633.

——— (1972). Orbital eccentricity of Mercury and the origin of the moon. *Nature* **240,** 299–300.

——— (1973a). Accumulation processes in the primitive solar nebula. *Icarus* **18,** 407–450.

——— (1973b). Formation of the outer planets. *Space Sci. Rev.* **14,** 383–391.

——— (1973c). The early evolution of the solar system. To appear in: "Evolutionary and Physical Problems of Meteoroids," Proc. Inter. Astron. Un. Symposium No. 13.

——— and M. R. PINE (1973). Numerical models of the primitive solar nebula. *Icarus* **18,** 377–406.

CLARK, S. P. (1963). The variation of density in the earth and the melting curve in the mantle. In: "*The Earth Sciences: Problems and Progress in Current Research.*" Univ. Chicago Press, Chicago.

———, K. TUREKIAN, and L. GROSSMAN (1972). Model for the early history of the earth. In: E. C. Robertson (ed.), "*The Nature of the Solid Earth,*" pp. 3-18. McGraw-Hill, New York.

DARWIN, G. H. (1908). "*Scientific Papers*," vol. 2, p. 318. Cambridge Univ. Press, London.

DODD, R. T. (1971). The petrology of chondrules in the Sharps meteorite. *Contr. Mineral. Petrol.* **31,** 201–227.

DYMOND, J. (1973). Rare gas abundance patterns in deep-sea basalts. (Abstract.) *Trans. Am. Geophys. Union* **54,** 485.

——— and L. HOGAN (1973). Noble gas abundance patterns in deep-sea basalts-primordial gases from the mantle. *Earth Planet Sci. Letters* **20,** 131–139.

ELSASSER, W. M. (1963). Early history of the earth. In: J. Geiss and E. Goldberg (eds.), "*Earth Science and Meteoritics*," pp. 1-30. North-Holland, Amsterdam.

EUCKEN, A. (1944). Physikalisch-chemische Betrachtungen über der früeste Entwicklungsgesichte der Erde. *Nachr. Akad. Wiss. Göttingen, Math-Physik Kl.* **1,** 1–25.

EZER, D., and A. G. W. CAMERON (1963). The early evolution of the sun. *Icarus* **1,** 422–441.

FISHER, D. E. (1973). Primordial rare gases in the deep earth. *Nature* **244,** 344–345.

FLASAR, F. M., and F. BIRCH (1973). Energetics of core formation: A correction. *J. Geophys. Res.* **78,** 6101–6103.

FREDRIKSSON, K. (1963). Chondrules and the meteoritic parent bodies. *Trans. N. Y. Acad. Sci.* (ser. 2) **25,** 756–769.

——— (1969). The Sharps chondrite—new evidence on the origin of chondrules and chondrites. In: P. M. Millman (ed.), "*Meteorite Research*," pp. 155–165. Reidel, Dordrecht, Holland.

———, A. NOONAN, and J. NELEN (1973). Meteoritic, lunar and Lonar impact chondrules. *The Moon* **7,** 475–482.

GANAPATHY, R., R. KEAYS, J. LAUL, and E. ANDERS (1970). Trace elements in Apollo 11 lunar rocks: implications for meteorite influx and origin of moon. In: A. Levinson (ed.), "*Proc. Apollo* 11 *Sci. Conference*," vol. 2, 1117-1142. Pergamon, New York.

GAST, P. W. (1960). Limitations on the composition of the upper mantle. *J. Geophys. Res.* **65,** 1287–1297

——— (1972). The chemical composition of the earth, the moon and chondritic meteorites. In: E. C. Robertson (ed.), "*The Nature of the Solid Earth*," pp. 19–40. McGraw-Hill, New York.

GOLDREICH, P. (1966). History of the lunar orbit. *Rev. Geophys. Space Phys.* **4,** 411–439.

——— (1972). Reply to discussion by J. O'Keefe. *Sci. Am.* **227,** 6.

GOLES, G. C. (1967). Trace elements in ultramafic rocks. In: P. J. Wyllie (ed.), "*Ultramafic and Related Rocks*," pp. 352–362. Wiley, New York.

GREEN, D. H., and A. E. RINGWOOD (1967). The stability fields of aluminous pyroxene peridotite and garnet peridotite and their relevance in upper mantle structure. *Earth Planet. Sci. Letters* **3,** 151–160.

——— and ——— (1973). Significance of a primitive lunar basaltic composition present in Apollo 15 soils and breccias. *Earth Planet. Sci. Letters* **19,** 1–8.

———, ———, N. G. WARE, W. O. HIBBERSON, A. MAJOR, and E. KISS (1971). Experimental petrology and petrogenesis of Apollo 12 basalts. "*Proc. Second Lunar Sci. Conference*," vol. 1, 601–615.

GROSSMAN, L. (1972a). Condensation in the primitive solar nebula. *Geochim. Cosmochim. Acta* **36,** 597–619.

——— (1972b). Condensation, chondrites and planets. Ph.D. Thesis, Yale University.

——— (1973). Refractory trace elements in Ca-Al inclusions in the Allende meteorite. *Geochim. Cosmochim. Acta* **37,** 1119–1140.

HALL, H., and V. R. MURTHY (1971). Early chemical history of the earth: some critical elemental fractionations. *Earth Planet. Sci. Letters* **11,** 239–244.

HANKS, T. C., and D. L. ANDERSON (1969). The early thermal history of the earth. *Phys. Earth Planet. Interiors* **2,** 19–29.

HANSEN, M., and K. ANDERKO (1958). "*Constitution of Binary Alloys*," 2d ed., p. 705. McGraw-Hill, New York.

HARRIS, P. G., and J. A. ROWELL (1960). Some geochemical aspects of the Mohorovicic discontinuity. *J. Geophys. Res.* **65,** 2443–2459.

HERBIG, G. H. (1962). The properties and problems of T-Tauri stars and related objects. In: Z. Kopal (ed.), "*Advances in Astronomy and Astrophysics*," pp. 47-103. Academic, New York.

——— (1970). Introductory remarks to Symposium: "Évolution Stellaire Avant La Séquence Principale." *Mem. Soc. Roy. Sci. Liège* Ser. 5, **19,** 13–26.

HIGGINS, G., and G. C. KENNEDY (1971). The adiabatic gradient and the melting point gradient in the core of the earth. *J. Geophys. Res.* **76,** 1870–1878.

HOLLAND, H. D. (1963). On the chemical evolution of the terrestrial and Cytherean atmospheres. In: P. Brancazio and A. G. W. Cameron (eds.), "*The Origin and Evolution of Atmospheres and Oceans*." pp. 86–101. Wiley, New York.

HOYLE, F. (1946). On the condensation of the planets. *Mon. Not. Roy. Astron. Soc.* **106,** 406–414.

——— (1960). On the origin of the solar nebula. *Quart. J. Roy. Astron. Soc.* **1,** 28–55.

——— and W. A. FOWLER (1964). On the abundances of uranium and thorium in solar system material. In: H. Craig, S. Miller, and G. Wasserburg (eds.), "*Isotopic and Cosmic Chemistry*," chap. 30, pp. 516–529. North-Holland, Amsterdam.

HUBBARD, N. J., and P. GAST (1971). Chemical composition and origin of nonmare lunar basalts. "*Proc. Second Lunar Sci. Conference*," vol. 2, 999–1620.

KAULA, W. M. (1968). "*An Introduction to Planetary Physics*," p. 423. Wiley, New York.

——— (1971). Dynamical aspects of lunar origin. *Rev. Geophys.* **9,** 217–238

KENNEDY, G. C., and G. H. HIGGINS (1973). Temperature gradients at the core-mantle interface. *The Moon* **7,** 14–21.

KIMURA, K., R. LEWIS, and E. ANDERS (1973). Siderophile elements on the earth, moon and the eucrite parent body. (Abstract.) Ann. Meeting Meteoritical Society, Davos.

KING, D. A., and T. J. AHRENS (1973). Shock compression of iron sulphide and possible sulphur content of the earth's core. *Trans. Am. Geophys. Union* **54,** 476.

KING, E. A., M. CARMAN, and J. BUTLER (1972). Chondrules in Apollo 14 samples: Implications for the origin of chondritic meteorites. *Science* **175,** 59–60.

KUHI, K. V. (1966). Mass loss from T-Tauri stars. *Astrophys. J.* **140,** 1409–1433.

KUIPER, G. P. (1952). In: G. P. Kuiper (ed.), "*The Atmospheres of the Earth and Planets*," 2d ed. pp. 306–405. Univ. Chicago Press, Chicago.

——— (1956). The origin of the satellites and the Trojans. In: A. Beer (ed.), "*Vistas in Astronomy*," vol. 2, pp. 1631–1666. Pergamon, New York.

——— (1957). Origin, age and possible ultimate fate of the earth. In: D. R. Bates (ed.), "*The Planet Earth*," pp. 12–30. Pergamon, New York.

KULLURUD, G., and H. S. YODER (1959). Pyrite stability relations in the Fe-S system. *Econ. Geol.* **54,** 533–572.

KURAT, G. (1970). Zür Genese der Ca-Al reichen Einschlüsse im Chondriten von Lance. *Earth PLanet Sci. Letters* **9,** 225–231.

———, K. KEIL, M. PRINZ, and C. NEHRU (1972). Chondrules of lunar origin. "*Proc. Third Lunar Sci. Conference*," vol. 1, 707–721.
LARIMER, J. W. (1967). Chemical fractionations in meteorites—I. Condensation of the elements. *Geochim. Cosmochim. Acta* **31,** 1215–1238.
——— and E. ANDERS (1967). Chemical fractionations in meteorites—II. Abundance patterns and their interpretation. *Geochim. Cosmochim. Acta* **31,** 1239–1270.
——— and ——— (1970). Chemical fractionations in meteorites—III. Major element fractionations in chondrites. *Geochim. Cosmochim. Acta* **34,** 367–387.
LATIMER, W. M. (1950). Astrochemical problems in the formation of the earth. *Science* **112,** 101–104.
LEVIN, B. J. (1972). Origin of the earth. *Tectonophysics* **13,** 7–29.
LEWIS, J. S. (1971). Consequences of the presence of sulphur in the core of the earth. *Earth Planet. Sci. Letters* **11,** 130–134.
——— (1972). Metal/silicate fractionation in the solar system. *Earth Planet. Sci. Letters* **15,** 286–290.
LORD, H. C. (1965). Molecular equilibria and condensation in a solar nebula and cool stellar atmospheres. *Icarus* **4,** 279–288.
LUBIMOVA, E. A. (1958). Thermal history of the earth. *Geophys. J.* **1,** 115–134.
MACDONALD, G. J. F. (1959). Calculations on the thermal history of the earth. *J. Geophys. Res.* **64,** 1967–2000.
MCCROSKY, R. E. (1971). Are meteors a tool for studying the asteroids or vice versa? In: T. Gehrels (ed.), "*Physical Studies of Minor Planets*," pp. 395–397. NASA SP-267, Washington, D. C.
MILLMAN, P. M. (1972). Cometary meteoroids. In: A. Elvius (ed.), "*From Plasma to Planet*," pp. 157–168. Interscience, a division of Wiley, New York.
MIZUTANI, H., T. MATSUI, and H. TAKEUCHI (1972). Accretion process of the moon. *The Moon* **4,** 476–489.
MUELLER, R. F. (1964). Phase equilibria and the crystallization of chondritic meteorites. *Geochim. Cosmochim. Acta* **28,** 189–207.
MUNK, W. H., and D. DAVIES (1964). The relationship between core accretion and the rotation rate of the earth. In: H. Craig, S. Miller, and G. Wasserburg (eds.), "*Isotopic and Cosmic Chemistry*," chap. 22, pp. 341–346. North-Holland, Amsterdam.
MURTHY, R. V., and H. HALL (1970). The chemical composition of the earth's core: Possibility of sulphur in the core. *Phys. Earth Planet. Interiors* **2,** 276–282.
——— and ——— (1972). The origin and composition of the earth's core. *Phys. Earth Planet. Interiors* **6,** 125–130.
O'KEEFE, J. A. (1972a). Discussion. *Sci. Am.* **207,** 6.
——— (1972b). The inclination of the Moon's orbit: The early history. *Irish Astron. J.* **10,** 241–250.
ÖPIK, E. J. (1955). The origin of the moon. *Irish Astron. J.* **3,** 245–248.
——— (1961). Tidal deformations and the origin of the moon. *Astron. J.* **66,** 60–67.
——— (1963a). Dissipation of the solar nebula. In: R. Jastrow and A. G. W. Cameron (eds.), "*Origin of the Solar System*," pp. 73–75. Academic, New York.
——— (1963b). Selective escape of gases. *Geophys. J.* **7,** 490–509.
——— (1969). The moon's surface. *Ann. Rev. Astron. Astrophys.* **7,** 473–526.
——— (1973) Comets and the formation of planets. *Astrophys. Space Science* **21,** 307–398.
OVERSBY, V. M., and A. E. RINGWOOD (1971). Time of formation of the earth's core. *Nature* **234,** 463-465.

——— and ——— (1972). Potassium distribution between metal and silicate and its bearing on the occurrence of potassium in the earth's core. *Earth Planet. Sci. Letters* **14,** 345-347.

——— and ——— (1973). Reply to comments by K. Goettel and J. S. Lewis. *Earth Planet. Sci. Letters* **18,** 151–152.

PRENTICE, A. J. R. (1974). The formation of planetary systems. In: J. P. Wild (ed.), "*In the Beginning*," chap. 2, pp. 15–47. Copernicus 500th Birthday Symposium. Austral. Acad. Aci. 133 pp.

RINGWOOD, A. E. (1959). On the chemical evolution and densities of the planets. *Geochim. Cosmochim, Acta* **15,** 257–283.

——— (1960). Some aspects of the thermal evolution of the earth. *Geochim. Cosmochim. Acta* **20,** 241-259.

——— (1961a). Chemical and genetic relationships among meteorites. *Geochim. Cosmochim. Acta* **24,** 159–197.

——— (1961b). Silicon in the metal phase of enstatite chondrites and some geochemical implications. *Geochim. Cosmochim. Acta* **25,** 1–13.

——— (1962). Present status of the chondritic earth model. In: C. B. Moore (ed.), "*Researches on Meteorites*," pp. 198–216. Wiley, New York.

——— (1966a). The chemical composition and origin of the earth. In: P. M. Hurley (ed.), "*Advances in Earth Science*," pp. 287–356. M. I. T. Press, Cambridge, Mass.

——— (1966b). Chemical evolution of the terrestrial planets. *Geochim. Cosmochim. Acta* **30,** 41–104.

——— (1966c). Genesis of chondritic meteorites. *Rev. Geophys.* **4,** 113–175.

——— (1970a). Origin of the moon: The precipitation hypothesis. *Earth Planet. Sci. Letters* **8,** 131–140.

——— (1970b). Petrogenesis of Apollo 11 basalts and implications for lunar origin. *J. Geophys. Res.* **75,** 6453–6479.

——— (1971). Core-mantle equilibrium: Comments on a paper by R. Brett. *Geochim. Cosmochim. Acta* **35,** 223–230.

——— (1972). Some comparative aspects of lunar origin. *Phys. Earth Planet. Interiors* **6,** 366–376.

——— (1974). The early chemical evolution of planets. In: J. P. Wild (ed.), "*In the Beginning*," chap. 3, pp. 48–85. Copernicus 500th Birthday Symposium Austral. Acad. Sci. 133 pp.

——— and S. P. CLARK (1971). Internal constitution of Mars. *Nature* **234,** 89–92.

——— and E. ESSENE (1970). Petrogenesis of lunar basalts, internal constitution and origin of the moon. In: A. Levinson (ed.), "*Proc. Apollo* 11 *Lunar Sci. Conference*," vol. 1, 769–799. Pergamon, New York.

——— and D. H. GREEN (1975). Petrogenesis of maria basalts. (In preparation.)

RUBEY, W. A. (1951). Geologic history of sea water. *Bull Geol. Soc. Am.* **62,** 111–1147.

RUNCORN, S. K. (1962). Palaeomagnetic evidence for continental drift and its geophysical cause. In: S. K. Runcorn (ed.), "*Continental Drift*," pp. 1–39. Academic, New York.

——— (1965). Changes in the convection pattern in the earth's mantle and continental drift: Evidence for a cold origin of the earth. *Phil. Trans. Roy. Soc. London* **A258,** 228–251.

SAFRONOV, V. S. (1959). On the primeval temperature of the earth. *Bull. Acad. Sci. USSR, Geophys. Ser.* No. 1, 139–143.

——— (1972). "*Evolution of the Protoplanetary Cloud and Formation of the Earth and Planets*," pp. 1–206. (English Transl.) IPST, Jerusalem.

SAGAN, C. (1967). Origins of the atmospheres of earth and planets. "*International Dictionary of Geophysics*," vol. 1, pp. 97-104. Pergamon, New York.

SCHATZMAN, E. (1967). Cosmogony of the solar system and origin of the deuterium. *Ann. Astrophysique* **30,** 963–973.

SCHMIDT, O. Y. (1950). "*Four Lectures on the Origin of the Earth*," 2d ed., pp. 65–66. (In Russian.)

——— (1958). "*A Theory of the Origin of the Earth: Four Lectures.*" Foreign Languages Publishing House, Moscow, 58–59; Lawrence and Wishart, London (1959).

SHOEMAKER, E. (1972). Cratering history and early evolution of the moon. In: C. Watkins (ed.), "*Lunar Science III*," pp. 696–698. Lunar Science Institute, Houston.

SINGER, F. (1971). Discussion of paper by A. E. Ringwood. *J. Geophys. Res.* **76,** 8071–8074.

STEWART, R. M. (1973). Composition and temperature of the outer core. *J. Geophys. Res.* **78,** 2586–2597.

STUDIER, M. H., R. HAYATSU, and E. ANDERS (1968). Origin of organic matter in early solar system, I. Hydrocarbons. *Geochim. Cosmochim. Acta* **32,** 151–173.

SUESS, H. E. (1949). Die Haufigkeit der Edelgas auf der Erde und im Kosmos. *J. Geol.* **57,** 600–607.

TER HAAR, D., and H. WERGELAND (1948). On the temperature of the earth's crust. *Kgl. Norske Vidensk, Selsk. Forh.* **20,** 52.

TOKSÖZ, N., S. SOLOMON, J. MINEAR, and D. JOHNSON (1972). Thermal evolution of the moon. *The Moon* **4,** 190–213.

TOZER, D. C. (1965). Thermal history of the earth. *Geophys. J.* **9,** 95–112.

TUREKIAN, K., and S. P. CLARK (1969). Inhomogeneous accumulation of the earth from the primitive solar nebula. *Earth Planet. Sci. Letters* **6,** 346–348.

UFFEN, R. J. (1952). A method of estimating the melting point gradient in the earth's mantle. *Trans. Am. Geophys. Union* **33,** 893–896.

UREY, H. C. (1952). "*The Planets.*" Yale Univ. Press, New Haven, Conn. 245 pp.

——— (1953). Discussion. In: "*Nuclear Processes in Geologic Settings*," p. 49. National Research Council (U. S. A.) Publ. 400. Washington, D. C.

——— (1954). On the dissipation of gas and volatilized elements from protoplanets. *Astrophys. J. Suppl.* **1,** 147–173.

——— (1956). Diamonds, meteorites and the origin of the solar system. *Astrophys. J.* **124,** 623 637.

——— (1957). Boundary conditions for the origin of the solar system. In: L. Ahrens, F. Press, K. Rankama, and S. K. Runcorn (eds.), "*Physics and Chemistry of the Earth*," vol. 2, pp. 46–76. Pergamon, London.

——— (1958). The early history of the solar system as indicated by the meteorites. *Proc. Chem. Soc.* 67–78.

——— (1960). On the chemical evolution and densities of the planets. *Geochim. Cosmochim. Acta* **18,** 151–153.

——— (1962). Evidence regarding the origin of the earth. *Geochim. Cosmochim. Acta* **26,** 1–13.

——— (1963). The origin and evolution of the solar system. In: D. Le Galley (ed.), "*Space Science*," pp. 123–168. Wiley, New York.

——— and H. CRAIG (1953). The composition of the stone meteorites and the origin of meteorites. *Geochim. Cosmochim. Acta* **4,** 36–82.

VERHOOGEN, J. (1956). Temperatures within the earth. In: L. Ahrens, K. Rankama, and S. Runcorn (eds.), *"Physics and Chemistry of the Earth,"* vol. 1, pp. 17–43. Pergamon, London.

——— (1973). Thermal regime of the earth's core. *Phys. Earth Planet, Interiors* **7,** 47–58.

VINOGRADOV, A. P. (1961). The origin of the material of the earth's crust. Communication 1. *Geochemistry USSR* (English Trans.), 1–32.

WIIK, H. B. (1956). The chemical composition of some stony meteorites. *Geochim. Cosmochim. Acta* **9,** 279–289.

WLOTZKA, F. (1969). On the formation of chondrules and metal particles by "shock melting." In: P. M. Millman (ed.), *"Meteorite Research,"* pp. 174-184. Reidel, Dordrecht, Holland.

WOOD, J. A. (1962). Chondrites and the origin of the terrestrial planets. *Nature* **194,** 127–130.

——— (1963). On the origin of chondrules and chondrites. *Icarus* **2,** 152–180.

——— (1970). Petrology of the lunar soil and geophysical implications. *J. Geophys. Res.* **75,** 6497–6513.

——— (1972). Thermal history and early magmatism in the moon. *Icarus* **16,** 229–240.

INDEX

INDEX